Undergraduate Lecture Notes in Physics

Undergraduate Lecture Notes in Physics (ULNP) publishes authoritative texts covering topics throughout pure and applied physics. Each title in the series is suitable as a basis for undergraduate instruction, typically containing practice problems, worked examples, chapter summaries, and suggestions for further reading.

ULNP titles must provide at least one of the following:

- An exceptionally clear and concise treatment of a standard undergraduate subject.
- A solid undergraduate-level introduction to a graduate, advanced, or non-standard subject.
- A novel perspective or an unusual approach to teaching a subject.

ULNP especially encourages new, original, and idiosyncratic approaches to physics teaching at the undergraduate level.

The purpose of ULNP is to provide intriguing, absorbing books that will continue to be the reader's preferred reference throughout their academic career.

More information about this series at http://www.springer.com/series/8917

Alessandro De Angelis
Mário João Martins Pimenta

Introduction to Particle and Astroparticle Physics

Questions to the Universe

Springer

Alessandro De Angelis
Department of Mathematics,
Physics and Computer Science
University of Udine and INFN Padova
Padova
Italy

Mário João Martins Pimenta
Laboratório de Instrumentação e Física
de Partículas
University of Lisboa
Lisboa
Portugal

ISSN 2192-4791 ISSN 2192-4805 (electronic)
Undergraduate Lecture Notes in Physics
ISBN 978-88-470-2687-2 ISBN 978-88-470-2688-9 (eBook)
DOI 10.1007/978-88-470-2688-9

Library of Congress Control Number: 2015946570

Springer Milan Heidelberg New York Dordrecht London

Printed on acid-free paper

Springer-Verlag Italia Srl. is part of Springer Science+Business Media (www.springer.com)

Preface

This book introduces particle physics, astrophysics, and cosmology starting from experiment. It provides a unified view of these fields, which is needed to answer in the best way our questions about the Universe—a unified view that has been lost somehow during the last years due to increasing specialization.

Particle physics has recently seen the incredible success of the so-called "standard model." A 50-year long search for the missing ingredient of the model, the Higgs particle, has just been concluded successfully, and some scientists claim that we are close to the limit of the physics humans may know. Also, astrophysics and cosmology have shown an impressive evolution, driven by experiments and complemented by theories and models. We have nowadays a "standard model of cosmology" which can describe the evolution of the Universe from a tiny time after its birth to any foreseeable future. The situation is similar to the one in the end of the nineteenth century, after the formulation of the Maxwell equations—and we know how the story went.

As in the end of the nineteenth century, there are some clouds that might hide a new revolution in physics. The main cloud is that experiments indicate that we are still missing the description of the main ingredients of the Universe from the point of view of its energy budget. We believe one of these ingredients to be a new particle, of which we know very little; and the other to be a new form of energy. The same experiments indicating the need for these new ingredients are probably not powerful enough to unveil them, and we must invent new experiments to do it.

The scientists who will solve this puzzle will base their project on a unified vision of physics, and this book helps providing such a vision.

This book is addressed primarily to advanced undergraduate or beginning graduate level students, since the reader is only assumed to know quantum physics and "classical" physics, in particular electromagnetism and analytical mechanics, at an introductory level; but it can also be useful for graduates and postgraduates, and postdoc researchers involved in high-energy physics or astrophysics research. It is also aimed at senior particle and astroparticle physicists as a consultation book. Exercises at the end of each chapter help the reader reviewing material from the chapter itself and synthesizing concepts from several chapters. A "further reading" list is also provided for readers who want to explore in more detail particular topics.

Our experience is based on research both at artificial particle accelerators (in our younger years) and in astroparticle physics after the late 1990s. We work as professors for more than 20 years, teaching courses on particle and/or astroparticle physics at undergraduate and graduate levels. We spent long time in several research institutions outside our countries, also teaching there and gaining experience with students from different backgrounds.

This book contains broad and interdisciplinary material, which is appropriate for a consultation book, but it can be too much for a textbook. In order to give coherence to the material for a course, one can think of at least three paths through the manuscript:

- For an "old-style" one-semester course on particle physics for students with a good mathematical background, one could select Chaps. 1–6, part of 7, and possibly (part of) 9.
- For a basic particle physics course centered in astroparticle physics, one could instead use Chaps. 1, 2 (possibly excluding Sect. 2.10.7), 3, 4 (excluding Sect. 4.4), Sects. 5.1, 5.2, 5.4 (excluding Sect. 5.4.1), 5.5 (possibly excluding Sect. 5.5.4), 5.6, 5.7, 6.1, 8.1, 8.4, part of Chap. 10, and if possible Chap. 11.
- A specialized half-semester course in high-energy astroparticle physics could be based on Sects. 4.3.2, 4.5 and Chaps. 8, 10, 11; if needed, an introduction to experimental techniques could be given based on Sects. 4.1 and 4.2.

This book would have not been possible without the help of friends and colleagues; we remind here (in alphabetical order) Pedro Abreu, Sofia Andringa, Stefano Ansoldi, Pedro Assis, Liliana Apolinario, Luca Baldini, Fernando Barão, Sandro Bettini, Barbara Biasuzzi, Giovanni Busetto, Per Carlson, Nuno Castro, Julian Chela-Flores, Ruben Conceiçao, Jim Cronin, Michela De Maria, Tristano di Girolamo, Jorge Dias de Deus, Anna Driutti, Catarina Espírito Santo, Fernando Ferroni, Giorgio Galanti, Riccardo Giannitrapani, Marco Laveder, Francesco Longo, José Maneira, Oriana Mansutti, Mauro Mezzetto, Alessandro Pascolini, Gianni Pauletta, Elena Pavan, Massimo Persic, Ignasi Reichardt, Jorge Romao, Marco Roncadelli, Sara Salvador, Ron Shellard, Franco Simonetto, Fabrizio Tavecchio, Bernardo Tomé, Ezio Torassa, Andrea Turcati, Robert Wagner, Alan Watson, and Jeff Wyss. We also thank all our students who where patiently listening and discussing with us during all the past years.

Lisbon
Padova
13 June 2015

Alessandro De Angelis
Mário João Martins Pimenta

Contents

1 Understanding the Universe: Cosmology, Astrophysics, Particles, and Their Interactions 1

1.1 Particle and Astroparticle Physics 1
1.2 Particles and Fields 3
1.3 A Quick Look at the Universe 7
1.4 Cosmic Rays 15

2 The Birth and the Basics of Particle Physics 21

2.1 The Atom 21
2.2 The Rutherford Experiment 22
2.3 β Decay and the Neutrino Hypothesis 24
2.4 Uncertainty Principle and the Scale of Measurements 26
2.5 Cross-Section and Interaction Length. 27
2.5.1 Total Cross-Section 27
2.5.2 Differential Cross-Sections 29
2.5.3 Cross-Sections at Colliders 29
2.5.4 Partial Cross-Sections 30
2.5.5 Interaction Length 31
2.6 Decay Width and Lifetime 32
2.7 The Fermi Golden Rule and the Rutherford Scattering 34
2.7.1 Transition Amplitude 35
2.7.2 Flux 37
2.7.3 Density of States 37
2.8 The Modern View of Scattering: Quantum Field Exchange . . . 38
2.8.1 Feynman Diagrams 39
2.9 Particle Scattering in Static Fields 40
2.9.1 Extended Charge Distributions (Non Relativistic) 40
2.9.2 Finite Range Interactions 41
2.9.3 Electron Scattering 42

2.10 Special Relativity 44
2.10.1 Lorentz Transformations 45
2.10.2 Space-Time Interval 49
2.10.3 Energy and Momentum 50
2.10.4 Examples of Relativistic Dynamics 52
2.10.5 Mandelstam Variables 53
2.10.6 Lorentz Invariant Fermi Rule 55
2.10.7 The Electromagnetic Tensor and the Covariant Formulation of Electromagnetism 57
2.11 Natural Units 61

3 Cosmic Rays and the Development of Particle Physics 71
3.1 The Puzzle of Atmospheric Ionization and the Discovery of Cosmic Rays 71
3.1.1 Experiments Underwater and in Height 73
3.1.2 The Nature of Cosmic Rays 78
3.2 Cosmic Rays and the Beginning of Particle Physics 78
3.2.1 Relativistic Quantum Mechanics and Antimatter 79
3.2.2 The Discovery of Antimatter 88
3.2.3 Cosmic Rays and the Progress of Particle Physics 89
3.2.4 The μ Lepton and the π Mesons 91
3.2.5 Strange Particles 93
3.2.6 Mountain-Top Laboratories 95
3.3 Particle Hunters Become Farmers 95
3.4 The Recent Years 97

4 Particle Detection 101
4.1 Interaction of Particles with Matter 101
4.1.1 Charged Particle Interactions 101
4.1.2 Range 110
4.1.3 Photon Interactions 111
4.1.4 Nuclear (Hadronic) Interactions 114
4.1.5 Interaction of Neutrinos 114
4.1.6 Electromagnetic Showers 115
4.1.7 Hadronic Showers 119
4.2 Particle Detectors 120
4.2.1 Track Detectors 120
4.2.2 Photosensors 129
4.2.3 Cherenkov Detectors 131
4.2.4 Transition Radiation Detectors 133
4.2.5 Calorimeters 133
4.3 High-Energy Particles 135
4.3.1 Artificial Accelerators 136
4.3.2 Cosmic Rays as Very-High-Energy Beams 139

4.4 Detector Systems and Experiments at Accelerators 140
4.4.1 Examples of Detectors for Fixed Target Experiments 142
4.4.2 Examples of Detectors for Colliders 145
4.5 Cosmic-Ray Detectors 153
4.5.1 Interaction of Cosmic Rays with the Atmosphere; Extensive Air Showers 154
4.5.2 Detectors of Charged Cosmic Rays 157
4.5.3 Photon Detection 164
4.5.4 Neutrino Detection 177
4.5.5 Detection of Gravitational Waves 182

5 Particles and Symmetries 189
5.1 A Zoo of Particles 189
5.2 Symmetries and Conservation Laws: The Noether Theorem . . . 191
5.3 Symmetries and Groups. 193
5.3.1 A Quantum Mechanical View of the Noether's Theorem 194
5.3.2 Some Fundamental Symmetries in Quantum Mechanics 196
5.3.3 Unitary Groups and Special Unitary Groups. 199
5.3.4 SU(2) 199
5.3.5 SU(3) 202
5.3.6 Discrete Symmetries: Parity, Charge Conjugation, and Time Reversal 204
5.3.7 Isospin 207
5.3.8 The Eightfold Way 211
5.4 The Quark Model 214
5.4.1 $SU(3)_{flavor}$ 214
5.4.2 The Color 216
5.4.3 Excited States (Nonzero Angular Momenta Between Quarks) 217
5.4.4 The Charm Quark. 218
5.4.5 Beauty and Top 221
5.5 Quarks and Partons. 223
5.5.1 Elastic Scattering 223
5.5.2 Inelastic Scattering Kinematics 224
5.5.3 Deep Inelastic Scattering 227
5.5.4 The Quark-Parton Model 230
5.5.5 The Number of Quark Colors. 236
5.6 Leptons 237
5.6.1 The Discovery of the τ Lepton 237
5.6.2 Three Neutrinos 239

5.7 The Particle Data Group and the Particle Data Book 240
5.7.1 PDG: Estimates of Physical Quantities 241
5.7.2 Averaging Procedures by the PDG 241

6 Interactions and Field Theories 245
6.1 The Lagrangian Representation of a Dynamical System 247
6.1.1 The Lagrangian and the Noether Theorem 248
6.1.2 Lagrangians and Fields; Lagrangian Density 249
6.1.3 Lagrangian Density and Mass 250
6.2 Quantum Electrodynamics (QED) 250
6.2.1 Electrodynamics 250
6.2.2 Minimal Coupling 253
6.2.3 Gauge Invariance 256
6.2.4 Dirac Equation Revisited 259
6.2.5 Klein–Gordon Equation Revisited 270
6.2.6 The Lagrangian for a Charged Fermion in an Electromagnetic Field: Electromagnetism as a Field Theory 272
6.2.7 An Introduction to Feynman Diagrams: Electromagnetic Interactions Between Charged Spinless Particles 275
6.2.8 Electron-Muon Elastic Scattering 279
6.2.9 Renormalization and Vacuum Polarization 281
6.3 Weak Interactions 285
6.3.1 The Fermi Model of Weak Interactions 285
6.3.2 Parity Violation 288
6.3.3 V-A Theory 290
6.3.4 “Left” and “Right” Chiral Particle States 292
6.3.5 Intermediate Vector Bosons 295
6.3.6 The Cabibbo Angle and the GIM Mechanism 304
6.3.7 Extension to Three Quark Families: The CKM Matrix 308
6.3.8 CP Violation 310
6.3.9 Matter–Antimatter Asymmetry 321
6.4 Strong Interactions and QCD 323
6.4.1 Yang–Mills Theories 324
6.4.2 The Lagrangian of QCD 326
6.4.3 Vertices in QCD; Color Factors 327
6.4.4 The Strong Coupling 328
6.4.5 Asymptotic Freedom and Confinement 331
6.4.6 Hadronization; Final States from Hadronic Interactions 331
6.4.7 Hadronic Cross Section 340

7 The Higgs Mechanism and the Standard Model of Particle Physics 361
7.1 The Higgs Mechanism and the Origin of Mass 363
7.1.1 Spontaneous Symmetry Breaking 363
7.1.2 An Example from Classical Mechanics 364
7.1.3 Application to Field Theory: Massless Fields Acquire Mass 365
7.1.4 From SSB to the Higgs Mechanism: Gauge Symmetries and the Mass of Gauge Bosons 367
7.2 Electroweak Unification 369
7.2.1 The Formalism of the Electroweak Theory 371
7.2.2 The Higgs Mechanism in the Electroweak Theory, and the Mass of the Electroweak Bosons 375
7.2.3 The Fermion Masses 378
7.2.4 Interactions Between Fermions and Gauge Bosons 379
7.2.5 Self-interactions of Gauge Bosons 382
7.2.6 Feynman Diagram Rules for the Electroweak Interaction 382
7.3 The Lagrangian of the Standard Model 383
7.3.1 The Higgs Particle in the Standard Model 384
7.3.2 Standard Model Parameters 384
7.3.3 Accidental Symmetries 386
7.4 Observables in the Standard Model 387
7.5 Experimental Tests of the SM at Accelerators 389
7.5.1 Data Versus Experiments: LEP (and the Tevatron) 390
7.5.2 LHC and the Discovery of the Higgs Boson 403
7.6 Beyond the Minimal SM of Particle Physics; Unification of Forces 408
7.6.1 GUT 410
7.6.2 Supersymmetry 412
7.6.3 Strings and Extra Dimensions; Superstrings 415
7.6.4 Compositeness 417

8 The Standard Model of Cosmology and the Dark Universe 421
8.1 Experimental Cosmology 422
8.1.1 The Universe Is Expanding 422
8.1.2 Cosmic Microwave Background 429
8.1.3 Primordial Nucleosynthesis 438
8.1.4 Astrophysical Evidence for Dark Matter 443
8.1.5 Age of the Universe 449
8.2 General Relativity 451
8.2.1 Flat and Curved Spaces 455
8.2.2 The Einstein Equations 458

8.2.3 The Friedmann–Lemaitre–Robertson–Walker Model . . . 461
8.2.4 Black Holes . . . 472
8.3 Past, Present, and Future of the Universe . . . 473
8.3.1 Early Universe . . . 473
8.3.2 Inflation and Large-Scale Structures . . . 481
8.3.3 The ΛCDM Model . . . 486
8.4 What Is Dark Matter Made Of, and How Can It Be Found? . . . 488
8.4.1 Baryonic Dark Matter . . . 488
8.4.2 Weakly Interacting Particles . . . 490
8.4.3 Other Nonbaryonic Candidates . . . 502

9 The Properties of Neutrinos . . . 505
9.1 Sources and Detectors; Evidence of the Transmutation of the Neutrino Flavor . . . 506
9.1.1 Solar Neutrinos, and the Solar Neutrino Problem . . . 506
9.1.2 Neutrino Oscillation in a Two-Flavor System . . . 511
9.1.3 Long-Baseline Reactor Experiments and the Estimate of $\nu_e \to \nu_\mu$ Oscillation Parameters . . . 515
9.1.4 Atmospheric Neutrinos, and the $\nu_\mu \to \nu_\tau$ Oscillation . . . 516
9.1.5 Phenomenology of Neutrino Oscillations: Extension to Three Families . . . 519
9.1.6 Short-Baseline Reactor Experiments, and the Determination of θ_{13} . . . 521
9.1.7 Accelerator Neutrinos . . . 522
9.1.8 Explicit Appearance Experiment . . . 524
9.1.9 A Gift from Nature: Geo-Neutrinos . . . 525
9.2 Neutrino Oscillations; Present Knowledge of the Oscillation Parameters . . . 526
9.3 Neutrino Masses . . . 528
9.3.1 The Constraints from Cosmological and Astrophysical Data . . . 528
9.3.2 Direct Measurements of the Electron Neutrino Mass: Beta Decays . . . 529
9.3.3 Direct Measurements of the Muon- and Tau-Neutrino Masses . . . 531
9.3.4 Incorporating Neutrino Masses in the Theory . . . 531
9.3.5 Majorana Neutrinos and the Neutrinoless Double Beta Decay . . . 533
9.3.6 Present Mass Limits and Prospects . . . 535

10 Messengers from the High-Energy Universe 537
10.1 The Data 537
10.1.1 Charged Cosmic Rays 538
10.1.2 Photons 549
10.1.3 Neutrinos 557
10.1.4 Gravitational Radiation 559
10.2 How Are High-Energy Cosmic Rays Produced? 561
10.2.1 Acceleration of Charged Cosmic Rays 561
10.2.2 Production of High-Energy Gamma Rays 566
10.2.3 Production of High-Energy Neutrinos 570
10.3 Possible Acceleration Sites and Sources 570
10.3.1 Stellar Endproducts 572
10.3.2 Other Galactic Sources 575
10.3.3 Extragalactic Sources 576
10.3.4 Gamma Rays and the Origin of Cosmic Rays 582
10.3.5 Sources of Neutrinos 585
10.3.6 Sources of Gravitational Waves 586
10.4 The Propagation Process 587
10.4.1 Propagation of Charged Cosmic Rays 587
10.4.2 Propagation of Photons 594
10.4.3 Propagation of Neutrinos 597
10.4.4 Propagation of Gravitational Waves 598
10.5 Frontier Physics and Open Questions 598
10.5.1 The Sources 598
10.5.2 Ultrahigh-Energy Phenomena 600
10.5.3 Top-Down Production Mechanisms for CR; WIMPs 603
10.5.4 Lorentz Symmetry Violation 607
10.5.5 Possible Anomalous Photon Propagation Effects 610

11 Astrobiology and the Relation of Fundamental Physics to Life 617
11.1 What is Life? 618
11.1.1 Schrödinger's Definition of Life 619
11.1.2 The Recipe of Life 619
11.1.3 Life in Extreme Environments 624
11.1.4 The Kickoff 625
11.2 Life on the Solar System, Outside Earth 626
11.2.1 Planets of the Solar System 628
11.2.2 Satellites of Giant Planets 629
11.3 Search for Life Outside the Solar System 631
11.3.1 The "Drake Equation" 631
11.3.2 The Search for Habitable Planets 632
11.3.3 The Fermi Paradox 635

11.3.4 Listening to Messages from Space 636
11.3.5 Sending Messages to the Universe 638
11.4 Conclusions 645

Appendix A: Periodic Table of the Elements 647

Appendix B: Properties of Materials 649

Appendix C: Physical and Astrophysical Constants 651

Index 653

Acronyms

ACE	Advanced composition explorer (astrophysical observatory orbiting the Earth)
AGASA	Akeno giant air shower array (experiment in Japan)
AGILE	Astro-rivelatore gamma a immagini leggero (gamma-ray telescope orbiting the Earth)
AGN	Active galactic nucleus
ALEPH	A LEP experiment (at CERN)
ALICE	A large ion collider experiment (at CERN)
ALLEGRO	A Louisiana low-temperature experimental gravitational radiation observatory (in the US)
ALP	Axion-like particle
ALPHA	Anti-hydrogen experiment at CERN
AMS	Alpha magnetic spectrometer (particle detector onboard the ISS)
ANTARES	Astronomy with a neutrino telescope and abyss environmental research (experiment in the Mediterranean Sea)
APD	Avalanche photodiode (detector)
ARGO-YBJ	Cosmic ray detector at the Yanbanjing Observatory (in Tibet)
a.s.l.	Above sea level (altitude)
ATIC	Advanced thin ionization calorimeter (balloon-borne experiment)
ATLAS	A toroidal LHC apparatus (experiment at CERN)
AU	Astronomical unit (a.u.)
AURIGA	An ultra cryogenic gravitational wave detector
BaBar	*B*—anti-*B* experiment at SLAC
BATSE	Burst and transient source experiment (in the CGRO)
BBN	Big bang nucleosynthesis
BEBC	Big European Bubble Chamber (experiment at CERN)
Belle	b physics experiment at KEK
BESS	Balloon-borne experiment with superconducting spectrometer
Bevatron	Billion electron volts synchrotron (accelerator in the USA)
BGO	$Bi_4\ Ge_3\ O_{12}$ (scintillating crystal)

BH	Black hole
BL Lac	Blazar Lacertae (an active galactic nucleus)
BNL	Brookhaven national laboratory (in Long Island, NY)
Borexino	Boron solar neutrino experiment (at the LNGS)
BR	Branching Ratio (in a decay process)
CANGAROO	Collaboration of Australia and Nippon (Japan) for a gamma-ray observatory in the outback (Cherenkov observatory)
CAST	CERN axion search telescope (experiment at CERN)
CDF	Collider detector at Fermilab (experiment)
CERN	European Organization for Nuclear Research, also European laboratory for particle physics
CGC	Color glass condensate
CGRO	Compton gamma-ray observatory (orbiting the Earth)
cgs	centimeter, gram, second (system of units)
CKM	Cabibbo, Kobayasha, Maskawa (matrix mixing the quark flavors.)
CMB	Cosmic microwave background (radiation)
CMS	Compact muon solenoid (experiment at CERN)
COBE	Cosmic background explorer (satellite orbiting the Earth)
CoGeNT	Coherent germanium neutrino telescope (experiment in the US)
COUPP	Chicagoland observatory for underground particle physics (experiment at Fermilab)
CP	Charge conjugation × Parity (product of symmetry operators)
CPT	Charge conjugation × Parity × Time reversal (product of symmetry operators)
CR	Cosmic rays
CRESST	Cryogenic rare event search with superconducting thermometers (experiment at LNGS)
CTA	Cherenkov telescope array (an international gamma-ray detector)
CUORE	Cryogenic underground observatory for rare events (experiment at LNGS)
D0	Experiment at Fermilab
DASI	Degree angular scale interferometer
DAMA	Dark matter experiment (at LNGS)
DAMPE	Dark matter particle explorer (astrophysical space observatory)
DAQ	Data acquisition (electronics system)
DARMa	De Angelis, Roncadelli, Mansutti (model of axion-photon mixing)
DAS	Data acquisition system
DELPHI	Detector with lepton, photon, and hadron identification (experiment at the CERN's LEP)
DESY	Deutsche synchrotron (laboratory in Germany)
DM	Dark matter
DNA	Desoxyribonucleic acid (the genetic base of life)
DONUT	Direct observation of the ν_τ (experiment at Fermilab)

DSA	Diffusive shock acceleration (of cosmic rays)
dSph	Dwarf spheroidal galaxy
EAS	Extensive air shower (cosmic rays)
EBL	Extragalactic background light
ECAL	Electromagnetic calorimeter (detector)
EGMF	Extragalactic magnetic field
EGO	European Gravitational Observatory (in Italy)
EGRET	Energetic gamma-ray experiment telescope (part of the CGRO)
EHE	Extremely high-energy
EHS	European hybrid spectrometer (experiment at CERN)
EJSM/Laplace	European Jupiter space mission—Laplace (ESA/NASA Mission)
ESA	European space agency
EUSO	Extreme universe space observatory
FCNC	Flavor changing neutral currents (hypothetical electroweak process)
FD	Fluorescence Detector
Fermilab	Fermi national accelerator laboratory (near Chicago, IL) also FNAL
Fermi-LAT	Large Area Tracker, a gamma-ray telescope orbiting the Earth
FLRW	Friedmann, Lemaitre, Robertson, Walker (metric model in general relativity)
FNAL	Fermi national accelerator laboratory (near Chicago, IL) also Fermilab
FoV	Field of View
FPGA	Field programmable gate array (processor)
FRI	Fanaroff and Riley class I (astrophysical sources)
FSRQ	Flat-spectrum radio quasars
GALLEX	Gallium experiment (at LNGS)
GAMMA-400	gamma-ray space observatory (space astrophysical observatory)
Gargamelle	Experiment at CERN
GBM	Gamma burst monitor (detector)
GC	Galactic center
GERDA	Germanium detector array (experiment at the LNGS)
GIM	Glashow, Iliopoulos, Maiani (mechanism)
GLAST	Gamma-ray large area space telescope, renamed Fermi after positioning in orbit
GPM	Gaseous photomultipliers
GPS	Global positioning system
GRB	Gamma-ray burst (astrophysical event)
GSW	Glashow-Salam-Weinberg model of electroweak unification
GUT	Grand unified theory
GZK	Greisen, Zatsepin, Kuz'min (energy cutoff for cosmic rays)
HAWC	High-altitude water Cherenkov (observatory in Mexico)
HBL	High-energy peaked BL Lac
HCAL	Hadron calorimeter (detector)

HE	High energy
HEGRA	High-energy gamma-ray astronomy (Cherenkov experiment in La Palma)
HERA	Hadron elektron ring anlage (particle accelerator at DESY)
H.E.S.S.	High-energy stereoscopic system (Cherenkov experiment in Namibia)
HPD	Hybrid photon detector
HST	Hubble space telescope (orbiting the Earth)
IACT	Imaging atmospheric Cherenkov telescope
IBL	Intermediate-energy peaked BL Lac
IC	Inverse Compton scattering (mechanism for the production of HE gamma-rays)
ICRR	Institute for Cosmic Ray Research (at the University of Tokyo, Japan)
IceCube	Neutrinos observatory in Antarctica
IDPASC	International doctorate on particle and astroparticle physics, astrophysics, and cosmology (doctoral network)
IMB	Irvine, Michigan, Brookhaven (experiment in the US)
INFN	Istituto Nazionale di Fisica Nucleare (in Italy)
IR	Infrared (radiation)
IRB	Infrared Background (photons)
ISS	International Space Station
IST	Instituto Superior Técnico (at the University of Lisboa, Portugal)
JEM	Japanese experimental module (onboard the ISS)
JEM-EUSO	Extreme universe space observatory at JEM
K2K	KEK to Kamioka experiment (Japan)
Kamiokande	Kamioka neutrino detector (experiment in Japan)
KamLAND	Kamioka liquid scintillator antineutrino detector (experiment in Japan)
KASCADE	Karlsruhe shower and cosmic array detector (experiment in Germany)
KATRIN	Karlsruhe tritium neutrino experiment (in Germany)
KEK	High-energy accelerator in Japan
Kepler	Mission to search for extraterrestrial planets (NASA)
KM	Parametrization of the CKM matrix in the original paper by Kobayasha and Maskawa
Km3NeT	kilometer cube neutrino telescope (experiment in the Mediterranean Sea)
kTeV	Experiment at Fermilab
L3	LEP 3rd (experiment at CERN)
LAr	Liquid argon
LAT	Large area telescope (detector in the Fermi Satellite)
LBL	Low-energy peaked BL Lac
ΛCDM	Lambda and Cold Dark Matter (model with cosmological constant Λ)

LEBC	LExan Bubble Chamber (experiment at CERN)
LEP	Large electron positron (collider at CERN)
LEP II	2nd phase of operation of LEP, at energies above the Z mass
LHC	Large hadron collider (at CERN)
LHCb	LHC beauty (experiment at CERN)
LHCf	LHC forward (experiment at CERN)
LIGO	Laser interferometer gravitational-wave observatory (in the US)
LISA	Laser interferometer space antenna (project for gravitational waves detection)
LIV	Lorentz invariance violation
LMC	Large Magellanic cloud (dwarf galaxy satellite of the Milky Way)
LNGS	Laboratorio Nazionale del Gran Sasso (Laboratory for particle and astroparticle physics in Italy)
LO	Leading order in perturbative expansions
LPHD	Local parton hadron duality (approximation in QCD predictions)
LPM	Landau–Pomeranchuk–Migdal (effect)
LSND	Liquid scintillator neutrino detector (experiment in the US)
LSP	Lightest supersymmetric particle
LST	Large-Size Telescope (Cherenkov telescope for CTA)
ly	Light-year
MACE	Major atmospheric Cherenkov experiment (Cherenkov experiment in India)
MACHO	Massive astronomical compact halo object
MAGIC	Major atmospheric gamma-ray imaging Cherenkov telescopes (Cherenkov experiment in Canary Islands)
MARE	Microcalorimeter arrays for a Rhenium experiment (in Italy)
MC	Monte Carlo (simulation technique)
MILAGRO	Cosmic ray (gamma in particular) experiment in the US
MINOS	Main injector neutrino oscillation search (experiment in Fermilab)
mip	minimum ionizing particle
MoEDAL	Monopole and exotics detector at the LHC (experiment at CERN)
MOND	Modified Newtonian dynamics
MSSM	Minimal supersymmetric standard model
MSW	Mikheyev–Smirnov–Wolfenstein (matter effect in neutrino oscillations)
NA#	North area # (experiment at CERN, # standing for its number)
NASA	National aeronautics and space agency (in the US)
NEMO	Neutrino Ettore Majorana observatory (in France)
NESTOR	Neutrino Extended Submarine Telescope with Oceanographic Research (experiment in the Mediterranean Sea)
NFW	Navarro, Frenk, and White (profile of dark matter distribution)
NIST	National Institute of Standards and Technology (US institute)

NKG	Nishimura-Kamata-Greisen (lateral density distribution function for showers)
NLO	Next-to-leading order in QCD perturbative expansions
NLSP	Next-to-lightest supersymmetric particle
NNLO	Next-to-next-to-leading order in perturbative expansions
NS	Neutron star
NT-200	Neutrino telescope (experiment in Russia)
NTP	Normal temperature and pressure
NU	Natural units (system of units)
OPAL	Omni-purpose apparatus for LEP (experiment at CERN)
OPERA	Oscillation project with emulsion-tracking apparatus (experiment at LNGS)
OZI	Okubo-Zweig-Iizuka (rule for transitions in particle processes)
PAMELA	Payload for antimatter-matter exploration and light-nuclei astrophysics (astrophysical observatory orbiting the Earth)
PAO	Pierre Auger Observatory (cosmic ray observatory in Argentina)
PDF	Parton density function
PDG	Particle data group
PHENIX	A physics experiment at RHIC
Planck	ESA mission for precise measurement of CMB anisotropy and other properties
PLATO	Planet transits and oscillations of stars (ESA mission to search for extraterrestrial planets)
PMNS	Pontecorvo-Maki-Nakagawa-Sakata (neutrino mixing matrix)
PMT	Photomultiplier tube (detector)
PSF	Point spread function (space or angular resolution)
PVLAS	Polarizzazione del vuoto con laser (experiment in Italy)
PWN	Pulsar wind nebula (astrophysical object)
QCD	Quantum chromodynamics
QED	Quantum electrodynamics
QG	Quantum gravity
QGP	Quark gluon plasma (state of matter)
QPM	Quark parton model
RF	Radiofrequency
RHIC	Relativistic heavy ion collider (at BNL)
RICH	Ring imaging Cherenkov (detector)
RMS	Root mean square
RPC	Resistive plate chamber (detector)
SAGE	Soviet american gallium experiment (in Russia)
SCT	Semiconductor tracker (detector)
SDP	Shower detector plane (cosmic rays)
SED	Spectral energy distribution
SETI	Seach for extraterrestrial intelligence
SI	International System (of units)
SiPM	Silicon photomultiplier (detector)

SK	Super-Kamiokande neutrino detector (experiment in Japan) also Super-K
SLAC	Stanford linear accelerator center (in the US)
SLD	SLAC large detector
SM	Standard model (of particle physics)
SMBH	Supermassive black hole
SMC	Small Magellanic cloud (dwarf galaxy satellite of the Milky Way)
SNO	Sudbury neutrino observatory (Canada)
SNR	Supernova remnant
SNU	Solar neutrino unit (of neutrino interactions)
SO(n)	Special orthogonal group of rank n
SPEAR	Stanford Positron Electron Asymmetric Rings (particle accelerator in the US)
S$p\bar{p}$S	Super proton antiproton synchrotron (collider at CERN)
SPS	Super proton synchrotron (particle accelerator at CERN)
SSB	Spontaneous symmetry breaking
SSC	Self-synchrotron Compton (mechanism for production of HE gamma-rays)
SSM	Standard solar model (of physics reactions in the Sun's core)
SU(n)	Special unitary group of rank n
Super-K	Super-Kamiokande neutrino detector (experiment in Japan) also SK
SUSY	Supersymmetry (model beyond the SM)
T2K	Tokai to Kamioka experiment (in Japan)
TA	Telescope Array (cosmic ray observatory in the US)
TDAQ	Trigger and data acquisition (electronics system)
Tevatron	Teraelectronvolt synchrotron (collider at Fermilab)
TGC	Triple gauge coupling (coupling between the electroweak gauge bosons – Z, W bosons, and the photon)
TeVCAT	Catalog of astrophysical VHE gamma-ray sources
Tibet-AS	Cosmic rays experiment
TMAE	Tetra dimethyl-amine ethylene
TNT	Trinitrotoluene (2-Methyl-1,3,5-trinitrobenzene, chemical explosive)
TOTEM	Total cross section, elastic scattering and diffraction dissociation at the LHC (experiment at CERN)
TPC	Time projection chamber (detector)
TRD	Transition radiation detector
TRT	Transition radiation tracker (detector)
U(n)	Unitary group of rank n
UA#	Underground area # (experiment at CERN, # standing for its number)
UHE	Ultra-high-energy (cosmic rays)
UHECR	Ultra-high-energy cosmic rays

UV	Ultraviolet (radiation)
V-A	Vector minus axial-vector relational aspect of a theory
VCV	Véron-Cetty Véron (catalogue of galaxies with active galactic nuclei)
VERITAS	Very energetic radiation imaging telescope array system (Cherenkov experiment in the US)
VHE	Very-high-energy (cosmic rays)
VIRGO	Italian-French laser interferometer collaboration at EGO (experiment in Italy)
VLBA	Very long baseline array (of radio telescopes, in the US)
WA#	West area # (experiment at CERN, # standing for its number)
WBF	Weak boson fusion (electroweak process)
WHIPPLE	Cherenkov telescope (in Arizona)
WIMP	Weakly interactive massive particle
WMAP	Wilkinson microwave anisotropy probe (satellite orbiting the Earth)
XCOM	Photon cross-sections database by NIST
XTR	X-ray transition radiation

About the Authors

Alessandro De Angelis is a high energy physicist and astrophysicist. Full Professor at the University of Udine and the Instituto Superior Técnico of the University of Lisbon, he is presently the Director of Research at INFN Padova and Chairman of the Board of the Collaboration Managing the MAGIC gamma-ray telescope at the Northern European Observatory, La Palma, Canary Islands. His main interest of research is fundamental physics, especially astrophysics and elementary particle physics at accelerators. Graduated in Padova, he had been employed at CERN in the 1990s, and has been later among the founding members of the NASA Fermi gamma-ray telescope. His original contributions have been mostly related to calorimetry, advanced trigger systems, QCD, artificial neural networks, and recently to the cosmological propagation of photons. He has been a teacher of electromagnetism and astroparticle physics in Italy and Portugal and a visiting professor in the ICRR of Tokyo, at the Max Planck Institute in Munich, and at the University of Paris VI.

Mário João Martins Pimenta is a high energy physicist and astrophysicist. Full Professor at the Instituto Superior Técnico of the University of Lisbon, he is presently the President of the Laboratório de Instrumentação e Partículas, Coordinator of the International PhD doctoral network IDPASC, and the representative for Portugal at the Pierre Auger Observatory in Argentina. Former member of the WA72, WA74, NA38, and DELPHI experiments at CERN and of the EUSO collaboration at ESA, his main interest of research is high-energy physics, especially cosmic rays of extremely high energy and development of innovative detectors for astroparticle physics. Graduated in Lisbon and in Paris VI, he has been employed at CERN in the late 1980s. His original contributions have been mostly related to advanced trigger systems, search for new particles, hadronic interactions at extremely high energies, and recently to innovative particle detectors. He has been a teacher of general physics and particle physics in Portugal, has lectured at the University of Udine, and has been a visiting professor at SISSA/ISAS in Trieste.

Chapter 1
Understanding the Universe: Cosmology, Astrophysics, Particles, and Their Interactions

Cosmology, astrophysics, and the physics of elementary particles and interactions are intimately connected. After reading this chapter, it will be clear that these subjects are part of the same field of investigation: this book will show you some of the connections, and maybe many more you will discover yourself in the future.

1.1 Particle and Astroparticle Physics

The Universe around us, the objects surrounding us, display an enormous diversity. Is this diversity built over small hidden structures? This interrogation started out, as it often happens, as a philosophical question, only to become, several thousand years later, a scientific one. In the sixth and fifth century BC in India and Greece the atomic concept was proposed: matter was formed by small, invisible, indivisible, and eternal particles: the atoms—a word invented by Leucippus (460 BC) and made popular by his disciple Democritos. In the late eighteenth and early nineteenth century, chemistry gave finally to atomism the status of a scientific theory (mass conservation law, Lavoisier 1789; ideal gas laws, Gay-Lussac 1802; multiple proportional law, Dalton 1805), which was strongly reinforced with the establishment of the periodic table of elements by Mendeleev in 1869—the chemical proprieties of an element depend on a "magic" number, its atomic number.

If atoms did exist, their shape and structure were to be discovered. For Dalton, who lived before the formalization of electromagnetism, atoms had to be able to establish mechanical links with each other. After Maxwell (who formulated the electromagnetic field equations) and J.J. Thomson (who discovered the electron) the binding force was supposed to be the electric one and in atoms an equal number of positive and negative electric charges had to be accommodated in stable configurations. Several solutions were proposed (Fig. 1.1), from the association of small electric dipoles by Philip Lenard (1903) to the Saturnian model of Hantora Nagaoka (1904), where the positive charges were surrounded by the negative ones like the planet Saturn and its rings. In the Anglo-Saxon world the most popular model was, however, the so-called

A. De Angelis and M.J.M. Pimenta, *Introduction to Particle and Astroparticle Physics*, Undergraduate Lecture Notes in Physics,
DOI 10.1007/978-88-470-2688-9_1

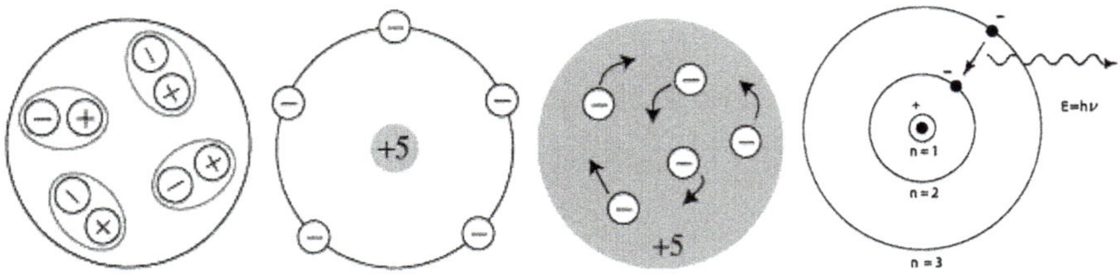

Fig. 1.1 Sketch of the atom according to atomic models by several scientists in the early twentieth century: from *left* to *right*, the Lenard model, the Nagaoka model, the Thomson model, and the Bohr model with the constraints from the Rutherford experiment. *Source* http://skullsinthestars.com/2008/05/27/the-gallery-of-failed-atomic-models-1903-1913/

plum pudding model of Thomson (1904), where the negative charges, the electrons, were immersed in a "soup" of positive charges. This model was clearly dismissed by Rutherford, who demonstrated in the beginning of the twentieth century that the positive charges had to be concentrated in a very small nucleus.

Natural radioactivity was the first way to investigate the intimate structure of matter; then people needed higher energy particles to access smaller distance scales. These particles came again from natural sources: it was discovered in the beginning of the twentieth century that the Earth is bombarded by very high energy particles coming from extraterrestrial sources. These particles were named "cosmic rays." A rich and not predicted spectrum of new particles was discovered. Particle physics, the study of the elementary structure of matter, also called "high energy physics," was born.

High energy physics is somehow synonymous with fundamental physics. The reason is that, due to Heisenberg's[1] principle, the minimum scale of distance Δx we can sample is inversely proportional to the momentum (which approximately equals the ratio of the energy E by the speed of light c for large energies) of the probe we are using for the investigation itself:

$$\Delta x \simeq \frac{\hbar}{p} .$$

[1]Werner Heisenberg (1901–1976) was a German theoretical physicist and was awarded the Nobel Prize in Physics for 1932 "for the creation of quantum mechanics." He also contributed to the theories of hydrodynamics, ferromagnetism, cosmic rays, and subatomic physics. During World War II he worked on atomic research, and after the end of the war he was arrested, then rehabilitated. Finally he organized the Max Planck Institute for Physics, which presently in Munich is entitled to him.

In the above equation, $\hbar = h/2\pi \simeq 10^{-34}$ J s is the so-called Planck[2] constant (sometimes the name of Planck constant is given to h). Accelerating machines, developed in the mid-twentieth century, provided higher and higher energy beams in optimal experimental conditions. The collision point was well-defined and multi-layer detectors could be built around it. Subnuclear particles (quarks) were discovered, and a "Standard Model of Particle Physics" was built, piece by piece, until its final consecration with the recent discovery of the Higgs boson. The energy scale of the TeV (that corresponds to distances down to 10^{-19}–10^{-20} m) is, for the time being, understood.

However, at the end of the twentieth century, the "end of fundamental physics research" announced once again by some, was dramatically dismissed by new and striking experimental evidence which led to the discovery of neutrino oscillations, which meant nonzero neutrino mass, and by the proof that the Universe is in a state of accelerated expansion and that we are immersed in a dark Universe composed mainly of Dark Matter and Dark Energy—whatever those entities, presently unknown to us, are. While the discovery that neutrinos have nonzero mass could be incorporated in the Standard Model by a simple extension, the problem of Dark Matter is still wide open.

The way to our final understanding of the fundamental constituents of the Universe, which we think will occur at energies of 10^{19} GeV (the so-called Planck scale), is hopelessly long. What is worse, despite the enormous progress made by particle acceleration technology, the energies we shall be able to reach at Earth will always be lower than those of the most energetic cosmic rays—particles reaching the Earth from not yet understood extraterrestrial accelerators. These high-energy beams from space may advance our knowledge of fundamental physics and interactions, and of astrophysical phenomena; last but not least, the messengers from space may advance our knowledge of the Universe on large scale, from cosmology to the ultimate quest on the origins of life, astrobiology. That is the domain and the ambition of the new recently born field of fundamental physics, called astroparticle physics. This book is addressed to this field.

Let us start from the fundamental entities: particles and their interactions.

1.2 Particles and Fields

The paradigm which is currently accepted by most researchers, and which is at the basis of the standard model, is that there is a set of elementary particles constituting matter. From a philosophical point of view, even the very issue of the existence of

---

[2]Max Planck (1858–1934) was the originator of quantum theory, and deeply influenced the human understanding of atomic and subatomic processes. Professor in Berlin, he was awarded the Nobel Prize in 1918 "in recognition of the services he rendered to the advancement of Physics by his discovery of energy quanta." Politically aligned with the German nationalistic positions during World War I, Planck was later opposing nazism. Planck's son, Erwin, was arrested attempting assassination of Hitler, and died at the hands of the Gestapo.

elementary particles is far from being established: the concept of elementarity may just depend on the energy scale at which matter is investigated—i.e., ultimately, on the experiment itself. And since we use finite energies, a limit exists to the scale one can probe. The mathematical description of particles, in the modern quantum mechanical view, is that of fields, i.e., of complex amplitudes associated to points in spacetime, to which a local probability can be associated.

Interactions between elementary particles are described by fields representing the forces; in the quantum theory of fields, these fields can be also seen as particles by themselves. In classical mechanics fields were just a mathematical abstraction; the real thing were the forces. The paradigmatic example was Newton's[3] instantaneous and universal gravitation law. Later, Maxwell gave to the electromagnetic field the status of a physical entity: it transports energy and momentum in the form of electromagnetic waves and propagates at a finite velocity—the speed of light. Then, Einstein[4] explained the photoelectric effect postulating the existence of photons—the interaction of the electromagnetic waves with free electrons, as discovered by Compton, was equivalent to elastic collisions between two particles: the photon and the electron. Finally with quantum mechanics the wave-particle duality was extended to all "field" and "matter" particles.

Field particles and matter particles have different behaviors. Whereas matter particles comply with the Pauli[5] exclusion principle—only one single particle can occupy a given quantum state (matter particles obey Fermi-Dirac statistics, and are called "fermions")—there is no limit to the number of identical and indistinguishable field

---

[3]Sir Isaac Newton (1642–1727) was an English physicist, mathematician, astronomer, alchemist, and theologian, who deeply influenced science and culture down to the present days. His monograph Philosophiae Naturalis Principia Mathematica (1687) provided the foundations for classical mechanics. Newton built the first reflecting telescope and developed theories of color and sound. In mathematics, Newton developed differential and integral calculus (independently from Leibnitz). Newton was also deeply involved in occult studies and interpretations of religion.

[4]Albert Einstein (1879–1955) was a German-born physicist who deeply changed the human representation of the Universe, and our concepts of space and time. Although he is best known by the general public for his theories of relativity and for his mass-energy equivalence formula $E = mc^2$ (the main articles on the special theory of relativity and the $E = mc^2$ articles were published in 1905), he received the 1921 Nobel Prize in Physics "especially for his discovery of the law of the photoelectric effect" (also published in 1905), which was fundamental for establishing quantum theory. The young Einstein noticed that Newtonian mechanics could not reconcile the laws of dynamics with the laws of electromagnetism; this led to the development of his special theory of relativity. He realized, however, that the principle of relativity could also be extended to accelerated frames of reference when one was including gravitational fields, which led to his general theory of relativity (1916). Professor in Berlin, he moved to the United States when Adolf Hitler came to power in 1933, becoming a US citizen in 1940. During World War II, he cooperated with the Manhattan Project, which led to the atomic bomb. Later, however, he took a position against nuclear weapons. In the US, Einstein was affiliated to the Institute for Advanced Study in Princeton.

[5]Wolfgang Ernst (the famous physicist Ernst Mach was his godfather) Pauli (Vienna, Austria, 1900—Zurich, Switzerland, 1958) was awarded the 1945 Nobel prize in physics "for the discovery of the exclusion principle, also called the Pauli principle." He also predicted the existence of neutrinos. Professor in ETH Zurich and in Princeton, he had a rich exchange of letters with psychologist Carl Gustav Jung. According to anecdotes, Pauli was a very bad experimentalist, and the ability to break experimental equipment simply by being in the vicinity was called the "Pauli effect".

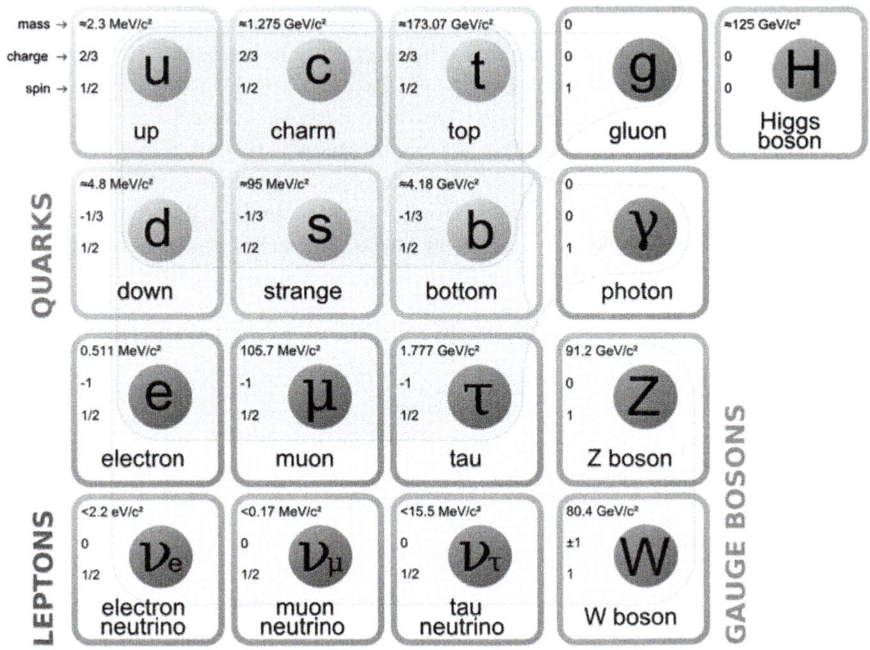

Fig. 1.2 Presently observed elementary particles. Fermions are listed in the first three columns; gauge bosons are listed in the fourth column. Adapted from MissMJ [CC BY 3.0 (http://creativecommons.org/licenses/by/3.0)], via Wikimedia Commons

particles that can occupy the same quantum state (field particles obey Bose–Einstein statistics, and are called "bosons"). Lasers (coherent streams of photons) and the electronic structure of atoms are thus justified. The spin of a particle and the statistics it obeys are connected by the spin-statistics theorem: according to this highly nontrivial theorem, demonstrated by Fierz (1939) and Pauli (1940), fermions have half-integer spins, whereas bosons have integer spins.

At the present energy scales and to our current knowledge, there are 12 elementary "matter" particles; they all have spin 1/2, hence they are fermions. The 12 "matter particles" currently known can be divided into two big families: 6 leptons (e.g., the electron, of charge $-e$, and the neutrino, neutral), and 6 quarks (a state of 3 bound quarks constitutes a nucleon, like the proton or the neutron). Each big family can be divided into three families of two particles each; generations have similar properties—but different masses. This is summarized in Fig. 1.2. A good scale for masses is one GeV/c^2, the proton mass being about 0.938 GeV/c^2. Quarks have fractional charges with respect to the absolute value of the electron charge, e: $\frac{2}{3}e$ for the up, charm, top quark, and $-\frac{1}{3}e$ for the down, strange, top. Quark names are just fantasy names.

The material constituting Earth can be basically explained by only three particles: the electron, the up quark, and the down quark (the proton being made of two up quarks and one down, *uud*, and the neutron by one up and two down, *udd*). To each known particle there is an antiparticle (antimatter) counterpart, with the same mass and opposite charge quantum numbers.

At the current energy scales of the Universe particles interact via four fundamental interactions. There are indications that this view is related to the present-day energy of the Universe: at higher energies—i.e., earlier epochs—some interactions would "unify" and the picture would become simpler. In fact, theorists think that these interactions might be the remnant of one single interaction that would occur at extreme energies—e.g., the energies typical of the beginning of the Universe. By increasing order of strength:

Quanta of the interaction fields	
Strong interaction	Eight gluons
Electromagnetic interaction	Photon (γ)
Weak interaction	Bosons W^+, W^-, Z
Gravitational interaction	Graviton (?)

1. The gravitational interaction, acting between whatever pair of bodies and dominant at macroscopic scales.
2. The weak interaction, also affecting all matter particles (with certain selection rules) and responsible, for instance, for the beta decay and thus for the energy production in the Sun.
3. The electromagnetic interaction, acting between pairs of electrically charged particles (i.e., all the matter particles, excluding neutrinos).
4. The color force, acting among quarks. The strong interaction,[6] responsible for binding the atomic nuclei (it ensures electromagnetic repulsion among protons in nuclei does not break them up) and for the interaction of cosmic protons with the atmosphere, is just a van der Waals shadow of the very strong interaction between quarks.

The relative intensity of such interactions spans many orders of magnitude. In a ^{2}H atom, in a scale where the intensity of strong interactions between the nucleons is 1, the intensity of electromagnetic interactions between electrons and the nucleus is 10^{-5}, the intensity of weak interactions is 10^{-13}, and the intensity of gravitational interactions between the electron and the nucleus is 10^{-45}. In the quantum mechanical view of interaction, interaction itself is mediated by quanta of the force field.

[6]This kind of interaction was first conjectured and named by Isaac Newton at the end of the seventeenth century: "There are therefore agents in Nature able to make the particles of bodies stick together by very strong attractions. And it is the business of experimental philosophy to find them out. Now the smallest particles of matter may cohere by the strongest attractions, and compose bigger particles of weaker virtue; and many of these may cohere and compose bigger particles whose virtue is still weaker, and so on for divers successions, until the progression ends in the biggest particles on which the operations in chemistry, and the colors of natural bodies depend." (I. Newton, Opticks)

According to most scientists, the gravitational interaction is mediated by the graviton, an electrically neutral boson of mass 0 and spin 2, yet undiscovered. The weak interaction is mediated by three vectors: two are charged, the W^+ (of mass ∼80.4 GeV/c^2) and its antiparticle, the W^-; one is neutral, the Z (with mass ∼91.2 GeV/c^2). The electromagnetic interaction is mediated by the well-known photon. The color interaction is exchanged by eight massless neutral particles called gluons. The couplings of each particle to the boson(s) associated to a given interaction are determined by the strength of the interaction and by "magic" numbers, called charges. The gravitational charge of a particle is proportional to its mass (energy); the weak charge is the weak isospin charge ($\pm 1/2$ for fermions, 0, ± 1 for bosons); the electrical charge is the well-known (positive and negative) charge; the strong charge comes in three types designated by color names (red, green, blue). Particles or combinations of particles can be neutral to the electromagnetic, weak or strong interaction, but not to the gravitational interaction. For instance, electrons have electric and weak charges but no color charge, and atoms are electrically neutral. At astrophysical scales, the dominant interaction is gravitation; at atomic scales, $\mathcal{O}(1\,\text{nm})$, it is the electromagnetic interaction; and at the scale of nuclei, $\mathcal{O}(1\text{fm})$, it is the strong interaction.

In quantum physics the vacuum is not empty at all. Heisenberg's uncertainty relations allow energy conservation violations by a quantity ΔE within small time intervals Δt such that $\Delta t \simeq \hbar/\Delta E$. Massive particles that live in such tiny time intervals are called "virtual." But, besides such particles which are at the origin of measurable effects (like the Casimir effect, see Chap. 6), we have just discovered that space is filled by an extra field to which is associated the Higgs boson, a neutral spinless particle with mass about 125 GeV/c^2. Particles in the present theory are intrinsically massless, and is their interaction with the Higgs field that originates their mass: the physical properties of particles are related to the properties of the quantum vacuum.

1.3 A Quick Look at the Universe

The origin and destiny of the Universe is, for most researchers, the fundamental question. Many answers were provided over the ages, a few of them built over scientific observations and reasoning. Over the last century enormous scientific theoretical and experimental breakthroughs have occurred: less than a century ago, people believed that the Milky Way, our own galaxy, was the only galaxy in the Universe; now we know that there are 10^{11} galaxies within the observable universe, each containing some 10^{11} stars. Most of them are so far, that we can not even hope to explore them.

Let us start an imaginary trip across the Universe from the Earth. The Earth, which has a radius of about 6400 km, is one of the planets orbiting around the Sun. The latter is a star with a mass of about 2×10^{30} kg located at a distance from us of about 150 million km (i.e., 500 light seconds). We call the average Earth-Sun distance the astronomical unit, AU. The ensemble of planets orbiting the Sun is called the Solar

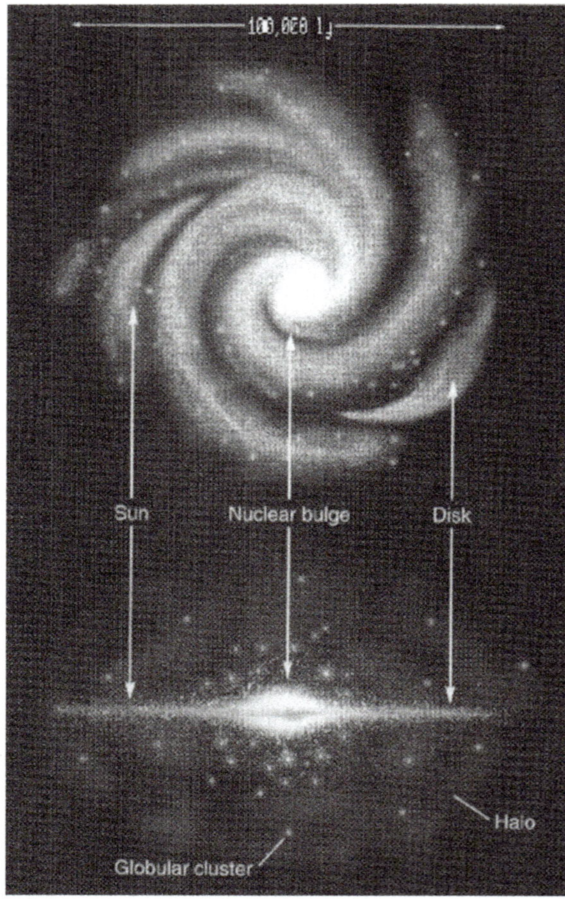

Fig. 1.3 The Milky Way seen from top and from side. *Source* http://casswww.ucsd.edu/archive/public/tutorial/mw.html

System. Looking to the aphelion of the orbit of the farthest acknowledged planet, Neptune, the Solar System has a diameter of 9 billion km (about 10 light hours, or 60 AU).

The Milky Way (Fig. 1.3) is the galaxy that contains our Solar System. Its name "milky" is derived from its appearance as a dim glowing band arching across the night sky in which the naked eye cannot distinguish individual stars. The ancient Romans named it "via lactea," which literally corresponds to the present name (being *lac* the latin word for milk)—the term "galaxy," too, descends from a Greek word indicating milk. Seen from Earth with the unaided eye, the Milky Way appears as a band because its disk-shaped structure is viewed edge-on from the periphery of the Galaxy itself.

Galilei[7] first resolved such band of light into individual stars with his telescope, in 1610.

The Milky Way is a spiral galaxy some 100 000 light-years (ly) across. The Solar System is located within the disk, about 30 000 light-years away from the Galactic Center, in the so-called Orion arm. The stars in the inner 10 000 light-years form a bulge and one or more bars that radiate from the bulge. The very center of the Galaxy, in the constellation of Sagittarius, hosts a supermassive black hole of some 4 million solar masses (this value has been determined precisely by studying the orbits of nearby stars).

The Milky Way is a relatively large galaxy. Teaming up with a similar-sized partner (called the Andromeda galaxy), it has gravitationally trapped many smaller galaxies: together, they all constitute the so-called Local Group. The Local Group comprises more than 50 galaxies, including numerous dwarf galaxies—some are just spherical collections of hundreds of stars that are called globular clusters. Its gravitational center is located somewhere between the Milky Way and the Andromeda galaxies. The Local Group covers a diameter of 10 million light-years, or 10 Mly (i.e., 3.1 megaparsec,[8] Mpc); it has a total mass of about 10^{12} solar masses.

Galaxies are not uniformly distributed; most of them are arranged into groups (containing some dozens of galaxies) and clusters (up to several thousand galaxies); groups and clusters and additional isolated galaxies form even larger structures called superclusters that may span up to 100 Mly.

This is how far our observations can go.

In 1929 the American astronomer Edwin Hubble, studying the emission of radiation from galaxies compared their speed (calculated from the Doppler shift of their emission lines) with the distance (Fig. 1.4), and discovered that objects in the Universe move away from us with velocity

---

[7]Galileo Galilei (1564–1642) was an Italian physicist, mathematician, astronomer, and philosopher who deeply influenced scientific thoughts down to the present days. He first formulated some of the fundamental laws of mechanics, like the principle of inertia and the law of accelerated motion; he formally proposed, with some influence from previous works by Giordano Bruno, the principle of relativity. Galilei was professor in Padua, nominated by the Republic of Venezia, and astronomer in Firenze. He built the first practical telescope (using lenses) and using this instrument he could perform astronomical observations which supported Copernicanism; in particular he discovered the phases of Venus, the four largest satellites of Jupiter (named the Galilean moons in his honor), and he observed and analyzed sunspots. Galilei also made major discoveries in military science and technology. He came into conflict with the catholic church, for his support of Copernican theories. In 1616 the Inquisition declared heliocentrism to be heretical, and Galilei was ordered to refrain from teaching heliocentric ideas. In 1616 Galilei argued that tides were an additional evidence for the motion of the Earth. In 1933 the Roman Inquisition found Galilei suspect of heresy, sentencing him to indefinite imprisonment; he was kept under house arrest in Arcetri near Firenze until his death.

[8]The parsec (symbol: pc, and meaning "parallax of one arcsecond") is often used in astronomy to measure distances to objects outside the Solar System. It is defined as the length of the longer leg of a right triangle, whose shorter leg corresponds to one astronomical unit, and the subtended angle of the vertex opposite to that leg is one arcsecond. It corresponds to approximately 3×10^{16} m, or about 3.26 light-years. Proxima Centauri, the nearest star, is about 1.3 pc from the Sun.

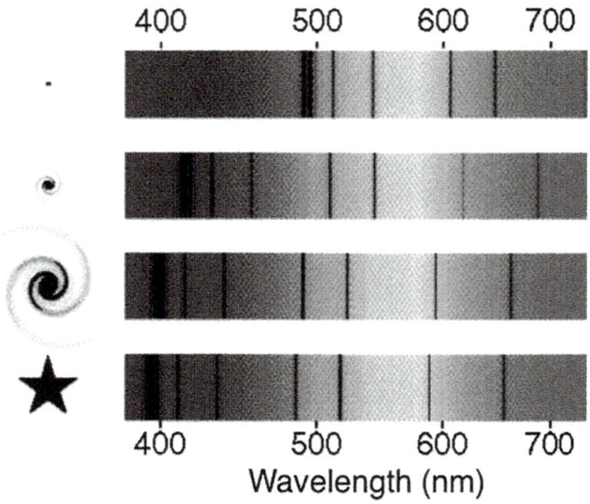

Fig. 1.4 Redshift of emission spectrum of stars and galaxies at different distances. A star in our galaxy is shown at the bottom left with its spectrum on the bottom right. The spectrum shows the dark absorption lines, which can be used to identify the chemical elements involved. The other three spectra and pictures from bottom to top show a nearby galaxy, a medium distance galaxy, and a distant galaxy. Using the redshift we can calculate the relative radial velocity between these objects and the Earth. From http://www.indiana.edu/~geol105

$$v = H_0 d\,, \tag{1.1}$$

where d is the distance between the objects, and H_0 is a parameter called the Hubble constant (whose value is known today to be about 68 km $s^{-1}Mpc^{-1}$, i.e., 21 km $s^{-1}Mly^{-1}$). The above relation is called Hubble's law (Fig. 1.5). Note that at that time galaxies beyond the Milky Way had just been discovered.

The Hubble law means that sources at cosmological distances (where local motions, often resulting from galaxies being in gravitationally bound states, are negligible) are observed to move away at speeds that are proportionally higher for larger distances. The Hubble constant describes the rate of increase of recession velocities for increasing distance. The Doppler redshift

$$z = \lambda'/\lambda - 1$$

can thus also be used as a metric of the distance of objects. To give an idea of what H_0 means, the speed of revolution of the Earth around the Sun is about 30 km/s. Andromeda, the large galaxy closest to the Milky Way, has a distance of about 2.5 Mly from us—however we and Andromeda are indeed approaching: this is an example of the effect of local motions.

Dimensionally, we note that H_0 is the inverse of a time: $H_0 \simeq (14 \times 10^9 \text{ years})^{-1}$. A simple interpretation of the Hubble law is that, if the Universe had always been expanding at a constant rate, about 14 billion years ago its volume was zero—naively,

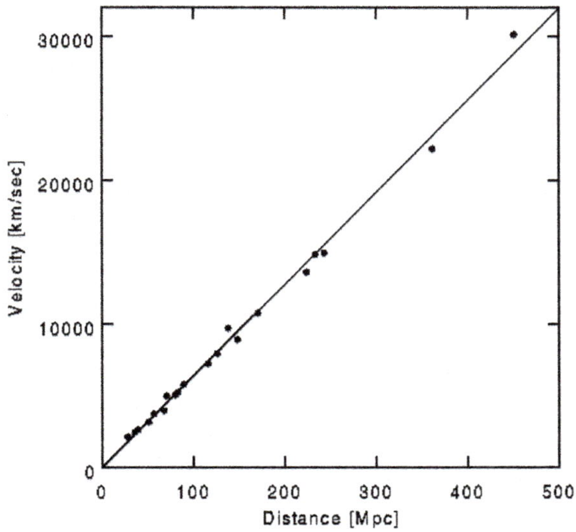

Fig. 1.5 Experimental plot of the relative velocity (in km/s) of known astrophysical objects as a function of distance from Earth (in Mpc). Several methods are used to determine the distances. Distances up to hundreds of parsecs are measured using stellar parallax (i.e., the difference between the angular positions from the Earth with a time delay of 6 months). Distances up to 50 Mpc are measured using cepheids, periodically pulsating stars for which the luminosity is related to the pulsation period (the distance can thus be inferred by comparing the intrinsic luminosity with the apparent luminosity). Finally, distances from 1 to 1000 Mpc can be measured with another type of standard candle, Type Ia supernova, a class of remnants of imploded stars. In between (from 15 to 200 Mpc), the Tully-Fisher relation, an empirical relationship between the intrinsic luminosity of a spiral galaxy and the width of its emission lines (a measure of its rotation velocity), can be used. The methods, having large superposition regions, can be cross-calibrated. The line is a Hubble-law fit to the data. From A. G. Riess, W. H. Press and R. P. Kirshner, Astrophys. J. 473 (1996) 88

we can think that it exploded through a quantum singularity, such an explosion being usually called the "Big Bang." This age is consistent with present estimates of the age of the Universe within gravitational theories, which we shall discuss later in this book, and slightly larger than the age of the oldest stars, which can be measured from the presence of heavy nuclei. The picture looks consistent.

The adiabatic expansion of the Universe entails a freezing with expansion time t, which in the nowadays quiet Universe can be summarized as a law for the evolution of the temperature T,

$$T \simeq \frac{30000\,\mathrm{K}}{\sqrt{t\ (\mathrm{years})}} .$$

The present temperature is slightly less than 3 K, and can be measured from the spectrum of the blackbody (microwave) radiation (the so-called cosmic microwave background, or CMB, permeating the Universe). The formula implies also that studying the ancient Universe in some sense means exploring the high energy world: subatomic physics and astrophysics are naturally connected.

At epochs corresponding to fractions of a second after the Big Bang, tiny quantum fluctuations in the distribution of cosmic matter led to galaxy formation. Density fluctuations grew with time into proto-structures which, after accreting enough mass from their surroundings, overcame the pull of the expanding universe and after the end of an initial era dominated by radiation collapsed into bound, stable structures. The average density of such structures was reminiscent of the average density of the Universe when they broke away from the Hubble expansion: so, earlier-forming structures have a higher mean density than later-forming structures. Proto-galaxies were initially dark. Only later, when enough gas had fallen into their potential well, stars started to form—again, by gravitational instability in the gas—and shine due to the nuclear fusion processes activated by the high temperatures cause by gravitational forces. The big picture of the process of galaxy formation is probably understood by now, but the details are not. The morphological difference between disk (i.e., spiral) galaxies and spheroidal (i.e., elliptical) galaxies are interpreted as due to the competition between the characteristic timescale of the infall of gas into the protogalaxy's gravitational well and the timescale of star formation: if the latter is shorter than the former, a spheroidal (i.e., three-dimensional) galaxy likely forms; if it is longer, a disk (i.e., two-dimensional) galaxy forms. A disk galaxy is rotation supported, whereas a spheroidal galaxy is pressure supported—stars behaving in this case like gas molecules. It is conjectured that the velocity dispersion ($\sim$200 km/s) among proto-galaxies in the earlier universe may have triggered rotation motions in disk galaxies, randomly among galaxies but orderly within individual galaxies.

Stars also formed by gravitational instabilities of the gas. For given conditions of density and temperature, gas (mostly hydrogen and helium) clouds collapse and, if their mass is suitable, eventually form stars. Stellar masses are limited by the conditions that (i) nuclear reactions can switch on in the stellar core ($>$0.1 solar masses), and (ii) the radiation drag of the produced luminosity on the plasma does not disrupt the star's structure ($<$100 solar masses). For a star of the mass of the Sun, formation takes 50 million years—the total lifetime is about 11 billion years before collapsing to a "white dwarf," and in the case of our Sun some 4.5 billion years are already gone.

Stars span a wide range of luminosities and colors, and can be classified according to these characteristics. The smallest stars, known as red dwarfs, may contain as little as 10 % the mass of the Sun and emit only 0.01 % as much energy, having typical surface temperatures of 3000 K. Red dwarfs are by far the most numerous stars in the Universe and have lifetimes of tens of billions of years, much larger than the age of the Universe. On the other hand, the most massive stars, known as hypergiants, may be 100 or more times more massive than the Sun, and have surface temperatures of more than 30 000 K. Hypergiants emit hundreds of thousands of times more energy than the Sun, but have lifetimes of only a few million years. They are thus extremely rare today and the Milky Way galaxy contains only a handful of them.

Luminosity and temperature of a star are in general linked. In the temperature-luminosity plane, stars populate a locus that can be described (in log scale) as a straight line: this is called the *main sequence* of stars. Our Sun is also found there—corresponding to very average temperature and luminosity.

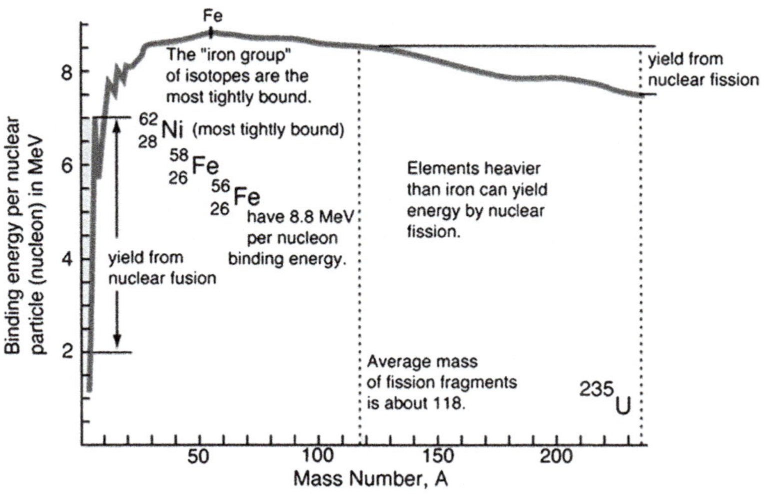

Fig. 1.6 Binding energy per nucleon for stable atoms. Iron (^{56}Fe) is the stable element for which the binding energy per nucleon is the largest (about 8.8 MeV); it is thus the natural endpoint of processes of fusion of lighter elements, and of fission of heavier elements (although ^{58}Fe and ^{56}Ni have a slightly higher binding energy, by less than 0.05 %, they are subject to nuclear photodisintegration). From http://hyperphysics.phy-astr.gsu.edu/

The fate of a star depends on its mass. The heavier is the star, the larger is its gravitational energy, the more effective are the nuclear processes powering it. In average stars like the Sun, the outer layers are supported against gravity until the stellar core stops producing fusion energy; then the star collapses into a "white dwarf"—an Earth-sized object. Main-sequence stars with mass over 8 solar masses die in a very energetic explosion called a supernova. In a supernova, the star's core, made by iron (which being the most stable atom, i.e., one whose mass defect per nucleon is maximum, is the endpoint of nuclear fusion processes, Fig. 1.6) collapses and the released gravitational energy goes into heating the overlying mass layers which, in an attempt to dissipate the sudden excess heat by increasing the star's radiating surface, expand at high speed (10 000 km/s and more) to the point that the star gets quickly disrupted—i.e., explodes. Supernovae release an enormous amount of energy, about 10^{46} J—mostly in neutrinos from the nuclear processes occurring in the core, and just 1 % in kinetic energies of the ejecta—in a few tens of seconds.[9] For a period of days to weeks, a supernova may outshine its entire host galaxy. Being the energy of the explosion large enough to generate hadronic interactions, basically any element and many subatomic particles are produced in these explosions. On average, in a typical galaxy (e.g., the Milky Way) supernova explosions occur just once or twice per century. Supernovae leave behind neutron stars or black holes.

[9]Note that frequently astrophysicist use as a unit of energy the old "cgs" (centimeter-gram-second) unit called *erg*; 1 erg = 10^{-7} J.

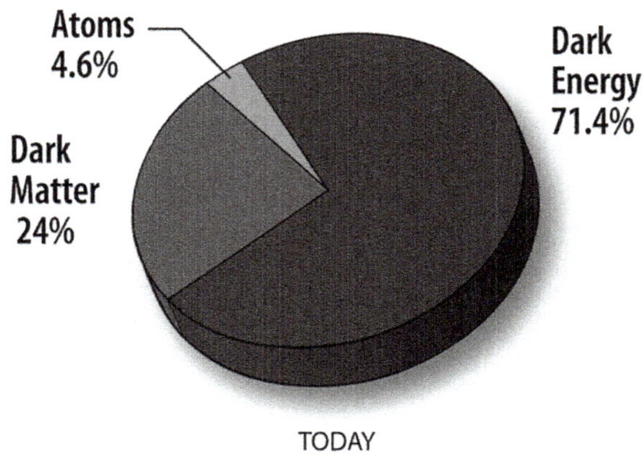

Fig. 1.7 Present energy budget of the Universe. Credit: NASA

The heavier the star, the most effective the fusion process, the shortest the lifetime. We need a star like our Sun, having a lifetime of a few tens of billion years, to both give enough time to life to develop and to guarantee large enough temperatures for humans. The Solar System is estimated to be some 4.6 billion years old and to have started from a molecular cloud. Most of the collapsing mass collected in the center, forming the Sun, while the rest flattened into a disk out of which the planets formed. The Sun is too young to have created heavy elements in such an abundance to justify carbon-based life on Earth. The carbon, nitrogen, and oxygen atoms in our bodies, as well as atoms of all other heavy elements, were created in previous generations of stars somewhere in the Universe.

The study of stellar motions in galaxies indicates the presence of a large amount of unseen mass in the Universe. This mass seems to be of a kind presently unknown to us; it neither emits nor absorbs electromagnetic radiation (including visible light) at any significant level. We call it *dark matter*: its abundance in the Universe amounts to an order of magnitude more than the conventional matter we are made of. Dark matter represents one of the greatest current mysteries of astroparticle physics. Indications exist also of a further form of energy, which we call *dark energy*. Dark energy contributes to the total energy budget of the Universe three times more than dark matter.

In summary, we live in a world that is mostly unknown even from the point of view of the nature of its main constituents (Fig. 1.7). The evolution of the Universe and our everyday life depend on this unknown external world. First of all, the ultimate destiny of the Universe—a perpetual expansion or a recollapse—depends on the amount of all the matter in the Universe. Moreover, every second high-energy particles (i.e., above 1 GeV) of extraterrestrial origin pass through each square centimeter on the Earth, and they are messengers from regions where highly energetic phenomena take place that we cannot directly explore. These are the so-called *cosmic rays*, discovered in the beginning of the nineteenth century (see Chap. 3). It is natural to try to use these messengers in order to obtain information on the highest energy events occurring in the Universe.

1.4 Cosmic Rays

The distribution in energy (the so-called spectrum) of cosmic rays is quite well described by a power law E^{-p} with p a positive number (Fig. 1.8). The spectral index p is around 3 in average. After the low energy region dominated by cosmic rays from the Sun (the solar wind), the spectrum becomes steeper for energy values of less than ~1000 TeV (150 times the maximum energy foreseen for the beams of the LHC collider at CERN): this is the energy region that we know to be dominated

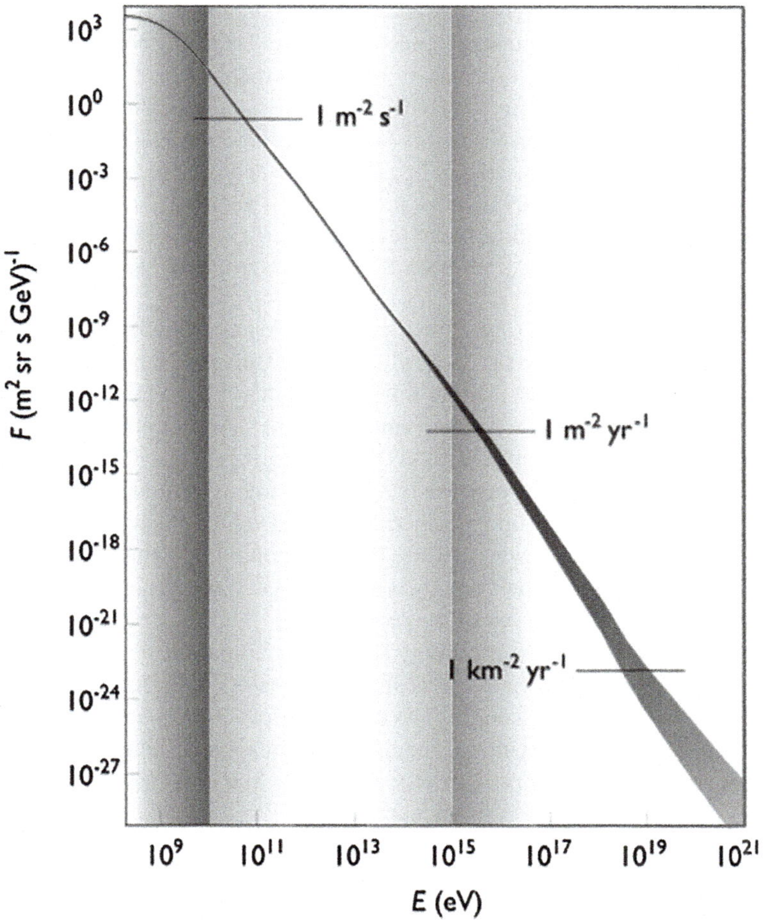

Fig. 1.8 The energy spectrum (number of incident particles per unit of energy, per second, per unit area, and per unit of solid angle) of the primary cosmic rays. The vertical band on the left indicates the energy region in which the emission from the Sun is supposed to be dominant; the central band the region in which most of the emission is presumably of galactic origin; the band on the right the region of extragalactic origin. By Sven Lafebre (own work) [GFDL http://www.gnu.org/copyleft/fdl.html], via Wikimedia Commons.

by cosmic rays produced by astrophysical sources in our galaxy, the Milky Way. For higher energies a further steepening occurs, the point at which this change of slope takes place being called the "knee"; we believe that the region above this energy is dominated by cosmic rays produced by extragalactic sources, mostly supermassive black holes growing at the centers of other galaxies. For even higher energies (more than one million TeV) the cosmic-ray spectrum becomes less steep, resulting in another change of slope, called the "ankle." About possible sources at such large energies, we have no clues. Finally, at the highest energies in the figure a drastic suppression is present—as expected from the interaction of long-traveling particles with the cosmic microwave background, remnant of the origin of the Universe.[10]

The majority of high-energy particles in cosmic rays are protons (hydrogen nuclei); about 10 % are helium nuclei (nuclear physicists usually call them alpha particles), and 1 % are neutrons or nuclei of heavier elements. Together, these account for 99 % of the cosmic rays, and electrons and photons make up the remaining 1 %. The number of neutrinos is estimated to be comparable to that of high-energy photons, but it is very high at low energy because of the nuclear processes that occur in the Sun: such processes involve a large production of neutrinos.

Cosmic rays hitting the atmosphere (called primary cosmic rays) generally produce secondary particles that can reach the Earth's surface, through multiplicative showers.

About once per minute, a single subatomic particle enters the Earth's atmosphere with an energy larger than 10 J. Somewhere in the Universe there are accelerators that can impart energies 100 million times larger than the energy reached by the most powerful accelerators on Earth to single protons. It is thought that the ultimate engine of the acceleration of cosmic rays is gravity. In gigantic gravitational collapses, such as those occurring in supernovae (stars imploding at the end of their lives, see Fig. 1.9, left) and in the accretion of supermassive black holes (equivalent to millions to billions of solar masses) at the expense of the surrounding matter (Fig. 1.9, right), part of the potential gravitational energy is transformed, through not fully understood mechanisms, into kinetic energy of the particles.

The reason why the maximum energy attained by human-made accelerators cannot compete with the still mysterious cosmic accelerators with the presently known acceleration technologies is simple. The most efficient way to accelerate particles requires their confinement within a radius R by a magnetic field B, and the final energy is proportional to the product $R \times B$. On Earth, it is difficult to imagine rea-

---

[10] A theoretical upper limit on the energy of cosmic rays from distant sources was computed in 1966 by Greisen, Kuzmin, and Zatsepin, and it is called today the GZK cutoff. Protons with energies above a threshold of about 10^{20} eV suffer a resonant interaction with the cosmic microwave background photons to produce pions through the formation of a short-lived particle (resonance) called Δ: $p + \gamma \to \Delta \to N + \pi$. This continues until their energy falls below the production threshold. Because of the mean path associated with the interaction, extragalactic cosmic rays from distances larger than 50 Mpc from the Earth and with energies greater than this threshold energy should be strongly suppressed on Earth, and there are no known sources within this distance that could produce them. A similar effect (nuclear photodisintegration) limits the mean free path for the propagation of nuclei heavier than the proton.

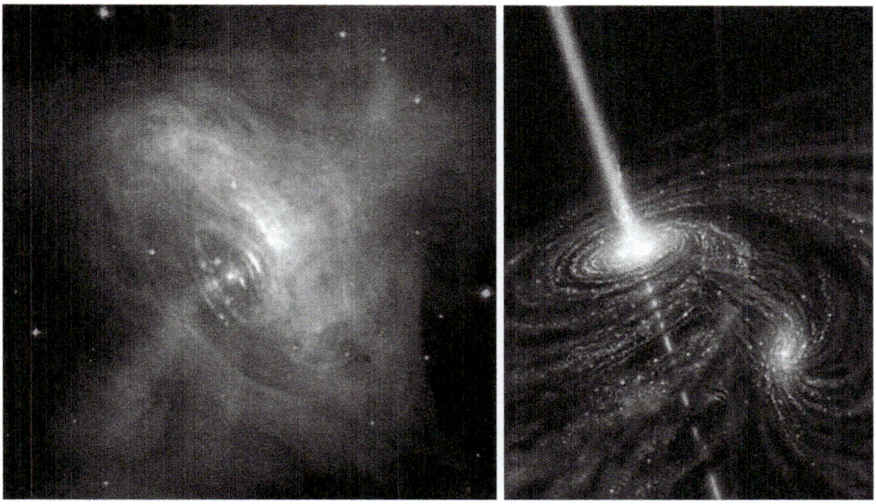

Fig. 1.9 *Left* The remnant of the supernova in the Crab region (Crab nebula), a powerful gamma emitter in our galaxy. The supernova exploded in 1054 and the phenomenon was recorded by Chinese astronomers. For 40 years, until 2010, most astronomers regarded the Crab as a standard candle for high energy photon emission, but recently it was discovered that Crab Nebula from time to time flickers. Anyway, most plots of sensitivity of detectors refer to a "standard Crab" as a reference unit. The vortex around the center is visible; a neutron star rapidly rotating (with a period of around 30 ms) and emitting pulsed gamma-ray streams (pulsar) powers the system. Some supernova remnants, seen by Earth, have an apparent dimension of a few tenths of degree—about the dimension of the Moon. *Right* A supermassive black hole accretes swallowing neighboring stellar bodies and molecular clouds, and emits jets of charged particles and gamma rays. Credits: NASA

sonable confinement radii greater than one hundred kilometers, and magnetic fields stronger than 10 tesla (i.e., one hundred thousand times the Earth's magnetic field). This combination can provide energies of a few tens of TeV, such as those of the LHC accelerator at CERN. In nature accelerators with much larger radii exist, such as supernova remnants (light-years) and active galactic nuclei (tens of thousands of light-years). Of course human-made accelerators have important advantages, as the flux and the possibility of knowing the initial conditions (cosmic ray researches do not know a-priori the initial conditions of their phenomena).

Among cosmic rays, photons are particularly important. The gamma photons (called gamma rays for historical reasons) are photons of very high energy, and occupy the most energetic part of the light spectrum; being neutral they can travel long distances without being deflected by galactic and extragalactic magnetic fields, hence they allow us to directly study their emission sources. These facts are now pushing us to study in particular the high-energy gamma rays and cosmic rays of hundreds of millions of TeV. However, gamma rays are less numerous than charged cosmic rays of the same energy, and the energy spectrum of charged cosmic rays is such that particles of hundreds of millions of TeV are very rare. The task of

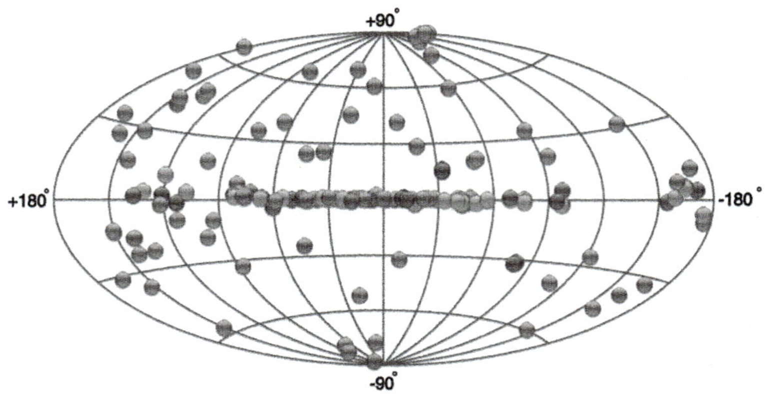

Fig. 1.10 Map of the emitters of photons above 100 GeV in the Universe, in galactic coordinates (from the TeVCAT catalog). The sources are indicated as circles—the colors represent different kinds of emitters which will be explained in Chap. 10. From http://tevcat.uchicago.edu/

experimental physics is, as usual, challenging, and often discoveries correspond to breakthroughs in detector techniques.

A sky map of the emitters of very-high energy photons in galactic coordinates[11] is shown in Fig. 1.10. One can identify both galactic emitters (in the equatorial plane) and extragalactic emitters.

[11]Usually the planar representations of maps of the Universe are done in galactic coordinates. To understand what this means, let us start from a celestial coordinate system in spherical coordinates, in which the Sun is at the center, the primary direction is the one joining the Sun with the center of the Milky Way, and the galactic plane is the fundamental plane. Coordinates are positive toward North and East in the fundamental plane.

We define as galactic longitude (l or λ) the angle between the projection of the object in the galactic plane and the primary direction. Latitude (symbol b or ϕ) is the angular distance between the object and the galactic plane. For example, the North galactic pole has a latitude of +90°.

Plots in galactic coordinates are then projected onto a plane, typically using an elliptical (Mollweide) projection preserving areas. This projection transforms latitude and longitude to plane coordinates x and y via the equations (angles are expressed in radiants):

$$x = R\frac{2\sqrt{2}}{\pi}\cos\theta$$
$$y = R\sqrt{2}\sin\theta\,,$$

where θ is defined by the equation

$$2\theta + \sin(2\theta) = \pi\sin\phi$$

and R is the radius of the sphere to be projected. The map has area $4\pi R^2$, obviously equal to the surface area of the generating globe. The x-coordinate has a range $[-2R\sqrt{2}, 2R\sqrt{2}]$, and the y-coordinate has a range $[-R\sqrt{2}, R\sqrt{2}]$. The Galactic center is located at (0,0).

Further Reading

[F1.1] A. Einstein and L. Infeld, "The Evolution of Physics," Touchstone. This inspiring book is about the main ideas in physics. With simplicity and a limited amount of formulas it gives an exciting account for the advancement of science down to the early quantum theory.
[F1.2] L. Lederman and D. Teresi, "The God Particle: If the Universe Is the Answer, What Is the Question?", Dell. This book provides a history of particle physics starting from Greek philosophers down to modern quantum physics.
[F1.3] G. Smoot and K. Davidson, "Wrinkles in time," Harper. This book discusses modern cosmology in a simple way.

Exercises

1. *The Universe.* Find a dark place close to where you live, and go there in the night. Try to locate the Milky Way and the Galactic center. Comment on your success (or insuccess).
2. *Telescopes.* Make a research on the differences between Newtonian and Galileian telescopes; discuss such a difference.
3. *Size of an atom.* Explain how you will be able to find the order of magnitude of an atom size using a drop of oil. Make the experiment and check the result.
4. *Thomson atom.* Consider the Thomson atom model applied to a helium atom (the two electrons are in equilibrium inside a homogeneous positive charged sphere of radius $r \sim 10^{-10}$ m).

   (a) Determine the distance of the electrons to the center of the sphere.
   (b) Determine the vibration frequency of the electrons in this model and compare it to the first line of the spectrum of hydrogen.

5. *Atom as a box.* Consider a simplified model where the hydrogen atom is described by a one dimensional box of length r with the proton at its center and where the electron is free to move around. Compute, taking into account the Heisenberg uncertainty principle, the total energy of the electron as a function of r and determine the value of r for which this energy is minimized.
6. *Cosmic ray fluxes and wavelength.* The most energetic particles ever observed at Earth are cosmic rays. Make an estimation of the number of such events with an energy between 3×10^{18} and 10^{19} eV that may be detected in one year by an experiment with a footprint of 1000 km^2. Evaluate the structure scale that can be probed by such particles.
7. *Energy from cosmic rays: Nikola Tesla's "free" energy generator.* "This new power for the driving of the world's machinery will be derived from the energy which operates the universe, the cosmic energy, whose central source for the Earth is the Sun and which is everywhere present in unlimited quantities." Immediately

after the discovery of natural radioactivity, in 1901, Nikola Tesla patented an engine using the energy involved (and expressed a conjecture about the origin of such radioactivity). As below, we show a drawing (made by Tesla himself) of Tesla's first radiant energy receiver. If an antenna (the higher the better: why?) is wired to one side of a capacitor (the other going to ground,) the potential difference will charge the capacitor. Suppose you can intercept all high-energy cosmic radiation (assume 1 particle per square centimeter per second with an average energy of 3 GeV); what is the power you could collect with a 1 m^2 antenna, and how does it compare with solar energy?

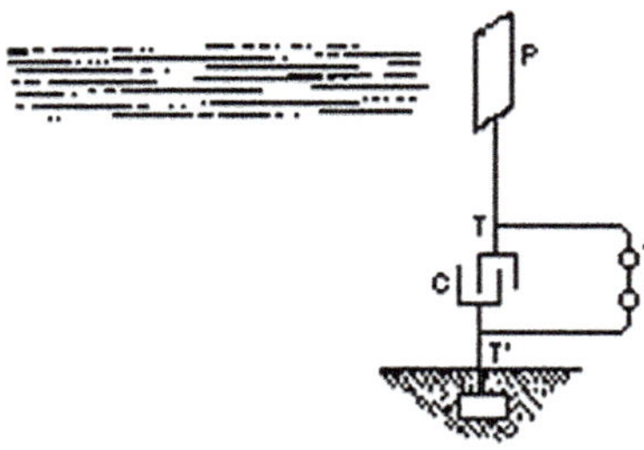

Chapter 2
The Birth and the Basics of Particle Physics

This chapter introduces the basics of the techniques for the study of the intimate structure of matter, described in a historical context. After reading this chapter, you should understand the basic tools which lead to the investigation and the description of the subatomic structure, and you should be able to compute the interaction probabilities of particles. A short reminder of the concepts of special relativity needed to understand astroparticle physics is also provided.

2.1 The Atom

In the second half of the nineteenth century, the work by Mendeleev[1] on the periodic table of the elements provided the paradigm that paved the way to the experimental demonstration of the atomic structure. The periodic table is an arrangement of the chemical elements. Mendeleev realized that the physical and chemical properties of elements are related to their atomic mass in a quasi-periodic way. He ordered the 63 elements known at his time according to their atomic mass and arranged them in a table so that elements with similar properties would be in the same column. Figure 2.1 shows this arrangement. Hydrogen, the lightest element, is isolated in the first row of the table. The following light elements are then disposed in octets. Mendeleev found some gaps in his table, and predicted that elements then unknown would be discovered which would fill these gaps: his predictions were successful.

Mendeleev's periodic table has been expanded and refined with the discovery of new elements and a better theoretical understanding of chemistry. The most important

[1]Dimitri Mendeleev (1834–1907) was a Russian chemist born in Tobolsk, Siberia. He studied science in St. Petersburg, where he graduated in 1856 and became full professor in 1863. Mendeleev is best known for his work on the periodic table, published in Principles of Chemistry in 1869, but also, according to a myth popular in Russia, for establishing that the minimum alcoholic fraction of vodka should be 40 %—this requirement was easy to verify, as this is the minimum content at which an alcoholic solution can be ignited at room temperature.

A. De Angelis and M.J.M. Pimenta, *Introduction to Particle and Astroparticle Physics*, Undergraduate Lecture Notes in Physics,
DOI 10.1007/978-88-470-2688-9_2

Reihen	Gruppe I. — R^2O	Gruppe II. — RO	Gruppe III. — R^2O^3	Gruppe IV. RH^4 RO^2	Gruppe V. RH^3 R^2O^5	Gruppe VI. RH^2 RO^3	Gruppe VII. RH R^2O^7	Gruppe VIII. — RO^4
1	H=1							
2	Li=7	Be=9,4	B=11	C=12	N=14	O=16	F=19	
3	Na=23	Mg=24	Al=27,3	Si=28	P=31	S=32	Cl=35,5	
4	K=39	Ca=40	—=44	Ti=48	V=51	Cr=52	Mn=55	Fe=56, Co=59, Ni=59, Cu=63
5	(Cu=63)	Zn=65	—=68	—=72	As=75	Se=78	Br=80	
6	Rb=85	Sr=87	?Yt=88	Zr=90	Nb=94	Mo=96	—=100	Ru=104, Rh=104, Pd=106, Ag=108
7	(Ag=108)	Cd=112	In=113	Sn=118	Sb=122	Te=125	J=127	
8	Cs=133	Ba=137	?Di=138	?Ce=140	—	—	—	— — — —
9	(—)	—	—	—	—	—	—	
10	—	—	?Er=178	?La=180	Ta=182	W=184	—	Os=195, Ir=197, Pt=198, Au=199
11	(Au=199)	Hg=200	Tl=204	Pb=207	Bi=208	—	—	
12	—	—	—	Th=231	—	U=240	—	— — — —

Fig. 2.1 Mendeleev's periodic table as published in Annalen der Chemie 1872 [public domain]. The noble gases had not yet been discovered, and are thus not displayed

modification was the use of atomic number (the number of electrons, which indeed characterizes an element) instead of atomic mass to order the elements. Since atoms are neutral, the same number of positive charges (protons) should be present. Starting from the element with atomic number 3, it had been conjectured by Mendeleev that electrons are disposed in shells. The n–th shell is complete with $2n^2$ electrons and the external shell alone dictates the chemical properties of an element. As we know, the quantum mechanical view is more complete but not as simple.

The present form of the periodic table (Appendix A) is a grid of elements with 18 columns and 7 rows, with an additional double row of elements. The rows are called periods; the columns, which define the chemical properties, are called groups; examples of groups are "halogens" and "noble gases."

Thanks to Mendeleev's table, a solid conjecture was formulated that atoms are composite states including protons and electrons, electrons being loosely bound. But how to understand experimentally the inner structure of the atom, i.e., how were protons and electrons arranged inside the atom? Were electrons "orbiting" around a positive nucleus, or were both protons and electrons embedded in a "plum pudding," with electrons (the "plums") more loosely bound? A technique invented around 1900 to answer this question pioneered the full history of particle physics.

2.2 The Rutherford Experiment

Collide a beam of particles with a target, observe what comes out, and try to infer the properties of the interacting objects and/or of the relevant interaction force. This is the paradigm of particle physics experiments. The first experiment was conducted

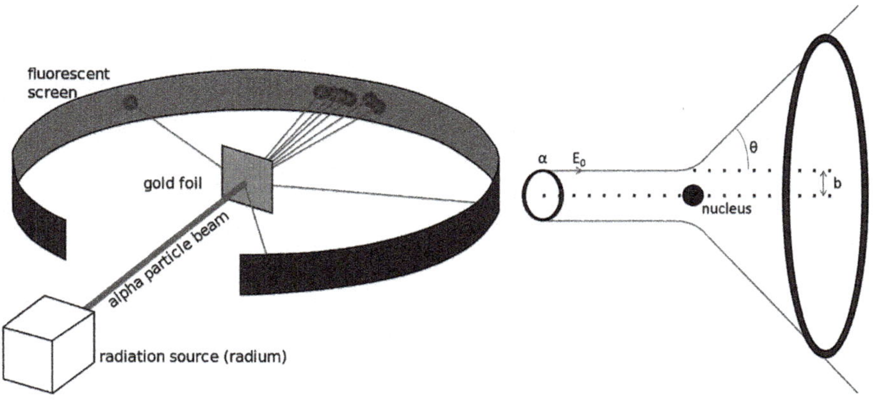

Fig. 2.2 *Left* Sketch of the Rutherford experiment (by Kurzon [own work, CC BY-SA 3.0], via Wikimedia Commons). *Right* Trajectories of the α particles

by Marsden and Geiger starting in 1908 and is remembered as the Rutherford[2] experiment. The beam consisted of α particles (helium nuclei); the target was a thin gold foil (some 400 nm); the detector, a scintillating screen which could be read by a microscope. The observation was that around 1 in 8000 α particles were deflected at very large angles (above 90°). A sketch of the experiment is shown in Fig. 2.2, left.

The interpretation of this result was given by Rutherford in 1911. It was based on a model in which the positive nucleus of the atom was a point fixed in the space and the scattering of the α particles was due to the Coulomb force and obeyed classical mechanics (quantum mechanics was yet to be born). The α particles were thus supposed to follow Keplerian trajectories. As energy and angular momentum are conserved, for a given impact parameter b (the perpendicular distance between the beam particle and the nucleus, see Fig. 2.2, right) there will be a well-defined scattering angle θ, and:

$$b = \left(\frac{1}{4\pi\epsilon_0}\right) \frac{Q_1 Q_2}{2E_0} \cot\left(\frac{\theta}{2}\right) \tag{2.1}$$

where ϵ_0 is the vacuum dielectric constant, Q_1 and Q_2 are the charges of the beam particle and of the target particle, and E_0 is the kinetic energy of the beam particle.

[2]Ernest Rutherford (1871–1937) was a New Zealand-born physicist. In early works at McGill University in Canada, he proved that radioactivity involved the transmutation of one chemical element into another; he differentiated and named the α (helium nuclei) and β (electrons) radiations. In 1907, Rutherford moved to Manchester, UK, where he discovered (and named) the proton. In 1908 he won the Nobel Prize in Chemistry "for his investigations into the disintegration of the elements, and the chemistry of radioactive substances." He became director of the Cavendish Laboratory at Cambridge University in 1919. Under his leadership, the neutron was discovered by James Chadwick in 1932. Also in 1932, his students John Cockcroft and Ernest Walton split for the first time the atom with a beam of particles. Rutherford was buried near Newton in Westminster Abbey, London. The chemical element rutherfordium—atomic number 104—was named after him in 1997.

If the number of beam particles per unit of transverse area n_{beam} is not a function of the transverse coordinates b and ϕ (the beam is uniform and wide with respect to the target size), the differential number of particles as a function of b is

$$\frac{dN}{db} = 2\pi b\, n_{\text{beam}}. \tag{2.2}$$

Expressing the differential number of particles as a function of the scattering angle θ:

$$\frac{dN}{d\theta} = \frac{dN}{db}\frac{db}{d\theta} \tag{2.3}$$

we obtain using Eq. 2.1:

$$\frac{dN}{d\theta} = \pi \left(\frac{1}{4\pi\epsilon_0}\frac{Q_1 Q_2}{2E_0}\right)^2 \frac{\cos\left(\frac{\theta}{2}\right)}{\sin^3\left(\frac{\theta}{2}\right)} n_{\text{beam}} \tag{2.4}$$

or, in terms of the solid angle Ω, ($d\Omega = 2\pi \sin\theta d\theta$):

$$\frac{dN}{d\Omega} = \left(\frac{1}{4\pi\epsilon_0}\frac{Q_1 Q_2}{4E_0}\right)^2 \frac{1}{\sin^4\left(\frac{\theta}{2}\right)} n_{\text{beam}}, \tag{2.5}$$

the well-known Rutherford formula. This equation explained the observation of scattering at large angles and became the paradigm for particle diffusion of nuclei.

2.3 β Decay and the Neutrino Hypothesis

Beta radioactivity, the spontaneous emission of electrons by some atoms, was discovered by Ernest Rutherford just a few years after the discovery by Henri Becquerel that uranium was able to impress photographic plates wrapped in black paper. It took then some years before James Chadwick in 1914 realized that the energy spectrum of the electrons originated in the beta decays was continuous and not discrete (Fig. 2.3). This was a unique feature in the new quantum world, in which decays were explained as transitions between well-defined energy levels. There was a missing energy problem and many explanations were tried along the years, but none was proved. In 1930, Niels Bohr even went so far as to suggest that the sacrosanct energy conservation law could be violated.

In December 1930, in a famous letter, Wolfgang Pauli proposed as "desperate remedy" the existence of a new neutral particle with spin one-half and low mass named neutron: "The continuous β spectrum would then become understandable from the assumption that in the β decay a neutron is emitted along with the electron,

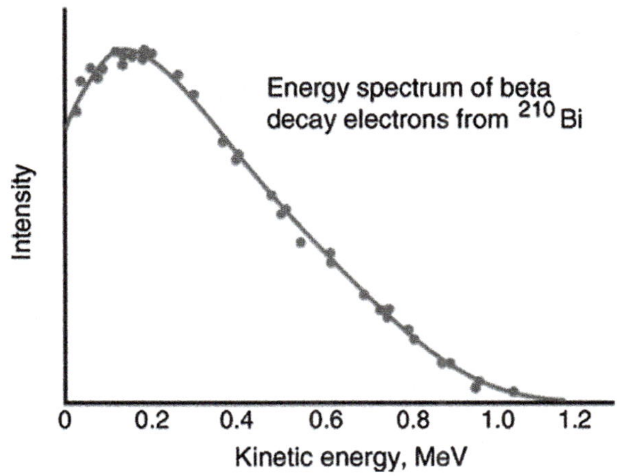

Fig. 2.3 Energy spectrum of electrons coming from the β decay of the $^{210)}$Bi (called historically "Radium E") to $^{210)}$Po (called historically "Radium F"). From http://hyperphysics.phy-astr.gsu.edu/; the measurements are from G. J. Neary, Roy. Phys. Soc. (London) A175 (1940) 71

in such way that the sum of the energies of the neutron and the electron is constant." This tiny new particle was later renamed "neutrino" by Enrico Fermi. The particle today known as neutron, constituent of the atomic nuclei, was discovered by James Chadwick in 1932, Nobel prize in Physics 1935, then at the University of Cambridge, Chadwick found a radiation consisting of uncharged particles of approximately the mass of the proton. His group leader Rutherford had conjectured the existence of the neutron already in 1920, in order to explain the difference between the atomic number of an atom and its atomic mass, and he modeled it as an electron orbiting a proton.

The atomic nuclei were thus composed (in the modern language) by protons and neutrons and the beta radioactive decays was explained by the decay of one of the nuclei neutrons into one proton, one electron, and one neutrino (in fact, as it will be discussed later, in one antineutrino):

$$n \rightarrow pe^{-}\bar{\nu}. \tag{2.6}$$

The β^{+} decay, i.e., the decay of one proton in the nucleus into one neutron, one positron (the antiparticle of the electron), and one neutrino

$$p \rightarrow ne^{+}\nu. \tag{2.7}$$

is also possible, although the neutron mass is larger than the proton mass—consider that nuclei are bound in the nucleus, and they are not free particles.

Neutrinos have almost no interaction with matter and therefore their experimental discovery was not an easy enterprise: intense sources and massive and performant detectors had to be made. Only in 1956 Reines and Cowan proved the existence of the neutrino placing a water tank near a nuclear reactor. Some of the antineutrinos produced in the reactor interacted with a proton in the water, giving rise to a neutron and a positron, the so-called inverse beta process:

$$\bar{\nu} p \rightarrow n e^{+}. \tag{2.8}$$

The positron then annihilates with an ordinary electron and the neutron is captured by cadmium chloride atoms dissolved in the water. Three photons were then detected (two from the annihilation and, 5 μs later, one from the de-excitement of the cadmium nucleus).

The mass of the neutrino is indeed very low (but not zero, as discovered by the end of the twentieth century with the observation of the oscillations between neutrinos of different families, a phenomenon that is possible only for massive neutrinos) and determines the maximum energy that the electron may have in the beta decay (the energy spectrum end-point). The present measurements are compatible with neutrino masses below a few eV.

2.4 Uncertainty Principle and the Scale of Measurements

If we want to investigate a structure below a length scale Δx, we are limited by the uncertainty principle (which indeed is a theorem). Since a wavelength

$$\lambda \simeq \frac{\hbar}{p} \tag{2.9}$$

can be associated to a particle of momentum p, this means that particles of energy (energy is close to momentum times c for high energy particles)

$$E > \frac{\hbar c}{\Delta x} \tag{2.10}$$

must be used. For example, a hard X-ray with an energy of 1 keV can investigate the structure of a target at a scale

$$\Delta x > \frac{\hbar c}{E} \simeq 2 \times 10^{-11}\,\text{m}, \tag{2.11}$$

an order of magnitude smaller than the atomic radius. A particle with an energy of 7 TeV (the running energy of the LHC accelerator at CERN, Geneva) can investigate the structure of a target at a scale

$$\Delta x > \frac{\hbar c}{E} \simeq 3 \times 10^{-20}\,\text{m}. \tag{2.12}$$

Since one can extract only a finite energy from finite regions of the Universe (and maybe the Universe itself has a finite energy), there is an intrinsic limit to the investigation of the structure of matter, below which the quest makes no more sense. However, as we shall see, there are practical limits much more stringent than that.

Does the concept of elementary particle have a meaning below this? The question is more philosophical than physical, since one cannot access infinite energies.

The maximum energies attainable by human-made accelerators are believed to be of the order of some 1000 TeV. However, Nature gives us for free beams of particles with much larger energies, hitting the Earth from extraterrestrial sources: cosmic rays.

2.5 Cross-Section and Interaction Length

Particle physicists observe and count particles, as pioneered by the Rutherford experiment. They count, for instance, the number of particles of a certain type with certain characteristics (energy, spin, scattering angle) that result from the interaction of a given particle beam at a given energy with a given target. It is then useful to express the results as quantities independent from the number of beam particles, or target particles. These quantities are the cross-sections, σ.

2.5.1 *Total Cross-Section*

The total cross-section measured in a collision of a beam with a single object (Fig. 2.4) is defined as

$$\sigma_{\text{tot}} = \frac{N_{\text{int}}}{n_{\text{beam}}} \tag{2.13}$$

where N_{int} is the total number of measured interactions and n_{beam} is, as previously defined, the number of beam particles per unit of transverse area.

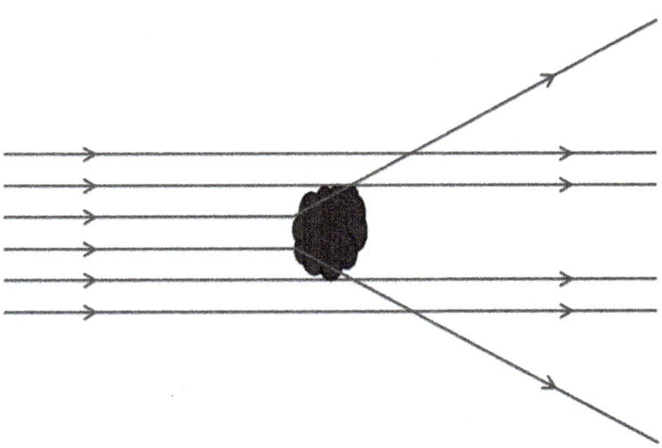

Fig. 2.4 Interaction of a particle beam with a single object target. *Lines* represent different particles in the beam

A cross-section has thus dimensions of area. It represents the effective area with which the interacting particles "see" each other. The usual unit for cross-section is the barn, b (1 b $= 10^{-24}$ cm^2) and its submultiples (millibarn—mb, microbarn—μb, nanobarn—nb, picobarn—pb, femtobarn—fb, etc.). To give an order of magnitude, the total cross-section for the interaction of two protons at a center-of-mass energy around 100 GeV is 40 mb (approximately the area of a circle with radius 1 fm).

We can write the total cross-section with a single target as

$$\sigma_{\text{tot}} = \frac{W_{\text{int}}}{J}, \tag{2.14}$$

in terms of the interaction rate W_{int} (number of interactions per unit time) and of the flux of incident particles J (number of beam particles that cross the unit of transverse area per unit of time). J is given as

$$J = \rho_{\text{beam}} v, \tag{2.15}$$

where ρ_{beam} is the density of particles in the beam and v is the beam particle velocity in the rest frame of the target.

In real life, most targets are composed of N_t small sub-targets (Fig. 2.5) within the beam incidence area. Considering as sub-targets the nuclei of the atoms of the target with the depth Δx and ignoring any shadowing between the nucleus, N_t is given as

$$N_t = \mathcal{N}\frac{\rho \Delta x}{w_a}, \tag{2.16}$$

where $\mathcal{N}$ is Avogadro's number, ρ is the specific mass of the target, w_a is its atomic weight. Note that N_t is a dimensionless number: it is just the number of sub-targets that are hit by one beam that has one unit of transverse area. In the case of several

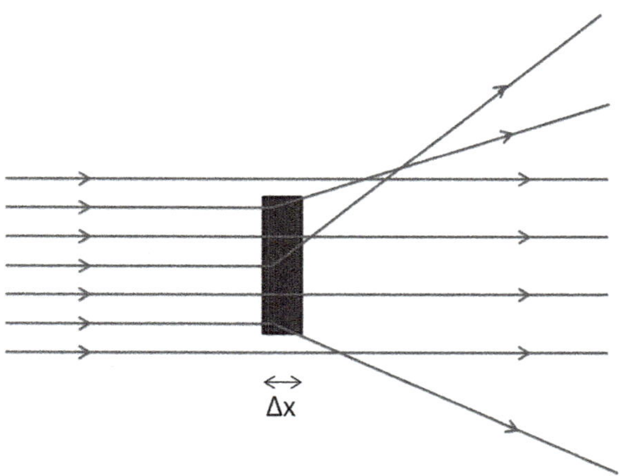

Fig. 2.5 Interaction of a particle beam with a target composed of many sub-targets

sub-targets, the total cross-section can thus be written as:

$$\sigma_{\text{tot}} = \frac{W_{\text{int}}}{J \cdot N_t} = \frac{W_{\text{int}}}{\mathcal{L}}, \tag{2.17}$$

where $\mathcal{L}$ is the luminosity.

The total number of interactions occurring in an experiment is then just the product of the total cross-section by the integral of the luminosity over the run time T of the experiment.

$$N_{\text{tot}} = \sigma_{\text{tot}} \int_T \mathcal{L} dt. \tag{2.18}$$

The units of integrated luminosity are therefore inverse barn, b^{-1}.

In this simplified model we are neglecting: the interactions between the scattered particles; the interactions between beam particles; the binding energies of the target particles; the absorption and the multi-scattering of the beam within the target.

2.5.2 *Differential Cross-Sections*

In practice, detectors often cover only a given angular region and we do not measure the total cross-section in the full solid angle. It is therefore useful to introduce the differential cross-section

$$\frac{d\sigma(\theta, \phi)}{d\Omega} = \frac{1}{\mathcal{L}} \frac{dW_{\text{int}}(\theta, \phi)}{d\Omega} \tag{2.19}$$

and

$$\sigma_{\text{tot}} = \int\int \frac{d\sigma(\theta, \phi)}{d\Omega} d\cos\theta. \tag{2.20}$$

The Rutherford formula (2.5) expressed as a differential cross-section is then

$$\frac{d\sigma}{d\Omega} = \left(\frac{1}{4\pi\epsilon_0} \frac{Q_1 Q_2}{4E_0}\right)^2 \frac{1}{\sin^4\left(\frac{\theta}{2}\right)}. \tag{2.21}$$

2.5.3 *Cross-Sections at Colliders*

In colliders, beam–target collisions are replaced by beam–beam collisions (Fig. 2.6). Particles in the beams come in bunches. The luminosity is thus defined as

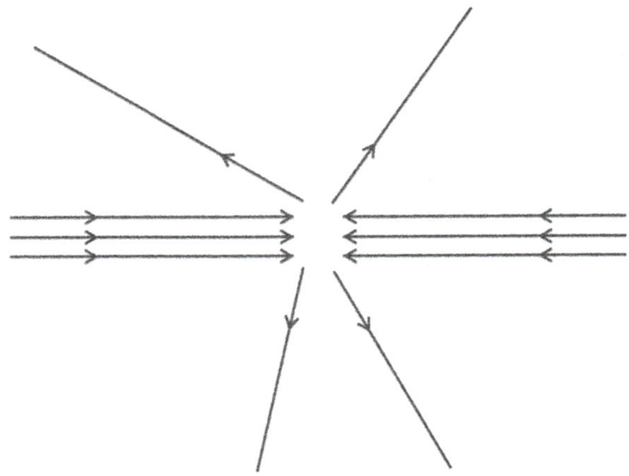

Fig. 2.6 Beam-beam interaction

$$\mathcal{L} = \frac{N_1 N_2}{A_T} N_b f \tag{2.22}$$

where N_1 and N_2 are the number of particles in the crossing bunches, N_b is the number of bunches per beam, A_T is the intersection transverse area, and f is the beam revolution frequency.

2.5.4 Partial Cross-Sections

When two particles collide, it is often the case that there are many possible outcomes. Quantum mechanics allows us to compute the occurrence probability for each specific final state. Total cross-section is thus a sum over all possible specific final states

$$\sigma_{\text{tot}} = \sum_i \sigma_i \tag{2.23}$$

where σ_i is defined as the partial cross-section for channel i.

A relevant partial cross-section is the elastic cross-section, σ_{el}. In an elastic process, the particles in the final state and in the initial state are the same—there is simply an exchange of energy-momentum. Whenever there is no available energy to create new particles, $\sigma_{\text{tot}} = \sigma_{el}$. This is shown in Fig. 2.7 for the case of proton–proton and antiproton–proton interactions.

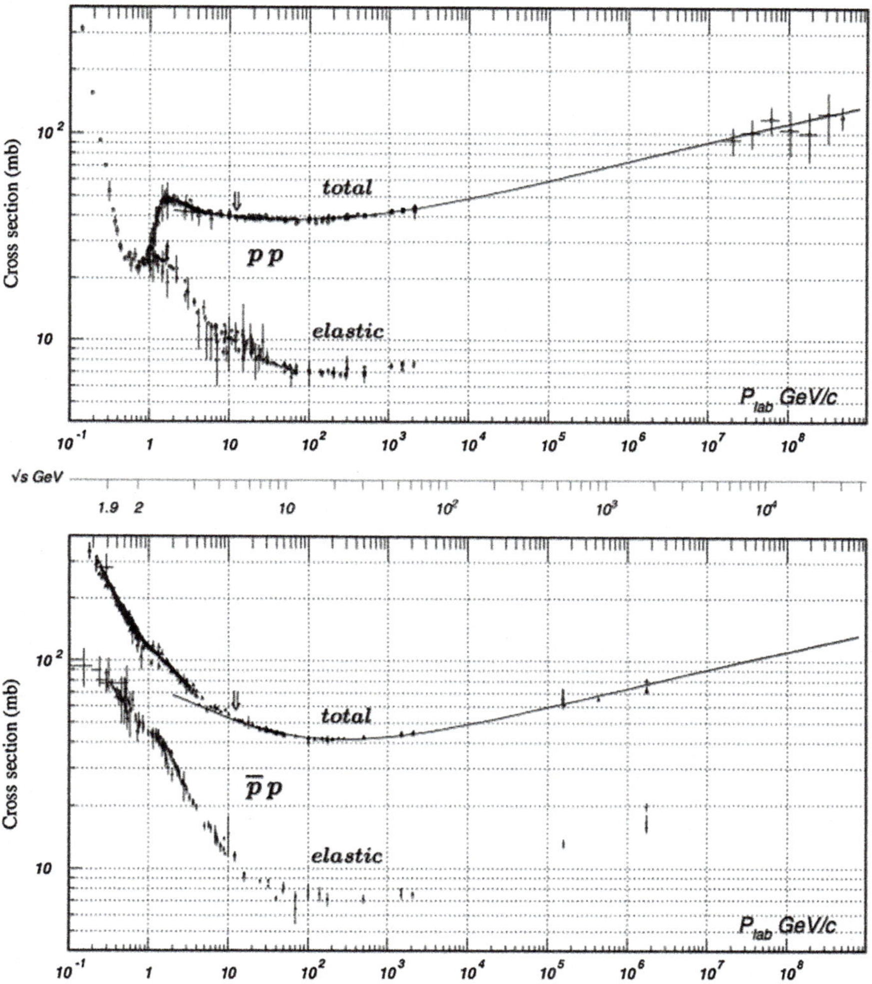

Fig. 2.7 Total and elastic cross sections for pp and $\bar{p}p$ collisions as a function of laboratory beam momentum and total center-of-mass energy. From the Review of Particle Physics, K.A. Olive et al. (Particle Data Group), Chin. Phys. C 38 (2014) 090001

2.5.5 Interaction Length

When a beam of particles crosses matter, its intensity is reduced. Using the definition of total cross-section 2.14 and Eq. (2.17), the reduction when crossing a slice of thickness Δx is:

$$\frac{\Delta N}{N} = \frac{W_{\text{int}}}{J} = \left(\mathcal{N}\frac{\rho}{w_A}\,\Delta x\right)\sigma_{\text{tot}} \tag{2.24}$$

where w_A is the atomic weight of the target. Defining the interaction length L_{int} as

$$L_{\text{int}} = \frac{w_A}{\sigma_{\text{tot}} \mathcal{N} \rho} \tag{2.25}$$

then

$$\frac{dN}{dx} = -\frac{1}{L_{\text{int}}} N \tag{2.26}$$

and

$$N = N_0 \, e^{-x/L_{\text{int}}} . \tag{2.27}$$

L_{int} has units of length (usually cm). However, this quantity is often redefined as

$$L'_{\text{int}} = L_{\text{int}} \, \rho = \frac{w_A}{\sigma_{\text{tot}} \, \mathcal{N}} \tag{2.28}$$

and its units will then be $\text{g}\,\text{cm}^{-2}$. This way of expressing L_{int} is widely used in cosmic ray physics. In fact, the density of the atmosphere has a strong variation with height (see next section). For this reason, to study the interaction of cosmic particles in their path in the atmosphere, the relevant quantity is not the path length but rather the amount of matter that has been traversed, $\int \rho dx$.

In a rough approximation, the atmosphere is isothermal; under this hypothesis, its depth x in g cm^{-2} varies exponentially with height h (km), according to the formula

$$x = X e^{-h/H} \tag{2.29}$$

where $H \simeq 6.5$ km, and $X \simeq 1030$ g·cm^{-2} is the total vertical atmospheric depth.

2.6 Decay Width and Lifetime

Stable particles like the proton and the electron (as much as we know) are the exception, not the rule. The lifetime of most particles is finite and its value may span over many orders of magnitude from, for instance, 10^{-25} s for the electroweak massive bosons (Z and W) to around 900 s for the neutron, depending on the strength of the relevant interaction and on the size of the decay phase space.

In order to describe decays we must use a genuine quantum mechanical language, being decay a genuine quantum process whose statistical nature cannot be described properly by classical physics. We shall use, thus, the language of wave functions. $|\Psi(x, y, z, t)|^2 \, dV$ is the probability density for finding a particle in a volume dV around point (x, y, z) at time t.

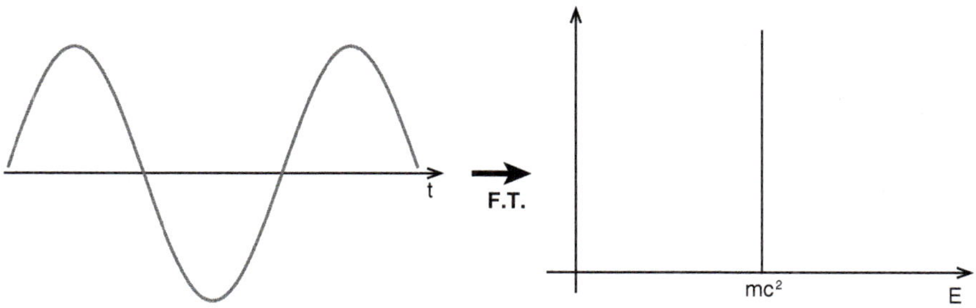

Fig. 2.8 The wave function of a stable particle and its energy spectrum

Stable particles are described by pure harmonic wave functions and their Fourier transforms are functions centered in well-defined proper energies—in the rest frame, $E = mc^2$ (Fig. 2.8):

$$\Psi(t) \propto \Psi(0)\, e^{-\imath \frac{E}{\hbar} t} \tag{2.30}$$

$$\Psi(E) \propto \delta(E - mc^2)\,. \tag{2.31}$$

Unstable particles are described by damped harmonic wave functions and therefore their proper energies are not well-defined (Fig. 2.9):

$$\Psi(t) \propto \Psi(0) e^{-\imath \frac{E}{\hbar} t} e^{-\frac{\Gamma}{2\hbar} t} \quad \Longrightarrow |\Psi(t)|^2 \propto |\Psi(0)|^2 e^{-t/\tau} \tag{2.32}$$

$$\Psi(E) \propto \frac{1}{(E - mc^2) + i\Gamma/2} \Longrightarrow |\Psi(E)|^2 \propto \frac{1}{(E - mc^2)^2 + \Gamma^2/4} \tag{2.33}$$

which is a Cauchy function (physicists call it a Breit-Wigner function) for which the width Γ is directly related to the particle lifetime τ:

$$\tau = \frac{\hbar}{\Gamma}. \tag{2.34}$$

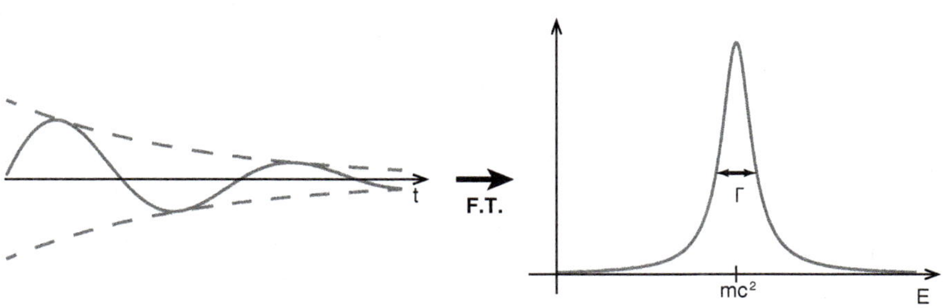

Fig. 2.9 The wave function of an unstable particle and its energy spectrum

If a particle can decay through different channels, its total width will be the sum of the partial widths Γ_i of each channel:

$$\Gamma_t = \sum \Gamma_i \,. \tag{2.35}$$

An unstable particle may thus have several specific decay rates, but it has just one lifetime:

$$\tau = \frac{\hbar}{\sum \Gamma_i} \,. \tag{2.36}$$

Therefore, all the Breit-Wigner functions related to the decays of the same particle have the same width Γ_t but different normalization factors which are proportional to the fraction of the decays in each specific channel, also called the branching ratio, BR_i (or Br_i), defined as

$$BR_i = \frac{\Gamma_i}{\Gamma_t} \,. \tag{2.37}$$

2.7 The Fermi Golden Rule and the Rutherford Scattering

Particles interact like corpuscles but propagate like waves. This was the turmoil created in physics in the early twentieth century by the photoelectric effect theory of Einstein. In the microscopic world, deterministic trajectories were no longer possible. Newton's laws had to be replaced by wave equations. Rutherford formulae, classically deduced, coincide anyway with calculations based on quantum mechanics.

In quantum mechanics, the scattering of a particle due to an interaction that acts only during a finite time interval can be described as the transition between an initial and a final stationary states characterized by well-defined momenta. The probability λ of such a transition is given, if the perturbation is small, by the Fermi[3] golden

[3]Enrico Fermi (Rome 1901–Chicago 1954) studied in Pisa and became full professor of Analytical Mechanics in Florence in 1925, and then of Theoretical Physics in Rome from 1926. Soon he surrounded himself by a group of brilliant young collaborators, the so-called "via Panisperna boys" (E. Amaldi, E. Majorana, B. Pontecorvo, F. Rasetti, E. Segré, O. D'Agostino). For Fermi, theory and experiment were inseparable. In 1934, he discovered that slow neutrons catalyzed a certain type of nuclear reactions, which made it possible to derive energy from nuclear fission. In 1938, Fermi went to Stockholm to receive the Nobel Prize, awarded for his fundamental work on neutrons, and from there he emigrated to the USA, where he became American citizen in open dispute with the Italian racial laws. He actively participated in the Manhattan Project for the use of nuclear power for the atomic bomb, but spoke out against the use of this weapon on civilian targets. Immediately after the end of World War II, he devoted himself to theoretical physics of elementary particles and to the origin of cosmic rays. Few scientists of the twentieth century impacted as profoundly as Fermi in different areas of physics: Fermi stands for elegance and power of thought in the group of immortal geniuses like Einstein, Landau, Heisenberg, and later Feynman.

rule: (see the book by Townsend among the recommended readings at the end of the chapter)

$$\lambda = \frac{2\pi}{\hbar}|H'_{fi}|^2\rho(E_i) \tag{2.38}$$

where H'_{fi} is the transition amplitude between states i and f ($H'_{fi} = \langle f|H'_{\text{int}}|i\rangle$ where H'_{int} is the interaction Hamiltonian) and $\rho(E_i)$ is the density of final states for a given energy $E_i = E_f$. The cross-section is, as it was seen above, the interaction rate per unit of flux J. Thus:

$$\sigma_{\text{tot}} = \frac{\lambda}{J}. \tag{2.39}$$

To compute the cross-section one needs then to determine the transition amplitude, the flux, and the density of final states.

2.7.1 Transition Amplitude

Rutherford scattering can be, in first approximation, treated as the nonrelativistic elastic scattering of a single particle by a fixed static Coulomb potential. The initial and final time-independent state amplitudes may be written as plane waves normalized in a box of volume L^3 (Fig. 2.10) and with linear momenta $\vec{p}_i = \hbar\,\vec{k}_i$ and $\vec{p}_f = \hbar\,\vec{k}_f$, respectively ($k = \left|\vec{k}_i\right| = \left|\vec{k}_f\right|$):

$$u_i = L^{-\frac{3}{2}}\exp(i\,\vec{k}_i\cdot\vec{r}) \tag{2.40}$$

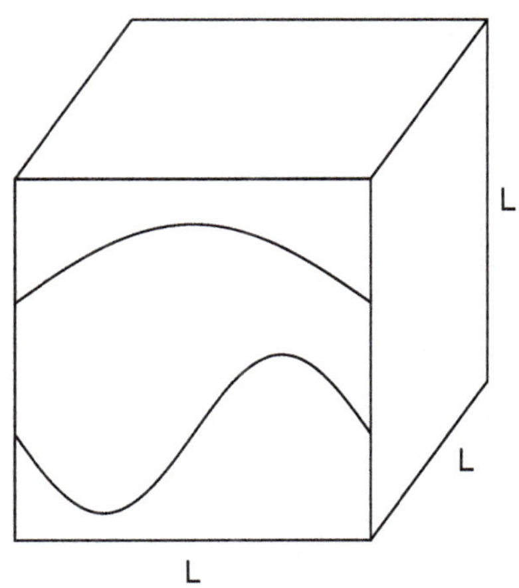

Fig. 2.10 Normalization box

and

$$u_f = L^{-\frac{3}{2}} \exp(i\,\vec{k_f} \cdot \vec{r})\,. \tag{2.41}$$

Assuming a scattering center at the origin of coordinates, the Coulomb potential is written as

$$V(r) = \frac{1}{4\pi\varepsilon_0} \frac{Q_1 Q_2}{r} \tag{2.42}$$

where ε_0 is the vacuum dielectric constant and Q_1 and Q_2 are the charges of the beam and of the target particles. The transition amplitude can thus be written as

$$H'_{fi} = L^{-3} \int \exp\left(-\,i\,\vec{k_f} \cdot \vec{r}\right) V(r) \exp\left(-\,i\,\vec{k_i} \cdot \vec{r}\right) d^3\,x. \tag{2.43}$$

Introducing the momentum transfer:

$$\vec{q} = \hbar\,\left(\vec{k_f} - \vec{k_i}\right) \tag{2.44}$$

the transition amplitude given by

$$H'_{fi} = L^{-3} \int V(r) \exp\left(-\frac{i}{\hbar}\,\vec{q}\,\cdot \vec{r}\right) d^3\,x \tag{2.45}$$

is just the Fourier transform of $V(r)$ and then

$$H'_{fi} = -\frac{4\,\pi\hbar^2}{L^3} \left(\frac{1}{4\pi\varepsilon_0} \frac{Q_1 Q_2}{|\vec{q}\,|^2}\right). \tag{2.46}$$

Expressing $|\vec{q}\,|^2$ as a function of the scattering angle θ as

$$|\vec{q}\,|^2 = 4\,\hbar^2\,k^2 \sin^2\left(\frac{\theta}{2}\right), \tag{2.47}$$

the transition amplitude may finally be written as

$$H'_{fi} = -\frac{4\pi\hbar^2}{L^3} \left(\frac{1}{4\pi\varepsilon_0} \frac{Q_1\,Q_2}{4\hbar^2\,k^2 \sin^2\left(\frac{\theta}{2}\right)}\right). \tag{2.48}$$

2.7.2 Flux

The flux, as seen in 2.15, is $J = \rho_{\text{beam}} v$, which in the present case may be written as

$$J = \frac{v}{L^3} = \frac{\hbar\, k}{m\, L^3}. \tag{2.49}$$

2.7.3 Density of States

The density of final states $\rho(E_i)$ is determined by the dimension of the normalization box. At the boundaries of the box, the wave function should be zero and so only harmonic waves are possible. Therefore, the projections of the wave number vector $\vec{\kappa}$ along each axis should also obey

$$k_x = \frac{2\pi\, n_x}{L}\ ;\ k_y = \frac{2\pi\, n_y}{L}\ ;\ k_z = \frac{2\pi\, n_z}{L}, \tag{2.50}$$

where n_x, n_y and n_z are the integer harmonic numbers.

Considering now the wave number $\vec{\kappa}$ in its vector space the volume associated to each possible state defined by a particular set of harmonic numbers is just

$$\frac{dk_x}{dn_x}\frac{dk_y}{dn_y}\frac{dk_z}{dn_z} = \left(\frac{2\pi}{L}\right)^3, \tag{2.51}$$

while the elementary volume d^3k in spherical coordinates is

$$d^3k = k^2\, dk\, d\Omega. \tag{2.52}$$

Then, the number of states dn in the volume d^3k is

$$dn = \left(\frac{L}{2\pi}\right)^3 k^2\, dk\, d\Omega. \tag{2.53}$$

Remembering that in nonrelativistic quantum mechanics

$$E = \frac{(\hbar k)^2}{2m}, \tag{2.54}$$

the density of states $\rho(E_i)$ is therefore given as

$$\rho(E_i) = \frac{dn}{dE} = \left(\frac{L}{2\pi\hbar}\right)^3 \frac{(\hbar k)^2}{v} d\Omega, \tag{2.55}$$

where v is the velocity of the particle.

Finally, replacing all the terms in 2.38 and 2.39:

$$\frac{d\sigma}{d\Omega} = \left(\frac{1}{4\pi\varepsilon_0}\frac{Q_1 Q_2}{4E_0}\right)^2 \frac{1}{\sin^4\left(\frac{\theta}{2}\right)} \tag{2.56}$$

and this is exactly the Rutherford formula.

In fact, the minimum distance at which a nonrelativistic beam particle with energy E_0 can approach the target nucleus is:

$$d_{\min} = \frac{Q_1 Q_2}{4\pi\varepsilon_0 E_0} \tag{2.57}$$

while the de Broglie wavelength associated to that particle is

$$\lambda = \frac{h}{\sqrt{2mE_0}}. \tag{2.58}$$

In the particular case of the Rutherford experiment (α particles with a kinetic energy of 7.7 MeV against a golden foil) $\lambda \ll d_{\min}$ and the classical approximation is, by chance, valid.

2.8 The Modern View of Scattering: Quantum Field Exchange

Quantum field theories, which provide in modern physics the description of interactions, describe nature in terms of fields. A force between two particles (described by "particle fields") is described in terms of the exchange of virtual force carrier particles (again described by appropriate fields) between them. For example, the electromagnetic force is mediated by the photon field, weak interactions are mediated by the Z and $W^{\pm}$ fields, while the mediators of the strong interaction are called gluons. "Virtual" means that these particles can be off-shell, i.e., they do not need to have the "right" relationship between mass, momentum, and energy—this is related to the virtual particles that we discussed in the previous chapter, which can violate energy-momentum conservation for short times.

2.8.1 Feynman Diagrams

Feynman diagrams are pictorial representations of interactions involving subatomic particles, introduced by Richard Feynman[4] in the late 1940s.

The orientation from left to right in a Feynman diagram represents normally time: a process begins on the left and ends on the right. Basic fermions are represented by straight lines with possibly an arrow to the right for particles, and to the left for antiparticles. Force carries are represented typically by wavy lines (photons), springs (gluons), dashed lines ($W^{\pm}$ and Z). Two important rules that the Feynman diagrams must satisfy clarify the meaning of such representation are as follows:

- conservation of energy and momentum is required at every vertex;
- lines entering or leaving the diagram represent real particles and must have $E^2 = p^2c^2 + m^2c^4$ (see later the discussion on Einstein's special relativity).

Associated with Feynman diagrams are mathematical rules (called the "Feynman rules") that enable calculation of the probability (quantum mechanically, the square of the absolute value of the amplitude) for a given reaction to occur; we shall describe the quantitative aspect in greater detail in Chaps. 6 and 7.

Figure 2.11, left, for example represents a simple Feynman diagram, in which an electron and a proton are mutually scattered as the result of an electromagnetic interaction (virtual photon exchange) between them. This process requires two vertices in which the photon interacts with the charged particle (one for each particle), and for this kind of scattering this is the minimum number of vertices—we say that this is the representation of the process at *leading order*.

The Feynman rules allow associating to each vertex a multiplication factor contributing to the total amplitude. For example, in the case of a photon coupling (two photon vertices) this factor is the "coupling parameter"

$$\frac{1}{4\pi\epsilon_0}\frac{e^2}{\hbar c} \simeq 1/137$$

for each photon, so the amplitudes for diagrams with many photons (see for example Fig. 2.11, right) are small, compared to those with only one.

[4]Richard Feynman (New York 1918—Los Angeles 1988), longtime professor at Caltech, is known for his work in quantum mechanics, in the theory of quantum electrodynamics, as well as in particle physics; he participated in the Manhattan project. In addition, he proposed quantum computing. He received the Nobel Prize in Physics in 1965 for his "fundamental work in quantum electrodynamics, with deep-ploughing consequences for the physics of elementary particles." His life was quite adventurous, and full of anecdotes. In the divorce file related to his second marriage, his wife complained that "He begins working calculus problems in his head as soon as he awakens. He did calculus while driving in his car, while sitting in the living room, and while lying in bed at night." He wrote several popular physics books, and an excellent general physics textbook now freely available at http://www.feynmanlectures.caltech.edu/.

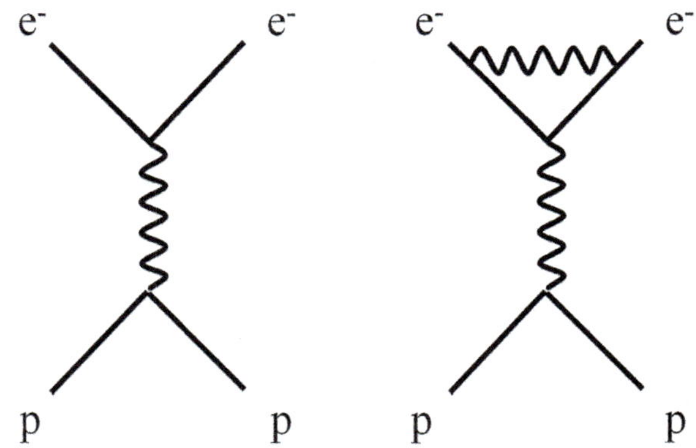

Fig. 2.11 Electromagnetic scattering: interaction between an electron and a proton. *Left* via the exchange of one virtual photon. *Right* the same process with one more virtual photon—the amplitude decreases by a factor of 1/137

Technically, the Feynman rules allow expressing the probability of a process as a power series expansion in the coupling parameter. One can draw all possible diagrams up to some number of mediators of the exchange, depending on the accuracy desired; then compute the amplitude for each diagram following the Feynman rules, sum all the amplitudes (note that the diagrams could display negative interference), and calculate the square of the modulus of the amplitude, which will give the probability. This perturbative technique is only of practical use when the coupling parameter is small, that is, as we shall see, for electromagnetic or weak interactions, but not for strong interactions, except at very high energies.

2.9 Particle Scattering in Static Fields

Rutherford formula was deduced assuming a static Coulomb field created by a fixed point charge. These assumptions can be either too crude or just not valid in many cases. Hereafter, some generalizations of the Rutherford formula are discussed.

2.9.1 Extended Charge Distributions (Non Relativistic)

Let us assume that the source of the static Coulomb field has some spatial extension $\rho(r')$ (Fig. 2.12) with

$$\int_0^\infty \rho\left(r'\right)\, dr' = 1\,. \tag{2.59}$$

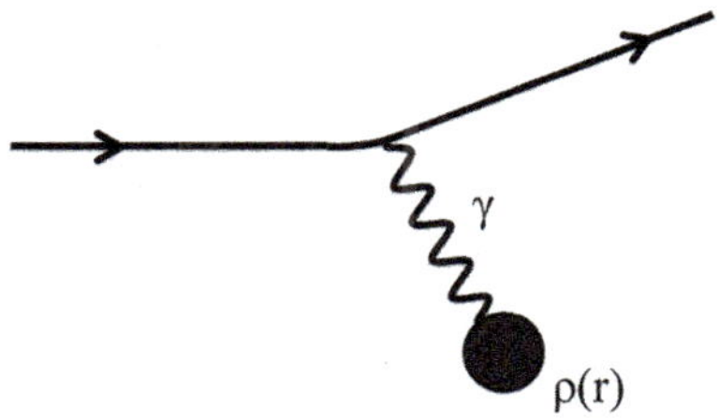

Fig. 2.12 Scattering by an extended source

Then

$$H'_{fi} = L^{-3} \int\int \left(\frac{1}{4\pi\varepsilon_0} \frac{Q_1 Q_2\, \rho(\vec{r}\,')}{|\vec{r}\,' - \vec{r}\,|} \right) \exp\left(-\frac{i}{\hbar}\, \vec{q} \cdot \vec{r} \right) d^3x\, d^3x' =$$

$$= L^{-3} \int\int \left(\frac{1}{4\pi\varepsilon_0} \frac{Q_1 Q_2\, \rho(\vec{r}\,')}{|\vec{r}\,' - \vec{r}\,|} \right) \exp\left(-\frac{i}{\hbar}\, \vec{q} \cdot (\vec{r}\,' - \vec{r}) \right) \exp\left(-\frac{i}{\hbar}\, \vec{q} \cdot \vec{r}\,' \right) d^3x\, d^3x' \tag{2.60}$$

and defining the electric form factor $F(q)$ as

$$F(q) = \int \rho(\vec{r}\,') \exp\left(-\frac{i}{\hbar}\, \vec{q} \cdot \vec{r}\,' \right) d^3x' \tag{2.61}$$

the modified scattering cross-section is

$$\frac{d\sigma}{d\Omega} = |F(\vec{q})|^2 \left(\frac{d\sigma}{d\Omega} \right)_0 \tag{2.62}$$

where $\left(\frac{d\sigma}{d\Omega}\right)_0$ is the Rutherford cross-section.

In the case of the proton the differential *ep* cross-section at low transverse momentum is described by such a formula, being the form factor given by the *dipole formula*

$$F(q) \propto \left(1 + \frac{|\vec{q}\,|^2}{\hbar^2 b^2} \right)^{-2} . \tag{2.63}$$

The charge distribution is the Fourier transform $\rho(r) \propto e^{-r/a}$, where $a = 1/b \simeq 0.2$ fm corresponds to a root mean square charge radius of 0.8 to 0.9 fm. The size of the proton is then determined to be at the scale of 1 fm.

2.9.2 Finite Range Interactions

The Coulomb field, as the Newton gravitational field, has an infinite range. Let us now consider a field with an exponential attenuation (Yukawa potential)

$$V(r) = \frac{g}{4\pi r} \exp\left(-\frac{r}{a}\right) \tag{2.64}$$

where a is some interaction range scale. Then,

$$H'_{fi} = L^{-3} \int \left(\frac{g}{4\pi r} \exp(-\frac{r}{a})\right) \exp\left(-\frac{i}{\hbar} \vec{q} \cdot \vec{r}\right) d^3 x \tag{2.65}$$

$$H'_{fi} = -\frac{\hbar^2}{L^3} \left(\frac{g}{\vec{q}^{\,2} + \frac{\hbar^2}{a^2}}\right) \tag{2.66}$$

and

$$\frac{d\sigma}{d\Omega} = g^2 \left(\frac{\vec{q}^{\,2}}{\vec{q}^{\,2} + M^2 c^2}\right)^2 \left(\frac{d\sigma}{d\Omega}\right)_0, \tag{2.67}$$

where $\left(\frac{d\sigma}{d\Omega}\right)_0$ is the Rutherford cross-section. $M = \hbar/(a \cdot c)$ was interpreted by Hideki Yukawa,[5] as it will be discussed in Sect. 3.2.4, as the mass of a particle exchanged between nucleons and responsible for the strong interaction which ensures the stability of nuclei. The relevant scale $a = 1$ fm corresponds to the size of nuclei, and the mass of the exchanged particle comes out to be $M \simeq 200$ MeV$/c^2$ (see Sect. 2.11 for the conversion).

2.9.3 Electron Scattering

Electrons have a nonzero spin $\left(S = \frac{1}{2}\hbar\right)$, and thus they have a nonzero magnetic momentum

$$\vec{\mu} = \frac{Q_e}{m_e} \vec{S} \tag{2.68}$$

where Q_e and m_e are, respectively, the charge and the mass of the electron.
The electron cross-scattering formula is given by the Mott cross-section (its derivation is out of the scope of the present chapter as it implies relativistic quantum mechanics):

$$\frac{d\sigma}{d\Omega} = \left(\frac{d\sigma}{d\Omega}\right)_0 \left(1 - \beta^2 \sin^2\left(\frac{\theta}{2}\right)\right). \tag{2.69}$$

[5] Hideki Yukawa (Tokyo, 1907–Kyoto, 1981), professor at the Kyoto University, gave fundamental contributions to quantum mechanics. For his research he won the prize Nobel Prize for Physics in 1949.

When $\beta \to 0$, the Rutherford scattering formula is recovered as

$$\frac{d\sigma}{d\Omega} = \left(\frac{d\sigma}{d\Omega}\right)_0 . \tag{2.70}$$

When $\beta \to 1$,

$$\frac{d\sigma}{d\Omega} = \left(\frac{d\sigma}{d\Omega}\right)_0 \cos^2\left(\frac{\theta}{2}\right) , \tag{2.71}$$

which translates the fact that, for massless particles (as it will be discussed in Sect. 6.2.4), the projection of the spin $\vec{S}$ over the direction of the linear momentum $\vec{p}$ is conserved (Fig. 2.13). The helicity quantum number h is defined as

$$\mathrm{h} = \vec{S} \cdot \frac{\vec{p}}{|\vec{p}|} . \tag{2.72}$$

A massless electron, thus, could not be backscattered.

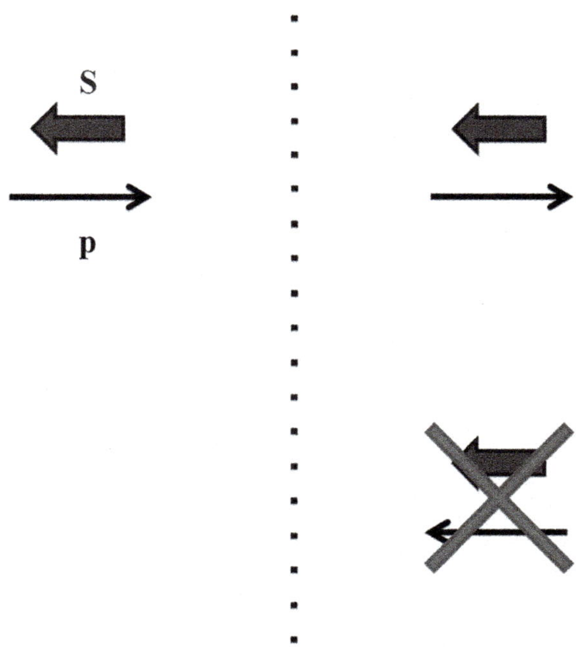

Fig. 2.13 Schematic representation of helicity conservation in the limit $\beta = 1$

2.10 Special Relativity

Physical laws are, since Galilei and Newton, believed to be the same in all inertial reference frames (i.e., in all frames moving with constant speed with respect to a frame in which they hold—classical mechanics postulates with the law of inertia the existence of at least one admissible frame). This is called the principle of special relativity, and it has been formulated in a quantitative way by Galilei. According to the laws of transformations of coordinates between inertial frames in classical physics (called Galilei transformations), accelerations are invariant with respect to a change of reference frame—while speeds are trivially non-invariant. Since the equations of classical physics (Newton's equations) are based on accelerations only, this automatically guarantees the principle of relativity.

Something revolutionary happened when Maxwell's equations[6] were formulated. The Maxwell's equations in vacuo

$$\vec{\nabla} \cdot \vec{\mathcal{E}} = \frac{\rho}{\epsilon_0} \tag{2.73}$$

$$\vec{\nabla} \times \vec{\mathcal{E}} = -\frac{\partial \vec{B}}{\partial t} \tag{2.74}$$

$$\vec{\nabla} \cdot \vec{B} = 0 \tag{2.75}$$

$$\vec{\nabla} \times \vec{B} = \frac{1}{c^2}\frac{\partial \vec{\mathcal{E}}}{\partial t} + \mu_0 \vec{j} \tag{2.76}$$

together with the equation describing the motion of a particle of electrical charge q in an electromagnetic field

$$\vec{F} = q(\vec{\mathcal{E}} + \vec{v} \times \vec{B}) \tag{2.77}$$

(Lorentz[7] force), provide a complete description of electromagnetic field and of its dynamical effects. Such laws contain explicitly a speed, the speed of light c, 299792458 m/s ($\sim$300 000 km/s, with a relative accuracy better than 10^{-3}). This speed is also present when the equations are written in vacuo, i.e., where neither charges ρ nor currents $\vec{j}$ are present: if they hold, thus, the classical formulation of relativity, based on the Galilei transformations, is not invariant in all inertial frames. One can easily create some paradoxes based on this (see the Exercises): electromagnetism is not consistent with classical mechanics.

---

[6]James Clerk Maxwell (1831–1879) was a Scottish physicist. His most prominent achievement was formulating classical electromagnetic theory. Maxwell's equations, published in 1865, demonstrate that electricity, magnetism, and light are all manifestations of the same phenomenon: the electromagnetic field. Maxwell also contributed to the Maxwell-Boltzmann distribution, which gives the statistical distribution of velocities in a classical perfect gas in equilibrium. Einstein had a photograph of Maxwell, one of Faraday and one of Newton in his office.

[7]Hendrik Antoon Lorentz (1853–1928) was a Dutch physicist who gave important contributions to electromagnetism. He also derived the equations subsequently used by Albert Einstein to describe the transformation of space and time coordinates in different inertial reference frames. He was awarded the 1902 Nobel Prize in Physics.

To solve the problem (i.e., maintaining the speed of light c a nature invariant, and guaranteeing the covariant formulation of the laws of mechanics) a deep change in our perception of space and time was needed: it was demonstrated that time and length intervals are no longer absolute. Two simultaneous events in one reference frame are not simultaneous in any other reference frame that moves with nonzero velocity with respect to the first one; the Galilean transformations had to be replaced by new ones, the Lorentz transformations. Another striking consequence of this revolution was that mass is just a particular form of energy; kinematics at velocities near c is quite different from the usual one and particle physics is the laboratory to test it.

2.10.1 Lorentz Transformations

Let S and S' be two inertial reference frames. S' moves at a constant velocity $\vec{v}$ directed along the common x axis with respect to S (Fig. 2.14). The coordinates in one reference frame transform into new coordinates in the other reference frame (Lorentz transformations) as:

$$\begin{aligned} ct &= \gamma(ct' + \beta x') \\ x &= \gamma(x' + \beta ct') \\ y &= y' \\ z &= z' \end{aligned}$$

where $\beta = v/c$ and $\gamma = 1/\sqrt{1-\beta^2}$.

It can be verified that applying the above transformations the speed of light is an invariant between S and S'.

A conceptually nontrivial consequence of these transformations is that for an observer in S the time interval ΔT is larger than the time measured by a clock in S' for two events happening at the same place, the so-called *proper time*, $\Delta T'$ (time dilation):

$$\Delta T = \gamma \Delta T' \tag{2.78}$$

Fig. 2.14 Inertial reference frames

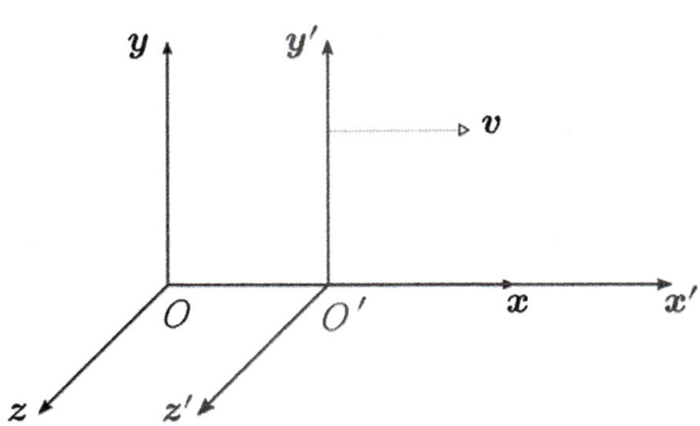

while the length of a ruler that is at rest in S' is shorter when measured in S (length contraction):

$$\Delta L = \Delta L'/\gamma . \tag{2.79}$$

Lorentz's transformations of coordinates guarantee automatically the invariance of the squared interval

$$ds^2 = c^2dt^2 - dx^2 - dy^2 - dz^2 \tag{2.80}$$

and now we want to extend the properties of the 4-ple (cdt, dx, dy, dz) to other 4-ple behaving in a similar way. We want to do something similar to what we did in classical physics: we want to introduce representations such that the equations become visually covariant with respect to transformations—i.e., the laws of physics are respected in different reference frames.

Let us introduce a simple convention: in the 4-ple (cdt, dx, dy, dz) the elements will be numbered from 0 to 3. Greek indices like μ will run from 0 to 3, and Roman symbols will run from 1 to 3 ($i = (1, 2, 3)$) as in the usual three-dimensional case.

We define four-vector a quadruple

$$A^\mu = \left(A^0, A^1, A^2, A^3\right) = \left(A^0, \vec{A}\right) \tag{2.81}$$

which transforms like (cdt, dx, dy, dz) for changes of reference systems. The A^μ (with high indices) is called *contravariant* representation of the four-vector.

Correspondingly, we define the 4-ple

$$A_\mu = (A_0, A_1, A_2, A_3) = \left(A^0, -A^1, -A^2, -A^3\right) = \left(A^0, -\vec{A}\right) \tag{2.82}$$

which is called *covariant* representation .

The coordinates of an event (ct, x, y, z) can be considered as the components of a four-dimensional radius vector in a four-dimensional space. So we shall denote its components by x^μ, where the index μ takes the values 0, 1, 2, 3 and

$$x^0 = ct \quad x^1 = x \quad x^2 = y \quad x^3 = z . \tag{2.83}$$

By our definition, the quantity $\equiv \sum_\mu A_\mu A^\mu \equiv A_\mu A^\mu$ is invariant. Omitting the sum sign when an index is repeated once in contravariant position and once in covariant position is called the Einstein's sum convention. Sometimes, when there is no ambiguity, this quantity is also indicated as A^2.

In analogy to the square of a four-vector, one forms the *scalar product* of two different four-vectors:

$$A^\mu B_\mu = A^0 B_0 + A^1 B_1 + A^2 B_2 + A^3 B_3 = A^0 B^0 - A^1 B^1 - A^2 B^2 - A^3 B^3 .$$

It is clear that it can be written either as $A^\mu B_\mu$ or $A_\mu B^\mu$—the result is the same.

The product $A^\mu B_\mu$ is a *four-scalar*: it is invariant under rotations of the four-dimensional coordinate system.

The component A^0 is called the *time component* and $(A^1,\ A^2,\ A^3)$ the *space components* of the four-vector. Under purely spatial rotations the three space components of the four-vector A^i form a three-dimensional vector $\vec{A}$.

The square of a four-vector can be positive, negative, or zero; accordingly, the four-vector is called *timelike-*, *spacelike-* and *null-vector* respectively.

We can write $A^\mu = g^{\mu\nu} A_\nu$, where

$$g^{\mu\nu} = \begin{pmatrix} 1 & 0 & 0 & 0 \\ 0 & -1 & 0 & 0 \\ 0 & 0 & -1 & 0 \\ 0 & 0 & 0 & -1 \end{pmatrix} \tag{2.84}$$

is called *metric tensor* (sometimes the Minkowski tensor or the Minkowski metric tensor), a symmetric matrix which transforms the contravariant A^μ in the covariant A_μ and vice versa.

Indeed we can also write $A_\mu = g_{\mu\nu} A^\mu$, where

$$g_{\mu\nu} = \begin{pmatrix} 1 & 0 & 0 & 0 \\ 0 & -1 & 0 & 0 \\ 0 & 0 & -1 & 0 \\ 0 & 0 & 0 & -1 \end{pmatrix} \tag{2.85}$$

is the covariant representation of the same metric tensor.

$g^{\mu\nu}$ is the completely contravariant metric tensor, $g_{\mu\nu}$ is the completely covariant metric tensor. The scalar product of two vectors can therefore be written in the form

$$A^\mu A_\mu = g_{\mu\nu} A^\mu A^\nu = g^{\mu\nu} A_\mu A_\nu\,. \tag{2.86}$$

Besides, we have that $g_{\mu\nu} g^{\mu\rho} = \delta^\rho_\mu = 1$. In this way we have enlarged the space adding a fourth dimension A^0: the time dimension.

The generic transformation between reference frames can be written expressing the Lorentz transformations by means of a 4 -matrix Λ:

$$A'_\mu = \Lambda^\nu_\mu A_\nu \tag{2.87}$$

where in the case of two frames moving along x

$$\Lambda^\nu_\mu = \begin{pmatrix} \gamma & -\gamma\,\beta & 0 & 0 \\ -\gamma\,\beta & \gamma & 0 & 0 \\ 0 & 0 & 1 & 0 \\ 0 & 0 & 0 & 1 \end{pmatrix}\,. \tag{2.88}$$

2.10.1.1 Tensors

A four-dimensional tensor of the second rank is a set of 16 quantities $A^{\mu\nu}$, which transforms like products of components of two four-vectors (i.e., they enter in covariant equations). We could similarly define four-tensor of higher rank.

For example, we could have the expression $A^{\mu}B_{\mu\sigma} \equiv C_{\sigma}$ where A^{μ} transforms as a vector, $B_{\mu\sigma}$ transforms as a product of vectors and C_{σ} is a four-vector.

A second-rank tensor can be written in three ways: covariant $A_{\mu\nu}$, contravariant $A^{\mu\nu}$ and mixed A^{μ}_{ν}. The connection between different types of components is determined from this general rule: raising or lowering a space index (1, 2, 3) changes the sign of the component, while raising or lowering the time index (0) does not.

The quantity $A^{\mu}_{\mu} = tr\left(A^{\nu}_{\mu}\right)$ is the trace of the tensor.

Our aim is now to rewrite the physical laws using these four-vectorial entities. To do that we introduce the *completely antisymmetric tensor* of rank 4. Like the tensors $g_{\mu\nu}$, $g^{\mu\nu}$, which are special because their components are the same in all coordinate systems, $\varepsilon^{\mu\nu\rho\sigma}$ has the same property.

The definition is

$$\varepsilon^{0123} = 1. \tag{2.89}$$

The components change sign under interchange of any pair of indices and clearly the nonzero components are those for which all four indices are different. Every permutation with an odd rank changes the sign. The number of component with nonzero value is $4! = 24$.

We have

$$\varepsilon_{\mu\nu\rho\sigma} = g_{\alpha\mu}g_{\beta\nu}g_{\gamma\rho}g_{\delta\sigma}\varepsilon^{\alpha\beta\gamma\delta} = -\varepsilon^{\mu\nu\rho\sigma}.$$

Thus, $\varepsilon^{\alpha\beta\gamma\delta}\varepsilon_{\alpha\beta\gamma\delta} = -24$ (number of nonzero elements changed of sign).

In fact with respect to rotations of the coordinate system, the quantities $\varepsilon^{\alpha\beta\gamma\delta}$ behave like the components of a tensor, but if we change the sign of one or three of the coordinates the components $^{\alpha\beta\gamma\delta}$, being defined as the same in all coordinate systems, do not change, whereas some of the components of a tensor should change sign.

2.10.1.2 An Example of Tensor: The Metric Tensor

The invariant interval can be written as

$$ds^2 = g_{\mu\nu}dx^{\mu}dx^{\nu}. \tag{2.90}$$

Under a Lorentz transformation, we get

$$ds^2 = g_{\mu\nu}dx'^{\mu}dx'^{\nu} = g_{\mu\nu}\Lambda^{\mu}{}_{\rho}\Lambda^{\nu}{}_{\sigma}dx^{\rho}dx^{\sigma}. \tag{2.91}$$

Since the interval is invariant, we get

$$g_{\mu\nu}\Lambda^{\mu}{}_{\rho}\Lambda^{\nu}{}_{\sigma}dx^{\rho}dx^{\sigma} = g_{\rho\sigma}dx^{\rho}dx^{\sigma} \Longrightarrow \left(g_{\mu\nu}\Lambda^{\mu}{}_{\rho}\Lambda^{\nu}{}_{\sigma} - g_{\rho\sigma}\right)dx^{\rho}dx^{\sigma} = 0. \tag{2.92}$$

Since the last equation must be true for any infinitesimal interval, the quantity in parentheses must be zero, so

$$g_{\rho\sigma} = g_{\mu\nu}\Lambda^{\mu}{}_{\rho}\Lambda^{\nu}{}_{\sigma}. \tag{2.93}$$

2.10.1.3 Covariant Derivatives

As a consequence of the total differential theorem, $(\partial s/\partial x^{\mu})dx^{\mu}$ is equal to the scalar ds. Thus $(\partial s/\partial x^{\mu})$ is a four-vector and since x^{μ} is contravariant, it is covariant, because ds is scalar. We call the operator $\partial_{\mu} = \dfrac{\partial}{\partial x^{\mu}}$ four-gradient.

We can write $\dfrac{\partial \phi}{\partial x^{\mu}} = \left(\dfrac{1}{c}\dfrac{\partial \phi}{\partial t}, \vec{\nabla}\phi\right)$. In general, the operators of differentiation with respect to the coordinates $x^{\mu} \equiv (ct, x, y, z)$, should be regarded as the covariant components of the operator four-gradient. For example,

$$\partial_{\mu} = \frac{\partial}{\partial x^{\mu}} = \left(\frac{1}{c}\frac{\partial}{\partial t}, \vec{\nabla}\right) \quad \partial^{\mu} = \frac{\partial}{\partial x_{\mu}} = \left(\frac{1}{c}\frac{\partial}{\partial t}, -\vec{\nabla}\right). \tag{2.94}$$

We can build from covariant quantities the operator

$$\partial_{\mu}\partial^{\mu} = \frac{1}{c^2}\frac{\partial^2}{\partial t^2} - \vec{\nabla}^2 \equiv \Box ; \tag{2.95}$$

this is called the D'Alembert operator.

2.10.2 Space-Time Interval

Two events in space-time can have a space-time difference such that

$$\begin{aligned} &\Delta s^2 > 0 \,(\text{time} - \text{like interval}) \\ &\Delta s^2 = 0 \\ &\Delta s^2 < 0 \,(\text{space} - \text{like interval}) \,. \end{aligned}$$

We remind that, due to the invariance of c, Δs^2 is an invariant.

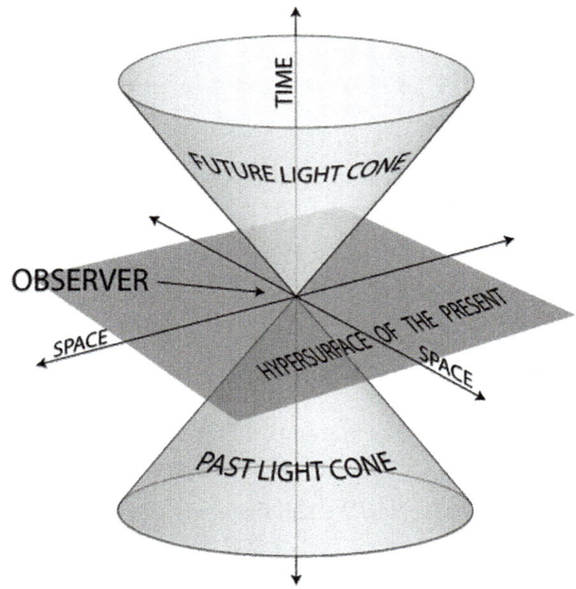

Fig. 2.15 Light cone. By Stib at en.wikipedia [GFDL http://www.gnu.org/copyleft/fdl.html], via Wikimedia Commons

The space-time is divided thus into two regions by the hypercone of the $\Delta s^2 = 0$ events (the so-called light cone, Fig. 2.15). If the interval between two causally connected events is "time-like" (no time travels, sorry) then the maximum speed at which information can be transmitted is c.

2.10.3 Energy and Momentum

Energy and momentum conservation have a deep meaning for physicists (they are closely connected to the invariance of physical laws under time and space translations). In order to keep these conservation laws in special relativity, new definitions have to be introduced:

$$E = \gamma m c^2$$
$$\vec{p} = \gamma m \vec{v} \, .$$

Expanding the new energy definition in powers of v:

$$E = mc^2 + \frac{1}{2} m v^2 + \frac{3}{8} m \frac{v^4}{c^2} + ... \qquad (2.96)$$

The classical expression for kinetic energy is recovered at first order but a constant term is now present. At rest, the energy of a body is $E_0 = mc^2$. In the words of

Einstein "mass and energy are both but different manifestations of the same thing." The dream of the alchemists is possible, but the recipe is very different.

The "Lorentz boost" γ of a particle and its velocity β normalized to the speed of light can be obtained as:

$$\gamma = E/mc^2$$
$$|\beta| = |pc|/E\,.$$

Energy and momentum form a four-vector p^μ with

$$p^0 = E/c\,;\ p^1 = p_x\,;\ p^2 = p_y\,;\ p^3 = p_z\,, \quad (2.97)$$

and the same Lorentz transformations, valid for any four-vector, apply to this object:

$$E/c = \gamma(E'/c + \beta\, p'_x)$$
$$p_x = \gamma(p'_x + \beta\, E'/c)$$
$$p_y = p'_y$$
$$p_z = p'_z.$$

The scalar product of p^μ with itself is by definition invariant and the result is

$$p^2 = p_\mu p^\mu = (E/c)^2 - |\vec{p}|^2 = m^2c^2 \quad (2.98)$$

and thus

$$E^2 = m^2c^4 + |\vec{p}|^2c^2\,. \quad (2.99)$$

While in classical mechanics you cannot have a particle of zero mass, this is possible in relativistic mechanics. The particle will have 4-momentum

$$p^\mu = (E/c, \vec{p}) \quad (2.100)$$

with

$$E^2 - p^2c^2 = 0\,, \quad (2.101)$$

and thus will move at the speed of light. The converse is also true: if a particle moves at the speed of light, its rest mass is zero—but the particle still carries a momentum E/c. The photon is such a particle.

2.10.4 Examples of Relativistic Dynamics

2.10.4.1 Decay

Let a body of mass M spontaneously decay into two parts with masses m_1 and m_2 respectively. In the frame of reference in which the body is at rest, energy conservation gives

$$Mc^2 = E_1 + E_2 \,. \tag{2.102}$$

where E_1 and E_2 are the energies of the final-state particles. Since for a particle of mass m $E \geq m$, this requires that $M \geq (m_1 + m_2)$: a body can decay spontaneously only into parts for which the sum of masses is smaller or equal to the mass of the body. If $M < (m_1 + m_2)$ the body is stable (with respect to that particular decay), and if we want to generate that process we have to supply from outside an amount of energy at least equal to its "binding energy" $(m_1 + m_2 - M)c^2$.

Momentum must be conserved as well in the decay: in the rest frame of the decaying particle $p_1 + p_2 = 0$. Consequently, $p_1^2 = p_2^2$ or

$$E_1^2 - m_1^2c^2 = E_2^2 - m_2^2c^2 \,. \tag{2.103}$$

Solving the two equations above, one gets

$$E_1 = \frac{M^2 + m_1^2 - m_2^2}{2M}c^2 \quad ; \quad E_2 = \frac{M^2 + m_2^2 - m_1^2}{2M}c^2 \,. \tag{2.104}$$

2.10.4.2 Elastic Scattering

Let us consider, from the point of view of relativistic mechanics, the elastic collision of particles. We denote the momenta and energies of the two colliding particles (with masses m_1 and m_2 by p_1^i and p_2^i respectively; we use primes for the corresponding quantities after collision. The laws of conservation of momentum and energy in the collision can be written together as the equation for conservation of the four-momentum:

$$p_1^\mu + p_2^\mu = p_1^{'\mu} + p_2^{'\mu} \,. \tag{2.105}$$

We rewrite it as $p_1^\mu + p_2^\mu - p_1^{'\mu} = p_2^{'\mu}$ and square:

$$p_1^\mu + p_2^\mu - p_1^{'\mu} = p_2^{'\mu} \Longrightarrow m_1^2c^4 + p_1^\mu p_{2\mu} - p_{1\mu}p_1^{'\mu} - p_{2\mu}p_1^{'\mu} = 0 \,. \tag{2.106}$$

Similarly,

$$p_1^\mu + p_2^\mu - p_2^{'\mu} = p_1^{'\mu} \Longrightarrow m_2^2c^4 + p_1^\mu p_{2\mu} - p_{2\mu}p_2^{'\mu} - p_{1\mu}p_2^{'\mu} = 0 \,. \tag{2.107}$$

Let us consider the collision in a reference frame in which one of the particles (m_2) was at rest before the collision. Then $\vec{p}_2 = 0$, and $p_1^{\mu} p_{2\mu} = E_1 m_2 c^2$, $p_{2\mu} p_1^{'\mu} = m_2 E_1' c^2$, $p_{1\mu} p_1^{'\mu} = E_1 E_1' - p_1 p_1' c^2 \cos\theta_1$ where $\cos\theta_1$ is the angle of scattering of the incident particle m_1. Substituting these expressions into Eq. (2.106) we get

$$\cos\theta_1 = \frac{E_1'(E_1 + m_2 c^2) - E_1 m_2 c^2 - m_1 c^4}{p_1 p_1' c^2} . \tag{2.108}$$

We note that if $m_1 > m_2$, i.e., if the incident particle is heavier than the target particle, the scattering angle θ_1 cannot exceed a certain maximum value. It is easy to find by elementary computations that this value is given by the equation

$$\sin\theta_{1\,\mathrm{max}} = m_2/m_1 \tag{2.109}$$

which coincides with the familiar classical result.

2.10.5 Mandelstam Variables

The kinematics of two-to-two particle scattering (two incoming and two outgoing particles, see Fig. 2.16) can be expressed in terms of Lorentz invariant scalars, the Mandelstam variables s, t, u, obtained as the square of the sum (or subtraction) of the four-vectors of two of the particles involved.

If p_1 and p_2 are the four-vectors of the incoming particles and p_3 and p_4 are the four-vectors of the outgoing particles, the Mandelstam variables are defined as

$$\begin{aligned} s &= (p_1 + p_2)^2 \\ t &= (p_1 - p_3)^2 \\ u &= (p_1 - p_4)^2 . \end{aligned}$$

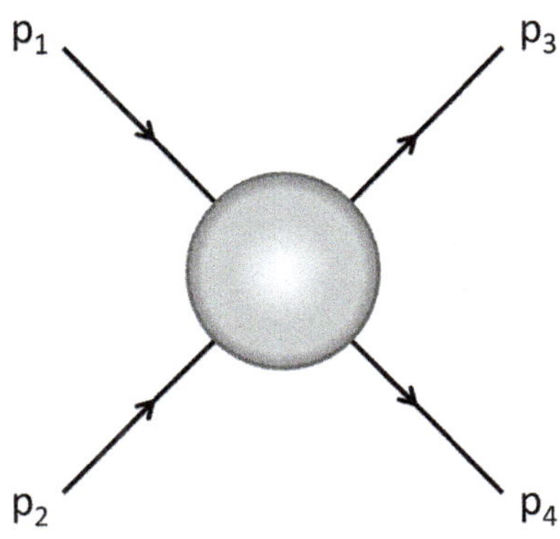

Fig. 2.16 Two-to-two particle scattering

The s variable is the square of the center-of-mass energy. In the center-of-mass reference frame S^*:

$$s = \left((E_1^*, \vec{p}^*) + (E_2^*, -\vec{p}^*)\right)^2 = E_{CM}^2 = (E_1^* + E_2^*)^2. \tag{2.110}$$

In the laboratory reference frame S:

$$\begin{aligned} s &= \left((E_{\text{beam}}, \vec{p}_{\text{beam}}) + (M_{\text{target}}c^2, 0)\right)^2 = E_{CM}^2 \\ &= M_{\text{beam}}^2 c^4 + M_{\text{target}}^2 c^4 + 2E_{\text{beam}} M_{\text{target}} c^2 . \end{aligned}$$

The center-of-mass energy is then proportional to the beam energy in a collider and (asymptotically for very high energies) to the square root of the beam energy in a fixed target experiment.

If the interaction is mediated by an intermediate particle X resulting from the "fusion" of particles 1 and 2 (s-channel, see Fig. 2.17 left),

$$1 + 2 \to X \to 3 + 4. \tag{2.111}$$

s is the square of the X particle four-vector, and one must have

$$s \geq M_X c^2 \tag{2.112}$$

so that particle X can live in our real world.

If the interaction is mediated by a particle X emitted by particle 1 and absorbed by particle 2 (t-channel, see Fig. 2.17 center):

$$t = ((E_1, \vec{p}_1) - (E_3, \vec{p}_3))^2 = \left((E_1 - E_3)^2 - (\vec{p}_1 - \vec{p}_3)^2\right)^2 . \tag{2.113}$$

If $mc^2 \ll E$

$$t = q^2 \simeq -4E_1 E_3 \sin^2\left(\frac{\theta}{2}\right) \tag{2.114}$$

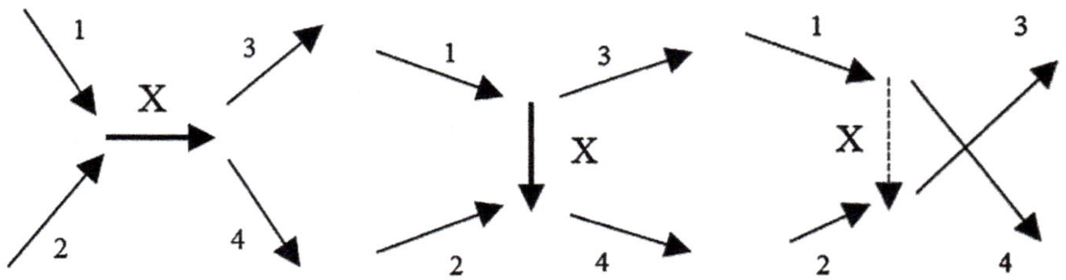

Fig. 2.17 Two-to-two particles interaction channels: *left* s-channel; *center* t-channel; *right* u-channel

where q, the four-vector of particle X, is the generalization in four dimensions of the momentum transfer previously introduced. Due to its space-like character q^2 is negative and then

$$q^2 \simeq -\vec{q}^{\,2}\,. \tag{2.115}$$

To avoid a negative sign a new variable Q^2 is defined as $Q^2 = -q^2$.

Finally, the u-channel is equivalent to the t-channel with the roles of particle 3 and 4 interchanged. It is relevant mainly in backward scattering processes.

In two-to-two scattering processes, there are eight outgoing variables (two four-vectors), four conservation equations (energy and momentum), and a relation between the energy of each outgoing particle and its momentum (see previous section). Then, there are only two independent outgoing variables and s, t, u must be related. In fact,

$$s + t + u = m_1c^2 + m_2c^2 + m_3c^2 + m_4c^2. \tag{2.116}$$

2.10.6 Lorentz Invariant Fermi Rule

In nonrelativistic quantum mechanics, the probability density $|\psi(\vec{r})|^2$ is usually normalized to 1 in some arbitrary box of volume V. However, V is not a Lorentz invariant and therefore the transition amplitude H'_{fi}, the density of final states $\rho(E_i)$, and the flux J as defined previously are not Lorentz invariant. The adopted convention is to normalize the density of probability to $2E$ (EV is a Lorentz invariant and the factor 2 is historical). The transition rate (2.38) is then redefined as

$$\lambda = \frac{2\pi}{\hbar} \frac{|\mathcal{M}|^2}{\prod_{i=1}^{n_i} 2E_i} \rho_{n_f}(E) \tag{2.117}$$

where the scattering amplitude is

$$|\mathcal{M}|^2 = |H'_{fi}|^2 \prod_{i=1}^{n_i} 2E_i \prod_{f=1}^{n_f} 2E_f \tag{2.118}$$

and the relativistic phase space is

$$\rho_{n_f}(E) = \frac{1}{(2\pi\hbar)^{3n_f}} \int \prod_{f=1}^{n_f} 2E_f \delta\left(\sum_{f=1}^{n_f} \vec{p}_f - \vec{p}_0\right) \delta\left(\sum_{f=1}^{n_f} E_f - E_0\right). \tag{2.119}$$

n_i and n_f are the number of particles in the initial and final states, respectively, and the δ functions ensure the conservation of linear momentum and energy.

In the case of a two-body final state, the phase space in the center-of-mass frame is simply

$$\rho_2(E^*) = \frac{\pi}{(2\pi\hbar)^6}\frac{|\vec{p}^*|}{E^*} \tag{2.120}$$

where $\vec{p}^*$ and E^* are the linear momentum of each final state particle and the total energy in the center-of-mass reference frame, respectively. The flux is now defined as

$$J = 2E_a 2E_b v_{ab} = 4F \tag{2.121}$$

where v_{ab} is the relative velocity of the two interacting particles and F is called the Möller's invariant flux factor. In terms of the four-vectors p_a and p_b of the incoming particles:

$$F = \sqrt{(p_a.p_b)^2 - {m_a}^2 {m_b}^2 c^4} \tag{2.122}$$

or, in terms of invariant variables,

$$F = \sqrt{\left(s - \left(m_a c^2 + m_b c^2\right)^2\right)\left(s - \left(m_a c^2 - m_b c^2\right)^2\right)}. \tag{2.123}$$

Putting together all the factors, the cross-section for the two-particle interaction is given as

$$\sigma_{a+b\to 1+2+\cdots+n_f} = \frac{1}{4F}\frac{S\hbar^2}{(2\pi)^{3n_f-4}}\int |\mathcal{M}|^2 \prod_{f=1}^{n_f}\frac{d^3p_f}{2E_f}\delta\left[\sum_{f=1}^{n_f}\vec{p_f} - \vec{p_0}\right]\delta\left[\sum_{f=1}^{n_f}E_f - E_0\right] \tag{2.124}$$

where S is a statistical factor that corrects for double counting whenever there are identical particles and also accounts for spin statistics.

In the special case of a two-to-two body scattering in the center-of-mass frame, a simple expression for the differential cross section $\frac{d\sigma}{d\Omega}$ can thus be obtained (if $|\mathcal{M}|^2$ is a function of the final momentum, the angular integration cannot be carried out):

$$\frac{d\sigma}{d\Omega} = \left(\frac{\hbar c}{8\pi}\right)^2 \frac{S\,|\mathcal{M}|^2}{s}\frac{|\vec{p_f}|}{|\vec{p_i}|}\,. \tag{2.125}$$

The partial width can be computed using again the relativistic version of the Fermi golden rule applied to the particular case of a particle at rest decaying into a final state with n_f particles

$$\Gamma_i = \frac{1}{2\hbar m_i}\frac{S}{(2\pi)^{(3n_f-4)}}\int |\mathcal{M}|^2 \prod_{f=1}^{n_f}\frac{d^3p_f}{2E_f}\delta\left(\sum_{f=1}^{n_f}\vec{p_f} - \vec{p_0}\right)\delta\left(\sum_{f=1}^{n_f}E_f - E_0\right). \tag{2.126}$$

In the particular case of only two particles in the final state, it simplifies to

$$\Gamma_i = \frac{S|\vec{p^*}|}{8\pi\hbar m_i^2 c}|\mathcal{M}|^2 \qquad (2.127)$$

where $\vec{p^*}$ is the linear momentum of each final state in the center-of-mass reference frame.

2.10.7 The Electromagnetic Tensor and the Covariant Formulation of Electromagnetism

In order to make the electromagnetism equations covariant, we introduced a new formalism, the quadrivectorial formalism. But the electromagnetic field is expressed in terms of 3-plets $\vec{\mathcal{E}}$ and $\vec{\mathcal{B}}$, which do not allow us to express the equations in a 4-vectorially covariant way.

We shall now write the equations of electromagnetism, Maxwell's equations (2.73–2.75, 2.77), in a completely covariant form.

First, let us express the electric and magnetic fields through the vector and scalar potential.

We begin examining the Maxwell equation $\vec{\nabla} \cdot \vec{\mathcal{B}} = 0$—the simplest of the equations. We know that it implies that $\vec{\mathcal{B}}$ can be expressed as the curl of a vector field. So, we write it in the form:

$$\vec{\mathcal{B}} = \vec{\nabla} \times \vec{A}\,. \qquad (2.128)$$

We take next the Faraday law, $\vec{\nabla} \times \vec{\mathcal{E}} = -\partial\vec{\mathcal{B}}/\partial t$. If we express $\vec{\mathcal{B}}$ as a function of the vector potential, and differentiate with respect to it, we can write Faraday's law in the form $\vec{\nabla} \times \vec{\mathcal{E}} + \partial(\vec{\nabla} \times \vec{A})/\partial t = 0$. Since we can differentiate either with respect to time or to space first, we can also write this equation as

$$\vec{\nabla} \times \left(\vec{\mathcal{E}} + \frac{\partial \vec{A}}{\partial t}\right) = 0\,. \qquad (2.129)$$

We see that $\vec{\mathcal{E}} + \partial\vec{A}/\partial t$ is a vector whose curl is equal to zero. Therefore that vector can be expressed as the gradient of a scalar field. In electrostatics, we take $\vec{\mathcal{E}}$ to be the gradient of $-\phi$. We do the same thing for $\vec{\mathcal{E}} + \partial\vec{A}/\partial t$; we set

$$\vec{\mathcal{E}} + \frac{\partial \vec{A}}{\partial t} = -\vec{\nabla}\phi\,. \qquad (2.130)$$

We use the same symbol ϕ, so that, in the electrostatic case where nothing changes, the relation $\vec{\mathcal{E}} = -\vec{\nabla}\phi$ still holds. Thus Faraday's law can be put in the form

$$\vec{\mathcal{E}} = -\vec{\nabla}\phi - \frac{\partial \vec{A}}{\partial t} . \tag{2.131}$$

We have solved two of Maxwell's equations already, and we have found that to describe the electromagnetic fields $\vec{\mathcal{E}}$ and $\vec{\mathcal{B}}$, we need four potential functions: a scalar potential ϕ, and a vector potential $\vec{A}$, which is, of course, three functions.

Now that $\vec{A}$ determines part of $\vec{\mathcal{E}}$, as well as $\vec{\mathcal{B}}$, what happens when we change $\vec{A}$ to $\vec{A}' = \vec{A} + \vec{\nabla}\psi$? Although $\vec{\mathcal{B}}$ does not change since $\vec{\nabla} \times \vec{\nabla}\psi = 0$, in general $\vec{\mathcal{E}}$ would change. We can, however, still allow $\vec{A}$ to be changed without affecting the electric and magnetic fields—that is, without changing the physics—if we always change $\vec{A}$ and ϕ together by the rules

$$\vec{A}' = \vec{A} + \vec{\nabla}\psi \; ; \; \phi' = \phi - \frac{\partial \psi}{\partial t} . \tag{2.132}$$

Now we return to the two remaining Maxwell equations which will give us relations between the potentials and the sources. Once we determine $\vec{A}$ and ϕ from the currents and charges, we can always get $\vec{\mathcal{E}}$ and $\vec{\mathcal{B}}$ from Eqs. (2.128) and (2.131), so we will have another form of Maxwell's equations.

We begin by substituting Eq. (2.131) into $\vec{\nabla} \cdot \vec{\mathcal{E}} = \rho/\epsilon_0$; we get

$$\vec{\nabla} \cdot \left(-\vec{\nabla}\phi - \frac{\partial \vec{A}}{\partial t} \right) = \frac{\rho}{\epsilon_0} \Longrightarrow -\nabla^2 \phi - \frac{\partial}{\partial t} \vec{\nabla} \cdot \vec{A} = \frac{\rho}{\epsilon_0} . \tag{2.133}$$

This is one equation relating ρ and $\vec{A}$ to the sources.

Our final equation will be the most complicated. Thanks to Eqs. (2.128) and (2.131), the fourth Maxwell equation

$$\vec{\nabla} \times \vec{\mathcal{B}} = \mu_0 \vec{j} + \frac{1}{c^2} \frac{\partial \vec{\mathcal{E}}}{\partial t} \tag{2.134}$$

can be written as

$$\vec{\nabla} \times (\vec{\nabla} \times \vec{A}) = \mu_0 \vec{j} + \frac{1}{c^2} \frac{\partial}{\partial t} \left(-\vec{\nabla}\phi - \frac{\partial \vec{A}}{\partial t} \right) \tag{2.135}$$

and since $\vec{\nabla} \times (\vec{\nabla} \times \vec{A}) = \vec{\nabla}(\vec{\nabla} \cdot \vec{A}) - \nabla^2 \vec{A}$ we can write

$$\begin{aligned} & -\nabla^2 \vec{A} + \left[\vec{\nabla}(\vec{\nabla} \cdot \vec{A}) + \frac{1}{c^2} \frac{\partial}{\partial t} \vec{\nabla}\phi \right] + \frac{1}{c^2} \frac{\partial^2 \vec{A}}{\partial t^2} = \mu_0 \vec{j} \\ \Longrightarrow & -\nabla^2 \vec{A} + \left[\vec{\nabla} \left(\vec{\nabla} \cdot \vec{A} + \frac{1}{c^2} \frac{\partial \phi}{\partial t} \right) \right] + \frac{1}{c^2} \frac{\partial^2 \vec{A}}{\partial t^2} = \mu_0 \vec{j} . \end{aligned} \tag{2.136}$$

Fortunately, we can now make use of our freedom to choose arbitrarily the divergence of $\vec{A}$, which is guaranteed by (2.132). What we are going to do is to use our choice to fix things so that the equations for $\vec{A}$ and for ϕ are separated but have the same form. We can do this by taking (this is called the Lorenz[8] gauge):

$$\vec{\nabla} \cdot \vec{A} = -\frac{1}{c^2}\frac{\partial \phi}{\partial t} . \tag{2.137}$$

When we do that, the two terms in square brackets in $\vec{A}$ and ϕ in Eq. (2.136) cancel, and that equation becomes much simpler:

$$\frac{1}{c^2}\frac{\partial^2 \vec{A}}{\partial t^2} - \vec{\nabla}^2 \vec{A} = \mu_0 \vec{j} \Longrightarrow \Box \vec{A} = \mu_0 \vec{j} \tag{2.138}$$

and also the equation for ϕ takes a similar form:

$$\frac{1}{c^2}\frac{\partial^2 \phi}{\partial t^2} - \vec{\nabla}^2 \phi = \frac{\rho}{\epsilon_0} \Longrightarrow \Box \phi = \frac{\rho}{\epsilon_0} . \tag{2.139}$$

These equations are particularly fascinating. We can easily obtain from Maxwell's equations the continuity equation for charge

$$\vec{\nabla} \cdot \vec{j} + \frac{\partial \rho}{\partial t} = 0 .$$

If there is a net electric current is flowing out of a region, then the charge in that region must be decreasing by the same amount. Charge is conserved. This provides a proof that $j^\mu = (\rho/c, \vec{j})$ is a four-vector, since we can write

$$\partial_\mu j^\mu = 0 . \tag{2.140}$$

If we define the 4-ple $A^\mu = (\phi/c, \vec{A})$, the Eqs. (2.138), (2.139) can be written all together as

$$\Box A^\mu = \mu_0 j^\mu . \tag{2.141}$$

Thus the 4-ple A^μ is also a four-vector; we call it the 4-potential of the electromagnetic field. Considering this fact, it appears clear that the Lorenz gauge (2.137) is covariant, and can be written as

$$\partial_\mu A^\mu = 0 . \tag{2.142}$$

[8]Ludvig Lorenz (1829–1891), not to be confused with Hendrik Antoon Lorentz, was a Danish mathematician and physicist, professor at the Military Academy in Copenhagen.

In regions where there are no longer any charges and currents, the solution of Eq. (2.141) is a 4-potential, which is changing in time but always moving out at the speed c. This 4-field travels onward through free space.

Since A^μ is a four-vector, it is thus a 4-tensor the antisymmetric matrix

$$F^{\mu\nu} = \partial^\mu A^\nu - \partial^\nu A^\mu . \tag{2.143}$$

Obviously the diagonal elements of this tensor are null. The 0-th row and column are, respectively,

$$F^{0i} = \partial^0 A^i - \partial^i A^0 = \frac{1}{c}\frac{\partial A^i}{\partial t} + \frac{\partial \phi}{\partial x^i} = -\mathcal{E}^i/c \tag{2.144}$$

$$F^{i0} = -F^{0i} = \mathcal{E}^i/c . \tag{2.145}$$

The 1...3 part or the matrix is

$$F^{12} = \partial^1 A^2 - \partial^2 A^1 = -(\vec{\nabla} \times \vec{A})_z = -B_z \tag{2.146}$$

$$F^{13} = \partial^1 A^3 - \partial^3 A^1 = (\vec{\nabla} \times \vec{A})_y = B_y \tag{2.147}$$

$$F^{23} = \partial^2 A^3 - \partial^3 A^2 = -(\vec{\nabla} \times \vec{A})_x = -B_x \tag{2.148}$$

and correspondingly for the symmetric components.

Finally, the *electromagnetic tensor is*

$$F^{\mu\nu} = \begin{pmatrix} 0 & -\mathcal{E}_x/c & -\mathcal{E}_y/c & -\mathcal{E}_z/c \\ \mathcal{E}_x/c & 0 & -B_z & B_y \\ \mathcal{E}_y/c & B_z & 0 & -B_x \\ \mathcal{E}_z/c & -B_y & B_x & 0 \end{pmatrix} . \tag{2.149}$$

The components of the electromagnetic field are thus elements of a tensor, the electromagnetic tensor.

The nonhomogeneous Maxwell equations have been written as (Eq. 2.141):

$$\Box A^\mu = (\partial_\nu \partial^\nu) A^\mu = j^\mu . \tag{2.150}$$

We can write

$$\partial_\nu F^{\nu\mu} = \partial_\nu(\partial^\nu A^\mu - \partial^\mu A^\nu) = (\partial_\nu \partial^\nu) A^\mu - \partial^\mu(\partial_\nu A^\nu) \tag{2.151}$$

and since $\partial_\nu A^\nu = 0$,

$$\partial_\nu F^{\nu\mu} = (\partial_\nu \partial^\nu) A^\mu = \Box A^\mu = j^\mu . \tag{2.152}$$

The covariant equation

$$\partial_\nu F^{\nu\mu} = j^\mu \tag{2.153}$$

is equivalent to the nonhomogeneous Maxwell equations.

In the same way, one has for the homogeneous equations:

$$\vec{\nabla} \times \vec{\mathcal{E}} + \frac{\partial \vec{\mathcal{B}}}{\partial t} = 0 \quad ; \quad \vec{\nabla} \cdot \vec{\mathcal{B}} = 0$$

the following result (4 equations):

$$\partial^2 F^{03} + \partial^3 F^{20} + \partial^0 F^{32} = 0 \quad \ldots \quad \partial^1 F^{23} + \partial^2 F^{31} + \partial^3 F^{12} = 0$$

and thus

$$\left(\vec{\nabla} \times \vec{\mathcal{E}} = -\frac{\partial \vec{\mathcal{B}}}{\partial t} \; \& \; \vec{\nabla} \cdot \vec{\mathcal{B}} = 0 \right) \Longleftrightarrow \epsilon_{\alpha\beta\gamma\delta} \partial^\beta F^{\gamma\delta} = 0 \quad (\alpha = 0, 1, 2, 3)\,.$$

Due to the tensor nature of $F_{\mu\nu}$, the two following quantities are invariant for transformations between inertial frames:

$$\frac{1}{2} F_{\mu\nu} F^{\mu\nu} = B^2 - \mathcal{E}^2/c^2$$
$$\frac{c}{8} \epsilon_{\alpha\beta\gamma\delta} F^{\alpha\beta} F^{\gamma\delta} = \vec{\mathcal{B}} \cdot \vec{\mathcal{E}}$$

where $\epsilon_{\alpha\beta\gamma\delta}$ is the completely antisymmetric unit tensor of rank 4.

2.11 Natural Units

The international system of units (SI) can be constructed on the basis of four fundamental units: a unit of length (the meter, m), a unit of time (the second, s), a unit of mass (the kilogram, kg) and a unit of charge (the coulomb, C).[9]

These units are inappropriate in the world of fundamental physics: the radius of a nucleus is of the order of 10^{-15} m (also called one femtometer or one fermi, fm); the mass of an electron is of the order of 10^{-30} kg; the charge of an electron is (in absolute value) of the order of 10^{-19} C. Using such units, we would carry along a lot of exponents. Thus, in particle physics, we rather use units like the electron charge

[9] For reasons related only to metrology (reproducibility and accuracy of the definition) in the standard SI the unit of electrical current, the ampere A, is used instead of the coulomb; the two definitions are however conceptually equivalent.

for the electrostatic charge, and the electron-volt eV and its multiples (keV, MeV, GeV, TeV) for the energy:

Length	1 fm	10^{-15} m
Mass	1 MeV/c^2	$\sim 1.78 \times 10^{-30}$ kg
Charge	$\lvert e \rvert$	$\sim 1.602 \times 10^{-19}$ C.

Note the unit of mass, in which the relation $E = mc^2$ is used implicitly: what one is doing here is to use 1 eV $\simeq 1.602 \times 10^{-19}$ J as the new fundamental unit of energy. In these new units, the mass of a proton is about 0.938 GeV/c^2, and the mass of the electron is about 0.511 MeV/c^2. The fundamental energy level of a hydrogen atom is about -13.6 eV.

In addition, nature provides us with two constants which are particularly appropriate in the world of fundamental physics: the speed of light $c \simeq 3.00 \times 10^8$ m/s $= 3.00 \times 10^{23}$ fm/s, and Planck's constant (over 2π) $\hbar \simeq 1.05 \times 10^{-34}$ J s $\simeq 6.58 \times 10^{-16}$ eV s.

It seems then natural to express speeds in terms of c, and angular momenta in terms of $\hbar$. We then switch to the so-called Natural Units (NU). The minimal set of natural units (not including electromagnetism) can then be chosen as

Speed	1 c	3.00×10^8 m/s
Angular momentum	1 $\hbar$	1.05×10^{-34} J · s
Energy	1 eV	1.602×10^{-19} J.

After the convention $\hbar = c = 1$, one single unit can be used to describe the mechanical Universe: we choose energy, and we can thus express all mechanical quantities in terms of eV and of its multiples. It is immediate to express momenta and masses directly in NU. To express 1 m and 1 s, we can write[10]

$$\begin{aligned} 1\text{ m} &= \frac{1\text{m}}{\hbar c} \simeq 5.10 \times 10^{12}\text{ MeV}^{-1} \\ 1\text{ s} &= \frac{1\text{s}}{\hbar} \simeq 1.52 \times 10^{21}\text{ MeV}^{-1} \\ 1\text{ kg} &= 1\text{J}/c^2 \simeq 5.62 \times 10^{29}\text{ MeV} \end{aligned}$$

Both length and time are thus, in natural units, expressed as inverse of energy. The first relation can also be written as 1fm $\simeq 5.10$ GeV^{-1}: note that when you have a quantity expressed in MeV^{-1}, in order to express it in GeV^{-1}, you must multiply (and not divide) by a factor of 1000.

[10] $\hbar c \simeq 1.97 \times 10^{-13}$MeV · m $= 3.15 \times 10^{-26}$J · m.

Let us now find a general rule to transform quantities expressed in natural units into SI, and vice versa. To express a quantity in NU back in SI, we first restore the $\hbar$ and c factors by dimensional arguments and then use the conversion factors $\hbar$ and c (or $\hbar c$) to evaluate the result. The dimensions of c are [m/s]; the dimensions of $\hbar$ are [kg m^2 s^{-1}].

The vice versa (from SI to NU) is also easy. A quantity with meter-kilogram-second [m · k · s] dimensions $M^p L^q T^r$ (where M represents mass, L length and T time) has the NU dimensions $[E^{p-q-r}]$, where E represents energy. Since $\hbar$ and c do not appear in NU, this is the only relevant dimension, and dimensional checks and estimates are very simple. The quantity Q in SI can be expressed in NU as

$$Q_{\rm NU} = Q_{\rm SI}\left(5.62\times 10^{29}\frac{\rm MeV}{\rm kg}\right)^p\left(5.10\times 10^{12}\frac{\rm MeV^{-1}}{\rm m}\right)^q \times\left(1.52\times 10^{21}\frac{\rm MeV^{-1}}{\rm s}\right)^r {\rm MeV}^{p-q-r}$$

The NU and SI dimensions are listed for some important quantities in Table 2.1.

Note that choosing natural units, all factors of $\hbar$ and c may be omitted from equations, which leads to considerable simplifications (we will profit from this in the next chapters). For example, the relativistic energy relation

$$E^2 = p^2c^2 + m^2c^4 \tag{2.154}$$

becomes

$$E^2 = p^2 + m^2 \tag{2.155}$$

Finally, let us discuss how to treat electromagnetism. To do so, we must introduce a new unit, charge for example. We can redefine the unit charge by observing that

Table 2.1 Dimensions of different physical quantities in SI and NU

	SI			NU
Quantity	p	q	r	n
Mass	1	0	0	1
Length	0	1	0	−1
Time	0	0	1	−1
Action ($\hbar$)	1	2	−1	0
Velocity (c)	0	1	−1	0
Momentum	1	1	−1	1
Energy	1	2	−2	1

$$\frac{e^2}{4\pi\epsilon_0} \tag{2.156}$$

has the dimension of [J · m], and thus is a pure, dimensionless, number in NU. Dividing by $\hbar c$ one has

$$\frac{e^2}{4\pi\epsilon_0 \hbar c} \simeq \frac{1}{137} . \tag{2.157}$$

Imposing that the electric permeability of vacuum $\epsilon_0 = 1$ (thus automatically $\mu_0 = 1$ for the magnetic permeability of vacuum, since from Maxwell's equations $\epsilon_0\mu_0 = 1/c^2$) we obtain the new definition of charge, and with this definition:

$$\alpha = \frac{e^2}{4\pi} \simeq \frac{1}{137} . \tag{2.158}$$

This is called the Lorentz-Heaviside convention. Elementary charge in NU becomes then a pure number

$$e \simeq 0.303. \tag{2.159}$$

Let us make now some applications.

The Thomson cross-section. Let us express a cross-section in NU. The cross-section for Compton scattering of a photon by a free electron is, for $E \ll m_e c^2$ (Thomson regime),

$$\sigma_T \simeq \frac{8\pi\alpha^2}{3m_e^2} . \tag{2.160}$$

The dimension of a cross-section is, in SI, [m^2]. Thus we can write

$$\sigma_T \simeq \frac{8\pi\alpha^2}{3m_e^2} \hbar^a c^b \tag{2.161}$$

and determine a and b such that the result has the dimension of a length squared. We find $a = 2$ and $b = -2$; thus

$$\sigma_T \simeq \frac{8\pi\alpha^2}{3(m_e c^2)^2} (\hbar c)^2 \tag{2.162}$$

and thus $\sigma_T \simeq 6.65 \times 10^{-29}$ m^2 = 665 mb.

The Planck mass, length, and time. According to quantum theory, a length called the Compton wavelength, λ_C, can be associated to any mass m. λ_C is defined as the wavelength of a photon with an energy equal to the rest mass of the particle:

$$\lambda_C = \frac{h}{mc} = 2\pi \frac{\hbar}{mc} . \tag{2.163}$$

The Compton wavelength sets the distance scale at which quantum field theory becomes crucial for understanding the behavior of a particle: wave and particle description become complementary at this scale.

On the other hand, we can compute for any mass m the associated Schwarzschild radius, R_S, such that compressing it to a size smaller than this radius we form a black hole. The Schwarzschild radius is the scale at which general relativity becomes crucial for understanding the behavior of the object, and

$$R_S = \frac{2G_N m}{c^2} \tag{2.164}$$

where G_N is the gravitational constant.

We call Planck mass the mass at which the Schwarzschild radius of a particle becomes equal to its Compton length, and Planck length their common value when this happens. The probe that could locate a particle within this distance would collapse to a black hole, something that would make measurements very weird. In NU, one can write

$$\frac{2\pi}{m_P} = 2G_N m_P \rightarrow m_P = \sqrt{\frac{\pi}{G_N}} \tag{2.165}$$

which can be converted into

$$m_P = \sqrt{\frac{\pi \hbar c}{G_N}} \simeq 3.86 \times 10^{-8}\text{kg} \simeq 2.16 \times 10^{19}\,\text{GeV/c}^2. \tag{2.166}$$

Since we are talking about orders of magnitude, the factor $\sqrt{\pi}$ is often neglected and we take as a definition:

$$m_P = \sqrt{\frac{\hbar c}{G_N}} \simeq 2.18 \times 10^{-8}\text{kg} \simeq 1.22 \times 10^{19}\,\text{GeV/c}^2. \tag{2.167}$$

Besides the Planck length ℓ_P, we can also define a Planck time $t_P = \ell_P/c$ (their value is equal in NU):

$$\ell_P = t_P = \frac{1}{m_P} = \sqrt{G_N} \tag{2.168}$$

(this corresponds to a length of about 1.6×10^{-20} fm, and to a time of about 5.4×10^{-44} s).

Both general relativity and quantum field theory are needed to understand the physics at mass scales about the Planck mass or distances about the Planck length, or times comparable to the Planck time. Traditional quantum physics and gravitation certainly fall short at this scale; since this failure should be independent of the reference frame, many scientists think that the Planck scale should be an invariant irrespective of the reference frame in which it is calculated (this fact would of course require important modifications to the theory of relativity).

Note that the shortest length you may probe with the energy of a particle accelerated by LHC is about 10^{15} times larger than the Planck length scale. Cosmic rays, whcih can reach c.m. energies beyond 100 TeV, are at the frontier of the exploration of fundamental scales.

Further Reading

[F2.1] J.S. Townsend, "A modern approach to quantum mechanics", Mc Graw-Hill 2012. An excellent quantum mechanics course at an advanced undergraduate level.

[F2.2] W. Rindler, "Introduction to Special Relativity", second edition, Oxford University Press 1991. A classical textbook on special relativity for undergraduates.

Exercises

1. *Rutherford formula.* Consider the Rutherford formula.
   (a) Determine the distance of closest approach of an α particle with an energy of 7.7 MeV to a gold target .
   (b) Determine the de Broglie wavelength of that α particle.
   (c) Explain why the classical Rutherford formula survived the revolution of quantum mechanics.

   You can find the numerical values of particle data and fundamental constants in the Appendices, or in your Particle Data Book(let).
2. *Cross-section at fixed target.* Consider a fixed target experiment with a monochromatic proton beam with an energy of 20 GeV and a 2 m long liquid hydrogen (H_2) target ($\rho = 60\,\mathrm{kg/m^3}$). In the detector placed just behind the target beam fluxes of 7×10^6 protons/s and 10^7 protons/s are measured, respectively, with the target full and empty. Determine the proton–proton total cross-section at this energy and its statistical error:
   (a) without taking into account the attenuation of the beam inside the target;
   (b) taking into account the attenuation of the beam inside the target.

3. *LHC collisions.* The LHC running parameters in 2012 were, for a c.m. energy $\sqrt{s} \simeq 8$ TeV: number of bunches = 1400; time interval between bunches $\simeq$ 50 ns; number protons per bunch $\simeq 1.1 \times 10^{11}$; beam width at the crossing point $\simeq 16\,\mu$m.

   (a) Determine the maximum instantaneous luminosity of the LHC in 2012.
   (b) Determine the number of interactions per collision ($\sigma_{pp} \sim 100$ mb).
   (c) As you probably heard, LHC found a particle called Higgs boson, which Leon Lederman called the "God particle" (a name the news like very much). If Higgs bosons are produced with a cross-section $\sigma_H \sim 21$ pb, determine the number of Higgs bosons decaying into 2 photons ($BR(H \to \gamma\gamma) \simeq 2.28 \times 10^{-3}$) which might have been produced in 2012 in the LHC, knowing that the integrated luminosity of the LHC (luminosity integrated over the time) during 2012 was around 20 fb^{-1}. Compare it to the real number of detected Higgs in this particular decay mode reported by the LHC collaborations (about 400). Discuss the difference.

4. *Classical electromagnetism is not a consistent theory.* Consider two electrons at rest, and let r be the distance between them. The (repulsive) force between the two electrons is the electrostatic force

$$F = \frac{1}{4\pi\epsilon_0}\frac{e^2}{r^2},$$

   where e is the charge of the electron; it is directed along the line joining the two charges. But an observer is moving with a velocity v perpendicular to the line joining the two charges will measure also a magnetic force (still directed as F)

$$F' = \frac{1}{4\pi\epsilon_0}\frac{e^2}{r^2} - \frac{\mu_0}{2\pi r}v^2e^2 \neq F\,.$$

   The expression of the force is thus different in the two frames of reference. But masses, charges, and accelerations are classically invariant. Comment.

5. *GZK threshold.* The Cosmic Microwave Background fills the Universe with photons with a peak energy of 0.37 meV and a density of $\rho \sim 400/\text{cm}^3$. Determine:

   (a) The minimal energy (known as the GZK threshold) that a proton should have in order that the reaction $p\gamma \to \Delta$ may occur.
   (b) The interaction length of such protons in the Universe considering a mean cross-section above the threshold of 0.6 mb.

6. *$\bar{p}$ production at the Bevatron.* The antiprotons were first produced in the laboratory in 1955, in proton–proton fixed target collisions at an accelerator called Bevatron (it was named for its ability to impart energies of billions of eV, i.e., Billions of eV Synchrotron), located at Lawrence Berkeley National Laboratory, US . The discovery resulted in the 1959 Nobel Prize in physics for Emilio Segrè and Owen Chamberlain.

(a) Describe the minimal reaction able to produce antiprotons in such collisions.
(b) When a proton is confined in a nucleus, it cannot have arbitrarily low momenta, as one can understand from the Heisenberg principle; the actual value of its momentum is called the "Fermi momentum." Determine the minimal energy that the proton beam must have in order that antiprotons were produced considering that the target protons have a Fermi momentum of around 150 MeV/c.

7. *Photon conversion.* Consider the conversion of one photon in one electron–positron pair. Determine the minimal energy that the photon has to have in order that this conversion would be possible if the photon is in presence of:

(a) one proton;
(b) one electron;
(c) when no charged particle is around.

8. π^-*decay.* Consider the decay of a flying π^- into $\mu^-\bar{\nu_\mu}$ and suppose that the μ^- was emitted along the flight line of flight of the π^-. Determine:

(a) The energy and momentum of the μ^- and of the $\bar{\nu_\mu}$ in the π^- frame.
(b) The energy and momentum of the μ^- and of the $\bar{\nu_\mu}$ in the laboratory frame, if the momentum $P_\pi^- = 100$ GeV/c.
(c) Same as the previous question but considering now that was the $\bar{\nu_\mu}$ that was emitted along the flight line of the π^-.

9. π^0 *decay.* Consider the decay of a π^0 into $\gamma\gamma$ (with pion momentum of 100 GeV/c). Determine:

(a) The minimal and the maximal angles between the two photons in the laboratory frame.
(b) The probability of having one of the photons with an energy smaller than an arbitrary value E_0 in the laboratory frame.
(c) Same as (a) but considering now that the decay of the π^0 is into e^+e^-.
(d) The maximum momentum that the π^0 may have in order that the maximal angle in its decay into $\gamma\gamma$ and in e^+e^- would be the same.

10. *Invariant flux.* In a collision between two particles a and b the incident flux is given by $F = 4|\vec{v_a} - \vec{v_b}|E_a E_b$ where $\vec{v_a}$, $\vec{v_b}$, E_a and E_b are, respectively, the vectorial speeds and the energies of particles a and b.

(a) Verify that the above formula is equivalent to: $F = 4\sqrt{(P_a P_b)^2 - (m_a m_b)^2}$ where P_a and P_b are, respectively, the four-vectors of particles a and b, and m_a and m_b their masses.
(b) Relate the expressions of the flux in the center-of-mass and in the laboratory reference frames.

11. *Units.* Determine in Natural Units:

    (a) Your own dimensions (height, weight, mass, age).
    (b) The mean lifetime of the muon ($\tau_\mu = 2.2\,\mu s$).

12. *Units.* In NU the expression of the muon life time is given as

$$\tau_\mu = \frac{192\pi^3}{G_F^2 m_\mu^5}$$

where G_F is the Fermi constant.

    (a) Is the Fermi constant dimensionless? If not compute its dimension in NU and in SI.
    (b) Obtain the conversion factor for transforming G_F from SI to NU.

Chapter 3
Cosmic Rays and the Development of Particle Physics

This chapter illustrates the path which led to the discovery that particles of extremely high energy, up to a few joule, come from extraterrestrial sources and collide with Earth's atmosphere. The history of this discovery started in the beginning of the twentieth century, but many of the techniques then introduced are still in use. A relevant part of the progress happened in recent years and has a large impact on the physics of elementary particles and fundamental interactions.

By 1785, Coulomb found that electroscopes (Fig. 3.1) can discharge spontaneously, and not simply due to defective insulation. The British physicist Crookes, in 1879, observed that the speed of discharge decreased when the pressure of the air inside the electroscope itself was reduced. The discharge was then likely due to the ionization of the atmosphere. But what was the cause of atmospheric ionization?

The explanation came in the early twentieth century and led to the revolutionary discovery of cosmic rays. We know today that cosmic rays are particles of extraterrestrial origin which can reach high energy (much larger than we shall ever be able to produce). They were the only source of high energy beams till the 1940s. World War II and the Cold War provided new technical and political resources for the study of elementary particles; technical resources included advances in microelectronics and the capability to produce high energy particles in human-made particle accelerators. By 1955, particle physics experiments would be largely dominated by accelerators, at least until the beginning of the 1990s, when explorations possible with the energies one can produce on Earth started showing signs of saturation, so that nowadays cosmic rays are again at the edge of physics.

3.1 The Puzzle of Atmospheric Ionization and the Discovery of Cosmic Rays

Spontaneous radioactivity (i.e., the emission of particles from nuclei as a result of nuclear instability) was discovered in 1896 by Becquerel. A few years later, Marie and Pierre Curie discovered that Polonium and Radium (the names Radium A,

A. De Angelis and M.J.M. Pimenta, *Introduction to Particle and Astroparticle Physics*, Undergraduate Lecture Notes in Physics,
DOI 10.1007/978-88-470-2688-9_3

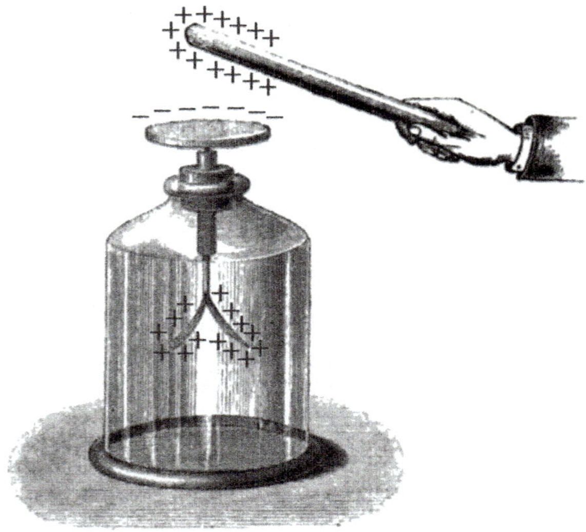

Fig. 3.1 The electroscope is a device for detecting electric charge. A typical electroscope (the configuration in the figure was invented at the end of the eighteenth century) consists of a vertical metal rod from the end of which two gold leaves hang. A disk or ball is attached to the top of the rod. The leaves are enclosed in a glass vessel, for protection against air movements. The test charge is applied to the top, charging the rod, and the gold leaves repel and diverge. By Sylvanus P. Thompson [public domain], via Wikimedia Commons

Radium B, ... indicated several isotopes of the element today called radon, and also some different elements) underwent transmutations by which they generated radioactivity; these processes were called "radioactive decays." A charged electroscope promptly
discharges in the presence of radioactive materials. It was concluded that the discharge was due to the emission of charged particles, which induce the formation of ions in the air, causing the discharge of electroscopes. The discharge rate of electroscopes was used to gauge the radioactivity level. During the first decade of the twentieth century, several researchers in Europe and in the New World presented progress on the study of ionization phenomena.

Around 1900, C.T.R. Wilson[1] in Britain and Elster and Geitel in Germany improved the sensitivity of the electroscope, by improving the technique for its insulation in a closed vessel (Fig. 3.2). This improvement allowed the quantitative measurement of the spontaneous discharge rate, and led to the conclusion that the radiation causing this discharge came from outside the vessel. Concerning the origin of such radiation, the simplest hypothesis was that its would be related to radioactive material in the surrounding of the apparatus. Terrestrial origin was thus a commonplace assumption, although experimental confirmation could not be achieved. Wilson did suggest that atmospheric ionization could be caused by a very penetrating radiation of extraterrestrial origin. His investigations in tunnels, with solid rock for

[1]Charles Thomson Rees Wilson, (1869–1959), a Scottish physicist and meteorologist, received the Nobel Prize in Physics for his invention of the cloud chamber; see the next chapter.

Fig. 3.2 *Left* The two friends Julius Elster and Hans Geitel, gymnasium teachers in Wolfenbuttel, around 1900. Credit http://www.elster-geitel.de. *Right* an electroscope developed by Elster and Geitel in the same period (private collection R. Fricke; photo by A. De Angelis)

shielding overhead, however could not support the idea, as no reduction in ionization was observed. The hypothesis of an extraterrestrial origin, though now and then discussed, was dropped for many years.

By 1909, measurements on the spontaneous discharge had proved that the discharging background radiation was also present in insulated environments and could penetrate metal shields. It was thus difficult to explain it in terms of α (He nuclei) or β (electron) radiation; it was thus assumed to be γ radiation, i.e., made of photons, which is the most penetrating among the three kinds of radiation known at the time. Three possible sources were then hypothesized for this radiation: it could be extraterrestrial (possibly from the Sun), it could be due to radioactivity from the Earth crust, or to radioactivity in the atmosphere. It was generally assumed that there had to be large contribution from radioactive materials in the crust, and calculations of its expected decrease with height were performed.

3.1.1 Experiments Underwater and in Height

Father Theodor Wulf, a German scientist and a Jesuit priest, thought of checking the variation of ionization with height to test its origin. In 1909, using an improved electroscope in which the two leaves had been replaced by metal coated silicon glass wires, making it easier to transport than previous instruments (Fig. 3.3), he measured the ionization rate at the top of the Eiffel tower in Paris, about 300 m high. Under the hypothesis that most of the radiation was of terrestrial origin, he expected the

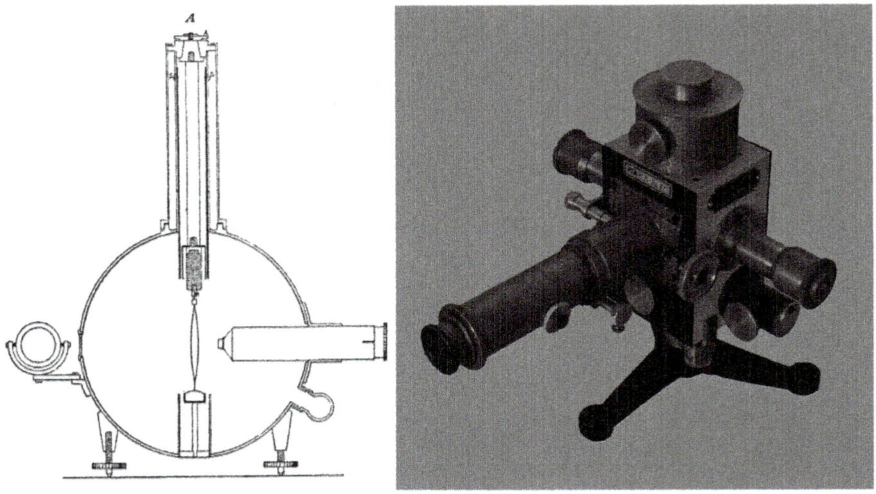

Fig. 3.3 *Left* Scheme of the Wulf electroscope (drawn by Wulf himself; reprinted from Z. Phys. [public domain]). The main cylinder was made of zinc, 17 cm in diameter and 13 cm deep. The distance between the two silicon glass wires (at the center) was measured using the microscope to the right. The wires were illuminated using the mirror to the left. According to Wulf, the sensitivity of the instrument was 1 V, as measured by the decrease of the inter-wire distance. *Right* an electroscope used by Wulf (private collection R. Fricke; photo by A. De Angelis)

ionization rate to be significantly smaller than the value at ground. The measured decrease was however too small to confirm the hypothesis: he observed that the radiation intensity "decrease at nearly 300 m [altitude] was not even to half of its ground value," while "just a few percent of the radiation" should remain if it did emerge from ground. Wulf's data, coming from experiments performed for many days at the same location and at different hours of the day, were of great value, and for long considered the most reliable source of information on the altitude variation of the ionization rate. However, his conclusion was that the most likely explanation for this unexpected result was still emission from ground.

The conclusion that atmospheric ionization was mostly due to radioactivity from the Earth's crust was challenged by the Italian physicist Domenico Pacini. Pacini developed a technique for underwater measurements and conducted experiments in the sea of the Gulf of Genova and in the Lake of Bracciano (Fig. 3.4). He found a significant decrease in the discharge rate in electroscopes placed three meters underwater. He wrote: "Observations carried out on the sea during the year 1910 led me to conclude that a significant proportion of the pervasive radiation that is found in air has an origin that is independent of direct action of the active substances in the upper layers of the Earth's surface. [...] [To prove this conclusion] the apparatus [...] was enclosed in a copper box so that it could immerse in depth. [...] Observations were performed with the instrument at the surface, and with the instrument immersed

Fig. 3.4 *Left* Pacini making a measurement in 1910. Courtesy of the Pacini family, edited by A. De Angelis [public domain, via Wikimedia Commons]. *Right* the instruments used by Pacini for the measurement of ionization. By D. Pacini (Ufficio Centrale di Meteorologia e Geodinamica), edited by A. De Angelis [public domain, via Wikimedia Commons]

in water, at a depth of 3 m." Pacini measured the discharge of the electroscope during 3 h, and repeated the measurement seven times. At the surface, the average ionization rate was 11.0 ions per cubic centimeter per second, while he measured 8.9 ions per cubic centimeter per second at a depth of 3 m in the 7 m deep sea (the depth of the water guaranteed that radioactivity from the soil was negligible). He concluded that the decrease of about 20 % was due to a radiation not coming from the Earth.

After Wulf's observations on the altitude effect, the need for balloon experiments (widely used for atmospheric electricity studies since 1885) became clear. The first high-altitude balloon with the purpose of studying the penetrating radiation was flown in Switzerland in December 1909 with a balloon from the Swiss aeroclub. Albert Gockel, professor at the University of Fribourg, ascended to 4500 m above sea level (a.s.l.). He made measurements up to 3000 m and found that ionization rate did not decrease with altitude as expected under the hypothesis of terrestrial origin. His conclusion was that "a non-negligible part of the penetrating radiation is independent of the direct action of the radioactive substances in the uppermost layers of the Earth."

In spite of Pacini's conclusions, and of Wulf's and Gockel's puzzling results on the altitude dependence, the issue of the origin of the penetrating radiation still raised doubts. A series of balloon flights by the Austrian physicist Victor Hess,[2] settled the issue, firmly establishing the extraterrestrial origin of at least part of the radiation causing the atmospheric ionization.

[2]Hess was born in 1883 in Steiermark, Austria, and graduated at Graz University in 1906 where he became professor of Experimental Physics in 1919. In 1936 Hess was awarded the Nobel Prize in Physics for the discovery of cosmic rays. He moved to the USA in 1938 as professor at Fordham University. Hess became an American citizen in 1944 and lived in New York until his death in 1964.

Fig. 3.5 *Left* Hess during the balloon flight in August 1912. [public domain], via Wikimedia Commons. *Right* one of the electrometers used by Hess during his flight. This instrument is a version of a commercial model of a Wulff electroscope especially modified by its manufacturer, Günther and Tegetmeyer, to operate under reduced pressure at high altitudes (Smithsonian National Air and Science Museum, Washington, DC). Photo by P. Carlson

Hess started by studying Wulf's results. He carefully checked the data on gamma-ray absorption coefficients (due to the large use of radioactive sources he will loose a thumb) and, after careful planning, he finalized his studies with balloon observations. The first ascensions took place in August 1911. From April 1912 to August 1912, he flew seven times, with three instruments (one of them with a thin wall to estimate the effect of β radiation, as for a given energy electrons have a shorter range than heavier particles). In the last flight, on August 7, 1912, he reached 5200 m (Fig. 3.5). The results clearly showed that the ionization rate first passied through a minimum and then increased considerably with height (Fig. 3.6). "(i) Immediately above ground the total radiation decreases a little. (ii) At altitudes of 1000–2000 m there occurs again a noticeable growth of penetrating radiation. (iii) The increase reaches, at altitudes of 3000–4000 m, already 50 % of the total radiation observed on the ground. (iv) At 4000–5200 m the radiation is stronger [more than 100 %] than on the ground."

Hess concluded that the increase in the ionization rate with altitude was due to radiation coming from above, and he thought that this radiation was of extraterrestrial origin. His observations during the day and during the night showed no variation and excluded the Sun as the direct source of this hypothetical radiation.

The results by Hess would later be confirmed by Kolhörster. In flights up to 9200 m, Kolhörster found an increase in the ionization rate up to ten times its value at sea level. The measured attenuation length of about 1 km in air at NTP came as a surprise, as it was eight times smaller than the absorption coefficient of air for γ rays as known at the time.

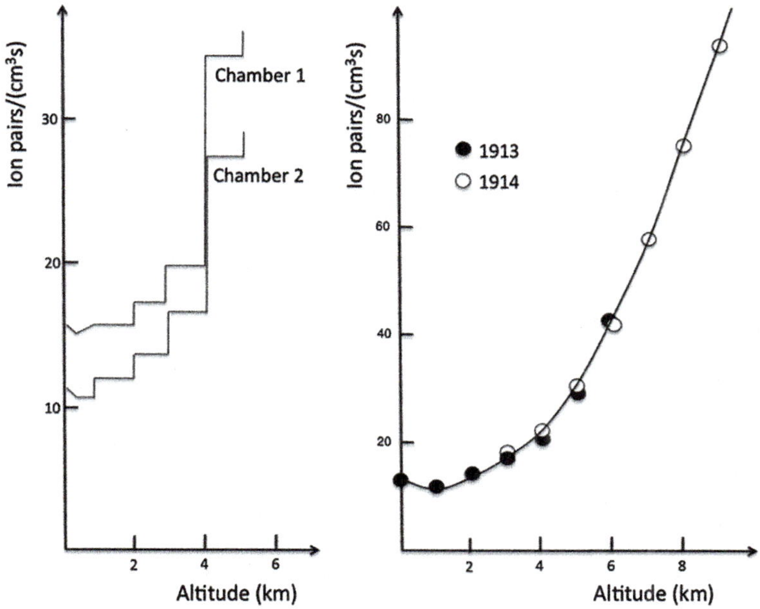

Fig. 3.6 Variation of ionization with altitude. *Left panel* Final ascent by Hess (1912), carrying two ion chambers. *Right panel* Ascents by Kolhörster (1913, 1914)

After the 1912 flights, Hess coined the name "Höhenstrahlung." Several other names were used for the extraterrestrial radiation: Ultrastrahlung, Ultra-X-Strahlung, kosmische Strahlung. The latter, used by Gockel and Wulf in 1909, inspired Millikan[3] who suggested the name "cosmic rays," which became generally accepted.

The idea of cosmic rays, despite the striking experimental evidence, was not immediately accepted (the Nobel prize for the discovery of cosmic rays will be assigned to Hess only in 1936). During the 1914–1918 war and the years that followed, very few investigations of the penetrating radiation were performed. In 1926, however, Millikan and Cameron performed absorption measurements of the radiation at different depths in lakes at high altitudes. They concluded that the radiation was made of high energy γ rays and that "these rays shoot through space equally in all directions," and called them "cosmic rays."

[3]Robert A.Millikan (Morrison 1868—Pasadena 1953) was an American experimental physicist, Nobel Prize in Physics in 1923 for his measurements of the electron charge and his work on the photoelectric effect. A scholar of classical literature before turning physics, he was president of the California Institute of Technology (Caltech) from 1921 to 1945. He was not famous for his deontology: a common saying at Caltech was "Jesus saves, and Millikan takes the credit."

3.1.2 The Nature of Cosmic Rays

Cosmic radiation was generally believed to be γ radiation because of its penetrating power (the penetrating power of relativistic charged particles was not known at the time). Millikan had launched the hypothesis that these γ rays were produced when protons and electrons formed helium nuclei in the interstellar space.

A key experiment on the nature of cosmic rays was the measurement of the intensity variation with geomagnetic latitude. During two voyages between Java and Genova in 1927 and 1928, the Dutch physicist Clay found that ionization increased with latitude; this proved that cosmic rays interacted with the geomagnetic field and, thus, they were mostly charged particles.

In 1928, the Geiger-Müller counter tube[4] was introduced, and soon confirmation came that cosmic radiation is indeed electrically charged. In 1933, three independent experiments by Alvarez and Compton, Johnson, and Rossi, discovered that close to the equator there were more cosmic rays coming from West than from East. This effect, due to the interaction with the geomagnetic field, showed that cosmic rays are mostly positively charged—and thus most probably protons, as some years later it was possible to demonstrate thanks to more powerful spectrometers.

3.2 Cosmic Rays and the Beginning of Particle Physics

With the development of cosmic ray physics, scientists knew that astrophysical sources provided high energy particles which entered the atmosphere. The obvious next step was to investigate the nature of such particles, and to use them to probe matter in detail, much in the same way as in the experiment conducted by Marsden and Geiger in 1909 (the Rutherford experiment, described in Chap. 2). Particle physics thus started with cosmic rays, and many of the fundamental discoveries were made thanks to cosmic rays.

In parallel, the theoretical understanding of the Universe was progressing quickly: at the end of the 1920s, scientists tried to put together relativity and quantum mechanics, and the discoveries following these attempts changed completely our view of nature. A new window was going to be opened: antimatter.

[4]The Geiger-Müller counter is a cylinder filled with a gas, with a charged metal wire inside. When a charged particle enters the detector, it ionizes the gas, and the ions and the electrons can be collected by the wire and by the walls. The electrical signal of the wire can be amplified and read by means of an amperometer. The tension V of the wire is large (a few thousand volts), in such a way that the gas is completely ionized; the signal is then a short pulse of height independent of the energy of the particle. Geiger-Müller tubes can be also appropriate for detecting γ radiation, since a photoelectron or a Compton-scattered electron can generate an avalanche.

3.2.1 Relativistic Quantum Mechanics and Antimatter

Schrödinger's[5] wave equation

$$i\hbar\frac{\partial\Psi}{\partial t} = -\frac{\hbar^2}{2m}\vec{\nabla}^2\Psi + V\Psi$$

can be seen as the translation into the wave language of the Hamiltonian equation of classical mechanics

$$H = \frac{p^2}{2m} + V,$$

where the Hamiltonian (essentially the energy of the system) is represented by the operator

$$\hat{H} = i\hbar\frac{\partial}{\partial t} + V$$

and momentum by

$$\hat{\vec{p}} = -i\hbar\vec{\nabla}.$$

The solutions of the equation are in general complex wavefunctions, which can be seen as probability amplitudes.

3.2.1.1 Some Properties of Schrödinger's Equation and of Its Solutions

In "classical" quantum mechanics physical states are represented by complex wavefunctions $\Psi(\vec{r}, t)$ which are solutions of the Schrödinger's equation.

Here we recall briefly some of the main characteristics of the solutions of the equation. They can be extended in general to any "good" hamiltonian equation; we shall not be anyway very formal here, but will just focus on the concepts.

The Meaning of Wavefunctions. What is a "wavefunction," and what can it tell us? In classical physics an elementary particle, by its nature, is localized at a point, whereas the wave function is spread out in space (for any given time t it is a function of x). How can such an object be said to describe the state of a particle? The answer

[5]Erwin Schrödinger was an Austrian physicist who obtained fundamental results in the fields of quantum theory, statistical mechanics and thermodynamics, physics of dielectrics, color theory, electrodynamics, cosmology, and cosmic-ray physics. He also paid great attention to the philosophical aspects of science, re-evaluating ancient and oriental philosophical concepts, and to biology and to the meaning of life. He formulated the famous paradox of the Schrödinger cat. He shared with P.A.M. Dirac the 1933 Nobel Prize for Physics "for the discovery of new productive forms of atomic theory."

is given by the so-called Born's statistical interpretation. If Ψ is normalized such that

$$\int dV\, \Psi^* \Psi = 1 \tag{3.1}$$

(the integral is extended over all the volume), the probability to find the particle in an infinitesimal volume dV around a point $\vec{r}$ at a time t is

$$dP = \Psi^*(\vec{r},t)\Psi(\vec{r},t)\, dV = |\Psi(\vec{r},t)|^2\, dV\, .$$

The left term in Eq. (3.1) is defined as the scalar product of the Ψ function by itself.

The statistical interpretation introduces a kind of uncertainty into quantum mechanics: even if you know everything the theory can tell you about the particle (its wavefunction), you cannot predict with certainty the outcome of a simple experiment to measure its position: all the theory gives is statistical information about the possible results.

Measurement: **Operators**. The expectation value of the measurement of, say, position along the x coordinate is thus given as

$$\langle x \rangle = \int dV\, \Psi^*\, x\, \Psi \tag{3.2}$$

and one can easily demonstrate (see for example [F3.4] in the Further Readings) that the expectation value of the momentum along x is

$$\langle p_x \rangle = \int dV\, \Psi^* \left[-i\hbar \frac{\partial}{\partial x} \right] \Psi\, . \tag{3.3}$$

In these two examples we saw that measurements are represented by operators acting on wavefunctions. The operator x represents position along x, and the operator $(-i\hbar\partial/\partial x)$ represents the x component of momentum, p_x. When ambiguity is possible we put a "hat" on top of the operator to distinguish it from the corresponding physical quantity.

To calculate expectation values, we put the appropriate operator in sandwich between Ψ^* and Ψ, and integrate. If A is a quantity and $\hat{A}$ the corresponding operator,

$$\langle A \rangle = \int dV\, \Psi^* \left(\hat{A} \right) \Psi\, . \tag{3.4}$$

Dirac Notation In the Dirac notation, a vector (an eigenstate in this case) is identified by the symbol $\mid \Phi\rangle$, and is called *ket*; the symbol $\langle \Psi \mid$ is called *bra*.

The *bracket* $\langle \Psi \mid \Phi \rangle$ is the scalar product of the two vectors:

$$\langle \Psi \mid \Phi \rangle = \int dV \Psi^* \Phi\, .$$

In this notation, an operator $\hat{A}$ acts on a ket $\mid \Phi >$, transforming it into a ket $\hat{A} \mid \Phi >$, and thus

$$< \Psi \mid \hat{A} \mid \Phi > = \int dV \, \Psi^*(\hat{A}\Phi) \, .$$

Good Operators Must be Hermitian. We define as hermitian conjugate or adjoint of $\hat{A}$ an operator $\hat{A}^\dagger$ such that for any $|\Psi >$

$$\langle \hat{A}^\dagger \Psi | \Psi \rangle \; = \; \langle \Psi | \hat{A} \Psi \rangle \, .$$

Let $\hat{A}$ represent an observable. One has for the expectation value

$$\langle \Psi \mid \hat{A} \mid \Psi \rangle \; = \; \langle \Psi \mid \hat{A}\Psi \rangle \; = \; \langle \hat{A}\Psi \mid \Psi \rangle^* == \langle \Psi \mid \hat{A}^+ \Psi \rangle^* = \langle \Psi \mid \hat{A}^+ \mid \Psi \rangle^*$$

and thus if we want all expectation values (and the results of any measurement) to be real, $\hat{A}$ must be a Hermitian operator (i.e., such that $\hat{A}^\dagger = \hat{A}$).

Now let us call Ψ_i the eigenvectors of $\hat{A}$ (which form a basis) and a_i the corresponding eigenvalues; for Ψ_m, Ψ_n such that $n \neq m$

$$\hat{A} \mid \Psi_m \rangle = a_m \mid \Psi_m \rangle \quad \hat{A} \mid \Psi_n \rangle = a_n \mid \Psi_n \rangle$$

and thus

$$\begin{aligned} a_n \langle \Psi_n \mid \Psi_m \rangle \; &= \langle \Psi_n \mid \hat{A} \mid \Psi_m \rangle = a_m \langle \Psi_n \mid \Psi_m \rangle \\ &\Rightarrow 0 = (a_n - a_m) \langle \Psi_n \mid \Psi_m \rangle \\ &\Rightarrow \; \langle \Psi_n \mid \Psi_m \rangle = 0 \quad \forall \, m \neq n \, . \end{aligned}$$

If the Ψ_i are properly normalized

$$\langle \Psi_n \mid \Psi_m \rangle = \delta_{mn} .$$

Hermitian operators are thus good operators for representing the measurement of physical quantities: their eigenvalues are real (and thus is the measurement of any quantity) and the solutions form an orthonormal basis.

When we measure a value, we obtain a well-defined value: thus the wavefunction "collapses" to an eigenfunction, and the measured value is one of the eigenvalues of the measurement operator.

Schrödinger's Equation: **Meaning of the Eigenvalues**. Schrödinger's equation is an equation for which the eigenvectors are eigenstates of definite energy. For a potential V not explicitly dependent on time, it can be split in general into two equations. One is a time-independent eigenvalue equation

$$\left(-\frac{\hbar^2}{2m} \vec{\nabla}^2 + V \right) \psi(\vec{r}) = E \psi(\vec{r})$$

and the other is an equation just involving time

$$\phi(t) = \exp(-iEt/\hbar)$$

with the complete solution being

$$\Psi(\vec{r}, t) = \psi(\vec{r})\phi(t) .$$

Spectral Decomposition of an Operator. Since the $\{|\Psi_i\rangle\}$ forms a basis, we can write

$$\sum_j |\Psi_j\rangle \langle\Psi_j| = I$$

where I is the unity operator.

This means that any wavefunction can be represented in this orthonormal basis by a unique combination:

$$| \Psi\rangle = \sum_m | \Psi_m\rangle\langle\Psi_m | \Psi\rangle = \sum_m c_m | \Psi_m\rangle$$

($\langle\Psi_m | \Psi\rangle = c_m$ are complex numbers).

The normalization of $|\Psi\rangle$ to 1 implies a relation on the c_m:

$$1 = \langle\Psi|\Psi\rangle = \sum_m \langle\Psi| \Psi_m\rangle\langle\Psi_m |\Psi\rangle = \sum_m |c_m|^2 ,$$

and the probability to obtain from a measurement the eigenvalue a_m is

$$P_m = |\langle\Psi_m|\Psi\rangle|^2 = |c_m|^2 .$$

In addition, we can determine coefficients $a_{mn} = \langle\Psi_m | \hat{A} | \Psi_n\rangle$ such that:

$$\hat{A} | \Psi\rangle = \sum_{mn} | \Psi_m\rangle\langle\Psi_m | \hat{A} | \Psi_n\rangle\langle\Psi_n | \Psi\rangle = \sum_{m,n} a_{mn}c_n | \Psi_m\rangle .$$

$[a_{mn}]$ is a square matrix representing $\hat{A}$ in the vector space defined by eigenvectors; the c_n are an n-tuple of components representing $|\Psi\rangle$.

Uncertainty Relations. We define commutator $[\hat{A}, \hat{B}]$ of two operators $\hat{A}$ e $\hat{B}$ the operator

$$[\hat{A}, \hat{B}] = (\hat{A}\hat{B} - \hat{B}\hat{A}) ,$$

and we say that the two operators commute when their commutator is zero. We can simultaneously measure observables whose operators commute, since such operators have a complete set of simultaneous eigenfunctions—thus one can have two definite measurements at the same time.

However, pairs of noncommuting operators cannot give rise to simultaneous measurements arbitrarily precise for the associated quantities (this is called usually Heisenberg's uncertainty principle, but indeed it is a theorem).

Let us define as spread of an operator the operator

$$\Delta \hat{A} = \hat{A} - \langle A \rangle \,.$$

Let $\hat{A}$ and $\hat{B}$ be two Hermitian oparators; we define $\hat{C}$ such that

$$[\hat{A}, \hat{B}] = i\hat{C}$$

($\hat{C}$ is Hermitian; you can demonstrate it). One has

$$\langle (\Delta A)^2 \rangle \langle (\Delta B)^2 \rangle \geq \frac{\langle C^2 \rangle}{4} \,. \tag{3.5}$$

In particular when a simultaneous measurement of position and momentum along an axis, say x, is performed, one has

$$\Delta x \Delta p_x \geq \frac{\hbar}{2} \sim \hbar \,.$$

Somehow linked to this is the fact that energy is not defined with absolute precision, but, if measured in a time Δt, has an uncertainty ΔE such that

$$\Delta E \Delta t \sim \hbar$$

(energy conservation can be violated for short times). The value of the Planck's constant $\hbar \simeq 6.58 \times 10^{-22}$ MeV · s is anyway small with respect to the value corresponding to the energies needed to create particles living for a reasonable (detectable) time.

Limits of the Schrödinger's Equation. Since Schrödinger's equation contains derivatives of different order with respect to space and time, it cannot be relativistically covariant, and thus, it cannot be the "final" equation. How can it be extended to be consistent with Lorentz invariance? We must translate relativistically covariant Hamiltonians in the quantum language, i.e., into equations using wavefunctions. We shall see in the following two approaches.

3.2.1.2 The Klein–Gordon Equation

In the case of a free particle ($V = 0$), the simplest way to extend the Schrödinger's equation to take into account relativity is to write the Hamiltonian equation

$$\hat{H}^2 = \hat{p}^2c^2 + m^2c^4$$
$$\Longrightarrow -\hbar^2\frac{\partial^2\Psi}{\partial t^2} = -\hbar^2c^2\vec{\nabla}^2\Psi + m^2c^4\Psi\,,$$

or, in natural units,

$$\left(-\frac{\partial^2}{\partial t^2} + \vec{\nabla}^2\right)\Psi = m^2\Psi\,.$$

This equation is known as Klein-Gordon equation,[6] but it was first considered as a quantum wave equation by Schrödinger; it was found in his notebooks from late 1925. Schrödinger had also prepared a manuscript applying it to the hydrogen atom; however, he could not solve some fundamental problems related to the form of the equation (which is not linear in energy, so that states are not easy to combine), and thus he went back to the equation today known by his name. In addition, the solutions of the Klein-Gordon equation do not allow for statistical interpretation of $|\Psi|^2$ as a probability density—its integral would in general not remain constant in time.

The Klein-Gordon equation displays one more interesting feature. Solutions of the associated eigenvalue equation

$$\left(-m^2 + \vec{\nabla}^2\right)\psi = E_p^2\psi$$

have both positive and negative eigenvalues for energy. For every plane wave solution of the form

$$\Psi(\vec{r}, t) = Ne^{i(\vec{p}\cdot\vec{r} - E_p t)}$$

with momentum $\vec{p}$ and positive energy

$$E_p = \sqrt{p^2 + m^2} \geq m,$$

there is a solution

$$\Psi^*(\vec{r}, t) = N^*e^{i(-\vec{p}\cdot\vec{r} + E_p t)}$$

[6]Oskar Klein (1894–1977) was a Swedish theoretical physicist; Walter Gordon (1893–1939) was a German theoretical physicist, former student of Max Planck.

with momentum $-\vec{p}$ and negative energy

$$E = -E_p = -\sqrt{p^2 + m^2} \leq -m\,.$$

Note that one cannot simply drop the solutions with negative energy as "unphysical": the full set of eigenstates is needed, because, if one starts from a given wave function, this could evolve with time into a wave function that, in general, has projections on all eigenstates (including those one would like to get rid of). We remind the reader that these are solutions of an equation describing a free particle.

A final comment about notation. The (classical) Schrödinger equation for a single particle in a time-independent potential can be decoupled into two equations: one (the so-called eigenvalue equation) just depending just on space, and the other depending just on time. The solution of the eigenvalue equation is normally indicated by a minuscule Greek symbol, $\psi(\vec{r})$ for example, while the time part has a solution independent of the potential, $e^{-(E/\hbar)t}$. The wavefunction is indicated by a capital letter:

$$\Psi(\vec{r}, t) = \psi(\vec{r}) e^{-i\frac{E}{\hbar}t}\,.$$

This distinction makes no sense for relativistically equations and in particular for the Klein-Gordon equation and for the Dirac equation which will be discussed later. Both $\Psi(x)$ and $\psi(x)$ are now valid notations for indicating a wavefunction which is function of the 4-vector $x = (ct, x, y, z)$.

3.2.1.3 The Dirac Equation

Dirac[7] in 1928 searched for an alternative relativistic equation starting from the generic form describing the evolution of a wave function, in the familiar form:

$$i\hbar \frac{\partial \Psi}{\partial t} = \hat{H}\Psi$$

with a Hamiltonian operator linear in $\hat{\vec{p}}$, t (Lorentz invariance requires that if the Hamiltonian has first derivatives with respect to time also the spatial derivatives should be of first order):

$$\hat{H} = c\vec{\alpha} \cdot \vec{p} + \beta mc^2.$$

[7]Paul Adrien Maurice Dirac (Bristol, UK, 1902—Tallahassee, US, 1984) was one of the founders of quantum physics. After graduating in engineering and later studying physics, he became professor of mathematics in Cambridge. In 1933 he shared the Nobel Prize with Schrödinger. He assigned to the concept of "beauty in mathematics" a prominent role among the basic aspects intrinsic to the nature so far as to argue that "a mathematically beautiful theory is more likely to be right and proper to an unpleasant as it is confirmed by the data."

This must be compatible with the Klein-Gordon equation, and thus

$$\alpha_i^2 = 1 \ ; \ \ \beta^2 = 1$$
$$\alpha_i \beta + \beta \alpha_i = 0$$
$$\alpha_i \alpha_j + \alpha_j \alpha_i = 0.$$

Therefore, parameters $\vec{\alpha}$ and β cannot be numbers. However, it works if they are matrices (and if these matrices are Hermitian it is guaranteed that the Hamiltonian is also Hermitian). It can be demonstrated that the lowest order is 4×4.

Using the explicit form of the momentum operator $\vec{p} = -i\hbar\vec{\nabla}$ the Dirac equation becomes

$$i\hbar \frac{\partial \Psi}{\partial t} = \left(i\vec{\alpha} \cdot \vec{\nabla} + \beta mc^2 \right) \Psi \, .$$

The wavefunctions Ψ must thus be four-component vectors:

$$\Psi(\vec{r}, t) = \begin{pmatrix} \Psi_1(\vec{r}, t) \\ \Psi_2(\vec{r}, t) \\ \Psi_3(\vec{r}, t) \\ \Psi_4(\vec{r}, t) \end{pmatrix} .$$

We arrived at an interpretation of the Dirac equation, as a four-dimensional matrix equation in which the solutions are four-component wavefunctions called spinors.[8] Plane wave solutions are

$$\Psi(\vec{r}, t) = u(\vec{p}) e^{i(\vec{p} \cdot \vec{r} - Et)}$$

where $u(\vec{p})$ is also a four-component spinor satisfying the eigenvalue equation

$$(c\vec{\alpha} \cdot \vec{p} + \beta m) \, u(\vec{p}) = E u(\vec{p}).$$

This equation has four solutions: two with positive energy $E = +E_p$ and two with negative energy $E = -E_p$. We discuss later the interpretation of the negative energy solutions.

Dirac's equation was a success. First, it accounted "for free" for the existence of two spin states (we remind that spin had to be inserted by hand in the Schrödinger equation of nonrelativistic quantum mechanics). In addition, since spin is embedded in the equation, the Dirac's equation:

---

[8]The term spinor indicates in general a vector which has definite transformation properties for a rotation in the proper angular momentum space—the spin space. The properties of rotation in spin space will be described in greater detail in Chap. 5 .

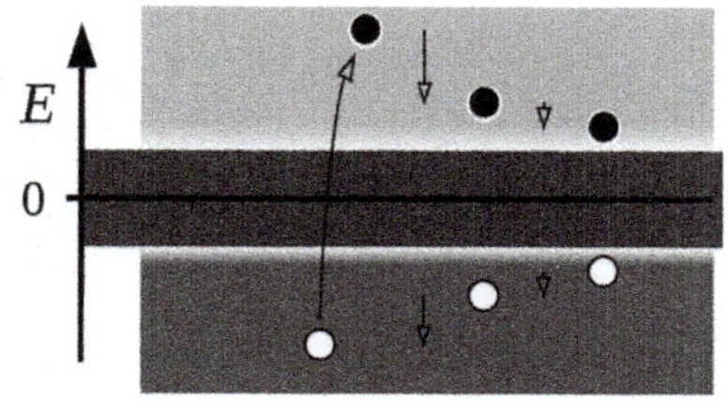

Fig. 3.7 Dirac picture of the vacuum. In normal conditions, the sea of negative energy states is totally occupied with two electrons in each level. By Incnis Mrsi [own work, public domain], via Wikimedia Commons.

- allows computing correctly the energy splitting of atomic levels with the same quantum numbers due to the spin–orbit interaction in atoms (fine and hyperfine splitting);
- explains the magnetic moment of point-like fermions.

The predictions on the values of the above quantities were incredibly precise and still resist to experimental tests.

3.2.1.4 Hole Theory and the Positron

Negative energy states must be occupied: if they were not, transitions from positive to negative energy states would occur, and matter would be unstable. Dirac postulated that the negative energy states are completely filled under normal conditions. In the case of electrons the Dirac picture of the vacuum is a "sea" of negative energy states, while the positive energy states are mostly free (Fig. 3.7). This condition cannot be distinguished from the usual vacuum.

If an electron is added to the vacuum, it finds in general place in the positive energy region since all the negative energy states are occupied. If a negative energy electron is removed from the vacuum, however, a new phenomenon happens: removing such an electron with $E < 0$, momentum $-\vec{p}$, spin $-\vec{S}$ and charge $-e$ leaves a "hole" indistinguishable from a particle with positive energy $E > 0$, momentum $\vec{p}$, spin $\vec{S}$ and charge $+e$. This is similar to the formation of holes in semiconductors. The two cases are equivalent descriptions of the same phenomena. Dirac's sea model thus predicts the existence of a new fermion with mass equal to the mass of the electron, but opposite charge. This particle, later called the positron, is the antiparticle of the electron, and is the prototype of a new family of particles: antimatter.

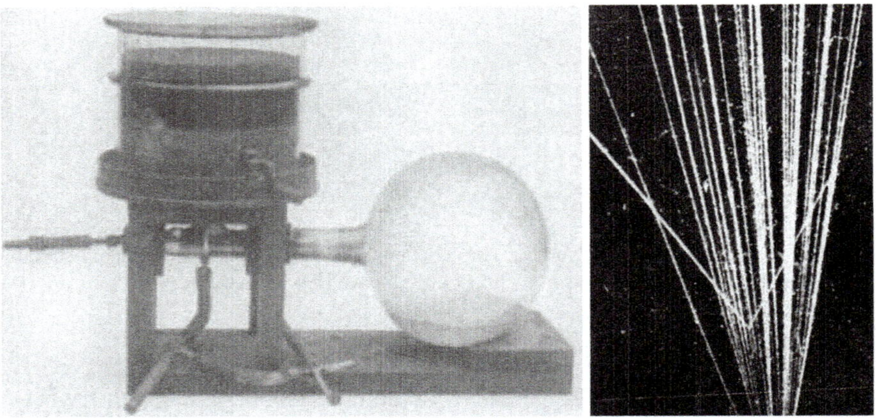

Fig. 3.8 *Left* A cloud chamber built by Wilson in 1911. By C.T.R. Wilson [public domain], via Wikimedia Commons. *Right* a picture of a collision in a cloud chamber [CC BY 4.0 http://creativecommons.org/licenses/by/4.0] via Wikimedia Commons

3.2.2 *The Discovery of Antimatter*

During his doctoral thesis (supervised by Millikan), Anderson was studying the tracks of cosmic rays passing through a cloud chamber[9] in a magnetic field (Fig. 3.8). In 1933 he discovered antimatter in the form of a positive particle of mass consistent with the electron mass, later called the positron (Fig. 3.9). Dirac's equation prediction was confirmed; this was a great achievement for cosmic ray physics. Anderson shared with Hess the Nobel Prize for Physics in 1936; they were nominated by Compton, with the following motivation:

> The time has now arrived, it seems to me, when we can say that the so-called cosmic rays definitely have their origin at such remote distances from the Earth that they may properly be called cosmic, and that the use of the rays has by now led to results of such importance that they may be considered a discovery of the first magnitude. [...] It is, I believe, correct to say that Hess was the first to establish the increase of the ionization observed in electroscopes with increasing altitude; and he was certainly the first to ascribe with confidence this increased ionization to radiation coming from outside the Earth.

---

[9]The cloud chamber (see also next chapter), invented by C.T.R. Wilson at the beginning of the twentieth century, was an instrument for reconstructing the trajectories of charged particles. The instrument is a container with a glass window, filled with air and saturated water vapor; the volume could be suddenly expanded, bringing the vapor to a supersaturated (metastable) state. A charged cosmic ray crossing the chamber produces ions, which act as seeds for the generation of droplets along the trajectory. One can record the trajectory by taking a photographic picture. If the chamber is immersed in a magnetic field, momentum and charge can be measured by the curvature. The working principle of bubble chambers is similar to that of the cloud chamber, but here the fluid is a liquid. Along the tracks' trajectories, a trail of gas bubbles condensates around the ions. Bubble and cloud chambers provide a complete information: the measurement of the bubble density and the range, i.e., the total track length before the particle eventually stops, provide an estimate for the energy and the mass; the angles of scattering provide an estimate for the momentum.

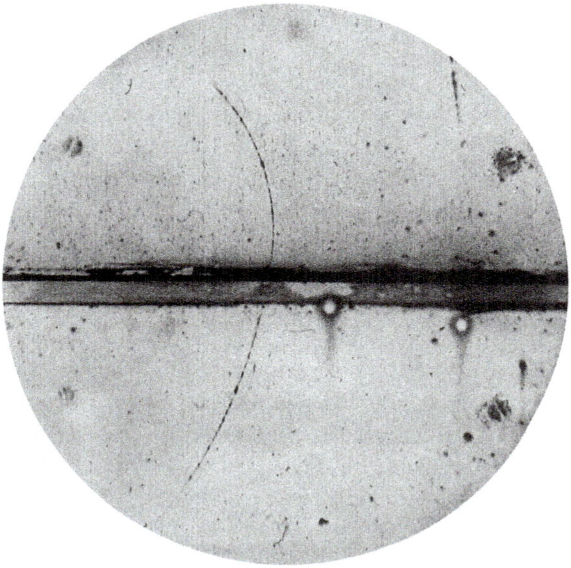

Fig. 3.9 The first picture by Anderson showing the passage of a cosmic antielectron, or positron, through a cloud chamber immersed in a magnetic field. One can understand that the particle comes from the *bottom* in the picture by the fact that, after passing through the sheet of material in the medium (and therefore losing energy), the radius of curvature decreases. The positive charge is inferred from the direction of bending in the magnetic field. The mass is measured by the bubble density (a proton would lose energy faster). Since most cosmic rays come from the *top*, the first evidence for antimatter comes thus from an unconventional event. From C.D. Anderson, "The Positive Electron," Physical Review 43 (1933) 491

Why so late a recognition to the discovery of cosmic rays? Compton writes:

> Before it was appropriate to award the Nobel Prize for the discovery of these rays, it was necessary to await more positive evidence regarding their unique characteristics and importance in various fields of physics.

3.2.3 Cosmic Rays and the Progress of Particle Physics

After Anderson's fundamental discovery of antimatter, new experimental results in the physics of elementary particles with cosmic rays were guided and accompanied by the improvement of the tools for detection, in particular by the improved design of the cloud chambers and by the introduction of the Geiger-Müller tube. According to Giuseppe Occhialini, one of the pioneers of the exploration of fundamental physics with cosmic rays, the Geiger-Müller counter was like the Colt in the Far West: a cheap instrument usable by everyone on one's way through a hard frontier.

At the end of the 1920s, Bothe and Kolhörster introduced the coincidence technique to study cosmic rays with the Geiger counter. A coincidence circuit activates the acquisition of data only when signals from predefined detectors are received within a given time window. The coincidence technique is widely used in particle physics experiments, but also in other areas of science and technology. Walther Bothe shared the Nobel Prize for Physics in 1954 with the motivation: "for the coincidence method and his discoveries made therewith." Coupling a cloud chamber to a system of Geiger counters and using the coincidence technique, it was possible to take photographs only when a cosmic ray traversed the cloud chamber (we call today such a system a "trigger"). This increased the chances of getting a significant photograph and thus the efficiency of cloud chambers.

Soon after the discovery of the positron by Anderson, a new important observation was made in 1933: the conversion of photons into pairs of electrons and positrons. Dirac's theory not only predicted the existence of antielectrons, but it also predicted that electron–positron pairs could be created from a single photon with energy large enough; the phenomenon was actually observed in cosmic rays by Blackett (Nobel Prize for Physics in 1948) and Occhialini, who further improved in Cambridge the coincidence technique. Electron–positron pair production is a simple and direct confirmation of the mass-energy equivalence and thus of what is predicted by the theory of relativity. It also demonstrates the behavior of light, confirming the quantum concept which was originally expressed as "wave-particle duality": the photon can behave as a particle.

In 1934, the Italian Bruno Rossi[10] reported the observation of the near-simultaneous discharge of two widely separated Geiger counters during a test of his equipment. In the report, he wrote: "[...] it seems that once in a while the recording equipment is struck by very extensive showers of particles, which causes coincidences between the counters, even placed at large distances from one another." In 1937 Pierre Auger, who was not aware of Rossi's report, made a similar observation and investigated the phenomenon in detail. He concluded that extensive showers originate when high energy primary cosmic rays interact with nuclei high in the atmosphere, leading to a series of interactions that ultimately yield a shower of particles that reach ground. This was the explanation of the spontaneous discharge of electroscopes due to cosmic rays.

[10]Bruno Rossi (Venice 1905—Cambridge, MA, 1993) graduated in Bologna, and then moved to Arcetri near Florence before becoming full professor of physics at the University of Padua in 1932. In Padua he was charged of overseeing the design and construction of the new Physics Institute, which was inaugurated in 1937. He was exiled in 1938, as a consequence the Italian racial laws, and he moved to Chicago and then to Cornell. In 1943 he joined the Manhattan project in Los Alamos, working to the development of the atomic bomb, and after the end of the second World War moved to MIT. At MIT Rossi started working on space missions as a scientfic consultant for the newborn NASA, and proposed the rocket experiment that discovered the first extra-solar source of X-rays. Many fundamental contributions to modern physics, for example the electronic coincidence circuit, the discovery and study of extensive air showers, the East–West effect, and the use of satellites for the exploration of the high-energy Universe, are due to Bruno Rossi.

3.2.4 The μ Lepton and the π Mesons

In 1935 the Japanese physicist Yukawa, 28 years old at that time, formulated his innovative theory explaining the "strong" interaction ultimately keeping together matter (strong interaction keeps together protons and neutrons in the atomic nuclei). This theory has been sketched in the previous chapter, and requires a "mediator" particle of intermediate mass between the electron and the proton, thus called meson—the word "meson" meaning "middle one."

To account for the strong force, Yukawa predicted that the meson must have a mass of about one-tenth of GeV, a mass that would explain the rapid weakening of the strong interaction with distance. The scientists studying cosmic rays started to discover new types of particles of intermediate masses. Anderson, who after the Nobel Prize had became a professor, and his student Neddermeyer, observed in 1937 a new particle, present in both positive and negative charge, more penetrating than any other particle known at the time. The new particle was heavier than the electron but lighter than the proton, and they suggested for it the name "mesotron." The mesotron mass, measured from ionization, was between 200 and 240 times the electron mass; this was fitting into Yukawa's prediction for the meson. Most researchers were convinced that these particles were the Yukawa's carrier of the strong nuclear force, and that they were created when primary cosmic rays collided with nuclei in the upper atmosphere, in the same way as electrons emit photons when colliding with a nucleus.

The lifetime of the mesotron was measured studying its flow at various altitudes, in particular by Rossi in Colorado (Rossi had been forced to emigrate to the United States to escape racial persecution); the result was of about two microseconds (a hundred times larger than predicted by Yukawa for the particle that transmits the strong interaction). Rossi found also that at the end of its life the mesotron decays into an electron and other neutral particles (neutrinos) that did not leave tracks in bubble chamber—the positive mesotron decays into a positive electron plus neutrinos.

Beyond the initial excitement, however, the picture did not work. In particular, the Yukawa particle is the carrier of strong interactions, and therefore it cannot be highly penetrating—the nuclei of the atmosphere would absorb it quickly. Many theorists tried to find complicated explanations to save the theory. The correct explanation was however the simplest one: the mesotron was not the Yukawa particle, as it was demonstrated in 1945/46 by three young Italian physicists, Conversi, Pancini, and Piccioni.

The experiment by Conversi, Pancini and Piccioni exploits the fact that slow negative Yukawa particles can be captured by nuclei in a time shorter than the typical lifetime of the mesotron, about 2 μs, and thus are absorbed before decaying; on the contrary, slow positive particles are likely to be repelled by the potential barrier of nuclei and thus have the time to decay. The setup is shown in Fig. 3.10; a magnetic lens focuses particles of a given charge, thus allowing charge selection. The Geiger counters A and B are in coincidence—i.e., a simultaneous signal is required; the C counters under the absorber are in "delayed coincidence," and it is requested that one of them fires after a time between 1 and 4.5 μs after the coincidence (AB). This

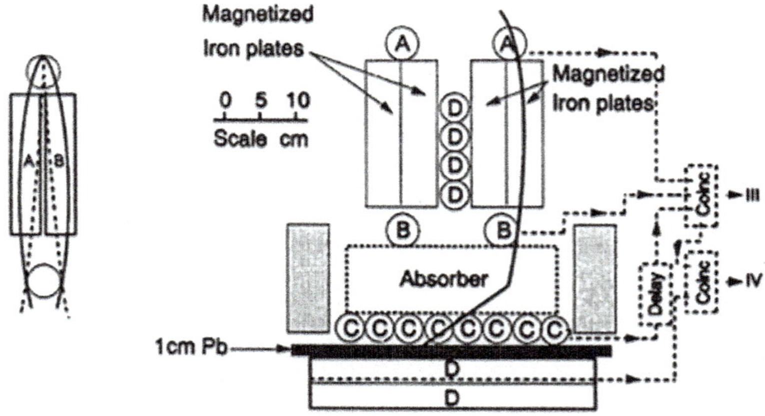

Fig. 3.10 *Left* a magnetic lens (invented by Rossi in 1930). *Right* setup of the Conversi, Pancini and Piccion experiment. From M. Conversi, E. Pancini, O. Piccioni, "On the disintegration of negative mesons", Physical Review 71 (1947) 209

guarantees that the particle selected is slow and, in case of decay, has a lifetime consistent with the mesotron. The result was that when carbon was used as absorber, a substantial fraction of the negative mesons decayed. The mesotron was not the Yukawa particle.

There were thus two particles of similar mass. One of them (with mass of about 140 MeV/c^2), corresponding to the particle predicted by Yukawa, was later called pion (or π meson); it was created in the interactions of cosmic protons with the atmosphere, and then interacted with the nuclei of the atmosphere, or decayed. Among its decay products there was the mesotron, since then called the muon (or μ lepton), which was insensitive to the strong force.

In 1947, Powell, Occhialini and Lattes, exposing nuclear emulsions (a kind of very sensitive photographic plates, with space resolutions of a few μm; we shall discuss them in the next chapter) to cosmic rays on Mount Chacaltaya in Bolivia, finally proved the existence of charged pions, positive and negative, while observing their decay into muons and allowing a precise determination of the masses. For this discovery Cecil Powell, the group leader, was awarded the Nobel Prize in 1950.

Many photographs of nuclear emulsions, especially in experiments on balloons, clearly showed traces of pions decaying into muons (the muon mass was reported to be about 106 MeV/c^2), decaying in turn into electrons. In the decay chain $\pi \rightarrow \mu \rightarrow e$ (Fig. 3.11) some energy is clearly missing, and can be attributed to neutrinos.

At this point, the distinction between pions and muons was clear. The muon looks like a "heavier brother" of the electron. After the discovery of the pion, the muon had no theoretical reason to exist (the physicist Isidor Rabi was attributed in the 1940s the famous quote: "Who ordered it?"). However, a new family was initiated: the family of leptons—including for the moment the electron and the muon, and their antiparticles.

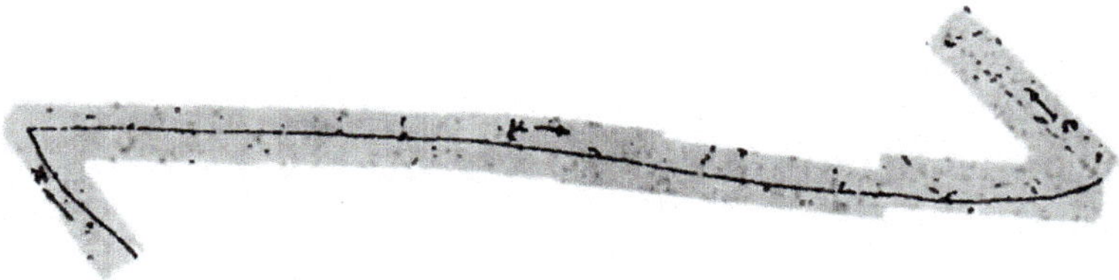

Fig. 3.11 The pion and the muon: the decay chain $\pi \rightarrow \mu \rightarrow e$. The pion travels from bottom to top on the left, the muon horizontally, and the electron from bottom to top on the right. The missing momentum is carried by neutrinos. From C.F. Powell, P.H. Fowler and D.H. Perkins, The Study of Elementary Particles by the Photographic Method (Pergamon Press 1959)

3.2.4.1 The Neutral Pion

Before it was even known that mesotrons were not the Yukawa particle, the theory of mesons had great development. In 1938, a theory of charge symmetry was formulated, conjecturing the fact that the forces between protons and neutrons, between protons and protons and between neutrons and neutrons are similar. This implies the existence of positive, negative and also neutral mesons.

The neutral pion was more difficult to detect than the charged one, due to the fact that neutral particles do not leave tracks in cloud chambers and nuclear emulsions—and also to the fact, discovered only later, that it lives only approximately 10^{-16} s before decaying mostly into two photons. However, between 1947 and 1950, the neutral pion was identified in cosmic rays by analyzing its decay products in showers of particles. So, after 15 years of research, the theory of Yukawa had finally complete confirmation.

3.2.5 Strange Particles

In 1947, after the thorny problem of the meson had been solved, particle physics seemed to be a complete science. Fourteen particles were known to physicists (some of them at the time were only postulated, and were going to be found experimentally later): the proton, the neutron (proton and neutron together belong to the family of baryons, the Greek etymology of the word referring to the concept of "heaviness") and the electron, and their antiparticles; the neutrino that was postulated because of an apparent violation of the principle of energy conservation; three pions; two muons; and the photon.

Apart from the muon, a particle that appeared unnecessary, all the others seemed to have a role in nature: the electron and the nucleons constitute the atom, the photon carries the electromagnetic force, and the pion the strong force; neutrinos are needed for energy conservation. But, once more in the history of science, when everything seemed understood a new revolution was just around the corner.

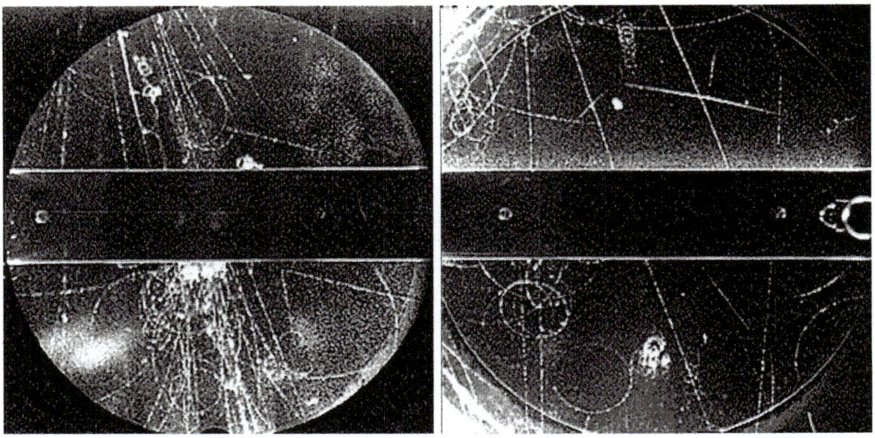

Fig. 3.12 The first images of the decay of particles known today as K mesons or kaons—the first examples of "strange" particles. The image on the *left* shows the decay of a neutral kaon. Being neutral it leaves no track, but when it decays into two lighter charged particles (just below the central bar to the *right*), one can see a "V". The picture on the *right* shows the decay of a charged kaon into a muon and a neutrino. The kaon reaches the *top right* corner of the chamber and the decay occurs where the track seems to bend sharply to the *left* (from G.D. Rochester, C.C. Butler, "Evidence for the Existence of New Unstable Elementary Particles" Nature 160 (1947) 855)

Since 1944, strange topologies of cosmic particles were photographed from time to time in cloud chambers. In 1947, G.D. Rochester and the C.C. Butler from the University of Manchester observed clearly in a photograph a couple of tracks from a single point with the shape of a "V"; the two tracks were deflected in opposite directions by an external magnetic field. The analysis showed that the parent neutral particle had a mass of about half a GeV (intermediate between the mass of the proton and that of the pion), and disintegrated into a pair of oppositely charged pions. A broken track in a second photograph showed the decay of a charged particle of about the same mass into a charged pion and at least one neutral particle (Fig. 3.12).

These particles, which were produced only in high energy interactions, were observed only every few hundred photographs. They are known today as K mesons (or kaons); kaons can be positive, negative, or neutral. A new family of particles had been discovered. The behavior of these particles was somehow strange: the cross section for their production could be understood in terms of strong interactions; however, their lifetime was inconsistent with strong interaction, being too long. These new particles were called "strange mesons". Later analyses indicated the presence of particles heavier than protons and neutrons. They decayed with a "V" topology into final states including protons, and they were called strange baryons, or hyperons (Λ, Σ, ...). Strange particles appear to be always produced in pairs, indicating the presence of a new conserved quantum number—thus called strangeness.

3.2.5.1 The τ-θ Puzzle

In the beginning, the discovery of strange mesons was made complicated by the so-called τ-θ puzzle. A strange meson was disintegrating into two pions, and was called the θ meson; another particle called the τ meson was disintegrating into three pions. Both particles disintegrated via the weak force and, apart from the decay mode, they turned out to be indistinguishable from each other, having identical masses within the experimental uncertainties. Were the two actually the same particle? It was concluded that they were (we are talking about the K meson); this opened a problem related to the so-called parity conservation law, and we will discuss it better in Chaps. 5 and 6.

3.2.6 Mountain-Top Laboratories

The discovery of mesons, which had put the physics world in turmoil after World War II, can be considered as the origin of the "modern" physics of elementary particles.

The following years showed a rapid development of the research groups dealing with cosmic rays, along with a progress of experimental techniques of detection, exploiting the complementarity of cloud and bubble chambers, nuclear emulsions, and electronic coincidence circuits. The low cost of emulsions allowed the spread of nuclear experiments and the establishment of international collaborations.

It became clear that it was appropriate to equip laboratories on top of the mountains to study cosmic rays. Physicists from all around the world were involved in a scientific challenge of enormous magnitude, taking place in small laboratories on the tops of the Alps, the Andes, the Rocky Mountains, the Caucasus.

Particle physicists used cosmic rays as the primary tool for their research until the advent of particle accelerators in the 1950s, so that the pioneering results in this field are due to cosmic rays. For the first 30 years cosmic rays allowed to gain information on the physics of elementary particles. With the advent of particle accelerators, in the years since 1950, most physicists went from hunting to farming.

3.3 Particle Hunters Become Farmers

In 1953, the Cosmic Ray Conference at Bagnères de Bigorre in the French Pyrenees was a turning point for high energy physics. The technology of artificial accelerators was progressing, and many cosmic ray physicists were moving to this new frontier. CERN, the European Laboratory for Particle Physics, was soon to be founded.

Also from the sociological point of view, important changes were in progress, and large international collaborations were formed. Only 10 years before, articles for which the preparation of the experiment and the data analysis had been performed by many scientists were signed only by the group leader. But the recent G-stack experiment, an international collaboration in which cosmic ray interactions were recorded in a series of balloon flights by means of a giant stack of nuclear emulsions, had introduced a new policy: all scientists contributing to the result were authors of the publications. At that time the number of signatures in one of the G-stack papers, 35, seemed enormous; in the twenty-first century things have further evolved, and the two articles announcing the discovery of the Higgs particle by the ATLAS and CMS collaborations have 2931 and 2899 signatures respectively.

In the 1953 Cosmic Ray Conference contributions coming from accelerator physics were not accepted: the two methods of investigation of the nature of elementary particles were kept separated. However, the French physicist Laprince-Ringuet, who was going to found CERN in 1954 together with scientists of the level of Bohr, Heisenberg, Powell, Auger, Edoardo Amaldi, and others, said in his concluding remarks:

> If we want to draw certain lessons from this congress, let's point out first that in the future we must use the particle accelerators. Let's point out for example the possibility that they will allow the measurement of certain fundamental curves (scattering, ionization, range) which will permit us to differentiate effects such as the existence of π mesons among the secondaries of K mesons. [...]
>
> I would like to finish with some words on a subject that is dear to my heart and is equally so to all the 'cosmicians', in particular the 'old timers'. [...] We have to face the grave question: what is the future of cosmic rays? Should we continue to struggle for a few new results or would it be better to turn to the machines? One can with no doubt say that the future of cosmic radiation in the domain of nuclear physics depends on the machines [...]. But probably this point of view should be tempered by the fact that we have the uniqueness of some phenomena, quite rare it is true, for which the energies are much larger.

It should be stressed that, despite the great advances of the technology of accelerators, the highest energies will always be reached by cosmic rays. The founding fathers of CERN, in their Constitution (Convention for the Establishment of a European Organization for Nuclear Research, 1953) explicitly stated that cosmic rays are one of the research items of the Laboratory.

A calculation made by Fermi about the maximum reasonably (and even *unreasonably*) achievable energy by terrestrial accelerators is interesting in this regard. In his speech "What can we learn from high energy accelerators") held at the American Physical Society in 1954 Fermi had considered a proton accelerator with a ring as large as the maximum circumference of the Earth (Fig. 3.13) as the maximum possible accelerator. Assuming a magnetic field of 2 tesla (Fermi assumed that this was the maximum field attainable in stable conditions and for long magnets; the conjecture is still true unless new technologies will appear), it is possible to obtain a maximum energy of 5000 TeV: this is the energy of cosmic rays just under the "knee," the typical energy of galactic accelerators. Fermi estimated with great optimism, extrapolating the rate of progress of the accelerator technology in the 1950s, that this accelerator could be constructed in 1994 and cost approximately 170 million dollars (the cost of LHC is some 100 times larger, and its energy is 700 times smaller).

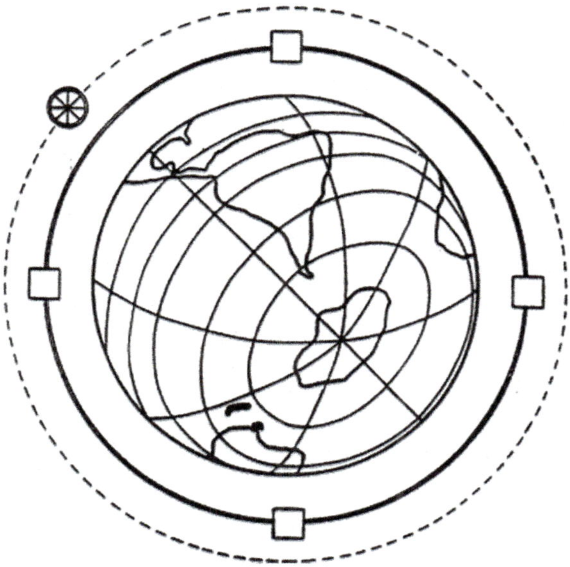

Fig. 3.13 The so-called "maximum accelerator" by Fermi (original drawing by Enrico Fermi reproduced from his 1954 speech at the annual meeting of the American Physical Society). Courtesy of Fermi National Laboratory, Batavia, Illinois

3.4 The Recent Years

Things went more or less as predicted by Leprince-Ringuet.

Between the 1950s and the 1990s most of the progress in fundamental physics was due to accelerating machines. Still, however, important experiments studying cosmic rays were alive, and were an important source of knowledge.

Cosmic rays are today central in the field of astroparticle physics, which has grown considerably in the last 20 years. Many large projects are active, with many different goals, including for example the search for dark matter in the Universe.

Gamma-ray space telescopes on satellites like the Fermi Large Area Telescope (Fermi-LAT) and AGILE, and the PAMELA and AMS-02 magnetic spectrometers, provided cutting-edge results; PAMELA in particular observed a yet unexplained anomalous yield of cosmic positrons, with a ratio between positrons and electrons growing with energy, which might point to new physics, in particular related to dark matter. The result was confirmed and extended to higher energies and with unprecedented accuracy by the AMS-02 detector onboard the International Space Station.

The study of very highest energy cosmic-ray showers, 80 years after the discovery of air showers by Rossi and Auger, is providing fundamental knowledge of the spectrum and sources of cosmic rays. In particular the region near the GZK cutoff is explored. The present day largest detector, the Pierre Auger Observatory, covers a surface of about 3000 km^2 in Argentina,

The ground-based very-high energy gamma telescopes HAWC, H.E.S.S., MAGIC and VERITAS are mapping the cosmic sources of gammas in the TeV and multi-TeV region. Together with the Fermi satellite, they are providing indications of a link between the photon accelerators and the cosmic ray accelerators in the Milky Way, in particular supernova remnants. Studying the propagation of very energetic photons traveling through cosmological distances, they are also sensitive to possible violations of the Lorentz invariance at very high energy, and to photon interactions with the quantum vacuum, which in turn are sensitive to the existence of yet unknown fields. A new detector, CTA, is planned and will outperform the present detectors by an order of magnitude at least.

Finally, the field of study of cosmic neutrinos registered impressive results. In the analysis of the fluxes of solar neutrinos and then of atmospheric neutrinos, studies performed using large neutrino detectors in Japan, US, Canada, China, and Italy have demonstrated that neutrinos can oscillate between different flavors; this phenomenon requires that neutrinos have nonzero mass—present indications favor masses of the order of tens of meV. Recently the IceCube South Pole Neutrino Observatory, a km^3 detector buried in the ice of Antarctica, has discovered the first solid evidence for astrophysical neutrinos from cosmic accelerators (some with energies above the PeV).

Cosmic rays and cosmological sources are thus again in the focus of very high energy particle and gravitational physics. This will be discussed in greater detail in Chap. 10.

Further Reading

[F3.1] P. Carlson, A. de Angelis, "Nationalism and internationalism in science: the case of the discovery of cosmic rays", The European Physical Journal H 35 (2010) 309.

[F3.2] A. de Angelis, "Atmospheric ionization and cosmic rays: studies and measurements before 1912", Astroparticle Physics 53 (2014) 19.

[F3.3] J.W. Cronin, "The 1953 Cosmic Ray Conference at Bagneres de Bigorre", The European Physical Journal H 36 (2011) 183.

[F3.4] D.H. Griffiths, "Introduction to Quantum Mechanics, 2nd edition", Addison-Wesley, Reading, MA, 2004.

[F3.5] J. Björken and S. Drell, "Relativistic Quantum Fields" McGraw-Hill, New York, 1969.

Exercises

1. *The measurement by Hess.* Discuss why radioactivity decreases with elevation up to some 1000 m, and then increases. Can you make a model? This was the subject of the thesis by Schrödinger in Wien in the beginning of twentieth century.

2. *Klein-Gordon equation.* Show that in the nonrelativistic limit $E \simeq mc^2$ the positive energy solutions Ψ of the Klein-Gordon equation can be written in the form

$$\Psi(\vec{r},t) \simeq \Phi(\vec{r},t)e^{-\frac{mc^2}{\hbar}} ,$$

where Φ satisfies the Schrödinger equation.
3. *Uncertainty relations.* Starting from Eq. (3.5), demonstrate the uncertainty principle.
4. *Antimatter.* The total number of nucleons minus the total number of anti nucleons is believed to be constant in a reaction—you can create nucleon–antinucleon pairs. What is the minimum energy of a proton hitting a proton at rest to generate an antiproton?
5. *Fermi maximum accelerator.* According to Enrico Fermi, the ultimate human accelerator, the "Globatron," would be built around 1994 encircling the entire Earth and attaining an energy of around 5000 TeV (with an estimated cost of 170 million US1954 dollars...). Discuss the parameters of such accelerator.
6. *Cosmic pions and muons.* Pions and muons are produced in the high atmosphere, at a height of some 10 km above sea level, as a result of hadronic interactions from the collisions of cosmic rays with atmospheric nuclei. Compute the energy at which charged pions and muons respectively must be produced to reach in average the Earth's surface.
You can find the masses of the lifetimes of pions and muons in your Particle Data Booklet.
7. *Very-high-energy cosmic rays.* Justify the sentence "About once per minute, a single subatomic particle enters the Earth's atmosphere with an energy larger than 10 J" in Chap. 1.
8. *Very-high-energy neutrinos.* The IceCube experiment in the South pole can detect neutrinos crossing the Earth from the North pole. If the cross section for neutrino interaction on a nucleon is $(6.7 \times 10^{-39} E)$ cm^2 with E expressed in GeV (note the linear increase with the neutrino energy E), what is the energy at which half of the neutrinos interact before reaching the the detector? Comment on the result.

Chapter 4
Particle Detection

After reading this chapter, you should be able to manage the basics of particle detection, and to understand the sections describing the detection technique in a modern article of high-energy particle or astroparticle physics.

Particle detectors measure physical quantities related to the result of a collision; they should ideally identify all the outcoming (and the incoming, if unknown) particles, and measure their kinematical characteristics (momentum, energy, velocity).

In order to detect a particle, one must make use of its interaction with a sensitive material. The interaction should possibly not destroy the particle that one wants to detect; however, for some particles this is the only way to obtain information.

In order to study the properties of detectors, we shall thus first need to review the characteristics of the interaction of particles with matter.

4.1 Interaction of Particles with Matter

4.1.1 Charged Particle Interactions

Charged particles interact basically with atoms, and the interaction is mostly electromagnetic: they might expel electrons (ionization), promote electrons to upper energy levels (excitation), or radiate photons (bremsstrahlung, Cherenkov radiation, transition radiation). High-energy particles may also interact directly with the atomic nuclei.

4.1.1.1 Ionization Energy Loss

This is one of the most important sources of energy loss by charged particles. The average value of the specific (i.e., calculated per unit length) energy loss due to ionization and excitation whenever a particle goes through a homogeneous material

A. De Angelis and M.J.M. Pimenta, *Introduction to Particle and Astroparticle Physics*, Undergraduate Lecture Notes in Physics,
DOI 10.1007/978-88-470-2688-9_4

of density ρ is described by the so-called Bethe formula.[1] This expression has an accuracy of a few % in the region $0.1 < \beta\gamma < 1000$ for materials with intermediate atomic number.

$$-\frac{dE}{dx} \simeq \rho D \left(\frac{Z}{A}\right) \frac{(z_p)^2}{\beta^2} \left[\frac{1}{2} \ln \left(\frac{2m_e c^2 \beta^2 \gamma^2}{I}\right) - \beta^2 - \frac{\delta(\beta, \rho)}{2}\right], \quad (4.1)$$

where

- ρ is the material density, in g/cm^3;
- Z and A are the atomic and mass number of the material, respectively;
- z_p is the charge of the incoming particle, in units of the electron charge;
- $D \simeq 0.307$ MeV cm^2/g;
- $m_e c^2$ is the energy corresponding to the electron mass, ~ 0.5 MeV;
- I is the mean excitation energy in the material; it can be approximated as $I \simeq 16 \text{ eV} \times Z^{0.9}$ for $Z > 1$;
- δ is a correction term that becomes important at high energies. It accounts for the reduction in energy loss due to the so-called *density effect*. As the incident particle velocity increases, media become polarized and their atoms can no longer be considered as isolated.

The energy loss by ionization (Fig. 4.1) in first approximation is:

- independent of the particle's mass;
- typically small for high-energy particles (about 2 MeV/cm in water; one can roughly assume a proportionality to the density of the material);
- proportional to $1/\beta^2$ for $\beta\gamma \leq 3$ (the minimum of ionization: minimum ionizing particle, often just called a "mip");
- basically constant for $\beta > 0.96$ (logarithmic increase after the minimum);
- proportional to Z/A (Z/A being about equal to 0.5 for all elements but hydrogen and the heaviest nuclei).

In practice, most relativistic particles (such as cosmic-ray muons) have mean energy loss rates close to the minimum; they can be considered within less than a factor of two as minimum ionizing particles. The loss from a minimum ionizing particle is well approximated as

$$\frac{1}{\rho}\frac{dE}{dx} \simeq -3.5 \left(\frac{Z}{A}\right) \text{MeV cm}^2/\text{g}.$$

---

[1] The 24-year-old Hans Bethe, Nobel prize in 1967 for his work on the theory of stellar nucleosynthesis, published this formula in 1930; the formula—not including the density term, added later by Fermi—was derived using quantum mechanical perturbation theory up to z_p^2. The description can be improved by considering corrections which correspond to higher powers of z_p: Felix Block obtained in 1933 a higher-order correction proportional to z_p^4, not reported in this text, and sometimes the formula is called "Bethe-Block energy loss"—although this naming convention has been discontinued by the Particle Data Group since 2008.

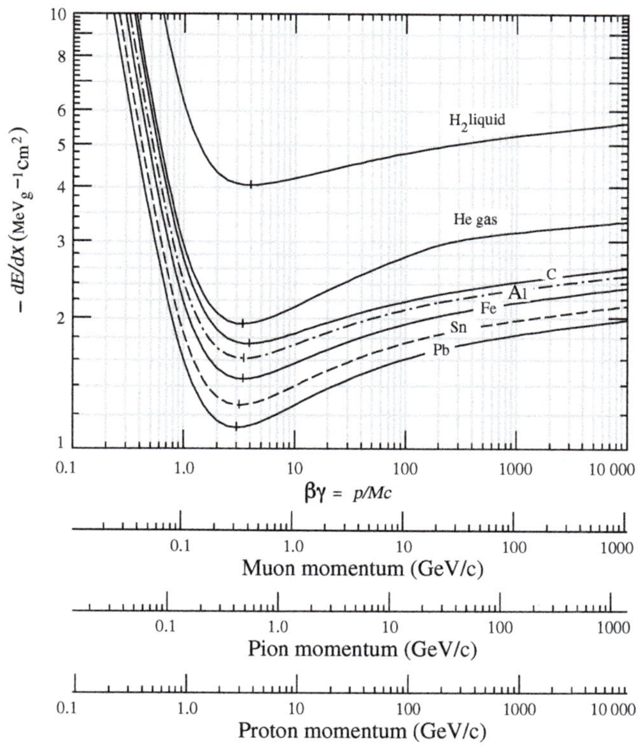

Fig. 4.1 Specific ionization energy loss for muons, pions, and protons in different materials. From K.A. Olive et al. (Particle Data Group), Chin. Phys. C 38 (2014) 090001

In any case, as we shall see later, the energy loss in the logarithmic increase region can be used by means of appropriate detectors for particle identification.

Due to the statistical nature of the ionization process, large fluctuations on the energy loss arise when fast charged particles pass through absorbers which are thin compared to the particle range. The energy loss is distributed around the most probable value according to an asymmetric distribution (named the Landau distribution). The average energy loss, represented by the Bethe formula, is larger than the most probable energy loss, since the Landau distribution has a long tail (as the width of the material increases, the most probable energy loss becomes however closer to the average, as one can see in Fig. 4.2).

Although its nature is quantum mechanical, the main characteristics of Eq. 4.1 can be derived classically, as it was first done by Bohr. Let us suppose a charged particle of mass m and charge z_p passes at a distance b from a target of mass M and charge Z. The momentum Δp transferred to the target depends on the electric field E produced by the charged traveling particle. Given the symmetry of the problem

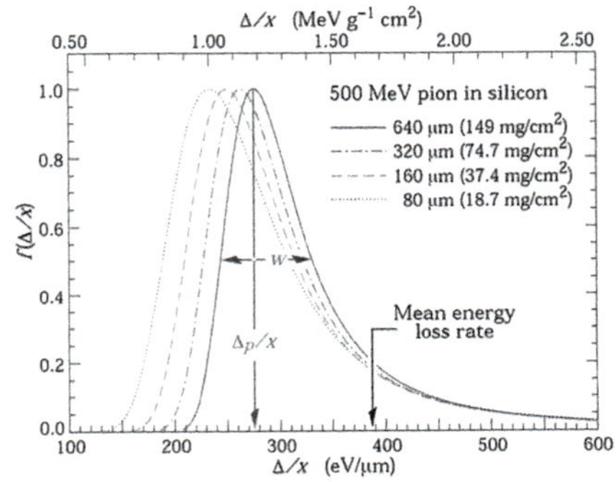

Fig. 4.2 Distribution of the energy loss (Landau distribution) in silicon for 500 MeV pions, normalized to unity at the most probable value. w is the full width at half maximum. From K.A. Olive et al. (Particle Data Group), Chin. Phys. C 38 (2014) 090001

only the transverse component of the electric field with the respect to the particle track $E_\perp$ matters. Relating the interaction time t with the velocity of the particle, $dt = dx/v$, one can write for the momentum transfer:

$$\Delta p = \int_{-\infty}^{+\infty} \vec{F}\, dt = \int_{-\infty}^{+\infty} e\, \vec{E}\, dt = \frac{e}{v} \int_{-\infty}^{+\infty} E_\perp\, dx\,.$$

The electric field integral can be calculated using the Gauss's law. In fact, the flux of the electric field passing through a cylinder of radius b is given by $\int E_\perp\, 2\pi b\, dx = z_p\, e/\varepsilon_0$. Therefore, the momentum transferred to the target particle can be written as

$$\Delta p = \frac{z_p\, e^2}{2\, \pi\, \varepsilon_0\, v\, b}$$

or still in terms of the energy and using the classical radius of the electron[2] $r_e = (e^2/4\pi\epsilon_0)/(m_e c^2) \simeq 0.003$ pm:

$$\Delta E = \frac{\Delta p^2}{2\, m} = \left(\frac{1}{4\pi\varepsilon_0}\right)^2 \frac{1}{m\, c^2}\, \frac{2\, z_p^2\, Z^2\, e^4}{b^2\, \beta^2} = \frac{(m_e\, c^2)^2}{m\, c^2}\, \frac{2\, z_p^2\, Z^2}{\beta^2}\, \left(\frac{r_e}{b}\right)^2 .$$

From this expression one can see that close collisions ($\Delta E \propto 1/b^2$) and low mass particles ($\Delta E \propto 1/m$) are the most important with respect to energy loss; thus one can neglect the effect of nuclei.

Photoluminescence. In some transparent media, part of the ionization energy loss goes into the emission of visible or near-visible light by the excitation of atoms and/or molecules. This phenomenon is called photoluminescence; often it results into a fast ($<100\,\mu$s) excitation/deexcitation—in this last case we talk of fluorescence, or scintillation. Specialists often use definitions which distinguish between fluorescence and scintillation; this separation is, however, not universally accepted.

---

[2] The classical electron radius is the size the electron would need to have for its mass to be completely due to its electrostatic potential energy, under the assumption that charge has a uniform volume density and that the electron is a sphere.

We shall discuss later fluorescence in the context of the detection of large showers induced in the atmosphere by high-energy cosmic rays.

4.1.1.2 High-Energy Radiation Effects

According to classical electromagnetism, a charged particle undergoing acceleration radiates electromagnetic waves. The intensity of the emitted radiation from a dipole is proportional to the square of the acceleration.

Particles deflected by the electric field of the material traversed, thus, also emit photons. We speak in this case of bremsstrahlung, or braking radiation.

At first order the emitted energy is (as in the classical case) proportional to the inverse of the square of the mass. On top of the ionization energy loss described by Eq. 4.1, above $\beta\gamma \sim 1000$ (which means an extremely high energy for a proton, $E \sim 1$ TeV, but just $E \sim 100$ GeV for a muon), such radiation effects become important (Fig. 4.3).

Bremsstrahlung is particularly relevant for electrons and positrons, particles for which the Bethe approximation starts to be inadequate at even lower energies. The average fractional energy loss by radiation for an electron of high energy ($E \gg m_e c^2$) is approximately independent of the energy itself, and can be described by

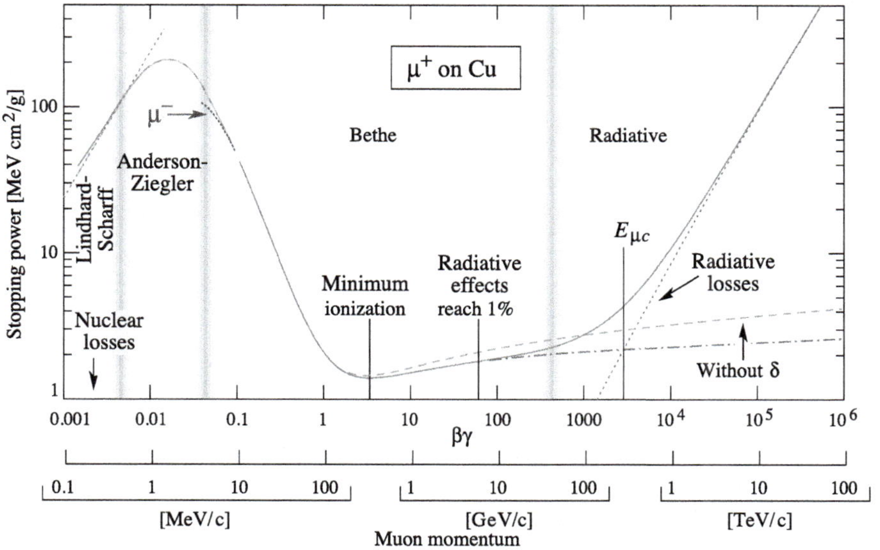

Fig. 4.3 The stopping power ($-dE/dx$) for positive muons in copper as a function of $\beta\gamma = p/Mc$ is shown over nine orders of magnitude in momentum (corresponding to 12 orders of magnitude in kinetic energy). From K.A. Olive et al. (Particle Data Group), Chin. Phys. C 38 (2014) 090001

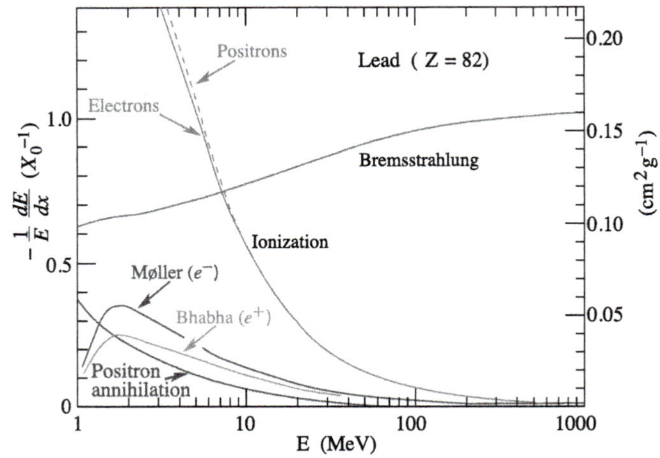

Fig. 4.4 Fractional energy loss per radiation length in lead as a function of the electron or positron energy. From K.A. Olive et al. (Particle Data Group), Chin. Phys. C 38 (2014) 090001

$$\frac{1}{E}\frac{dE}{dx} \simeq -\frac{1}{X_0} \tag{4.2}$$

where X_0 is called the radiation length, and is characteristic of the material—for example it is about 300 m for air at Normal Temperature and Pressure (NTP),[3] about 36 cm for water, about 0.5 cm for lead.

The radiation length has been tabulated for the different elements in Appendix B; a good approximation (for $Z > 4$) is

$$\frac{1}{X_0} = 4\left(\frac{\hbar}{m_e c}\right)^2 Z(Z+1)\alpha^3 n_a \ln\left(\frac{183}{Z^{1/3}}\right), \tag{4.3}$$

where n_a is the density of atoms per cubic centimeter in the medium, or more simply

$$\frac{1}{\rho}X_0 \simeq 180\frac{A}{Z^2}\text{cm} \quad \left(\frac{\Delta X_0}{X_0} < \pm 20\,\% \text{ for } 12 < Z < 93\right). \tag{4.4}$$

The total average energy loss by radiation increases rapidly (linearly in the approximation of the equation (4.2)) with energy, while the average energy loss by collision is practically constant. At high energies, radiation losses are thus much more important than collision losses (Fig. 4.4).

[3]NTP is commonly used as a standard condition; it is defined as air at 20 °C (293.15 K) and 1 atm (101.325 kPa). Density is 1.204 kg/m^3. Standard Temperature and Pressure STP, another condition frequently used in physics, is defined by IUPAC (International Union of Pure and Applied Chemistry) as air at 0°C (273.15 K) and 100 kPa.

The energy at which the radiation energy loss overtakes the collision energy loss (called the critical energy, E_c) decreases with increasing atomic number:

$$E_c \simeq \frac{550\,\text{MeV}}{Z} \quad \left(\frac{\Delta E_c}{E_c} < \pm 10\,\% \text{ for } 12 < Z < 93\right) . \tag{4.5}$$

Critical energy for air at NTP is about 84 MeV; for water is about 74 MeV.

The photons radiated by bremsstrahlung are distributed at leading order in such a way that the energy loss per unit energy is constant, i.e.,

$$N_\gamma \propto \frac{1}{E_\gamma}$$

between 0 and E. This results in a divergence for $E_\gamma \to 0$, which anyway does not contradict energy conservation: the integral of the energy released for each energy bin is constant.

The emitted photons are collimated: the typical angle of emission is $\sim m_e c^2/E$.

4.1.1.3 Cherenkov Radiation

The Vavilov–Cherenkov[4] radiation (commonly called just Cherenkov radiation) occurs when a charged particle moves through a medium faster than the speed of light in that medium. The total energy loss due to this process is negligible; however, Cherenkov radiation is important related to the possibility of use in detectors.

The light is emitted in a coherent cone (Fig. 4.5) at an angle such that

$$\cos\theta_c = \frac{1}{n\beta} \tag{4.6}$$

from the direction of the emitting particle. The threshold velocity is thus $\beta = 1/n$, where n is the refractive index of the medium. The presence of a coherent wavefront can be easily derived by using the Huygens–Fresnel principle.

The number of photons produced per unit path length and per unit energy interval of the photons by a particle with charge $z_p e$ at the maximum (limiting) angle is

$$\frac{d^2N}{dEdx} \simeq \frac{\alpha z_p^2}{\hbar c} \sin^2\theta_c \simeq 370 \sin^2\theta_c \,\text{eV}^{-1}\text{cm}^{-1} \tag{4.7}$$

or equivalently

---

[4]Pavel Cherenkov (1904–1990) was a Soviet physicist who shared the Nobel prize in physics in 1958 with con-nationals Ilya Frank (1908–1990) and Igor Tamm (1895–1971) for the discovery of Cherenkov radiation, made in 1934. The work was done under the supervision of Sergey Vavilov, who died before the recognition of the discovery by the Nobel committee.

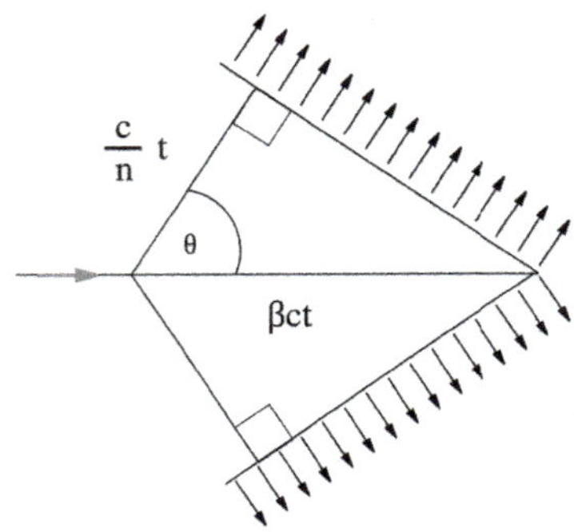

Fig. 4.5 The emission of Cherenkov radiation by a charged particle. User Harp, Creative Commons Attribution-Share Alike 2.5 Generic license via Wikimedia Commons

$$\frac{d^2N}{d\lambda dx} \simeq \frac{2\pi\alpha z_p^2}{\lambda^2}\sin^2\theta_c \tag{4.8}$$

(the index of refraction n is in general a function of photon energy E; Cherenkov radiation is relevant when $n > 1$ and the medium is transparent, and this happens close to the range of visible light).

The total energy radiated is small, some 10^{-4} times the energy lost by ionization. In the visible range (300–700 nm), the total number of emitted photons is about 40/m in air, about 500/cm in water. Due to the dependence on λ, it is important that Cherenkov detectors be sensitive close to the ultraviolet region.

Dense media can be transparent not only to visible light, but also to radio waves. The development of Cherenkov radiation in the radiowave region due to the interactions with electrons in the medium is often referred to as the Askar'yan effect. This effect has been experimentally confirmed for different media (namely sand, rock salt and ice) in accelerator experiments at SLAC; presently attempts are in progress to use this effect for in particle detectors.

4.1.1.4 Transition Radiation

X-ray transition radiation (XTR) occurs when a relativistic charged particle crosses from one medium to another with different dielectric permittivity.

The energy radiated when a particle with charge $z_p e$ and $\gamma \simeq 1000$ crosses the boundary between vacuum and a different transparent medium is typically concentrated in the soft X-ray range 2–40 keV.

The process is closely related to Cherenkov radiation, and also in this case the total energy emitted is low (typically the expected number of photons per transition is smaller than unity; one thus needs several layers to build a detector).

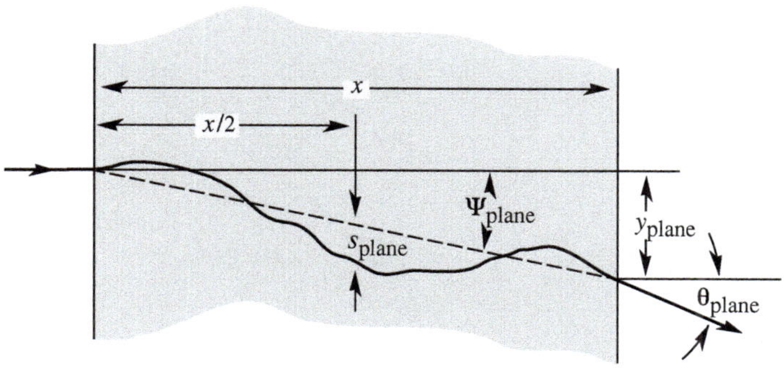

Fig. 4.6 Multiple Coulomb scattering. From K.A. Olive et al. (Particle Data Group), Chin. Phys. C 38 (2014) 090001

4.1.1.5 Multiple Scattering

A charged particle passing near a nucleus undergoes deflection, with an energy loss that is in most cases negligible (approximately zero). This phenomenon is called elastic scattering and is caused by the interaction between the particle and the Coulomb field of the nucleus. The global effect is that the path of the particle becomes a random walk (Fig. 4.6), and information on the original direction is partly lost—this fact can create problems for the reconstruction of direction in tracking detectors. For very-high-energy hadrons, also hadronic cross section can contribute to the effect.

Summing up many relatively small random changes on the direction of flight of a particle of unit charge traversing a thin layer of material, the distribution of its projected scattering angle can be approximated by a Gaussian distribution of standard deviation projected on a plane (one has to multiply by $\sqrt{2}$ to determine the standard deviation in space):

$$\theta_0 \simeq \frac{13.6\,\text{MeV}}{\beta cp} z_p \sqrt{\frac{x}{X_0}} \left[1 + 0.038 \ln \frac{x}{X_0} \right] .$$

X_0 is the radiation length defined before.

The above expression comes from the so-called Molière theory, and is accurate to some 10 % or better for $10^{-3} < x/X_0 < 100$.

This approximation is crude, mainly due to the underlying assumption of a Gaussian distribution. In particular, the occurrence of scattering at large angles is underestimated, and some 2 % of the particles can suffer more important deflections due to Rutherford scattering and contribute to a sizable tail.

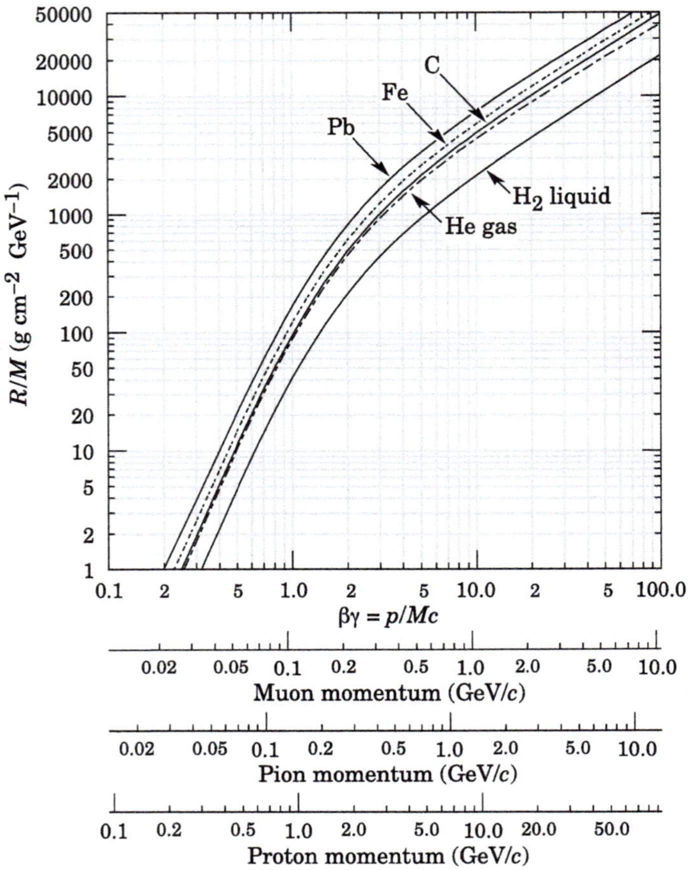

Fig. 4.7 Range per unit of density and of mass for heavy charged particles in liquid (bubble chamber) hydrogen, helium gas, carbon, iron, and lead. Example: a K^+ with momentum 700 MeV/c, $\beta\gamma \simeq 1.42$, and we read $R/M \simeq 396$, in lead, corresponding to a range of 195 g/cm^2. From K.A. Olive et al. (Particle Data Group), Chin. Phys. C 38 (2014) 090001

4.1.2 Range

From the specific energy loss as a function of energy, we can calculate the fraction of energy lost as a function of the distance x traveled in the medium. This is known as the Bragg curve. For charged particles, most of the ionization loss occurs near the end of the path, where the particle speed is low. The Bragg curve has a pronounced peak close to the end of the path length, where it then falls rapidly to zero. The range R for a particle of energy E is the average distance traveled before reaching the energy at which the particle is absorbed (Fig. 4.7):

$$R(E') = \int_E^{Mc^2} \frac{1}{\frac{dE}{dx}} dE \, .$$

4.1.3 Photon Interactions

Photons mostly interact with matter via photoelectric effect, Compton scattering, and electron–positron pair production. Other processes, like Rayleigh scattering and photonuclear interactions, have in general much smaller cross sections.

4.1.3.1 Photoelectric Effect

The photoelectric effect is the ejection of an electron from a material that has just absorbed a photon. The ejected electron is called a *photoelectron*.

The photoelectric effect was pivotal in the development of quantum physics (for the explanation of this effect Albert Einstein was awarded the Nobel prize). Due to photoelectric effect, a photon of angular frequency $\omega > V/e$ can eject from a metal an electron which pops up with a kinetic energy $\hbar\omega - V$, where V is the minimum gap of energy of electrons trapped in the metal (V is frequently called the *work function* of the metal).

No simple relationship between the attenuation of the incident electromagnetic wave and the photon energy E can be derived, since the process is characterized by the interaction with the (quantized) orbitals. The plot of the attenuation coefficient (the distance per unit density at which intensity is reduced by a factor $1/e$) as a function of the photon energy displays sharp peaks at the binding energies of the different orbital shells and depends strongly on the atomic number. A reasonable approximation for the cross section σ is

$$\sigma \propto \frac{Z^{\nu}}{E^3} ,$$

with the exponent ν varying between 4 and 5 depending on the energy. The cross section rapidly decreases with energy above the typical electron binding energies (Fig. 4.8).

The photoelectric effect can be used for detecting photons below the MeV; a photosensor (see later) sensitive to such energies can "read" the signal generated by a photoelectron, possibly amplified by an avalanche process.

4.1.3.2 Compton Scattering

Compton scattering is the collision between a photon and an electron. Let E be the energy of the primary photon (corresponding to a wavelength λ) and suppose the electron is initially free and at rest. After the collision, the photon is scattered at an angle θ and comes out with a reduced energy E', corresponding to a wavelength λ'; the electron acquires an energy $E - E'$. The conservation laws of energy and momentum yield the following relation (Compton formula):

$$\lambda' - \lambda = \lambda_C(1 - \cos\theta) \longrightarrow E' = \frac{E}{1 + \frac{E}{m_e c^2}(1 - \cos\theta)}$$

where θ is the scattering angle of the emitted photon; $\lambda_C = h/m_e c \simeq 2.4$ pm is the Compton wavelength of the electron.

It should be noted that, in case the target electron is not at rest, the energy of the scattered photon can be larger than the energy of the incoming one. This regime is called *inverse Compton,* and it has great importance in the emission of high-energy photons by astrophysical sources: in practice, thanks to inverse Compton, photons can be "accelerated."

The differential cross section for Compton scattering was calculated by Klein and Nishina around 1930. If the photon energy is much below $m_e c^2$ (so the scattered electrons are non-relativistic) then the total cross section is given by the Thomson cross section. This is known as the Thomson limit. The cross section for $E \ll m_e c^2$ (Thomson regime) is about

$$\sigma_T \simeq \frac{8\pi\alpha^2}{3m_e^2} = \frac{8\pi r_e^2}{3}, \tag{4.9}$$

where $r_e = (e^2/4\pi\epsilon_0)/(m_e c^2) \simeq 0.003$ pm is the classical radius of the electron. If the photon energy is $E \gg m_e c^2$, we are in the so-called Klein-Nishina regime and the total cross section falls off rapidly with increasing energy (Fig. 4.8):

$$\sigma_{KN} \simeq \frac{3\sigma_T}{8}\frac{\ln 2E}{E}. \tag{4.10}$$

As in the case of the photoelectric effect, the ejected electron can be detected (possibly after multiplication) by an appropriate sensor.

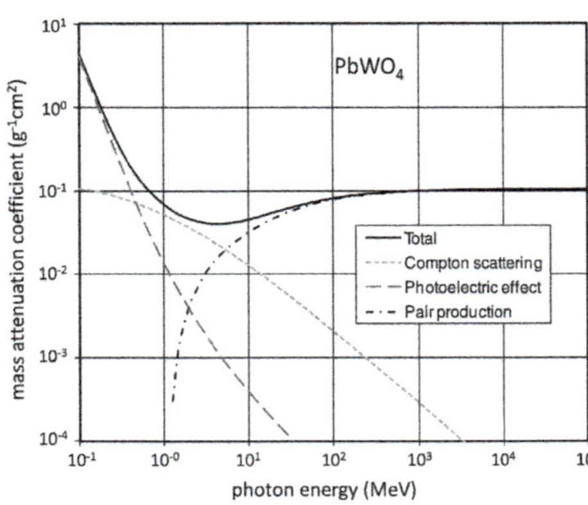

Fig. 4.8 Photon mass attenuation coefficient (cross section per gram of material) as a function of energy in lead tungstate (data from the NIST XCOM database)

4.1.3.3 Pair Production

Pair production is the most important interaction process for a photon above an energy of a few tenth of MeV. In the electric field in the neighborhood of a nucleus, a high-energy photon has a non-negligible probability of transforming itself into a negative and a positive electron—the process being kinematically forbidden unless an external field, regardless of how little, is present.

Energy conservation yields the following relation between the energy E of the primary photon and the total energies U and U' of the electrons:

$$E = U + U' .$$

With reasonable approximation, for 1 TeV> $E > 100\,\text{MeV}$ the fraction of energy $u = U/E$ taken by the secondary electron/positron is uniformly distributed between 0 and 1 (becoming peaked at the extremes as the energy increases to values above 1 PeV).

The cross section grows quickly from the kinematical threshold of about 1 MeV to its asymptotic value reached at some 100 MeV:

$$\sigma \simeq \frac{7}{9} \frac{1}{n_a X_0} ,$$

where n_a is the density of atomic nuclei per unit volume, in such a way that the interaction length is

$$\lambda \simeq \frac{9}{7} X_0 .$$

The angle of emission for the particles in the pair is typically $\sim m_e c^2/E$.

4.1.3.4 Rayleigh Scattering and Photonuclear Interactions

Rayleigh scattering (the dispersion of electromagnetic radiation by particles with radii less than or in the order of 1/10 the wavelength of the radiation) is usually of minor importance for the conditions of high-energy particle and astroparticle physics, but it can be important for light in the atmosphere, and thus for the design of instruments detecting visible light. The photonuclear effect, i.e., the excitation of nuclei by photons, is mostly restricted to the region around 10 MeV, and it may amount to as much as 10 percent of the total cross section due to electrodynamic effects.

4.1.3.5 Comparison Between Different Processes for Photons

The total Compton scattering probability decreases rapidly when the photon energy increases. Conversely, the total pair production probability is a slowly increasing function of energy. At large energies, most photons are thus absorbed by pair production, while photon absorption by Compton effect dominates at small energies (being the photoelectric effect characteristic of even smaller energies). The absorption of photons by pair production, Compton, and photoelectric effect is compared in Fig. 4.8.

As a matter of fact, above about 30 MeV the dominant process is pair production, and the interaction length of a photon is with extremely good approximation equal to $9X_0/7$.

At extremely high matter densities and/or at extremely high energies (typically above 10^{16}–10^{18} eV, depending on the medium composition and density) collisions cannot be treated independently, and the result of the collective quantum mechanical treatment is a reduction of the cross section. The result is the so-called Landau-Pomeranchuk-Migdal effect, or simply LPM effect, which entails a reduction of the pair production cross section, as well as of bremsstrahlung.

4.1.4 Nuclear (Hadronic) Interactions

The nuclear force is felt by hadrons, charged and neutral; at high energies (above a few GeV) the inelastic cross section for hadrons is dominated by nuclear interaction. High-energy nuclear interactions can be characterized by an inelastic interaction length λ_H. Values for $\rho\lambda_H$ are typically of the order of 100 g/cm^2; a listing for some common materials is provided in Appendix B—where the inelastic length λ_I and the total length λ_T are separately listed, and the rule for the composition is $1/\lambda_T = 1/\lambda_H + 1/\lambda_I$.

The final state products of inelastic high-energy hadronic collisions are mostly pions, since these are the lightest hadrons. The rate of positive, negative, and neutral pions is more or less equal—as we shall see, this fact is due to a fundamental symmetry of hadronic interactions, called the *isospin* symmetry.

4.1.5 Interaction of Neutrinos

The case of neutrinos is a special one. Neutrinos have a very low interaction cross section, which can be parameterized (on a single nucleon) for intermediate energies (Fig. 4.9) as

$$\sigma_{\nu N} \simeq (6.7 \times 10^{-39} E)\,\mathrm{cm}^2\,, \tag{4.11}$$

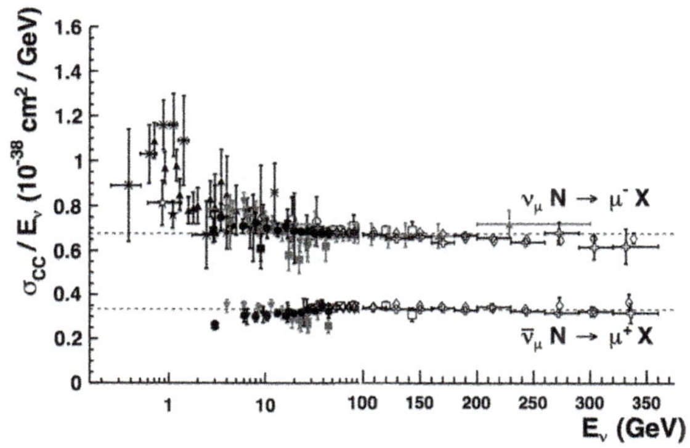

Fig. 4.9 Measurements of muon neutrino and antineutrino inclusive scattering cross sections divided by neutrino energy as a function of neutrino energy; different symbols represent experimental measurements by different experiments. Note the transition between logarithmic and linear scales at 100 GeV. From K.A. Olive et al. (Particle Data Group), Chin. Phys. C 38 (2014) 090001

E being the neutrino energy in GeV. Solar neutrinos, which have MeV energies, typically cross the Earth undisturbed (see a more complete discussion in Chap. 9).

The low value of the cross section makes the detection of neutrinos very difficult.

4.1.6 Electromagnetic Showers

High-energy electrons lose most of their energy by radiation. Thus, in their interaction with matter, most of the energy is spent in the production of high-energy photons and only a small fraction is dissipated. The secondary photons, in turn, undergo pair production (or, at lower energies, Compton scattering); in the first case, electrons and positrons can in turn radiate. This phenomenon continues generating cascades (showers) of electromagnetic particles; at each step the number of particles increases while the average energy decreases, until the energy falls below the critical energy.

Given the characteristics of the interactions of electrons/positrons and of photons with matter, it is natural to describe the process of electromagnetic cascades in terms of the scaled distance

$$t = \frac{x}{X_0}$$

(where X_0 is the radiation length), and of the scaled energy

$$\epsilon = \frac{E}{E_c}$$

(where E_c is the critical energy); the radiation length and the critical energy have been defined in Sect. 4.1.1.2. Since the opening angles for bremsstrahlung and pair production are small, the process can be in first approximation (above the critical energy) considered as one-dimensional (the lateral spread will be discussed at the end of this section).

A simple approximation (a "toy model"), proposed by Heitler in the late 1930s, assumes that

- the incoming charged particle has an initial energy E_0 much larger than the critical energy E_c;
- each electron travels one radiation length and then gives half of its energy to a bremsstrahlung photon;
- each photon travels one radiation length and then creates an electron–positron pair; the electron and the positron each carry half of the energy of the original photon.

In the above model, asymptotic formulas for radiation and pair production are assumed to be valid; the Compton effect and the collision processes are neglected. The branching stops abruptly when $E = E_c$, and then electrons and positrons lose their energy by ionization.

This simple branching model is schematically shown in Fig. 4.10, left. It implies that after t radiation lengths the shower will contain 2^t particles and there will be roughly the same number of electrons, positrons, and photons, each with an average energy

$$E(t) = E_0/2^t \, .$$

The cascading process will stop when $E(t) = E_c$, at a thickness of absorber $t_{\max}$, that can be written in terms of the initial and critical energies as

$$t_{\max} = \log_2(E_0/E_c) \, ,$$

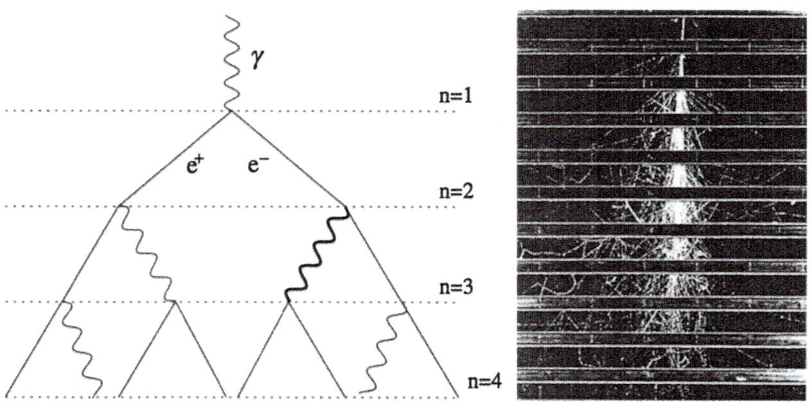

Fig. 4.10 *Left* Scheme of the Heitler approximation for the development of an electromagnetic shower. From J. Matthews, Astropart. Phys. 22 (2005) 387. *Right* Image of an electromagnetic shower developing through a number of brass plates 1.25 cm thick placed across a cloud chamber (from B. Rossi, "Cosmic rays", McGraw-Hill 1964)

with the number of particles at this point given by

$$N_{\max} = \frac{E_0}{E_c} \equiv y\,.$$

The model suggests that the shower depth at its maximum varies as the logarithm of the primary energy. This emerges also from more sophisticated shower models and is observed experimentally. A real image of an electromagnetic shower in a cloud chamber is shown in Fig. 4.10, right.

An improved model was formulated by Rossi in the beginning of the 1940s. Rossi (see for example reference [F4.1]) computed analytically the development of a shower in the so-called "approximation B," in which: electrons lose energy by ionization and bremsstrahlung (described by asymptotical formulae); photons undergo pair production, also described by asymptotical formulae. All the process is one-dimensional. The results of the "Rossi approximation B" are summarized in Table 4.1. Under this approximation, the number of particles grows exponentially in the beginning up to the maximum, and then decreases as shown in Figs. 4.11 and 4.12.

Table 4.1 Shower parameters for a particle on energy E_0 according to Rossi approximation B (y = E_0/E_c)

	Incident electron	Incident photon
Peak of shower $t_{\max}$	$1.0 \times (\ln y - 1)$	$1.0 \times (\ln y - 0.5)$
Center of gravity t_{med}	$t_{\max} + 1.4$	$t_{\max} + 1.7$
Number of e^+ and e^- at peak	$0.3y/\sqrt{\ln y - 0.37}$	$0.3y/\sqrt{\ln y - 0.31}$
Total track length	y	y

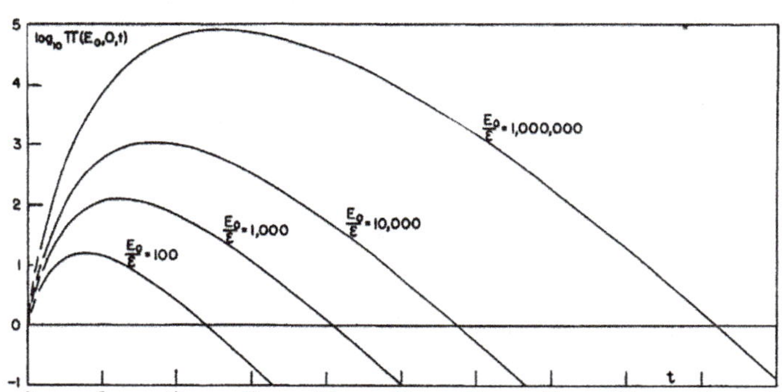

Fig. 4.11 Logarithm of the number of electrons for electron-initiated showers, calculated under Rossi approximation B, as a function of the number of radiation lengths traversed. Multiplication by E_c/I (E_c is called ε in the Figure) yields the specific ionization energy loss [F4.1]

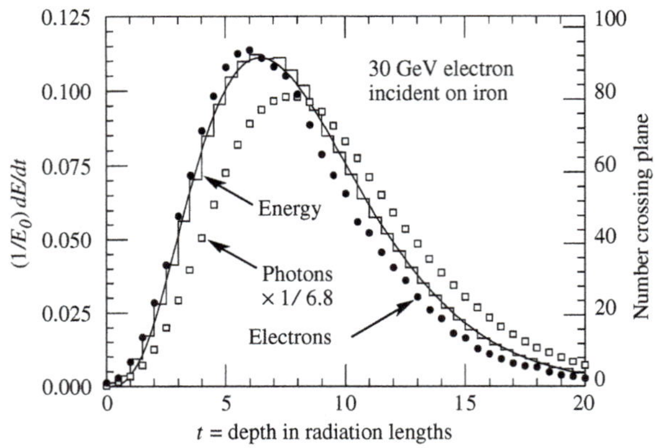

Fig. 4.12 A Monte Carlo simulation of a 30 GeV electron-induced cascade in iron. The histogram shows the fractional energy deposition per radiation length, and the *curve* is a fit to the distribution using Eq. (4.12). The circles indicate the number of electrons with total energy greater than 1.5 MeV crossing planes at $X_0/2$ intervals (scale on the *right*) and the squares the number of photons above the same energy crossing the planes (scaled down to have the same area as the electron distribution). From K.A. Olive et al. (Particle Data Group), Chin. Phys. C 38 (2014) 090001

A common parameterization of the longitudinal profile for a shower of initial energy E_0 is

$$\frac{dE}{dt} = E_0 \frac{\beta}{\Gamma(\alpha)} (\beta t)^{\alpha-1} e^{-\beta t} , \tag{4.12}$$

where Γ is the Euler's gamma function $\Gamma(z) = \int_0^{+\infty} t^{z-1}\, e^{-t}\, dt$. In the above approximation, $t_{\max} = (\alpha - 1)/\beta$, which should be thus equal to $\ln(E_0/E_c) - C$ with $C = 1$ for an electron and $C = 0.5$ for a photon.

Fluctuations on the total track length are dominated by the fluctuations on the total number of particles, and thus they grow as $\sqrt{E_0}$. An incomplete longitudinal containment of the shower badly increases fluctuations on the deposited energy. A rule of thumb for the longitudinal containment of 95 % of the shower is

$$T(95\,\%) = (t_{\max} + 0.08Z + 9.6) ,$$

expressed in radiation lengths.

Despite the elegance of Rossi's calculations, one can do better using computers, and most calculations are performed nowadays by Monte Carlo methods.[5] Monte Carlo calculations of electromagnetic cascades have the advantages of using accurate cross sections for bremsstrahlung and pair production, the correct energy dependence of ionization loss, and including all electromagnetic interactions. Monte Carlo calculations, in addition, give correct account for the fluctuations in the shower devel-

[5] Monte Carlo methods are computational algorithms based on repeated random sampling. The name is due to its resemblance to the act of playing in a gambling casino.

opment, as well as for the angular and lateral distribution of the shower particles. Rossi approximation B, however, is faster and represents a rather accurate model.

The description of the transverse development of a shower is more complicated. Usually the normalized lateral density distribution of electrons is approximated by the Nishimura-Kamata-Greisen (NKG) function, which depends on the "shower age" s, being 0 at the first interaction, 1 at the maximum and 3 at the death [F4.1]:

$$s = \frac{3t}{t + 2t_{\max}} \tag{4.13}$$

but not on its energy. The NKG function:

$$\rho_{\text{NKG}}(r, s, N_e) = \frac{N_e}{R_M^2} \frac{\Gamma(4.5 - s)}{2\pi\Gamma(s)\Gamma(4.5 - 2s)} \left(\frac{r}{R_M}\right)^{s-2} \left(1 + \frac{r}{R_M}\right)^{s-4.5} \tag{4.14}$$

where N_e is the electron shower size, r is the distance from the shower axis, and R_M is a transverse scale called the Molière radius described below, is accurate for a shower age $0.5 < s < 1.5$. A variety of transverse distribution functions can be found in the literature (Greisen, Greisen–Linsley, etc.) and are mostly specific modifications of the NKG function.

In a crude approximation, one can assume the transverse dimension of the shower to be dictated by the Molière radius:

$$R_M \simeq \frac{21\,\text{MeV}}{E_c} X_0 \,.$$

About 90 % of the shower energy is deposited in a cylinder of radius R_M; about 95 % is contained in a radius of $2R_M$, and about 99 % in a radius of $3R_M$. In air at NTP, $R_M \simeq 80$ m; in water $R_M \simeq 9$ cm.

4.1.7 Hadronic Showers

The concept of hadronic showers is similar to the concept of electromagnetic showers: primary hadrons can undergo a sequence of interactions and decays creating a cascade. However, on top of electromagnetic interactions one has now nuclear reactions. In addition, in hadronic collisions with the nuclei of the material, a significant part of the primary energy is consumed in the nuclear processes (excitation, emission of low-energy nucleons, etc.). One thus needs ad-hoc Monte Carlo corrections to account for the energy lost, and fluctuations are larger. The development of appropriate Monte Carlo codes for hadronic interactions has been a problem in itself, and still the calculation requires huge computational "loads." At the end of a hadronic cascade, most of the particles are pions, and one-third of the pions are neutral and decay almost instantaneously ($\tau \sim 10^{-16}$ s) into a pair of photons; thus in average

Fig. 4.13 Image of a hadronic shower developing through a number of brass plates 1.25 cm thick placed across a cloud chamber (from B. Rossi, "Cosmic rays," McGraw-Hill 1964). To be compared to Fig. 4.10, *right*

one-third of the hadronic cascade is indeed electromagnetic (but it can be more or less, subject to fluctuations).

In the first approximation, the development of the shower can be described by the inelastic hadronic interaction length λ_H; however, the approximation is less accurate than the one we did when we assumed electromagnetic reactions to scale with the radiation length X_0, and intrinsic fluctuations are larger.

As an example, the image of a hadronic shower in a cloud chamber is shown in Fig. 4.13.

4.2 Particle Detectors

The aim of a particle detector is to measure the momenta and to identify the particles that pass through it after being produced in a collision or a decay; this is called an "event." The position in space where the event occurs is known as the interaction point.

In order to identify every particle produced by the collision, and plot the paths they have taken—i.e., to "completely reconstruct the event"—it is necessary to know the masses and momenta of the particles themselves. The mass can be computed by measuring the momentum and either the velocity or the energy.

The characteristics of the different instruments that allow these measurements are presented in what follows.

4.2.1 Track Detectors

A tracking detector reveals the path taken by a particle by measurements of sampled points (hits). Momentum measurements can be made by measuring the curvature of the track in a magnetic field, which causes the particle to curve into a spiral orbit with a radius proportional to the momentum of the particle. This requires the determination of the best fit to a helix of the hits (particle fit). For a particle of unit charge

$$p \simeq 0.3 B_{\perp} R \, ,$$

where $B_{\perp}$ is the component of the magnetic field perpendicular to the particle velocity, expressed in tesla (which is the order of magnitude of typical fields in detectors), the momentum p is expressed in GeV, and R is the radius of curvature of the helix in meters.

A source of uncertainty for this determination is given by the errors in the measurement of the hits; another (intrinsic) noise is given by multiple scattering. In what follows we shall review some detectors used to determine the trajectory of charged tracks.

4.2.1.1 Cloud Chamber and Bubble Chamber

The cloud chamber was invented by C.T.R. Wilson in the beginning of XX century, and was used as a detector for reconstructing the trajectories of charged cosmic rays. The instrument, already discussed in the previous chapter, is a container with a glass window, filled with air and saturated water vapor (Fig. 3.8); the volume can be suddenly expanded, and the adiabatic expansion causes the temperature to decrease, bringing the vapor to a supersaturated (metastable) state. A charged particle crossing the chamber produces ions, which act as seeds for the generation of droplets along the trajectory. One can record the trajectory by taking a photographic picture. If the chamber is immersed in a magnetic field B, momentum and charge can be measured by the curvature.

The working principle of bubble chambers (Fig. 4.14) is similar to that of the cloud chamber, but here the fluid is a liquid. Along the trajectory of the particle, a trail of gas bubbles evaporates around the ions.

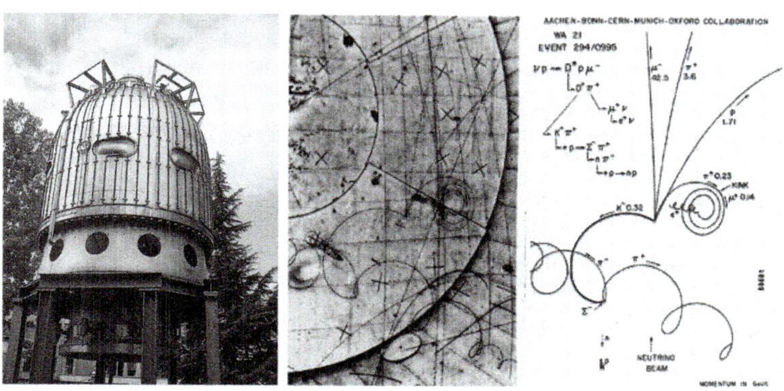

Fig. 4.14 *Left* The BEBC bubble chamber. *Center* A picture taken in BEBC, and *right* its interpretation. Credits: CERN

Due to the higher density of liquids compared with gases, the interaction probability is larger for bubble chambers than for gas chambers, and bubble chambers act at the same time both as an effective target and as a detector. Different liquids can be used, depending on the type of experiment: hydrogen to have protons as a target nucleus, deuterium to study interactions on neutrons, etc. From 1950 to the mid 1980s, before the advent of electronic detectors, bubble chambers were the reference tracking detectors. Very large chambers were built (the Big European Bubble Chamber BEBC now displayed at the entrance of the CERN exhibition is a cylinder with an active volume of 35 cubic meters), and wonderful pictures have been recorded.

Bubble and cloud chambers provide a complete information: the measurement of the bubble density (their number per unit length) provides an estimate of the specific ionization energy loss dE/dx, hence $\beta\gamma = p/Mc$; the range, i.e., the total track length before the particle eventually stops (if the stopping point is recorded), provides an estimate for the initial energy; the multiple scattering (see below) provides an estimate for the momentum.

A weak point of cloud and bubble chambers is their dead time: after an expansion, the fluid must be re-compressed. This might take a time ranging from about 50 ms for small chambers (LEBC, the LExan Bubble Chamber, used in the beginning of the 1980s for the study of the production and decay of particles containing the quark charm, had an active volume of less than a liter) to several seconds. Due to this limitation and to the labor-consuming visual scanning of the photographs, bubble chambers have been abandoned in the mid 1980s—cloud chambers had been abandoned much earlier.

4.2.1.2 Nuclear Emulsions

A nuclear emulsion is a photographic plate with a thick emulsion layer and very uniform grain size. Like bubble chambers and cloud chambers they record the tracks of charged particles passing through, by changing the chemical status of grains that have absorbed photons (which makes them visible after photographic processing). They are compact, have high density, but have the disadvantages that the plates must be developed before the tracks can be observed, and they must be visually examined.

Nuclear emulsion have very good space resolution of the order of about 1 μm. They had great importance in the beginning of cosmic-ray physics, and they are still used in neutrino experiments (where interactions are rare) due to the lower cost per unit of volume compared to semiconductor detectors and to the fact that they are unsurpassed for what concerns to the single-point space resolution. They recently had a revival with the OPERA experiment at the LNGS underground laboratory in Gran Sasso, Italy, detecting the interactions of a beam of muon neutrinos sent from the CERN SPS in Geneva, 730 km away.

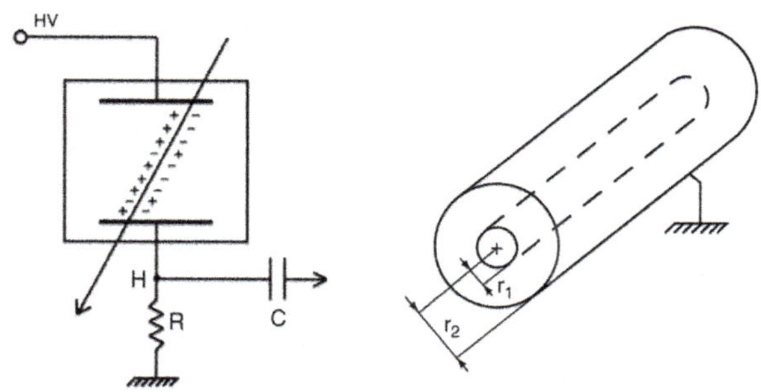

Fig. 4.15 *Left* Operational scheme of a ionization chamber. *Right* A chamber made in a "tube" shape, using coaxial cylindrical electrodes. From Braibant, Giacomelli and Spurio, "Particles and fundamental interactions," Springer 2014

4.2.1.3 Ionization Counter, Proportional Counter and Geiger-Müller Counter

These three kinds of detectors have the same principle of operation: they consist of a tube filled with a gas, with a charged metal wire inside (Fig. 4.15). When a charged particle enters the detector, it ionizes the gas, and the ions and the electrons can be collected by the wire and by the walls (the mobility of electrons being larger than the mobility of ions, it is convenient that the wire's potential is positive). The electrical signal of the wire can be amplified and read by means of an amperometer. The tension V of the wire must be larger than a threshold below which ions and electrons spontaneously recombine.

Depending on the tension V of the wire, one can have three different regimes (Fig. 4.16):

- The ionization chamber regime when $V < I/e$ (I is the ionization energy of the gas, and e the electron charge). The primary ions produced by the track are collected by the wire, and the signal is then proportional to the energy released by the particle.
- The proportional counter regime when $V > I/e$, but V is smaller than a breakdown potential V_{GM} (see below). The ions and the electrons are then accelerated at an energy such that they can ionize the gas. The signal is thus amplified and it generates an avalanche of electrons around the anode. The signal is then proportional to the wire tension.
- Above a potential V_{GM}, the gas is completely ionized; the signal is then a short pulse of height independent of the energy of the particle (Geiger-Müller regime). Geiger-Müller tubes are also appropriate for detecting gamma radiation, since a photoelectron can generate an avalanche.

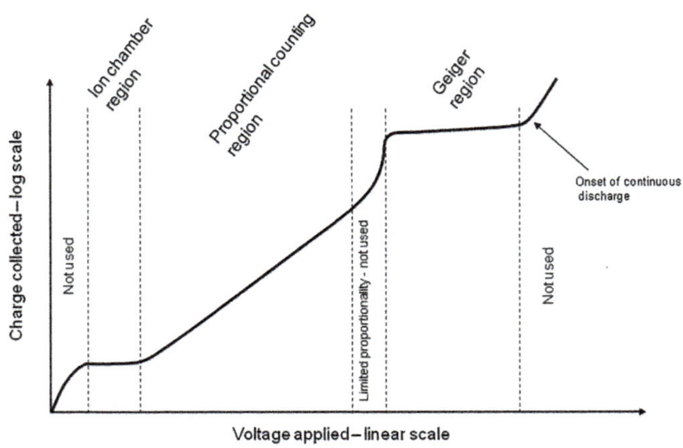

Fig. 4.16 Practical gaseous ionization detector regions: variation of the ion charge with applied voltage in a counter, for a constant incident radiation. By Doug Sim (own work) [CC BY-SA 3.0 http://creativecommons.org/licenses/by-sa/3.0], via Wikimedia Commons

Fig. 4.17 Scheme of a multiwire chamber. By Michael Schmid (own work) [GFDL http://www.gnu.org/copyleft/fdl.html], via Wikimedia Commons

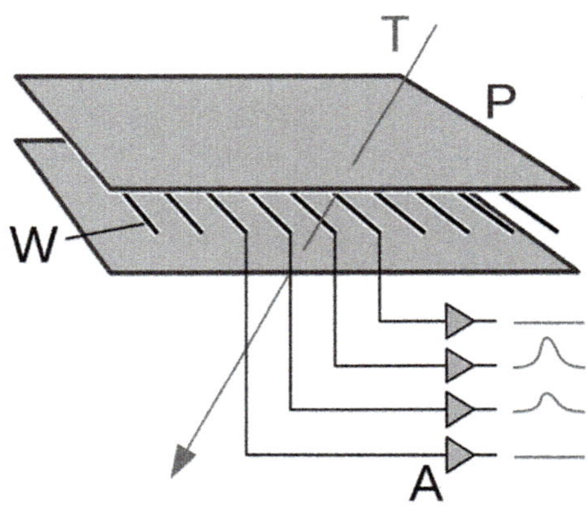

4.2.1.4 Wire Chamber

The multiwire chamber[6] is basically a sequence of proportional counters. Tubes are replaced by two parallel cathodic planes; the typical distance between the planes is 1–2 cm and the typical distance between the anodic wires is 1 mm (Fig. 4.17). A charged particle deposits the ionization charge on the closest wire, inducing an electric current; by a sequence of two parallel detectors with the wires aligned per-

[6]Jerzy ("Georges") Charpak (1924–2010) was awarded the Nobel prize in Physics in 1992 "for his invention and development of particle detectors, in particular the multiwire proportional chamber." Charpak was a Polish born, French physicist. Coming from a Jewish family, he was deported in the Nazi concentration camp in Dachau; after the liberation he studied in Paris, and then worked since 1959 at CERN, Geneva.

Fig. 4.18 The flash chamber built by the Laboratório de Instrumentação e Partículas (LIP) in Lisbon for didactical purposes records a shower of cosmic rays

pendicularly one can determine the position of a particle. The typical response time is of the order of 30 ns.

4.2.1.5 Streamer Chamber and Flash Chamber

These are typically multianode (can be multiwire) chambers operating in the Geiger-Müller regime. Short electric pulses of the order of 10 kV/cm are sent between subsequent planes; when a particle passes in the chamber, it can generate a series of discharges which can be visible—a sequence of flashes along the trajectory, Fig. 4.18.

4.2.1.6 Drift Chamber

The drift chamber is a multiwire chamber in which spatial resolution is achieved by measuring the time electrons need to reach the anode wire. This results in wider wire spacing with respect to what can be used in multiwire proportional chambers. Fewer channels have to be equipped with electronics in order to obtain a comparable overall space resolution; in addition, drift chambers are often coupled to high-precision space measurement devices like silicon detectors (see below).

Drift chambers use longer drift distances than multiwire chambers, hence can be slower. Since the drift distance can be long and drift velocity needs to be well known, the shape and constancy of the electric field need to be carefully adjusted and controlled. To do this, besides the anode wires (also called "signal" or "sense" wires), thick field-shaping cathode wires called "field wires" are often used.

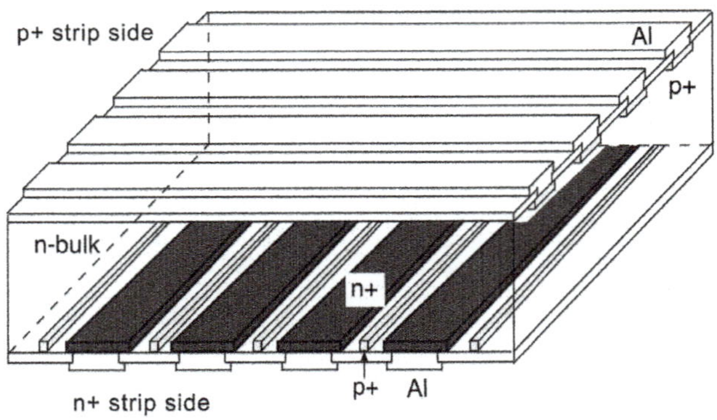

Fig. 4.19 Scheme of a silicon microstrip detector, arranged in a double-side geometry (strips are perpendicular). *Source* http://spie.org/x20060.xml

An extreme case is time projection chambers (TPC), for which drift lengths can be very large (up to 2 m), and the sense wires are arranged at one end; signals in pads or strips near the signal wire plane are used to obtain three-dimensional information.

4.2.1.7 Semiconductor Detectors

Silicon detectors are solid-state particle detectors, whose principle of operation is similar to that of a ionization chamber: the passage of ionizing particles produces in them a number of electron–hole pairs proportional to the energy released. The detector is like a diode (p-n junction) with reverse polarization, the active area being the depleted region. The electron–hole pairs are collected thanks to the electric field, and generate an electrical signal.

The main feature of silicon detectors is the small energy required to create a electron–hole pair—about 3.6 eV, compared with about 30 eV necessary to ionize an atom in an Ar gas ionization chamber.

Furthermore, compared to gaseous detectors, they are characterized by a high density and a high stopping power, much greater than that of the gaseous detectors: they can thus be very thin, typically about 300 μm.

An arrangement of silicon detectors is the so-called microstrip arrangement. A microstrip is a conducting strip separated from a ground plane by a dielectric layer known as the substrate. The general pattern of a silicon microstrip detector is shown in Fig. 4.19. The distance between two adjacent strips, called the pitch, can be of the order of 100 μm, as the width of each strip.

From the signal collected on the strip one can tell if a particle has passed through the detector. The accuracy can be smaller than the size and the pitch: the charge sharing between adjacent strips improves the resolution to some 10 μm. As in the case of mutiwire chambers, the usual geometry involves adjacent parallel planes of mutually perpendicular strips.

Fig. 4.20 A scintillator. From http://www.tnw.tudelft.nl/fileadmin/Faculteit/TNW/Over_de_faculteit/Afdelingen/Radiation_Radionuclides_Reactors/Research/Research_Groups/RDM/Home/img/scintillators.jpg

A recent implementation of semiconductor detectors is the silicon pixel detector. Wafers of silicon are segmented into little squares (pixels) that are as small as 100 μm on a side. Electronics is more expensive (however with modern technology it can be bonded to the sensors themselves); the advantage is that one can measure directly the hits without ambiguities.

4.2.1.8 Scintillators

Scintillators are among the oldest particle detectors. They are slabs of transparent material, organic or inorganic; the ionization induces fluorescence, and light is conveyed towards a photosensor (photosensors will be described later). The light yield is large (can be as large as 10^4 photons per MeV of energy deposited), and the time of formation of the signal is very fast (typically less than 1 ns): they are appropriate for trigger[7] systems.

To make the light travel efficiently towards the photosensor (photomultiplier), light guides are frequently used (Fig. 4.20). Sometimes the fluorescence is dominated by low wavelengths; in this case it is appropriate to match the photosensor optical efficiency with a *wavelength shifter* (a material which induces absorption of light and re-emission in an appropriate wavelength).

The scintillators can be used as tracking devices, in the so-called "hodoscope" configuration (from the Greek "hodos" for path, and "skope" for observation) as in the case of silicon strips.

[7] A trigger is an electronic system that uses simple criteria to rapidly decide which events in a particle detector to keep in cases where only a small fraction of the total number of events can be recorded.

The experimenter can use two segments shaped like strips, arranged in two layers. The strips of the two layers should be arranged in perpendicular directions (let us call them horizontal and vertical). A particle passing through hits a strip in each layer; the vertical scintillator strip reveals the horizontal position of the particle, and the horizontal strip indicates its vertical position (as in the case of two wire chambers with perpendicular orientation of the wires, but with poorer resolution). Scintillator hodoscopes are among the cheapest detectors for tracking charged particles.

Among scintillators, some are polymeric (plastic); plastic scintillators are particularly important due to their good performance at low price, to their high light output and relatively quick (few ns) signal, and in particular to their ability to be shaped into almost any desired form.

4.2.1.9 Resistive-Plate Chambers

The resistive-plate chamber (RPC) is a lower-cost alternative to large scintillator planes. A RPC is usually constructed from two parallel high-resistivity glass or melaminic plates with a gap of a few millimeters between them, which is filled with gas at atmospheric pressure. A high potential (of the order of 10 kV) is maintained between the plates.

A charged particle passing through the chamber initiates an electric discharge, which size and duration are limited by the fact that the current brings the local potential below the minimum required to maintain it. The signal induced is read by metallic strips on both sides of the detector and outside the gas chamber, which are separated from the high voltage coatings by thin insulating sheets.

RPC detectors combine high efficiency (larger than 95 %) with excellent time resolution (about 1 ns), and they are therefore a good choice for trigger systems.

4.2.1.10 Comparison of the Performance of Tracking Detectors

The main characteristics of tracking detectors are summarized in Table 4.2.

Table 4.2 Typical characteristics of different kinds of tracking detectors (data come from K.A. Olive et al. (Particle Data Group), Chin. Phys. C 38 (2014) 090001)

Detector type	Spatial resolution	Time resolution	Dead time
RPC	$\leq$10mm	$\sim$1ns (down to $\sim$50ps)	–
Scintillation counter	10 mm	0.1 ns	10 ns
Emulsion	1 μm	–	–
Bubble chamber	10–100 μm	1 ms	50 ms–1 s
Proportional chamber	50–100 μm	2 ns	20–200 ns
Drift chamber	50–100 μm	few ns	20–200 ns
Silicon strip	Pitch/5 (few μm)	few ns	50 ns
Silicon pixel	10 μm	few ns	50 ns

4.2.2 Photosensors

Most detectors in particle physics and astrophysics rely on the detection of photons near the visible range, i.e., in the eV energy range. This range covers scintillation and Cherenkov radiation as well as the light detected in many astronomical observations.

Essentially, one needs to extract a measurable signal from a (usually very small) number of incident photons. This goal can be achieved by generating a primary photoelectron or electron–hole pair by an incident photon (typically by photoelectric effect), amplifying the signal to a detectable level (usually by a sequence of avalanche processes), and collecting the secondary charges to form the electrical signal.

The important characteristics of a photodetector include as follows:

- the quantum efficiency QE, namely the probability that a primary photon generates a photoelectron;
- the collection efficiency C related to the overall acceptance;
- the gain G, i.e., the number of electrons collected for each photoelectron generated;
- the dark noise DN, i.e. the electrical signal when there is no incoming photon; and
- the intrinsic response time of the detector.

Several kinds of photosensor are used in experiments.

4.2.2.1 Photomultiplier Tubes

Photomultiplier tubes (photomultipliers or PMT) are detectors of light in the ultraviolet, visible, and near-infrared ranges of the electromagnetic spectrum; they are the oldest photon detectors used in high-energy particle and astroparticle physics.

They are constructed (Fig. 4.21) from a glass envelope with a high vacuum inside, housing a photocathode, several intermediate electrodes called dynodes, and an anode. As incident photons hit the photocathode material (a thin deposit on the entry window of the device) electrons are produced by photoelectric effect and directed by

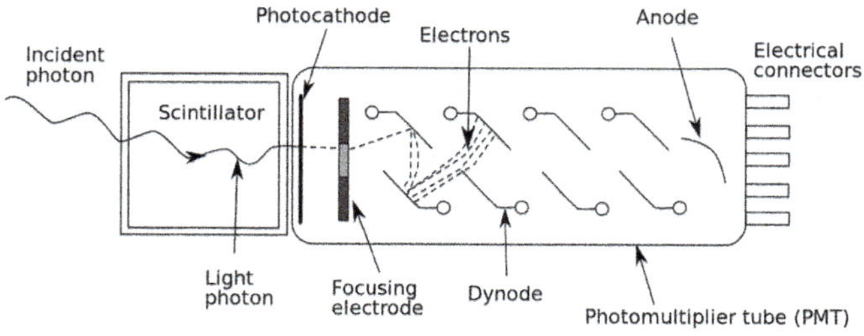

Fig. 4.21 Scheme of a photomultiplier attached to a scintillator. *Source* Colin Eberhardt [public domain], via Wikimedia Commons

the focusing electrode toward the electron multiplier chain, where they are multiplied by secondary emission.

The electron multiplier consists of several dynodes, each held at a higher positive voltage than the previous one (the typical total voltage in the avalanche process being of 1–2 kV). The electrons produced in the photocathode have the energy of the incoming photon (minus the work function of the photocathode, i.e., the energy needed to extract the electron itself from the metal, which typically amounts to a few eV). As the electrons enter the multiplier chain, they are accelerated by the electric field. They hit the first dynode with an already much higher energy. Low-energy electrons are then emitted, which in turn are accelerated towards the second dynode. The dynode chain is arranged in such a way that an increasing number of electrons are produced at each stage. When the electrons finally reach the anode, the accumulation of charge results in a sharp current pulse. This is the result of the arrival of a photon at the photocathode.

Photocathodes can be made of a variety of materials with different properties. Typically materials with a low work function are chosen.

The typical quantum efficiency of a photomultiplier is about 30 % in the range from 300 to 800 nm of wavelength for the light, and the gain G is in the range 10^5–10^6.

A recent improvement to the photomultiplier was obtained thanks to hybrid photon detectors (HPD), in which a vacuum PMT is coupled to a silicon sensor. A photoelectron ejected from the photocathode is accelerated through a potential difference of about $V \simeq 20\,\text{kV}$ before it hits a silicon sensor/anode. The number of electron–hole pairs that can be created in a single acceleration step is $G \sim V/(3.6\,\text{V})$, the denominator being the mean voltage required to create an electron–hole pair. The linear behavior of the gain is helpful because, unlike exponential gain devices, high voltage stability translates in gain stability. HPD detectors can work as single-photon counters.

4.2.2.2 Gaseous Photon Detectors

In gaseous photomultipliers (GPM) a photoelectron entering a suitably chosen gas mixture (a gas with low photoionization work function, like the tetra dimethyl-amine ethylene TMAE) starts an avalanche in a high-field region. Similarly to what happens in gaseous tracking detectors, a large number of secondary ionization electrons are produced and collected.

Since GPMs can have a good space resolution and can be made into flat panels to cover large areas, they are often used as position-sensitive photon detectors. Many of the ring imaging Cherenkov (RICH) detectors (see later) use GPM as sensors.

4.2.2.3 Solid-State Photon Detectors

Semiconductor photodiodes were developed during World War II, approximately at the same time photomultiplier tube became a commercial product. Only in recent years, however, a technique which allows the Geiger-mode avalanche in silicon was engineered, and the semiconductor photodetectors reached sensitivities comparable to photomultiplier tubes. Solid-state photodetectors (often called SiPM) are more compact, lightweight, and they might become cheaper than traditional PMT in the near future. They also allow fine pixelization, of the order of 1 mm $\times$ 1 mm, are easy to integrate in large systems and can operate at low-electric potentials.

One of the recent developments in the field was the construction of large arrays of tiny avalanche photodiodes (APD) packed over a small area and operated in Geiger mode.

The main advantages of SiPM with respect to the standard PMT are as follows: compact size, low power consumption, low operating voltage (less than 100 V), and immunity to electromagnetic field. The main disadvantages of SiPM are dark current caused by thermally generated avalanches (avalanches generated even if a photon was not detected), cross-talk between different channels, and the dependence of gain on temperature (temperature needs thus to be stabilized, or at least monitored).

4.2.3 Cherenkov Detectors

The main ingredients of Cherenkov detectors are a medium to produce Cherenkov radiation (usually called the radiator) and a system of photodetectors to detect Cherenkov photons. The yield of Cherenkov radiation is usually generous so to make these detectors performant.

If one does not need particle identification, a cheap medium (radiator) with large n can be used so to have a threshold for the emission as low as possible. A typical radiator is water, with $n \simeq 1.33$. The IceCube detector in Antarctica uses ice as a radiator (the photomultipliers are embedded in the ice).

Since the photon yield and the emission angle depend on the mass of the particle, some Cherenkov detectors are also used for particle identification.

Threshold Cherenkov detectors make a yes/no decision based on whether a particle velocity is or not above the Cherenkov threshold velocity c/n—this depends exclusively on the velocity and, if the momentum has been measured, is in practice a threshold measurement on the value of the mass. A more advanced version uses the number of detected photoelectrons to discriminate between particle species.

Imaging Cherenkov detectors measure the ring-correlated angles of emission of the individual Cherenkov photons. Low-energy photon detectors measure the position (and sometimes the arrival time) of each photon. These must then be "imaged" onto a detector so that their angles can be derived. Typically the optics maps the Cherenkov cone onto (a portion of) a conical section at the photodetector.

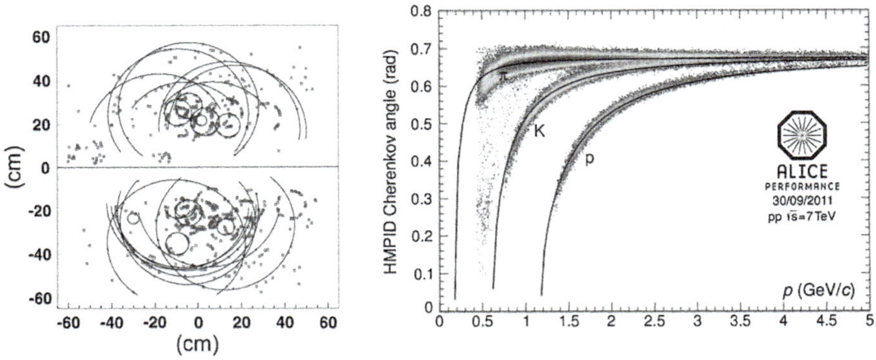

Fig. 4.22 *Left* Image of the hits on the photon detectors of the RICHs of the LHCb experiment at CERN with superimposed rings. Credit: LHCb collaboration. *Right* Dependence of the Cherenkov angle measured by the RICH of the ALICE experiment at CERN on the particle momentum; the angle can be used to measure the mass through Eq. 4.6 ($\beta = p/E$). Credit: ALICE Collaboration

Among imaging detectors, in the so-called Ring Imaging Cherenkov (RICH) detectors, the Cherenkov light cone produced by the passage of a high-speed charged particle in a suitable gaseous or liquid radiator is detected on a position-sensitive planar photon detector. This allows for the reconstruction of a conical section (can be a ring), and its parameters give a measurement for the Cherenkov emission angle (Fig. 4.22). Both focusing and proximity-focusing detectors are used. In focusing detectors, photons are collected by a parabolic mirror and focused onto a photon detector at the focal plane. The result is a conic section (a circle for normal incidence); it can be demonstrated that the radius of the circle is independent of the emission point along the particle track. This scheme is suitable for low refractive index radiators such as gases, due to the large radiator length needed to accumulate enough photons. In proximity-focusing detectors, more compact, the Cherenkov light emitted in a thin volume traverses a short distance (the proximity gap) and is detected in the photon detector plane. The image is a ring of light, with radius defined by the Cherenkov emission angle and the proximity gap.

Atmospheric Cherenkov telescopes for high-energy γ astrophysics are also in use. If one uses a parabolic telescope, again the projection of the emission by a particle along its trajectory is a conical section in the focal plane. If the particle has generated through a multiplicative cascade a shower of secondary particles (see later), the projection is a spot, whose shape can allow distinguishing if the primary particle was a hadron or an electromagnetic particle (electron, positron, or photon).

4.2.4 Transition Radiation Detectors

Similar to Cherenkov detectors, transition radiation detectors (TRD) couple interfaces between different media (used as radiators) to photon detectors. Thin foils of lithium, polyethylene, or carbon are common. Randomly spaced radiators are also in use, like foams. The main problem in the TRD is the low number of photons. In order to intensify the photon flux, periodic arrangements of a large number of foils are used, interleaved with X-ray detectors such as multiwire proportional chambers filled with xenon or a Xe/CO_2 mixture.

4.2.5 Calorimeters

Once entering an absorbing medium, particles undergo successive interactions and decays, until their energy is degraded, as we have seen in Sect. 4.1.6. Calorimeters are blocks of matter in which the energy of a particle is measured through the absorption to the level of detectable atomic ionizations and excitations. Such detectors can be used to measure not only the energy, but also the position in space, the direction, and in some cases the nature of the particle.

4.2.5.1 Electromagnetic Calorimeters

An ideal material used for an electromagnetic calorimeter should have a short radiation length, so that one can contain the electromagnetic shower in a compact detector, and the signal should travel unimpeded through the absorber (homogeneous calorimeters). However, sometimes materials which can be good converters and conductors of the signals are very expensive: one then uses sampling calorimeters, where the degraded energy is measured in a number of sensitive layers separated by passive absorbers.

The performance of calorimeters is limited both by the unavoidable fluctuations of the elementary phenomena through which the energy is degraded and by the technique chosen to measure the final products of the cascade processes.

Homogeneous Calorimeters. Homogeneous calorimeters may be built with heavy (high density, high Z) scintillating crystals, i.e., crystals in which ionization energy loss results in the emission of visible light, or Cherenkov radiators such as lead glass and lead fluoride. The material acts as a medium for the development of the shower, as a transducer of the electron signal into photons, and as a light guide towards the photodetector. Scintillation light and/or ionization can be detected also in noble liquids.

Sampling Calorimeters. Layers of absorbers are typically alternated to layers of active material (sandwich geometry). The absorber helps the development of the electromagnetic shower, while the active material transforms part of the energy into photons, which are guided towards the photodetector. Different geometries can be used: for example sometimes rods of active material cross the absorber (spaghetti geometry).

Converters have high density, short radiation length. Typical materials are iron (Fe), lead (Pb), uranium. Typical active materials are plastic scintillator, silicon, liquid ionization chamber gas detectors.

Disadvantages of sampling calorimeters are that only part of the deposited particle energy is detected in the active layers, typically a few percent (and even one or two orders of magnitude less in the case of gaseous detectors). These sampling fluctuations typically result in a worse energy resolution for sampling calorimeters.

Electromagnetic Calorimeters: Comparison of the Performance. The fractional energy resolution $\Delta E/E$ of a calorimeter can be parameterized as

$$\frac{\Delta E}{E} = \frac{a}{\sqrt{E}} \oplus b \oplus \frac{c}{E},$$

where $\oplus$ represents addition in quadrature. The stochastic term a originates from statistics-related effects such as the intrinsic fluctuations in the shower, number of photoelectrons, dead material in front of the calorimeter, and sampling fluctuations—we remind that the number of particles is roughly proportional to the energy, and thus the Poisson statistics gives fluctuations $\propto \sqrt{E}$. The a term is at a few percent level for a homogeneous calorimeter and typically 10 % for sampling calorimeters. The systematic or constant b term represents contributions from the detector non-uniformity and calibration uncertainty. In the case of hadronic cascades (discussed below), the different response of the instrument to hadrons and leptons, called non-compensation, also contributes to the constant term. The constant term b can be reduced to below one percent. The c term is due to electronic noise. Some of the above terms can be negligible in calorimeters.

The best energy resolution for electromagnetic shower measurement is obtained with total absorption, homogeneous calorimeters, such as those built with heavy crystal scintillators like $Bi_4Ge_3O_{12}$, called BGO. They are used when optimal performance is required. A relatively cheap scintillator with relatively short X_0 is the caesium iodide CsI, which becomes more luminescent when activated with thallium, and is called CsI(Tl); this is frequently used for dosimetry in medical applications.

Energy resolutions for some homogeneous and sampling calorimeters are listed in Table 4.3.

4.2.5.2 Hadronic Calorimeters

We have examined the main characteristics of hadronic showers in Sect. 4.1.7.

Detectors capable to absorb hadrons and detect a signal related to it were developed around 1950 for the measurement of the energy of cosmic rays. It can be assumed

Table 4.3 Main characteristics of some electromagnetic calorimeters (data from K.A. Olive et al. (Particle Data Group), Chin. Phys. C 38 (2014) 090001)

Technology (experiment)	Depth (X_0)	Energy resolution (relative)
BGO (L3)	22	$2\,\%/\sqrt{E} \oplus 0.7\,\%$
CsI (kTeV)	27	$2\,\%/\sqrt{E} \oplus 0.45\,\%$
$PbWO_4$ (CMS)	25	$3\,\%/\sqrt{E} \oplus 0.5\,\% \oplus 0.2\,\%/E$
Lead glass (DELPHI, OPAL)	20	$5\,\%/\sqrt{E}$
Scintillator/Pb (CDF)	18	$18.5\,\%/\sqrt{E}$
Liquid Ar/Pb (SLD)	21	$12\,\%/\sqrt{E}$

that the energy of the incident particle is proportional to the multiplicity of charged particles.

Most large hadron calorimeters are sampling calorimeters installed as part of complex detectors at accelerator experiments. The basic structure typically consists of absorber plates (Fe, Pb, Cu, or occasionally U or W) alternating with plastic scintillators (shaped as plates, tiles, bars), liquid argon (LAr) chambers, or gaseous detectors (Fig. 4.23). The ionization is measured directly, as in LAr calorimeters, or via scintillation light observed in photodetectors (usually photomultipliers).

The fluctuations in the invisible energy and in the hadronic component of a shower affect the resolution of hadron calorimeters.

A hadron with energy E generates a cascade in which there are repeated hadronic collisions. In each of these, neutral pions are produced which immediately decay to photons. A fraction of the energy is converted to a potentially observable signal with an efficiency which is in general different, usually larger, than the hadronic detection efficiency. The response to hadrons is thus not compensated with respect to the response to electromagnetic particles (or to the electromagnetic part of the hadronic shower).

Due to all these problems, typical fractional energy resolutions are in the order of 30–50 $\%/\sqrt{E}$.

4.3 High-Energy Particles

We have seen that when we use a beam of particles as a microscope, like Rutherford did in his experiment, the minimum distance we can sample (for example, to probe a possible substructure in matter) decreases with increasing energy. According to de Broglie's equation, the relation between the momentum p and the wavelength λ of a wave packet is given by

$$\lambda = \frac{h}{p} \, .$$

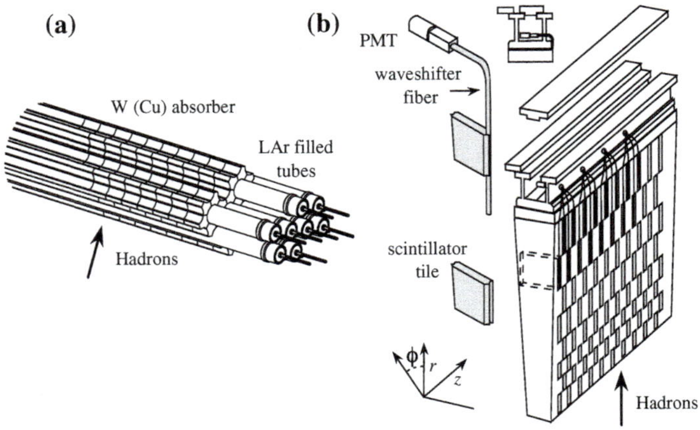

Fig. 4.23 The hadronic calorimeters of the ATLAS experiments at LHC. Credit: CERN

Therefore, larger momenta correspond to shorter wavelengths, and allow accessing to smaller structures. Particle acceleration is thus a fundamental tool for research in physics.

In addition, we might be interested in using high-energy particles to produce new particles in collisions. This requires the more energy the more massive the particles we want to produce are.

4.3.1 Artificial Accelerators

A particle accelerator is an instrument using electromagnetic fields to accelerate charged particles at high energies.

There are two schemes of collision:

- collision with a fixed target (fixed-target experiments);
- collision of a beam with another beam running in opposite direction (collider experiments).

We also distinguish two main categories of accelerators depending on the geometry: linear accelerators and circular accelerators. In linear accelerators the bremsstrahlung energy loss is much reduced since there is no centripetal acceleration, but particles are wasted after a collision, while in circular accelerators the particles which did not interact can be re-used.

The center of-mass energy E_{CM} sets the scale for the maximum mass of the particles we can produce (the actual value being in general smaller due to constraints related to conservation laws). We want now to compare fixed target and colliding beam experiments concerning the available energy.

In the case of beam–target collisions between a particles of energy E much larger than its mass, and a target of mass m,

$$E_{\mathrm{CM}} \simeq \sqrt{2mE}\,.$$

This means that, in a fixed target experiment, the centre-of-mass energy grows only with the square root of E. In beam–beam collisions in the centre-of-mass, instead,

$$E_{\mathrm{CM}} = 2E\,.$$

It is therefore much more efficient to use two beams in opposite directions. As a result, most of the recent experiments at accelerators are done at colliders.

Making two beams to collide, however, is not trivial: one must control the fact that the beams tend to defocus due to mutual repulsion. In addition, the Liouville theorem states that the phase space volume (the product of the spread in terms of the space coordinates times the spread in the momentum coordinate) of an isolated beam is constant: reducing the momentum dispersion is done at the expense of the space dispersion—and one needs small space dispersion in order that the particles in the beam actually collide. Beating the Liouville theorem requires feedback on the beam itself.[8]

Since beams are circulated for several hours, circular accelerators are based on beams of stable particles and antiparticles, such as electrons, protons, and their antiparticles. In the future, muon colliders are an interesting candidate: as clean as electrons, muons have a lower energy dissipation (due to synchrotron radiation and bremsstrahlung) thanks to their mass 200 times higher than electrons.

The accelerators and detectors are often situated underground in order to provide the maximal shielding possible from natural radiation such as cosmic rays that would otherwise mask the events taking place inside the detector.

4.3.1.1 Acceleration Methods

A particle of charge q and speed $\vec{v}$ in an electric field $\vec{E}$ and a magnetic field $\vec{B}$ feels a force

$$\vec{F} = q(\vec{E} + \vec{v} \times \vec{B})\,.$$

The electric field can thus accelerate the particle; the work by the magnetic field is zero, nevertheless the magnetic field can be used to control the particle's trajectory.

[8]The Nobel prize for physics in 1984 was assigned to the Italian physicist Carlo Rubbia (1984 –) and to the Dutch Simon van der Meer (1925–2011) "for their decisive contributions to the large project, which led to the discovery of the field particles W and Z, communicators of weak interaction." In short, Rubbia and van der Meer used feedback signals sent in the ring to reduce the entropy of the beam; this technique allowed the accumulation of focalized particles with unprecedented efficiency, and is at the basis of all modern accelerators.

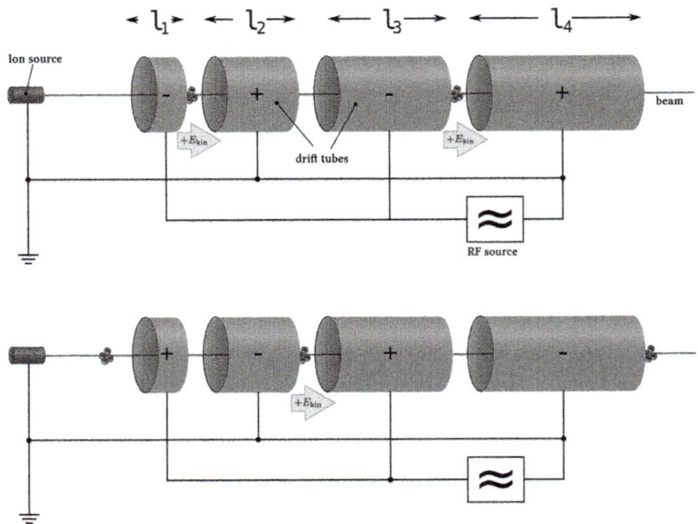

Fig. 4.24 Scheme of an acceleration line displayed at two different times. By Sgbeer (own work) [GFDL http://www.gnu.org/copyleft/fdl.html], via Wikimedia Commons

For example, a magnetic field perpendicular to $\vec{v}$ can constrain the particle along a circular trajectory perpendicular to $\vec{B}$.

If a single potential were applied, increasing energy would demand increasing tensions. The solution is to apply multiple times a limited potential.

An acceleration line (which corresponds roughly to a linear accelerator) works as follows. In a beam pipe (a cylindrical tube in which vacuum has been made) cylindrical electrodes are aligned. A pulsed radiofrequency (RF) source of electromotive force V is applied. Thus particles are accelerated when passing in the RF cavity (Fig. 4.24); the period is adjusted in such a way that half of the period corresponds of the time needed to the particle to cross the cavity. The potential between the cylinders is reversed when the particle is located within them.

To have a large number of collisions, it is useful that particles are accelerated in bunches. This introduces an additional problem, since the particles tend to diverge due to mutual electrostatic repulsion. Divergence can be compensated thanks to focusing magnets (for example quadrupoles, which squeeze beams in a plane).

A collider consists of two circular or almost circular accelerator structures with vacuum pipes, magnets and accelerating cavities, in which two beams of particles travel in opposite directions. The particles may be both protons, or protons and antiprotons, or electrons and positrons, or electrons and protons, or also nuclei and nuclei. The two rings intercept each other at a few positions along the circumference, where bunches can cross and particles can interact. In a particle–antiparticle collider (electron–positron or proton-antiproton), as particles and antiparticles have opposite charges and the same mass, a single magnetic structure is sufficient to keep the two beams circulating in opposite directions.

4.3.1.2 Parameters of an Accelerator

An important parameter for an accelerator is the maximum center of-mass (c.m.) energy $\sqrt{s}$ available, since this sets the maximum mass of new particles that can be produced.

Another important parameter is *luminosity,* already discussed in Chap. 2. Imagine a physical process has a cross section σ_{proc}; the number of outcomes of this process per unit time can be expressed as

$$\frac{dN_{\text{proc}}}{dt} = \frac{dL}{dt}\sigma_{\text{proc}} \, .$$

dL/dt is called *differential luminosity* of the accelerator, and is measured in cm^{-2} s^{-1}; however, for practical reasons it is customary to use "inverse barns" and its multiples instead of cm^{-2} (careful: due to the definition, 1 mbarn^{-1} = 1000 barn^{-1}).

The integrated luminosity can be obtained by integrating the differential luminosity over the time of operation of an accelerator:

$$L = \int_{\text{time of operation}} \frac{dL(t)}{dt} dt \, .$$

In a collider, the luminosity is proportional to the product of the numbers of particles, n_1 and n_2, in the two beams. Notice that in a proton–antiproton collider the number of antiprotons is smaller than that of protons, due to the "cost" of the antiprotons (antiprotons are difficult to store and to accumulate, since they easily annihilate). The luminosity is also proportional to the number of crossings in a second f and inversely proportional to the transverse section $\mathcal{A}$ at the intersection point

$$\frac{dL}{dt} = f\frac{n_1 n_2}{\mathcal{A}} \, .$$

4.3.2 *Cosmic Rays as Very-High-Energy Beams*

As we have already shown, cosmic rays can attain energies much larger than particles produced at human-made accelerators. The main characteristics of cosmic rays have been explained in Sect. 1.4 and in Chap. 3.

We just recall here that the distribution in energy (the so-called spectrum) of cosmic rays is quite well described by a power law E^{-p}, with the so-called spectral index p around 3 in average (Fig. 1.8), extending up to about 10^{21} eV (above this energy the GZK cutoff, explained in the previous chapters, stops the cosmic travel of particles; a similar mechanism works for heavier nuclei, which undergo photo-disintegration during their cosmic travel). The majority of the high-energy particles in cosmic rays are protons (hydrogen nuclei); about 10 % are helium nuclei (nuclear physicists call them usually “alpha particles”), and 1 % are neutrons or nuclei of heavier elements. These together account for 99 % of the cosmic rays, and electrons and photons make up the remaining 1 %. The number of neutrinos is estimated to be comparable to that of high-energy photons, but it is very high at low energy because of the nuclear processes that occur in the Sun: such processes involve a large production of neutrinos. Cosmic rays hitting the atmosphere (called primary cosmic rays) generally produce secondary particles that can reach the Earth’s surface, through multiplicative showers.

The reason why human-made accelerators cannot compete with cosmic accelerators from the point of view of the maximum attainable energy is that with the present technologies acceleration requires confinement within a radius R by a magnetic field B, and the final energy is proportional to the product of R times B. On Earth, it is difficult to imagine reasonable radii of confinement larger than one hundred kilometers and magnetic fields stronger than ten tesla (one hundred thousand times the Earth’s magnetic field). This combination can provide energies of a few tens of TeV, such as those of the LHC accelerator at CERN. In nature there are accelerators with much larger radii, as the remnants of supernovae (hundreds of light years) and active galactic nuclei of galaxies (tens of thousands of light years): one can thus reach energies as large as 10^{21} eV (the so-called Extremely-High-Energy, EHE, cosmic rays; cosmic rays above 10^{18} eV are instead called Ultra-High-Energy, UHE). Of course terrestrial accelerators have great advantages like luminosity and possibility to know the initial conditions.

The conditions are synthetically illustrated in the so-called Hillas plot (Fig. 10.32), a scatter plot in which different cosmic objects are grouped according to their sizes and magnetic fields; this will be discussed in larger detail in Chap. 10. EHE can be reached in the surroundings of active galactic nuclei, or in gamma-ray bursts. The product of R times B in supernova remnants are such that particles can reach energies of some PeV.

4.4 Detector Systems and Experiments at Accelerators

Detectors at experimental facilities are in general hybrid, i.e., they combine many of the detectors discussed so far, such as the drift chambers, Cherenkov detectors, electromagnetic, and hadronic calorimeters. They are built up in a sequence of layers, each one designed to measure a specific aspect of the particles produced after the collision.

Starting with the innermost layer the successive layers are typically as follows:

- A tracking system: this is designed to track all the charged particles and allow for complete event reconstruction. It is in general the first layer crossed by the particles, in such a way that their properties have not yet been deteriorated by the interaction with the material of the detector. It should have as little material as possible, so to preserve the particles for the subsequent layer.
- A layer devoted to electromagnetic calorimetry.
- A layer devoted to hadronic calorimetry.

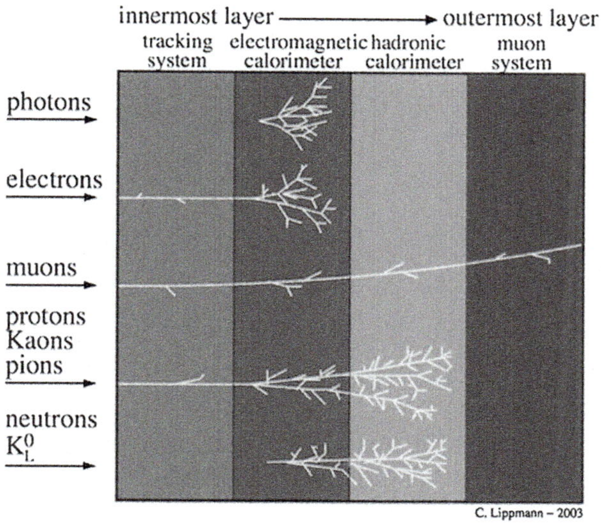

Fig. 4.25 Overview of the signatures by a particle in a multilayer hybrid detector. Credit: CERN

- A layer of muon tracking chambers: any particle releasing signal on these tracking detectors (often drift chambers) has necessarily traveled through all the other layers and is very likely a muon (neutrinos have extremely low interaction cross sections, and most probably they cross also the muon chambers without leaving any signal).

A layer containing a solenoid can be inserted after the tracking system, or after the calorimeter. Tracking in magnetic field will allow momentum measurement.

The particle species can be identified by energy loss, curvature in magnetic field, and Cherenkov radiation. However, the search for the identity of a particle can be significantly narrowed down by simply examining which parts of the detector it deposits energy in:

- Photons leave no tracks in the tracking detectors (unless they undergo pair production) but produce a shower in the electromagnetic calorimeter.
- Electrons and positrons leave a track in the tracking detectors and produce a shower in the electromagnetic calorimeter.
- Muons leave tracks in all the detectors (likely as a minimum ionizing particle in the calorimeters).
- Long-lived charged hadrons (protons for example) leave tracks in all the detectors up to the hadronic calorimeter where they shower and deposit all their energy.
- Neutrinos are identified by missing energy-momentum when the relevant conservation law is applied to the event.

These signatures are summarized in Fig. 4.25.

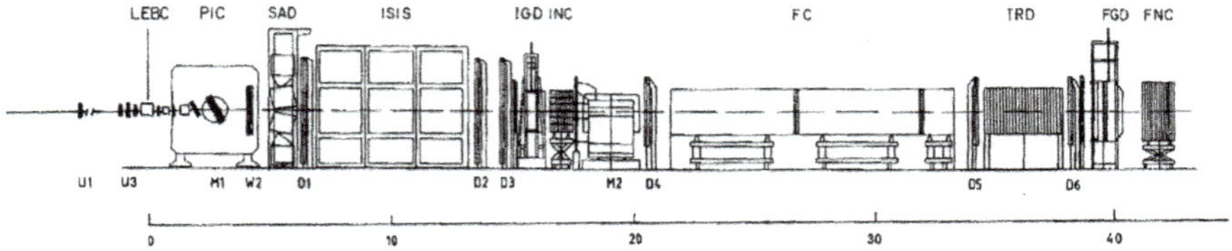

Fig. 4.26 A configuration of the European Hybrid Spectrometer (a fixed target detector at the CERN Super-proton Synchrotron). From M. Aguilar-Benitez et al., "The European hybrid spectrometer," Nucl. Instr. Methods 258 (1987) 26

4.4.1 Examples of Detectors for Fixed Target Experiments

In a fixed target experiment, relativistic effects make the interaction products highly collimated. In such experiments then, in order to enhance the possibility of detection in the small-x_T ($x_T = p_T/\sqrt{s}$, where p_T is the momentum component perpendicular to the beam direction), different stages are separated by magnets opening up the charged particles in the final state (lever arms).

The first detectors along the beam line should be non-destructive; at the end of the beam line, one can have calorimeters. Two examples is given in the following; the first is a fixed target experiment from the past, while the second is an *almost* fixed target detector presently operating.

4.4.1.1 The European Hybrid Spectrometer at the SPS

The European Hybrid Spectrometer EHS has been operational during the years 1970s and the beginning of the 1980s at the North Area of CERN, where beams of protons were extracted from the SPS (Super-proton Synchrotron)[9] accelerator at energies ranging from 300 to 400 GeV. Such particles might possibly generate secondary beams of charged pions of slightly smaller energies by a beam-dump and a velocity selector based on magnetic field. EHS was a multi-stage detector serving different experiments (NA16, NA22, NA23, NA27). Here we describe a typical configuration; Fig. 4.26 shows a schematic drawing of the EHS set-up.

In the Figure, the beam particles come in from the left. Their direction is determined by the two small wire chambers U1 and U3. From the collision point inside a rapid cycling bubble chamber (RCBC; the previously described LEBC is an example, with a space resolution of 10 μm) most of the produced particles will enter the downstream part of the spectrometer. The fast ones (typically with momentum $p > 30$ GeV/c) will go through the aperture of the magnet M2 to the so-called second lever arm.

---

[9]A synchrotron is a particle accelerator ring, in which the guiding magnetic field (bending the particles into a closed path) is time dependent and synchronized to a particle beam of increasing kinetic energy. The concept was developed by the Soviet physicist Vladimir Veksler in 1944.

The RCBC acts both as a target and as a vertex detector. If an event is triggered, stereoscopic pictures are taken with 3 cameras and recorded on film.

The momentum resolution of the secondary tracks depends on the number of detector element hits available for the fits. For low momentum tracks, typically $p < 3$ GeV/c, length and direction of the momentum vector at the collision-point can be well determined from RCBC. On the other hand, tracks with $p < 3$ GeV/c have a very good chance to enter the so-called first lever arm. This is defined by the group of four wire chambers W2, D1, D2, and D3 placed between the two magnets M1 and M2.

Fast tracks, typically $p > 50$ GeV/c, have a good chance to go in addition through the gap of the magnet M2 and enter the second lever arm, consisting of the three drift chambers D4, D5, and D6.

To detect gammas, two electromagnetic calorimeters are used in EHS, the intermediate gamma detector (IGD) and the forward gamma detector (FGD). IGD is placed before the magnet M2. It has a central hole to allow fast particles to pass into the second lever arm. FGD covers this hole at the end of the spectrometer. The IGD has been designed to measure both the position and the energy of a shower in a two-dimensional matrix of lead-glass counters 5 cm × 5 cm in size, each of them connected to a PMT. The FGD consists of three separate sections. The first section is the converter (a lead glass wall), to initiate the electromagnetic shower. The second section (the position detector) is a three-plane scintillator hodoscope. The third section is the absorber, a lead-glass matrix deep enough (60 radiation length) to totally absorb showers up to the highest available energies. For both calorimeters, the relative accuracy on energy reconstruction is $\Delta E/E \simeq 0.1/\sqrt{E} \oplus 0.02$.

The spectrometer included also three detectors devoted to particle identification: the silica-aerogel Cherenkov detector SAD, the ISIS chamber measuring specific ionization, and the transition radiation detector TRD.

4.4.1.2 LHCb at LHC

LHCb ("Large Hadron Collider beauty") is a detector at the Large Hadron Collider accelerator at CERN. LHCb is specialized in the detection of $b-$hadrons (hadrons containing a bottom quark). A sketch of the detector is shown in Fig. 4.27.

Although in strict terms, LHCb is a colliding beam experiment, it is done as a fixed target one: the strongly boosted b-hadrons fly along the beam direction, and one side is instrumented.

At the heart of the detector is the vertex detector, recording the decays of the b particles, which have typical lifetimes of about 1 ps and will travel only about 10 mm before decaying. It has 17 planes of silicon (radius 6 cm) spaced over a meter and consisting of two disks (in order to measure radial and polar coordinates) and provides a hit resolution of about 10 and 40 μm for the impact parameter of high momentum tracks.

Downstream of the vertex detector, the tracking system (made of 11 tracking chambers) reconstructs the trajectories of emerging particles. LHCb's 1.1 T super-

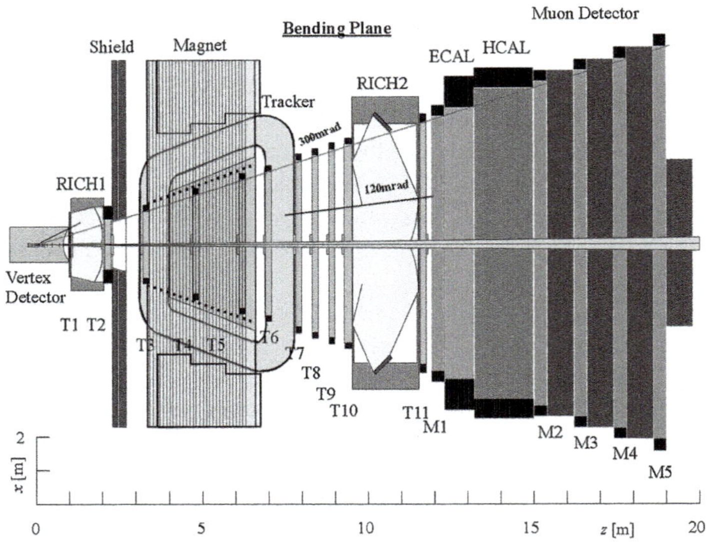

Fig. 4.27 Sketch of the LHCb detector. Credit: CERN

conducting dipole spectrometer magnet (inherited from the DELPHI detector at LEP, see later) opens up the tracks.

Particle identification is performed by two ring-imaging Cherenkov (RICH) detector stations. The first RICH is located just behind the vertex detector and equipped with a 5 cm silica aerogel and 1 m C_4F_{10} gas radiators, while the second one consists of 2 m of CF_4 gas radiator behind the tracker. Cherenkov photons are picked up by a hybrid photodiode array.

The electromagnetic calorimeter, installed following the second RICH, is a "shashlik" structure of scintillator and lead read out by wavelength-shifting fibers. It has three annular regions with different granularities in order to optimize readout. A lead-scintillator preshower detector improves electromagnetic particle identification.

The hadron calorimeter is made of scintillator tiles embedded in iron. Like the electromagnetic calorimeter upstream, it has three zones of granularity. Downstream, shielded by the calorimetry, is four layers of muon detectors. These are multigap resistive plate chambers and cathode pad chambers embedded in iron, with an additional layer of cathode pad chambers mounted before the calorimeters. Besides muon identification, this provides important input for triggering.

There are four levels of triggering. The initial (level 0) decisions are based on a high transverse-momentum particle and use the calorimeters and muon detectors. This reduces by a factor of 40 the 40 MHz input rate. The next trigger level (level 1) is based on vertex detector (to look for secondary vertices) and tracking information, and reduces the data by a factor of 25 to an output rate of 40 kHz. Level 2, suppressing fake secondary decay vertices, achieves further eightfold compression. Level 3 reconstructs B decays to select specific decay channels, achieving another compression factor of 25. Data are written to tape at 200 Hz.

4.4.2 *Examples of Detectors for Colliders*

The modern particle detectors in use today at colliders are as much as possible hermetic detectors. They are designed to cover most of the solid angle around the interaction point (a limitation being given by the presence of the beam pipe). The typical detector consists of a cylindrical section covering the "barrel" region and two endcaps covering the "forward" regions.

In the standard coordinate system, the z axis is along the beam direction, the x axis points towards the center of the ring, and the y axis points upwards. The polar angle to the z axis is called θ and the azimuthal angle is called ϕ; the radial coordinate is $R = \sqrt{x^2 + y^2}$.

Frequently the polar angle is replaced by a coordinate called *pseudorapidity* η and defined as

$$\eta = \ln \left[\tan \left(\frac{\theta}{2} \right) \right] ;$$

the region $\eta \simeq 0$ corresponds to $\theta \simeq \pi/2$, and is called the central region. When in Chap. 6 we shall discuss the theory of hadronic interactions, quantum chromodynamics, we shall clarify the physical significance of this variable.

The detector has the typical onion-like structure described in the previous section: a sequence of layers, the innermost being the most precise for tracking.

The configuration of the endcaps is similar to that in a fixed-target experiment except for the necessary presence of a beam pipe, which makes it impossible to detect particles at very small polar angles, and entails the possible production of secondary particles in the pipe wall.

In the following we shall shortly describe three generations of collider detectors operating at the European Organization for Particle Physics, CERN: UA1 at the S$p\bar{p}$S $p\bar{p}$ accelerator, DELPHI at the LEP e^+e^- accelerator, and both the main detectors at the LHC pp accelerator: CMS and ATLAS. We shall see how much the technology developed and the required labor increased; the basic ideas are anyway still common to the prototype detector, UA1.

4.4.2.1 UA1 at the S$p\bar{p}$S

The UA1 experiment, named as the first experiment in the CERN Underground Area (UA), runs at CERN's S$p\bar{p}$S (Super-proton-antiproton-Synchrotron) accelerator-collider from 1981 till 1993. The discovery of the W and Z bosons, mediators of the weak interaction, by this experiment in 1983, led to the Nobel prize for physics to Carlo Rubbia and Simon van der Meer in 1984 (the motivation of the prize being more related to the development of the collider technology). The S$p\bar{p}$S was colliding protons and antiprotons at a typical c.m. energy of 540 GeV; three bunches of protons and three bunches of antiprotons, 10^{11} particles per bunch, were present in the ring at the same time, and the luminosity was about 5×10^{27} cm^{-2}/s (5 inverse millibarn per second).

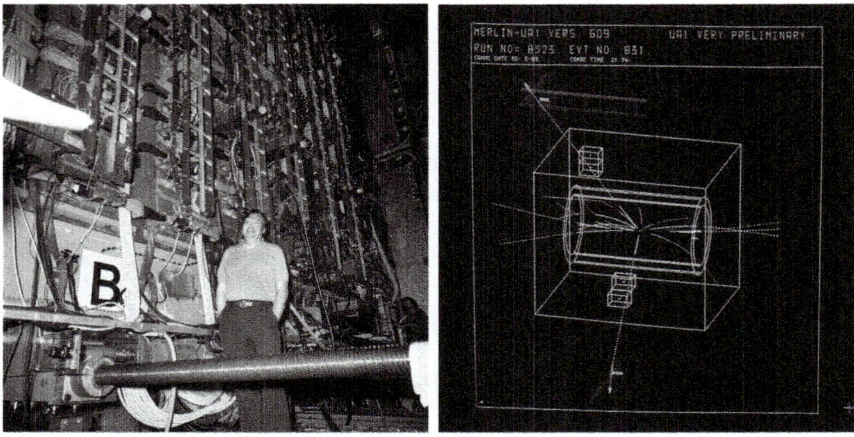

Fig. 4.28 *Left* The UA1 detector, and Carlo Rubbia. *Right* A Z boson decaying into a muon–antimuon pair as seen at the event display of UA1 (*Source* CERN)

UA1 was a huge and complex detector for its days, and it was and still is the prototype of collider detectors. The collaboration constructing and managing the detector included approximately 130 scientists from all around the world.

UA1 was a general purpose detector. The central tracking system was an assembly of six drift chambers 5.8 m long and 2.3 m in diameter. It recorded the tracks of charged particles curving in a 0.7 T magnetic field, measuring their momenta with typical accuracy $\delta p/p \simeq 0.01\, p_T$ (where p_T is the momentum component transverse to the beam axis, also called the transverse momentum[10]; p is expressed in GeV/c) and possibly identifying them by the specific energy loss dE/dx. The geometrical arrangement of the about 17000 field wires and 6125 sense wires allowed a three-dimensional reconstruction of events. UA1 introduced also the concept of event display (Fig. 4.28).

After the tracking chamber and an air gap of 0.5 m, the particles next encountered the calorimeter plus 60 cm of additional iron shielding, including the magnet yoke. The calorimeter started with an electromagnetic calorimeter made of a sandwich of lead and scintillator, with a total relative energy resolution about $0.2/\sqrt{E}$. The iron shielding was partially instrumented with streamer tubes[11] measuring the position and the number of minimum ionizing particles, and thus, acting as a hadronic calorimeter with relative energy resolution about $0.8/\sqrt{E}$. Together, the calorimeter and the shielding corresponded to more than eight interaction lengths of material, which almost completely absorbed strongly interacting particles. Finally, muons

[10]Being proportional to $(1/p_T)$ the fitted quantity by means of the radius of curvature, the accuracy on the momentum measurement can be parameterized as $\delta(1/p_T) = \delta(p)/(p_T\, p) \sim$ constant.

[11]Limited streamer tubes, often simply called streamer tubes, are made of a resistive cathode in the form of a round or square tube, with a thick (0.1 mm) anode wire in its axis; they operate at voltages close to the breakdown (see Sect. 4.2.1.3). Such detectors can be produced with 1–2 cm diameter, and they are cheap and robust.

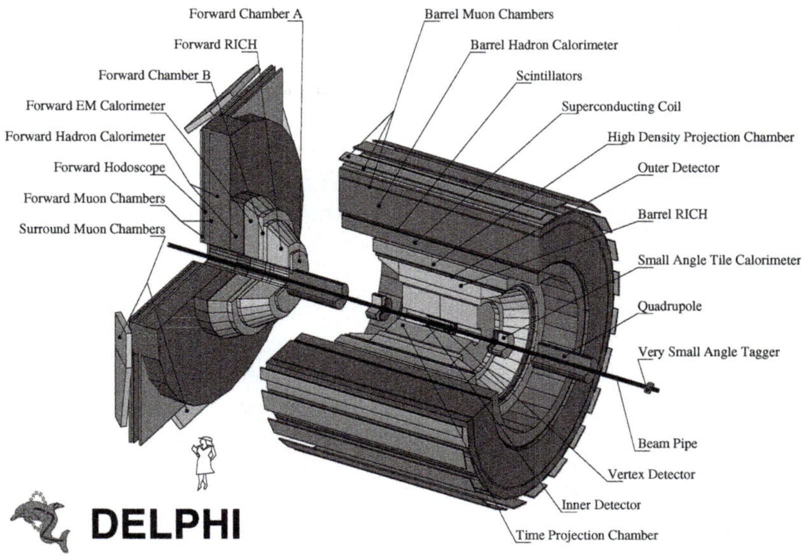

Fig. 4.29 The DELPHI detector at LEP (*Source* CERN)

were detected in the outer muon chambers, which covered about 75 % of the solid angle in the pseudorapidity range $|\eta| < 2.3$. Muon trigger processors required tracks in the muon chambers pointing back to the interaction region to retain an event as significant.

4.4.2.2 DELPHI at LEP

DELPHI (DEtector with Lepton Photon and Hadron Identification, Fig. 4.29) was one of the four experiments built for the LEP (Large Electron–Positron) collider at CERN. The main aim of the experiment was the verification of the theory known as the Standard Model of particle physics. DELPHI started collecting data in 1989 and ran about 8 months/year, 24 h a day, until the end of 2000; it recorded the products of collisions of electrons and positrons at c.m. energies from 80 to 209 GeV (most of the data being taken at the Z peak, around 91.2 GeV). Typical luminosity was 2×10^{31} cm^{-2}/s (20 inverse microbarn per second). DELPHI was built and operated by approximately 600 scientists from 50 laboratories all over the world.

DELPHI consisted of a central cylindrical section and two endcaps, in a solenoidal magnetic field of 1.2 T provided by a superconducting coil. The overall length and the diameter were over 10 meters and the total weight was 2500 tons. The electron–positron collisions took place inside the vacuum pipe at the center of DELPHI and the products of the annihilations would fly radially outwards, tracked by several detection layers and read out via about 200,000 electronic channels. A typical event was about one million bits of information.

The DELPHI detector was composed of subdetectors as described in Fig. 4.29. In the barrel part of the detector (covering approximately the region of polar angle θ between 40° and 140° and there was an onion-like structure of tracking detectors, the ones closest to the collision point being characterized by best resolution: the silicon Vertex Detector (VD), a cylinder of proportional counters called the Inner Detector (ID), the Time Projection Chamber (TPC), another cylinder of proportional counters called the Outer Detector (OD) and the Barrel Muon Chambers (MUB). The Time Projection Chamber (TPC), shown as the big cylinder in the Figure, was the principal tracking device of DELPHI, helping as well in charged particle identification by measuring the specific ionization energy loss dE/dx. The detector provided points per particle trajectory at radii from 40 to 110 cm.

Also in the forward part, a sequence of tracking chambers was present: the Forward Silicon Detector, the Forward Chambers A and B (FCA and FCB), the Forward Muon Chambers (MUF) were devoted to precise measurement of the trajectories of charged particles, and hence to the precise determination of the directions and momenta of the charged particles.

Electron and photon identification was provided primarily by the electromagnetic calorimetry system. This was composed of a barrel calorimeter (the High-density Projection Chamber, HPC) and a forward calorimeter (FEMC); a smaller calorimeter in the forward region was used mainly for luminosity measurement. The HPC was installed as a cylindrical layer outside the Outer Detector, inside the solenoid; it was an accordion of lead with gas as sensitive detector. The Forward ElectroMagnetic Calorimeter (FEMC) consisted of two 5 m diameter disks made of lead-glass, with the front faces placed at $|z| = 284$ cm; it detected the Cherenkov light emitted by charged particles in the shower.

The hadron calorimeter (HCAL) was a sampling gas detector incorporated in the magnet yoke (made mainly of iron), covering both the barrel and the endcap regions. It provided calorimetric energy measurements of charged and neutral hadrons (strongly interacting particles).

The identification of charged hadrons in DELPHI relied also on the specific ionization energy loss per unit length in the TPC, and on two Ring Imaging Cherenkov (RICH) detectors, respectively, in the barrel and in the forward region.

The overall accuracy in momentum can be parameterized as

$$\frac{\delta p}{p} \simeq 0.6\,\%\, p_T\,, \tag{4.15}$$

where p is expressed in GeV/c, and the typical calorimetric resolution in the barrel part is

$$\frac{\delta E}{E} \simeq \frac{7\,\%}{\sqrt{E}} \oplus 1\,\%\,, \tag{4.16}$$

with E expressed in GeV.

Two reconstructed events are shown in Fig. 4.30.

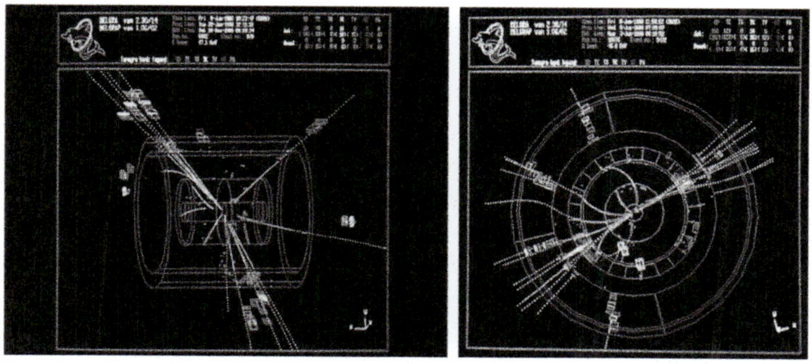

Fig. 4.30 Two events reconstructed by DELPHI, projected on the xz plane (*left*) and on the xy plane (*right*). Credits: CERN

4.4.2.3 CMS at LHC

The Compact Muon Solenoid (CMS) experiment is one of the two large general purpose particle physics detectors built on the proton–proton Large Hadron Collider (LHC) at CERN, the other being called ATLAS.[12] Approximately 3000 scientists, representing 183 research institutes and 38 countries, form the CMS collaboration who built and now since 2008 operate the detector (the size of the ATLAS collaboration is similar). The detector is shown in Fig. 4.31. Proton–proton collisions at c.m. energies of 8 TeV are recorded; typical luminosity is $7 \times 10^{34}\,\text{cm}^{-2}/\text{s}$ (70 inverse nanobarn per second).

As customary for collider detectors, CMS in structured in layers arranged in an onion-like structure.

Layer 1 is devoted to tracking. The inner silicon tracker is located immediately around the interaction point. It is used to identify the tracks of individual particles and match them to the vertices from which they originated. The curvature of charged particle tracks in the magnetic field allows their charge and momentum to be measured. The CMS silicon tracker consists of 13 layers in the central region and 14 layers in the endcaps. The three innermost layers (up to 11 cm radius) are made of $100 \times 150\,\mu\text{m}$ pixels (a total of 66 million pixels) and the next four (up to 55 cm radius) are silicon strips (9.6 million strip channels in total). The CMS silicon tracker is the world's largest silicon detector, with $205\,\text{m}^2$ of silicon sensors (approximately the area of a tennis court) and 76 million channels.

[12] Seven detectors have been constructed at the LHC, located underground in large caverns excavated at the LHC's intersection points. ATLAS and CMS are large, general purpose particle detectors. A Large Ion Collider Experiment (ALICE) and LHCb have more specific roles, respectively, the study of collisions of heavy ions and the study of the physics of the b quark. The remaining three are much smaller; two of them, TOTEM and LHCf, study the cross section in the forward region (which dominates the total hadronic cross section, as we shall see in Chap. 6); finally, MoEDAL searches for exotic particles, magnetic monopoles in particular.

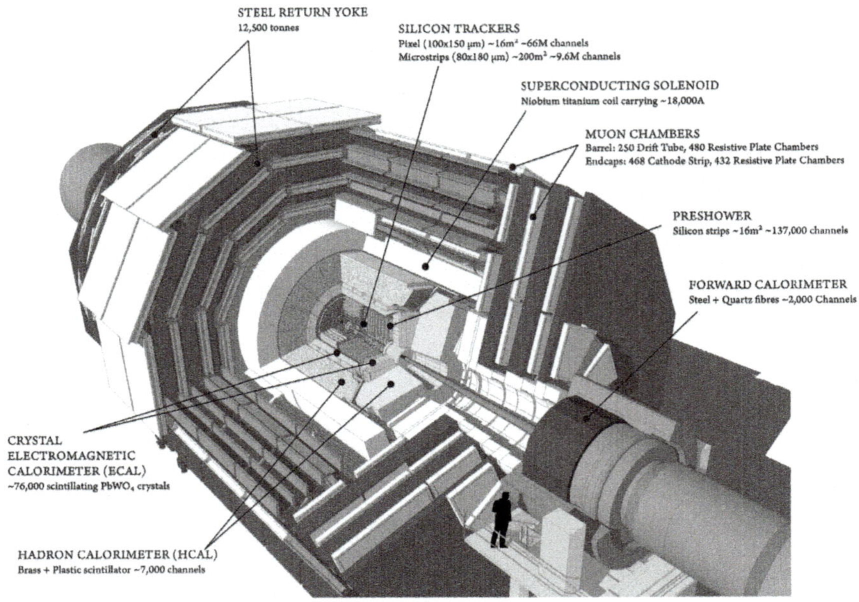

Fig. 4.31 The CMS detector at LHC (*Source* CERN)

Layer 2 is devoted to electromagnetic calorimetry. The Electromagnetic Calorimeter (ECAL) is constructed from crystals of lead tungstate, $PbWO_4$, a very dense but optically clear material. It has a radiation length of 0.89 cm, and a rapid light yield (80 % of light yield within one crossing time of 25 ns) of about 30 photons per MeV of incident energy. The crystals front size is 22 mm × 22 mm, with a depth of 23 cm. They are readout by silicon avalanche photodiodes and sit in a matrix of carbon fiber that ensures optical isolation. The barrel region consists of about 60 thousand crystals, with a further about seven thousand in each of the endcaps.

Layer 3 is devoted to Hadronic Calorimetry. The Hadronic Calorimeter (HCAL) consists of layers of dense material (brass or steel) interleaved with tiles of plastic scintillators, read out via wavelength-shifting fibers by hybrid photodiodes. This combination was optimized to guarantee the maximum amount of absorbing material inside the magnet coil.

Layer 4 is the magnet. It is 13 m long and 6 m in diameter, and it is a refrigerated superconducting niobium-titanium coil providing a solenoidal field of 3.8 T (the current being about 18000 A, giving a total stored energy of about 2.5 GJ, equivalent to about 500 kg of TNT: dump circuits to safely dissipate this energy are in place, should the magnet quench).

Layer 5 is occupied by the muon detectors and the return yoke. To identify muons and measure their momenta, CMS uses mostly drift tubes and resistive plate chambers. The drift tubes provide precise trajectory measurements in the central barrel region. The RPC provide an accurate timing signal at the passage of a muon.

The amount of raw data from each crossing is approximately 1 MB, which at the 40 MHz crossing rate would result in 40 TB of data per second, A multi-stage trigger system reduces the rate of interesting events down to about 100/s. At the first stage, the data from each crossing are held in buffers within the detector and some key information is used to identify interesting features (such as large transverse momentum particles, high energy jets, muons or missing energy). This task is completed in around 1 μs, and the event rate is reduced by a factor of about thousand down to 50 kHz. The data corresponding to the selected events are sent over fiber-optic links to a higher level trigger stage, which is a software trigger. The lower rate allows for a much more detailed analysis of the event, and the event rate is again reduced by a further factor of about a thousand, down to around 100 events per second. In a high-luminosity colliders as the LHC, one single bunch crossing may produce several separate events, so-called pile-up events. Trigger systems must thus be very effective.

The overall accuracy in momentum can be parameterized as

$$\frac{\delta p}{p} \simeq 0.015\,\%\, p_T \oplus 0.5\,\% \,, \tag{4.17}$$

where p is expressed in GeV/c, and the typical calorimetric resolution in the barrel part is

$$\frac{\delta E}{E} \simeq \frac{3\,\%}{\sqrt{E}} \oplus 0.3\,\% \,, \tag{4.18}$$

with E expressed in GeV.

A reconstructed event is seen in Fig. 4.32.

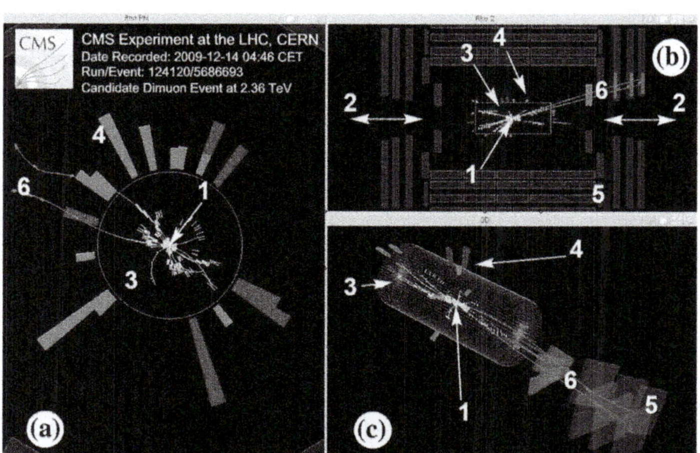

Fig. 4.32 An event reconstructed by CMS as shown in different projections by the CMS event display (*Source* CERN)

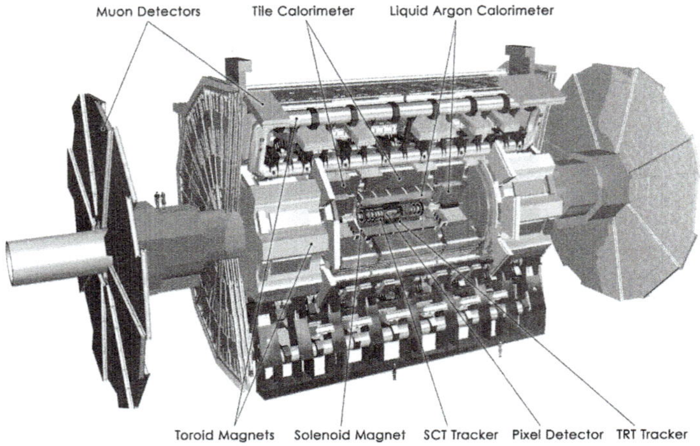

Fig. 4.33 Sketch of the ATLAS detector. Credit: CERN

4.4.2.4 ATLAS at LHC

ATLAS (A Toroidal LHC ApparatuS, Fig. 4.33) is 46 m long, 25 m in diameter, and weighs about 7000 tons. It consists of a series of ever-larger concentric cylinders around the interaction point:

- An inner tracking system: operating inside an axial magnetic field of 2 T, it is based on three types of tracking devices—an innermost silicon pixel detector is followed by a silicon strip detector and finally by straw tubes with particle identification capabilities based on transition radiation in the outer tracker.
- A hybrid calorimeter system: liquid argon with different types of absorber materials is used for the electromagnetic part, the hadronic end-cap and the forward calorimeter. The central hadronic calorimeter is a sampling calorimeter with steel as the absorber material and scintillator as the active medium.
- A large muon spectrometer: an air-core toroid system generates an average field of 0.5 T (1 T), in the barrel (end-cap) region of this spectrometer, resulting in a bending power between 2.0 and 7.5 Tm. Tracks are measured by monitored drift tubes and cathode strip chambers. Trigger information is provided by Thin Gap Chambers (TGC) in the end-cap and RPC in the barrel.

An electromagnetic calorimeter and a Cherenkov counter instrument the end-cap region. Two scintillator wheels were mounted in front of the electromagnetic end-caps to provide trigger signals with minimum bias.

The ATLAS trigger and data acquisition is a multi-level system with buffering at all levels. Trigger decisions are based on calculations done at three consecutive trigger levels.

Table 4.4 Comparison of the main design parameters of CMS and ATLAS

Parameter	ATLAS	CMS
Total weight (tons)	7000	12,500
Overall diameter (m)	22	15
Overall length (m)	46	20
Magnetic field for tracking (T)	2	4
Solid angel for precision measurement ($\Delta\phi \times \Delta\eta$)	$2\pi \times 5.0$	$2\pi \times 5.0$
Solid angel for energy measurement ($\Delta\phi \times \Delta\eta$)	$2\pi \times 9.6$	$2\pi \times 9.6$
Total cost (million Swiss francs)	550	550

The main design parameters of ATLAS and CMS are compared in Table 4.4. The overall accuracy in momentum can be parameterized as

$$\frac{\delta p}{p} \simeq 0.05\,\%\, p_T \oplus 1\,\%\,, \tag{4.19}$$

where p is expressed in GeV/c, and the typical calorimetric resolution in the barrel part is

$$\frac{\delta E}{E} \simeq \frac{2.8\,\%}{\sqrt{E}} \oplus 0.3\,\%\,, \tag{4.20}$$

with E expressed in GeV.

4.5 Cosmic-Ray Detectors

The strong decrease in the flux Φ of cosmic rays with energy, in first approximation $\Phi \propto E^{-3}$, poses a big challenge to the dimensions and the running times of the experimental installations when large energies are studied. Among cosmic rays, a small fraction about 10^{-3} are photons, which are particularly interesting since they are not deflected by intergalactic magnetic fields, and thus point directly to their sources; the large background from charged cosmic rays makes the detection even more complicated. Neutrinos are expected to be even less numerous than photons, and their detection is even more complicated due to the small cross section.

We shall examine first the detectors of cosmic rays which have a sizable probability of interactions with the atmosphere: nuclei, electrons/positrons, and photons.

Balloon and satellite born detectors operate at an altitude above 15 km where they can detect the interaction of the primary particle inside the detector, but they are limited in detection area and therefore also limited in the energy range they can measure (the maximum primary energy that can be measured by means of direct observations is of the order of 1 PeV; above this energy the observations are performed by exploiting the cascades induced in atmosphere by the interactions of cosmic rays).

4.5.1 Interaction of Cosmic Rays with the Atmosphere; Extensive Air Showers

The physics of electromagnetic and hadronic showers has been described before; here we particularize the results obtained to the development of the showers due to the interaction of high-energy particles with the atmosphere. These are called extensive air showers (EAS).

High-energy hadrons, photons, and electrons interact in the high atmosphere. As we have seen, the process characterizing hadronic and electromagnetic showers is conceptually similar (Fig. 4.34).

For photons and electrons above a few hundred MeV, the cascade process is dominated by the pair production and the bremsstrahlung mechanisms: an energetic photon scatters on an atmospheric nucleus and produces a pair, which emits secondary photons via bremsstrahlung; such photons produce in turn an e^+e^- pair, and so on, giving rise to a shower of charged particles and photons.

The longitudinal development of typical photon-induced extensive air showers is shown in Fig. 4.35 for different values of the primary energies. The maximum shower size occurs approximately $\ln(E/\epsilon_0)$ radiation lengths, the radiation length for air being about 37 g/cm^2 (approximately 300 m at sea level and NTP). The critical

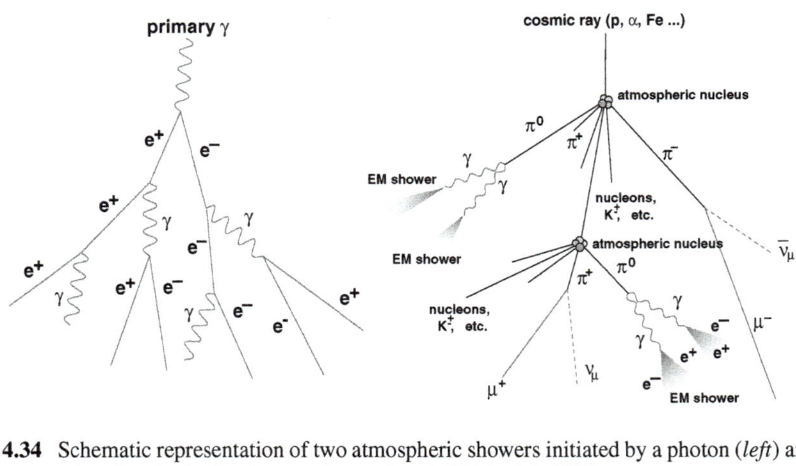

Fig. 4.34 Schematic representation of two atmospheric showers initiated by a photon (*left*) and by a proton (*right*). Credit: R.M. Wagner, dissertation, MPI Munich 2007

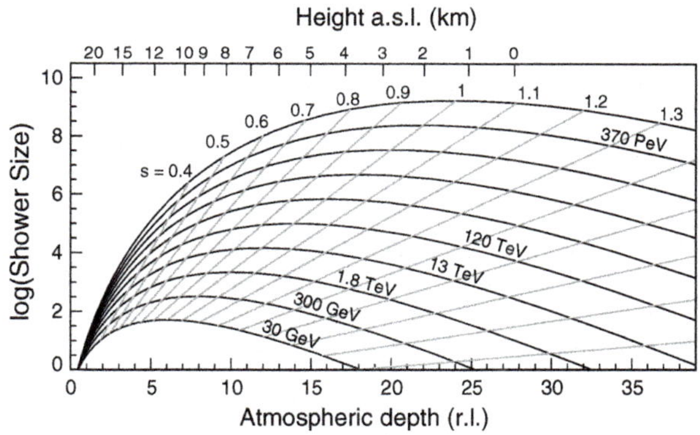

Fig. 4.35 Longitudinal shower development from a photon-initiated cascade. The parameter s describes the shower age. Credit: R.M. Wagner, dissertation, MPI Munich 2007; adapted from reference [F4.1] in the "Further reading"

energy ϵ_0, the energy at which the ionization energy loss starts dominating the energy loss by bremsstrahlung, is about 80 MeV in air.[13]

The hadronic interaction length in air is about 61 g/cm^2 for protons (500 meters for air at NTP), being shorter for heavier nuclei—the dependence of the cross section on the mass number A is approximately $A^{2/3}$. The transverse profile of hadronic showers is in general wider than for electromagnetic showers, and fluctuations are larger.

Particles release energy in the atmosphere, which acts like a calorimeter, through different mechanisms—which give rise to a measurable signal. We have discussed these mechanisms in Sect. 4.1.1; now we reexamine them in relation to their use in detectors.

4.5.1.1 Fluorescence

As the charged particles in an extensive air shower go through the atmosphere, they ionize and excite the gas molecules (mostly nitrogen). In the de-excitation processes that follow, visible and UV radiations are emitted. This is the so-called fluorescence light associated to the shower.

The number of emitted fluorescence photons is small—of the order of a few photons per electron per meter in air. This entails the fact that fluorescence technique can be used only at high energies. However, it is not directional as in the case of Cherenkov photons (see below), and thus it can be used in serendipitous observations.

---

[13] In the isothermal approximation, the depth x of the atmosphere at a height h (i.e., the amount of atmosphere above h) can be approximated as

$$x \simeq X e^{-h/7\,\mathrm{km}} ,$$

with $X \simeq 1030$ g/cm^2.

4.5.1.2 Cherenkov Emission

Many secondary particles in the EAS are superluminal, and they thus emit Cherenkov light that can be used for the detection. The properties of the Cherenkov emission have been discussed in Sect. 4.2.

At sea level, the value of the Cherenkov angle θ_C in air for $\beta = 1$ is about 1.3°, while at 8 km a.s.l. it is about 1°. The energy threshold at sea level is 21 MeV for a primary electron and 44 GeV for a primary muon.

Half of the emission occurs within 20 m of the shower axis (about 70 m for a proton shower).

Since the intrinsic angular spread of the charged particles in an electromagnetic shower is about 0.5°, the opening of the light cone is dominated by the Cherenkov angle. As a consequence, the ground area illuminated by Cherenkov photons from a shower of 1 TeV (the so-called "light pool" of the shower) has a radius of about 120 m. The height of maximal emission for a primary of 1 TeV of energy is approximately 8 km a.s.l., and about 150 photons per m^2 arrive at 2000 m a.s.l. (where typically Cherenkov detectors are located) in the visible frequencies. This dependence is not linear, being the yield of about 10 photons per square meter at 100 GeV.

The atmospheric extinction of light drastically changes the Cherenkov light spectrum (originally proportional to $1/\lambda^2$) arriving at the detectors, in particular suppressing the UV component (Fig. 4.36) which is still, anyway, dominant. There are several sources of extinction: absorption bands of several molecules, molecular (Rayleigh), scattering from aerosol (Mie).

Radio Emission. Cosmic-ray air showers also emit radio waves in the frequency range from a few to a few hundred MHz, an effect that opens many interesting possibilities in the study of UHE and EHE extensive air showers. At present, however, open questions still remain concerning both the emission mechanism and its strength.

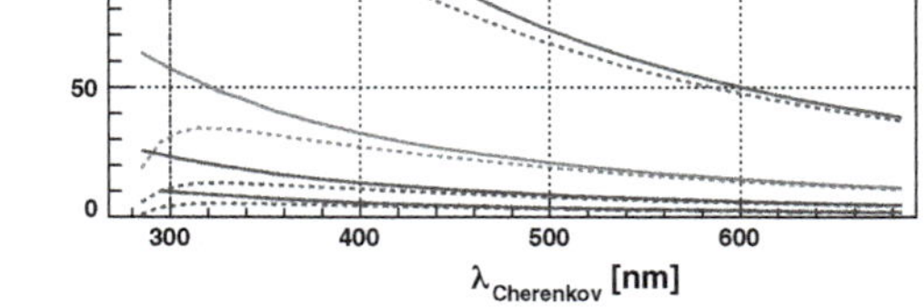

Fig. 4.36 Spectrum of the Cherenkov radiation emitted by gamma showers at different energies initiated at 10 km a.s.l. (*solid curves*) and the corresponding spectra detected at 2200 meters a.s.l. (*lower curve*). Credit: R.M. Wagner, dissertation, MPI Munich 2007

4.5.2 Detectors of Charged Cosmic Rays

The detection of charged cosmic rays may be done at top of the Earth atmosphere in balloon or satellite-based experiments whenever the fluxes are high enough (typically below tens or hundreds of GeV) and otherwise in an indirect way by the observation of the extensive air showers produced in their interaction with the atmosphere (see Sect. 4.5.1).

In the last thirty years, several experiments to detect charged cosmic rays in the space or at the top of the Earth atmosphere were designed and a few were successfully performed. In particular:

- The Advanced Composition Explorer (ACE) launched in 1997 and still in operation (enough propellant until $\sim$2024) has been producing a large set of measurements on the composition (from H to Ni) of solar and Galactic Cosmic rays covering energies from the 1 KeV/nucleon to 500 MeV/nucleon. ACE have several instruments which are able to identify the particle charge and mass using different types of detectors (for example silicon detectors, gas proportional counters, fiber-optics hodoscopes) and techniques (for example the specific energy loss dE/dx, the time-of-flight, electrostatic deflection). The total mass at launch (including fuel) was about 800 kg.
- The Balloon-borne Experiment with Superconducting Spectrometer (BESS) performed successive flights starting in 1993 with the main aim to measure the low-energy anti-proton spectrum and to search for anti-matter namely anti-helium. The last two flights (BESS-Polar) were over Antarctica and had a long duration (8.5 days in 2004 and 29.5 days in 2007/2008). The instrument, improved before every flight, had to ensure a good charge separation and good particle identification. It had a horizontal cylindrical configuration and its main components were as follows: a thin-wall superconducting magnet; a central tracker composed by drift chambers; time-of-flight scintillation counters hodoscopes; an aerogel (an ultra-light porous material derived from a gel by replacing its liquid component with a gas) Cherenkov counter.
- The PAMELA experiment launched in June 2006 measured charged particle and anti-particles out of the Earth atmosphere during a long (six years)-time period. A permanent magnet of 0.43 T and a microstrip silicon tracking system ensured a good charge separation between electrons and positrons up to energies of the order of the hundred of GeV measured by a silicon/tungsten electromagnetic calorimeter complemented by a neutron counter to enhance the electromagnetic/hadronic discrimination power. The trigger was provided by a system of plastic scintillators which were also used to measure the time-of-flight and an estimation of the specific ionization energy loss (dE/dX).
- The Alpha Magnetic Spectrometer (AMS-02) was installed in May 2011 on the International Space Station. Its layout is similar to PAMELA but with a much larger acceptance and a more complete set of sophisticated and higher performing detectors. Apart from the permanent magnet and the precision silicon tracker

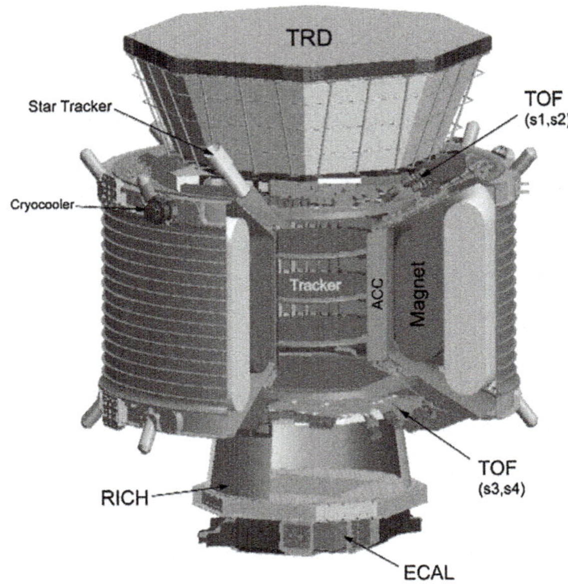

Fig. 4.37 The AMS-02 detector layout. Credit: AMS Collaboration

it consists of a transition radiation detector, time-of-flight and anti-coincidence counters, a ring imaging Cherenkov detector, and an electromagnetic calorimeter (Fig. 4.37). Its total weight is 8500 kg and its cost was over 2 billion euro.

Extensive air showers produced by high-energy cosmic rays in their interaction with the atmosphere are detected using three different techniques:

- The measurement of a fraction of the EAS particles arriving at the Earth surface through an array of surface detectors (SD);
- The measurement in moonless nights of the fluorescence light emitted mainly by the excitation of the atmosphere nitrogen molecules by the shower low energetic electrons through an array of ultra-violet fluorescence detectors (FD) placed on the Earth surface or even in satellites; and
- The measurement of the Cherenkov light emitted by the ultra-relativistic air shower particles in a narrow cone around the shower axis, through dedicated telescopes as the Imaging Atmosphere Cherenkov Telescopes (IACTs), which will be discussed in the next section.

Other possible techniques (radio detection for example) might be exploited in the future.

Surface detectors measure at specific space locations the time of arrival of individual particles. The most widely used surface detectors are scintillation counters and water Cherenkov counters. More sophisticated tracking detectors as resistive plate chambers, drift chambers, and streamer tube detectors have been also used or proposed.

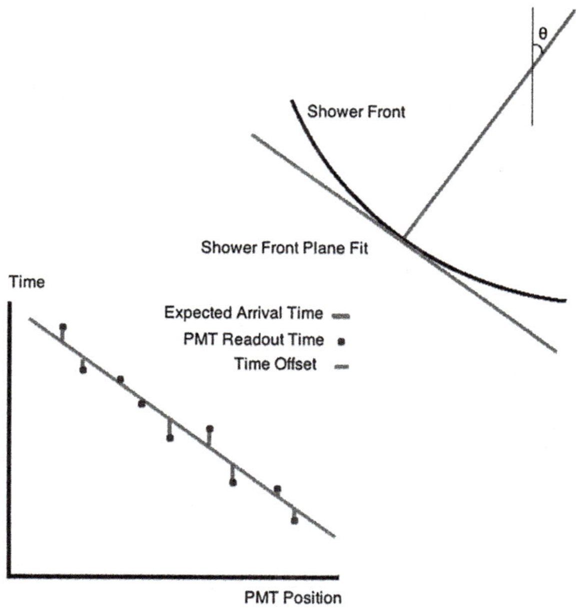

Fig. 4.38 Air shower front arriving at the Earth surface; times of arrival are measured by surface detectors and allow the determination of the direction. From HAWC Collaboration, A.U. Abeysekara et al. arXiv:1310.0074

The arrival direction of an air shower is determined from the arrival time at the different surface detectors of the shower front. In a first approximation, the front can be described by a thin disk that propagates with the speed of light; second-order corrections can be applied to improve the measurement (Fig. 4.38).

The impact point of the air shower axis at the Earth surface (the air shower core) is defined as the point of maximum particle density and is determined from the measured densities at the different surface detectors using, in a first approximation, a modified center-of-mass algorithm. In Fig. 4.39, the particle density pattern of the highest energy event at the AGASA array experiment[14] is shown. The energy of the event was estimated to be about 2×10^{20} eV. The measured densities show a fast decrease with the distance of the core and are usually parameterized by empirical or phenomenological inspired formulae—the most popular being the NKG function, introduced in Sect. 4.1.6—which depend also on the shower age (the level of development of the shower in the moment when it arrive at the Earth surface). Such functions allow for a better determination of the shower core and for the extrapolation of the particle density to a reference distance of the core which is then used as an estimator of the shower size and thus of the shower energy. The exact function as well as the reference distance depends on the particular experiment setup.

The fluorescence telescopes record the intensity and arrival time of the light emitted in the atmosphere in specific solid angle regions and thus are able to reconstruct

[14]The Akeno Giant Air-Shower Array (AGASA) is a very large surface array covering an area of 100 km^2 in Japan and consisting of 111 surface detectors (scintillators) and 27 muon detectors (proportional chambers shielded by Fe/concrete).

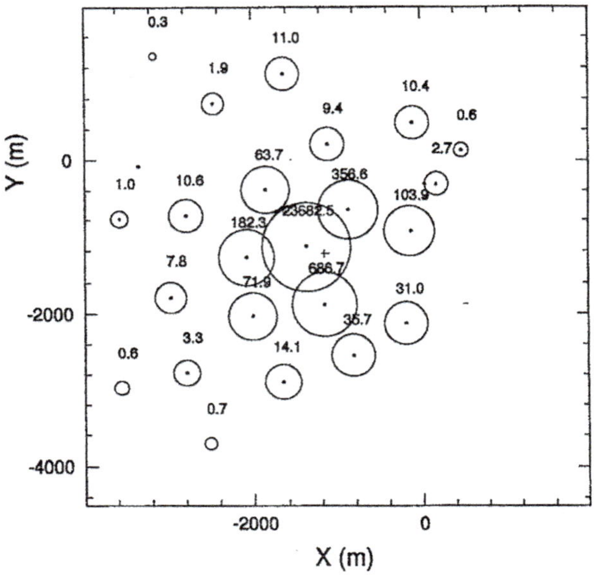

Fig. 4.39 Map of the observed particle density pattern of the highest energy event at the AGASA array experiment. The cross corresponds to the fitted position of the shower core. From http://www.icrr.u-tokyo.ac.jp

the shower axis geometry and the shower longitudinal profile. In Fig. 4.40 the image of a shower in the focal plane of one of the Pierre Auger fluorescence telescopes (see later) is shown. The third dimension, time, is represented in a color code.

The geometry of the shower (Fig. 4.41) is then reconstructed in two steps: first the shower detector plane (SDP) is found by minimizing the direction of the SDP normal to the mean directions of the triggered pixels, and then the shower axis parameters within the SDP are found from the measured arrival time of the light in each pixel assuming that the shower develops along a line at the speed of light.

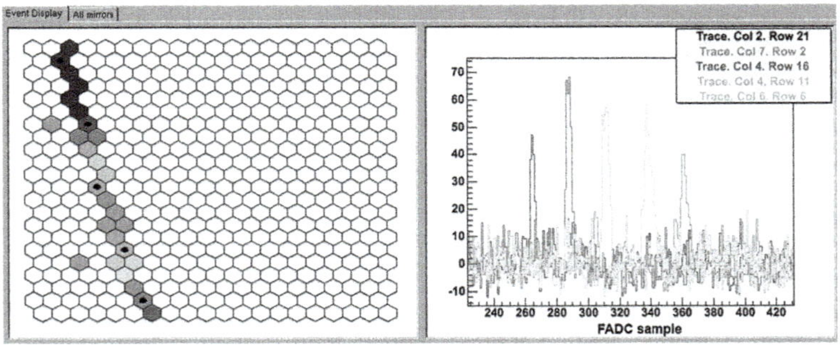

Fig. 4.40 Display of one shower in the focal plane of one of the Pierre Auger fluorescence telescopes. *Left* Pattern of the pixels with signal; *right* response (signal versus time, with a time bin of 100 ns) of the selected pixels (marked with a *black dot* in the *left* panel). The development of the shower in the atmosphere can be qualitatively pictured. From https://www.auger.org

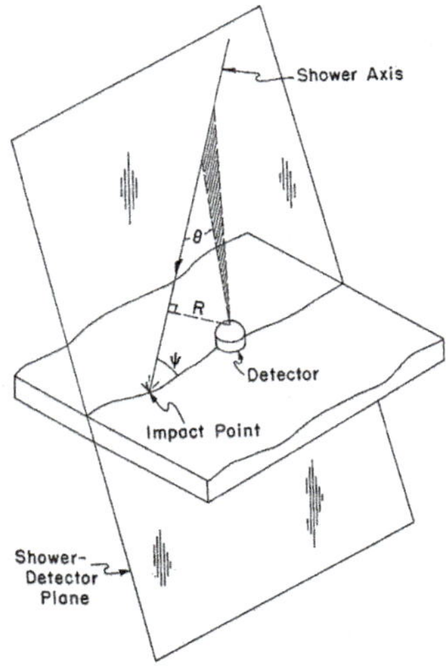

Fig. 4.41 The shower geometry as seen by a fluorescence telescope. From K.-H. Kampert and A. Watson, "Extensive Air Showers and Ultra High Energy Cosmic Rays: A Historical Review," EPJ-H 37 (2012) 359

The simultaneous observation of the shower by two (stereo) or more fluorescence detectors or by a surface detector array (hybrid detection) provides further geometric constrains which improve considerably the resolution of the shower geometric reconstruction.

The intensity of collected light along the shower axis is corrected for the detector efficiency, the solid angle seen by each detector pixel, the attenuation in the atmosphere, the night sky background, and the contribution of fluorescence (dominant unless the shower axis points in the direction of the telescope) and of the Cherenkov light (direct and scattered) are estimated. Finally, the shower longitudinal profile (Fig. 4.42) is obtained assuming the proportionality between the fluorescence

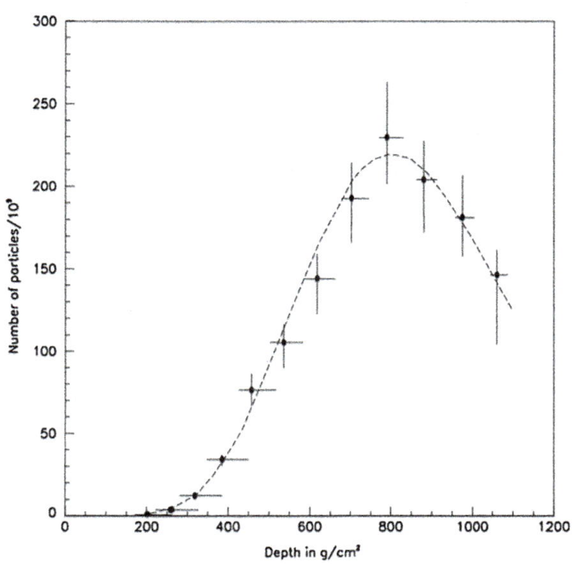

Fig. 4.42 Shower longitudinal profile of the most energetic event observed by the Fly's Eye experiment. From D.J. Bird et al., Astrophys. J. 441 (1995) 144

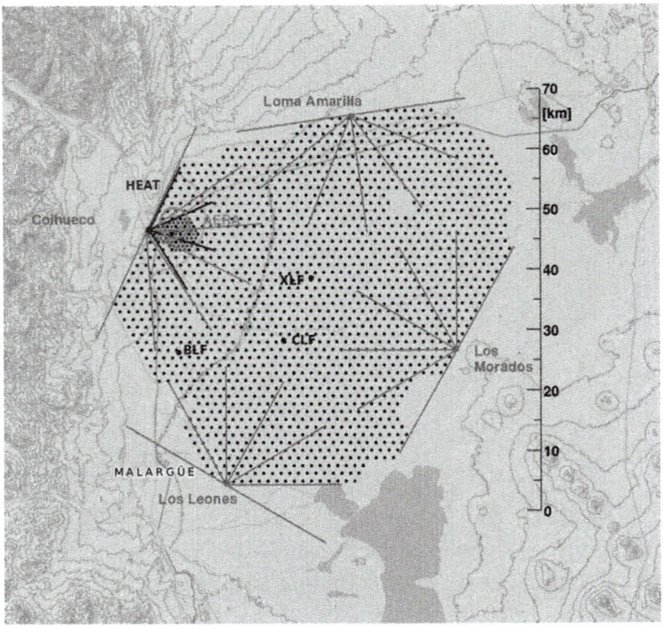

Fig. 4.43 The Pierre Auger Observatory near Malargue, Argentina. The radial lines point to the fluorescence detectors (FD, $4 \times 6 = 24$). The *black dots* are the 1600 ground stations (SD). Sites with specialized equipment are also indicated. By Darko Veberic [GFDL http://www.gnu.org/copyleft/fdl.html], via Wikimedia Commons

light emitted and the number of particles in the shower. The integral of such profile is a good estimator of the energy of the shower (small "missing energy" corrections due to low interacting particles in the atmosphere, like muons and neutrinos, have to be taken into account).

The Pierre Auger Observatory. The Pierre Auger Observatory in Malargue, Argentina, is the largest cosmic-ray detector ever built. It covers a surface of about 3000 square kilometers with 1600 surface detector stations (Cherenkov water tanks) arranged in a triangular grid of 1.5 km side complemented by 24 fluorescence telescopes, grouped into four locations to cover the atmosphere above the detector area (Fig. 4.43).

Each water tank is a cylinder of 10 m^2 base by 1.5 m height filled with 12 tons of water (Fig. 4.44). The inner walls of the tank are covered with a high reflectivity material. The Cherenkov light, produced by the charged particles crossing the tank, is collected by three PMT placed on the top of the tank. Each tank is autonomous being the time given by a GPS unit and the power provided by a solar panel; it communicates via radio with the central data acquisition system.

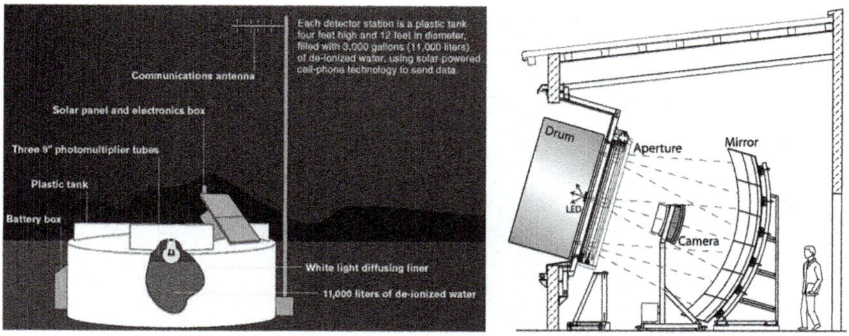

Fig. 4.44 Sketch of one of the Pierre Auger surface detectors (*left*); a fluorescence telescope (*right*). From https://www.auger.org

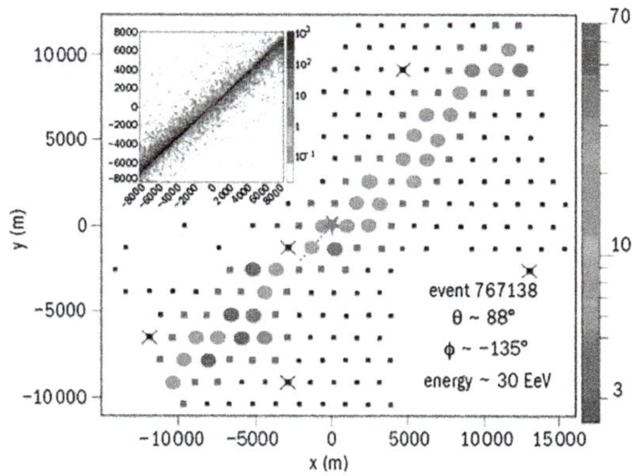

Fig. 4.45 A 30 EeV event at a zenith angle of about 88° recorded by the Auger detector. The *inset* shows a simulation of an event of the same energy and angle. From https://www.auger.org

Each fluorescence detector is a Schmidt telescope[15] with a field of view of 30° in azimuth and 29° in elevation (Fig. 4.44). The light enters the telescope through an ultra-violet filter installed over the telescope circular diaphragm cross and is collected in a 3.5 m diameter spherical mirror which focus it in a 440 PMT camera.

The signal by an event of extremely high energy is shown in Fig. 4.45.

The Telescope Array. The largest cosmic-ray detector in the northern hemisphere is the Telescope Array (TA) in Utah, USA. Similar to Auger, it is also an hybrid detector composed by a surface array of 507 scintillator detectors, each 3 m in size, located on a 1.2 km square grid, plus three fluorescence stations each one with a dozen of telescopes each instrumented with a 256 PMT camera covering 3–33 ° in elevation. The total surface covered is about 800 km^2.

Future Prospects: Detection from Space. An innovative approach is proposed by the JEM-EUSO collaboration (which succeeded to the EUSO—Extreme Uni-

[15] In a Schmidt telescope, a spherical mirror receives light that passed through a thin aspherical lens that compensates for the image distortions that will occur in the mirror. The light is then reflected in the mirror into a detector that records the image.

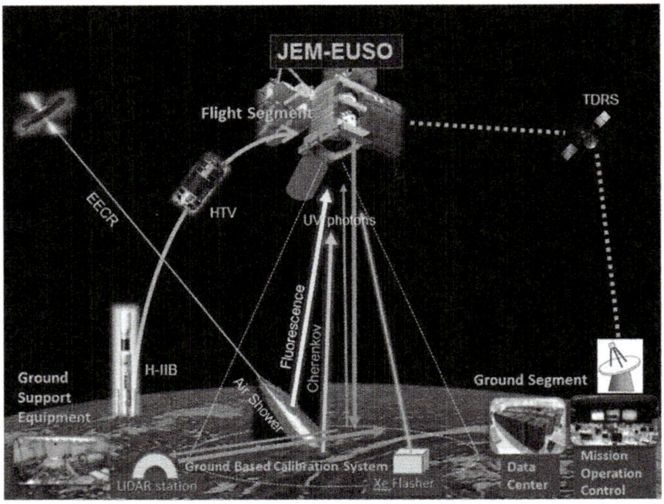

Fig. 4.46 JEM-EUSO observational principle. From http://jemeuso.riken.jp/en/

verse Space Observatory, discontinued in 2004). JEM-EUSO intends to install in the Japanese Experimental Module of the International Space Station a large and sophisticated double-sided Fresnel lens telescope able to detect the fluorescence light emitted by the extended air showers (Fig. 4.46). Observing the Earth from 400 km above and having a large field of view ($\pm 30°$) the covered Earth surface is huge (above 1.9 $\times$ 10^5 km^2) but the energy threshold is high (around 3 $\times$ 10^{19} eV). The camera is composed by about 5000 multi-anode PMT with a total of 3.2 $\times$ 10^5 pixels.

4.5.3 Photon Detection

High-energy photons are expected to be produced in astrophysical objects by leptonic acceleration mechanisms (see Chap. 10), and by the decays of neutral pions made in cosmic-ray interactions with radiation or gas. As these pions decay, they produce photons with typical energies one order of magnitude smaller than those of the cosmic-ray nucleons.

The detection of high-energy photons is complicated by the absorption in the atmosphere, and by the faintness of the signal, in particular when compared to the corresponding charged particles of similar energy—being the latter three–four orders of magnitude more frequent.

Photons above the ultra-violet (UV) region are shielded by the Earth's atmosphere (see Fig. 4.47). They interact with matter mostly due to the Compton mechanism and to the photoelectric effect at energies up to about 20 MeV, while e^+e^- pair production dominates above about 20 MeV.

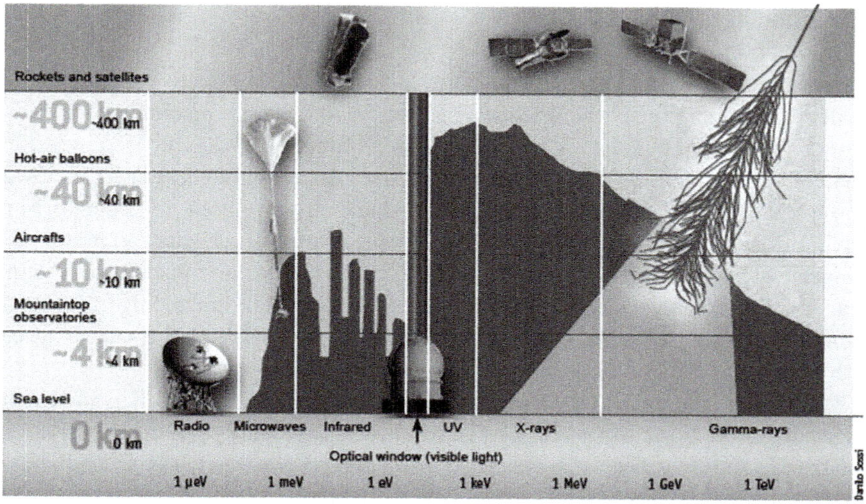

Fig. 4.47 Transparency of the atmosphere for different photon energies and possible detection techniques. *Source* A. De Angelis and L. Peruzzo, "Le magie del telescopio MAGIC," Le Scienze, April 2007

Although arbitrary, a classification of gamma rays as a function of their energy defines as high-energy (HE) the photons above 30 MeV—i.e., the threshold for the production of e^+e^- pairs, plus some additional energy to allow the phase space for the pair production process to become large; very-high-energy (VHE) the photons above 30 GeV. This classification, in particular, corresponds to different detection techniques, as we shall see later. The HE—and VHE in particular—regions are especially important related to the physics of cosmic rays and to fundamental physics.

Due to the conversion probability in the atmosphere (whose thickness is about 28 radiation length at sea level) only satellite-based detectors can detect primary X/γ-rays. Satellites are anyway small, being of about 1 m^2 their maximum dimensions because of the cost of space technology. If the energy of the primary particle is large enough, some of the products of the shower generated by the interaction with the atmosphere, acting as a calorimeter, can reach ground-based detectors. It is appropriate to build such detectors at high altitudes, where atmospheric dimming is lower. The area sampled by ground-based detectors can be much larger than the typical areas of satellites. Since the fluxes of high-energy photons are low and decrease rapidly with increasing energy, VHE and UHE gammas can be detected only from the atmospheric showers they produce, i.e., by means of ground-based detectors. This fact clarifies another meaning of the division between HE and VHE photons: HE photons are detected using satellites, while for VHE photons the detection using ground-based detectors becomes possible.

4.5.3.1 Satellites

Main figures of merit for a satellite-borne detector are its effective area (i.e., the product of the area times the detection efficiency), the energy resolution, the space or angular resolution (called as well point-spread function, or PSF).

Satellite HE gamma telescopes can detect the primary photons at energies lower than ground-based telescopes. They have a small effective area, of order of $1\,\mathrm{m}^2$ maximum, which limits the sensitivity. They have a large duty cycle, and they suffer a low rate of background events, since they can be coupled to anticoincidence systems rejecting the charged cosmic rays. They have a large cost, dominated by the costs of launch and by the strong requirements of instruments which must be sent in space, with little or no possibility of intervention to fix possible bugs.

Two modern gamma-ray telescopes are in orbit; they are called AGILE and Fermi (Fig. 4.48).

Their technology has been inherited from the smaller and less technological EGRET instrument, operational in the years 1991–2000 on the Compton Gamma-Ray Observatory, and from particle physics. The direction of an incident photon is determined through the geometry of its conversion into an e^+e^- pair in foils of heavy materials which compose the instrument, and detected by planes of silicon detectors. The presence of an anticoincidence apparatus realizes a veto against unwanted incoming charged particles. The principle of operation is illustrated in Fig. 4.48, right.

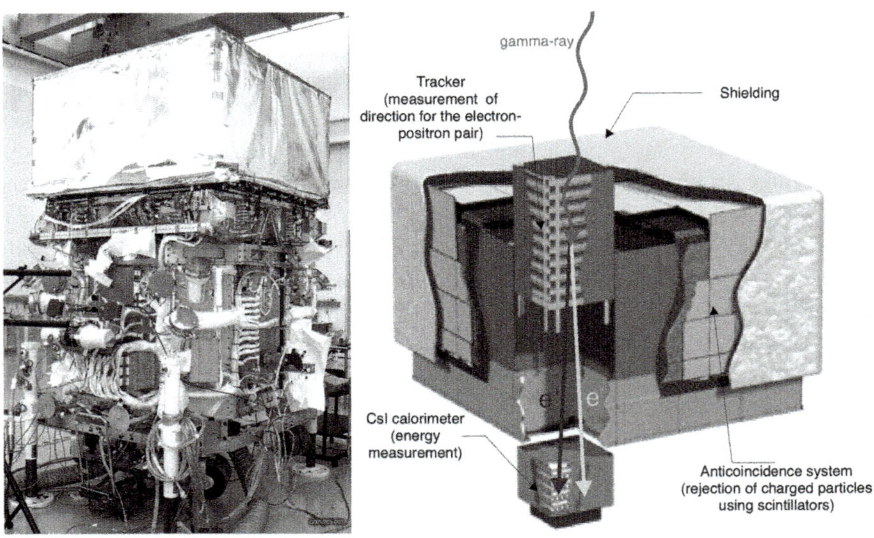

Fig. 4.48 On the *left*, the Fermi satellite. On the *right*, the layout of the Large Area Telescope (LAT), and principle of operation. Credits: NASA

The angular resolution of these telescopes is limited by the opening angle of the e^+e^- pair, approximately $(m_ec^2)/E \ \ln(E/(m_ec^2))$, and especially by the effect of multiple scattering. To achieve a good energy resolution, in this kind of detector, a calorimeter in the bottom of the tracker is possibly used, depending on the weight that the payload is planned to comply with. Due to weight limitations, however, it is difficult to fit in a calorimeter that completely contains the showers; this leakage downgrades the energy resolution. Since at low energies multiple scattering is the dominant process, the optimal detector design is a tradeoff between small radiation length (which decreases the conversion efficiency) and large number of samplings (which increases the power consumption, limited by the problems of heat dissipation in space).

The largest gamma-ray space-based detector ever built is the Fermi observatory, launched in June 2008—and called GLAST before the successful positioning in orbit. It is composed by the spacecraft and by two instruments: the Large Area Telescope (LAT) and the Fermi Gamma Burst Monitor (GBM); the two instruments are integrated and they work as a single observatory.

The structure of the LAT consists mainly in a tracker, an anticoincidence apparatus and a calorimeter (see Fig. 4.48). Its energy range goes from 20 MeV to about 300 GeV and above, while the energy range explored by the GBM is 10–25 MeV. Fermi was built and it is operated by an international collaboration with contributions from space agencies, high-energy particle physics institutes, and universities in France, Italy, Japan, Sweden, and the United States; it involves about 600 scientists. After the first year, data are public, i.e., every scientist in the world can in principle analyze them.

The scientific objectives of the LAT include the understanding of the nature of unidentified gamma-ray sources and origins of diffuse galactic emission; of particle acceleration mechanisms at the sources, particularly in active galactic nuclei, pulsars, supernovae remnants, and the Sun; of the high-energy behavior of gamma-ray burst and transient sources. The observations will also be used to probe dark matter and, at high-energy, the early universe and the cosmic evolution of high-energy sources to $z \geq 6$.

The characteristics and performance of the LAT will enable significant progress in the understanding of the high-energy sky. In particular, it has good angular resolution for source localization and multi-wavelength study, high sensitivity in a broad field-of-view to detect transients and monitor variability, good calorimetry over an extended energy band for detailed emission spectrum studies, and good calibration and stability for absolute, long term flux measurements.

The LAT tracker is done by 16 planes of high-Z material (W) in which incident γ-rays can convert to an e^+e^- pair. The converter planes are interleaved with 18 silicon detectors two-layer planes that measures the tracks of the particles resulting from pair-conversion. This information is used to reconstruct the directions of the incident γ-rays.

After the tracker, a calorimeter can measure the energy. It is made of CsI(Tl) crystals with a total depth of 8.6 radiation lengths, arranged in a hodoscope configuration in order to provide longitudinal and transverse information on the energy deposition. The depth and the segmentation of the calorimeter enable the high-energy reach of the LAT and significantly contribute to background rejection. The aspect ratio of the tracker (height/width) is 0.4 (the width being about 1.7 m), resulting in a large field-of-view (2.4 sr) and ensuring that most pair-conversion showers initiated in the tracker will reach the calorimeter for energy measurement.

Around the tracker, an anticoincidence detector (ACD) made of plastic scintillator provides charged particle background rejection.

The overall performance of Fermi can be summarized as follows in the region of main interest (30 MeV–30 GeV):

- Effective area of about 1 m^2;
- Relative energy resolution decreasing between 10 % at 100 MeV and 5 % at 1 GeV, increasing again to 10 % at 30 GeV; and
- Angular resolution of 0.1° at 10 GeV, and approximately varying as $1/\sqrt{E}$.

AGILE, the precursor of Fermi, is a completely Italian satellite launched in April 2007. Its structure is very similar to Fermi, but its effective area is about one order of magnitude smaller. However, many remarkable physics results were obtained thanks to the AGILE data.

4.5.3.2 Ground-Based Gamma Detectors

Ground-based VHE telescopes such as HAWC, H.E.S.S., MAGIC, and VERITAS, detect the atmospheric showers produced by primary photons and cosmic rays of energy higher than those observed by satellites.

The two kinds of detectors (on satellite and at ground) are complementary. At energies below 1 GeV or so, the showers generated by photons do not have the time to develop properly, and thus the only way to detect such photons below this energy is related to the use of satellites. At TeV energies, however, the flux is too low to be detected with satellite-based detectors: due to their cost, and in particular to the cost of the launch, the satellites have areas of the order of 1 m^2 at most, and at these energies even the most luminous gamma sources have a flux smaller than one photon per square meter every ten hours. Ground-based detectors have a huge effective area, so their sensitivity is high; they detect a huge amount of background events, but they have low cost.

There are two main classes of ground-based VHE gamma detectors: the EAS arrays and the Cherenkov telescopes (see Fig. 4.49).

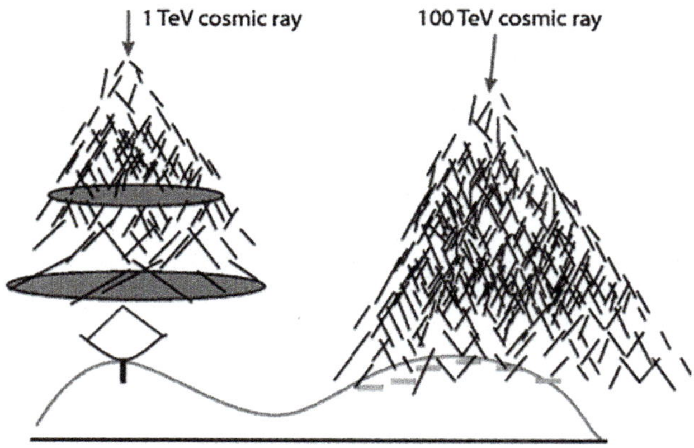

Fig. 4.49 Sketch of the operation of Cherenkov telescopes and of EAS detectors

EAS Detectors. The EAS detectors, such as MILAGRO, Tibet-AS and ARGO in the past, and HAWC which is presently in operation, are large arrays of detectors sensitive to charged secondary particles generated in the atmospheric showers. They have a high duty cycle and a large field-of-view, but a relatively low sensitivity. The energy threshold of such detectors is rather large—a shower initiated by a 1 TeV photon typically has its maximum at about 8 km a.s.l.

The principle of operation is the same as the one for the UHE cosmic rays detectors like Auger, i.e., direct sampling of the charged particles in the shower. This can be achieved:

- either using a sparse array of scintillator-based detectors, as for example in Tibet-AS (located at 4100 m a.s.l. to reduce the threshold; for an energy of 100 TeV there are about 50,000 electrons at mountain-top altitudes);
- or by effective covering of the ground, to ensure efficient collection and hence lower the energy threshold.
  - The ARGO-YBJ detector at the Tibet site followed this approach. It was of an array of resistive plate counters. Its energy threshold was in the 0.5 TeV-1 TeV range. The Crab Nebula with a significance of about 5 standard deviations (σ) in 50 days of observation.
  - MILAGRO was a water-Cherenkov instrument located in New Mexico (at an altitude of about 2600 m a.s.l.). It detected the Cherenkov light produced by the secondary particles of the shower when they enter the water pool instrumented with photomultipliers. MILAGRO could detect the Crab Nebula with a significance of about 5σ in 100 days of observation, at a median energy of about 20 TeV.

The energy threshold of EAS detectors is at best in the 0.5–1 TeV range, so they are built to detect UHE photons as well as the most energetic VHE gammas. At such energies fluxes are small and large surfaces of order of 10^4 m^2 are required.

Fig. 4.50 *Left* Layout of the HAWC detector. The location of the different water tanks is shown. *Right* A water tank. Credit: HAWC Collaboration

Concerning the discrimination from the charged cosmic ray background, muon detectors devoted to hadron rejection may be present. Otherwise, it is based on the reconstructed shower shape. The direction of the detected primary particles is computed from the arrival times with an angular precision of about 1°. The calibration can be performed by studying the shadow in the reconstructed directions caused by the Moon. Energy resolution is poor.

Somehow, the past generation EAS detectors were not sensitive enough and just detected a handful of sources. This lesson lead to a new EAS observatory with much larger sensitivity: the High Altitude Water Cherenkov detector HAWC, inaugurated in 2015.

HAWC. HAWC (Fig. 4.50) is a very high-energy gamma-ray observatory located in Mexico at an altitude of 4100 m. It consists of 300 steel tanks of 7.3 m diameter and 4.5 m deep, covering an instrumented area of about 22,000 m^2. Each tank is filled with purified water and contains three PMT of 20 cm diameter, which observe the Cherenkov light emitted in water by superluminal particles in atmospheric air showers. Photons traveling through the water typically undergo Compton scattering or produce an electron–positron pair, also resulting in Cherenkov light emission. This is an advantage of the water Cherenkov technique, as photons constitute a large fraction of the electromagnetic component of an air shower at ground.

HAWC improves the sensitivity for a Crab-like point spectrum by a factor of 15 in comparison to MILAGRO. The sensitivity should be also such to possibly detect gamma-ray burst emissions at high energy.

A future installation in the Northern hemisphere, a hybrid detector called LHAASO to be deployed in China, is currently under discussion.

Cherenkov Telescopes. Most of the experimental results on VHE photons are presently due to Imaging Atmospheric Cherenkov Telescopes (IACTs).

IACTs such as the first successful cosmic detectors, called WHIPPLE, and then the second generation instruments HEGRA and CANGAROO, and presently the third generation H.E.S.S., MAGIC and VERITAS, detect the Cherenkov photons

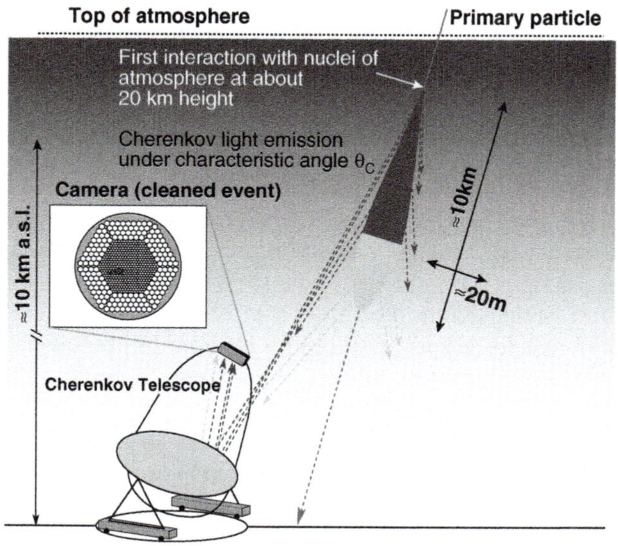

Fig. 4.51 The observational technique adopted by Cherenkov telescopes. Credit: R.M. Wagner, dissertation, MPI Munich 2007

produced in air by charged, locally superluminal particles in atmospheric showers. For reasons explained below, they have a low duty cycle (about 1000–1500 h/year) and a small field-of-view (FoV), but they have a high sensitivity and a low energy threshold.

The observational technique used by the IACTs is to project the Cherenkov light collected in a large optical reflecting surface onto a camera which is basically an array of photomultipliers, with a typical quantum efficiency of about 30 %, in the focal plane of the reflector (see Fig. 4.51). The camera has a typical diameter of about 1 m, and covers a FoV of $5° \times 5°$. The signal collected by the camera is analogically transmitted to trigger systems, similar to the ones used in high-energy physics. The events which pass the trigger levels are sent to the data acquisition system, which typically operates at a frequency of a few hundreds Hz. The typical resolution on the arrival time of a signal on a photomultiplier is better than 1 ns.

The shower has a duration of a few ns (about 2–3) at ground; this duration can be maintained by an isochronous (parabolic) reflector.

Since, as discussed above, about 10 photons per square meter arrive in the light pool for a primary photon of 100 GeV, a light collector of area 100 m^2 is sufficient to detect gamma-ray showers if placed at mountain-top altitudes. Due to the faintness of the signal, data can typically be taken only in moonless time, or with moderate moonlight, and without clouds, which limits the total observation time to some 1200 h/year.

In the GeV-TeV region, the background from charged particles is three orders of magnitude larger than the signal. Hadronic showers, however, have a different topology, being larger and more subject to fluctuations than electromagnetic showers. Showers induced by gamma-rays can thus be separated from the hadronic ones on the basis of the shower shape.

Most of the present identification techniques rely on a technique pioneered by Hillas in the 1980s; the discriminating variables are called "Hillas parameters." The intensity (and area) of the image produced is an indication of the shower energy, while the image orientation is related to the shower direction (photons "point" to emission sources, while hadrons are in first approximation isotropic). The shape of the image is different for events produced by photons and by other particles; this characteristic can be used to reject the background from charged particles (Figs. 4.52 and 4.53).

The time structure of Cherenkov images provides an additional discriminator against the hadronic background, which can be used by isochronous detectors (with parabolic shape) and with a signal integration time smaller than the duration of the shower (i.e., better than 1–2 GHz).

Systems of more than one Cherenkov telescope provide a better background rejection, and a better angular and energy resolution than a single telescope.

There are three large operating IACTs: H.E.S.S., MAGIC, and VERITAS; the first located in the southern hemisphere and the last two in the northern hemisphere.

- The H.E.S.S. observatory (Fig. 4.54) in Namibia is composed by four telescopes with a diameter of 12 m each, working since early 2003. A fifth large telescope, a surface of about 600 m^2, is located in the center; it has been inaugurated in 2012.
- The MAGIC observatory (Fig. 4.55) in the Canary Island of La Palma is a twin telescope system; each parabola has a diameter of 17 m and a reflecting surface of 236 m^2, and it is the largest single-dish Cherenkov telescope in operation.
- VERITAS is constituted by an array of four telescopes with a diameter of 12 m and is located near Tucson, Arizona. It is operational since April 2007.

These instruments are managed by international collaborations of some 150 scientists.

Typical sensitivities of H.E.S.S., MAGIC, and VERITAS are such that a source as luminous as less than 1 % of Crab can be detected at a 5σ significance in 50 hours of observation.

An overlap in the regions of the sky explored by the IACTs allows an almost continuous observation of sources placed at mid-latitude; there is however space for two more installations, one in South America and one (MACE, already scheduled for construction) in India.

Agreements between the Cherenkov telescopes and Fermi allow a balance between competition and cooperation.

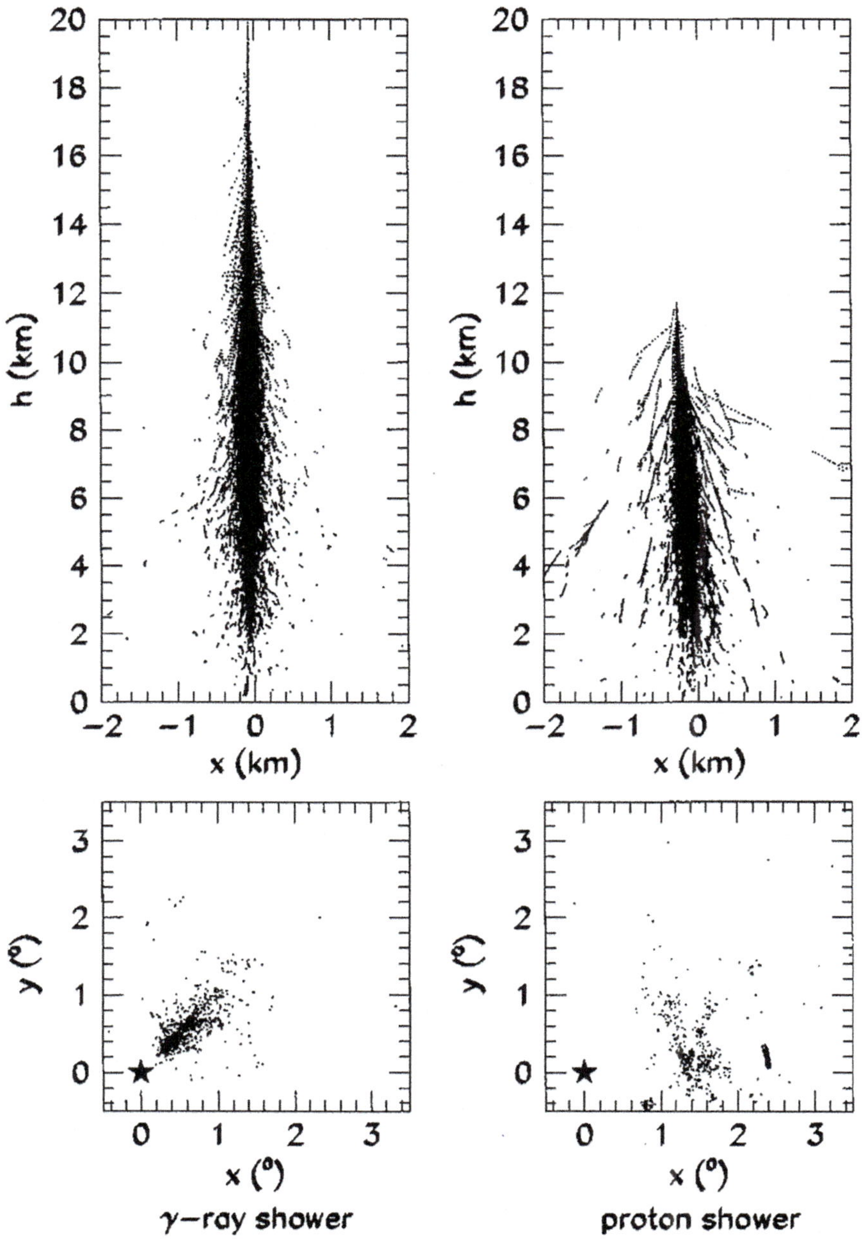

Fig. 4.52 Development of vertical 1 TeV photon (*left*) and proton (*right*) showers in the atmosphere. The *upper panels* show the positions in the atmosphere of all shower electrons above the Cherenkov threshold; the *lower panels* show the resulting Cherenkov images in the focal plane of a 10 m reflecting mirror when the showers fall 100 m from the detector (the center of the focal plane is indicated by a *star*). From C.M. Hoffmann et al., "Gamma-ray astronomy at high energies," Reviews of Modern Physics 71 (1999) 897

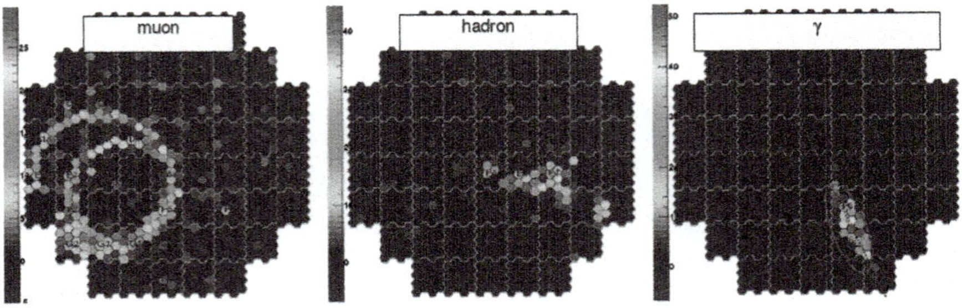

Fig. 4.53 Images from the focal camera of a Cherenkov telescope. The electromagnetic events differ from the hadronic events by several features: the envelope of the electromagnetic shower can be quite well described by an ellipse whereas the important fraction of large transverse momentum particles in hadronic showers will result in a more scattered reconstructed image. Muons are characterized by a conical section. From http://www.isdc.unige.ch/cta/images/outreach/

Fig. 4.54 The H.E.S.S. telescopes. Credit: H.E.S.S. Collaboration

4.5.3.3 Summary of the Performance of Gamma Detectors

A simplified comparison of the characteristics of the Fermi LAT satellite detector, of the IACTs and of the EAS detectors (ground-based), is given in Table 4.5. The sensitivities of the above described high-energy detectors are shown in Fig. 4.56.

A Cherenkov Telescope: MAGIC. We shall now describe in larger detail one of the Cherenkov telescopes: MAGIC. The MAGIC experiment, located at a height of 2200 m a.s.l. on the Canary island of La Palma, is composed by two 17 m diameter IACTs devoted to the observation of very-high-energy (VHE, >30 GeV) gamma rays. The first of the MAGIC telescopes started operations in 2004; the second was built some years later allowing stereo observations since autumn 2009.

Fig. 4.55 One of the MAGIC telescopes. Credit: Robert Wagner, University of Stockholm

Table 4.5 A comparison of the characteristics of Fermi, the IACTs and of the EAS particle detector arrays. Sensitivity computed over one year for Fermi and the EAS, and over 50 h for the IACTs

Quantity	Fermi	IACTs	EAS
Energy range	20 MeV–200 GeV	100 GeV–50 TeV	400 GeV–100 TeV
Energy res.	5–10 %	15–20 %	$\sim$ 50 %
Duty cycle	80 %	15 %	> 90 %
FoV	$4\pi/5$	5 deg $\times$ 5 deg	$4\pi/6$
PSF (deg)	0.1	0.07	0.5
Sensitivity	1 % Crab (1 GeV)	1 % Crab (0.5 TeV)	0.5 Crab (5 TeV)

MAGIC II was constructed like a copy of MAGIC I with a few improvements. Both telescopes are built using the same light-weight carbon-fiber structure. The size of the mirror dish (17 m diameter) and the camera field of view (3.5°) are the same in both telescopes. Each MAGIC camera is composed of 1039 0.1° hexagonal pixels (a hexagonal reflecting cone, called Winston cone, collecting the light onto a photomultiplier).

The reflectors are made of many square mirrors with a curved surface; each mirror is 1 m in size. Their position is corrected thanks to an automatic mirror control (AMC) in such a way that they point to the focal camera.

In both telescopes the signals from the PMT in each pixel are optically transmitted to the counting house where trigger and digitization of the signals take place. The signals of both telescopes are digitized using a frequency of 2 GSample/s.

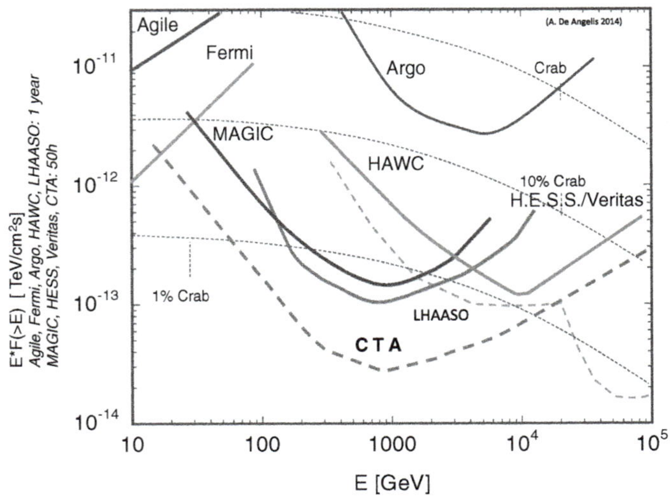

Fig. 4.56 Sensitivities of some present and future HE gamma detectors, measured as the minimum intensity source detectable at 5 σ. The performance for EAS and satellite detector is based on one year of data taking; for Cherenkov telescopes it is based on 50 h of data

Regular observations are performed in stereoscopic mode. Only events that trigger both telescopes are recorded. The trigger condition for the individual telescope is that at least 3 neighboring pixels must be above their pixel threshold (level-0 trigger). The stereo trigger makes a tight time coincidence between both telescopes taking into account the delay due to the relative position of the telescopes and their pointing direction. Although the individual telescope trigger rates are of several kHz, the stereo trigger rate is in the range of 150–200 Hz with just a few Hz being accidental triggers. The lower observational threshold can be reduced to 30 GeV thanks to a dedicated low-energy trigger.

4.5.3.4 Future Detectors for High-Energy Photons

It is difficult to think for this century of an instrument for GeV photons improving substantially the performance of the Fermi LAT: the cost of space missions is such that the size of Fermi cannot be reasonably overcome with present technologies. New satellites in construction (like the Russian-Italian mission GAMMA400 and the Chinese-Italian mission DAMPE) improve some of the aspects of Fermi, e.g., calorimetry.

Improvements are however possible in the sector of VHE gamma astrophysics. VHE gamma astrophysics in the current era has been dominated by Cherenkov telescopes. We know today that the previous generation EAS telescopes were under-dimensioned in relation to the strength of the sources.

The research in the future will push both on EAS and IACT, which have mutual advantages and disadvantages. The sensitivities of the main present and future detectors are illustrated in Fig. 4.56. We are already seen the characteristics of HAWC, which is under completion; a very large Cherenkov Telescope Array (CTA) is also in construction.

CTA. The CTA is a future instrument for VHE gamma astrophysics that is expected to provide an order of magnitude improvement in sensitivity over existing instruments.

An array of tens of telescopes will detect gamma-ray-induced showers over a large area on the ground, increasing the efficiency and the sensitivity, while providing a much larger number of views of each cascade. This will result in both improved angular resolution and better suppression of charged cosmic-ray background events.

In the present design scenario, CTA will be deployed in two sites. The southern hemisphere array will consist of three types of telescopes with different mirror sizes, in order to cover the full energy range. In the northern hemisphere array, the two larger telescope types would be deployed.

- The low energy (the goal is to detect showers starting from an energy of 20 GeV) instrumentation will consist of four 23 m telescopes with a FoV of about 4–5 degrees.
- The medium energy range, from around 100 GeV–1 TeV, will be covered by some 20 telescopes of the 12 m class with a FoV of 6–8 degrees.
- The high-energy instruments, operating above 10 TeV, will consist of a large number (30–40) of small (4–6 m diameter) telescopes with a FoV of around 10 degrees.

The southern CTA will cover about three square kilometers of land with around 60 telescopes that will monitor all the energy ranges in the center of the Milky Way's galactic plane. The northern CTA will cover one square kilometer and be composed of some 30 telescopes. These telescopes will be mostly targeted at extragalactic astronomy.

The telescopes of different sizes will be disposed in concentric circles, the largest in the center (Fig. 4.57). Different modes of operation will be possible: deep field observation; pointing mode; scanning mode.

4.5.4 Neutrino Detection

The energy spectrum of neutrinos interesting for particle and astroparticle physics spans more than 20 orders of magnitude, from the 2 eV of relic neutrinos from the big bang, to the few MeV of the solar neutrinos, to the few GeV of the neutrinos produced by the interaction of cosmic rays with the atmosphere (atmospheric neutrinos), to the region of extremely high energy where the production from astrophysical sources is dominant. We concentrate here on the detection of neutrinos of at least some MeV, and we present some of the most important neutrino detectors operating.

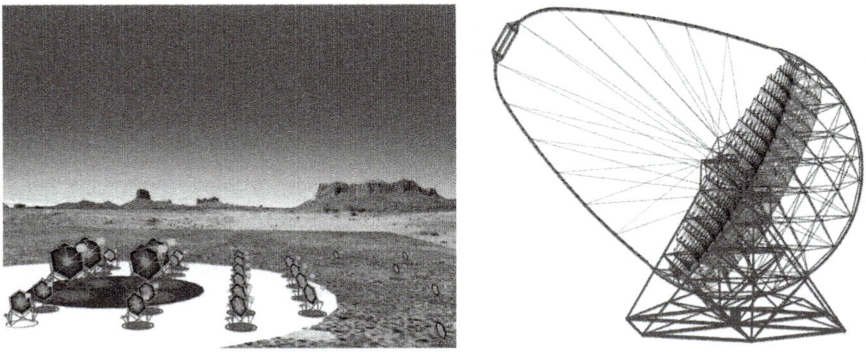

Fig. 4.57 *Left* Possible layout of the CTA. *Right* Project of the large telescope (LST). Credit: CTA Collaboration

Since neutrino cross section is small, it is important that neutrino detectors be located underground or under water to shield from cosmic rays.

4.5.4.1 MeV Neutrinos

Neutrino detectors mostly use the detection of the products of induced β decays. The first setups used a solution of cadmium chloride in water and two scintillation detectors. Antineutrinos with an energy above the 1.8 MeV threshold cause charged inverse beta-decay interactions with the protons in the water, producing a positron which in turn annihilated, generating pairs of photons that could be detected.

Radiochemical chlorine detectors consist instead of a tank filled with a chlorine solution in a fluid. A neutrino converts a ^{37}Cl atom into one of ^{37}Ar; the threshold neutrino energy for this reaction is 0.814 MeV. From time to time the argon atoms are counted to measure the number of radioactive decays.

The first detection of solar neutrinos was achieved using a chlorine detector containing 470 tons of fluid in the former Homestake Mine near Lead, South Dakota. This was first measurement of the deficit of electron neutrinos. For this discovery the leader of the experiment, Ray Davis, won the Nobel prize for physics.[16] A similar detector design, with a lower detection threshold of 0.233 MeV, uses the $Ga \to Ge$ transition.

---

[16]One half of the Nobel prize in physics 2002 was assigned jointly to the US physicist Raymond Davis Jr. (Washington 1914—New York 2006) and to the leader of the Kamiokande collaboration (see later) Masatoshi Koshiba (Aichi, Japan, 1926) "for pioneering contributions to astrophysics, in particular for the detection of cosmic neutrinos."

4.5.4.2 MeV and GeV Neutrinos

Probably the most important results in the sector of MeV to GeV neutrinos in the recent years are due to a Cherenkov-based neutrino detector, Kamiokande, in Japan. We give here a short description of this detector in its present version, called Super-Kamiokande.

The Super-Kamiokande Detector. Super-Kamiokande (often abbreviated to Super-K or SK) is a neutrino observatory located in a mine 1000 m underground under Mount Kamioka near the city of Hida, Japan. The observatory was initially designed to search for proton decay, predicted by several unification theories (see Sect. 7.6.1).

Super-K (Fig. 4.58) consists of a cylindrical tank about 40 m tall and 40 m in diameter containing 50,000 tons of ultra-pure water. The volume is divided by a stainless steel structure into an inner detector (ID) region (33.8 m in diameter and

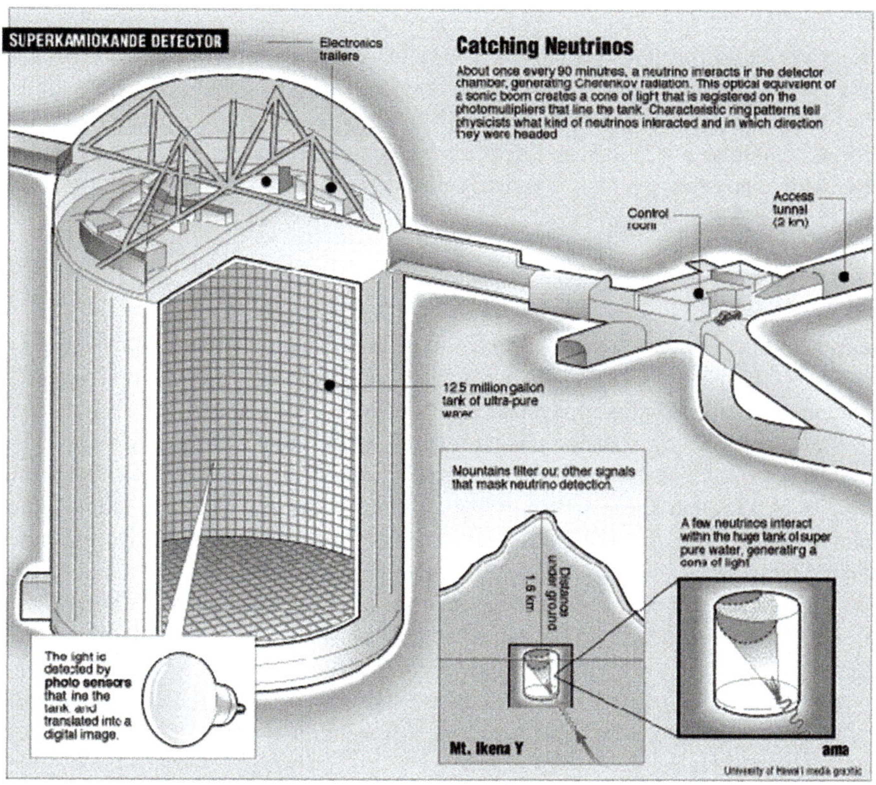

Fig. 4.58 The Super-Kamiokande detector (from the Super-Kamiokande web site)

36.2 m in height) and an outer detector (OD) consisting of the remaining tank volume. Mounted on the structure are about 11000 PMT 50 cm in diameter that face the ID and 2000 20 cm PMT facing the OD.

The interaction of a neutrino with the electrons or nuclei in the water can produce a superluminal charged particle producing Cherenkov radiation, which is projected as a ring on the wall of the detector and recorded by the PMT. The information recorded is the timing and charge information by each PMT, from which one can reconstruct the interaction vertex, the direction and the size of the cone.

The typical threshold for the detection of electron neutrinos is of about 6 MeV. Electrons lose quickly their energy, and, if generated in the ID, are likely to be fully contained (not penetrating inside the OD). Muons instead can penetrate, and the muon events can be partially contained in the detector (or not contained). The threshold for the detection of muon neutrinos is about 2 GeV.

A new detector called Hyper-Kamiokande is envisaged, with a volume 20 times larger than Super-Kamiokande. Construction is expected to start in 2018, and observations might start in 2025.

The SNO Detector. The Sudbury Neutrino Observatory (SNO) uses 1000 tons of heavy water (D_2O) contained in a 12 m diameter vessel surrounded by a cylinder of ordinary water, 22 m in diameter and 34 m high. In addition to the neutrino interactions visible in a detector as SK, the presence of deuterium allows the reaction producing a neutron, which is captured releasing a gamma-ray that can be detected.

4.5.4.3 Very-High-Energy Neutrinos

Very-high-energy neutrinos are expected to be produced in astrophysical objects by the decays of charged pions made in primary cosmic ray interactions with radiation or molecular clouds in astrophysical objects (the so-called hadronic mechanism). As these pions decay, they produce neutrinos with typical energies one order of magnitude smaller than those of the cosmic-ray nucleons—more or less the same energies as photons. These neutrinos can travel long distances undisturbed by either the absorption experienced by high-energy photons or the magnetic deflection experienced by charged particles, making them a unique tracer of cosmic-ray acceleration. Additional sources can be the interactions of cosmic rays with the atmosphere (atmospheric neutrinos), and decays of heavier particles formed by the interaction of cosmic rays with the CMB, or decay of new, heavy particles.

Above an energy of 100 TeV, the expected atmospheric neutrino background falls to the level of one event per year per cubic kilometer, and any harder astrophysical flux can be clearly seen.

The challenge in the field of UHE neutrinos is to build telescopes with good enough sensitivity to see events, since the flux is expected to be lower than the photon flux (the main mechanism for the production of neutrinos, i.e., the hadronic mechanism, is common to photons). This requires instrumenting very large volumes. Efforts to use large quantities of water and ice as detectors are ongoing. Several experiments are completed, operating, or in development using Antarctic ice, the oceans, and lakes, with detection methods including optical and coherent radio detection as well as particle production.

Among the experiments in operation, the largest sensitivity detectors are Baikal NT-200 and IceCube.

Baikal NT-200 Detector. The underwater neutrino telescope NT-200 is located in the Siberian lake Baikal at a depth of approximately 1 km and is taking data since 1998. It consists of 192 optical sensors deployed in eight strings, with a total active volume of 5 million cubic meters. Deployment and maintenance are carried out during winter, when the lake is covered with a thick ice sheet and the sensors can easily be lowered into the water underneath. Data are collected over the whole year and permanently transmitted to the shore over electrical cables.

The IceCube Experiment. IceCube, a cube of 1 km^3 instrumented in the Antarctica ices, has been in operation at the South Pole since 2010 (Fig. 4.59). The telescope views the ice through approximately 5160 optical sensors, deployed in 80 sparse and 6 dense vertical strings, at a depth of 1450–2450 m. At the surface, an air-shower array is coupled to the detector. As the Earth is opaque to UHE neutrinos, detection of extremely high-energy neutrinos must come from neutrinos incident at or above the horizon, while intermediate energy neutrinos are more likely to be seen from below.

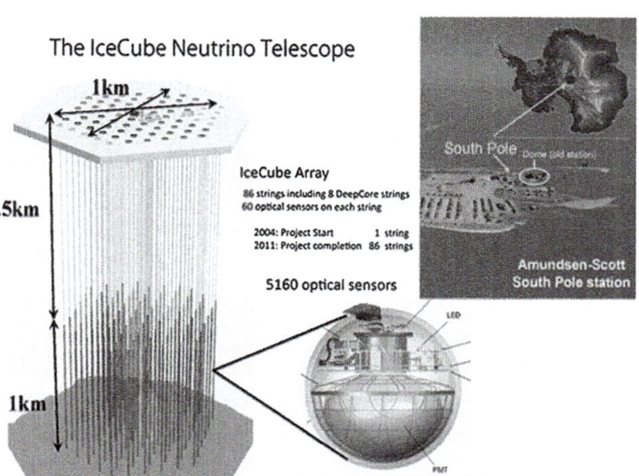

Fig. 4.59 The IceCube detector. Credit: http://www.icehap.chiba-u.jp/en/frontier/

To date some possible high-energy events have been seen, consistent with cosmogenic sources (astrophysical interactions of cosmic rays or production from a cosmic accelerator), with energies most likely between 1 and 10 PeV. It is not excluded, however, although unlikely, that these events are atmospheric neutrinos. IceCube sensitivity has reached the upper range of the high-energy neutrino fluxes predicted in cosmogenic neutrino models.

Km3NeT. A large underwater neutrino detector, Km3NeT, is planned. KM3NeT will host a neutrino telescope with a volume of several cubic kilometers at the bottom of the Mediterranean sea.

This telescope is foreseen to contain of the order of 12,000 pressure-resistant glass spheres attached to about 300 detection units—vertical structures with nearly one kilometer in height. Each glass sphere will contain 31 photomultipliers and connected to shore via a high-bandwidth optical link. At shore, a computer farm will perform the first data filtering in the search for the signal of cosmic neutrinos. KM3NeT builds on the experience of three pilot projects in the Mediterranean sea: the ANTARES detector near Marseille, the NEMO project in Sicily, and the NESTOR project in Greece. ANTARES was completed in 2008 and is the largest neutrino telescope in the northern hemisphere.

4.5.5 Detection of Gravitational Waves

Gravitational waves are predicted to be generated by aspherical concentrations of moving matter, to propagate unimpeded at the speed of light, and to change locally the metric of space-time and, thus, the distance between two free masses, located at fixed spatial coordinates. This effect is however small, and even the most violent astrophysical phenomena (e.g., colliding black holes or neutron stars, collapsing stars) emit gravitational waves which, given the typical distance to the event, are expected to cause relative shifts on distances of only 10^{-20} on Earth. In fact, gravitational waves have been predicted by Albert Einstein in his theory of general relativity over 80 years ago, but only recently technology enabled us to start addressing the problem of their detection.

The main figure of merit for a detector is the space strain ℓ:

$$\Delta L/L \sim \ell$$

where L is the distance between the two masses and ΔL is its variation.

The idea explored first to detect gravitational waves was to detect the elastic energy induced by the compression/relaxation of a metal bar due to the compression/relaxation of distance. Detectors were metal cylinders, and the energy converted to longitudinal oscillations of the bar was measured by piezoelectric transducers. The first large gravitational wave detector, built by Joseph Weber in the early 1960s, was a 1.2 ton aluminum cylindrical bar of 1.5 m length and 61 cm diameter (Fig. 4.60)

Fig. 4.60 Joseph Weber working on his gravitational antenna (1965). From http://www.physics.umd.edu/

working at room temperature and isolated as much as possible from acoustic and ground vibrations. The mechanical oscillation of the bar was translated into electric signals by piezoelectric sensors placed in its surface close to the central region. The detector behaved as a narrow band high$-Q$ (quality factor) mechanical resonator with a central frequency of about 1600 Hz. The attenuation of the oscillations is, in such devices, very small and therefore the bar should oscillate for long periods well after the excitation induced by the gravitational waves. The sensitivity of Weber's gravitational antenna was of the order of $\ell \sim 10^{-16}$ over timescales of 10^{-3} s.

The present bar detectors (ALLEGRO, AURIGA, Nautilus, Explorer, Niobe) reach sensitivities of $\ell \sim 10^{-21}$, thanks to the introduction of cryogenic techniques which allow for a substantial reduction in the thermal noise as well as the use of very performing superconducting sensors. However, their frequency bandwidths remain very narrow ($\sim$tens of Hz) and the resonant frequencies ($\sim$1 kHz) correspond typically to acoustic wavelengths of the order of the detector length. A further increase in sensitivity implies a particular attention to the quantum noise, and thus a considerable increase of the detector mass (bars with hundred tons of mass are being considered).

Nowadays the most sensitive gravitational wave detectors are Michelson-type interferometers with kilometer-long arms and very stable laser beams (see Fig. 4.61). The lengths of the perpendicular arms of the interferometer will be differently modified by the incoming gravitational wave and the interference pattern will change according to the difference induced in the round-trip time between the two arms. These detectors are per nature broadband, being their sensitivity limited only by the smallest time difference they are able to measure. The present and the aimed sensitivities ($\ell \sim 10^{-22} - 10^{-24}$) correspond to interferences over distances many orders of magnitude ($\sim 10^{14} - 10^{16}$) smaller than the dimension of an atom, and thus both the stability of the laser beam and the control of all possible noise sources are critical.

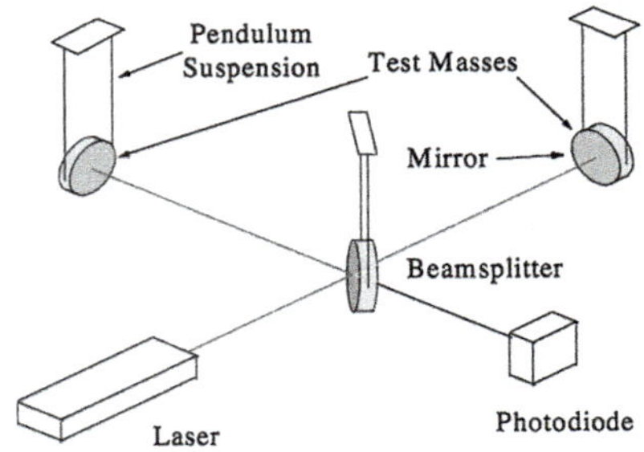

Fig. 4.61 Sketch of a laser interferometer gravitational wave detector. From M. Pitkin et al., "Gravitational Wave Detection by Interferometry (Ground and Space)," http://www.livingreviews.org/lrr-2011-5

These noise sources may be classified as thermal, readout, and seismic:

- The thermal noise is associated both to the Brownian motion of the test masses due to the impact of the surrounding air molecules and to their internal vibrations. To minimize such effects the rest masses should be placed in a high vacuum environment and the frequencies of the intrinsic resonances of the system should be set as far as possible from the target signal frequency band.
- The intrinsic readout noise is due to the fluctuations induced by the quantum nature of the interaction of the laser light beams with the mirrors. The light beams may be modeled as discrete set of photons obeying in their arrival time to the mirror to Poisson statistics. The number of photons measured in a time window has a statistical intrinsic fluctuation ("shot noise") which decreases with the increase of the laser power. On the other hand, the increase of the laser power increases the momentum transfer to the mirrors ("radiation pressure"), which will change the phase of the beams. To minimize such contradictory effects, the tests masses should be as heavy as possible and the Heisenberg uncertainties relations (the quantum limit) carefully handled.
- The seismic noise accounts for all the natural or human-made perturbations comprising a large range of diversified phenomena like earthquakes, environment perturbations or nearby automobile traffic. The measured spectrum of such noise, in a quiet location, decreases with the frequency and imposes already an important sensitivity constraint to the next generation of laser interferometer detectors. To minimize such effects, the test masses are isolated from ground through several attenuation stages characterized by resonance frequencies much lower than the expected signal frequencies. To access lower frequencies (1–10 Hz), the possibility to build large interferometers underground in a low seismic region are being studied. To go further into the 10^{-4}–10^{-1} Hz region it will be necessary to build large arm interferometer in space.

Presently the largest gravitational wave observatories are LIGO and VIRGO. LIGO is built over two sites in the US (at Hanford, Washington, and at Livingston, Louisiana), each one with a 4 km arm interferometer (with another 2 km arm interferometer installed at Hanford), while VIRGO is installed near Pisa, Italy, and consists of a 3 km arm interferometer. In all these interferometers, resonant Fabry-Perot cavities were installed along their arms in a way that the light beams suffer multiple reflections increasing by a large factor the effective arm lengths.

Both LIGO and VIRGO are carrying out extensive upgrade programs (advanced LIGO and advanced VIRGO) and a close collaboration among all of the gravitational waves observatories is now in place. The first direct detection of gravitational waves may be near in the 10–1000 Hz range.

In order to further extend the lever arms, one needs to go to space. In a more distant future (one/two decades) a space observatory may be built extending the detection sensitivity to a much lower frequency range (10^{-4}–10^{-1} Hz). The LISA concept project, comprising three satellite detectors spaced by more than five million kilometers (Fig. 4.62), is under study by ESA and NASA.

The present and expected sensitivities of gravitational wave detectors are summarized in Fig. 4.63.

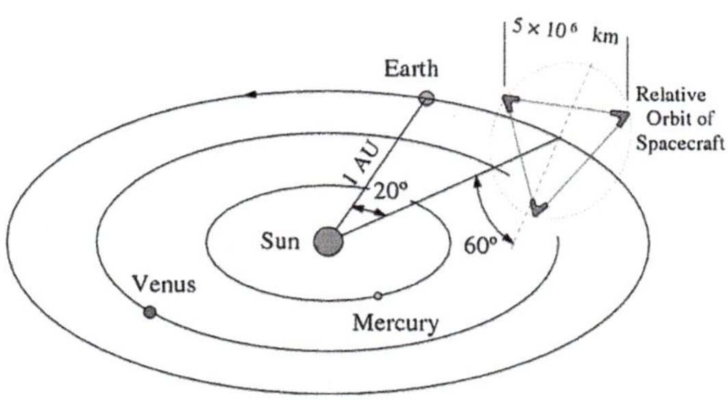

Fig. 4.62 The proposed LISA detector (the size is increased by a factor of 10). From M. Pitkin et al., "Gravitational Wave Detection by Interferometry (Ground and Space)," http://www.livingreviews.org/lrr-2011-5

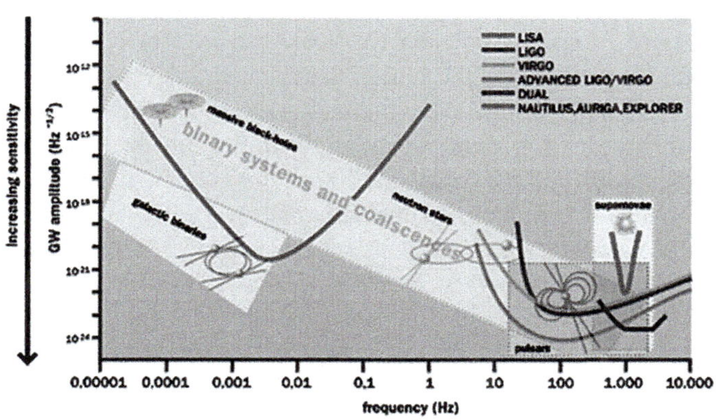

Fig. 4.63 Present and expected sensitivities of gravitational wave detectors (http://www.aspera-eu.org/)

Further Reading

[F4.1] B. Rossi, "High-Energy Particles," Prentice-Hall, New York 1952. Still a fundamental book on particle detection, in particular related to the interaction of particles with matter and to multiplicative showers.
[F4.2] K. Kleinknecht, "Detectors for Particle Radiation" Cambridge University Press 1986.
[F4.3] W.R. Leo, "Techniques for Nuclear and Particle Physics Experiments," Springer Verlag 1994.

Exercises

1. *Compton scattering.* A photon of wavelength λ is scattered off a free electron initially at rest. Let λ' be the wavelength of the photon scattered in the direction θ. Compute: (a) λ' as a function of λ, θ and universal parameters; (b) the kinetic energy of the recoiling electron.
2. *Cherenkov radiation.* A proton with momentum 1.0 GeV/c passes through a gas at high pressure. The index of refraction of the gas can be changed by changing the pressure. Compute: (a) the minimum index of refraction at which the proton will emit Cherenkov radiation; (b) the Cherenkov radiation emission angle when the index of refraction of the gas is 1.6.
3. *Nuclear reactions.* The mean free path of fast neutrons in lead is of the order of 5 cm. What is the total fast neutron cross section in lead?
4. *Photodetectors.* What gain would be required from a photomultiplier in order to resolve the signal produced by three photoelectrons from that due to two or four photoelectrons? Assume that the fluctuations in the signal are described by Poisson statistics, and consider that two peaks can be resolved when their centers are separated by more than the sum of their standard deviations.
5. *Cherenkov counters.* Estimate the minimum length of a gas Cherenkov counter used in the threshold mode to be able to distinguish between pions and kaons with momentum 20 GeV. Assume that 200 photons need to be radiated to ensure a high probability of detection and that radiation covers the whole visible spectrum (neglect the variation with wavelength of the refractive index of the gas).
6. *Electromagnetic calorimeters.* Electromagnetic calorimeters have usually 20 radiation lengths of material. Calculate the thickness (in cm) for a calorimeters made of of BGO, $PbWO_4$ (as in the CMS experiment at LHC), uranium, iron, and lead. Take the radiation lengths from Appendix B or from the Particle Data Book.
7. *Cherenkov telescopes.* Suppose you have a Cherenkov telescope with 7 m diameter, and your camera can detect a signal only when you collect 100 photons from a source. Assuming a global efficiency of 0.1 for the acquisition system (including reflectivity of the surface and quantum efficiency of the PMT), what is

the minimum energy (neglecting the background) that such a system can detect at a height of 2 km a.s.l.?

8. *Cherenkov telescopes*. If a shower is generated by a gamma ray of $E = 1$ TeV penetrating the atmosphere vertically, considering that the radiation length X_0 of air is approximately 37 g/cm^2 and its critical energy E_c is about 88 MeV, calculate the height h_M of the maximum of the shower in the Heitler model and in the Rossi approximation B.
9. *Cherenkov telescopes*. Show that the image of the Cherenkov emission from a muon in the focal plane of a parabolic IACT is a conical section (approximate the Cherenkov angle as a constant).
10. *Colliders*. The luminosity at the Large Electron–Positron Collider (LEP) was determined by measuring the elastic e^+e^- scattering (Bhabha scattering) as its cross section at low angles is well known from QED. In fact, assuming small polar angles, the Bhabha scattering cross section integrated between a polar angle $\theta_{\min}$ and $\theta_{\max}$ is given at first order by

$$\sigma \approx \frac{1040\,\text{nb}}{s\ [\text{GeV}^2]} \left(\frac{1}{\theta_{\max}^2 - \theta_{\min}^2} \right).$$

Determine the luminosity of a run of LEP knowing that this run lasted 4 hours, and the number of identified Bhabha scattering events was 1200 in the polar range of $\theta \in [29; 185]$ mrad. Take into account a detection efficiency of 95 % and a background of 10 % at $\sqrt{s} = m_Z$.
11. *Initial state radiation*. The effective energy of the elastic e^+e^- scattering can be changed by the radiation of a photon by the particles of the beam (initial radiation), which is peaked at very small angles. Supposing that a measured e^+e^- pair has the following transverse momenta: $p_1^t = p_2^t = 5$ GeV, and the radiated photon is collinear with the beam and has an energy of 10 GeV, determine the effective energy of the interaction of the electron and positron in the center of mass, $\sqrt{s_{e^+e^-}}$. Consider that the beam was tuned for $\sqrt{s} = m_Z$.
12. *Energy loss*. In the Pierre Auger Observatory the surface detectors are composed by water Cherenkov tanks 1.2 m high, each containing 12 tons of water. These detectors are able to measure the light produced by charged particles crossing them. Consider one tank crossed by a single vertical muon with an energy of 5 GeV. The refraction index of water is $n \simeq 1.33$ and can be in good approximation considered constant for all the relevant photon wavelengths. Determine the energy lost by ionization and compare it with the energy lost by Cherenkov emission.
13. *Synchrotron radiation*. Consider a circular synchrotron of radius R_0 which is capable of accelerating charged particles up to an energy E_0. Compare the radiation emitted by a proton and an electron and discuss the difficulties to accelerate these particles with this technology.

Chapter 5
Particles and Symmetries

Symmetry simplifies the description of physical phenomena, in such a way that humans can understand them: the Latin word for "understanding," capire, *means "to contain"; and as we are a part of it we cannot contain the full Universe, unless we find a way to reduce its complexity–this is the meaning of symmetry. Symmetry plays a particularly important role in particle physics. The key mathematical framework for symmetry is group theory: symmetry transformations form groups. Although the symmetries of a physical system are not sufficient to fully describe its behavior—for this purpose, one needs a complete dynamical theory—it is possible to use symmetry to discover fundamental properties of a system. Examples of symmetries include space-time symmetries, internal symmetries of particles, and the so-called gage symmetries of field theories.*

5.1 A Zoo of Particles

In the beginning of the thirties of last century, just a few particles were known: the proton and the electron, the charged constituents of the atom; the neutron, which was ensuring the stability of the atomic nuclei; the neutrino (predicted but by then not discovered yet), whose existence was conjectured to guarantee the conservation of energy and momentum in beta decays; and the photon, i.e., the quantum of the electromagnetic field. Then, as we discussed in Chap. 3, new and mostly unexpected particles were discovered in cosmic rays: the positron, the antiparticle of the electron; the muon, the heavy brother of the electron; the charged pion, identified as the particle postulated by Yukawa as the mediator of the strong interaction; the strange "V" particles K^0 and Λ (called "V" from the topology of their decays in bubble chambers).

Human-made accelerators were meanwhile developed (the first linear accelerator by R. Wideroe in 1928; the first Van de Graaf by R.J. Van de Graaf in 1929; the first cyclotron by E.O. Lawrence in 1929; the first multistage voltage multiplier by J.D. Cockcroft and E.T.S. Walton in 1932). The impressive exponential increase of the energy of the beams produced by accelerators, from a few hundred keV in the 1930s

A. De Angelis and M.J.M. Pimenta, *Introduction to Particle and Astroparticle Physics*, Undergraduate Lecture Notes in Physics,
DOI 10.1007/978-88-470-2688-9_5

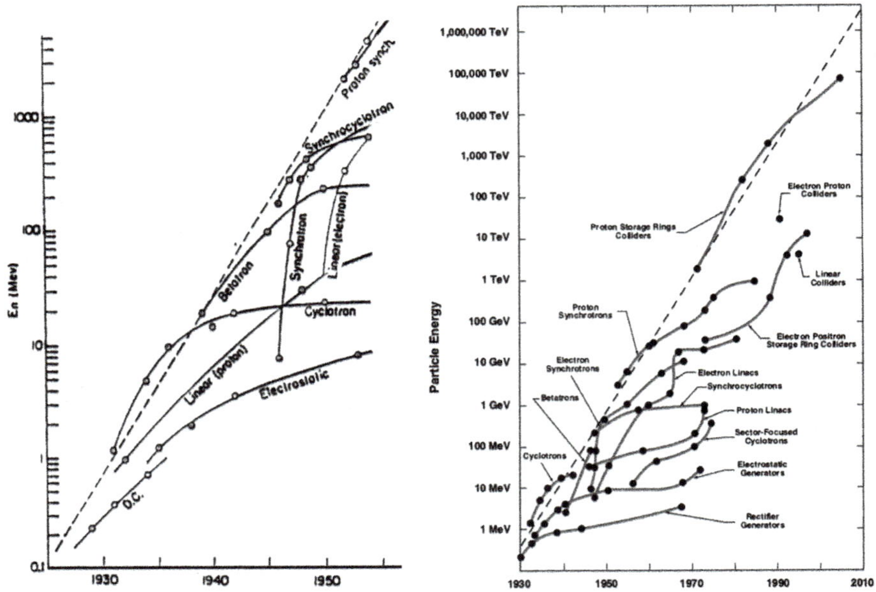

Fig. 5.1 The Livingston plot, representing the maximum energies attained by accelerators as a function of the year: original (*left*) and updated (*right*). For colliders, energies are translated into the laboratory system. Original figures from M.S. Livingston & J.P. Blewett, "Particle Accelerators," MacGraw Hill 1962; A. Chao et al. "2001 Snowmass Accelerator R&D Report," eConf C010630 (2001) MT100

to the GeV in the beginning of the 1950s, was summarized by M. Stanley Livingston in 1954, in the so-called *Livingston plot* (Fig. 5.1). This increase went on along the last fifty years of the twentieth century and just in the most recent years it may have reached a limit ~14 TeV with the Large Hadron Collider (LHC) at CERN.

Accelerators provide the possibility to explore in a systematic way the energy scale up to a few TeV, and thanks to this a huge number of new particles have been discovered. Already in 1950s, many discoveries of particles with different masses, spins, charges, properties took place. Their names almost exhausted the Greek alphabet: these particles were called π, ρ, η, η', ϕ, ω, Δ, Λ, Σ, Ξ ...

Classifications had to be devised to put an order in such a complex zoology. Particles were first classified according to their masses in classes with names inspired, once again, to Greek words: heavy particles like the proton were called *baryons*; light particles like the electron were called *leptons*; and particles with intermediate masses were called *mesons*. The strict meaning of the class names was soon lost, and now we know a lepton, the tau, heavier than the proton. According to the present definition, leptons are those fermions (particles with half-integer spins) that do not interact strongly with the nuclei of atoms, while baryons are the fermions that do. Mesons are bosons (particles with integer spins) subject to strong interactions. Baryons and mesons interact thus mainly via the strong nuclear force and have a common designation of *hadrons*.

The detailed study of these particles shows that there are several conserved quantities in their interactions and decays. The total electric charge, for example, is always conserved, but also the total number of baryons appears to be conserved, and thus the proton, being the lightest baryon, cannot decay (the present experimental limit for the proton lifetime is of about 10^{34} years). Strange particles if they decay by strong interactions give always birth to a lighter strange particle, but the same is not true when they decay via weak interactions. To each (totally or partially) conserved quantity, a new quantum number was associated: baryons, for instance, have "baryonic quantum number" $+1$ (anti-baryons have baryonic quantum number -1, and mesons have baryonic quantum number 0).

As a consequence of baryon number conservation, for example, the most economic way to produce an antiproton in a proton–proton collision is the reaction $pp \rightarrow ppp\bar{p}$. A proton beam with energy above the corresponding kinematic threshold is thus needed to make this process possible. The Bevatron, a proton synchrotron at the Lawrence Berkeley National Laboratory providing beams with energy of 6.2 GeV, was designed for this purpose and started operation in 1954. In the next year, Chamberlain, Segrè, Wiegand, and Ypsilantis announced the discovery of the antiproton; the new particle was identified by measuring its momentum and mass using a spectrometer with a known magnetic field, a Cherenkov detector, and a time-of-flight system. This discovery confirmed that indeed, as predicted by the Dirac equation, to each particle an oppositely charged particle with the same mass and spin corresponds.

The existence of particles and antiparticles is an example of symmetry, and symmetries became more and more present in the characterization of particles and of their interactions. Particle physicists had to study or reinvent group theory in order to profit of the possible simplifications guaranteed by the existence of symmetries.

5.2 Symmetries and Conservation Laws: The Noether Theorem

Being a part of the Universe, it is difficult to imagine how humans can expect to understand it. But we can simplify the representation of the Universe if we find that its description is symmetrical with respect to some transformations. For example, if the representation of the physical world is invariant with respect to translation in space, we can say that the laws of physics are the same everywhere, and this simplifies our description of Nature—quite a lot.

The dynamical description of a system of particles can be classically expressed by the positions q_j of the particles themselves, by their momenta p_j and by the potentials of the interactions within the system. One way to express this is to use the so-called Hamiltonian function

$$H = K + V \tag{5.1}$$

which represents the total energy of the system (K is the term corresponding to the kinetic energies, while V corresponds to the potentials). An equivalent description, using the Lagrangian function, will be discussed in the next chapter.

From the Hamiltonian, the time evolution of the system is obtained by the Hamilton's equations:

$$\frac{dp_j}{dt} = -\frac{\partial H}{\partial q_j} \tag{5.2}$$

$$\frac{dq_j}{dt} = \frac{\partial H}{\partial p_j} \tag{5.3}$$

where

$$p_j = \frac{\partial H}{\partial \dot{q}_j} \,. \tag{5.4}$$

For example, in the case of a single particle in a conservative field in one dimension,

$$H = \frac{p^2}{2m} + V \tag{5.5}$$

and Hamilton's equations become

$$\frac{dp}{dt} = -\frac{dV}{dx} = F \; ; \; \frac{dx}{dt} = \frac{p}{m} \,. \tag{5.6}$$

To the Hamiltonian, there corresponds a quantum mechanical operator, which in the non-relativistic theory can be written as

$$\hat{H} = \frac{\hat{p}^2}{2m} + V \,. \tag{5.7}$$

We shall expand this concept in this chapter and in the next one.

Symmetries of a Hamiltonian with respect to given operations entail conservation laws: this fact is demonstrated by the famous Noether theorem.[1] In the opinion of the authors, this is one of the most elegant theorems in physics.

Let us consider an invariance of the Hamiltonian with respect to a certain transformation—for example, a translation along x. One can write

[1]Emmy Noether (1882–1935) was a German mathematician. After dismissing her original plan to become a teacher in foreign languages, she studied mathematics at the University of Erlangen, where her father was a professor. After graduating in 1907, she worked for seven years as an unpaid assistant (at the time women could not apply for academic positions). In 1915, she joined the University of Göttingen, thanks to an invitation by David Hilbert and Felix Klein, but the faculty did not allow her to receive a salary and she worked four years unpaid. In that time, she published her famous theorem. Finally, Noether moved to the US to take up a college professorship in Philadelphia, where she died at the age of 53.

$$0 = dH = dx\frac{\partial H}{\partial x} = -dx\frac{dp_x}{dt} \Longrightarrow \frac{dp_x}{dt} = 0\,.$$

If the Hamiltonian is invariant with respect to a translation along a coordinate, the momentum associated to this coordinate is constant. And the Hamiltonian of the world should be invariant with respect to translations if the laws of physics are the same everywhere. In a similar way, we could demonstrate that the invariance of an Hamiltonian with respect to time entails energy conservation, and the rotational invariance entails the conservation of angular momentum.

5.3 Symmetries and Groups

A set (*a*,*b*,*c*, ... finite or infinite) of objects or transformations (called hereafter elements of the group) form a group $\mathcal{G}$ if there is an operation (called hereafter product and represented by the symbol $\odot$) between any two of its elements and if:

1. It is closed: the product of any of two elements *a*, *b* is an element *c* of the group

$$c = a \odot b\,.$$

2. There is one and only one identity element: the product of any element *a* by the identity element *e* is the proper element *a*

$$a = a \odot e = e \odot a\,.$$

3. Each element has an inverse: the product of an element *a* by its inverse element *b* (designated also as a^{-1}) is the identity *e*

$$e = a \odot b = b \odot a\,.$$

4. The associativity law holds: the product between three elements *a*, *b*, *c* can be carried out as the product of one element by the product of the other two or as the product of the product of two elements by the other element, keeping however the order of the elements:

$$a \odot b \odot c = a \odot (b \odot c) = (a \odot b) \odot c\,.$$

The commutativity law in the product of any two elements *a*, *b*

$$a \odot b = b \odot a$$

can hold or not. If it does, the group is called *Abelian*.

A symmetry is an invariance over a transformation or a group of transformations. In physics, there are well-known symmetries, some related to fundamental properties of space and time. For instance in mechanics, the description of isolated systems is invariant with respect to space and time translations as well as to space rotations. We have seen that a fundamental theorem, the Noether's theorem, grants that for each symmetry of a system there is a corresponding conservation law and therefore a conserved quantity. The formulation of Noether's theorem in quantum mechanics is particularly elegant.

5.3.1 A Quantum Mechanical View of the Noether's Theorem

Suppose that a physical system is invariant under some transformation U (it can be, for example, the rotation of the coordinate axes). This means that invariance holds when the wave function is subject to the transformation

$$\psi \to \psi' = U\psi .$$

A minimum requirement for U to not change the physical laws is unitarity, since normalization of the wave function should be kept

$$\langle\psi'|\psi'\rangle = \langle U\psi|U\psi\rangle = \langle\psi|U^\dagger U|\psi\rangle = \langle\psi|\psi\rangle \Longrightarrow U^\dagger U = I$$

where I represents the unit operator (which can be, for example, the identity matrix) and $U^\dagger$ is the Hermitian conjugate of U. We shall use in what follows without distinction the terms Hermitian conjugate, conjugate transpose, Hermitian transpose, or adjoint of an $m \times n$ complex matrix A to indicate the $n \times m$ matrix obtained from A by taking the transpose and then taking the complex conjugate of each entry.

For physical predictions to be unchanged by the symmetry transformation, the eigenvalues of the Hamiltonian should be unchanged, i.e., if

$$\hat{H}\psi_i = E_i\psi_i$$

then

$$\hat{H}\psi_i' = E_i\psi_i' .$$

The last equation implies

$$\hat{H}U\psi_i = E_iU\psi_i = UE_i\psi_i = U\hat{H}\psi_i$$

and since the $\{\psi_i\}$, eigenstates of the Hamiltonian, are a complete basis, U commutes with the Hamiltonian:

$$[\hat{H}, U] = \hat{H}U - U\hat{H} = 0 .$$

Thus for every symmetry of a system there is a unitary operator that commutes with the Hamiltonian.

As a consequence, the expectation value of U is constant, since

$$\frac{d}{dt}\langle\psi|U|\psi\rangle = -i\hbar\langle\psi|[U, H]|\psi\rangle = 0. \tag{5.8}$$

5.3.1.1 Continuum Symmetries

Suppose that U is continuous, and consider the infinitesimal transformation

$$U(\epsilon) = I + i\epsilon G$$

(G is called the generator of the transformation U). We shall have, to first order,

$$U^\dagger U = (I - i\epsilon G^\dagger)(I + i\epsilon G) = I + i\epsilon(G - G^\dagger) = I$$

i.e.,

$$G^\dagger = G .$$

The generator of the unitary group is thus Hermitian, and it is thus associated with an observable quantity (its eigenvalues are real). Moreover, it commutes with the Hamiltonian:

$$[H, I + i\epsilon G] = 0 \Longrightarrow [H, G] = 0$$

(trivially $[H, I] = 0$), and since the time evolution of the expectation value of G is given by the equation

$$\frac{d}{dt}\langle G\rangle = i\langle[H, G]\rangle = 0$$

the quantity $\langle G\rangle$ is conserved.

Continuum symmetries in quantum mechanics are thus associated to conservation laws related to the group generators. In the next section, we shall make examples; in particular, we shall see how translational invariance entails momentum conservation (the momentum operator being the generator of space translations).

Let us see now how Noether's theorem can be extended to discrete symmetries.

5.3.1.2 Discrete Symmetries

In case one has a discrete Hermitian operator $\hat{P}$ which commutes with the Hamiltonian

$$[\hat{H}, \hat{P}] = 0$$

and a system is in an eigenstate of the operator itself, whatever transformation due to the Hamiltonian cannot change the eigenvalue.

Let us take for example the parity operator, which will be discussed later. The parity operator, reversing the sign of the space coordinates, is such that

$$\hat{P}^2 = I$$

and thus its eigenvalues are ± 1.

Parity-invariant Hamiltonians represent interaction which conserve parity.

Let us examine now some examples of symmetries.

5.3.2 Some Fundamental Symmetries in Quantum Mechanics

5.3.2.1 Phase Shift Invariance

In quantum mechanics, a system is described by wave functions which are complex functions but only the square of their amplitude has physical meaning: it represents the probability density of the system in a point of the space. A global change of the phase of the wave function leaves the system invariant. Indeed if

$$\psi'(x) = \exp(i\alpha)\,\psi(x)$$

where α is a real number, then

$$\int \psi'^*(x)\,\psi'(x)\,dx = \int \psi^*(x)\exp(-i\alpha)\exp(i\alpha)\,\psi(x)\,dx = \int \psi^*(x)\,\psi(x)\,dx\,.$$

The operator $U = \exp(i\alpha)$ associated to this transformation is a unitary operator: this means that its Hermitian conjugate $U^\dagger = \exp(-i\alpha)$ is equal to its inverse operator U^{-1}.

5.3.2.2 Space Translation Invariance

Due to fundamental properties of space and time, a generic system is invariant with respect to space translations. We can consider, without loss of generality, a translation along x:

$$\psi'(x) = \psi(x + \Delta x) = \psi(x) + \Delta x \frac{\partial}{\partial x}\psi(x) + \frac{1}{2}\Delta x^2 \frac{\partial^2}{\partial x^2}\psi(x) + \cdots \qquad (5.9)$$

which can be written in a symbolic way as

$$\psi'(x) = \exp\left(\Delta x \frac{\partial}{\partial x}\right)\psi(x)\,.$$

The linear momentum operator along x can be expressed as

$$\hat{p}_x = -i\hbar \frac{\partial}{\partial x}$$

and thus

$$\psi'(x) = \exp\left(\frac{i}{\hbar}\Delta x\, \hat{p}_x\right)\; \psi(x)\,.$$

The operator associated to finite space translation Δx along x

$$U_x(\Delta x) = \exp\left(\frac{i}{\hbar}\Delta x\, \hat{p}_x\right)$$

is unitary and it is said to be generated by the momentum operator $\hat{p}_x$.

$\hat{p}_x$ commutes with the Hamiltonian of an isolated system and the associated conserved quantity is the linear momentum p_x.

For an infinitesimal translation δx, just the first terms can be retained and

$$U_x(\delta x) \simeq \left(1 + \frac{i}{\hbar}\delta x\, \hat{p}_x\right)\,. \tag{5.10}$$

5.3.2.3 Rotational Invariance

The same exercise can be done for other transformations which leave a physical system invariant like for instance rotations about an arbitrary axis (this invariance is due to isotropy of space). In the case of a rotation about the z axis, the rotation operator will be

$$R_z(\theta_z) = \exp\left(\frac{i}{\hbar}\theta_z \hat{L}_z\right)$$

where $\hat{L}_z$, the angular momentum operator about the z axis, is the generator of the rotation:

$$\hat{L}_z = -i\hbar\left(x\frac{\partial}{\partial y} - y\frac{\partial}{\partial x}\right)\,.$$

The infinitesimal rotation operator about the z axis will be then

$$R_z(\delta\theta_z) = \left(1 + \frac{i}{\hbar}\delta\theta_z \hat{L}_z\right)\,.$$

It can be shown that in rectangular coordinates $(x,\ y,\ z)$ angular momentum can be replaced in the perturbative development of the rotation by the matrix

$$\hat{L}_z = \hbar\begin{pmatrix} 0 & -i & 0 \\ i & 0 & 0 \\ 0 & 0 & 0 \end{pmatrix} \tag{5.11}$$

which in fact ensures that, expanding the exponential, the usual rotation matrix is recovered:

$$R_z(\theta_z) = \exp\left(\frac{i}{\hbar}\theta_z \hat{L}_z\right) = \begin{pmatrix} \cos\theta_z & \sin\theta_z & 0 \\ -\sin\theta_z & \cos\theta_z & 0 \\ 0 & 0 & 1 \end{pmatrix} . \tag{5.12}$$

A sequence of rotations about axes x, y, and z is described by the product of the corresponding operators:

$$R_x(\theta_x)\ R_y(\theta_y)\ R_z(\theta_z) = \exp\left(\frac{i}{\hbar}\theta_x \hat{L}_x\right) \exp\left(\frac{i}{\hbar}\theta_y \hat{L}_y\right) \exp\left(\frac{i}{\hbar}\theta_z \hat{L}_z\right)$$

which is, as the angular operators do not commute, different from the exponential of the sum of the exponents

$$R_x(\theta_x)\ R_y(\theta_y)\ R_z(\theta_z) \neq \exp\left[\frac{i}{\hbar}\left(\theta_x \hat{L}_x + \theta_y \hat{L}_y + \theta_z \hat{L}_z\right)\right]:$$

the result of a sequence of rotations depends on the order in which the rotations were done.

In fact being $\hat{A}$ and $\hat{B}$ two operators, the following relation holds:

$$\exp\left(\hat{A}+\hat{B}\right) = \exp\left(\frac{1}{2}\left[\hat{A}, \hat{B}\right]\right) \exp(\hat{A})\ \exp(\hat{B}) \tag{5.13}$$

where $\left[\hat{A}, \hat{B}\right]$ is the commutator of the two operators.

The commutators of the angular momentum operators are indeed not zero and given by

$$\left[\hat{L}_x, \hat{L}_y\right] = i\hbar\hat{L}_z\ ;\ \left[\hat{L}_y, \hat{L}_z\right] = i\hbar\hat{L}_x\ ;\ \left[\hat{L}_z, \hat{L}_x\right] = i\hbar\hat{L}_y\ . \tag{5.14}$$

The commutation relations between the generators determine thus the product of the elements of the rotation group and are known as the Lie algebra of the group.

Once a basis is defined, operators can in most cases be associated to matrices; there is a isomorphism between vectors and states, matrices and operators.[2] In the following, whenever there is no ambiguity, we shall identify operators and matrices, and we shall omit when there is no ambiguity the "hat" associated to operators.

[2]Here, we are indeed cutting a long story short; we address the interested readers to a textbook in quantum physics to learn what is behind this fundamental point.

5.3.3 *Unitary Groups and Special Unitary Groups*

Unitary groups U(n) and Special Unitary groups SU(n) of a generic rank n play a central role in particle physics both related to the classification of the elementary particles and to the theories of fundamental interactions.

The unitary group U(n) is the group of unitary complex square matrices with n rows and n columns. A complex $n \times n$ matrix has $2n^2$ parameters, but the unitarity condition ($U^\dagger U = UU^\dagger = 1$) imposes n^2 constrains and thus the number of free parameters is n^2. A particularly important group is the group U(1) which has just one free parameter and so one generator. It corresponds to a phase transformation:

$$U = \exp\left(\frac{i}{\hbar}\alpha\hat{A}\right)$$

where α is a real number and $\hat{A}$ is a Hermitian operator. Relevant cases are $\hat{A}$ being the identity operator (like a global change of the phase of the wave function as discussed above) or an operator associated to a single measurable quantity (like the electric charge or the baryonic number). Noether's theorem ensures that the invariance of the Hamiltonian with respect to such transformation entails the conservation of a corresponding measurable quantity.

The Special Unitary group SU(n) is the group of unitary complex matrices of dimension $n \times n$ and with determinant equal to 1. The number of free parameters and generators of the group is thus $(n^2 - 1)$. Particularly important groups will be the groups SU(2) and SU(3).

5.3.4 *SU(2)*

SU(2) is the group of the spin rotations. The generic SU(2) matrix can be written as

$$\begin{pmatrix} a & b \\ -b^* & a^* \end{pmatrix} \tag{5.15}$$

where a and b are complex numbers and $|a|^2 + |b|^2 = 1$. This group has three free parameters and thus three generators. SU(2) operates in an internal space of particle *spinors*, which in this context are complex two-dimensional vectors introduced to describe the spin $\frac{1}{2}$ (as the electron) polarization states. For instance in a ($|z\rangle$, $|-z\rangle$) basis, the polarization states along z, x, and y can be written as

$$|+z\rangle = \begin{pmatrix} 1 \\ 0 \end{pmatrix}, \quad |-z\rangle = \begin{pmatrix} 0 \\ 1 \end{pmatrix} \tag{5.16}$$

$$|+x\rangle = \frac{1}{\sqrt{2}}\begin{pmatrix}1\\1\end{pmatrix},\ |-x\rangle = \frac{1}{\sqrt{2}}\begin{pmatrix}1\\-1\end{pmatrix} \tag{5.17}$$

$$|+y\rangle = \frac{1}{\sqrt{2}}\begin{pmatrix}1\\i\end{pmatrix},\ |-x\rangle = \frac{1}{\sqrt{2}}\begin{pmatrix}1\\-i\end{pmatrix}. \tag{5.18}$$

The generators, of the group, which are the spin $\frac{1}{2}$ angular momentum operators, can be for example (it is not a unique choice) the Pauli matrices σ_z, σ_x e σ_y

$$\sigma_z = \begin{pmatrix}1 & 0\\0 & -1\end{pmatrix};\ \sigma_x = \begin{pmatrix}0 & 1\\1 & 0\end{pmatrix};\ \sigma_y = \begin{pmatrix}0 & -i\\i & 0\end{pmatrix}, \tag{5.19}$$

being

$$\hat{S}_z = \frac{\hbar}{2}\sigma_z\ ;\ \hat{S}_x = \frac{\hbar}{2}\sigma_x\ ;\ \hat{S}_y = \frac{\hbar}{2}\sigma_y\,. \tag{5.20}$$

The following commutation relations hold:

$$\left[\hat{S}_i, \hat{S}_j\right] = i\hbar\,\varepsilon_{i,j,k}\hat{S}_k \tag{5.21}$$

where $\varepsilon_{i,j,k}$, the Levi-Civita symbol, is the completely antisymmetric tensor which takes the value 1 if i, j, k, is obtained by an even number of permutations of x, y, z, the value -1 if i, j, k, is obtained by a odd number for permutations of x, y, z, and is zero otherwise.

These commutation relations are identical to those discussed above for the generators of space rotations (the normal angular momentum operators) in three dimensions, which form a SO(3) group (group of real orthogonal matrices of dimension 3×3 with determinant equal to 1). SU(2) and SO(3) have thus the same algebra and there is a mapping between the elements of SU(2) and elements of SO(3) which respect the respective group operations. But, while in our example SO(3) operates in the real space transforming particle wave functions, SU(2) operates in the internal space of particle spinors.

The rotation operator in this spinor space around a generic axis i can then be written as

$$U = \exp\left(i\ \frac{\theta_i}{2}\sigma_i\right) \tag{5.22}$$

and in general, defining $\vec{\sigma} = \sigma_x\vec{e}_x + \sigma_y\vec{e}_y + \sigma_z\vec{e}_z$ where the $\vec{e}_{x,y,z}$ are the unit vectors of the coordinate axes, and aligning the rotation axis to a unit vector $\vec{n}$:

$$U = \exp\left(i\ \frac{\theta}{2}\ \vec{n}\cdot\vec{\sigma}\right) = \cos\frac{\theta}{2} + i\sin\frac{\theta}{2}\ \vec{n}\cdot\vec{\sigma} \tag{5.23}$$

where the cosine term is implicitly multiplied by the identity 2×2 matrix.

Spin projection operators do not commute and thus the Heisenberg theorem tells us that the projection of the spin along different axis cannot be measured simultaneously with arbitrary precision. However, $(n-1)$ operators (called the Casimir operators) exist which do commute with all the SU(n) generators. In case of SU(2), there is then just one, and it can be chosen as the square of the total spin:

$$\hat{S}^2 = \hat{S}_x^2 + \hat{S}_y^2 + \hat{S}_z^2$$

which has eigenvalues $s\ (s+1)$ where s is the total spin:

$$\hat{S}^2 \left|s, m_s\right\rangle = \hbar^2\, s\ (s+1) \left|s, m_s\right\rangle .$$

If $\hat{S}_z$ is chosen as the projection operator

$$\hat{S}_z\ |s, m_s\rangle = \hbar\, m_s\ |s, m_s\rangle$$

spin eigenstates can be thus labeled by the eigenvalues m_s of the projection operator along a given axis and by the total spin s. The two other operators $\hat{S}_x$ e $\hat{S}_y$ can be combined forming the so-called raising and lowering operators $\hat{S}_+$ e $\hat{S}_-$.

$$\hat{S}_+ = \hat{S}_x + i\hat{S}_y \tag{5.24}$$

$$\hat{S}_- = \hat{S}_x - i\hat{S}_y \tag{5.25}$$

the names "raising" and "lowering" being justified by the facts that

$$\hat{S}_z\ \hat{S}_+\ |s, m_s\rangle = \hbar\ (m_s + 1)\ \ \hat{S}_+|s, m_s\rangle \tag{5.26}$$

$$\hat{S}_z\ \hat{S}_-\ |s, m_s\rangle = \hbar\ (m_s - 1)\ \ \hat{S}_+|s, m_s\rangle . \tag{5.27}$$

Particles with spins higher than $\frac{1}{2}$ have to be accommodated in SU(n) representations of higher order. For example, for spin 1 particles, the spin projection operator has three eigenvalues $(1, 0, -1)$ and therefore the spin states are described by a three-component vector. In this case, the spin operators are 3×3 matrices. For instance, in a $|z\rangle$ basis:

$$\hat{S}_z = \begin{pmatrix} 1 & 0 & 0 \\ 0 & 0 & 0 \\ 0 & 0 & -1 \end{pmatrix} ; \ \hat{S}_x = \frac{1}{\sqrt{2}} \begin{pmatrix} 0 & -1 & 0 \\ -1 & 0 & 1 \\ 0 & 1 & 0 \end{pmatrix} ; \ \hat{S}_y = \frac{i}{\sqrt{2}} \begin{pmatrix} 0 & 1 & 0 \\ -1 & 0 & -1 \\ 0 & 1 & 0 \end{pmatrix} \tag{5.28}$$

which have the same commutation relation as the 2×2 fundamental representation.

A graphical representation of a spin s multiplet can be done as $(2s+1)$ nodes aligned along an axis m_s (Fig. 5.2).

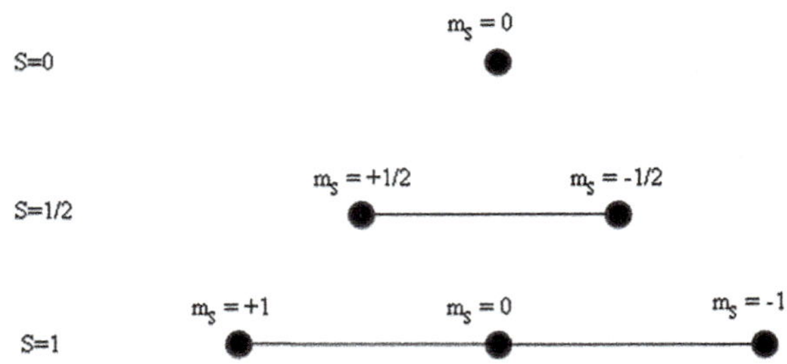

Fig. 5.2 Graphical representation of SU(2) multiplets

5.3.5 *SU(3)*

SU(3) is the group behind the so-called "eightfold way" (the organization of baryons and mesons in octets and decuplets, which was the first successful attempt of classification of hadrons) and behind QCD (quantum chromodynamics, the modern theory of strong interactions). Indeed, SU(3) operates in an internal space of three-dimensional complex vectors and thus can accommodate at the same level rotations among three different elements (flavors u, d, s or colors Red, Green, Blue). The eightfold way will be discussed later in this Chapter, while QCD will be discussed in Chap. 6; here, we present the basics of SU(3).

The elements of SU(3) generalizing SU(2) can be written as

$$U_j = \exp\left(i\frac{\theta_j}{2}\lambda_j\right)$$

where the 3×3 matrices λ_j are the generators. Since for a generic matrix A

$$\det(e^A) = e^{tr(A)} ,$$

the $\{\lambda_j\}$ matrices should be traceless. SU(3) has thus $3^2 - 1 = 8$ traceless and Hermitian generators that in analogy with SU(2) Pauli matrices can be defined as

$$t_i = \frac{\hbar}{2}\lambda_i$$

where λ_i are the Gell-Mann matrices:

$$\lambda_1 = \begin{pmatrix} 0 & 1 & 0 \\ 1 & 0 & 0 \\ 0 & 0 & 0 \end{pmatrix} ; \ \lambda_2 = \begin{pmatrix} 0 & -i & 0 \\ i & 0 & 0 \\ 0 & 0 & 0 \end{pmatrix} ; \ \lambda_3 = \begin{pmatrix} 1 & 0 & 0 \\ 0 & -1 & 0 \\ 0 & 0 & 0 \end{pmatrix}$$

$$\lambda_4 = \begin{pmatrix} 0 & 0 & 1 \\ 0 & 0 & 0 \\ 1 & 0 & 0 \end{pmatrix} ; \ \lambda_5 = \begin{pmatrix} 0 & 0 & -i \\ 0 & 0 & 0 \\ i & 0 & 0 \end{pmatrix}$$

$$\lambda_6 = \begin{pmatrix} 0 & 0 & 0 \\ 0 & 0 & 1 \\ 0 & 1 & 0 \end{pmatrix} \; ; \; \lambda_7 = \begin{pmatrix} 0 & 0 & 0 \\ 0 & 0 & -i \\ 0 & i & 0 \end{pmatrix} \; ; \; \lambda_8 = \frac{1}{\sqrt{3}} \begin{pmatrix} 1 & 0 & 0 \\ 0 & 1 & 0 \\ 0 & 0 & -2 \end{pmatrix}$$

and have the following commutation relations:

$$\left(\lambda_i, \lambda_j\right) = 2i \sum_k f_{ijk} \lambda_k$$

where the nonzero structure constants f_{ijk} are

ijk	123	147	156	246	257	345	367	458	678
f_{ijk}	1	$1/2$	$-1/2$	$1/2$	$1/2$	$1/2$	$-1/2$	$\sqrt{3}/2$	$\sqrt{3}/2$

SU(3) contains three SU(2) subgroups corresponding to the different rotations between any pair of the three group elements. Note for instance that the first three λ matrices are built as the extension of the SU(2) generators we have discussed before.

The generators λ_3 and λ_8 commute and thus they have common eigenstates; their eigenvalues can thus be used to label the eigenstates. We call the corresponding quantum numbers "third isospin component" I_3 and "hypercharge" Y quantum numbers. The other operators can be, in a similar way as it was done for SU(2), combined two by two to form raising and lowering (step) operators. Then the *standard* SU(3) generators will be defined as:

$$\hat{I}_3 = t_3 \; ; \; \hat{Y} = \frac{2}{\sqrt{3}} t_8$$

$$\begin{aligned} \hat{I}_\pm &= t_1 \pm i\, t_2 \\ \hat{V}_\pm &= t_4 \pm i\, t_5 \\ \hat{U}_\pm &= t_6 \pm i\, t_7\,. \end{aligned}$$

The step operators act as follows:

- $\hat{I}_\pm$ leaves Y unchanged and changes $\hat{I}_3$ by ± 1,
- $\hat{V}_\pm$ changes Y by $+1$ and changes $\hat{I}_3$ by $\pm 1/2$,
- $\hat{U}_\pm$ changes Y by -1 and changes $\hat{I}_3$ by $\pm 1/2$.

Graphically, these operations can be represented as vectors in a (I_3,Y) space (Fig. 5.3). SU(3) multiplets form then in this space plane figures (triangles, octagons, ...) as it will be discussed later on (Figs. 5.7 and 5.8).

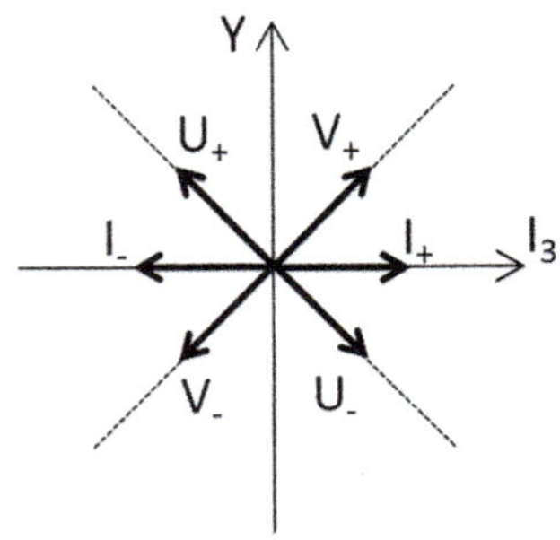

Fig. 5.3 Graphical representation of the SU(3) step operators

5.3.6 *Discrete Symmetries: Parity, Charge Conjugation, and Time Reversal*

Let us examine now in larger detail three fundamental discrete symmetries: parity, charge conjugation, and time reversal.

5.3.6.1 Parity

We have already introduced the parity transformation (sometimes called wrongly "mirror reflection"), which reverses all spatial coordinates:

$$\vec{x} \rightarrow \vec{x'} = -\vec{x} \ .$$

A *vector* (for instance, the position vector, the linear momentum or the electric field) will be inverted under parity transformation, while the cross product of two vectors (like the angular momentum or the magnetic field) will not be changed. The latter is called *pseudo—(or axial) vector*. The internal product of two vectors is a *scalar* and is invariant under parity transformation but the internal product of a vector and a pseudo-vector changes sign under parity transformation and it is called a *pseudo-scalar*.

The application of the parity operator $\hat{P}$ once and twice to a wave function leads to

$$\hat{P}\,\psi\left(\vec{x}\right) = \psi\left(-\vec{x}\right) = \lambda_P\,\psi\left(\vec{x}\right)$$

$$\hat{P}^2\,\psi\left(\vec{x}\right) = {\lambda_P}^2\,\psi\left(\vec{x}\right) = \psi\left(\vec{x}\right)$$

implying that the eigenvalues of the $\hat{P}$ operator are $\lambda_P = \pm 1$. The parity group has just two elements: $\hat{P}$ and the identity operator $\hat{I}$. It is thus Hermitian, and a measurable quantity, *parity*, can be associated to its eigenvectors, which is a legal quantum number.

Electromagnetic and strong interactions appear to be invariant under parity transformations (and therefore the parity quantum number is conserved in these interactions) but, as it will be discussed in the next chapter, weak interactions, with the surprise of most of physicists in the 1950s, are not.

For any system bound by a central potential, $V(r)$, the spatial part of the wave function can be written as the product of a radial and an angular part, with the angular part described by spherical harmonics:

$$\psi(r, \theta, \phi) = R(r) Y^l_m(\theta, \phi) .$$

The parity operator in polar coordinates changes from θ to $\pi - \theta$ and ϕ to $\pi + \phi$. One can easily demonstrate that

$$P Y^l_m = (-1)^l Y^l_m$$

Elementary particles are (with good approximation) eigenstates of $\hat{P}$, since a generic free-particle Hamiltonian is with good approximation parity invariant, and an "intrinsic" parity is assigned to each particle. Fermions and antifermions have opposite parities; bosons and anti-bosons have the same parity.

The photon has a negative parity: this can be seen by the fact that the basic atomic transition is characterized by the emission of a photon and a change of orbital angular momentum by one unit.

Example: Experimental Determination of Parity for the Pion. By convention, we define that protons and neutrons have positive intrinsic parity. The negative parity of pions can be determined by assuming parity and angular momentum conservation in the capture at rest of a π^- by a deuterium nucleus producing two neutrons in the final state ($\pi^- d \rightarrow nn$). The parity of a system of two particles is the product of the parities of the two particles multiplied by a $(-1)^l$ factor where l is the orbital angular momentum of the system ($l = 0$ is the ground state). In the case of the nn system discussed above, $l = 1$ and thus the final state parity is -1. All the vector bosons have a negative parity, while the axial vector bosons have a positive parity.

Combining Parity in a set of Particles. Parity is a multiplicative quantum number contrary to, for instance, electric charge which is additive. In fact while discrete symmetry groups are usually defined directly by the corresponding operators, continuous symmetry groups are, as it was seen, associated to the exponentiation of generators,

$$U = \exp\left(\frac{i}{\hbar}\alpha \hat{Q}\right) .$$

5.3.6.2 Charge Conjugation

Charge conjugation reverses the sign of all "internal" quantum numbers (electric charge, baryon number, strangeness, ...) keeping the values of mass, momentum, energy, and spin. It transforms a particle in its own antiparticle. Applying the charge conjugation operator $\hat{C}$ twice brings the state back to its original state, as in the case of parity. The eigenvalues of $\hat{C}$ are again $\lambda_C = \pm 1$ but most of the elementary particles are not eigenstates of $\hat{C}$ (particle and antiparticle are not usually the same); this is trivial for electrically charged particles.

Once again electromagnetic and strong interactions appear to be invariant under charge conjugation but weak interactions are not (they are "almost" invariant to the product $\hat{C}\,\hat{P}$ as it will discussed later on).

Electric charge changes sign under charge conjugation and so do electric and magnetic field. The photon has thus a negative $\hat{C}$ eigenvalue. The neutral pion π^0 decays into two photons: its $\hat{C}$ eigenvalue is positive. Now you should be able to answer the question: is the decay $\pi^0 \rightarrow \gamma\gamma\gamma$ possible?

5.3.6.3 Time Reversal and CPT

Time reversal inverts the time coordinate:

$$t \rightarrow t' = -t \,.$$

Physical laws that are invariant to such transformation have no preferred time direction. Going back in the past would be as possible as going further in the future. Although we have compulsory evidence in our lives that this is not generally the case, the Hamiltonians of fundamental interactions were believed to exhibit such invariance. On the other hand, in relativistic quantum field theory in flat space-time geometry, it has been demonstrated (CPT theorem), under very generic assumptions, that any quantum theory is invariant under the combined action of charge conjugation (C) , space reversal (parity P), and time reversal (T). This is the case of the Standard Model of particle physics. As a consequence of the CPT theorem, particles and antiparticles must have identical masses and lifetimes. Stringent tests have been performed being the best the limit at 90 % CL on the mass difference between the K^0 and the $\overline{K}^0$:

$$\left| \frac{m_{K^0} - m_{\overline{K}^0}}{m_{K^0}} \right| < 0.6 \times 10^{-18} \,. \tag{5.29}$$

So far CPT remains both experimentally and theoretically an exact symmetry. This implies that any violation of one of the individual symmetries (C, P, T) must be compensated by corresponding violation(s) in at least one of the others symmetries. In the late 1950s, and early 1960s, it was found that P and C are individually violated in weak interactions and that, beyond all the expectations, their combined action (CP) is also violated in particular particle systems (see Sect. 6.3.8). Therefore, T should be also violated in such systems. Indeed, the T violation parity violation has been recently detected in the B meson sector. The arrow of time is also manifest at the level of fundamental particles.

5.3.7 *Isospin*

In 1932, J. Chadwick discovered the neutron after more than 10 years of intensive experimental searches following the observation by Rutherford that to explain the mass and charges of all atoms, and apart hydrogen, the nucleus should consist of protons and of neutral bound states of electrons and protons. The particle discovered was not a bound state of electron and proton—meanwhile, the uncertainty relations demonstrated by Heisenberg in 1927 had indeed forbidden it. The neutron was indeed a particle like the proton with almost the same mass ($m_p \simeq 939.57$ MeV/c^2, $m_n \simeq 938.28$ MeV/c^2), the same behavior with respect to nuclear interaction, but with no electric charge. It was the neutral "brother" of the proton.

Soon after neutron discovery, Heisenberg proposed to regard proton and neutron as two states of a single particle later on called the nucleon. The formalism was borrowed from the Pauli spin theory and Wigner, in 1937, called "isospin" symmetry this new internal symmetry with respect to rotations in the space defined by the vectors $(p, 0)$ and $(0, n)$. Strong interactions should be invariant with respect to an internal SU(2) symmetry group, the nucleons would have isospin $I = 1/2$, and their states would be described by isospin spinors. By convention, the proton is identified with the isospin-up ($I_3 = +1/2$) projection and the neutron with the isospin-down ($I_3 = -1/2$) projection.

As we have discussed in Chap. 3, Yukawa postulated in 1935 that the short-range nuclear interaction might be explained by the exchange of a massive meson, the pion. The charge independence of nuclear interactions suggested later on that three pions (π^+, π^-, π^0) should exist. Nuclear interaction could thus be seen as an interaction between the nucleon isospin doublet ($I = 1/2$) and a isovector ($I = 1$) triplet of pions. Three spin 0 and isospin 1 pions (π^+ with $I_3 = +1$, π^0 with $I_3 = 0$, π^- with $I_3 = -1$) with almost the same masses ($m_{\pi^\pm} \simeq 139.6$ MeV/c^2, $m_{\pi^0} \simeq 135.0$ MeV/c^2) were indeed discovered in the late 1940s and in the beginning of the 1950s. The isospin theory of nuclear interactions was established.

5.3.7.1 The Isospin of a System of Particles

The isospin of a system of particles can be expressed as a function of the isospin of the individual particles using the same addition rules valid for the sum of ordinary spins or angular momenta.

In a system of two particles, the projection of the total spin of the system on an axis (conventionally assumed to be the z axis) is just the sum of the projections of the individual spins, $m_s = m_{s1} + m_{s2}$, while the total spin s can take values from $|s_1 - s_2|$ to $s_1 + s_2$. The weights of the different final states in the total amplitude are given by the squares of the so-called Clebsch–Gordan coefficients $C_{sm,s1m1,s2m2}$. The calculation of these coefficients is not relevant for the purpose of this book; they can be found in tables, an example being given in Fig. 5.4.

For example, the addition of two spin $\frac{1}{2}$ particles gives

$$
\begin{aligned}
|\frac{1}{2},\frac{1}{2}\rangle + |\frac{1}{2},\frac{1}{2}\rangle &= |1,1\rangle \\
|\frac{1}{2},\frac{1}{2}\rangle + |\frac{1}{2},-\frac{1}{2}\rangle &= \frac{1}{\sqrt{2}}|1,0\rangle + \frac{1}{\sqrt{2}}|0,0\rangle \\
|\frac{1}{2},-\frac{1}{2}\rangle + |\frac{1}{2},\frac{1}{2}\rangle &= \frac{1}{\sqrt{2}}|1,0\rangle - \frac{1}{\sqrt{2}}|0,0\rangle \\
|\frac{1}{2},-\frac{1}{2}\rangle + |\frac{1}{2},-\frac{1}{2}\rangle &= |1,-1\rangle
\end{aligned}
\tag{5.30}
$$

The final states can thus be organized in a symmetric triplet of total spin 1

$$
\begin{aligned}
|1,1\rangle &= |\frac{1}{2},\frac{1}{2}\rangle|\frac{1}{2},\frac{1}{2}\rangle \\
|1,0\rangle &= \frac{1}{\sqrt{2}}|\frac{1}{2},\frac{1}{2}\rangle|\frac{1}{2},-\frac{1}{2}\rangle + \frac{1}{\sqrt{2}}|\frac{1}{2},-\frac{1}{2}\rangle|\frac{1}{2},-\frac{1}{2}\rangle \\
|1,-1\rangle &= |\frac{1}{2},-\frac{1}{2}\rangle|\frac{1}{2},-\frac{1}{2}\rangle
\end{aligned}
\tag{5.31}
$$

and in an antisymmetric singlet of total spin 0

$$
|0,0\rangle = \frac{1}{\sqrt{2}}|\frac{1}{2},\frac{1}{2}\rangle|\frac{1}{2},-\frac{1}{2}\rangle - \frac{1}{\sqrt{2}}|\frac{1}{2},-\frac{1}{2}\rangle|\frac{1}{2},-\frac{1}{2}\rangle . \tag{5.32}
$$

In the language of group theory, the cross product of two SU(2) doublets gives a triplet and a singlet:

$$
2 \otimes 2 = 3 \oplus 1 . \tag{5.33}
$$

5.3.7.2 Isospin and Cross Section

Strong interactions are invariant under SU(2) rotations in the internal isospin space and according to Noether‘s theorem, total isospin I is conserved in such interactions. The transition amplitudes between initial and final states are a function of the isospin I and can be labeled as $\mathcal{M}_I$.

Let us consider the inelastic collision of two nucleons giving a deuterium nucleus and a pion. Three channels are considered:

1. $p + p \to d + \pi^+$
2. $p + n \to d + \pi^0$
3. $n + n \to d + \pi^-$.

The deuteron is a *pn* bound state and must have isospin I = 0; otherwise, the bound states *pp* and *nn* should exist (experimentally, they do not exist). The isospin quan-

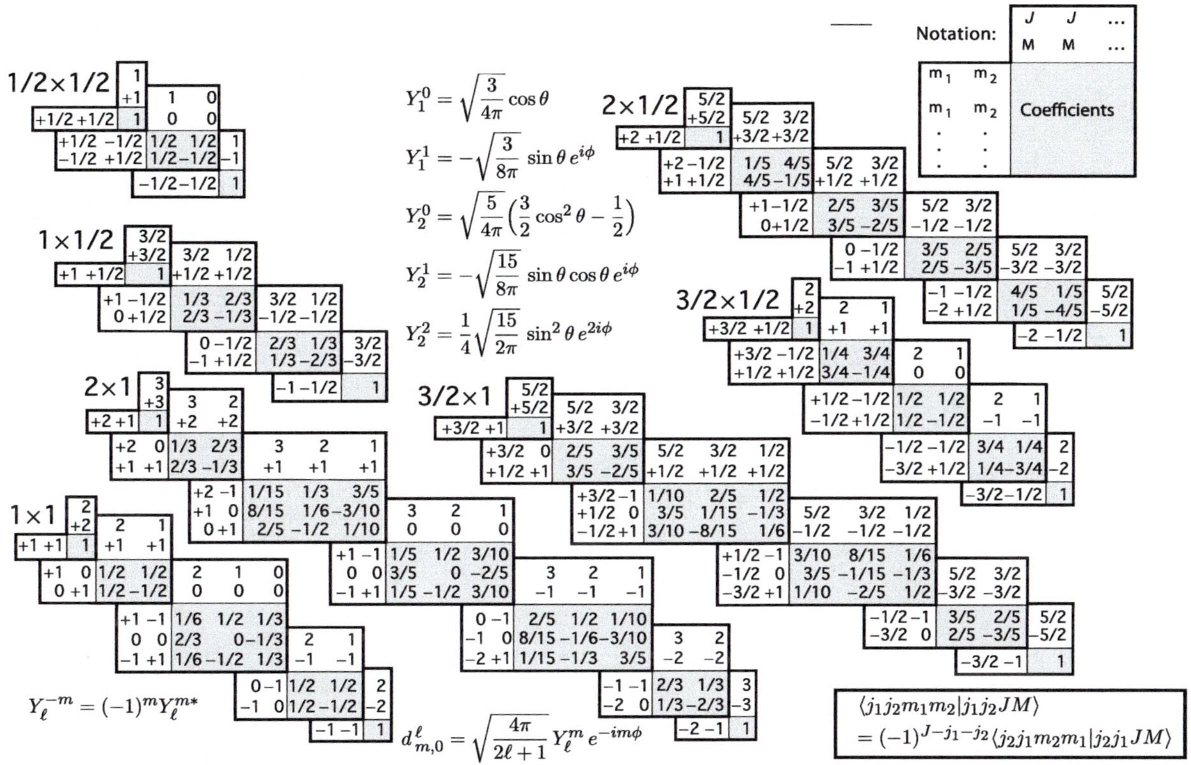

Fig. 5.4 Clebsch–Gordan coefficients and spherical harmonics. From K.A. Olive et al. (Particle Data Group), Chin. Phys. C 38 (2014) 090001. *Note* A square-root sign is to be understood over *every* coefficient, *e.g.*, for $-8/15$ read $-\sqrt{8/15}$

tum numbers $|I,\ I_3\rangle$ of the final states are thus those of the π, which means $|1,1\rangle$, $|1,0\rangle$, $|1,-1\rangle$, respectively. The isospins of the initial states follow the rules of the addition of two isospin 1/2 states discussed above, and are, respectively, $|1,1\rangle$, $\frac{1}{\sqrt{2}}|1,0\rangle + \frac{1}{\sqrt{2}}|0,0\rangle$ and $|1,-1\rangle$. As the final state is a pure I = 1 state, only the transition amplitude corresponding to I = 1 is possible. The cross section (proportional to the square of the sum of the scattering amplitudes) for the reaction $p+n \to d+\pi^0$ should then be half of each of the cross sections of any of the other reactions.

Let us consider now the $\pi^+ p$ and $\pi^- p$ collisions:

1. $\pi^+ + p \to \pi^+ + p$
2. $\pi^- + p \to \pi^- + p$
3. $\pi^- + p \to \pi^0 + n$.

Using the Clebsch–Gordan tables, the isospin decomposition of the initial and final states are

$$\pi^+ + p : |1,1\rangle + |\frac{1}{2},\frac{1}{2}\rangle = |\frac{3}{2},\frac{3}{2}\rangle$$

$$\pi^- + p : |1,-1\rangle + |\frac{1}{2},\frac{1}{2}\rangle = \sqrt{\frac{1}{3}}|\frac{3}{2},-\frac{1}{2}\rangle - \sqrt{\frac{2}{3}}|\frac{1}{2},-\frac{1}{2}\rangle$$

$$\pi^0 + n : |1,1\rangle + |\frac{1}{2},\frac{1}{2}\rangle = \sqrt{\frac{2}{3}}|\frac{3}{2},-\frac{1}{2}\rangle + \sqrt{\frac{1}{3}}|\frac{1}{2},-\frac{1}{2}\rangle .$$

Therefore, there are two possible transition amplitudes $\mathcal{M}_{1/2}$ e $\mathcal{M}_{3/2}$ corresponding to $I = \frac{1}{2}$ and $I = \frac{3}{2}$, respectively, and

$$M(\pi^+ p \to \pi^+ p) \propto M_{3/2}$$

$$M(\pi^- p \to \pi^- p) \propto \frac{1}{3}M_{3/2} + \frac{2}{3}M_{1/2}$$

$$M(\pi^- p \to \pi^0 p) \propto \frac{\sqrt{2}}{3}M_{3/2} - \frac{\sqrt{2}}{3}M_{1/2}$$

Experimentally, in 1951, the group led by Fermi in Chicago discovered in the $\pi^+ p$ elastic scattering channel an unexpected and dramatic increase at center-of-mass energies of 1232 MeV (Fig. 5.5). Such increase was soon after interpreted by Keith Brueckner (Fermi was not convinced) as evidence that the pion and the proton form at that energy a short-lived bound state with isospin number $I = \frac{3}{2}$. Indeed, whenever $M_{3/2} \gg M_{1/2}$,

$$\frac{\sigma(\pi^+ p \to \pi^+ p)}{\sigma(\pi^- p \to \pi^- p) + \sigma(\pi^- p \to \pi^0 p)} \sim \frac{(M_{3/2})^2}{\frac{1}{9}(M_{3/2})^2 + \frac{2}{9}(M_{3/2})^2} \sim 3$$

in agreement with the measured value of such ratio as shown in Fig. 5.5.

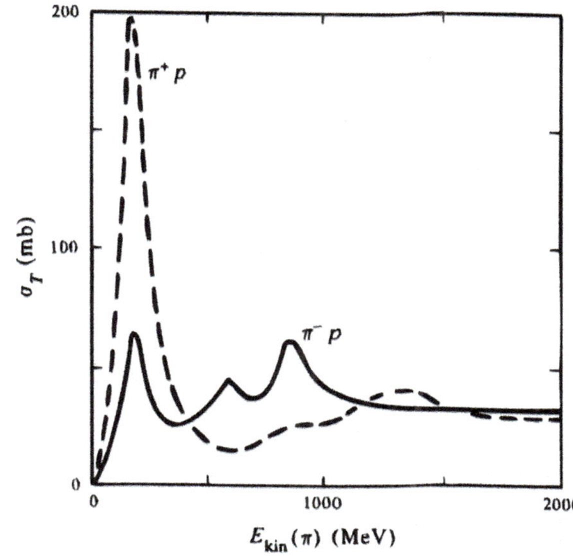

Fig. 5.5 Total cross section for the collision of positive and negative pions with protons as a function of the pion kinetic energy. Credit: E.M Henley and A. Garcia, Subatomic physics, World Scientific 2007

Fig. 5.6 Dharma wheel used to represent the Buddhist "eightfold path." By Shazz, Esteban (own work) [GFDL http://www.gnu.org/copyleft/fdl.html], via Wikimedia Commons

This resonance is now called the Δ; being a $I = \frac{3}{2}$ state it has, as expected, four projections $\left(\text{called } \Delta^{++}, \Delta^{+}, \Delta^{0}, \Delta^{-}\right)$.

5.3.8 The Eightfold Way

The "eightfold way" is the name Murray Gell-Mann, inspired by the noble Eightfold Path from the Buddhism (Fig. 5.6), gave to the classification of mesons and baryons proposed independently by him and by Yuval Ne'eman in the early 1960s.

As discussed in Chap. 3, strange particles had been discovered in the late 1940s in cosmic rays, and later abundantly produced in early accelerator experiments in the beginning of the 1950s. These particles were indeed strange considering the knowledge at that time: they have large masses and they are produced in pairs with large cross sections but they have large lifetimes as compared with what expected

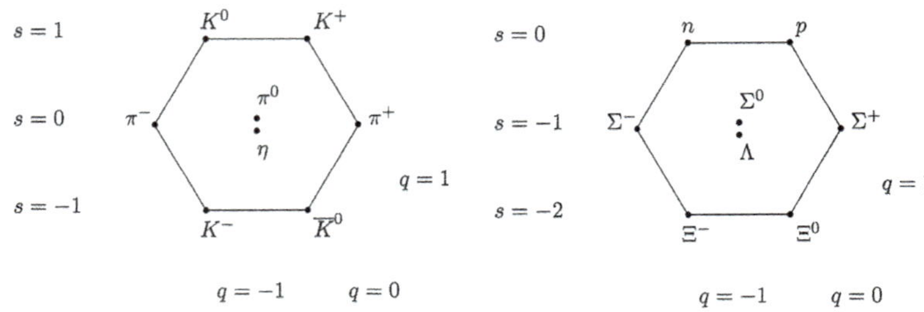

Fig. 5.7 Fundamental meson and baryon octets: on the *left* spin 0, parity −1 (pseudo-scalar mesons); on the *right* the spin 1/2 baryons. The I_3 axis is the abscissa, while the Y axis is the ordinate

for nuclear resonances. Their production is ruled by strong interactions while they decay weakly. A new quantum number, the *strangeness*, was assigned in 1953 to these particles by Nakano and Nishijima and, independently, by Gell-Mann. By convention, the positive K mesons (kaons) have strangeness $+1$, while the Λ baryons have strangeness -1. "Ordinary" (non-strange) particles (proton, neutron, pions, ...) have strangeness 0.

Strangeness is conserved in the associated production of kaons and lambdas, as for instance, in

$$\pi^+ \, n \; \rightarrow \; K^+ \, \Lambda \; ; \; \pi^- \, p \; \rightarrow \; K^0 \, \Lambda \tag{5.34}$$

but not conserved in strange particle decays, e.g.,

$$\Lambda \; \rightarrow \; \pi^- p \; ; \;\; K^0 \; \rightarrow \pi^- \, \pi^+ \, . \tag{5.35}$$

Strange particles can also be grouped in isospin multiplets but the analogy with strangeless particles is not straightforward. Pions are grouped in an isospin triplet being the π^+ the antiparticle of the π^- and the π^0 its own antiparticle. For kaons, the existence of the strangeness quantum number implies that there are four different states which are organized in two isospin doublets: $\left(K^+, K^0\right)$ and $\left(K^-, \overline{K}^0\right)$ having, respectively, strangeness S $= +1$ and S $= -1$ and being the antiparticles of each other.

Gell-Mann and Nishijima noticed also that there is an empirical relation between the electric charge, the third component of isospin I_3, and a new quantum number, the hypercharge $Y = B + S$, defined as the sum of the baryonic number B (being $B = 0$ for mesons, $B = 1$ for baryons and $B = -1$ for anti-baryons) and strangeness S:

$$Q = I_3 + \frac{1}{2} \, Y \, .$$

Mesons and baryons with the same spin and parity and with similar masses can be grouped forming geometrical hexagons and triangles in the $(I_3, \, Y)$ plane (Fig. 5.7).

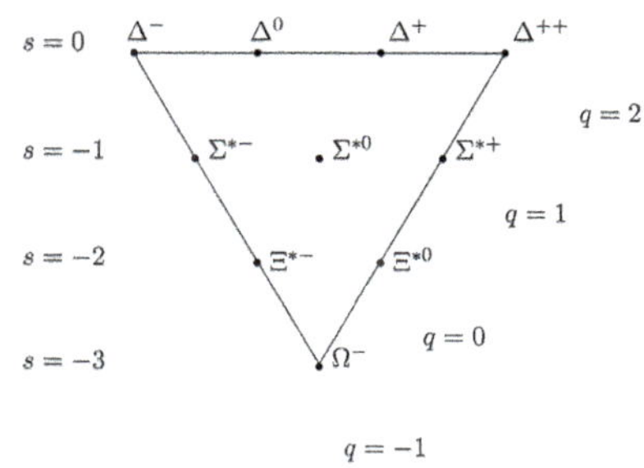

Fig. 5.8 The spin 3/2, parity 1 baryon decuplet. The I_3 axis is the abscissa, while the Y axis is the ordinate. The Ω^- has $Y = 0$, and the Σs have $I_3 = 0$

In the middle of each hexagon, there are two particles with $I_3 = 0$, $Y = 0$: one with $I = 0$ (η^0, Λ) and one with $I = 1$ (π^0, Σ^0).

The masses of the particles in each multiplet are not equal (they would be if the symmetry were perfect). Indeed, while particles lying on the horizontal lines with the same isospin have almost equal masses, the masses of the particles in consecutive horizontal lines differ by 130–150 MeV/c^2.

Besides the two basic octets, a decuplet of 3/2 baryons was also present (Fig. 5.8).

There was however an empty spot: a baryon with $Q = -1$, I $= 0$, $Y = -2$, $S = -3$ and a mass around 1670 MeV/c^2 was missing. This particle, which we call now the Ω^-, was indeed discovered in the Brookhaven National Laboratory 2-meter hydrogen bubble chamber in 1964 (Fig. 5.9). A K^- meson interacts with a proton in the liquid hydrogen of bubble chamber producing a Ω^-, a K^0, and a K^+, which then decay in other particles according to the following scheme:

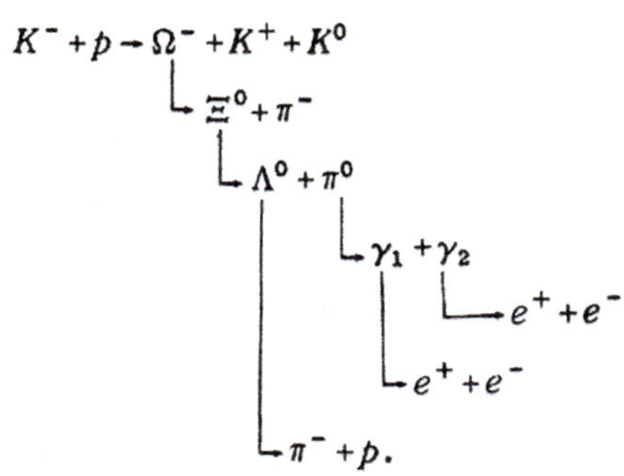

Measuring the final state charged particles and applying energy-momentum conservation, the mass of the Ω^- was reconstructed with a value of (1686 ± 12) MeV/c^2, in agreement with the prediction of Gell-Mann and Ne'eman.

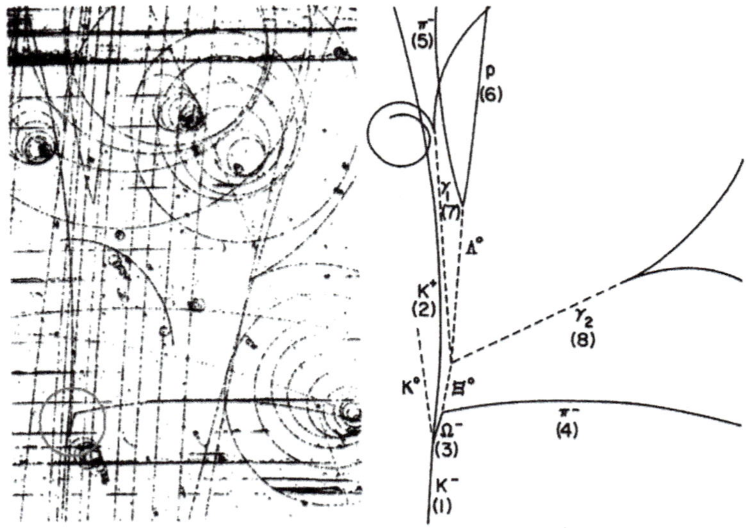

Fig. 5.9 Bubble chamber picture of the first Ω^-. From V.E. Barnes et al., "Observation of a Hyperon with Strangeness Minus Three," Physical Review Letters 12 (1964) 204

5.4 The Quark Model

5.4.1 SU(3)$_{flavor}$

The meson and baryon multiplets were recognized as representations of SU(3) but it was soon realized that they are not the fundamental ones, and they could be generated by the combination of more fundamental representations. In 1964, Gell-Mann and Zweig proposed as fundamental representation a triplet (3) of hypothetical spin 1/2 particles named *quarks*. Its conjugate representation $(\bar{3})$ is the triplet of *antiquarks*. Two of the quarks (named *up, u,* and *down, d,* quarks) form a isospin duplet and the other (named *strange, s*), which has strangeness quantum number different from zero, a isospin singlet. This classification of quarks into u, d, and s states introduces a new symmetry called *flavor* symmetry and the corresponding SU(3) group is labeled as SU(3)$_{flavor}$ or shortly SU(3)$_f$, whenever confusion is possible with the group SU(3)$_c$ (SU(3)$_{\text{color}}$) of strong interactions that will be discussed in the next chapter.

Mesons are quark/antiquark bound states, whereas baryons are composed by three quarks (anti-baryons by three antiquarks). To reproduce the hadrons quantum numbers, quarks must have fractional electric charge and fractional baryonic number. Their quantum numbers are as follows:

The fundamental representations of quarks and antiquarks form triangles in the (I_3, Y) plane (Fig. 5.10).

	Q	I	I_3	S	B	Y
u	$+2/3$	1/2	$+1/2$	0	1/3	1/3
d	$-1/3$	1/2	$-1/2$	0	1/3	1/3
s	$-1/3$	0	0	-1	1/3	$-2/3$

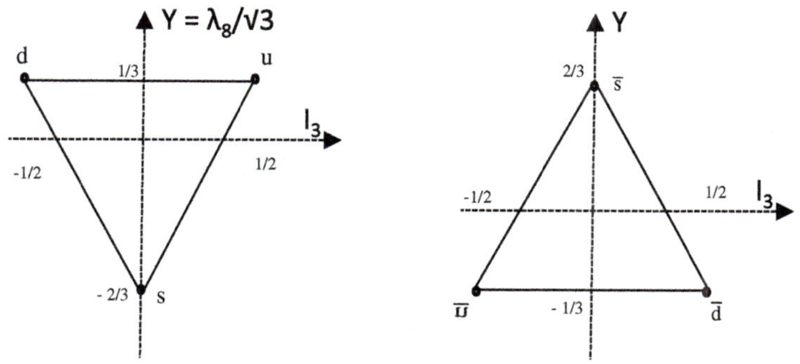

Fig. 5.10 Fundamental representations (3) and $(\overline{3})$ of SU(3)

The mesons multiplets are thus obtained by the cross product of the (3) and $(\overline{3})$ SU(3) representations, which gives an octet and a singlet:

$$3 \otimes \overline{3} = 8 \oplus 1 . \tag{5.36}$$

Graphically, the octet can be drawn centering in each quark vertex the inverse anti-quark triangle (Fig. 5.11).

Fig. 5.11 The "nonet" (octet + singlet) of pseudo-scalars mesons

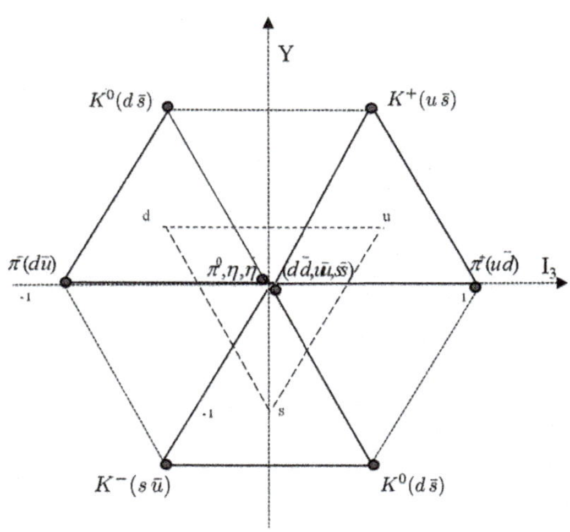

There are three states with $I_3 = 0$ and $Y = 0$ both for pseudo-scalars (π^0, η, η') and vectors (ρ^0, ω, ϕ). The π^0 and the ρ^0 are the already well-known states with

$I = 1,\ I_3 = 0\ \left(\frac{1}{\sqrt{2}}\ (u\overline{u} - d\overline{d})\right)$. The other states should then correspond to the SU(3) symmetric singlet $\left(\frac{1}{\sqrt{3}}\ (u\overline{u} + d\overline{d} + s\overline{s})\right)$ and to the octet isospin singlet, orthogonal to the SU(3) singlet, $\left(\frac{1}{\sqrt{6}}\ (u\overline{u} + d\overline{d} - 2\,s\overline{s})\right)$; however, the physically observed states are mixtures of these two "mathematical" singlets. Due to these mixings, there is in fact a combination of the SU(3) octet and singlet which is commonly designated as the "nonet."

The quark content of a meson can be accessed studying its decay modes.

Baryon multiplets are obtained by the triple cross product of the (3) SU(3) representations. The 27 possible three quark combinations are then organized in a decuplet, two octets, and a singlet:

$$3 \otimes 3 \otimes 3 = 10 \oplus 8 \oplus 8 \oplus 1\,. \tag{5.37}$$

In terms of the exchange of the quark flavor, it can be shown that the decuplet state wave functions are completely symmetric, while the singlet state wave function is completely antisymmetric. The octet state wave functions have mixed symmetry. The total wave function of each state is however not restricted to the flavor component. It must include also a spatial component (corresponding to spin and to the orbital angular momentum) and a color component which will be discussed in the next section.

5.4.2 The Color

Color is at the basis of the present theory of strong interactions, QCD (see Sect. 6.4), and its introduction solves the so-called Δ^{++} puzzle. The Δ^{++} is formed by three u quarks with orbital angular momentum $l = 0$ (it is a ground state) and in the same spin projection state (the total spin is $J = 3/2$). Therefore, its flavor, spin, and orbital wave functions are symmetric, while the Pauli exclusion principle imposes that the total wave functions of states of identical fermions (as it is the case) should be antisymmetric.

In color space, quarks are represented by complex three-vectors (the generalization of the two-dimensional spinors). The number of colors in QCD is $N_c = 3$, as we shall see later in this Chapter; the quark colors are usually designated as *Red*, *Blue* and *Green*, having the antiquark the corresponding anti-colors.

Quarks interact via the emission or absorption of color field bosons, the gluons. There are eight gluons corresponding to the eight generators of the SU(3) group (see Sect. 5.3.5). Gluons are in turn colored, and the emission of a gluon changes the color.

(Anti-)baryons are singlet states obtained by the combination of three (anti)quarks; mesons are singlet states obtained by the combination of one quark and one antiquark. In summary, all stable hadrons are color singlets, i.e., they are neutral in color. This

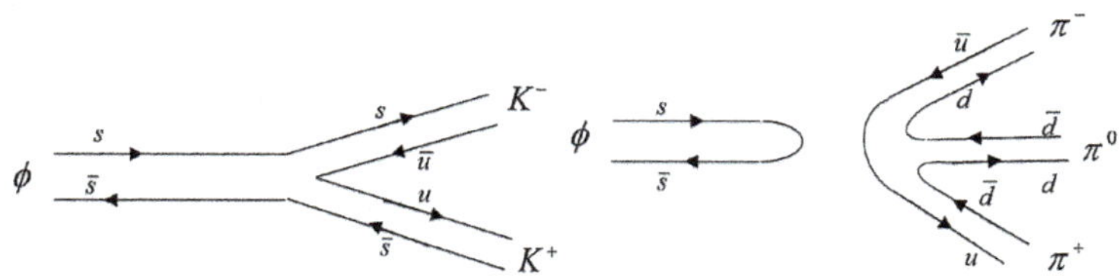

Fig. 5.12 OZI favored (*left*) and suppressed (*right*) ϕ decay diagrams

is the main reason behind the so-called OZI (Okubo–Zweig–Iizuka) rule, which can be seen, for example, in the case of the ϕ the decay into a pair of kaons which is experimentally favored (86 % branching ratio) in relation to the decay into three pions which however has a much larger phase space. The suppression of the 3π mode can be seen as a consequence of the fact that "decays with disconnected quark lines are suppressed" (Fig. 5.12). Being the $s\bar{s}$ state a color singlet, the initial and the final state in the right plot cannot be connected by a single gluon, being the gluon a colored object (see Sect. 6.4). Indeed, one can prove that the "disconnected" decay would need the exchange of at least three gluons.

In color space, the physical states are antisymmetric singlets (the total color charge of hadrons is zero, the strong interactions are short range, confined to the interior of hadrons and nuclei). The product of the spin wave function and the flavor wave function in ground states (angular orbital momentum $L = 0$) must then be symmetric. The net result is that the ground-state baryons are organized in a symmetric spin 1/2 octet and in a symmetric spin 3/2 decuplet (Fig. 5.13).

5.4.3 Excited States (Nonzero Angular Momenta Between Quarks)

Hundreds of excited states have been discovered with nonzero orbital angular momentum; they can be organized in successive SU(3) multiplets.

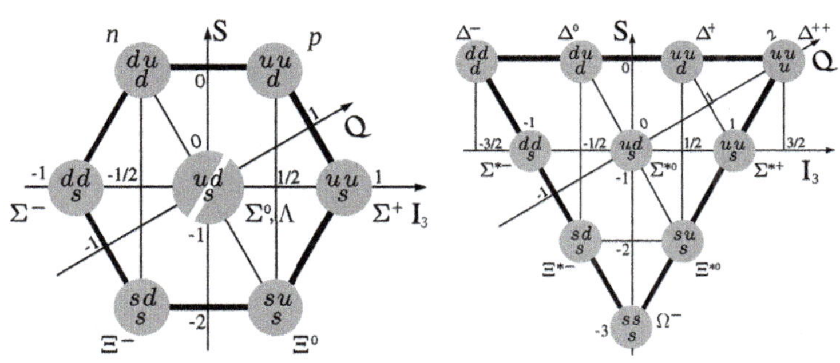

Fig. 5.13 Baryon ground states in the quark model: the spin 1/2 octet (*left*), and the spin 3/2 decuplet (*right*). The vertical (S) axis corresponds to the Y axis, shifted by 1 ($Y = 0$ corresponds to $S = -1$). By Trassiorf [own work, public domain], via Wikimedia Commons

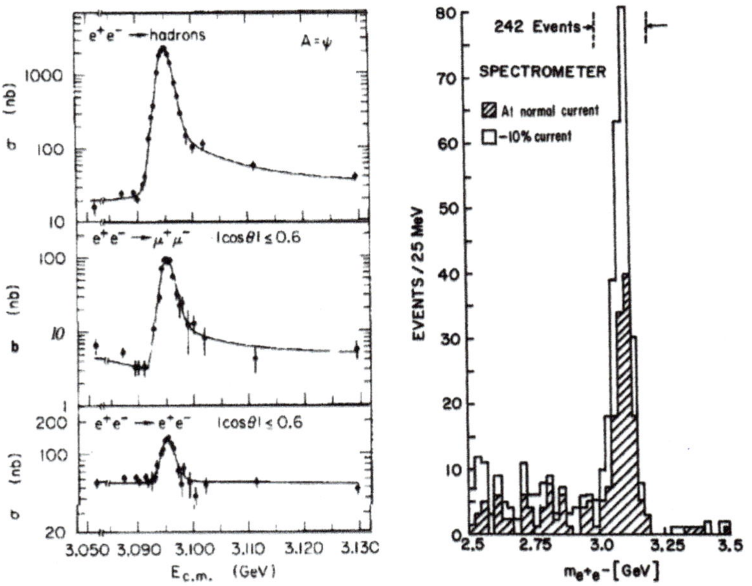

Fig. 5.14 J/ψ invariant mass plot in e^+e^- annihilations (*left*) and in proton–beryllium interactions (*right*). Credits: Nobel foundation

In the case of mesons, these states are labeled using the notation of atomic physics. In the case of baryons, two independent orbital angular momenta can be defined (between for instance two of the quarks, said to form a diquark state, and between this diquark and the third quark), and the picture is more complex.

5.4.4 The Charm Quark

In November 1974, there was a revolution in particle physics: the simultaneous discovery by two groups[3] of a heavy and narrow (implying with relatively long lifetime) resonance (Fig. 5.14). One group was lead by Burton Richter and was studying electron–positron annihilations at SLAC (Stanford Linear Accelerator Center), and the other was lead by Samuel Ting and studied proton–beryllium interactions at BNL (Brookhaven National Laboratory). The BNL group named the particle "J," while the SLAC group called it "ψ"; it was finally decided to name it "J/ψ." The resonance was too narrow to be an excited state—in this case, a hadronic decay would have been expected, and thus a width of $\sim$150 MeV/c^2.

[3]K. Niu and collaborators had already published candidates for charm (no such name was ascribed to this new quark at that time) in emulsion as early as 1971. These results, taken seriously in Japan, were not accepted as evidence for the discovery of charm by the majority of the US and European scientific communities. Once again, cosmic-ray physics was the pathfinder.

In terms of the quark model, the J/ψ can be interpreted as a $c\bar{c}$ (where c stands for a new quark flavor, called the *charm*, which has an electric charge of 2/3) vector (being produced in $e^+e^- \to \gamma^* \to J/\psi$) meson. The possibility of the existence of a fourth flavor was suggested in 1964 by Bjorken and Glashow for symmetry reasons: at that time four leptons—the electron, the muon, and their respective neutrinos—were known, and just three quarks. Later, in 1970, Glashow, Iliopoulos, and Maiani demonstrated that a fourth quark was indeed needed to explain the suppression of some *neutral current* weak processes—this is the so-called GIM mechanism that will be discussed in Sect. 6.3.6.

With the existence of a fourth flavor, the flavor symmetry group changes from SU(3) to SU(4) giving rise to more complex multiplets which were named "super-multiplets." Super-multiplets can be visualized (Fig. 5.15) as solids in a three-dimensional space I_3, Y, C, where C is the new charm quantum number.

A rich spectroscopy of charmed hadrons was open. For instance, the pseudo-scalar SU(3) octet becomes a 15-particle SU(4) multiplet with seven new mesons with at least a c quark (D^0 $(c\bar{u})$, D^+ $\left(c\bar{d}\right)$, D_s $(c\bar{s})$, η_c $(c\bar{c})$, $\overline{D}^0$ $(\bar{c}\,u)$, $\overline{D}^+$ $(\bar{c}\,d)$, $\overline{D}_s$ $(\bar{c}\,s)$), and the spin 3/2 decuplet of baryons becomes a 20-particle multiplet.

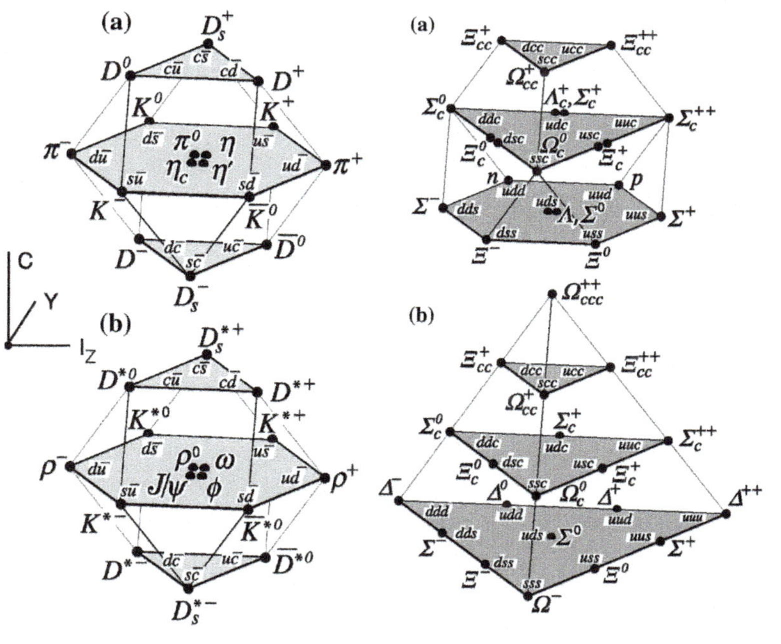

Fig. 5.15 *Left* The 16-plets for the pseudo-scalar (**a**) and vector (**b**) mesons made of the u, d, s, and c quarks as a function of isospin I_3, charm C, and hypercharge $Y = B + S + C$. The nonets of light mesons occupy the central planes to which the $c\bar{c}$ states have been added. *Right* SU(4) multiplets of baryons made of u, d, s, and c quarks: (**a**) The 20-plets including an SU(3) octet; (**b**) The 20-plets with an SU(3) decuplet. From K.A. Olive et al. (Particle Data Group), Chin. Phys. C 38 (2014) 090001

5.4.4.1 Quarkonia: The Charmonium

The $c\bar{c}$ states, named *charmonium* states, are particularly interesting; they correspond to non-relativistic particle/antiparticle bound states. Charmonium states have thus a structure of excited states similar to positronium (an e^+e^- bound state) but their energy levels can be explained by a potential in which a linear term is added to the Coulomb-like potential, which ensures that an infinite energy is needed to separate the quark and the antiquark (no free quarks have been observed up to now, and we found a clever explanation for this fact, as we shall see):

$$V(r) \simeq -\frac{4}{3}\frac{\alpha_s}{r} + k\,r \tag{5.38}$$

where r is the radius of the bound state, α_s is the equivalent for the strong interactions of the fine structure constant α, k is a positive constant, and the coefficient 4/3 has to do with the color structure of strong interactions; we shall discuss it in larger detail in Sect. 6.4.6.

The linear term dominates for large distances. Whenever the pair is stretched, a field string is formed storing potential energy; at some point, (part of) the stored energy can be converted, by tunnel effect, into mass and a new quark/antiquark pair can be created transforming the original meson into two new mesons (Fig. 5.16). This process is named quark hadronization and plays a central role in high-energy hadronic interactions.

In positronium spectroscopy, one can obtain the energy levels by solving the Schrödinger equation with a potential $V_{em} = -\alpha/r$; the result is

$$E_{p;n} = -\frac{\alpha m_e c^2}{4n^2} \,.$$

Note that these levels are equal to the energy levels of the hydrogen atom, divided by two: this is due to the fact that the mass entering in the Schrödinger equation is the *reduced* mass m_r of the system, which in the case of hydrogen is approximately equal to the electron mass ($m_r = m_e m_p/(m_e + m_p)$, while in the case of positronium, it is exactly $m_e/2$. The spin–orbit interaction splits the energy levels (fine splitting), and a further splitting (hyperfine splitting) is provided by the spin–spin interactions.

The left plot of Fig. 5.17 shows the energy levels of positronium. They are indicated by the symbols $n^{2S+1}\,L\,s$ (n is the principal quantum number, S is the

Fig. 5.16 Hadronization: the string mechanism

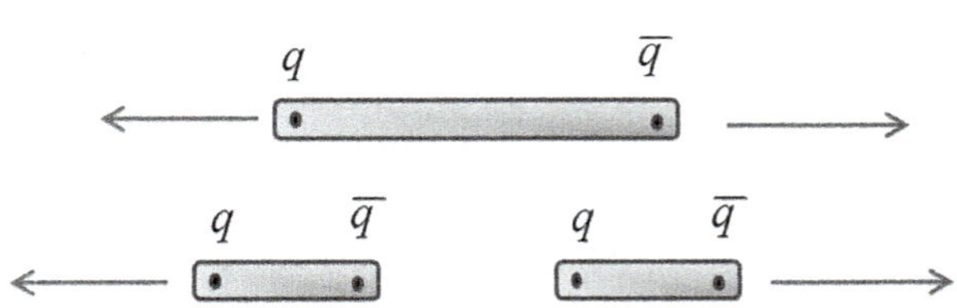

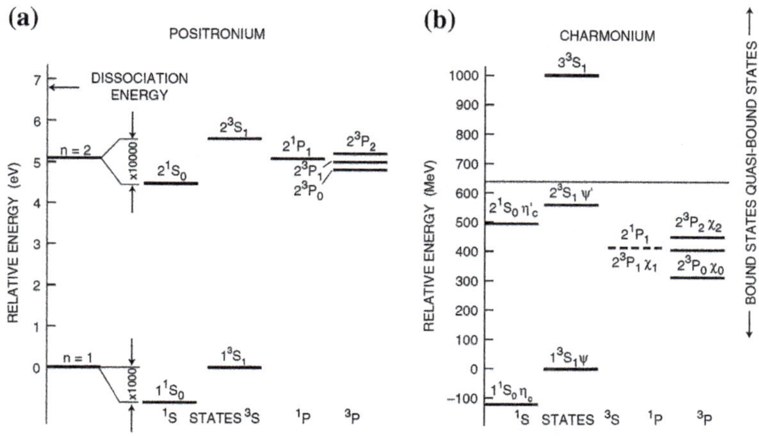

Fig. 5.17 Energy levels for (**a**) the positronium and (**b**) the charmonium states. From S. Braibant, G. Giacomelli, and M. Spurio, "Particles and fundamental interactions," Springer 2012

total spin, L indicates the orbital angular momentum in the spectroscopic notation (S being the $\ell = 0$ state), and s is the spin projection.

The right plot of Fig. 5.17 shows the energy levels of charmonium; they can be obtained inserting the potential (5.38) into the Schrödinger equation. One obtains $k \sim 1$ GeV, and $\alpha_s \sim 0.3$.

The quark bottom, which will be introduced in the next section, has an even larger mass, and it gives rise a similar spectroscopy of quarkonia.

5.4.5 Beauty and Top

A fifth quark was discovered a few years later. In 1977, in Fermilab, an experiment led by Leon Lederman studied the mass spectrum of $\mu^-\mu^+$ pairs produced in the interaction of a 400 GeV proton beam on copper and platinum targets. A new heavy and narrow resonance, named the upsilon Υ, was found, with a mass of around 9.46 GeV/c^2.

The Υ was interpreted as a $b\bar{b}$ vector meson where b stands for a new quark flavor, the *bottom* or *beauty*, which has, like the d and the s, an electric charge of $-1/3$. Several hadrons containing at least a b quark were discovered. A family of $b\bar{b}$ mesons, called the bottomium and indicated by the letter Υ, was there, as well as mesons and baryons resulting from the combination of b quarks with lighter quarks: pseudo-scalar mesons like the B^+ $(u\bar{b})$, B_c^+ $(c\bar{b})$, the B^0 $(d\bar{b})$, and the B_s^0 $(s\bar{b})$; bottom baryons like $\Lambda_b^0\,(udb)$, $\Xi_b^0\,(usb)$, $\Xi_b^-\,(dsb)$, $\Omega_b^-\,(ssb)$. Heavy mesons and baryons with a single heavy quark are very interesting. The light quarks surround the heavy quark in a strong force analogy of electrons around the proton in the electromagnetically bound hydrogen atom.

The discovery of the b quark inaugurated a third generation or family of quarks. Each family is formed by two quarks, one with electric charge $+2/3$ and the other with electric charge $-1/3$ (the first family is formed by quarks u and d; the second by quarks c and s). In the lepton sector, as will be discussed at the end of this chapter, there were at that time already five known leptons (the electron, the muon, their corresponding neutrinos, and a recently discovered heavy charged lepton, the tau). With the bottom quark, the symmetry was restored between quarks and leptons but the sixth partners both in the quark and in lepton (the tau neutrino) sector were missing. The existence of a third family of quarks had indeed been predicted already in 1973, before the discovery of the J/ψ, by Makoto Kobayashi and Toshihide Maskawa, to accommodate in the quark model the CP violation observed in the $K^0\overline{K}^0$ system (this will be discussed in Chap. 6). The hypothetical sixth quark was named before its discovery the "top" quark.

The top quark was missing and for many years, and many people looked for it in many laboratories (in the United States and in Germany, Japan, CERN), both at electron–positron and at proton–(anti)proton colliders. A strong indication of a top with a mass around 40 GeV/c^2 was even announced in 1984 but soon dismissed. Lower limits on the top mass were later established and indications on the value of the mass were derived from the standard model (see Chap. 7); finally, in 1995, the discovery of the top quark, with a measured mass of (175 ± 20) GeV/c^2 (the present world average is (173.2 ± 0.9) GeV/c^2) was published by the CDF experiment—and soon after by the D0 experiment—at Fermilab. The top is heavier than a gold nucleus; with such a large mass, its decay phase space is huge and its lifetime is very short, even if the decay is mediated by the weak force. It is so short (the estimated value is around 5×10^{-25} s) that the top does not live long enough to hadronize: there are no top hadrons.

At present six quark flavors (u, d, c, s, t, b) are known and they are organized into three families:

$$\begin{pmatrix} u \\ d \end{pmatrix} \begin{pmatrix} c \\ s \end{pmatrix} \begin{pmatrix} t \\ b \end{pmatrix}.$$

Their masses cover an enormous range, from the tens of MeV/c^2 for the u and the d quarks,[4] to the almost 200 GeV/c^2 for the t quark. The flavor symmetry that was the clue to organize the many discovered hadrons is strongly violated. Why? Is there a fourth, a fifth (...), family to be discovered? Are quarks really elementary? These are questions we hope to answer during this century.

[4]The problem of the determination of the quark masses is not trivial. We can define as a "current" quark mass the mass entering in the Lagrangian (or Hamiltonian) representation of a hadron; this comes out to be of the order of some MeV/c^2 for u, d quarks, and $\sim$0.2 GeV/c^2 for s quarks. However, the strong field surrounds the quarks in such a way that they acquire a "constituent "(effective) quark mass including the equivalent of the color field; this comes out to be of the order of some 300 MeV/c^2 for u, d quarks, and $\sim$0.5 GeV/c^2 for s quarks. Current quark masses are almost the same as constituent quark mass for the heavy quarks.

5.5 Quarks and Partons

In the words of Murray Gell-Mann in 1967, quarks seemed to be just *mathematical entities*. This picture was deeply changed in a few years by the results of deep inelastic scattering experiments .

Indeed in the 1950s, Robert Hofstadter,[5] in a series of Rutherford-like experiments, using a beam of electrons instead of α particles (electrons have no strong interactions) showed departures from the expected elastic point cross section. Nucleons (protons and neutrons) are not point-like particles. The proton must have a structure and the quarks could be thought as its constituents.

5.5.1 Elastic Scattering

The electron–proton cross section, in case the target proton is thought as a of point-like spin $1/2$ particle with a mass m_p, was calculated by Rosenbluth (Sect. 6.2.8):

$$\frac{d\sigma}{d\Omega} = \frac{\alpha^2 \cos^2\left(\frac{\theta}{2}\right)}{4\,E^2 \sin^4\left(\frac{\theta}{2}\right)} \frac{E'}{E} \left[1 + \frac{Q^2}{2M_p^2} \tan^2\left(\frac{\theta}{2}\right)\right] \tag{5.39}$$

where the first factor is the Mott cross section (scattering of an electron in a Coulomb field, see Chap. 2); the second $\left(E'/E\right)$ takes into account the energy lost in the recoil of the proton; and the third factor is the spin/spin interaction. Note that for a given energy of the incident electron there is just one independent variable, which is usually chosen by experimentalists to be the scattering angle, θ. In fact, the energy of the scattered electron, E', can be expressed as a function of θ as

$$E' = \frac{E}{1 + E(1 - \cos\,\theta)/m_p}\,. \tag{5.40}$$

[5]Robert Hofstadter (1915–1990) was an American physicist. He was awarded the 1961 Nobel Prize in Physics “for his pioneering studies of electron scattering in atomic nuclei and for his consequent discoveries concerning the structure of nucleons.” He worked at Princeton before joining Stanford University, where he taught from 1950 to 1985. In 1948, Hofstadter patented the thallium activated NaI gamma-ray detector, still one of the most used radiation detectors. He coined the name “fermi,” symbol fm, for the scale of 10^{-15} m. During his last years, Hofstadter became interested in astrophysics and participated to the design of the EGRET gamma-ray space telescope (see Chap. 10).

The measured cross section had however a stronger Q^2 dependence as would be expected in the case of a finite size proton. This cross section can be parameterized as

$$\frac{d\sigma}{d\Omega} = \frac{\alpha^2 \cos^2\left(\frac{\theta}{2}\right)}{4\, E^2 \sin^4\left(\frac{\theta}{2}\right)} \frac{E'}{E} \left[\frac{G_E^2\left(Q^2\right) + \frac{Q^2}{4M_p^2} G_M^2\left(Q^2\right)}{1 + \frac{Q^2}{4M_p^2}} + 2\frac{Q^2}{4M_p^2}\, G_M^2\left(Q^2\right) \tan^2\left(\frac{\theta}{2}\right) \right] \tag{5.41}$$

where $G_E^2\left(Q^2\right)$ and $G_M^2\left(Q^2\right)$ are called, respectively, the electric and the magnetic form factors (if $G_E = G_M = 1$, the Rosenbluth formula 5.47 is recovered).

At low Q^2, $G_E\left(Q^2\right)$ and $G_M\left(Q^2\right)$ can be interpreted as the Fourier transforms of the electric charge and of the magnetisation current density inside the proton. In the limit $Q^2 \to 0$ ($\lambda \to \infty$ for the exchanged virtual photon), the electron "sees" the entire proton and it could be expected that $G_E\,(0) = G_M\,(0) = 1$. This is what the experiment tells for G_E, but it is not the case for G_M. The measured value is $G_M\,(0) = \mu_p \simeq 2.79$. The proton has an anomalous magnetic moment μ_p which reveals already that the proton is not a Dirac point-like particle. The same is observed for the neutron which has $\mu_n \simeq -1.91$.

In fact, at low Q^2 ($Q^2 < 1 - 2\ \mathrm{GeV}^2$), the experimental data on G_E and G_M are well described by the dipole formula

$$G_E\left(Q^2\right) \simeq \frac{G_M\left(Q^2\right)}{\mu_p} \simeq \left(\frac{1}{1 + Q^2/0.71\mathrm{GeV}^2}\right)^2 \tag{5.42}$$

suggesting similar spatial distributions for charges and currents. However, recent data at higher Q^2 using polarized beams showed a much richer picture reflecting a complex structure of constituents and their interactions.

5.5.2 *Inelastic Scattering Kinematics*

The scattering of an electron on a proton may, if the electron energy is high enough, show the substructure of the proton. At first order, such scattering (Fig. 5.18) can be seen as the exchange of a virtual photon (γ^*) with four-momentum:

$$q^2 = -Q^2 = (p_1 - p_3)^2 = (p_4 - p_2)^2.$$

where p_1 and p_3 are, respectively, the four-momentum of the incoming and outcoming electron, p_2 is the target four-momentum, and p_4 is the four-momentum of the final hadronic state which has an invariant mass $M = \sqrt{p_4^2}$ (see Fig. 5.18). In case of elastic scattering, $M = m_p$.

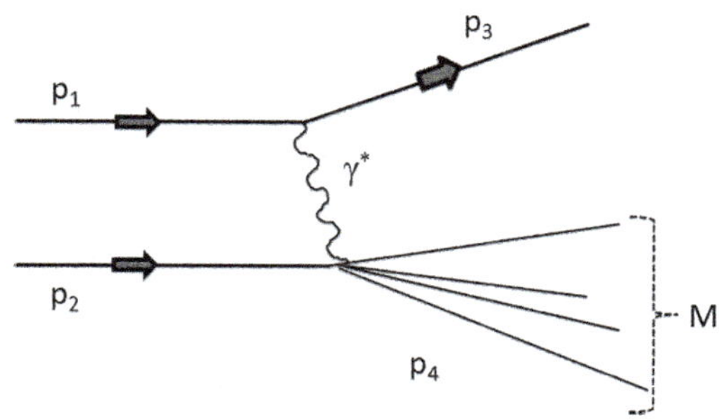

Fig. 5.18 Deep inelastic scattering kinematics

In the laboratory reference frame:

$$p_1 = \left(E, \vec{K}\right); \; p_2 = \left(M_p, 0\right); \; p_3 = \left(E', \vec{K'}\right); \; p_4 = \left(\sqrt{M^2 + \vec{p}_4^{\,2}}, \vec{p}_4\right). \tag{5.43}$$

The center-of-mass energy is, as it was seen in Chap. 2, just the square root of the Mandelstam variable s:

$$s = (p_1 + p_2)^2$$

which in the laboratory reference frame is given (neglecting $m_e^2 \sim 0$) by

$$s \simeq m_p(2E + m_p).$$

It is then useful to construct, using internal products of the above four-vectors, other Lorentz invariant variables, namely:

- the energy lost ν

$$\nu = \frac{q p_2}{m_p}$$

which in the laboratory reference frame is just the energy lost by the electron:

$$\nu = E - E';$$

- the inelasticity y

$$y = \frac{q p_2}{p_1 p_2}$$

which in the laboratory frame is just the fraction of the energy lost by the electron:

$$y = \frac{\nu}{E}$$

(y is thus dimensionless and it is limited to the interval $0 \leq y \leq 1$);

- the Bjorken variable, x

$$x = \frac{Q^2}{2p_2 q}$$

which can be expressed, using the definition of ν, as

$$x = \frac{Q^2}{2m_p \nu}$$

or, imposing energy and momentum conservation at the hadronic vertex:

$$x = \frac{Q^2}{Q^2 + M^2 - m_p^2}$$

(x is therefore also dimensionless and limited to the interval $0 \leq x \leq 1$; if $x = 1$, $M = m_p$ and the elastic scattering formula is recovered).

At a fixed center-of-mass energy, $\sqrt{s}$, the inelastic scattering final state can the be characterized by the Lorentz invariant variables, Q^2, M^2, x, y, as well as by the scattered electron energy E' and scattered angle θ in the laboratory reference frame. However, from all those variables, only two are independent. The experimental choice is usually the directly measured variables E' and θ, while the theoretical interpretation is usually done in terms of Q^2 and ν or Q^2 and x.

Many relations can be built connecting all these variables. The following are particularly useful:

$$Q^2 \simeq 4\, E\, E' \sin^2\left(\frac{\theta}{2}\right)$$

$$Q^2 \simeq 2\, M\nu$$

$$Q^2 = x\, y\, \left(s^2 - {M_p}^2\right)$$

$$M^2 = {M_p}^2 + 2M_p\, \nu - Q^2$$

$$\frac{2m_p x^2}{Q^2}\frac{y^2}{2} = \frac{1}{m_p}\frac{E'}{E}\sin^2\left(\frac{\theta}{2}\right)$$

$$1 - y - \frac{m_p^2 x^2 y^2}{Q^2} = \frac{E'}{E}\cos^2\left(\frac{\theta}{2}\right) .$$

5.5.3 Deep Inelastic Scattering

The differential electron–proton inelastic cross section can be parameterized, similar to what was done in the case of the electron–proton elastic cross section, introducing two independent functions. These functions, called the structure functions W_1 and W_2, can be expressed as a function of any two of kinematic variables discussed in the previous section. Hereafter, the choice will be Q^2 and ν. Hence,

$$\frac{d\sigma}{d\Omega dE'} = \frac{\alpha^2 \cos^2\left(\frac{\theta}{2}\right)}{4\ E^2 \sin^4\left(\frac{\theta}{2}\right)} \frac{E'}{E} \left[W_2\left(Q^2, \nu\right) + 2W_1\left(Q^2, \nu\right)\ \tan^2\left(\frac{\theta}{2}\right)\right]. \quad (5.44)$$

$W_1\left(Q^2, \nu\right)$ describes the interaction between the electron and the proton magnetic moments and can be neglected for low Q^2.

In the limit of electron–proton elastic scattering ($x \to 1$, $\nu = Q^2/2m_p$), these structure functions should reproduce the elastic cross-section formula discussed above:

$$W_1\left(Q^2, \nu\right) = \frac{Q^2}{4M_p^2} G_M^2\left(Q^2\right) \delta\left(\nu - \frac{Q^2}{2\,m_p}\right); \quad (5.45)$$

$$W_2\left(Q^2, \nu\right) = \frac{G_E^2\left(Q^2\right) + \frac{Q^2}{4M_p^2} G_M^2\left(Q^2\right)}{1 + \frac{Q^2}{4M_p^2}} \delta\left(\nu - \frac{Q^2}{2\,M_p}\right). \quad (5.46)$$

In the case of the elastic scattering of electrons on a point 1/2 spin particle with mass m_p and charge e, the Rosenbluth formula (5.47) should be reproduced:

$$W_1\left(Q^2, \nu\right) = \frac{Q^2}{4M_p^2} \delta\left(\nu - \frac{Q^2}{2\,m_p}\right);\ W_2\left(Q^2, \nu\right) = \delta\left(\nu - \frac{Q^2}{2\,m_p}\right). \quad (5.47)$$

The difference between scattering on point-like or finite size particles is thus translated into the form factors G_E and G_M. In the case of a scattering over point-like particles, the exchanged virtual photon "sees" always the same charge whatever the Q^2, and $G_E = G_M = 1$. In the case of a finite size particle, the photon wavelength ($\lambda \sim 1/\sqrt{Q^2}$) limits the *observed* volume inside the target. For smooth charge distributions inside the target, it is therefore trivial to predict that when $Q^2 \to \infty$:

$$W_1\left(Q^2, \nu\right) \to 0;\ W_2\left(Q^2, \nu\right) \to 0.$$

On the contrary if the charge distribution is "concentrated" in a few space points, some kind of point-like behavior should be recovered. Such behavior was predicted by James Bjorken in 1967 who postulated, for high Q^2 and ν, the scaling of the structure functions:

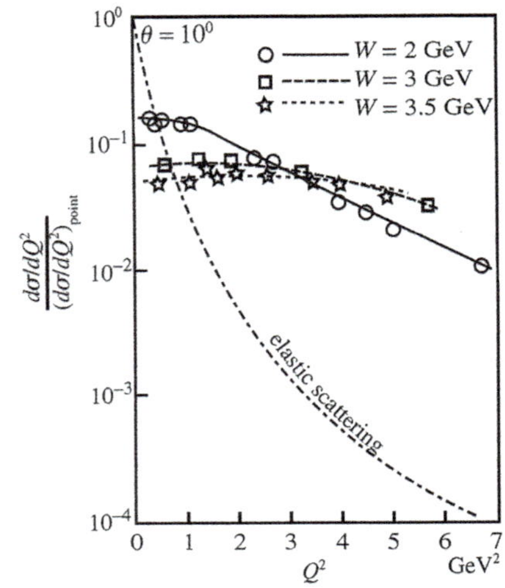

Fig. 5.19 Deep inelastic electron–proton differential cross section normalized to the Mott cross section, as measured by SLAC. *Source* Ref. [F5.2] in the "Further readings." Adapted from Nobel Foundation, 1990

$$W_1\left(Q^2,\nu\right) \rightarrow \frac{1}{m_p} F_1(w)\ ;\ W_2\left(Q^2,\nu\right) \rightarrow \frac{1}{\nu} F_2(w) \tag{5.48}$$

where w, the Bjorken scaling variable, is the inverse of the x variable:

$$w = \frac{1}{x} = \frac{2M_p\nu}{Q^2}\,. \tag{5.49}$$

According to the above definitions, F_1 and F_2 are dimensionless functions, while W_1 and W_2 have dimensions E^{-1}.

While Bjorken was suggesting the scaling hypothesis, the groups lead by J. Friedman, H. Kendall and R. Taylor[6] designed and built at SLAC electron spectrometers able to measure energies up to 20 GeV for different scattering angles. The strong Q^2 dependence of the elastic form factors was then confirmed up to $Q^2 \simeq 30\ \text{GeV}^2$ while, surprisingly, in the region $M > 2$ GeV, the inelastic cross section showed a very mild Q^2 dependence (Fig. 5.19) and the $W_2\left(Q^2,\nu\right)$ structure function showed, at these relative small energies, already an approximate Bjorken scaling. $F_2(w) = \nu\, W_2\left(Q^2,\nu\right)$ was indeed found to be a universal function of $w = \frac{1}{x}$ as demonstrated by measurements at different energies and angles of the scattered particle (Fig. 5.20) or at different Q^2 but keeping w constant (Fig. 5.21).

In 1968, Richard Feynman, just a few months after the presentation of the SLAC results, worked out a simple and elegant model which could explain the results: the electron–nucleon scattering at high energy might be seen as the scattering of the

[6]The Nobel Prize in Physics 1990 was assigned to Jerome I. Friedman, Henry W. Kendall and Richard E. Taylor "for their pioneering investigations concerning deep inelastic scattering of electrons on protons and bound neutrons, which have been of essential importance for the development of the quark model in particle physics."

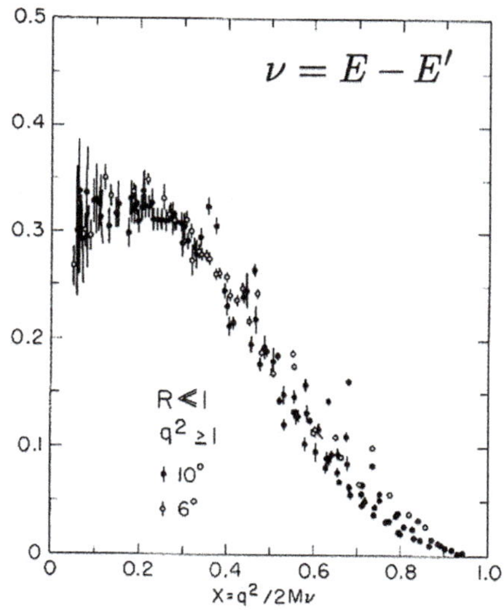

Fig. 5.20 νW_2 scaling: measurements at different scattered energies and angles. *Source* W.B. Atwood, "Lepton Nucleon Scattering," Proc. 1979 SLAC Summer Institute (SLAC-R-224)

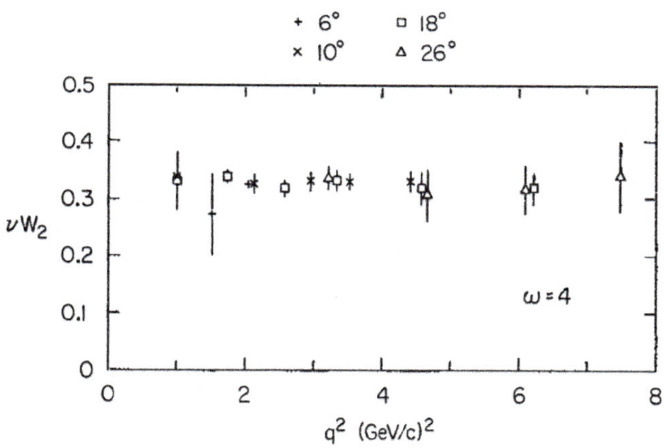

Fig. 5.21 νW_2 scaling for the proton: measurements at fixed x and at different Q^2. *Source* J.I. Friedman and H.W. Kendall, Annual Rev. Nucl. Science 22 (1972) 203

electron into free point-like charged particles in the nucleon: the *partons* (Fig. 5.22). This is the so-called quark-parton model (QPM).

The nucleon has point-like constituents but, despite the strong interaction that holds the nucleon, they appear to be basically free. The partons are, however, confined inside the hadrons: nobody has ever observed a parton out of a hadron.

The Feynman partons were soon identified as the Gell-Mann and Ne'eman quarks; nowadays, the term "parton" is used, however, to denominate all the nucleon constituents, i.e., the quarks and the gluons.

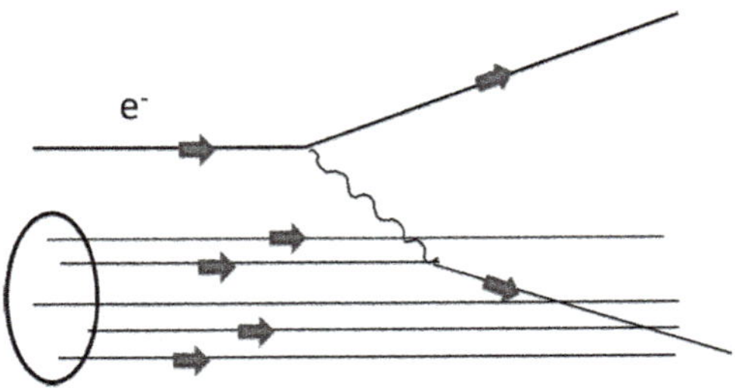

Fig. 5.22 Representation of electron–parton scattering in Feynman's Quark-Parton Model

5.5.4 The Quark-Parton Model

In a first approximation, the transverse momentum of the partons can be neglected and each parton shares a fraction Z_i of the nucleon momentum and energy:

$$E_i = Z_i\, E$$

$$\vec{p}_i = Z_i\, \vec{p}$$

and thus the parton mass is also a fraction Z_i of the nucleon mass:

$$m_i = Z_i\, M_N\,.$$

These assumptions are exact in a frame where the parton reverses its linear momentum keeping constant its energy (collision against a "wall"). In such frame (called the Breit frame, or also the infinitum momentum frame), the energy of the virtual photon is zero ($q_{\gamma^*} = (0, 0, 0, -Q)$) and the proton moves with a very high momentum toward the photon. However, even if the parton model was built in such an extreme frame, its results are valid whenever $Q^2 \gg m_N$, where m_N is the nucleon mass.

Remembering the previous section, the elastic form factor of the scattering electron on a point-like spin 1/2 particle with electric charge e_i and mass m_i can be written as

$$W_1\left(Q^2, \nu\right) = \frac{Q^2}{4m_i^2}\, e_i^2 \delta\left(\nu - \frac{Q^2}{2\, m_i}\right)\; ;\; W_2\left(Q^2, \nu\right) = e_i^2\, \delta\left(\nu - \frac{Q^2}{2\, m_i}\right)\,.$$

Using the property of the δ function $\delta\,(ax) = \frac{1}{|a|}\delta\,(x)$, and remembering that $m_i = Z_i\, M_N$ and $x = \frac{Q^2}{2M_N\nu}$:

$$W_1\left(Q^2, \nu\right) = e_i^2\, \frac{x}{2\, M_N Z_i}\, \delta\,(\, Z_i - x)\; ;\; W_2\left(Q^2, \nu\right) = e_i^2 \frac{Z_i}{\nu}\, \delta\,(\, Z_i - x) \quad (5.50)$$

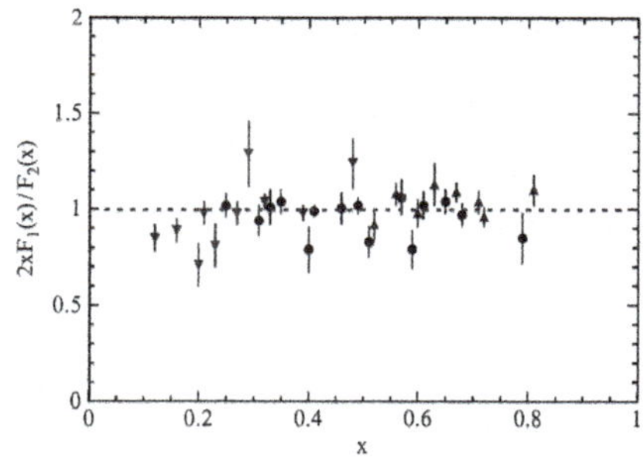

Fig. 5.23 Validation of the Callan–Gross relation. Adapted from S. Braibant, G. Giacomelli and M. Spurio, "Particles and fundamental interactions," Springer 2012

or, in terms of F_1 and F_2:

$$F_1\left(Q^2,\nu\right) = e_i^2\,\frac{x}{2\,Z_i}\,\delta\left(Z_i - x\right)\ ;\ F_2\left(Q^2,\nu\right) = e_i^2 Z_i\,\delta\left(Z_i - x\right)\,. \qquad (5.51)$$

The δ function imposes that $Z_i \equiv x$, meaning that to comply with the elastic kinematics constraints the exchanged virtual photon has to pick up a parton with precisely a fraction x of the nucleon momentum.

Comparing F_1 and F_2 , the so-called Callan–Gross relation is established:

$$F_2\left(Q^2,\nu\right) = 2x\,F_1\left(Q^2,\nu\right)\,. \qquad (5.52)$$

This relation derives directly from the assumption that partons are spin $1/2$ particles (if their spin would be 0 then $F_1\left(Q^2,\nu\right) = 0$) and was well verified experimentally (Fig. 5.23).

Let us now define $f_i\,(Z_i)$ as the density probability function to find a parton carrying a fraction of momentum Z_i. The total electron–nucleon form factors can be obtained integrating over all the entire Z_i range and summing up all the partons.

$$F_1\left(Q^2,x\right) = \sum_i \int_0^1 e_i^2\,\frac{x}{2\,Z_i}\,f_i\,(Z_i)\,\delta\left(Z_i - x\right) dZ_i$$

$$F_1\left(Q^2,x\right) = \sum_i e_i^2\,\frac{1}{2}\,f_i\,(x)$$

and

$$F_2\left(Q^2,x\right) = \sum_i \int_0^1 e_i^2 Z_i\,f_i\,(Z_i)\,\delta\left(Z_i - x\right) dZ_i$$

$$F_2\left(Q^2,x\right) = x\sum_i e_i^2\,f_i\,(x)\,.$$

The functions $f_i\ (x)$ are called the parton density functions (PDFs).

The sum of all parton momentum fractions should be 1, if the partons are the only constituents of the nucleon:

$$\sum_i \int_0^1 e_i^2\ f_i\ (x)\ dx = 1.$$

Experimentally, however, the partons (the charged constituents of the nucleon) carry only around 50 % of the total nucleon momentum. The rest of the momentum is carried by neutral particles, the gluons, which are, as it will be discussed in the next chapter, the bosons associated with the strong field that holds together the nucleon.

The real picture is more complex: instead of a nucleon composed by three quarks, there are inside the nucleon an infinite number of quarks. In fact, as it happens in the case of the electromagnetic field where electron–positron pairs can be created even in the vacuum (the Casimir effect being a spectacular demonstration), virtual quark/antiquark pairs can be created inside the nucleon. These pairs are formed in time scales allowed by the Heisenberg uncertainties relations. In an artistic picture (Fig. 5.24), the nucleon is formed by three quarks which determine the nucleon quantum numbers and carry a large fraction of the nucleon momentum (the valence quarks) surrounded by clouds of virtual quark/antiquark pairs (the "sea" quarks) and everything is embedded in a background of strong field bosons (the gluons).

The quarks of the hadrons may have different flavors and thus different charges and masses. The corresponding PDFs are denominated according to the corresponding flavor: $u\,(x)\,,\ d\,(x)\,,\ s\,(x)\,,\ c\,(x)\,,\ b\,(x)\,,\ t\,(x)$ for the quarks; $\overline{u}\,(x)\,,\ \overline{d}\,(x)\,,$ $\overline{s}\,(x)\,,\ \ldots$ for the antiquarks.

The form factor F_2 for the electron–proton scattering can now be written as a function of the specific quarks PDFs:

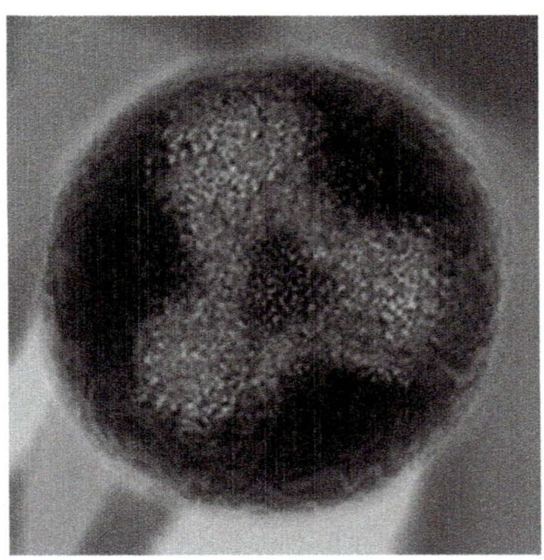

Fig. 5.24 Artistic picture of a nucleon. *Source* http://hendrix2.uoregon.edu/~imamura/

$$F_2{}^{ep}\left(Q^2, x\right) \simeq x\left[\frac{4}{9}\left(u\left(x\right)+\overline{u}\left(x\right)\right)+\frac{1}{9}\left(d\left(x\right)+\overline{d}\left(x\right)+s\left(x\right)+\overline{s}\left(x\right)\right)\right].$$

The small contributions from heavier quarks and antiquarks can be usually neglected (due to their high masses, they are strongly suppressed). The PDFs can still be divided into valence and sea. To specify if a given quark PDF refers to valence or sea, a subscript V or S is used. For instance, the total u quark proton PDF is the sum of two PDFs:

$$u\left(x\right)=\ u_V\left(x\right)+u_S\left(x\right)\ .$$

For the $\overline{u}$ antiquark PDF, we should remember that in the proton there are no valence antiquarks, just sea antiquarks. Moreover, as the sea quarks and antiquarks are produced in pairs, the sea quarks and antiquarks PDFs with the same flavor should be similar. Therefore, the $\overline{u}$ component in the proton can be expressed as

$$\overline{u}\left(x\right)=\overline{u}_S\left(x\right)=u_S\left(x\right)\ . \tag{5.53}$$

There are thus several new functions (the specific quarks PDFs) to be determined from the data. A large program of experiments has been carried out and in particular deep inelastic scattering experiments with electron, muon, neutrino, and antineutrino beams. The use of neutrinos and antineutrinos is particularly interesting since, as it will be discussed in the next chapter, their interactions with the quarks arises through the weak force and having a well-defined helicity (neutrinos have left helicity, antineutrinos right helicity) they "choose" between quarks and antiquarks. The results of all experiments are globally analyzed and PDFs for quarks but also for gluons ($g\left(x\right)$), are obtained, like these shown in Fig. 5.25.

At low x, the PDFs of sea quarks and gluons behave as $1/x$, and therefore their number inside the proton becomes extremely large at $x \to 0$. However, the physical observable $xf\left(x\right)$ (the carried momentum) is well behaved.

The valence quark PDFs can then be obtained subtracting the relevant quark and antiquark PDFs:

$$u_V\left(x\right)=u\left(x\right)-\overline{u}\left(x\right)\ ;\ d_V\left(x\right)=d\left(x\right)-\overline{d}\left(x\right).$$

Their integration over the full x range is consistent with the quark model. In fact for the proton,

$$\int_0^1 u_V\ \left(x\right)\ dx \simeq 2\ ;\ \int_0^1 d_V\ \left(x\right)\ dx \simeq 1. \tag{5.54}$$

The $x\ u_V\left(x\right)$ and $x\ d_V\left(x\right)$ distributions have a maximum around $1/3$ as expected but the sum of the momenta carried out by the valence quarks is (as it was discussed before) smaller than the total momentum:

$$\int_0^1 x\ u_V\ \left(x\right)\ dx \simeq 0.36\ ;\ \int_0^1 x\ d_V\ \left(x\right)\ dx \simeq 0.18\,. \tag{5.55}$$

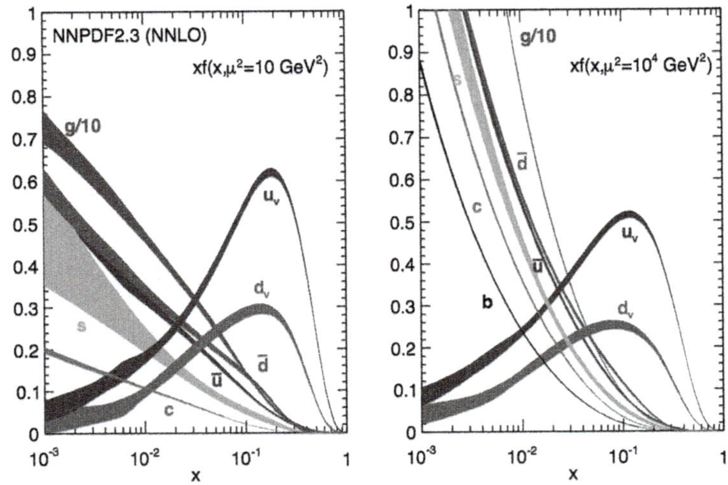

Fig. 5.25 Parton distribution functions at $Q^2 = 10\,\text{GeV}^2$ (*left*) and $Q^2 = 10000\ \text{GeV}^2$ (*right*). The gluon and sea distributions are scaled down by a factor of 10. The experimental, model, and parameterization uncertainties are shown. From K.A. Olive et al. (Particle Data Group), Chin. Phys. C 38 (2014) 090001

Many tests can be done by combining the measured form factors. An interesting quantity, for instance, is the difference of the form factor functions F_2 for electron–proton and electron–neutron scattering.

Assuming isospin invariance:

$$u^p(x) = d^n(x) \ ; \ d^p(x) = u^n(x)$$

$$\overline{u}^p(x) = \overline{u}^n(x) = \overline{d}^p(x) = \overline{d}^n(x)$$

$$s^p(x) = \overline{s}^p(x) = s^n(x) = \overline{s}^n(x) \ .$$

Then

$$F_2^{ep}\left(Q^2, x\right) \simeq x\left[\frac{4}{9}u_v^p(x) + \frac{1}{9}d_v^p(x) + \frac{10}{9}\overline{u}^p(x) + \frac{2}{9}\overline{s}^p(x)\right]$$

$$F_2^{en}\left(Q^2, x\right) \simeq x\left[\frac{1}{9}u_v^p(x) + \frac{4}{9}d_v^p(x) + \frac{10}{9}\overline{u}^p(x) + \frac{2}{9}\overline{s}^p(x)\right]$$

and

$$F_2^{ep}\left(Q^2, x\right) - F_2^{en}\left(Q^2, x\right) \sim \frac{1}{3}x\left(u_V{}^p(x) - d_V{}^p(x)\right) .$$

Integrating over the full x range, one has

$$\int_0^1 \frac{1}{x}\left\{F_2{}^{ep}\left(Q^2, x\right) - F_2{}^{en}\left(Q^2, x\right)\right\} dx \simeq \frac{1}{3}. \tag{5.56}$$

This is the so-called Gottfried sum rule. This rule is, however, strongly violated in experimental data (the measured value is 0.235 ± 0.026) showing the limits of the naïve quark-parton model. There is probably an isospin violation in the sea quark distributions.

The Q^2 dependence of the structure functions (Fig. 5.26) was systematically measured by several experiments, in particular, at the HERA electron–proton collider, where a wide Q^2 and x range was covered ($2.7 < Q^2 < 30000$ GeV^2; $6\ 10^{-5} < x < 0.65$). For $x > 0.1$, the scaling is reasonably satisfied but for small x, the F_2 structure function clearly increases with Q^2. This behavior is well predicted by the theory of strong interactions and basically reflects the resolution power of the exchanged virtual photon. A higher Q^2 corresponds to a smaller wavelength ($\lambda \sim 1/\sqrt{Q^2}$), and therefore a much larger number of sea quarks with a very small x can be seen.

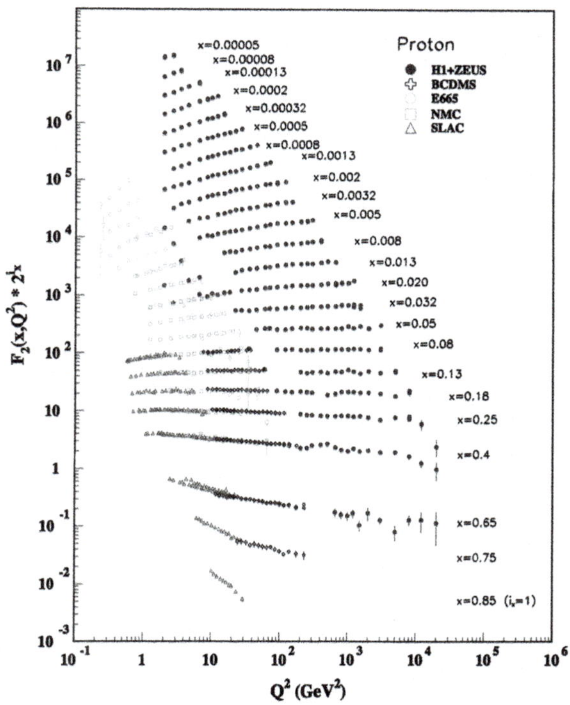

Fig. 5.26 Q^2dependence of $F_2{}^{ep}$. From K.A. Olive et al. (Particle Data Group), Chin. Phys. C 38 (2014) 090001

5.5.5 The Number of Quark Colors

A direct experimental test of the number of colors, N_c, comes from the measurement of the R ratio of the hadronic cross section in e^+e^- annihilations to the $\mu^+\mu^-$ cross section, defined as

$$R = \frac{\sigma\left(e^+e^- \to q\overline{q}\right)}{\sigma(e^+e^- \to \mu^+\mu^-)} .$$

At low energies $\left(\sqrt{s} < m_Z\right)$, these processes are basically electromagnetic and are mediated at the first order by one virtual photon (γ^*). The cross sections are thus proportional to the squares of the electric charges of the final state particles q_i^2. A rule-of-thumb that can frequently be helpful (note the analogies with the Rutherford cross section) is

$$\sigma(e^+e^- \to \mu^+\mu^-) \simeq \frac{4\pi\alpha}{3s} q_i^2 \simeq \frac{86.8\text{nb}}{s[\text{GeV}^2]} q_i^2 . \tag{5.57}$$

When considering more than one flavor (for example in the case of hadronic final states), a sum over all the possible final states has to be performed. Thus

$$R = \frac{\sigma(e^+e^- \to hadrons)}{\sigma(e^+e^- \to \mu^+\mu^-)} \simeq N_c \sum_i q_i^2 . \tag{5.58}$$

The sum runs over all the quark flavors with mass $m_i < \frac{1}{2}\sqrt{s}$, and over all colors. For $\sqrt{s} \lesssim 3$ GeV, just the u, d and s quarks can contribute and then

$$R = \frac{2}{3} N_c .$$

For 3 GeV $\lesssim \sqrt{s} \lesssim 5$ GeV, there is also the contribution of the c quark and

$$R = \frac{10}{9} N_c .$$

Finally, for $\sqrt{s} \gtrsim 5$ GeV, the b quark contributes

$$R = \frac{11}{9} N_c .$$

The mass of the top quark is too high for the $t\overline{t}$ pair production to be accessible at the past and present e^+e^- colliders.

The measurements for $\sqrt{s} \lesssim 40$ GeV, summarized in Fig. 5.27, show, apart from regions close to the resonances, a fair agreement between the data and this naïve prediction provided $N_c = 3$. Above $\sqrt{s} \gtrsim 40$ GeV, the annihilation via the exchange of a Z boson starts to be non-negligible and the interference between the two channels is visible, the calculation in Eq. 5.58 being no more valid (see Chap. 7).

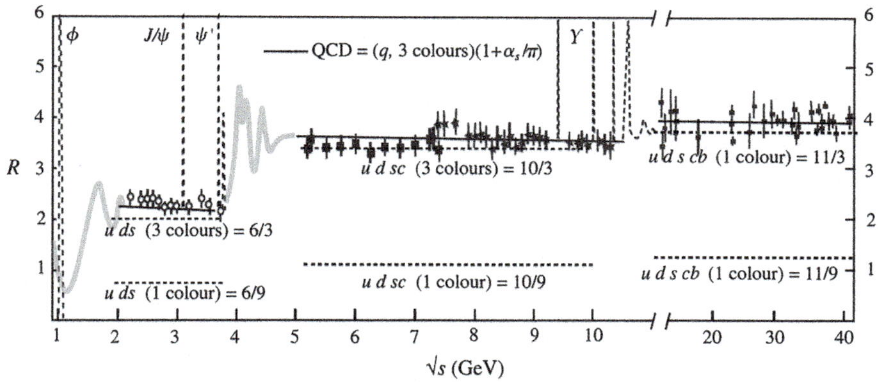

Fig. 5.27 Measurements of $R\left(\sqrt{s}\right)$. From [F5.2] in the "further readings"; the data are taken from K.A. Olive et al. (Particle Data Group), Chin. Phys. C 38 (2014) 090001

5.6 Leptons

The existence of particles not interacting strongly (the leptons) is indispensable to the architecture of the Universe. The first such particle discovered, the electron, has electromagnetic charge -1 and it is one of the fundamental constituents of the atoms. Later on, the neutral neutrino had to be postulated to save the energy-momentum conservation law, as seen in Chap. 2. Finally, three families each made by one charged and one neutral leptons (and their corresponding antiparticles) were discovered, symmetrically with the structure of the three quark families:

$$\begin{pmatrix} \nu_e \\ e^- \end{pmatrix} \begin{pmatrix} \nu_\mu \\ \mu^- \end{pmatrix} \begin{pmatrix} \nu_\tau \\ \tau^- \end{pmatrix} .$$

Electrons, muons, and the experimental proof of the existence of neutrinos were already discussed in Chaps. 2 and 3. Neutrino oscillations and neutrino masses will be discussed in Chap. 9.

The τ (tau) lepton, with its neutrino, was the last discovered.

5.6.1 The Discovery of the τ Lepton

The third charged lepton, the τ (tau), was discovered in a series of experiments lead by Martin Perl, using the Mark I detector at the SPEAR e^+e^-storage ring in the years 1974–1976. The first evidence was the observation of events with only two charged particles in the final state: an electron or a positron and an opposite sign muon, with missing energy and momentum (Fig. 5.28). The conservation of energy and momentum indicated the existence in such events of at least two undetected particles (neutrinos).

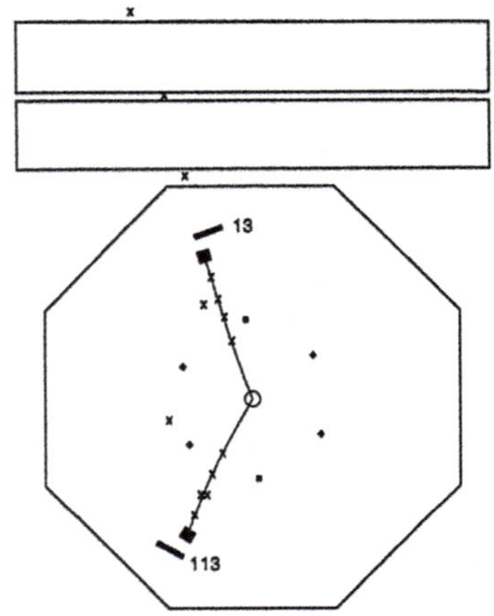

Fig. 5.28 An $e\mu$ event observed at Mark I. The muon moves upward and the electron downward. The numbers 13 and 113 give the relative amount of the electromagnetic energy deposited. Credit: Martin Perl et al., M. L. Perl et al., "Evidence for Anomalous Lepton Production in e^+e^- Annihilation," Phys. Rev. Lett. 35 (1975) 1489

There was no conventional explanation for those events: one had to assume the existence of a new heavy lepton, the τ. In this case, a $\tau^+\tau^-$ pair could have been produced,

$$e^+e^- \rightarrow \tau^+\tau^-$$

followed by the weak decay of each τ into its (anti)neutrino plus a W boson (Fig. 5.29); the W boson, as it will be explained in the next chapter, can then decay in one of the known charged leptons ($l = e, \mu$) plus the corresponding neutrino and antineutrino:

$$\tau^- \rightarrow \nu_\tau\, W^- \rightarrow \nu_\tau\, l^-\, \overline{\nu}_l \;\; ; \;\; \tau^+ \rightarrow \overline{\nu}_\tau\, W^+ \rightarrow \overline{\nu}_\tau\, l^+ \nu_l \; .$$

A confirmation came two years later with the observation of the τ hadronic decay modes:

$$\tau^- \rightarrow \nu_\tau \; + W^- \rightarrow \nu_\tau \; + hadrons \, .$$

Indeed, the τ is massive enough ($m_\tau \simeq$ 1.8 GeV) to allow such decays (the W^- being virtual).

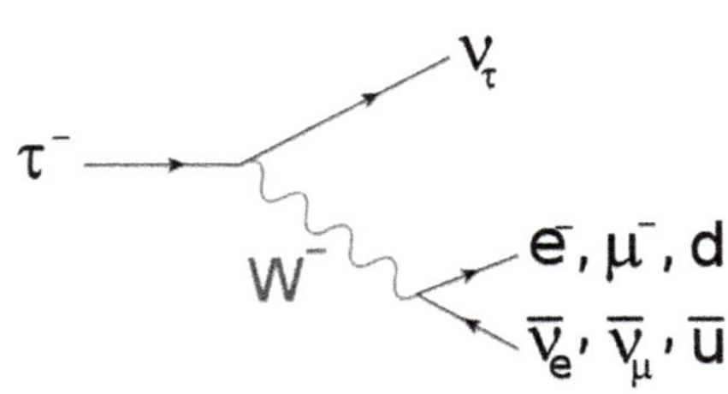

Fig. 5.29 Feynman diagram for the τ^- decay in $\nu_\tau\, e^-\, \overline{\nu}_e$, $\nu_\tau\, \mu^-\, \overline{\nu}_\mu$, $\nu_\tau\, d\, \overline{u}$. By en:User:JabberWok and Time 3000 [GFDL http://www.gnu.org/copyleft/fdl.html], via Wikimedia Commons

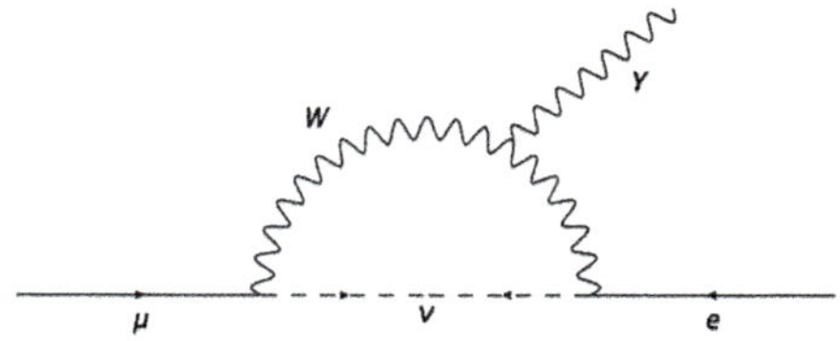

Fig. 5.30 Feynman diagram for the decay $\mu^- \to e^- \gamma$ in the case where just one neutrino species exists

5.6.2 Three Neutrinos

Muons decay into electrons and the electron energy spectrum is continuous. Once again, this excludes a two-body decay and thus at least two neutral "invisible" particles should be present in the final state:

$$\mu^- \to e^- \nu_1 \overline{\nu}_2 \, .$$

Is $\overline{\nu}_2$ the antiparticle of ν_1? Lee and Yang were convinced, in 1960, that it should not be so (otherwise the Feynman diagram represented in Fig. 5.30, would be possible and then the branching fraction for $\mu^- \longrightarrow e^- \gamma$ would be large). At least two different species of neutrinos should exist.

Around the same time, the possibility to produce a neutrino beam from the decay of pions created in the collision of GeV protons on a target was intensively discussed in the cafeteria of the Columbia University. In 1962, a kind of a neutrino beam was finally available at BNL (Brookhaven National Laboratory): the idea was to put an internal target in a long straight section of the proton accelerator and to drive with a magnet the proton beam on it; the pions coming from the proton interactions were then decaying into pions. An experiment led by Leon Lederman, Melvin Schwartz, and Jack Steinberger was set to observe the neutrino reaction within a 10-ton spark chamber. Hundreds of millions of neutrinos were produced mostly accompanied by a muon ($BR(\pi \longrightarrow \mu\, \nu) \gg BR(\pi \longrightarrow e\, \nu)$ as it will be discussed in the next chapter). Forty neutrino interactions in the detector were clearly identified; in six of them, the final state was an electron, and in 34, the final state was a muon. The ν_μ and ν_e are, thus, different particles, otherwise the same number of events with one electron and with one muon in the final state should have been observed.

Only in the last year of the 20th century, the direct evidence for the third neutrino was established. The DONUT experience at Fermilab found four events in six millions where a τ lepton was clearly identified. In these events, the reconstruction of the charged tracks in the iron/emulsion target showed a characteristic kink indicating at least one invisible particle produced in the decay of a heavy particle into a muon (Fig. 5.31).

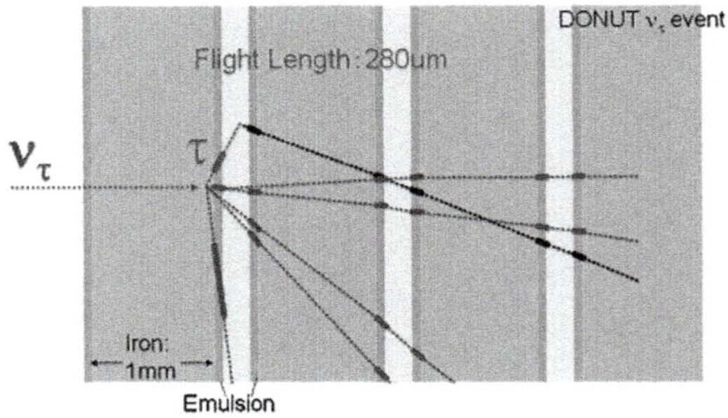

Fig. 5.31 Tau-neutrino event in DONUT. A tau-neutrino produces several charged particles. Among them a tau particle, which decays to another charged particle with missing energy (at least one neutrino). From K. Kodama et al., DONUT Collaboration, "Observation of tau neutrino interactions," Phys. Lett. B 504 (2001) 218

5.7 The Particle Data Group and the Particle Data Book

How can one manage all this information about so many particles? How can one remember all these names? The explosion of particle discoveries has been so great, that Fermi said, "If I could remember the names of all these particles, I'd be a botanist."

Fortunately, a book, called the Review of Particle Physics (also known as the Particle Data Book), can help us. It is edited by the Particle Data Group (in short PDG), an international collaboration of about 150 particle physicists, helped by some 500 consultants, that compiles and organizes published results related to the properties of particles and fundamental interactions, reviewing in addition theoretical advancements relevant for experimental physics. The PDG publishes the Review of Particle Physics and its pocket version, the Particle Physics Booklet, which are printed biennially in paper, and updated annually in the Web. The PDG also maintains the standard numbering scheme for particles in event generators (Monte Carlo simulations).

The Review of Particle Physics is a voluminous reference work (more than one thousand pages); it is currently the most referenced article in high energy physics, being cited more than 2,000 times per year in the scientific literature. It is divided into 3 sections:

- Particle physics summary tables—Brief tables with the properties of of particles.
- Reviews, tables, and plots—Review of fundamental concepts from mathematics and statistics, tables related to the chemical and physical properties of materials, review of current status in the fields of Standard Model, Cosmology, and experimental methods of particle physics, tables of fundamental physical and astronomical constants, summaries of relevant theoretical subjects.
- Particle listings—Extended version of the Particle Physics Summary Tables, with reference to the experimental measurements.

The Particle Physics Booklet (about 300 pages) is a condensed version of the Review, including the summary tables, and a shortened section of reviews, tables, and plots.

The publication of the Review of Particle Physics in its present form started in 1970; formally, it is a journal publication, appearing in different journals depending on the year.

5.7.1 PDG: Estimates of Physical Quantities

The "particle listings" (from which the "summary tables" are extracted) contain all relevant data known to the PDG team that are published in journals. From these data, "world averages" are calculated.

Sometimes a measurement might be excluded from a world average. Among the reasons of exclusion are the following (as reported by the PDG itself, K.A. Olive et al. (Particle Data Group), Chin. Phys. C 38 (2014) 090001):

- it is superseded by or included in later results;
- no error is given;
- it involves questionable assumptions;
- it has a poor signal-to-noise ratio, low statistical significance, or is otherwise of poorer quality than other data available;
- it is clearly inconsistent with other results that appear to be more reliable.

Several kinds of "world average" are provided:

- OUR AVERAGE—From a weighted average of selected data.
- OUR FIT—From a constrained or overdetermined multiparameter fit of data.
- OUR EVALUATION—Not from a direct measurement, but evaluated from measurements of related quantities.
- OUR ESTIMATE—Based on the observed range of the data, not from a formal statistical procedure.

5.7.2 Averaging Procedures by the PDG

The average is computed by a χ^2 minimization. There is an attempt to use uncorrelated variables as much as possible, but correlations are taken into account.

When the error for a measurement x is asymmetric, the error used is a continuous function of the errors δx_+ and δx_-. When the resultant average x is less than $x - \delta x_-$, δx_- is used; when it is greater than $x + \delta x_+$, δx_+ is used; in between, the error is a linear function of x.

Sometimes measurements are inconsistent. Possible inconsistencies are evaluated on the basis of the χ^2, as follows. The PDG calculates a weighted average and error as

$$\bar{x} \pm \delta\bar{x} = \frac{\sum_i w_i x_i}{\sum_i w_i} \quad \text{with} \quad w_i = \frac{1}{(\delta x_i)^2}\,. \tag{5.59}$$

Then $\chi^2 = \sum_i w_i (x_i - \bar{x})^2$.

- If $\chi^2/(N-1)$ is less than or equal to 1, and there are no known problems with the data, the results are accepted.
- If $\chi^2/(N-1)$ is very large, the PDG may
  - not to use the average at all, or
  - quote the calculated average, making an educated (conservative) guess of the error.
- If $\chi^2/(N-1)$ is greater than 1, but not so much, the PDG still averages the data, but then also increases the error by $S = \sqrt{\chi^2/(N-1)}$. This scaling procedure for errors does not affect central values.

If the number of experiments is at least three, and $\chi^2/(N-1)$ is greater than 1.25, an ideogram of the data is shown in the Particle Listings. Figure 5.32 is an example. Each measurement is shown as a Gaussian with a central value x_i, error δx_i, and area proportional to $1/\delta x_i$.

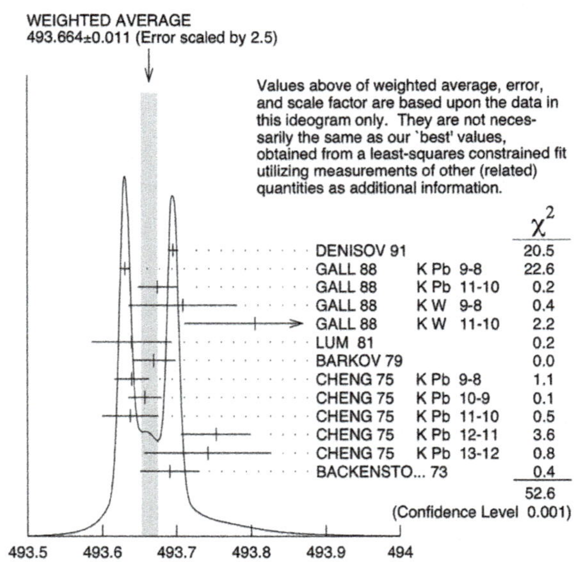

Fig. 5.32 An ideogram representing clearly inconsistent data—the measurements of the mass of the charged kaon. The *arrow* at the *top* shows the position of the weighted average, while the width of the *shaded pattern* shows the error in the average after scaling by the factor S. From K.A. Olive et al. (Particle Data Group), Chin. Phys. C 38 (2014) 090001

Further Reading

[F5.1] S. Haywood, "Symmetries and Conservation laws in Particle Physics: an introduction to group theory for experimental physicists" Imperial College Press 2011. An excellent introduction to group theory and its application in particle physics.

[F5.2] A. Bettini, "Introduction to Elementary Particle Physics," Cambridge University Press 2014. A good introduction to Particle Physics at the undergraduate level putting together experimental and theoretical aspects and discussing basic and relevant experiments

[F5.3] M. Thomson, "Modern Particle Physics," Cambridge University Press, 2013. A recent, pedagogical and rigorous book covering the main aspects of Particle Physics at advanced undergraduate and early graduate level.

[F5.4] PDG 2014, K.A. Olive et al. (Particle Data Group), Chin. Phys. C, 38, 090001 (2014). URL: http://pdg.lbl.gov/. This is also the most quoted reference in this book.

Exercises

1. Compute the kinematic threshold of the reaction $pp \to ppp\bar{p}$ in a fixed target experiment.
2. Why is the hypothesis that the neutron is a bound state of electron and proton is inconsistent with Heisenberg's uncertainty principle?
3. Demonstrate that if $\hat{A}$ and $\hat{B}$ are two operators the relation 5.13 holds:

$$\exp\left(\hat{A}+\hat{B}\right) = \exp\left(\frac{1}{2}\left[\hat{A}, \hat{B}\right]\right) \exp(\hat{A})\ \exp(\hat{B}) .$$

4. *Cross sections and isospin.* Determine the ratio of the following interactions cross sections at the Δ^{++} resonance:
   $\pi^- p \to K^0\Sigma^0$; $\pi^- p \to K^+\Sigma^-$; $\pi^+ p \to K^+\Sigma^+$.
5. *Decay branching ratios and isospin.* Consider the decays of the Σ^{*0} into $\Sigma^+\pi^-$, $\Sigma^0\pi^0$ and $\Sigma^-\pi^+$. Determine the ratios between the decay rates in these decay channels.
6. *Quantum numbers.* Verify if the following reactions/decays are possible and if not say why:

   (a) $pp \to \pi^+\pi^-\pi^0$,
   (b) $pp \to ppn$,
   (c) $pp \to ppp\bar{p}$,
   (d) $p\bar{p} \to \gamma$,
   (e) $\pi^- p \to K^0\Lambda$,
   (f) $n \to pe^-\nu$,

(g) $\Lambda \to \pi^- p$,
(h) $e^- \to \nu_e \gamma$.

7. Ω^- mass. Verify the relations between the masses of all the particles lying in the fundamental baryon decuplet but the $\Omega^-$and predict the mass of this one. Compare your prediction with the measured value.
8. *Experimental resolution in deep inelastic scattering.* Consider an $e^- p$ deep inelastic scattering experiment where the electron scattering angle is $\sim 6°$. Make an estimation of the experimental resolution in the measurement of the energy of the scattered electron that is needed to distinguish the elastic channel ($e^- p \to e^- p$) from the first inelastic channel ($e^- p \to e^- p\pi^0$).
9. *$e^- p$ deep inelastic scattering kinematics.* Consider the $e^- p$ deep inelastic scattering and deduce the following formula:

$$Q^2 = 4EE' \sin^2(\theta/2)$$

$$Q^2 = 2M_p \nu$$

$$Q^2 = xy(s^2 - y^2)\,.$$

10. *Gottfried sum rule.* Deduce in the framework of the quark-parton model the Gottfried sum rule

$$\int \frac{1}{x}\left(F_2^{ep}(x) - F_2^{ep}(x)\right) dx = \frac{1}{3} + \frac{2}{3}\int \left(\bar{u}(x) - \bar{d}(x)\right) dx$$

and comment the fact that the value measured in $e^- p$ and $e^- d$ deep inelastic elastic scattering experiments is approximately 1/4.

Chapter 6
Interactions and Field Theories

Quantum field theories are a theoretical framework for constructing models describing particles and their interactions. The dynamics of a system can be determined starting from the Lagrangian of an interaction through canonical equations. This chapter introduces the basic formalism, illustrates the relation between symmetries of the Lagrangian and conserved quantities, and finally describes the Lagrangian for the most relevant interactions at the particle level: the electromagnetic interaction (QED), the weak interaction, and the strong interaction.

The structure and the dynamics of the Universe are determined by the so-called fundamental interactions: gravitational, electromagnetic, weak, and strong. In their absence, the Universe would be an immense space filled with ideal gases of structureless particles. Interactions between "matter" particles (fermions) are in relativistic quantum physics associated to the exchange of "wave" particles (bosons)—note that bosons can also interact among themselves. Such a picture can be visualized (and observables related to the process can be computed) using the schematic diagrams invented in 1948 by Richard Feynman: the Feynman diagrams (Fig. 6.1), that we have shortly presented in Chap. 2.

Each diagram corresponds to a specific term of a perturbative expansion of the scattering amplitude. It is a symbolic graph, where initial and final state particles are represented by incoming and outgoing lines (which are not space–time trajectories) and the internal lines represent the exchange of virtual particles (the term "virtual" meaning that their energy and momentum do not have necessarily to be related through the relativistic equation $E^2 = p^2 + M^2$; if they are not, they are said to be off the mass shell). Solid straight lines are associated to fermions while wavy, curly, or broken lines are associated to bosons. Arrows indicate the time flow of the external particles and antiparticles (in the plot time runs usually from left to right, but having it running from bottom to top is also a possible convention). A particle (anti-particle) moving backwards in time is equivalent to its anti-particle (particle) moving forwards in time.

At the lowest order, the two initial state particles exchange only a particle mediating the interaction (for instance a photon). Associated to each vertex (a point where at least three lines meet) is a number, the coupling constant (in the case of

A. De Angelis and M.J.M. Pimenta, *Introduction to Particle and Astroparticle Physics*, Undergraduate Lecture Notes in Physics,
DOI 10.1007/978-88-470-2688-9_6

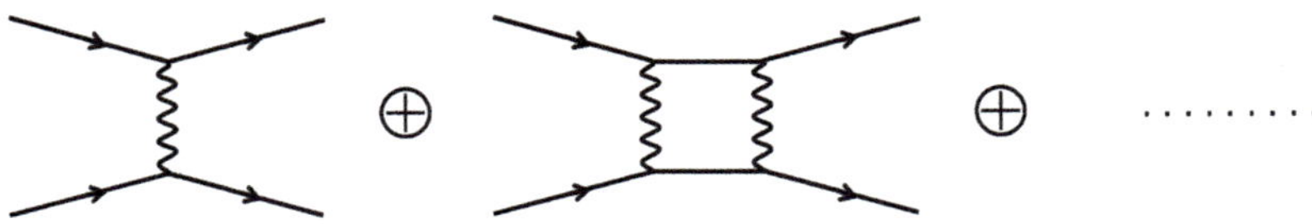

Fig. 6.1 Feynman diagrams

electromagnetic interaction $z\sqrt{\alpha} = ze/\sqrt{4\pi}$ for a particle with electrical charge z), which indicates the probability of the emission/absorption of the field particle and thus the strength of the interaction. Energy–momentum, as well as all the quantum numbers, is conserved in each vertex.

At higher orders, more than one field particle can be exchanged (second diagram from the left in the Figure) and there is an infinite number of possibilities (terms in the perturbative expansion) for which amplitudes and probabilities are proportional to increasing powers of the coupling constants. Although the scattering amplitude is proportional to the square of the sum of all the terms, if the coupling constants are small enough, just the first diagrams will be relevant. However, even low-order diagrams can give an infinite contribution. Indeed in the second diagram, there is a loop of internal particles and an integration over the exchanged energy-momentum has to be carried out. Since this integration is performed in a virtual space, it is not bound and therefore it might, in principle, diverge. Curing divergent integrals (or, in jargon, "canceling infinities") became the central problem of quantum field theory in the middle of the twentieth century (classically the electrostatic self-energy of a point charged particle is also infinite) and it was successfully solved in the case of electromagnetic interaction, as it will be briefly discussed in Sect. 6.2.9, within the renormalization scheme.

The quantum equations for "matter" (Schrödinger, Klein-Gordon, Dirac equations) must be modified to incorporate explicitly the couplings with the interaction fields. The introduction of these new terms makes the equations invariant to a combined *local* (space–time dependent) transformation of the matter and of the interactions fields (the fermion wave phase and the four-momentum potential degree of freedom in case of the electromagnetic interactions). Conversely requiring that the "matter" quantum equations should be invariant to local transformation of some internal symmetry groups implies the existence of well-defined interaction fields, the gauge fields. These ideas, developed in particular by Feynman and by Yang and Mills in the 1950s, were applied to the electromagnetic, weak and strong interactions field theories; they provided the framework for the unification of the electromagnetic and weak interactions (electroweak interactions) which has been extensively tested with an impressive success (see next Chapter) and may lead to further unification involving strong interaction (GUTs—Grand Unified Theories) and even gravity (ToE—Theories of Everything). One could think that we are close to the "end of physics." However, the experimental discovery that most of the energy of the Universe cannot be explained by the known physical objects quickly dismissed such claim—in fact Dark Matter and Dark Energy represent around 95 % of the total energy budget of the Universe, and they are not explained by present theories.

6.1 The Lagrangian Representation of a Dynamical System

In the quantum world, we usually find it convenient to use the Lagrangian or the Hamiltonian representation of a system to compute the equations of motion. The Lagrangian L of a system of particles is defined as

$$L = K - V \tag{6.1}$$

where is the K is the total kinetic energy of the system, and V the total potential energy.

Any system with n degrees of freedom, is fully described by n generalized coordinates q_i and n generalized velocities $\dot{q}_i$. The equations of motion of the system are the so-called Euler–Lagrange equations

$$\frac{d}{dt}\left(\frac{\partial L}{\partial \dot{q}_j}\right) = \frac{\partial L}{\partial q_j} \tag{6.2}$$

where the index $j = 1, 2, \ldots, n$ runs over the degrees of freedom. For example, in the case of a single particle in a conservative field in one dimension, x, one can write

$$L = \frac{1}{2}mv^2 - V(x) \tag{6.3}$$

and applying the Euler–Lagrange equations

$$\frac{d}{dt}mv = -\frac{d}{dx}V = F \Longrightarrow F = ma$$

(Newton's law).

Although the mathematics required for Lagrange's equations might seem more complicated than Newton's law, Lagrange equations make often the solution easier, since the generalized coordinates can be conveniently chosen to exploit symmetries in the system, and constraint forces are incorporated in the geometry of the problem.

The Lagrangian is of course not unique: you can multiply it by a constant factor, for example, or add a constant, and the equations will not change. You can also add the four-divergence of an arbitrary vector function: it will cancel when you apply the Euler–Lagrange equations, and thus the dynamical equations are not affected.

The so-called Hamiltonian representation uses instead the Hamiltonian function $H(p_j, q_j, t)$:

$$H = K + V\,. \tag{6.4}$$

We have already shortly discussed in the previous chapter this function, which represents the total energy in terms of generalized coordinates q_j and of generalized momenta

$$p_j = \frac{\partial H}{\partial \dot{q}_j} \, . \tag{6.5}$$

The time evolution of the system is obtained by the Hamilton's equations:

$$\frac{dp_j}{dt} = -\frac{\partial H}{\partial q_j} \tag{6.6}$$

$$\frac{dq_j}{dt} = \frac{\partial H}{\partial p_j} \, . \tag{6.7}$$

The two representations, Lagrangian and Hamiltionian, are equivalent. For example, in the case of a single particle in a conservative field in one dimension

$$H = \frac{p^2}{2m} + V \tag{6.8}$$

and Hamilton's equations become

$$\frac{dp}{dt} = -\frac{dV}{dx} = F \; ; \; \frac{dx}{dt} = \frac{p}{m} \, . \tag{6.9}$$

We shall use more frequently Lagrangian mechanics. Let us now see how Lagrangian mechanics simplifies the description of a complex system.

6.1.1 The Lagrangian and the Noether Theorem

Noether's theorem is particularly simple when the Lagrangian representation is used. If the Lagrangian does not depend on the variable q_i, the Euler–Lagrange equation related to this coordinate becomes

$$\frac{d}{dt}\left(\frac{\partial L}{\partial \dot{q}_i}\right) = 0 \tag{6.10}$$

and thus the quantity

$$\left(\frac{\partial L}{\partial \dot{q}_i}\right) = p_i$$

is conserved. For example the invariance to space translation implies that linear momentum is conserved. By a similar approach, we could see that the invariance to rotational translation implies that angular momentum is conserved.

6.1.2 Lagrangians and Fields; Lagrangian Density

The Euler–Lagrange equations are derived imposing the stationarity of an action S defined as $S = \int dt\, L$; such a form, giving a special role to time, does not allow a relativistically covariant Lagrangian L.

We can recover relativistic covariance using instead of the Lagrangian a "Lagrangian density" $\mathcal{L}$, such that the Lagrangian will be the integral of $\mathcal{L}$ over all space,

$$L = \int d^3x\, \mathcal{L}\,. \tag{6.11}$$

Now we can write

$$S = \int dt\, L = \int d^4x\, \mathcal{L}\,. \tag{6.12}$$

In a quantum mechanical world, $\mathcal{L}$ can depend, instead than on coordinates and velocities, on fields, $\phi(\vec{r}, t) = \phi(x^\mu)$, which are meaningful quantities in the four-dimensional space of relativity. Quantum mechanics guarantees the invariance of physics with respect to a *global* rotation of the wave function in complex space, i.e., the multiplication for a constant phase: $\phi \to \phi e^{i\theta}$. This means that, in general, a Lagrangian will be the combination of functions $|\phi|^2$ or $|\partial\phi|^2$. The latter are called, with obvious meaning, kinetic terms.

The same argument leading to the Euler–Lagrange equations leads now to generalized Euler–Lagrange equations

$$\partial_\mu \left(\frac{\partial \mathcal{L}}{\partial(\partial_\mu \phi_i)} \right) - \frac{\partial \mathcal{L}}{\partial \phi_i} = 0 \tag{6.13}$$

for fields ϕ_i $(i = 1, \ldots, n)$.

Noether's theorem guarantees that, if the Lagrangian density does not depend explicitly on the field ϕ, we have a four-current

$$j^\mu \equiv \frac{\partial \mathcal{L}}{\partial(\partial_\mu \phi)} \tag{6.14}$$

subject to the continuity condition

$$\partial_\mu j^\mu = 0 \Rightarrow -\frac{\partial j^0}{\partial t} + \vec{\nabla} \cdot \vec{j} = 0\,, \tag{6.15}$$

where j^0 is the charge density, and $\vec{j}$ is the current density. The total (conserved) charge will be

$$Q = \int_{\text{all space}} d^3x\, j^0\,. \tag{6.16}$$

Hamilton's formalism can be also extended to relativistic quantum fields.

In the rest of the book, we shall in general make use of Lagrangian densities $\mathcal{L}$, but unless and otherwise specified we shall refer to the Lagrangian densities simply as Lagrangians.

6.1.3 Lagrangian Density and Mass

A Lagrangian is in general composed of generalized coordinates and of their derivatives (or of fields and their derivatives).

We shall show later that a nonzero mass—i.e., a positive energy for a state at rest—is associated in field theory to an expression quadratic in the field; for instance, in the case of a scalar field,

$$\mathcal{L}_K = \frac{1}{2} m^2 |\phi|^2 \,. \tag{6.17}$$

The dimension of the Lagrangian density is [energy^4] since the action (6.12) is dimensionless; the scalar field ϕ has thus the dimension of an energy.

6.2 Quantum Electrodynamics (QED)

Electromagnetic effects were known since the antiquity, but just during the nineteenth century the (classical) theory of electromagnetic interactions was firmly established. In the twentieth century, the marriage between electrodynamics and quantum mechanics (Maxwell's equations were already relativistic even before the formulation of Einstein's relativity) gave birth to the theory of Quantum Electrodynamics (QED), which is the most accurate theory ever formulated. QED describes the interactions between charged electrical particles mediated by a quantized electromagnetic field.

6.2.1 Electrodynamics

In 1864, James Clerk Maxwell accomplished the "second great unification in Physics" (the first one was realized by Isaac Newton) formulating the theory of electromagnetic field and summarizing it in a set of coupled differential equations. The Maxwell's equations can be written using the vector notation introduced by Heaviside and following the Lorentz–Heaviside convention for units (see Chap. 2) as

$$\vec{\nabla} \cdot \vec{E} = \rho$$

$$\vec{\nabla} \cdot \vec{B} = 0$$

$$\vec{\nabla} \times \vec{E} = -\frac{\partial \vec{B}}{\partial t}$$

$$\vec{\nabla} \times \vec{B} = \vec{j} + \frac{\partial \vec{E}}{\partial t}\,.$$

A scalar potential ϕ and a vector potential $\vec{A}$ can be introduced such that

$$\vec{E} = -\vec{\nabla}\phi - \frac{\partial \vec{A}}{\partial t}$$

$$\vec{B} = \vec{\nabla} \times \vec{A}\,.$$

Then two of the Maxwell equations are automatically satisfied:

$$\vec{\nabla} \cdot \vec{B} = \vec{\nabla} \cdot \left(\vec{\nabla} \times \vec{A}\right) = 0$$

$$\vec{\nabla} \times \vec{E} = \vec{\nabla} \times \left(-\vec{\nabla}\phi - \frac{\partial \vec{A}}{\partial t}\right) = -\frac{\partial \vec{B}}{\partial t}$$

and the other two can be written as:

$$\vec{\nabla} \cdot \vec{E} = \vec{\nabla} \cdot \left(-\vec{\nabla}\phi - \frac{\partial \vec{A}}{\partial t}\right) = \rho$$

$$\vec{\nabla} \times \vec{B} = \vec{\nabla} \times \left(\vec{\nabla} \times \vec{A}\right) = \vec{j} + \frac{\partial}{\partial t}\left(-\vec{\nabla} \cdot \phi - \frac{\partial \vec{A}}{\partial t}\right).$$

However, the potential fields $(\phi, \vec{A})$ are not totally determined, having a local degree of freedom. In fact, if $\chi(t, \vec{x})$ is a scalar function of the time and space coordinates, then the potentials $(\phi, \vec{A})$ defined as

$$\phi' = \phi - \frac{\partial \chi}{\partial t}$$

$$\vec{A}' = \vec{A} + \vec{\nabla}\chi$$

give origin to the same $\vec{E}$ and $\vec{B}$ fields. These transformations are designated as gauge transformations and generalize the freedom that exist in electrostatics in the definition of the space points where the electric potential is zero (the electrostatic field is invariant under a global transformation of the electrostatic potential, but the electromagnetic field is invariant under a joint local transformation of the scalar and vector potential).

The arbitrariness of these transformations can be used to write the Maxwell equations in a simpler way. What we are going to do is to use our choice to fix things so that the equations for $\vec{A}$ and for ϕ are separated but have the same form. We can do this by taking (this is called the Lorenz gauge):

$$\vec{\nabla} \cdot \vec{A} = -\frac{\partial \phi}{\partial t} . \tag{6.18}$$

Thus

$$\frac{\partial^2 \phi}{\partial t^2} - \nabla^2 \phi = \rho$$

$$\frac{\partial^2 \vec{A}}{\partial t^2} - \nabla^2 \vec{A} = \vec{j} .$$

These two equations can be written in an extremely compact way if four-vectors A^μ and j^μ are introduced and if the D'Alembert operator $\Box \equiv \partial^\mu \partial_\mu$ is used. Defining

$$A^\mu = \left(\phi, \vec{A}\right) \; ; \; j^\mu = \left(\rho, \vec{j}\right) \tag{6.19}$$

(notice that the Lorenz gauge $\partial_\mu A^\mu = 0$ is covariant), the two equations are summarized in

$$\Box A^\mu = j^\mu . \tag{6.20}$$

In the absence of charges and currents (free electromagnetic field)

$$\Box A^\mu = 0 . \tag{6.21}$$

This equation is similar to the Klein–Gordon equation for one particle with $m = 0$ (see Sects. 5.7.2 and 6.2.5) but with spin 1. A^μ is identified with the wave function of a free photon, and the solution of the above equation is, up to some normalization factor:

$$A^\mu = \epsilon^\mu (q) \, e^{-iqx} \tag{6.22}$$

where q is the four-momentum of the photon and ϵ^μ its the polarization four-vector. The four components of ϵ^μ are not independent. The Lorenz condition imposes one constraint, reducing the number of independent component to three. However, even after imposing the Lorenz condition, there is still the possibility, if $\partial^2 \chi = 0$, of a further gauge transformation

$$A^\mu \to A^\mu + \partial^\mu \chi . \tag{6.23}$$

This extra gauge transformation can be used to set the time component of the polarization four-vector to zero $\left(\epsilon^0 = 0\right)$ and thus converting the Lorenz condition into

$$\vec{\epsilon} \cdot \vec{q} = 0 \,.$$

This choice is known as the Coulomb gauge and it makes clear that there are just two degrees of freedom left for the polarization which is the case of mass zero spin 1 particles ($m_s = \pm 1$).

6.2.1.1 Modification for a Nonzero Mass: The Proca Equation

In the case of a photon with a tiny mass μ_γ:

$$\left(\Box - {\mu_\gamma}^2\right) A^\mu = j^\mu \,, \tag{6.24}$$

the Maxwell equations would be transformed into the Proca[1] equations:

$$\vec{\nabla} \cdot \vec{E} = \rho - {\mu_\gamma}^2 \phi$$

$$\vec{\nabla} \cdot \vec{B} = 0$$

$$\vec{\nabla} \times \vec{E} = -\frac{\partial \vec{B}}{\partial t}$$

$$\vec{\nabla} \times \vec{B} = \vec{j} + \frac{\partial \vec{E}}{\partial t} - {\mu_\gamma}^2 \vec{A}.$$

In this scenario, the electrostatic field would show a Yukawa-type exponential attenuation, $e^{-\mu_\gamma r}$. Experimental tests of the validity of the Coulomb inverse square law have been performed since many years in experiments using different techniques, leading to stringent limits: $\mu_\gamma < 10^{-18}\,\mathrm{eV} \sim 10^{-51}\,\mathrm{g}$. Stronger limits ($\mu_\gamma < 10^{-26}\,\mathrm{eV}$) are reported from the analyses of astronomical data, but are model dependent.

6.2.2 *Minimal Coupling*

Classically, the coupling between a particle with charge e and the electromagnetic field is given by the Lorentz force:

---

[1] Alexandru Proca (1897–1955) was a Romanian physicist who studied and worked in France (he was a student of Louis de Broglie). He developed the vector meson theory of nuclear forces and worked on relativistic quantum field equations.

$$\vec{F} = e\left(\vec{E} + \vec{v} \times \vec{B}\right)$$

which can be written in terms of scalar and vector potential as

$$\frac{d\vec{p}}{dt} = e\left(-\vec{\nabla}\phi - \frac{\partial \vec{A}}{\partial t} + \vec{v} \times \left(\vec{\nabla} \times \vec{A}\right)\right) = e\left(-\vec{\nabla}\phi - \frac{\partial \vec{A}}{\partial t} + \vec{\nabla}\left(\vec{v} \cdot \vec{A}\right) - \left(\vec{v} \cdot \vec{\nabla}\right)\vec{A}\right)$$

$$= e\left(-\vec{\nabla}\left(\phi - \vec{v} \cdot \vec{A}\right) - \frac{\partial \vec{A}}{\partial t} - \left(\vec{v} \cdot \vec{\nabla}\right)\vec{A}\right) = e\left(-\vec{\nabla}\left(\phi - \vec{v} \cdot \vec{A}\right) - \frac{d\vec{A}}{dt}\right)$$

$$\Longrightarrow \frac{d}{dt}(\vec{p} + e\vec{A}) = e\left(-\vec{\nabla}\left(\phi - \vec{v} \cdot \vec{A}\right)\right).$$

Referring to the Euler–Lagrange equations:

$$\frac{\partial}{\partial t}\frac{\partial L}{\partial \dot{x}_i} = -\frac{\partial L}{\partial x_i}$$

with the nonrelativistic Lagrangian L defined as

$$L = \sum_i \frac{1}{2} m \dot{x}_i^2 - U\left(x_i, \dot{x}_i, t\right)$$

a generalized potential $U(x_i, \dot{x}_i, t)$ for this dynamics is

$$U = e\left(\phi - \vec{\dot{x}}_i \cdot \vec{A}\right).$$

The momentum being given by

$$p_i = \frac{\partial L}{\partial \dot{x}_i}$$

one has for $\vec{p}$ and for the Hamiltonian H

$$\vec{p} = m\vec{\dot{x}}_i + e\vec{A}$$

$$H = \frac{1}{2m}\left(\vec{p} - e\vec{A}\right)^2 + e\phi.$$

Then the free-particle equation

$$E = \frac{\vec{p}^{\,2}}{2m}$$

is transformed in the case of the coupling with the electromagnetic field in:

$$E - e\phi = \frac{1}{2m}\left(\vec{p} - e\vec{A}\right)^2 . \tag{6.25}$$

This is equivalent to the following replacements for the free-particle energy and momentum:

$$E \to E - e\phi \,;\; \vec{p} \to \vec{p} - e\vec{A} \tag{6.26}$$

i.e., in terms of the relativistic energy-momentum four-vector:

$$p^\mu \to p^\mu - eA^\mu \tag{6.27}$$

or, in the operator view ($p^\mu \to i\hbar\partial^\mu$):

$$\partial^\mu \to D^\mu \equiv \partial^\mu + ieA^\mu . \tag{6.28}$$

The operator D^μ is designated the *covariant derivative.*

The replacement $\partial^\mu \to D^\mu$ is called the *minimal coupling prescription*. This prescription involves only the charge distribution and is able to account for all electromagnetic interactions.

Wave equations can now be generalized to account for the coupling with the electromagnetic field using the minimal coupling prescription.

For instance, the free-particle Schrödinger equation

$$i\hbar\frac{\partial}{\partial t}\Psi = -\frac{1}{2m}\left(-i\hbar\vec{\nabla}\right)^2\Psi \tag{6.29}$$

becomes under such a replacement

$$\left(i\hbar\frac{\partial}{\partial t} - e\phi\right)\Psi = -\frac{1}{2m}\left(-i\hbar\vec{\nabla} - e\vec{A}\right)^2\Psi . \tag{6.30}$$

The Schrödinger equation couples directly to the scalar and vector potential and not to the force, and quantum effects not foreseen in classic physics do exist. One of them is the well-known Bohm–Aharonov effect predicted in 1959 by David Bohm and his student Yakir Aharonov.[2] Whenever a particle is confined in a region where the electric and the magnetic field are zero but the potential four-vector is not, its wave function changes the phase.

This is the case of particles crossing a region outside an infinite thin solenoid (Fig. 6.2, left).

---

[2]David Bohm (1917–1992) was an American scientist who contributed innovative and unorthodox ideas to quantum theory, neuropsychology, and the philosophy of mind. Yakir Aharonov (1932) is an Israeli physicist specialized in quantum physics, interested in quantum field theories and interpretations of quantum mechanics.

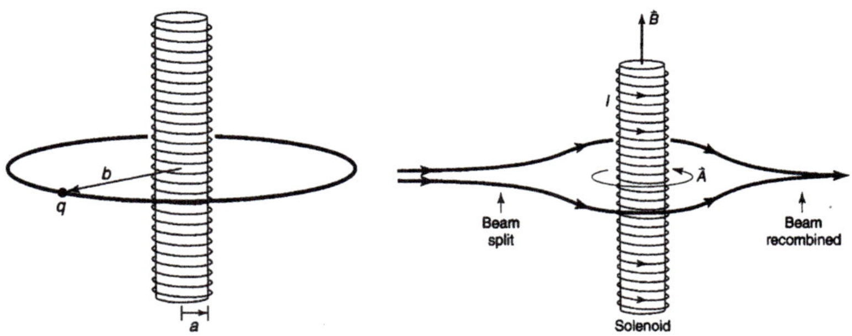

Fig. 6.2 *Left* Vector potential in the region outside an infinite solenoid. *Right* Double-slit experiment demonstrating the Bohm–Aharonov effect. From D. Griffiths, "Introduction to quantum mechanics," second edition, Pearson 2004

In this region, the magnetic field $\vec{B}$ is zero but the vector potential vector $\vec{A}$ is not

$$\vec{\nabla} \times \vec{A} = \vec{B}$$

$$\oint \vec{A} \cdot d\vec{l} = \int_S \vec{B} \cdot d\vec{s}\,.$$

The line integral of the vector potential $\vec{A}$ around a closed loop is equal to the magnetic flux through the area enclosed by the loop. As $\vec{B}$ inside the solenoid is not zero, the flux is also not zero and therefore $\vec{A}$ is not null.

This effect was experimentally verified observing shifts in an interference pattern whether or not the current in a microscopic solenoid placed in between the two fringes is turned on (Fig. 6.2, right).

6.2.3 Gauge Invariance

We have seen that physical observables connected to a wave function Ψ are invariant to global change in the phase of the wave function itself

$$\Psi(\vec{x}, t) \rightarrow \Psi(\vec{x}, t)\; e^{iq\alpha} \tag{6.31}$$

where α is a real number.

The free-particle Schrödinger equation in particular is invariant with respect to a global change in the phase of the wave function. It is easy however to verify that this does not apply, in general, to a local change

$$\Psi(\vec{x}, t) \rightarrow \Psi(\vec{x}, t)\; e^{iq\alpha(\vec{x},t)}\,. \tag{6.32}$$

On the other hand, the electromagnetic field is, as it was discussed in Sect. 6.2.1, invariant under a combined local transformation of the scalar and vector potential:

$$\phi \rightarrow \phi - \frac{\partial \chi}{\partial t} \tag{6.33}$$

$$\vec{A} \rightarrow \vec{A} + \vec{\nabla} \chi \tag{6.34}$$

where $\chi\,(t, \vec{x})$ is a scalar function of the time and space coordinates.

Remarkably the Schrödinger equation modified using the minimal coupling prescription is invariant under a joint local transformation both of the phase of the wave function and of the electromagnetic four-potential:

$$\Psi\,(\vec{x}, t) \rightarrow \Psi\,(\vec{x}, t)\; e^{ie\alpha(\vec{x})} \tag{6.35}$$

$$A^{\mu} \rightarrow A^{\mu} - \partial^{\mu}\,\alpha\,(\vec{x})\;. \tag{6.36}$$

Applying the minimal coupling prescription to the relativistic wave equations (Klein-Gordon and Dirac equations) these equations become also invariant under local gauge transformations, as we shall verify later.

Conversely, imposing the invariance under a local gauge transformation of the free-particle wave equations implies the introduction of a gauge field.

The gauge transformation of the wave functions can be written in a more general form as

$$\Psi\,(\vec{x}, t) \rightarrow \Psi\,(\vec{x}, t)\;\exp\left(i\alpha\,(\vec{x})\,\hat{A}\right) \tag{6.37}$$

where $\alpha\,(\vec{x})$ is a real function of the space coordinates and $\hat{A}$ a unitary operator (see Sect. 5.3.3).

In the case of QED, Herman Weyl, Vladmir Foch, and Fritz London found in the late 1920s that the invariance of a Lagrangian including fermion and field terms with respect to transformations associated to the U(1) group, corresponding to local rotations by $\alpha\,(\vec{x})$ of the wave function phase, requires (and provides) the interaction term with the electromagnetic field, whose quantum is the photon.

The generalization of this symmetry to non-Abelian groups was introduced in 1954 by Chen Yang and Robert Mills.[3] Indeed we shall see that

- The weak interaction is modeled by a "weak isospin" symmetry linking "weak isospin up" particles (identified for example with the u-type quarks and with the

[3]Chen Yang (1922) is a Chinese-born American physicist who works on statistical mechanics and particle physics. He shared the 1957 Nobel prize in physics with T.D. Lee for their work on parity non-conservation in weak interactions. While working with the US physicist Robert Mills (1927–1999) at Brookhaven National Laboratory, in 1954 he proposed a tensor equation for what are now called Yang–Mills fields.

neutrinos) and "weak isospin down" particles (identified for example with the d-type quarks and with the leptons). We have seen that SU(2) is the minimal representation for such a symmetry. If $\hat{A}$ is chosen to be one of the generators of the SU(2) group then the associated gauge transformation corresponds to a local rotation in a spinor space. The needed gauge field to assure the invariance of the wave equations under such transformations are the weak fields whose quanta are the $W^{\pm}$ and Z (see Sect. 6.3).

- The strong interaction is modeled by QCD, a theory exploiting the invariance of the strong interaction with respect to a rotation in color space. We shall see that SU(3) is the minimal representation for such a symmetry. If $\hat{A}$ is chosen to be one of the generators of the SU(3) group then the associated gauge transformation corresponds to a local rotation in a complex three-dimensional vector space, which represents the color space. The gauge fields needed to assure the invariance of the

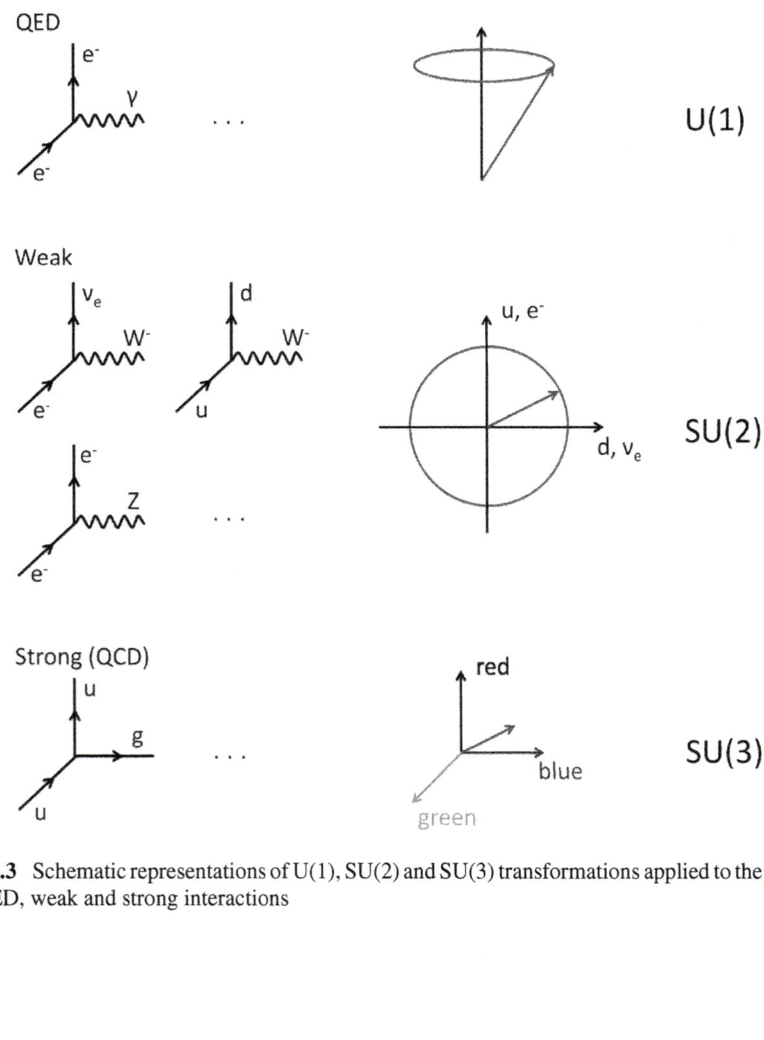

Fig. 6.3 Schematic representations of U(1), SU(2) and SU(3) transformations applied to the models of QED, weak and strong interactions

wave equations under such transformations are the strong fields whose quanta are called gluons (see Sect. 6.4).

Figure 6.3 shows schematic representations of such transformations.

6.2.4 Dirac Equation Revisited

Dirac equation was briefly introduced in Sect. 3.2.1. It is a *linear* equation describing free relativistic particles with spin 1/2 (electrons and positrons for instance); linearity allows overcoming some difficulties coming from the nonlinearity of the Klein-Gordon equation, which was the translation in quantum mechanical form of the relativistic Hamiltonian

$$H^2 = p^2 + m^2$$

replacing the Hamiltonian itself and the momentum with the appropriate operators:

$$\hat{H}^2 = \hat{p}^2 + m^2 \Longrightarrow -\frac{\partial^2 \psi}{\partial t^2} = -\vec{\nabla}^2 \psi + m^2 \psi \,. \tag{6.38}$$

Dirac searched for an alternative relativistic equation starting from the generic form describing the evolution of a wave function, in the familiar form:

$$i\frac{\partial \Psi}{\partial t} = \hat{H}\psi$$

with a Hamiltonian operator linear in $\hat{\vec{p}}$, t (Lorentz invariance requires that if the Hamiltonian has first derivatives with respect to time also the spatial derivatives should be of first order):

$$\hat{H} = \vec{\alpha} \cdot \vec{p} + \beta m \,.$$

This must be compatible with the Klein-Gordon equation, and thus

$$\begin{aligned} \alpha_i^2 = 1 \quad &; \quad \beta^2 = 1 \\ \alpha_i \beta + \beta \alpha_i &= 0 \\ \alpha_i \alpha_j + \alpha_j \alpha_i &= 0 \,. \end{aligned} \tag{6.39}$$

Therefore, the parameters $\vec{\alpha}$ and β cannot be numbers. However, things work if they are matrices (and if these matrices are hermitian it is guaranteed that the hamiltonian is also hermitian). It can be demonstrated that their lowest possible rank is 4.

Using the explicit form of the momentum operator $\vec{p} = -i\vec{\nabla}$, the Dirac equation can be written as

$$i\frac{\partial \psi}{\partial t} = \left(i\vec{\alpha} \cdot \vec{\nabla} + \beta m\right)\psi \,. \tag{6.40}$$

The wave functions ψ must thus be four-component vectors:

$$\psi(\vec{r}, t) = \begin{pmatrix} \psi_1(x) \\ \psi_2(x) \\ \psi_3(x) \\ \psi_4(x) \end{pmatrix}.$$

We arrived at an interpretation of the Dirac equation as a four-dimensional matrix equation in which the solutions are four-component wave functions called spinors. Plane wave solutions are

$$\psi(x) = u(\vec{p})e^{i(\vec{p}\cdot\vec{r} - Et)}$$

where $u(\vec{p})$ is also a four-component spinor satisfying the eigenvalue equation

$$(c\vec{\alpha} \cdot \vec{p} + \beta m)\, u(\vec{p}) = E u(\vec{p})\,.$$

This equation has four solutions: two with positive energy $E = +E_p$ and two with negative energy $E = -E_p$. We will discuss later the interpretation of the negative energy solutions. The Dirac equation accounts "for free" for the existence of two spin states, which had to be inserted by hand in the Schrödinger equation of nonrelativistic quantum mechanics, and therefore explains the magnetic moment of point-like fermions. In addition, since spin is embedded in the equation, the Dirac's equation allows computing correctly the energy splitting of atomic levels with the same quantum numbers due to the spin-orbit and spin–spin interactions in atoms (fine and hyperfine splitting).

We shall now write the free-particle Dirac equation in a more compact form, from which relativistic covariance is immediately visible. This requires the introduction of a new set of important 4×4 matrices, the γ^μ matrices, which replace the α_i and β matrices discussed before. To account for electromagnetic interactions, the minimal coupling prescription can once again be used.

One possible, choice for α_i and β satisfying the conditions (6.39) is the set of matrices:

$$\alpha_i = \begin{pmatrix} 0 & \sigma_i \\ \sigma_i & 0 \end{pmatrix} \tag{6.41}$$

$$\beta = \begin{pmatrix} I & 0 \\ 0 & -I \end{pmatrix} \tag{6.42}$$

being σ_i the 2×2 Pauli matrices (see Sect. 5.7.2) and I the unit 2×2 matrix.

Multiplying the Dirac equation (6.40) by β one has

$$i\beta\frac{\partial\psi}{\partial t} = \left(i\beta\vec{\alpha} \cdot \vec{\nabla} + m\right)\, \psi\,.$$

and introducing the Pauli-Dirac γ^μ matrices defined as

$$\gamma^0 = \beta \; ; \; \vec{\gamma} = \beta\vec{\alpha} \tag{6.43}$$

$$\gamma^0 = \begin{pmatrix} 1 & 0 & 0 & 0 \\ 0 & 1 & 0 & 0 \\ 0 & 0 & -1 & 0 \\ 0 & 0 & 0 & -1 \end{pmatrix}, \gamma^1 = \begin{pmatrix} 0 & 0 & 0 & 1 \\ 0 & 0 & 1 & 0 \\ 0 & -1 & 0 & 0 \\ -1 & 0 & 0 & 0 \end{pmatrix}, \gamma^2 = \begin{pmatrix} 0 & 0 & 0 & -i \\ 0 & 0 & i & 0 \\ 0 & i & 0 & 0 \\ -i & 0 & 0 & 0 \end{pmatrix}, \gamma^3 = \begin{pmatrix} 0 & 0 & 1 & 0 \\ 0 & 0 & 0 & -1 \\ -1 & 0 & 0 & 0 \\ 0 & 1 & 0 & 0 \end{pmatrix}$$

then:

$$\left[i \left(\gamma^0 \frac{\partial}{\partial x_0} + \gamma^i \frac{\partial}{\partial x_i} \right) - m \right] \psi = 0 .$$

If we use a four-vector notation

$$\gamma^\mu = (\beta, \beta\vec{\alpha})$$

taking into account that

$$\partial_\mu = \left(\frac{\partial}{\partial t}, \vec{\nabla} \right) ,$$

the Dirac equation can be finally written as:

$$(i\gamma^\mu \partial_\mu - m)\psi = 0 . \tag{6.44}$$

This is an extremely compact form of writing a set of four differential equations applied to a four-component vector ψ (a spinor, often called in this particular case bi-spinor). We call it the covariant form of the Dirac equation (its form is preserved in all the inertial frames).

Let us examine now the solutions of the Dirac equation in some particular cases.

6.2.4.1 Particle at Rest

Particles at rest have $\vec{p} = 0$ and thus

$$\left(i\gamma^0 \frac{\partial}{\partial t} - m \right) \psi = 0$$

$$\begin{pmatrix} I & 0 \\ 0 & -I \end{pmatrix} \begin{pmatrix} \frac{\partial}{\partial t}\psi_A \\ \frac{\partial}{\partial t}\psi_B \end{pmatrix} = -im \begin{pmatrix} \psi_A \\ \psi_B \end{pmatrix}$$

being ψ_A and ψ_B spinors:

$$\psi_A = \begin{pmatrix} \psi_1 \\ \psi_2 \end{pmatrix}$$

$$\psi_B = \begin{pmatrix} \psi_3 \\ \psi_4 \end{pmatrix} .$$

In this simple case, the two spinors are subject to two independent differential equations:

$$\frac{\partial}{\partial t}\psi_A = -im\psi_A$$

$$\frac{\partial}{\partial t}\psi_B = im\psi_B$$

which have as solution (up to some normalization factor):

- $\psi_A = e^{-imt}\psi_A(0)$ with energy $E = m > 0$;
- $\psi_B = e^{imt}\psi_B(0)$ with energy $E = -m < 0$

or in terms of each component of the wave function vector

$$\psi_1 = e^{-imt}\begin{pmatrix} 1 \\ 0 \\ 0 \\ 0 \end{pmatrix} \quad \psi_2 = e^{-imt}\begin{pmatrix} 0 \\ 1 \\ 0 \\ 0 \end{pmatrix}$$

$$\psi_3 = e^{imt}\begin{pmatrix} 0 \\ 0 \\ 1 \\ 0 \end{pmatrix} \quad \psi_4 = e^{imt}\begin{pmatrix} 0 \\ 0 \\ 0 \\ 1 \end{pmatrix} .$$

There are then four solutions which can accommodate a spin $1/2$ particle or antiparticle. The positive energy solutions ψ_1 and ψ_2 correspond to fermions (electrons for instance) with spin up and down, respectively, while the negative energy solutions ψ_3 and ψ_4 correspond to antifermions (positrons for instance) with spin up and down.

6.2.4.2 Free Particle

Free particles have $\vec{p}$ = constant and their wave function is be a plane wave of the form:

$$\psi(\vec{x}, t) = u(p_0, \vec{p})\, e^{-i(p_0 t - \vec{p}\cdot\vec{x})}$$

where

$$u(p_0, \vec{p}) = N \begin{pmatrix} \phi \\ \chi \end{pmatrix}$$

is a bi-spinor (ϕ, χ are spinors) and N a normalization factor.

The Dirac equation can be written as a function of the energy-momentum operators as

$$\left(\left(\gamma^0 p_0 - \vec{\gamma}\cdot\vec{p}\right) - m\right)\psi = 0$$

Inserting the equation of a plane wave as a trial solution and using the Pauli–Dirac representation of the γ matrices:

$$\begin{pmatrix} (p_0 - m)I & -\vec{\sigma}\cdot\vec{p} \\ \vec{\sigma}\cdot\vec{p} & (-p_0 - m)I \end{pmatrix}\begin{pmatrix} \phi \\ \chi \end{pmatrix} = 0\,.$$

I is again the 2×2 unity matrix which is often omitted writing the equations and

$$\vec{\sigma}\cdot\vec{p} = \begin{pmatrix} p_z & p_x - ip_y \\ p_x + ip_x & -p_z \end{pmatrix}.$$

For $\vec{p} = 0$, the "particle at rest" solution discussed above is recovered. Otherwise, there are two coupled equations for the spinors ϕ and χ:

$$\phi = \frac{\vec{\sigma}\cdot\vec{p}}{E - m}\chi$$

$$\chi = \frac{\vec{\sigma}\cdot\vec{p}}{E + m}\phi$$

and then the u bi-spinor can be written either in terms of the spinor ϕ or in term of the spinor χ:

$$u_1 = N\begin{pmatrix} \phi \\ \frac{\vec{\sigma}\cdot\vec{p}}{E+m}\phi \end{pmatrix}$$

$$u_2 = N\begin{pmatrix} \frac{-\vec{\sigma}\cdot\vec{p}}{-E+m}\chi \\ \chi \end{pmatrix}.$$

The first solution corresponds to states with $E > 0$ (particles) and the second to states with $E < 0$ (antiparticles) as can be seen by going to the $\vec{p} = 0$ limit. These last states can be rewritten changing the sign of E and $\vec{p}$ and labeling the bi-spinor u_2 as v (u_1 is then labeled just as u).

$$v = N\begin{pmatrix} \frac{\vec{\sigma}\cdot\vec{p}}{E+m}\chi \\ \chi \end{pmatrix}.$$

Both ϕ and χ can be written in a base of unit vectors χ_s with

$$\chi_{s=1} = \begin{pmatrix} 1 \\ 0 \end{pmatrix}$$

$$\chi_{s=2} = \begin{pmatrix} 0 \\ 1 \end{pmatrix} .$$

Finally, we have then again four solutions: two for the particle states and two for the antiparticle states.

The normalization factor N is often defined as

$$N = \frac{\sqrt{E+m}}{\sqrt{V}}$$

ensuring a standard relativistic normalization convention of $2E$ particles per box of volume V. In fact introducing the bi-spinors transpose conjugate $u^\dagger$ and $v^\dagger$

$$u^\dagger u = v^\dagger v = 2E/V .$$

6.2.4.3 Helicity

The spin operator $\vec{S}$ introduced in Sect. 5.7.2 can now be generalized in this bi-spinor space as

$$\vec{S} = \frac{1}{2}\vec{\Sigma} \tag{6.45}$$

where

$$\vec{\Sigma} = \begin{pmatrix} \vec{\sigma} & 0 \\ 0 & \vec{\sigma} \end{pmatrix} . \tag{6.46}$$

More generally, defining the helicity operator h as the projection of the spin over the momentum direction:

$$\mathrm{h} = \frac{1}{2}\frac{\vec{\sigma}\cdot\vec{p}}{|\vec{p}|} \tag{6.47}$$

there are always four eigenstates of this operator. Indeed, using spherical polar coordinates (θ, ϕ):

$$\vec{p} = |\vec{p}|(\sin\theta\cos\phi\vec{e}_x + \sin\theta\sin\phi\vec{e}_y + \cos\theta\vec{e}_z) \tag{6.48}$$

the helicity operator is given by

$$\mathrm{h} = \begin{pmatrix} \cos\theta & \sin\theta e^{-i\phi} \\ \sin\theta e^{i\phi} & -\cos\theta \end{pmatrix} . \tag{6.49}$$

The eigenstates of the operator h can also be written as

$$u_{\uparrow} = \sqrt{E+m}\begin{pmatrix} \cos\left(\frac{\theta}{2}\right) \\ \sin\left(\frac{\theta}{2}\right)e^{i\phi} \\ \frac{p}{E+m}\cos\left(\frac{\theta}{2}\right) \\ \frac{p}{E+m}\sin\left(\frac{\theta}{2}\right)e^{i\phi} \end{pmatrix} ; u_{\downarrow} = \sqrt{E+m}\begin{pmatrix} -\sin\left(\frac{\theta}{2}\right) \\ \cos\left(\frac{\theta}{2}\right)e^{i\phi} \\ \frac{p}{E+m}\sin\left(\frac{\theta}{2}\right) \\ -\frac{p}{E+m}\cos\left(\frac{\theta}{2}\right)e^{i\phi} \end{pmatrix}$$

$$v_{\uparrow} = \sqrt{E+m}\begin{pmatrix} \frac{p}{E+m}\sin\left(\frac{\theta}{2}\right) \\ -\frac{p}{E+m}\cos\left(\frac{\theta}{2}\right)e^{i\phi} \\ -\sin\left(\frac{\theta}{2}\right) \\ \cos\left(\frac{\theta}{2}\right)e^{i\phi} \end{pmatrix} ; v_{\downarrow} = \sqrt{E+m}\begin{pmatrix} \frac{p}{E+m}\cos\left(\frac{\theta}{2}\right) \\ \frac{p}{E+m}\sin\left(\frac{\theta}{2}\right)e^{i\phi} \\ \cos\left(\frac{\theta}{2}\right) \\ \sin\left(\frac{\theta}{2}\right)e^{i\phi} \end{pmatrix}.$$

Note that helicity is Lorentz invariant only in the case of massless particles (otherwise the direction of $\vec{p}$ can be inverted choosing an appropriate reference frame).

6.2.4.4 Dirac Adjoint, the γ^5 Matrix and Bilinear Covariants

The Dirac bi-spinors are not real four-vectors and it can be shown that the product $\psi^{\dagger}\psi$ is not a Lorentz invariant (a scalar). On the contrary the product $\overline{\psi}\psi$ is a Lorentz invariant being $\overline{\psi}$ named the *adjoint Dirac spinor* and defined as:

$$\overline{\psi} = \psi^{\dagger}\gamma^0 =$$

$$= \left(\psi_1^*, \psi_2^*, \psi_3^*, \psi_4^*\right)\begin{pmatrix} 1 & 0 & 0 & 0 \\ 0 & 1 & 0 & 0 \\ 0 & 0 & -1 & 0 \\ 0 & 0 & 0 & -1 \end{pmatrix} = \left(\psi_1^*, \psi_2^*, -\psi_3^*, -\psi_4^*\right).$$

The parity operator P in the Dirac bi-spinor space is just the matrix γ^0 (it reverts the sign of the terms which are function of $\vec{p}$), and

$$P\left(\overline{\psi}\psi\right) = \psi^{\dagger}\gamma^0\gamma^0\gamma^0\psi = \overline{\psi}\psi$$

as $\left(\gamma^0\right)^2 = 1$.

Other quantities can be constructed using ψ and $\overline{\psi}$ (bilinear covariants). In particular introducing γ^5 as

$$\gamma^5 = i\gamma^0\gamma^1\gamma^2\gamma^3 = \begin{pmatrix} 0 & 0 & 0 & 1 \\ 0 & 0 & 1 & 0 \\ 0 & 1 & 0 & 0 \\ 1 & 0 & 0 & 0 \end{pmatrix} :$$

- $\overline{\psi}\gamma^5\psi$ is a pseudoscalar.
- $\overline{\psi}\gamma^\mu\psi$ is a four-vector.
- $\overline{\psi}\gamma^\mu\gamma^5\psi$ is a pseudo four-vector.
- $\left(\overline{\psi}\,\sigma^{\mu\nu}\psi\right)$, where $\sigma^{\mu\nu} = \frac{i}{2}\left(\gamma^\mu\gamma^\nu - \gamma^\nu\gamma^\mu\right)$, is an antisymmetric tensor.

6.2.4.5 Dirac Equation in the Presence of an Electromagnetic Field

The Dirac equation in the presence of an electromagnetic field can be obtained applying the minimal coupling prescription discussed in Sect. 6.2.2. In practice this is obtained by replacing the ∂^μ derivatives by the covariant derivative D^μ:

$$\partial^\mu \to D^\mu \equiv \partial^\mu + ieA^\mu .$$

Then

$$\left(i\gamma^\mu D_\mu - m\right)\,\psi = 0$$

$$\left(i\gamma^\mu\partial_\mu - e\,\gamma^\mu A_\mu - m\right)\,\psi = 0 .$$

The interaction with a magnetic field can be then described introducing the two spinors ϕ and χ and using the Pauli–Dirac representation of the γ matrices:

$$\begin{pmatrix} p_0 - m & -\vec{\sigma}\cdot\left(-i\vec{\nabla} - e\vec{A}\right) \\ \vec{\sigma}\cdot\left(-i\vec{\nabla} - e\vec{A}\right) & -p_0 - m \end{pmatrix}\begin{pmatrix}\phi \\ \chi\end{pmatrix} = 0 .$$

In the nonrelativistic limit ($E \approx m$), the Dirac equation reduces to

$$\frac{1}{2m}\left|\vec{p} - e\vec{A}\right|^2\psi - \frac{e\vec{B}\cdot\vec{\Sigma}}{2m}\psi = 0$$

where the magnetic field $\vec{B} = \vec{\nabla}\times\vec{A}$ has been reintroduced.

There is thus a coupling of the form $-\vec{\mu}\cdot\vec{B}$ between the magnetic field and the spin of a point-like charged particle (the electron or the muon for instance), and the quantity

$$\vec{\mu} = \vec{\mu}_S = \frac{e}{m}\frac{1}{2}\vec{\Sigma} = \frac{e}{m}\vec{S}$$

can be identified with the intrinsic magnetic moment of a charged particle with spin $\vec{S}$.

Defining the *gyromagnetic ratio* g as the ratio between $\vec{\mu}_S$ and the classical magnetic moment $\vec{\mu}_L$ of a charged particle with an angular momentum $\vec{L} = \vec{S}$:

$$g = \frac{\vec{\mu}_S}{\vec{\mu}_L} = 2 .$$

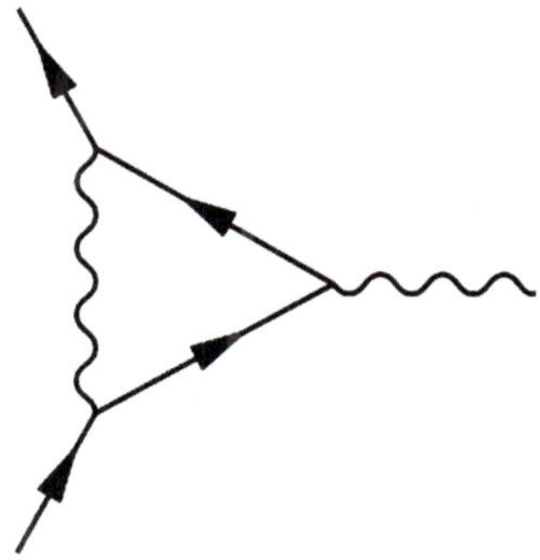

Fig. 6.4 Second-order vertex correction to g

6.2.4.6 $g - 2$

The value of the coupling between the magnetic field and the spin of the point charged particle is however modified by higher order corrections which can be translated in successive Feynman diagrams, as the ones we have seen in Fig. 6.1. In second order, the main correction is introduced by a vertex correction, described by the diagram represented in Fig. 6.4 computed in 1948 by Schwinger, leading to deviation of g from 2 with the magnitude:

$$a_e = \frac{g-2}{2} \simeq \frac{\alpha}{2\pi} \simeq 0.0011614\,.$$

Nowadays, the theoretical corrections are completely computed up to the eighth-order (891 diagrams) and the most significant tenth-order terms as well as electroweak and hadronic corrections are also computed. There is a remarkable agreement with the present experimental value of:

$$a_e^{\exp} = 0.00115965218076 \pm 0.00000000000027\,.$$

Historically, the first high precision $g - 2$ measurements were accomplished by H. Richard Crane and his group in the years 1950–1967 at the University of Michigan, USA. A beam of electrons is first polarized and then trapped in a magnetic bottle for a (long) time T. After this time, the beam is extracted and the polarization is measured (Fig. 6.5).

Under the influence of the magnetic field B in the box, the spin of the electron precesses with angular velocity

$$\omega_p = \frac{g\, e\, B}{2\, m}$$

while the electron follows a helicoidal trajectory with an angular velocity of

$$\omega_{rot} = \frac{e\, B}{m}\,.$$

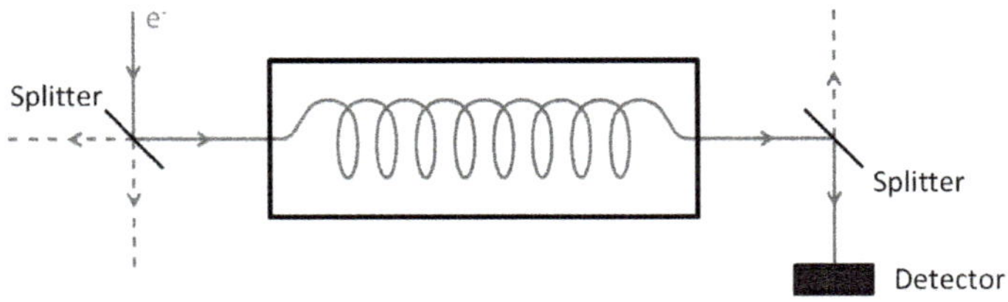

Fig. 6.5 Schematic drawing of the $g-2$ experiment from H. Richard Crane

The polarization of the outgoing beam is thus proportional to the ratio

$$\frac{w_p}{w_{rot}} = \frac{g}{2}.$$

Nowadays, Penning traps are used to keep electrons (and positrons) confined for months at a time. Such a device, invented by H. Dehmelt in the 1980s uses a homogeneous static magnetic field and a spatially inhomogeneous static electric field to trap charged particles (Fig. 6.6).

The muon and electron magnetic moments should be at first-order equal. However, the loop corrections are proportional to the square of the respective masses and thus those of the muon are much larger $\left(m_\mu{}^2/m_e{}^2 \sim 4 \cdot 10^4\right)$. In particular, the sensitivity to loops involving hypothetical new particles (see Chap. 7 for a survey) is much higher and a precise measurement of the muon anomalous magnetic moment a_μ may be used as a test of the Standard Model.

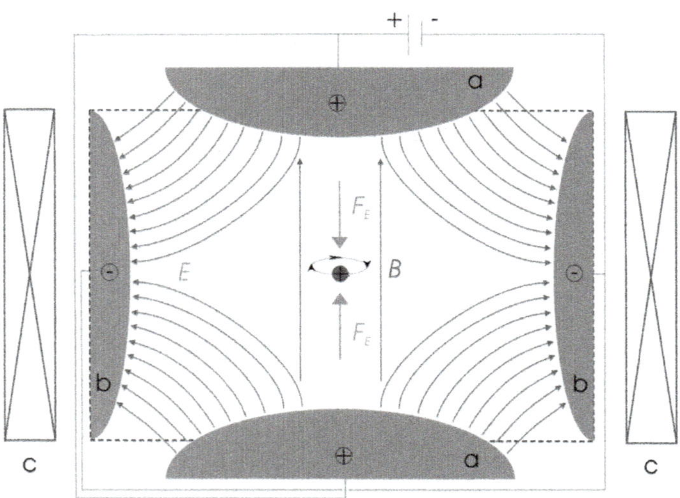

Fig. 6.6 Schematic representation of the electric and magnetic fields inside a Penning trap. By Arian Kriesch Akriesch 23:40, [own work, GFDL http://www.gnu.org/copyleft/fdl.html, CC-BY-SA-3.0], via Wikimedia Commons

Fig. 6.7 The E821 storage ring. From Brookhaven National Laboratory

The most precise measurement of a_μ so far was done by the experiment E821 at Brookhaven National Laboratory (BNL). A beam of polarized muons circulates in a storage ring with a diameter of $\sim$14 m under the influence of an uniform magnetic field (Fig. 6.7). The muon spin precess and the polarization of the beam is a function of time. After many turns, muons decay to electron (and neutrinos) whose momentum is basically aligned with the direction of the muon spin (see Sect. 6.3). The measured value is

$$a_\mu^{\text{exp}} = 0.00116592083 \pm 0.00000000063\,.$$

This result is more than 3 σ away from the expected one which leads to a wide discussion both on the accuracy of the theoretical computation (in particular in the hadronic contribution) and the possibility of an indication of new physics. This discussion is quite interesting but the statistical significance of the result is still marginal.

6.2.4.7 The Lagrangian Density Corresponding to the Dirac Equation

Consider the Lagrangian density

$$\mathcal{L} = i\bar{\psi}\gamma^\mu \partial_\mu \psi - m\bar{\psi}\psi \tag{6.50}$$

and apply the Euler–Lagrange equations to $\bar{\psi}$. We find

$$\frac{\partial \mathcal{L}}{\partial(\partial_\mu \bar{\psi})} = i\gamma^\mu \partial_\mu \psi - m\psi = 0\,,$$

which is indeed the Dirac equation for a free particle.

Notice that:

- the mass (i.e., the energy associated to rest—whatever this can mean in quantum mechanics) is associated to a term quadratic in the field

$$m\bar{\psi}\psi\,;$$

- the dimension of the field ψ is [energy$^{3/2}$] ($m\psi^2 d^4x$ is a scalar).

6.2.5 *Klein–Gordon Equation Revisited*

The Klein–Gordon equation was briefly introduced in Sect. 3.2.1. It describes free relativistic particles with spin 0 (scalars or pseudoscalars). With the introduction of the four-vector notation, it can be written in a covariant form. To account for electromagnetic interactions, the minimal coupling prescription can be used.

6.2.5.1 Covariant Form of the Klein–Gordon Equation

In Sect. 5.7.2, the Klein–Gordon equation was written as

$$\left(\frac{\partial^2}{\partial t^2} - \vec{\nabla}^2 + m^2\right)\phi\,(x) = 0$$

where $\phi\,(x)$ is a scalar wave function.

Remembering that

$$\Box = \partial_\mu \partial^\mu = \frac{\partial^2}{\partial t^2} - \vec{\nabla}^2$$

the Klein–Gordon equation can be written in a covariant form:

$$\left(\partial_\mu \partial^\mu + m^2\right)\phi\,(x) = 0\,.$$

The solutions are, as it was discussed before, plane waves

$$\phi\,(x) = N\;e^{i(\vec{p}\cdot\vec{r} - Et)}$$

with

$$E = \pm\sqrt{\vec{p}^2 + m^2}$$

(the positive solutions correspond to particles and the negative ones to antiparticles).

Doing some arithmetic with the Klein–Gordon equation and its conjugate a continuity equation can also be obtained for a particle with charge e:

$$\vec{\nabla} \cdot \vec{j} = -\frac{\partial \rho}{\partial t}$$

where

$$\rho(x) = ie\left(\phi^* \partial t \phi - \phi \partial t \phi^*\right) \; ; \; \vec{j}(x) = -ie\left(\phi^* \nabla \phi - \phi \nabla \phi^*\right)$$

or in terms of four-vectors:

$$\partial^\mu j_\mu = 0$$

where

$$j_\mu(x) = ie\left(\phi^* \partial_\mu \phi - \phi \partial_\mu \phi^*\right) .$$

In the case of plane waves:

$$j_\mu(x) = 2e|N|^2 p_\mu .$$

6.2.5.2 Klein–Gordon Equation in Presence of an Electromagnetic Field

In the presence of an electromagnetic field, the Klein–Gordon equation can be modified applying, as it was done previously for the Schrödinger and the Dirac equations, the minimal coupling prescription. The normal derivatives are replaced by the covariant derivatives:

$$\partial^\mu \rightarrow D^\mu \equiv \partial^\mu + ieA^\mu$$

and thus

$$\left(\left(\partial_\mu + ieA_\mu\right)\left(\partial^\mu + ieA^\mu\right) + m^2\right)\phi(x) = 0$$

$$\left(\partial_\mu \partial^\mu + m^2 + ie\left(\partial_\mu A^\mu + A_\mu \partial^\mu\right) - e^2 A_\mu A^\mu\right)\phi(x) = 0 .$$

The e^2 term is of second order and can be neglected. Then the Klein-Gordon equation in presence of an electromagnetic field can be written at first order as

$$\left(\partial_\mu \partial^\mu + V(x) + m^2\right)\phi(x) = 0$$

where

$$V(x) = ie\left(\partial_\mu A^\mu + A_\mu \partial^\mu\right)$$

is the potential.

6.2.5.3 The Lagrangian Density Corresponding to the Klein-Gordon Equation

Consider the Lagrangian density

$$\mathcal{L} = \frac{1}{2}(\partial_\mu \phi)(\partial^\mu \phi) - \frac{1}{2}m^2\phi^2 \tag{6.51}$$

and apply the Euler–Lagrange equations to ϕ. We find

$$\partial_\mu \left(\frac{\partial \mathcal{L}}{\partial(\partial_\mu \phi)} \right) - \frac{\partial \mathcal{L}}{\partial \phi} = \partial_\mu \partial^\mu \phi + m^2 \phi = 0 \, ,$$

which is indeed the Klein-Gordon equation for a free scalar field.

Notice that

- the mass (i.e., the energy associated to rest—or better, in a quantum mechanical language, to the ground state) is associated to a term quadratic in the field

$$\frac{1}{2}m^2\phi^2 \, ;$$

- the dimension of the field ϕ is [energy] ($m^2\phi^2 d^4x$ is a scalar).

6.2.6 The Lagrangian for a Charged Fermion in an Electromagnetic Field: Electromagnetism as a Field Theory

Let us draw a field theory which is equivalent to the Dirac equations in the presence of an external field.

We already wrote a Lagrangian density equivalent to the Dirac equation for a free particle:

$$\mathcal{L} = \bar{\psi}(i\gamma^\mu \partial_\mu - m)\psi \, . \tag{6.52}$$

Electromagnetism can be translated into the quantum world by assuming a Lagrangian density

$$\mathcal{L} = \bar{\psi}(i\gamma^\mu D_\mu - m)\psi - \frac{1}{4}F_{\mu\nu}F^{\mu\nu} \tag{6.53}$$

where $D_\mu \equiv \partial_\mu + ieA_\mu$ is called the covariant derivative (remind the "minimal prescription"), and A_μ is the four-potential of the electromagnetic field; $F_{\mu\nu} = \partial_\mu A_\nu - \partial_\nu A_\mu$ is the electromagnetic field tensor (see Sect. 2.10.7).

If the field A^μ transforms under a local gauge transformation as

$$A^\mu \to A^\mu - \partial^\mu \theta(x)$$

the Lagrangian is invariant with respect to a *local* U(1) gauge transformation $\psi \to \psi e^{i\theta(x)}$.

Substituting the definition of D into the Lagrangian gives us

$$\mathcal{L} = i\bar{\psi}\gamma^\mu \partial_\mu \psi - e\bar{\psi}\gamma_\mu A^\mu \psi - m\bar{\psi}\psi - \frac{1}{4}F_{\mu\nu}F^{\mu\nu} . \qquad (6.54)$$

To find the equations of motion, we can apply to this Lagrangian the Euler–Lagrange equations for a field:

$$\partial_\mu \left(\frac{\partial \mathcal{L}}{\partial(\partial_\mu \psi)} \right) - \frac{\partial \mathcal{L}}{\partial \psi} = 0 . \qquad (6.55)$$

One has

$$\partial_\mu \left(\frac{\partial \mathcal{L}}{\partial(\partial_\mu \psi)} \right) = \partial_\mu \left(i\bar{\psi}\gamma^\mu \right) \; ; \; \frac{\partial \mathcal{L}}{\partial \psi} = -e\bar{\psi}\gamma_\mu A^\mu - m\bar{\psi} .$$

Substituting these two expressions back into the Euler–Lagrange equation results in

$$\partial_\mu \bar{\psi}\gamma^\mu + e\bar{\psi}\gamma_\mu A^\mu + m\bar{\psi} = 0 ,$$

and taking the complex conjugate and moving the term in the potential to the right side:

$$\gamma^\mu \partial_\mu \psi - m\psi = e\gamma_\mu A^\mu \psi .$$

This is the Dirac equation including electrodynamics, as we have seen when discussing the minimal coupling prescription.

Let us now apply the Euler–Lagrange equations this time to the field A_μ in the Lagrangian (6.53):

$$\partial_\nu \left(\frac{\partial \mathcal{L}}{\partial(\partial_\nu A_\mu)} \right) - \frac{\partial \mathcal{L}}{\partial A_\mu} = 0 . \qquad (6.56)$$

We find

$$\partial_\nu \left(\frac{\partial \mathcal{L}}{\partial(\partial_\nu A_\mu)} \right) = \partial_\nu \left(\partial^\mu A^\nu - \partial^\nu A^\mu \right) \; ; \; \frac{\partial \mathcal{L}}{\partial A_\mu} = -e\bar{\psi}\gamma^\mu \psi$$

and substituting these two terms into (6.56) gives:

$$\partial_\nu F^{\nu\mu} = e\bar{\psi}\gamma^\mu \psi . \qquad (6.57)$$

For the spinor matter fields, the current takes the simple form:

$$j^\mu(x) = \sum_i q_i \bar{\psi}_i(x)\gamma^\mu \psi_i(x) \tag{6.58}$$

where q_i is the charge of the field ψ_i in units of e. The equation

$$\partial_\nu F^{\nu\mu} = j^\mu \tag{6.59}$$

is equivalent, as we discussed in Chap. 2, to the nonhomogeneous Maxwell equations. Notice that the two homogeneous Maxwell equations

$$\epsilon_{\mu\nu\rho\sigma} F^{\mu\nu} F^{\rho\sigma} = 0$$

are automatically satisfied due to the definition of the tensor $F^{\mu\nu} = \partial_\mu A_\nu - \partial_\nu A_\mu$ when we impose the Lorenz gauge $\partial_\mu A^\mu = 0$.

Again, if we impose the Lorenz gauge $\partial_\mu A^\mu = 0$,

$$\Box A^\mu = e\bar{\psi}\gamma^\mu\psi$$

which is a wave equation for the four-potential—the QED version of the classical Maxwell equations in the Lorenz gauge.

Notice that the Lagrangian (6.53) of QED, based on a local gauge invariance, contains all the physics of electromagnetism. It reflects also some remarkable properties, confirmed by the experiments:

- The interaction conserves separately P, C, and T.
- The current is diagonal in flavor space (i.e., it does not change the flavors of the particles).

We can see how the massless electromagnetic field A^μ "appears" thanks the gauge invariance. This is the basis of QED, quantum electrodynamics.

If a mass $m \neq 0$ would be associated to A, this new field would enter in the Lagrangian with a Proca term

$$-\frac{1}{2}F^{\mu\nu}F_{\mu\nu} + mA^\mu A_\mu$$

which is not invariant under local phase transformation. The field must, thus, be massless.

Summarizing, the requirement of local phase invariance under U(1), applied to the free Dirac Lagrangian, generates all of electrodynamics and specifies the electromagnetic current associated to Dirac particles; moreover, it introduces a massless field which can be interpreted as the photon. This is QED.

Notice that introducing local phase transformations just implies a simple difference in the calculation of the derivatives: we pick up an extra piece involving A^μ.

We replace the derivative with the covariant derivative

$$\partial^{\mu} \rightarrow D^{\mu} = \partial^{\mu} + iqA^{\mu}$$

and the invariance of the Lagrangian is restored. Substituting ∂^{μ} with D^{μ} transforms a globally invariant Lagrangian into a locally invariant one.

6.2.7 *An Introduction to Feynman Diagrams: Electromagnetic Interactions Between Charged Spinless Particles*

Electrons and muons have spin 1/2; but, for a moment, let us see how to compute transition probabilities in QED in the case of spinless particles, since the computation of the electromagnetic scattering amplitudes between charged spinless particles is much simpler.

6.2.7.1 Spinless Particles in an Electromagnetic Field

The scattering of a particle due to an interaction that acts only in a finite time interval can be described, as it was discussed in Sect. 2.7, as the transition between an initial and a final stationary states characterized by well-defined momentum. The first-order amplitude for such transition is written, in relativistic perturbative quantum mechanics, as (see Fig. 6.8, left):

$$H'_{if} = -i \int \phi_f^* (x) V(x) \; \phi_i (x) \, d^4x \, .$$

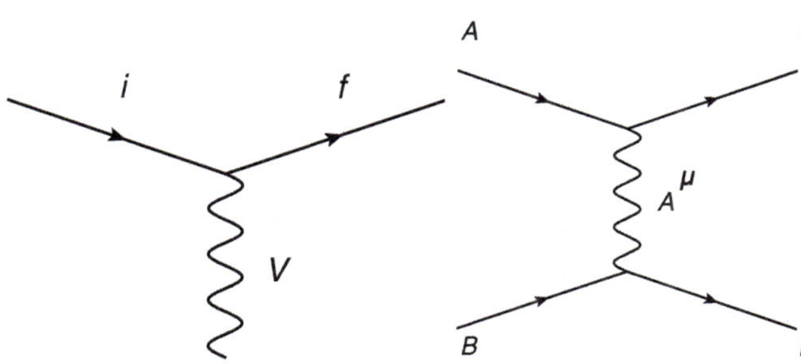

Fig. 6.8 *Left* Schematic representation of the first-order interaction of a particle in a field. *Right* Schematic representation (Feynman diagram) of the first-order elastic scattering of two charged non-identical particles

In the case of the electromagnetic field

$$H'_{if} = e\int \phi_f^*(x)\left(\partial_\mu A^\mu + A_\mu \partial^\mu\right)\phi_i(x)\,d^4x$$

and noting that integrating by parts assuming that the potential vanishes at $t \to \pm\infty$ or $x \to \pm\infty$

$$\int \phi_f^*(x)\,\partial_\mu\left(A^\mu \phi_i\right)d^4x = -\int \partial_\mu\left(\phi_f^*\right)A^\mu \phi_i d^4x$$

this amplitude can be transformed into:

$$H'_{if} = -i\int j_\mu^{if}\,A^\mu\,d^4\,x\,,$$

introducing a "transition" current j_μ^{if} between the initial and final state defined as:

$$j_\mu^{if} = ie\left(\phi_f^*\left(\partial_\mu \phi_i\right) - \left(\partial_\mu \phi_f^*\right)\phi_i\right).$$

In the case of plane waves describing particles with charge e the current can be written as:

$$j_\mu^{if} = eN_i N_f\left(p_i + p_f\right)_\mu e^{i(p_f - p_i)x}\,. \tag{6.60}$$

6.2.7.2 Elastic Scattering of Two Non-identical Charged Spinless Particles

The interaction of two charged particles can be treated as the interaction of one of the particles with the field created by the other (which thus acts as the source of the field).

The initial and final states of particle 1 are labeled as the states A and C, respectively, while for the particle 2 (taken as the source of the field) the corresponding labels are B and D (see Fig. 6.8, right). Let us assume that particles 1 and 2 are not of the same type (otherwise they would be indistinguishable) and have charge e. Then:

$$H'_{if} = e\int j_\mu^{AC}\,A^\mu\,d^4\,x$$

with

$$j_\mu^{AC} = eN_A N_C(p_A + p_C)_\mu e^{i(p_C - p_A)x}\,.$$

Being A^μ generated by the current associated to particle 2 (see Sect. 6.2.1)

$$\Box A^\mu = j_{BD}^\mu$$

with

$$j^{\mu}_{BD} = eN_B N_D (p_B + p_D)^{\mu} e^{i(p_D - p_B)x} \,.$$

Defining

$$q = (p_D - p_B) \;=\; (p_A - p_C)$$

then

$$A^{\mu} = -\frac{1}{q^2} j^{\mu}_{BD}$$

and

$$H'_{if} = -i \int j^{AC}_{\mu} \left(-\frac{1}{q^2}\right) j^{\mu}_{BD} \, d^4 x \,.$$

Solving the integral:

$$H'_{if} = \; -i \; N_A N_B N_C N_D (2\pi)^4 \, \delta^4 \, (p_A + p_B - p_C + p_D) \; \mathcal{M} \,,$$

where $\delta^4()$ ensures the conservation of energy-momentum, and the amplitude $\mathcal{M}$ is defined as

$$i\mathcal{M} = \left(ie(p_A + p_C)^{\mu}\right) \left(\frac{-ig_{\mu\nu}}{q^2}\right) \left(ie(p_B + p_D)^{\nu}\right) \,.$$

The different factors of this amplitude can be associated directly to the several elements of the Feynman diagrams (Fig. 6.8, right). In particular, the factors $(ie(p_A + p_C)^{\mu})$ and $(ie(p_B + p_D)^{\nu})$ can be associated, respectively, to the vertex of the particle 1 and 2 with the photon and the factor $\left(-ig_{\mu\nu}/q^2\right)$, called the "propagator" since it is the mathematical term accounting for the interaction, to the exchanged photon. There is energy-momentum conservation at each vertex.

Being θ the scattering angle in the center-of-mass reference frame (see Fig. 6.9) and p the module of center-of-mass momentum, the four-vectors of the initial and final states at high-energy ($E \gg m$) can be written as

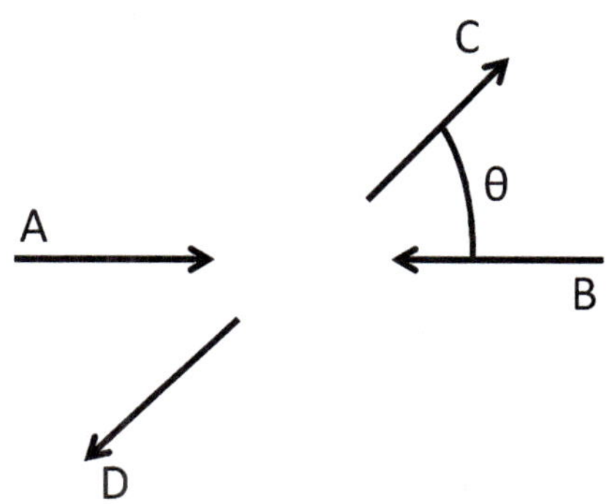

Fig. 6.9 Scattering of two charged particles in the center-of-mass reference frame

$$p_A = (p, p, 0, 0)$$

$$p_B = (p, -p, 0, 0)$$

$$p_C = (p, p\ \cos\theta, p\sin\theta, 0)$$

$$p_D = (p, -p\ \cos\theta,\ -p\sin\theta, 0)\,.$$

Then:

$$(p_A + p_C) = (2p, p\ (1+\cos\theta)\,,\, p\sin\theta,\ 0)$$

$$(p_B + p_D) = (2p, -p\ (1+\cos\theta)\,,\, -p\sin\theta,\ 0)$$

$$q = (p_D - p_B) = (0, p\ (1-\cos\theta)\,,\, -p\sin\theta,\ 0)$$

and

$$\mathcal{M} = -e^2\frac{1}{q^2}\left((p_A+p_C)^0\,(p_B+p_D)^0 - \sum_{i=1}^{3}(p_A+p_C)^i\,(p_B+p_D)^i\right)$$

$$\mathcal{M} = -e^2\frac{1}{p^2\,(1-\cos\theta\,)^2 + p^2\sin\theta^{\,2}}\left(4p^2 + p^2\,(1+\cos\theta\,)^2 + p^2\sin\theta^{\,2}\right)$$

$$\mathcal{M} = e^2\frac{(3+\cos\theta)}{(1-\cos\theta)}\,.$$

The differential cross section of an elastic two-body scattering between spinless nonidentical particles in the center-of-mass frame is given by (see Sect. 5.7.2):

$$\frac{d\sigma}{d\Omega} = \frac{|\mathcal{M}|^2}{64\pi^2 s}$$

where $s = (E_A + E_B)^2$ is the square of the center-of-mass energy (s is one of the Mandelstam variables, see Sect. 5.7.2).

Finally

$$\frac{d\sigma}{d\Omega} = \frac{\alpha^2}{4s}\frac{(3+\cos\theta)^2}{(1-\cos\theta)^2}$$

where

$$\alpha = \frac{e^2}{4\pi} \tag{6.61}$$

is the fine structure constant.

Note that when $\cos\theta \to 1$ the cross section diverges. This fact is a consequence of the infinite range of the electromagnetic interactions, translated into the fact that photons are massless.

6.2.8 Electron-Muon Elastic Scattering

Electron and muons have spin 1/2 and are thus described by Dirac bi-spinors (see Sect. 6.2.4). The computation of the scattering amplitudes is more complex than the one discussed in the previous subsection for the case of spinless particles but the main steps, summarized hereafter, are similar.

The Dirac equation in presence of an electromagnetic field is written as

$$\left(i\gamma^{\mu}\partial_{\mu} - e\,\gamma^{\mu}A_{\mu} - m\right)\,\psi = 0.$$

The corresponding current is

$$j_{\mu}\,(x) = \; -e\overline{\psi}\,\gamma_{\mu}\psi\,.$$

The transition amplitude for the electron (states A and C)/muon (states B and D) scattering can then be written as (Fig. 6.10):

$$H'_{if} = -i\int j_{\mu}^{\text{elect}}\left(-\frac{1}{q^2}\right) j^{\mu}_{\text{muon}}\; d^4\,x$$

where

$$j_{\mu}^{\text{elect}} = -e\left(\overline{u}_C\gamma_{\mu}u_A\right)e^{-iqx}$$

$$j^{\mu}_{\text{muon}} = -e\left(\overline{u}_D\gamma^{\mu}u_A\right)e^{iqx}$$

with

$$q = (p_D - p_B) \;=\; (p_A - p_C)\;.$$

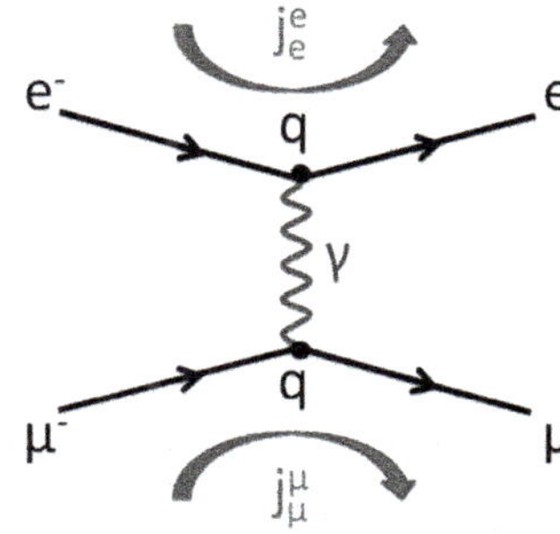

Fig. 6.10 Lowest order Feynman diagram for electron-muon scattering

Solving the integral,

$$H'_{if} = -i\ N_A N_B N_C N_D (2\pi)^4\ \delta^4\left(p_A + p_B - p_C + p_D\right)\ \mathcal{M}$$

where the amplitude $\mathcal{M}$ is defined as

$$-i\mathcal{M} = \left(ie\left(\overline{u}_C \gamma^\mu u_A\right)\right)\left(\frac{-ig_{\mu\nu}}{q^2}\right)\left(ie\left(\overline{u}_D \gamma^\upsilon u_B\right)\right).$$

Once again there is a strict correspondence between the different factors of this amplitude and the several elements of the Feynman diagrams. Now to each incoming or outgoing external line one spinor (u_A, u_B) or one adjoint spinor ($\overline{u}_C$, $\overline{u}_D$) is, respectively, associated, while the factor associated to the photon internal line (photon propagator) is $\left(-ig_{\mu\nu}/q^2\right)$. The vertex factor is now given by $(ie\gamma^\mu)$ and again energy-momentum is conserved at each vertex.

Higher order terms correspond to more complex diagrams which may have internal loops and fermion internal lines (see Fig. 6.11). In this case, the factor associated to each internal fermion line is

$$\left(\frac{i\left(\gamma^\mu p_\mu + m\right)}{p^2 - m^2}\right)$$

and one should not forget that each internal four-momentum loop has to be integrated over the full momentum range.

The cross sections can be computed once again using the Fermi golden rule. This requires the computation of the square of the transition amplitude $|\mathcal{M}|^2$ and is, even at the first order, out of the scope of the present text (it involves the sum over all internal indices and products of γ matrices which can be considerably simplified using the so-called trace theorems). If several Feynman diagrams contribute to the same process (same initial and final states) the corresponding amplitudes have to be summed before squaring. On the other hand, if several different final states (different spin configurations for instance) correspond to the same initial state, the sum must be

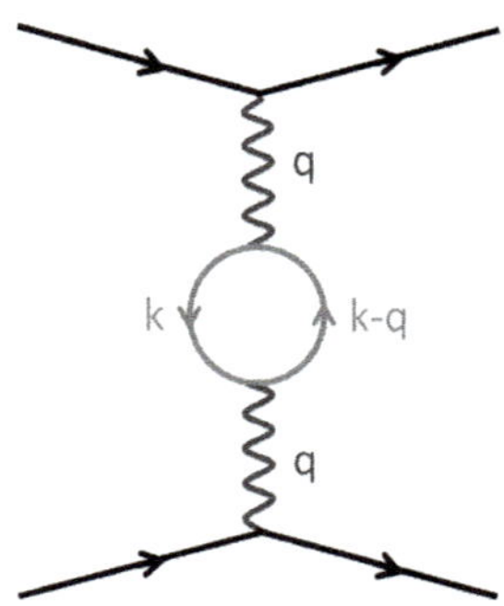

Fig. 6.11 A higher order diagram with a fermion loop

performed over the squares of the individual amplitudes. Finally, if the beam and/or the target are not polarized (if they are not in spin eigenstates) an average over the possible spin configuration has to be calculated.

The first-order elastic electron/muon scattering cross section at high-energy ($E \gg m$) in the center-of-mass reference frame is given by

$$\frac{d\sigma}{d\Omega} = \frac{\alpha^2}{2s} \frac{1+\cos^4(\theta/2)}{\sin^4(\theta/2)} \tag{6.62}$$

while in the laboratory reference frame (muon at rest) is

$$\frac{d\sigma}{d\Omega} = \frac{\alpha^2 \cos^2\left(\frac{\theta}{2}\right)}{4\, E^2 \sin^4\left(\frac{\theta}{2}\right)} \frac{E'}{E} \left(1 - \frac{q^2}{2m_\mu^2} \tan^2\left(\frac{\theta}{2}\right)\right). \tag{6.63}$$

This is the Rosenbluth formula referred in Sect. 5.5.1.

6.2.9 Renormalization and Vacuum Polarization

High-order diagrams involve often closed loops where integration over all possible momentum should be performed (see Fig. 6.11). As these loops are virtual, they represent phenomena that occur in time scales compatible with the Heisenberg uncertainty relations. Since there is no limit on the range of the integration and on the number of diagrams, the probabilities may a-priori diverge to infinity. We shall see, however, that the effect of higher order diagrams is the redefinition of some quantities; for example, the "bare" charge of the electron becomes a new quantity e that we measure in experiments. A theory with such characteristics—i.e., a theory for which the series of the contributions form all diagrams converges—is said to be renormalizable.

To avoid confusion in what follows, shall call now g_e the "pure" electromagnetic coupling.

Following the example of the amplitude corresponding to the diagram represented in Fig. 6.10, the photon propagator is modified by the introduction of the integration over the virtual fermion/antifermion loop leading to

$$\mathcal{M}_2 \sim \frac{-g_e^4}{q^4} \left(\left(\overline{u}_C \gamma^\mu u_A\right)\right) \left(\left(\overline{u}_D \gamma^\mu u_B\right)\right) \left(\int_0^\infty \frac{(\ldots)}{(k^2 - m^2)\left((k-q)^2 - m^2\right)} d^4k \right)$$

where g_e is the "bare" coupling constant ($g_e = \sqrt{4\pi\alpha_0}$, in the case of QED; α_0 refers to the "bare" coupling, without renormalization).

The integral can be computed setting some energy cut-off M and making $M \to \infty$ in the end of the calculation. Then it can be shown that

$$\lim(M \to \infty)\left(\int_0^M \frac{(\ldots)}{(k^2 - m^2)\left((k-q)^2 - m^2\right)} d^4k\right) \sim \frac{q^2}{12\pi^2}\left\{\ln\left(\frac{M^2}{m^2}\right) - f\left(\frac{-q^2}{m^2}\right)\right\}$$

having $(\ldots)$ dimensions of m^2, and

$$\mathcal{M}_2 \sim \frac{-g_e^2}{q^2}(\ldots)(\ldots)\left(1 - \frac{g_e^2}{12\pi^2}\left\{\ln\left(\frac{M^2}{m^2}\right) - f\left(\frac{-q^2}{m^2}\right)\right\}\right).$$

The divergence is now logarithmic but it is still there.

The "renormalization miracle" consists in absorbing the infinity in the definition of the coupling constant. Defining

$$g_R \equiv g_e \sqrt{1 - \frac{g_e^2}{12\pi^2} \ln\left(\frac{M^2}{m^2}\right)} \tag{6.64}$$

and neglecting g_e^6 terms (for that many other diagrams have to be summed up, but the associated probability is expected to become negligible)

$$\mathcal{M}_2 \sim \frac{-{g_R}^2}{q^2}(\ldots)(\ldots)\left[1 + \frac{{g_R}^2}{12\pi^2} f\left(\frac{-q^2}{m^2}\right)\right].$$

The $\mathcal{M}_2$ is no more divergent but the coupling constant g_R (the electric charge) is now a function of q^2:

$$g_R\left(q^2\right) = g_R\left(q_0^2\right)\sqrt{1 + \frac{g_R\left(q_0^2\right)}{12\pi^2} f\left(\frac{-q^2}{m^2}\right)}. \tag{6.65}$$

Other diagrams as those represented Fig. 6.12 lead to the renormalization of fundamental constants. In the left diagram, "emission" and "absorption" of a virtual photon by one of the fermion external lines contributes to the renormalization of the fermion mass, while in the one on the right, "emission" and "absorption" of a virtual photon between the fermion external lines from a same vertex contribute to the renormalization of the fermion magnetic moment and thus are central in the

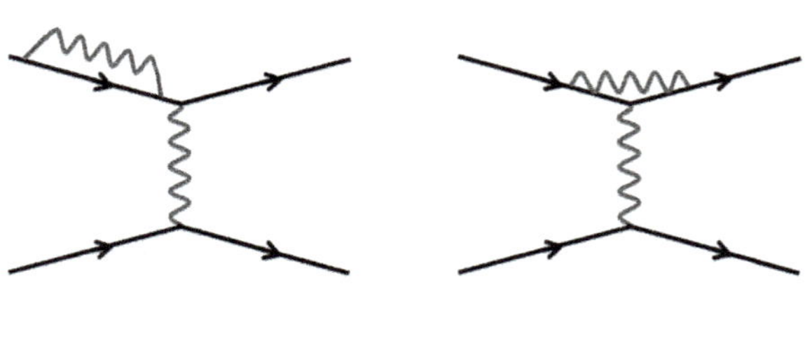

Fig. 6.12 Higher order diagrams with a fermion loop leading to the renormalization of the fermion mass (*left*) and of the magnetic moment (*right*)

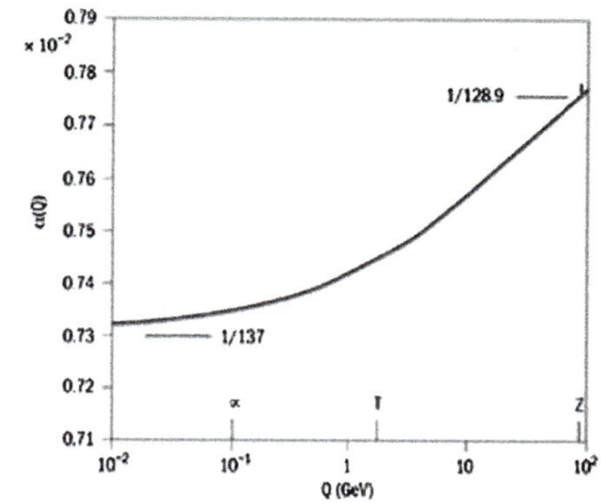

Fig. 6.13 Evolution of the QED effective coupling constant with momentum transfer. The theoretical *curve* is compared with measurements at the Z mass at CERN's LEP e^+e^- collider. From CERN Courier, August 2001

calculation of $(g-2)$ as discussed in Sect. 6.2.4. The contribution of these kinds of diagrams to the renormalization of the charge cancels out, ensuring that the electron and the muon charges remain the same.

The result in 6.65 can be written at first order as

$$\alpha(q^2) \simeq \alpha(\mu^2)\frac{1}{1-\frac{\alpha(\mu^2)}{3\pi}\ln\frac{q^2}{\mu^2}}\,. \tag{6.66}$$

The electromagnetic coupling can be obtained by an appropriate renormalization of the electron charge defined at an arbitrary scale μ^2. The electric charge, and the electromagnetic coupling constant, "run" and increase with q^2. We know that at momentum transfers close to the electron mass $\alpha \simeq 1/137$, while close to the Z mass $\alpha \sim 1/128$. The "running" behavior of the coupling constants is not a mathematical artifact: it is experimentally well established that the strength of the electromagnetic interaction between two charged particles increases as the center-of-mass energy of the collision increases (Fig. 6.13).

Such an effect can be qualitatively described by the polarization of the cloud of the virtual fermion/antifermions pairs (mainly electron/positrons) by the "bare" charge

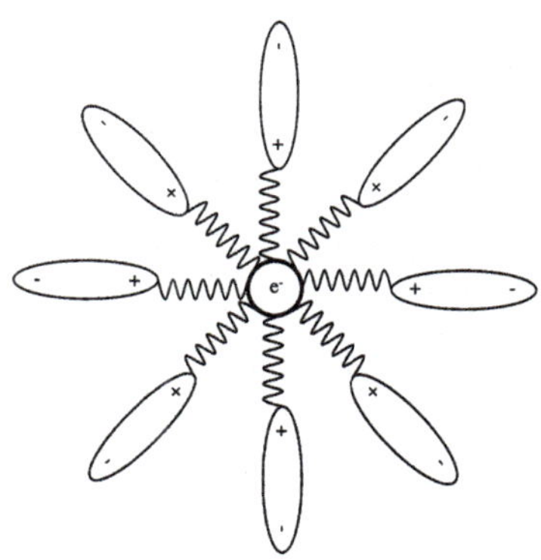

Fig. 6.14 Artistic representation of the screening of a charge by its own cloud of virtual charged particle–antiparticle pairs

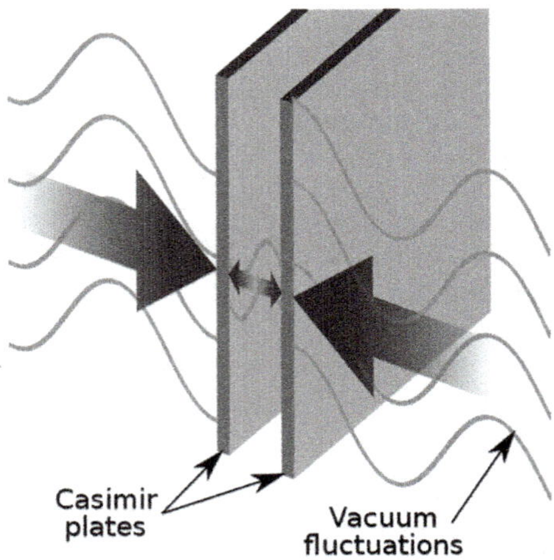

Fig. 6.15 Artistic view of the Casimir effect. From the Scientific American blog of Jennifer Ouellette, April 19, 2012

that is at the same time the source of the electromagnetic field (Fig. 6.14). This bare charge is screened by this polarized medium and its intensity decreases with the distance to the charge (increases with the square of the transferred momentum).

Even in the absence of any "real" matter particle (i.e., in the vacuum) there is no empty space in quantum field theory. A rich spectrum of virtual wave particles (like the photons) can be created and destroyed under the protection of the Heisenberg incertitude relations and within its limits be transfigurated into fermion/antifermion pairs. The space is thus full of electromagnetic waves and the energy of its ground state (the *zero point energy*) is, like the ground state of any harmonic oscillator, different from zero. The integral over all the space of this ground-state energy will be infinite, which leads once again to an enormous challenge to theoretical physicists (what is the relation with a nonzero cosmological constant which may explain the accelerated expansion of the Universe observed in the last years as discussed in Sect. 8.1).

A spectacular consequence is the attraction experimented by two neutral planes of conductor when placed face to face at very short distances, typically of the order of the micrometer (see Fig. 6.15). This effect is known as the Casimir effect, since it was predicted by Hendrick Casimir[4] in 1948 and later on experimentally demonstrated. The two plates impose boundary conditions to the electromagnetic waves originated by the vacuum fluctuations, and the energy decreases with the distance in such a way that the net result is a very small but measurable attractive force.

A theory is said to be renormalizable if (as in QED) all the divergences at all orders can be absorbed into running physical constants; corrections are then finite at any order of the perturbative expansion. The present theory of the so-called Standard Model of Particle Physics was proven to be renormalizable. In contrast, the quantization of general relativity leads easily to non-renormalizable terms and this is one

---

[4]Hendrick Casimir (1909–2000) was a Dutch physicist mostly known for his works on superconductivity.

of the strong motivations for alternative theories (see Chap. 7). Nevertheless, the fact that a theory is not renormalizable does not mean that it is useless: it might just be an effective theory that works only up to some physical scale.

6.3 Weak Interactions

Weak interactions are, of course, weak; they have short range and contrary to the other interactions do not bind particles together. Their existence was first revealed in β decay and their universality was the object of many controversies until being finally established in the second half of the twentieth century. All fermions have weak charges and are thus subject to their subtle or dramatic effects. The structure of the weak interactions was found to be similar to the structure of QED, and this fact is at the basis of one of the most important and beautiful pieces of theoretical work in the 20th century: the Glashow–Weinberg–Salam model of electroweak interactions, which, together with the theory of strong interactions (QCD), constitutes the Standard Model (SM) of particle physics, that will be discussed in the next chapter.

There are however striking differences between QED and weak interactions: parity is conserved, as it was expected, in QED, but not in weak interactions; the relevant symmetry group in weak interactions is SU(2) (fermions are grouped in left doublets and right singlets) while in QED the symmetry group is U(1); in QED there is only one massless vector boson, the photon, while weak interaction are mediated by three massive vector bosons, the $W^{\pm}$ and the Z.

6.3.1 *The Fermi Model of Weak Interactions*

The β decay was known since long time when Enrico Fermi in 1933 realized that the associate transition amplitude could be written in a way similar to QED (see Sect. 6.2.8). Assuming time reversal symmetry, one can see that the transition amplitude for β decay,

$$n \rightarrow \; p\, e^-\, \overline{\nu}_e\,,$$

is, for instance, the same as:

$$\nu_e\, n \rightarrow \; p\, e^-\,;$$

$$e^-\; p \;\rightarrow\; n\; \nu_e \text{ (K capture)};$$

$$\overline{\nu}_e\; p \;\rightarrow\; n\; e^+ \text{ (inverse } \beta \text{ decay)}\,.$$

The transition amplitude can then be seen as the interaction of one hadronic and one leptonic current (Fig. 6.16) and may be written, in analogy to the electron-muon elastic scattering discussed before (Fig. 6.10) as

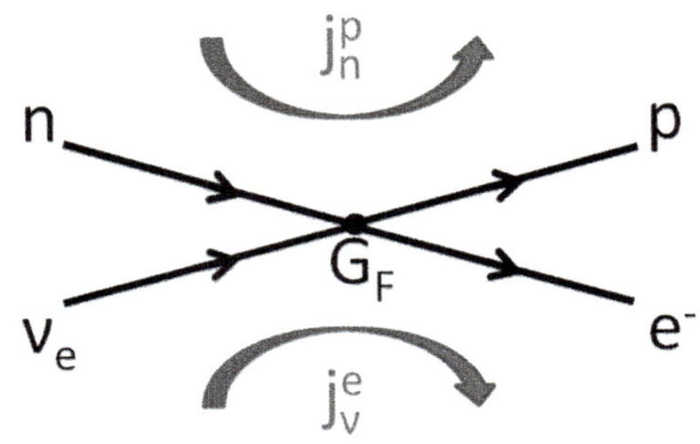

Fig. 6.16 Current–current description of the β decay in the Fermi model

$$\mathcal{M} = G_F \left((\overline{u}_p \gamma^\mu u_n) \right) \left((\overline{u}_e \gamma_\mu u_{\nu_e}) \right) . \tag{6.67}$$

Contrary to QED the fermions change their identity in the interaction ($n \rightarrow p$; $\nu_e \rightarrow e^-$), currents mix different charges (the electric charges of the initial states are not the same as those of the final states) and there is no propagator (the currents meet at a single point: we are in front of a contact interaction).

The coupling constant G_F, known nowadays as the Fermi constant, replaces the e^2/q^2 factor present in the QED amplitudes and thus has dimensions E^{-2} (GeV^{-2} in natural units). Its order of magnitude, deduced from the measurements of the β decay rates, is $G_F \sim (300\,\text{GeV})^{-2} \sim 10^{-5}\,\text{GeV}^{-2}$ (see Sect. 6.3.3). Assuming point-like interactions has striking consequences: the Fermi weak interaction cross sections diverge at high energies.

On a dimensional basis, one can deduce for instance that the neutrino-nucleon cross section behaves like:

$$\sigma \sim {G_F}^2 \ E^2 . \tag{6.68}$$

The cross section grows with the square of the center-of-mass energy and this behavior is indeed observed in low-energy neutrino scattering experiments.

However, from quantum mechanics, it is well known that a cross section can be decomposed in a sum over all the possible angular momenta l and then

$$\sigma \leq \frac{4\pi}{k^2} \sum_{l=0}^{\infty} (2l+1) . \tag{6.69}$$

Being $\lambda = 1/k$, this relation just means that contribution of each partial wave is bound and its scale is given by the area ($\pi\lambda^2$) "seen" by the incident particle. In a contact interaction, the impact parameter is zero and so the only possible contribution is the S wave ($l = 0$). Thus, the neutrino–nucleon cross section cannot increase forever. Given the magnitude of the Fermi constant G_F, the Fermi model of weak interactions cannot be valid for center-of-mass energies above a few hundreds of GeV (this bound is commonly said to be imposed by unitarity in the sense that the probability of an interaction cannot be >1).

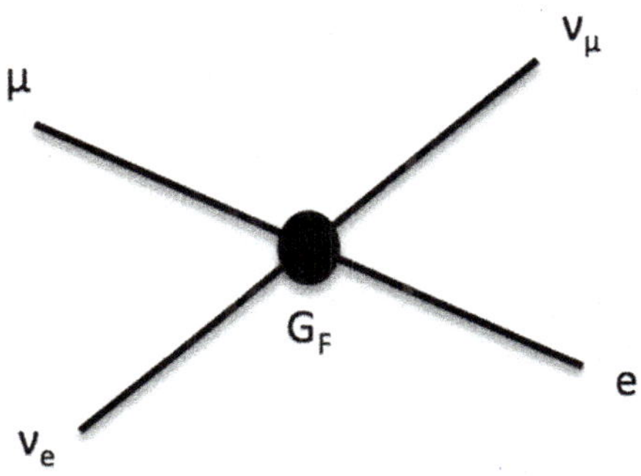

Fig. 6.17 Current–current description of the muon decay in the Fermi model

The solution of such a problem was found associating the weak interactions to a new field of short range, the weak field, whose massive charged bosons (the $W^{\pm}$) will act as propagators. In practice (see Sect. 6.3.5),

$$G_F \rightarrow \frac{g_w^2}{q^2 - M_W^2}\,, \tag{6.70}$$

and therefore the Fermi model is a low-energy approximation which is valid whenever the center-of-mass energy $\sqrt{s} \ll m_W$ ($m_W \sim 80$ GeV).

The discovery of the muon extended the applicability of the Fermi model of weak interactions. Bruno Pontecorvo realized in the late 1940s that the capture of a muon by a nucleus,

$$\mu^- p \rightarrow n\, \nu_\mu$$

as well as its weak decay (Fig. 6.17)

$$\mu^- \rightarrow e^-\, \nu_\mu \overline{\nu}_e$$

may be described by the Fermi model[5] as

$$\mathcal{M} = G_F \left(\left(\overline{u}_{\nu_\mu} \gamma^\rho u_\mu \right) \right) \left(\left(\overline{u}_e \gamma_\rho u_{\nu_e} \right) \right) . \tag{6.71}$$

Although β and μ decays are due to the same type of interaction their phenomenology is completely different: the neutron lifetime is ∼900 s while the muon lifetime is ∼2.2 μs; the energy spectrum of the decay electron is in both cases continuum (three-body decay) but its shape is quite different (Fig. 6.18). While in β decay it vanishes at the endpoint, in the case of μ is clearly nonzero. These striking differences are basically a reflection of the decay kinematics.

[5] The electromagnetic decay $\mu^- \rightarrow e^- \gamma$ violates the lepton family number and was never observed: $\Gamma_{\mu^- \rightarrow e^- \gamma} / \Gamma_{\text{tot}} < 5.7 \cdot 10^{-13}$.

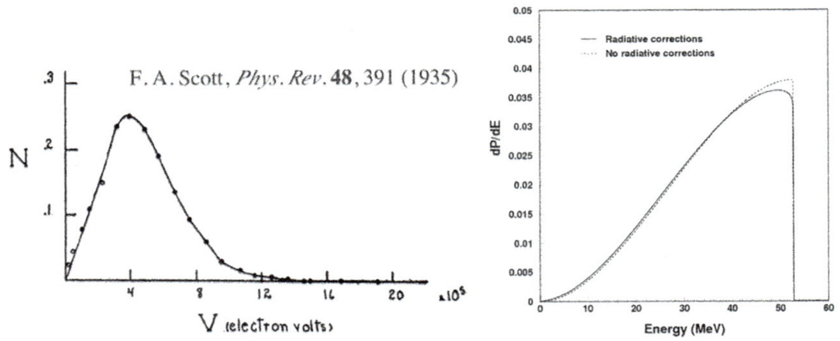

Fig. 6.18 Electron energy spectrum in β decay of thallium 206 (*left*) and in μ decay (*right*). *Sources* F.A. Scott, Phys. Rev. 48 (1935) 391; ICARUS Collaboration (S. Amoruso et al.), Eur. Phys. J. C33 (2004) 233

In fact using once more dimensional arguments the decay width of these particles should behave as

$$\Gamma \sim {G_F}^2 \ \Delta E^5 \ ,$$

where ΔE is the energy released in the decay. In the case of the β decay:

$$\Delta E_n \sim \left(m_n - m_p\right) \sim 1.29 \text{ MeV}$$

while in the μ decay

$$\Delta E_\mu \sim m_\mu \sim 105 \text{ MeV}$$

and therefore

$$\Delta E_n^5 \ll \Delta E_\mu^5 \, .$$

On the other hand, the shape of the electron energy spectrum at the endpoint is determined by the available phase space. At the endpoint, the electron is aligned against the other two decay products but, while in the β decay the proton is basically at rest (or remains "imprisoned" inside the nucleus) and there is only one possible configuration in the final state, in the case of μ decay, as neutrinos have negligible mass, the number of endpoint configurations is quite large reflecting the different ways to share the remaining energy between the neutrino and the anti-neutrino.

6.3.2 Parity Violation

The conservation of parity (see Sect. 5.3.6) was a dogma for physicists until the 1950s. Then, a puzzle appeared: apparently two strange mesons (denominated as θ^+

and τ^+) had the same mass, the same lifetime but different parities according to their decay modes:

$$\theta^+ \rightarrow \pi^+\pi^0 \qquad \text{(even parity)} \tag{6.72}$$

$$\tau^+ \rightarrow \pi^+\pi^+\pi^- \quad \text{(odd parity).} \tag{6.73}$$

In the 1956 Rochester conference, the conservation of parity in weak decays was questioned by Feynman reporting a suggestion of Martin Block. Few months later, Lee and Yang reviewed all the past experimental data and found that there was no evidence of parity conservation in weak interactions and they proposed new experimental tests based in the measurement of observables depending on axial vectors.

C.S. Wu (known as Madame Wu) was able, in a few months, to design and execute a β decay experiment where nuclei of ^{60}Co (with total angular momentum J = 5) decay into an excited state ^{60}Ni** (with total angular momentum J = 4):

$$^{60}\text{Co} \rightarrow {}^{60}\text{Ni}^{**}\text{e}^-\overline{\nu}_\text{e}$$

The ^{60}Co was polarized (a strong magnetic field was used, and the temperatures were as low as a few mK) and the number of decay electrons emitted in the direction (or opposite to) of the polarization field was measured (Fig. 6.19). The observed angle θ between the electron and the polarization direction followed a distribution of the form:

$$N(\theta) \sim 1 - P\,\beta \cos\theta$$

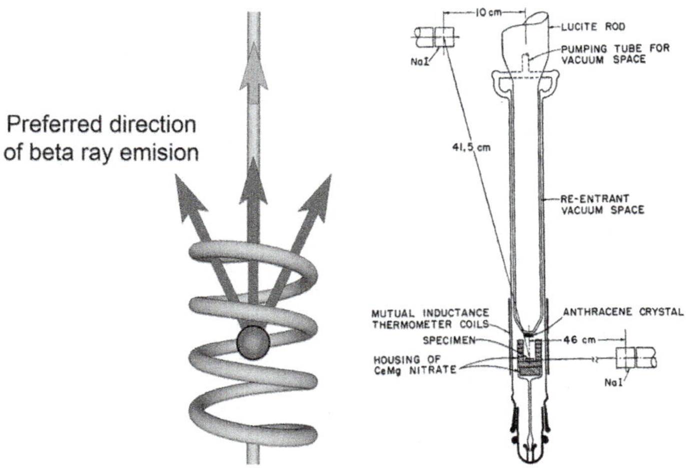

Fig. 6.19 Conceptual (*left*) and schematic diagram (*right*) of the experimental apparatus used by Wu et al. (1957) to detect the violation of the parity symmetry in β decay. The *left plot* comes from Wikimedia commons; the *right plot* from the original article by Wu et al., Physical Review 105 (1957) 1413

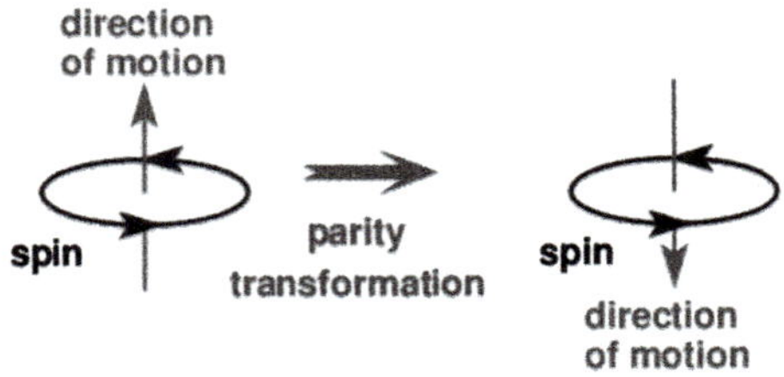

Fig. 6.20 Parity transformation of electron and magnetic field direction. The Wu experiment preferred the *left side* of the mirror to the *right* one

where P is the degree of polarization of the nuclei and β is the speed of the electron normalized to the speed of light.

The electrons were emitted preferentially in the direction opposite to the polarization of the nuclei, thus violating parity conservation. In fact under a parity transformation the momentum of the electron (a vector) reverses its direction while the magnetic field (an axial vector) does not (Fig. 6.20). Pauli placed a bet: "I don't believe that the Lord is a weak left-hander, and I am ready to bet a very high sum that the experiment will give a symmetric angular distributions of electrons"—and lost.

6.3.3 V-A Theory

The universality of the Fermi model of weak interactions was questioned long before the Wu experiment. In fact in the original Fermi model, only β decays in which there was no angular momentum change in the nuclei (Fermi transitions) were allowed, while the existence of β decays where the spin of the nuclei changed by one unity (the Gamow–Teller transitions) was already well established. The Fermi model had to be generalized.

In the most general way, the currents involved in the weak interactions could be written as a sum of Scalar (S), Pseudoscalar (P), Vector (V), Axial (A), or Tensor (T) terms following the Dirac bilinear forms referred in Sect. 6.2.4:

$$J_{1,2} = \sum_i C_i \left(\overline{u}_1 \Gamma_i u_2\right)$$

where C_i are arbitrary complex constants and the Γ_i are S, P, V, A, T operators.

At the end of 1956 George Sudarshan, a young Indian Ph.D. student working in Rochester University under the supervision of Robert Marshak, realized that the results on the electron-neutrino angular correlation reported by several experiments were not consistent. Sudarshan, contrary to the current opinion, suggested that the weak interaction had a V-A structure. This structure was (in the own words of Feynman) "*publicized by Feynman and Gell-Mann*" in 1958 in a widely cited article.

Each vectorial current in the Fermi model is, in the (V-A) theory, replaced by a vectorial minus an axial-vectorial current. For instance, the neutrino-electron vectorial current present in the β decay and in the muon decay amplitudes (Eqs. 6.67, 6.71, and Fig. 6.16, respectively):

$$\left(\overline{u}_e \gamma_\mu u_{\nu_e}\right) \tag{6.74}$$

is replaced by

$$\left(\overline{u}_e \gamma_\mu (1-\gamma^5) u_{\nu_e}\right). \tag{6.75}$$

In terms of the Feynman diagrams, the factor associated to the vertex becomes

$$\gamma^\mu (1-\gamma^5)\,. \tag{6.76}$$

Within the (V-A) theory, the transition amplitude of the muon decay, which is a golden example of a leptonic weak interaction, can be written as

$$\mathcal{M} = \frac{G_F}{\sqrt{2}} \left(\overline{u}_{\nu_\mu} \gamma^\mu (1-\gamma^5) u_\mu\right) \left(\overline{u}_e \gamma_\mu (1-\gamma^5) u_{\nu_e}\right). \tag{6.77}$$

The factor $\sqrt{2}$ is just a convention to keep the same definition for G_F and the only relevant change in relation to the Fermi model is the replacement:

$$\gamma^\mu \rightarrow \gamma^\mu (1-\gamma^5)\,.$$

The muon lifetime can now be computed using the Fermi golden rule. This detailed computation, which is beyond the scope of the present text, leads to:

$$\tau_\mu = \frac{192\ \pi^3}{{G_F}^2\ {m_\mu}^5}\,, \tag{6.78}$$

showing the $\mathrm{m}_\mu{}^{-5}$ dependence anticipated in Sect. 6.3.1—based just on dimensional arguments.

In practice, it is the measurement of the muon lifetime which is used to derive the value of the Fermi constant:

$$G_F = 1.166\,378\,7\,(6)\ \ 10^{-5}\ \mathrm{GeV}^{-2} \simeq \frac{1}{(300\ \mathrm{GeV})^2}\,.$$

The transition amplitude of the β decay can, in analogous way, be written as

$$\mathcal{M} = \frac{G_F^*}{\sqrt{2}} \left(\overline{u}_p \gamma^\mu (C_V - C_A \gamma^5) u_n\right) \left(\overline{u}_e \gamma_\mu (1-\gamma^5) u_{\nu_e}\right). \tag{6.79}$$

The C_V and C_A constants reflect the fact that the neutron and the proton are not point-like particles and thus form factors may lead to a change on their weak charges. Experimentally, the measurement of many nuclear β decays is compatible with the preservation of the value of the "vector weak charge" and a 25 % change in the axial charge:

$$C_V = 1.000$$

$$C_A = 1.255 \pm 0.006\,.$$

The value of G_F^* was found to be slightly lower (2 %) than the one found from the muon decay. This "discrepancy" was cured with the introduction of the Cabibbo angle as will be discussed in Sect. 6.3.6.

6.3.4 "Left" and "Right" Chiral Particle States

The violation of parity in weak interactions observed in the Wu experiment and embedded in the (V-A) structure can be translated in terms of interactions between particles with well-defined states of chirality.

"Chiral" states are eigenstates of γ^5 and they coincide with the helicity states for massless particles; however, no such particles (massless 4-spinors) appear to exist, to our present knowledge—neutrinos have very tiny mass. The operators $\frac{1}{2}\left(1+\gamma^5\right)$ and $\frac{1}{2}(1-\gamma^5)$, when applied to a generic particle bi-spinor u, (Sect. 6.2.4) project, respectively, on eigenstates with chirality +1 (R—Right) and -1 (L—Left). Chiral particle spinors can thus be defined as

$$u_L = \frac{1}{2}\left(1-\gamma^5\right)\, u \;;\; u_R = \frac{1}{2}\left(1+\gamma^5\right)\, u \tag{6.80}$$

with $u = u_L + u_R$. The adjoint spinors are given by

$$\overline{u}_L = \overline{u}\frac{1}{2}\left(1+\gamma^5\right) \;;\; \overline{u}_R = \overline{u}\frac{1}{2}\left(1-\gamma^5\right). \tag{6.81}$$

For antiparticles

$$v_L = \frac{1}{2}\left(1+\gamma^5\right)\, v \;;\; v_R = \frac{1}{2}\left(1-\gamma^5\right)\, v \tag{6.82}$$

$$\overline{v}_L = \overline{v}\frac{1}{2}\left(1-\gamma^5\right) \;;\; \overline{v}_R = \overline{v}\frac{1}{2}\left(1+\gamma^5\right). \tag{6.83}$$

Chiral states are closely related to helicity states but they are not identical. In fact, applying the chiral projection operators defined above to the helicity eigenstates (Sect. 6.2.4) one obtains, for instance, for the right helicity eigenstate:

$$u_\uparrow = \left(\frac{1}{2}(1-\gamma^5) + \frac{1}{2}(1+\gamma^5)\right) u_\uparrow$$

$$u_\uparrow = \frac{1}{2}\left(1 + \frac{p}{E+m}\right) u_R + \frac{1}{2}\left(1 - \frac{p}{E+m}\right) u_L \,. \tag{6.84}$$

In the limit $m \to 0$ or $p \to \infty$ right helicity and right chiral eigenstates coincide, otherwise not.

There is also a subtle but important difference: helicity is not Lorentz invariant but it is time invariant ($[h, H] = 0$), while chirality is Lorentz invariant but it is not time invariant ($[\gamma^5, H] \propto m$). The above relation is basically valid for $t \sim 0$.

Now, since

$$\gamma_\mu \left(\frac{1-\gamma^5}{2}\right) = \left(\frac{1+\gamma^5}{2}\right) \gamma_\mu \left(\frac{1-\gamma^5}{2}\right)$$

the weak (V-A) neutrino-electron current (Eq. 6.75) can be written as:

$$\overline{u}_e \gamma_\mu (1 - \gamma^5) u_{\nu_e} = 2 \left[\overline{u}_e \left(\frac{1+\gamma^5}{2}\right) \gamma_\mu \left(\frac{1-\gamma^5}{2}\right) u_{\nu_e} \right] = 2 \left(\overline{u}_{eL} \gamma_\mu u_{\nu_{eL}}\right) : \tag{6.85}$$

the weak charged leptonic current involves then only chiral left particles (and right chiral anti-particles).

In the case of the ^{60}Co β decay (the Wu experiment), the electron and anti-neutrino masses can be neglected and so the anti-neutrino must have right helicity and the electron left helicity. Thus, as the electron and anti-neutrino have to add up their spin to compensate the change by one unity in the spin of the nucleus, the electron is preferentially emitted in the direction opposite to the polarization of the nucleus (Fig. 6.21).

The confirmation of the negative helicity of neutrinos came from a sophisticated and elegant experiment by M. Goldhaber, L. Grodzins and A. Sunyar in 1957, studying neutrinos produced in a K capture process ($e^- \; p \to n \; \nu_e$). The helicity of the emitted neutrino was translated in the longitudinal polarization of a photon resulting from the de-excitation of a nucleus produced in the K capture:

$$^{152}\text{Eu} \; e^- \to \; ^{152}\text{Sm}^* \; \nu_e$$

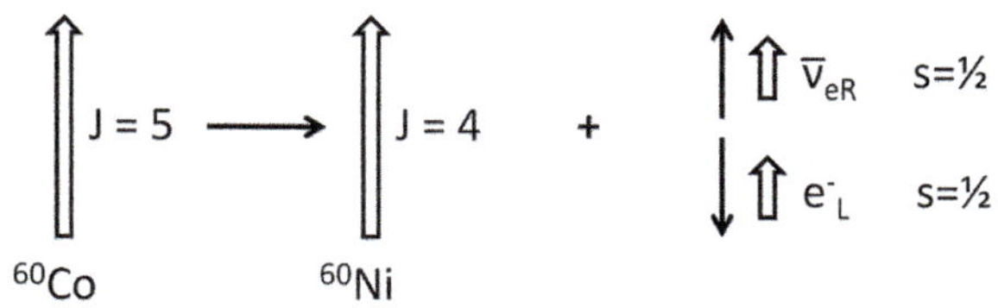

Fig. 6.21 Schematic representation of the spin alignment in the ^{60}Co β decay

$$^{152}\text{Sm}^* \rightarrow {}^{152}\text{Sm}\ \gamma\,.$$

The result was conclusive: neutrinos were indeed left-handed particles.

The accurate calculation of the ratio of the decay width of charged $\pi^{\pm}$ mesons into electron neutrinos with respect to muon neutrinos was also one of the successes of the (V-A) theory. According to (V-A) theory at first order:

$$\frac{BR\left(\pi^- \rightarrow e^-\,\overline{\nu}_e\right)}{BR\left(\pi^- \rightarrow \mu^-\,\mu\right)} = \frac{m_e^2\left(m_\pi^2 - m_e^2\right)^2}{m_\mu^2\left(m_\pi^2 - m_\mu^2\right)^2} \simeq 1.28 \times 10^{-4}\,,$$

while at the time this ratio was first computed the experimental limit was wrongly much smaller $\left(<10^{-6}\right)$. In fact the (V-A) theoretical prediction is confirmed by the present experimental determination:

$$\frac{BR\left(\pi^- \rightarrow e^-\,\overline{\nu}_e\right)}{BR\left(\pi^- \rightarrow \mu^-\,\mu\right)} \simeq 1.2 \times 10^{-4}\,.$$

In the framework of (V-A) theory, if the charged leptons were massless these weak decays would be forbidden. In fact the pion has spin 0, the anti-neutrino is a right-handed particle and thus to conserve angular momentum the helicity of the electron should be positive (Fig. 6.22) which is impossible for a massless left electron. However, the suppression of the decay into electron neutrino face to the decay into muon neutrino, contrary to what would be expected from the available decay phase space, is not a proof of the (V-A) theory. It can be shown that a theory with V or A couplings (or any combination of them) would also imply a suppression factor of the order m_e^2/m_μ^2.

As a last example, the neutrino and anti-neutrino handedness is revealed in the observed ratio of cross sections for neutrino and anti-neutrino in isoscalar nuclei (with an equal number of protons and neutrons) N at GeV energies:

$$\frac{\sigma\left(\overline{\nu}_\mu\,N \rightarrow \mu^+ X\right)}{\sigma\left(\nu_\mu\,N \rightarrow \mu^- X\right)} \sim \frac{1}{3}\,.$$

Note that at these energies, the neutrinos and the anti-neutrinos interact directly with the quarks and antiquarks the protons and neutrons are made of (similarly to the electrons in the deep inelastic scattering discussed in Sect. 5.5.3).

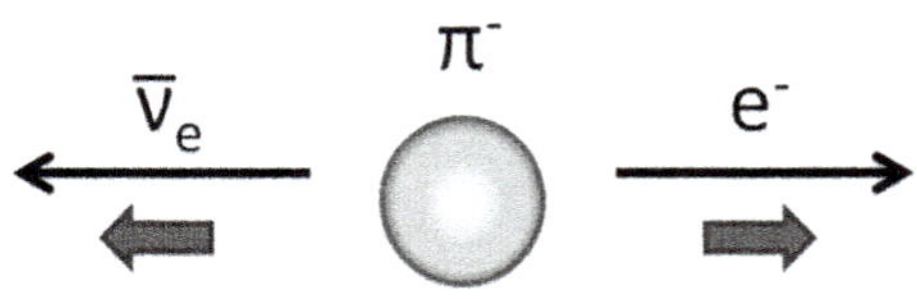

Fig. 6.22 Schematic representation of the spin alignment in the π^- decay

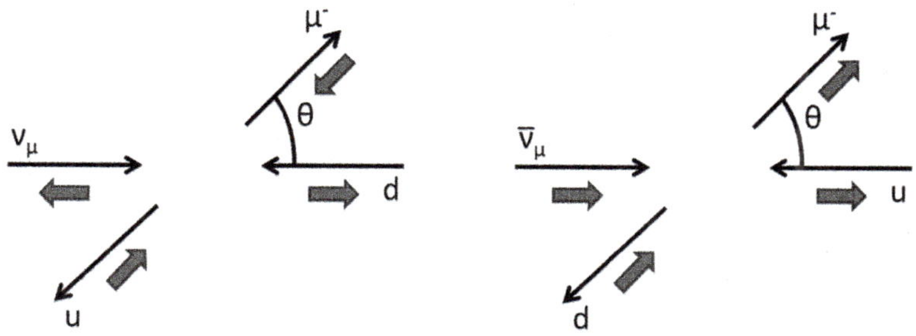

Fig. 6.23 Schematic representation of the spin alignments in $\nu_\mu\, d \to \mu^- u$ (*left*) and in $\overline{\nu}_\mu\, u \to \mu^+ d$ (*right*) interactions

Let us now consider just valence quarks in a first approximation. As electric charge and leptonic number are conserved, a neutrino can just pick up a d quark transforming it into a u quark and emitting a μ^-. Anti-neutrinos will do the opposite. In these conditions, neglecting masses, all fermions have negative helicity and all antifermions have positive helicity. The total angular momentum is therefore 0 for neutrino interactions and 1 for anti-neutrino interactions (Fig. 6.23). Thus, the former interaction will be isotropic while the amplitude of the latter will be weighted by a factor $1/2\,(1+\cos\theta)$. Then

$$\frac{d\sigma\left(\overline{\nu}_\mu\, u \to \mu^+ d\right)}{d\Omega} = \frac{d\sigma\left(\nu_\mu\, d \to \mu^- u\right)}{d\Omega}\,\frac{(1+\cos\theta\,)^2}{4}$$

and integrating over the solid angle

$$\frac{\sigma\left(\overline{\nu}_\mu\, u \to \mu^+ d\right)}{\sigma\left(\nu_\mu\, d \to \mu^- u\right)} = \frac{1}{3}\,.$$

6.3.5 *Intermediate Vector Bosons*

Four-fermion interaction theories (like Fermi model—see Sect. 6.3.1) violate unitarity at high energy and are not renormalizable (all infinities cannot be absorbed into running physical constants—see Sect. 6.2.9). The path to solve such problem was to construct, in analogy with QED, a gauge theory of weak interactions which lead to the introduction of intermediate vector bosons, with spin 1: the $W^\pm$ and the Z. However, in order to model the short range of the weak interactions, such bosons could not have zero mass, and thus would violate the gauge symmetry. The problem was solved by the introduction of spontaneously broken symmetries, which then led to the prediction of the existence of the so-called Higgs boson.

In this section, the modification introduced on the structure of the charged weak currents as well as the discovery of the neutral currents and of the $W^{\pm}$ and the Z bosons will be briefly reviewed. The overall discussion on the electroweak unification and its experimental tests will be the object of the next chapter.

6.3.5.1 Charged Weak Currents

The structure of the weak charged and of the electromagnetic interactions became similar with the introduction of the $W^{\pm}$ bosons, with the relevant difference that weak-charged interactions couple left-handed fermions (right-handed antifermions) belonging to SU(2) doublets, while electromagnetic interactions couple fermions belonging to U(1) singlets irrespective of chirality.

The muon decay amplitude deduced in (V-A) theory (Eq. 6.77) can now, introducing the massive $W^{\pm}$ propagator (Fig. 6.24), be written as:

$$\mathcal{M} = \frac{g_W{}^2}{8}\left(\overline{u}_{\nu_\mu}\gamma^\mu(1-\gamma^5)u_\mu\right)\frac{-i\left(g_{\mu\nu}-q_\mu q_\nu/M_W{}^2\right)}{\left(q^2-M_W{}^2\right)}\left(\overline{u}_e\gamma^\nu(1-\gamma^5)u_{\nu_e}\right)$$

or, introducing explicitly the left and right spinors:

$$\mathcal{M} = \frac{g_W{}^2}{2}\left(\overline{u}_{\nu_{\mu L}}\gamma_\mu u_{\mu L}\right)\frac{i\left(g_{\mu\nu}-q_\mu q_\nu/M_W{}^2\right)}{\left(q^2-M_W{}^2\right)}\left(\overline{u}_{e_L}\gamma_\mu u_{\nu_{eL}}\right).$$

The derivation of the expression of the propagator for massive spin 1 boson is based on the Proca equation (Sect. 6.2.1) and it is out of the scope of the present text. But as the second term can be neglected a Yukawa-type expression, $g_{\mu\nu}/\left(q^2-M_W{}^2\right)$, is recovered.

In the low-energy limit, $\left(q^2 \ll M_W{}^2\right)$ the two coupling constants are thus related by:

$$G_F = \frac{\sqrt{2}}{8}\frac{g_W{}^2}{M_W{}^2}.$$

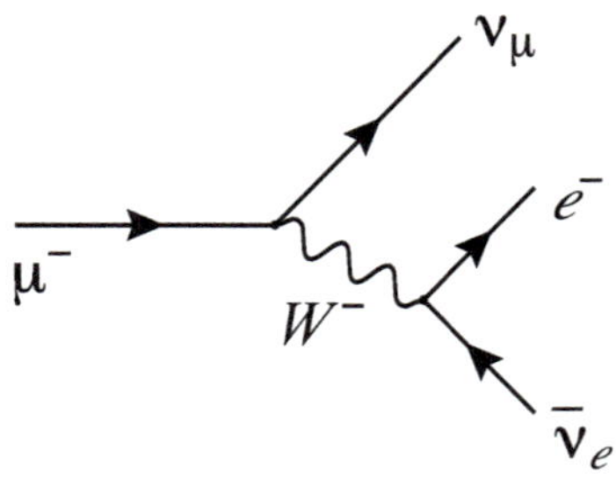

Fig. 6.24 First-order Feynman diagram for muon decay

6.3.5.2 Neutral Weak Currents

Neutral weak currents were predicted long before their discovery at CERN in 1973 (N. Kemmer 1937, O. Klein 1938, S. A. Bludman 1958). Indeed the SU(2) structure of charged interactions (leptons organized in weak isospin doublets) suggested the existence of a triplet of weak bosons similarly to the pion triplet responsible for the proton–neutron strong isospin rotations.

However, if the charged components would be the $W^{\pm}$, the neutral one could not be the γ, which has no weak charge. Furthermore, in the 1960s it was discovered that strangeness-changing neutral currents (for instance $K^{+} \rightarrow \pi^{+} \nu \overline{\nu}$) were highly suppressed and thus some thought that neutral weak interactions may not exist. Many theorists however became enthusiastic about neutral currents around the 1970s since they were embedded in the work by Glashow, Salam, and Weinberg on electroweak unification (see Sect. 7.2). From the experimental point of view, it was clearly a very difficult issue and the previous experimental searches on neutral weak processes lead just to upper limits.

Neutrino beams were the key to such searches. In fact, as neutrinos do not have electromagnetic and strong charges, their only possible interaction is the weak one. Neutrino beams are produced in laboratory (Fig. 6.25) by the decay of secondary pions and kaons coming from a primary high-energy proton interaction on a fixed target. The charge and the momentum range of the pions and kaons can be selected using a sophisticated focusing magnetic optics system (narrow-band beam) or just loosely selected maximizing the beam intensity (wide-band beam). The energy spectra of such beams are quite different (Fig. 6.26). While the narrow-band beam has an almost flat energy spectrum, the wide band is normally peaked at low energies.

In the 1960s a large heavy liquid bubble chamber (18 tons of freon under a pressure of 10–15 atmospheres, within a magnetic field of 2 T) called Gargamelle was designed at CERN. Gargamelle could collect a significant number (one order of

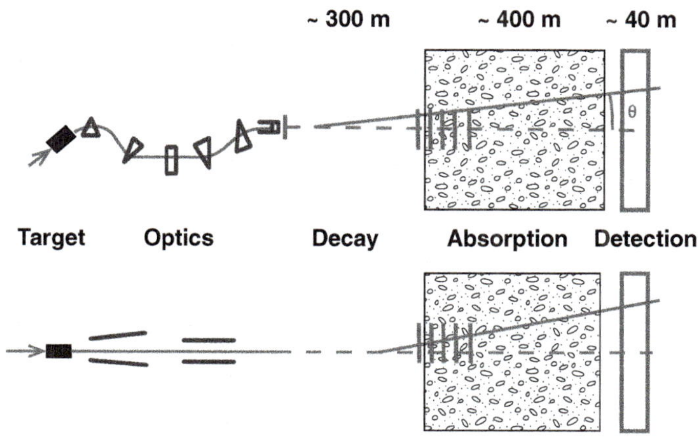

Fig. 6.25 Neutrino narrow-band beam (*top*) and wide-bam beam (*bottom*) production

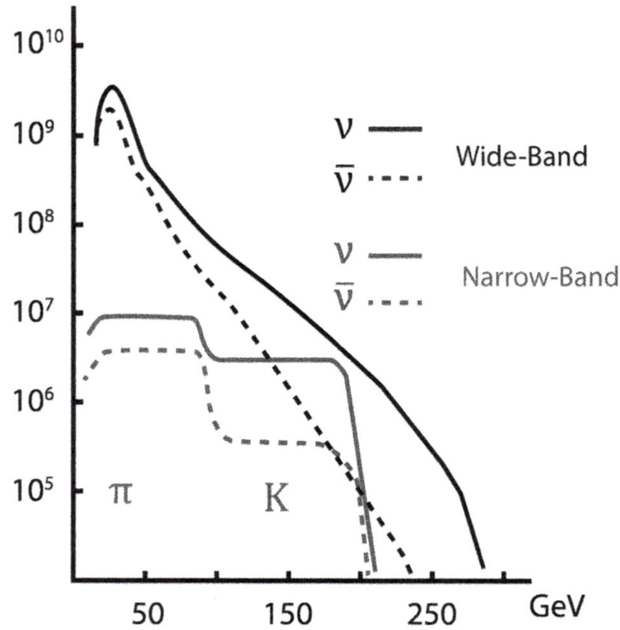

Fig. 6.26 Narrow-band and wide-band neutrino energy spectra. The y axis represents the number of particles per bunch

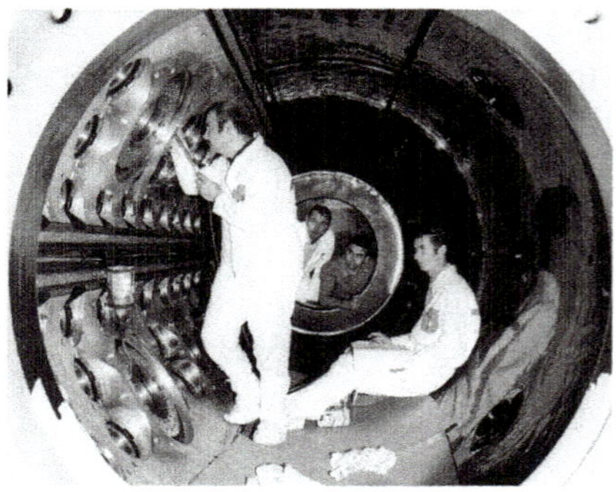

Fig. 6.27 Technicians at work in the Gargamelle bubble chamber at CERN. *Source* CERN

magnitude above the previous experiments) of neutrino interactions (Fig. 6.27). Its first physics priority was, in the beginning of the 1970s, the test of the structure of the protons and neutrons just revealed in the deep inelastic scattering experiment at SLAC (Sect. 5.5.3).

In a batch of about 700,000 photos of neutrino interactions, one event emerged as anomalous. In that photo (Fig. 6.28), taken with an anti-neutrino beam, just an electron was visible (giving rise to a small electromagnetic cascade). This event is a perfect candidate for a $\overline{\nu}_\mu \; e^- \to \overline{\nu}_\mu \; e^-$ interaction (Fig. 6.29). The background in the anti-neutrino beam was estimated to be negligible.

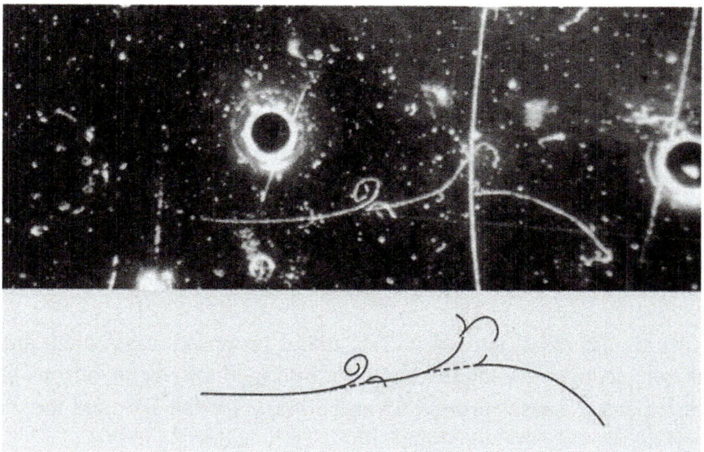

Fig. 6.28 Gargamelle picture (*top*) and sketch (*bottom*) of the first observed neutral current process $\overline{\nu}_\mu\ e^- \rightarrow \overline{\nu}_\mu\ e^-$. A muon anti-neutrino coming from the left knocks an electron forward, creating a small shower of electron–positron pairs. *Source* CERN

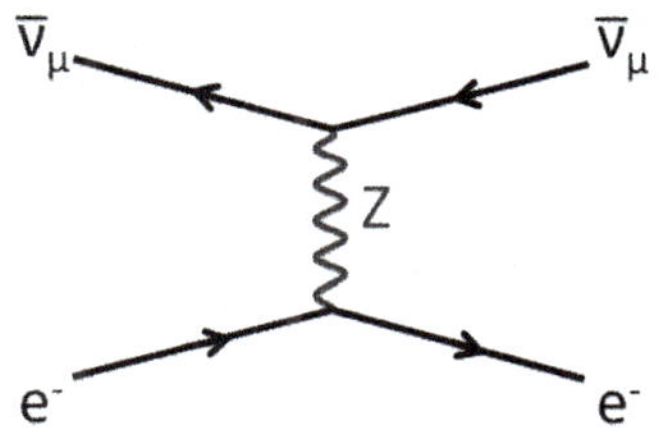

Fig. 6.29 First-order Feynman diagram for the neutral leptonic weak interactions $\overline{\nu}_\mu\ e^- \rightarrow \overline{\nu}_\mu\ e^-$

If neutral current interactions did exist, they should be even more visible in the semileptonic channel. The signature should be clear: in charged semileptonic weak interactions an isolated muon and several hadrons should be produced in the final state, while in the interactions mediated by the neutral current there could be no muon (Fig. 6.30).

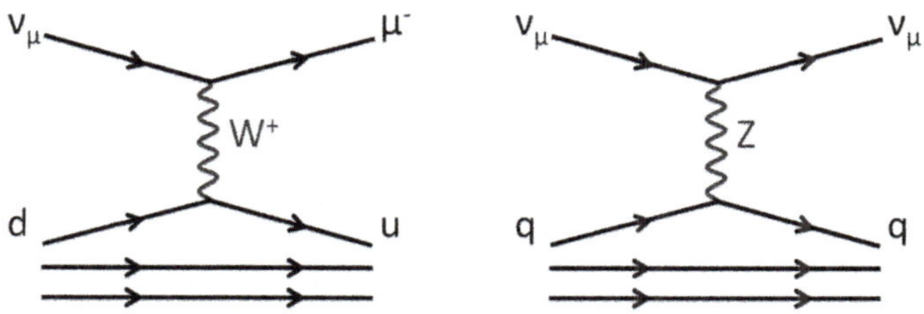

Fig. 6.30 First-order Feynman diagrams for the charged (*left*) and neutral (*right*) semileptonic weak interactions

However, the background resulting from neutron interactions in the chamber, being the neutrons produced in neutrino interactions upstream the detector, is not negligible. Careful background estimation had to be performed. The final result, after several months of work and public discussions, was that the number of events without a muon was clearly above the expected number of background events. The existence of the weak neutral currents was finally firmly established.

6.3.5.3 The Discovery of the W and Z Bosons

Neutral currents did exist, and the GSW model proposed a complete and unified framework for electroweak interactions: the intermediate vector bosons should be there (with expected masses around 65 and 80 GeV for the $W^{\pm}$ and the Z, respectively, based on the data known at that time). They had to be found.

In 1976 Carlo Rubbia pushed the idea to convert the existing Super Proton Synchrotron accelerator at CERN (or the equivalent machine at Fermilab) into a proton/antiproton collider. It was not necessary to build a new accelerator (protons and antiprotons would travel in opposite directions within the same vacuum tube) but antiprotons had to be produced and kept alive during many hours to be accumulated in an auxiliary storage ring. Another big challenge was to keep the beam focused. Simon van der Meer made this possible developing an ingenious strategy of beam cooling, to decrease the angular dispersion while maintaining monochromaticity. In beginning of the 1980s, the CERN SPS collider operating at a center-of-mass energy of 540 GeV was able to produce the first $W^{\pm}$ and Z (Fig. 6.31) by quark/antiquark annihilation ($u\,\overline{u} \to Z; d\,\overline{d} \to Z; u\,\overline{d} \to W^{+}; d\,\overline{u} \to W^{-}$).

The leptonic decay channels with electrons and muons in the final state were the most obvious signatures to detect the so awaited bosons. The hadronic decay channels as well as final states with tau leptons suffer from a huge hadronic background due

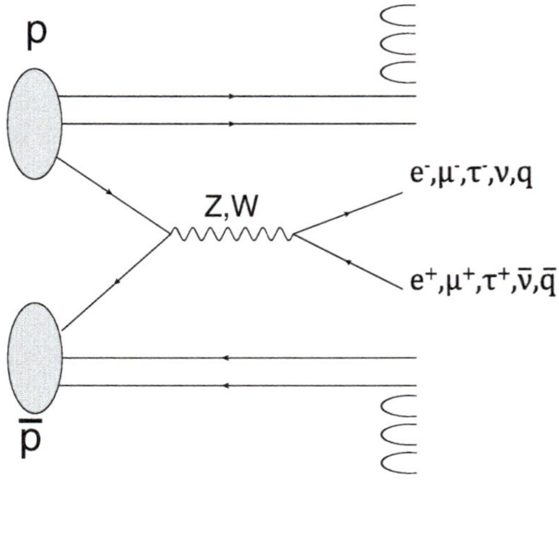

Fig. 6.31 $W^{\pm}$ and Z production in proton/antiproton colliders

to the "normal" quark and gluons strong interactions. Priority was then given to searches into the channels:

$$p\,\overline{p} \to ZX \to e^- e^+ X$$

$$p\,\overline{p} \to ZX \to \mu^- \mu^+ X$$

and

$$p\,\overline{p} \to W^\pm\,X \to e^\pm \nu_e\,X$$

$$p\,\overline{p} \to W^\pm\,X \to \mu^\pm \nu_e\,X\,.$$

Two general purpose experiments, UA1 and UA2, were built having the usual "onion" structure (a tracking detector surrounded by electromagnetic and hadronic calorimeters, surrounded by an exterior layer of muons detectors). In the case of UA1, the central detector (tracking and electromagnetic calorimeter) was immersed in a 0.7 T magnetic field, perpendicular to the beam line, produced by a magnetic coil (Fig. 6.32); the iron return yoke of the field was instrumented to operate as a hadronic calorimeter. UA1 was designed to be as hermetic as possible.

The first $W^\pm$ and Z events were recorded in 1983. $Z \to e^- e^+$ events were characterized by two isolated high-energy deposits in the cells of the electromagnetic calorimeter (Fig. 6.33 left) while $W^\pm\,X \to e^\pm \nu_e$ events were characterized by one isolated high-energy deposit in the cells of the electromagnetic calorimeter and an important transverse missing energy (Fig. 6.33 right).

The Z mass in this type of events can be reconstructed just computing the invariant mass of the final state electron and positron:

$$m_Z^2 = (E_1 + E_2)^2 - \left(\vec{P}_1 + \vec{P}_2\right)^2$$

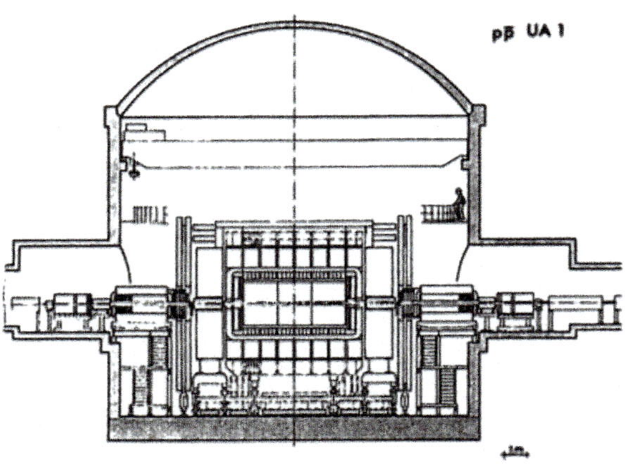

Fig. 6.32 Longitudinal cross section of the UA1 detector. From CERN http://cern-discoveries.web.cern.ch/cern-discoveries/courier/experiments/page-20.jpg

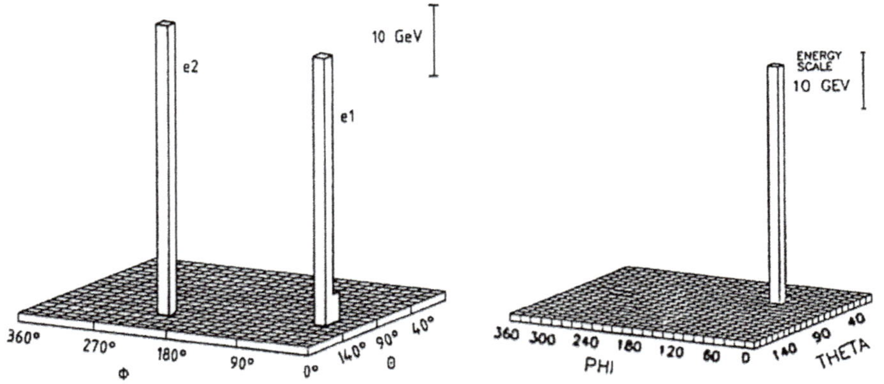

Fig. 6.33 *Left* Two high-energy deposits from a $Z \to e^-e^+$ event seen in the electromagnetic calorimeter of the UA2 experiment. *Right* One high-energy deposit and the missing transverse momentum from a $W^\pm X \to e^\pm \nu_e$ event. From http://cern-discoveries.web.cern.ch

$$m_Z^2 \cong 4\, E_1 E_2 \sin^2(\alpha/2)$$

where α is the angle between the electron and positron.

The distribution of the measured m_Z for the first $Z \to e^-e^+$ and $Z \to \mu^+\mu^-$ candidate events by UA1 and UA2 is represented in Fig. 6.34. The best-fit value presented by Carlo Rubbia in his Nobel lecture (1984) was of $m_Z = (95.6 \pm 1.4 \pm 2.9)$ GeV—the present value, after LEP, is 91.1876 ± 0.0021 GeV.

The reconstruction of the $W^\pm$ mass is more subtle—the missing energy does not allow a full kinematical constraint. The best way is to take it from the shape of the differential $W^\pm$ cross section as a function of the transverse momentum (the so-called "Jacobian peak" method). In fact neglecting the electron and neutrino masses, the

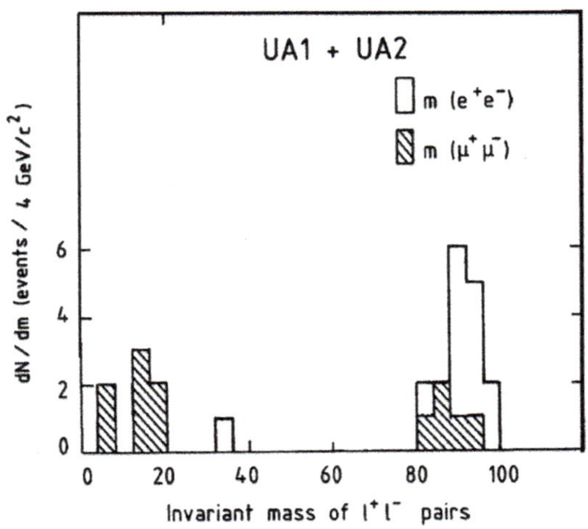

Fig. 6.34 Invariant mass distribution for the first candidate $Z \to e^+e^-$ and $Z \to \mu^+\mu^-$ events recorded by UA1 and UA2 (from the Nobel lecture of Carlo Rubbia, ©The Nobel Foundation). A clear peak of 17 events is visible around 95 GeV

transverse momentum of the $W^{\pm}$ is given by

$$P_T \cong \frac{m_W}{2} \sin\left(\theta^*\right)$$

where θ^* is the $W^{\pm}$ production angle in the center-of-mass reference frame.

Then

$$\cos\left(\theta^*\right) = \sqrt{1 - 4\,\frac{P_T^2}{m_W^2}}$$

and

$$\frac{d\cos\left(\theta^*\right)}{dP_T} = \frac{4P_T/m_W}{\sqrt{\left(1 - 4\,\frac{P_T^2}{m_W^2}\right)}}\,.$$

Writing the differential cross section as

$$\frac{d\sigma}{dP_T} = \frac{d\sigma}{d\cos\theta^*}\,\frac{d\cos\theta^*}{dP_T}$$

it is clear (Fig. 6.35) that a peak is present at

$$P_T = \frac{m_W}{2}\,.$$

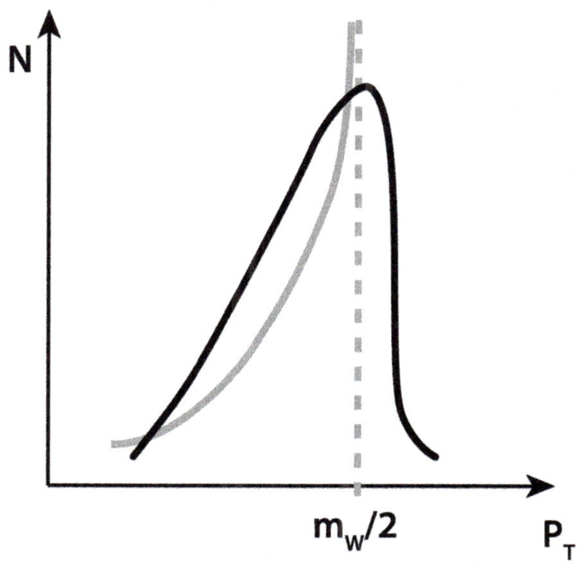

Fig. 6.35 Differential $W^{\pm}$ cross section as a function of transverse momentum. The *gray* (*black*) *line* refers to a measurement with an ideal (real) detector

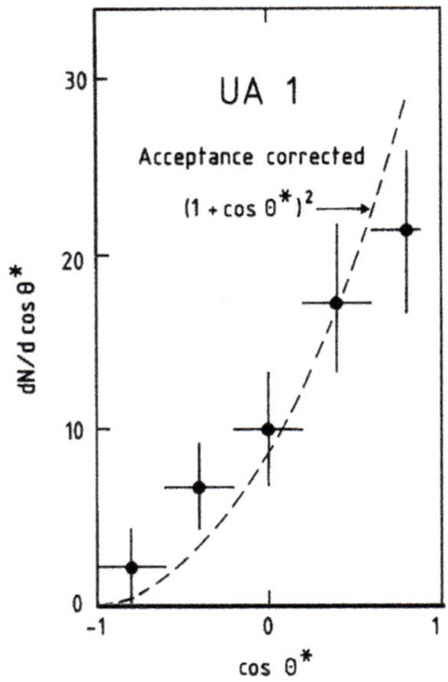

Fig. 6.36 The angular distribution of the electron emission angle θ^* in the rest frame of the W after correction for experimental acceptance, as detected by the UA1 detector (from the Nobel lecture of Carlo Rubbia, ©The Nobel Foundation)

The measured value for m_W by UA1 and UA2 was, respectively, $m_W = (82.7 \pm 1.0 \pm 2.7)$ GeV and $m_W = (80.2 \pm 0.6 \pm 0.5)$ GeV—the present world average is (80.385 ± 0.015) GeV.

Finally the V-A character of the charged weak interactions, as well as the fact that the W has spin 1, is revealed by the differential cross section as a function of $\cos\theta^*$ for the electron produced in the W semileptonic decay, which displays a $(1+\cos\theta^*)^2$ dependence (Fig. 6.36).

In fact, at CERN collider energies, neglecting the masses of the quarks and leptons and considering that $W^{\pm}$ are mainly produced by the interaction of valence quarks (from the proton) and valence antiquarks (from the antiproton), the helicity of the third component of spin of the $W^{\pm}$ is along the antiproton beam direction and thus the electron (positron) is emitted preferentially in the proton (antiproton) beam direction (Fig. 6.37).

6.3.6 The Cabibbo Angle and the GIM Mechanism

The universality of weak interactions established in the end 1940s (see Sect. 6.3.1) was questioned when it was discovered that some strange particle decays (as for instance $K^- \to \mu^- \, \overline{\nu}_\mu$ or $\Lambda \to p \, e^- \, \overline{\nu}_e$) were suppressed by a factor around 20 in relation to what expected.

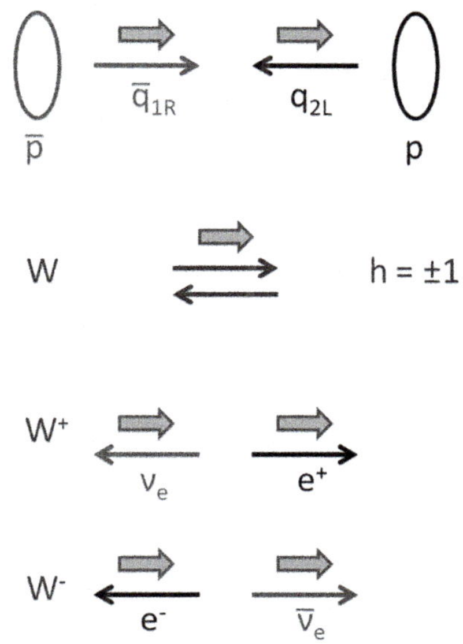

Fig. 6.37 Helicity in the $W^{\pm}$ production and decay

The problem was solved in 1963 by Nicola Cabibbo,[6] who suggested that the quark weak and strong eigenstates may be not the same. At that time only the u, d and s quarks were known (Sect. 5.7.2) and Cabibbo conjectured that the two quarks with electromagnetic charge $-1/3$ (d and s) mixed into a weak eigenstate d' such as

$$d' = d\cos\theta_c + s\sin\theta_c$$

where θ_c is a mixing angle, then designated as the Cabibbo angle.

Then the W–quark couplings involved in the μ, n and Λ are, respectively g_w, $g_w\cos\theta_c$ and $g_w\sin\theta_c$ (Fig. 6.38). The value of the Cabibbo angle is not predicted in the theory of electroweak interactions. Its present experimental value is $\sin\theta_c = 0.2253 \pm 0.0014$, which corresponds to an angle of about 13°.

A-priori, transitions between the s and d quarks could happen both via neutral currents (coupling through the Z) or charged currents (coupling through $W^{\pm}$) as shown in Fig. 6.39. Decays like $K^0 \to \mu^-\mu^+$ could then be allowed (Fig. 6.40), both at leading order and at one loop. However, the experimental branching ratio of the $K^0 \to \mu^-\mu^+$ process is of the order of 10^{-9}: flavor changing neutral currents (FCNC) appear to be strongly suppressed, even below what predicted by the diagram involving double W exchange.

It appears evident that the neutral current does not change flavor; however, this is not enough: the branching fraction is smaller than calculated from double W exchange. Glashow, Iliopoulos, and Maiani proposed in 1970 the introduction of a

[6]Nicola Cabibbo (1935–2010) was a professor in Rome, and president of the Italian Institute for Nuclear Physics (INFN). He gave fundamental contributions to the development of the standard model.

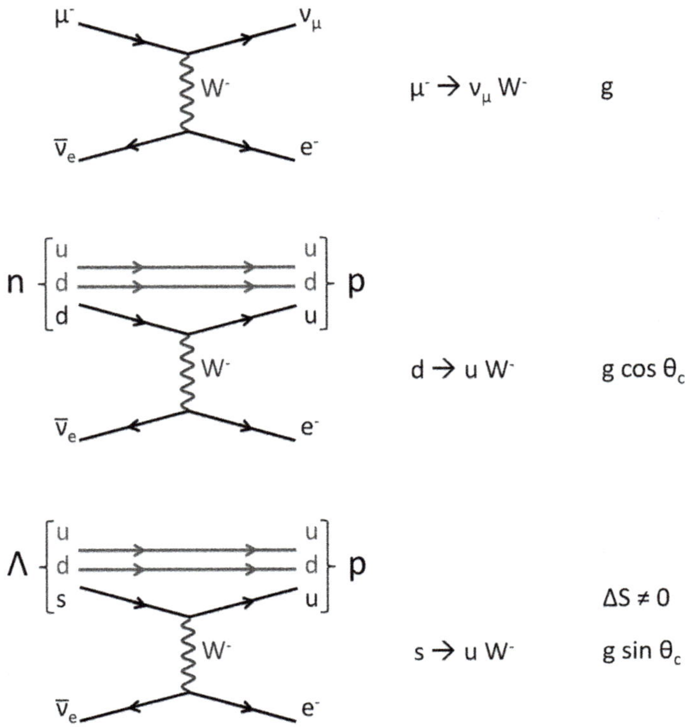

Fig. 6.38 Weak decay coupling: Leptonic (*top*), semileptonic with (*bottom*) and without (*middle*) involving strange quarks

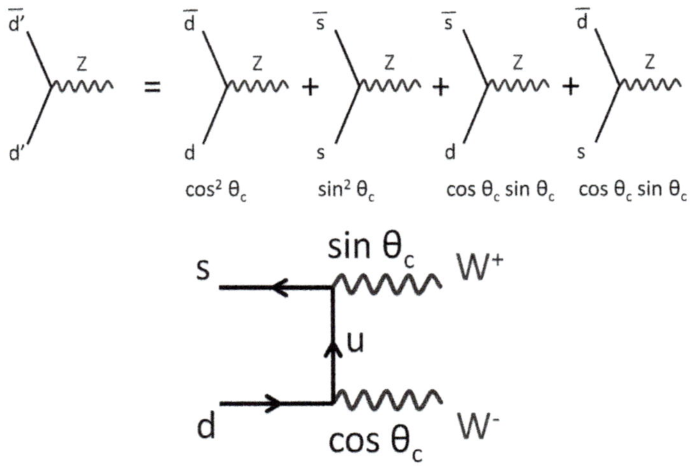

Fig. 6.39 Possible s and d quarks transitions generated by Z (*top*) and $W^{\pm}$ (*bottom*) couplings (three families)

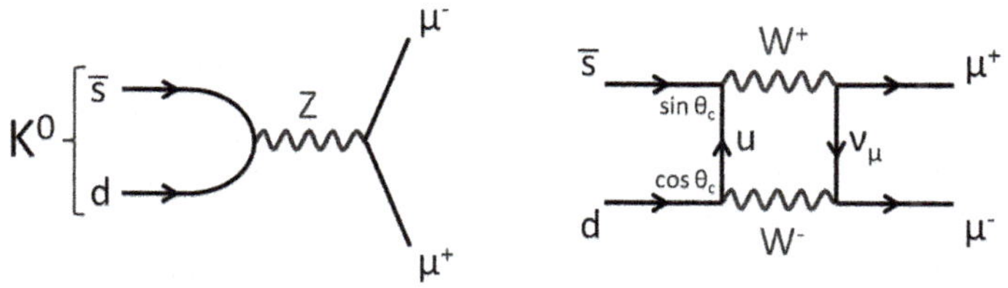

Fig. 6.40 $K^0 \to \mu^-\mu^+$ decay diagrams

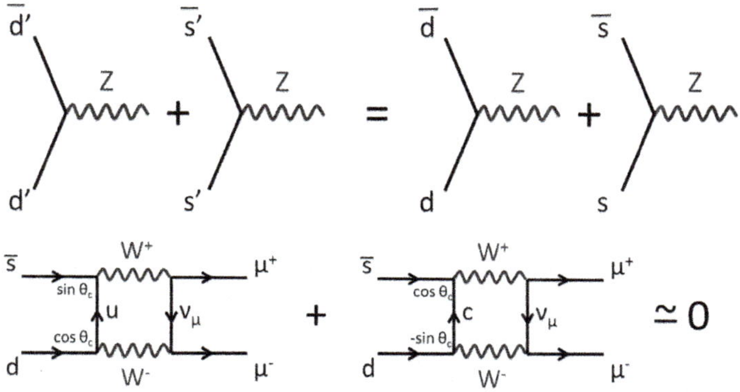

Fig. 6.41 FCNC suppression by diagram cancellation

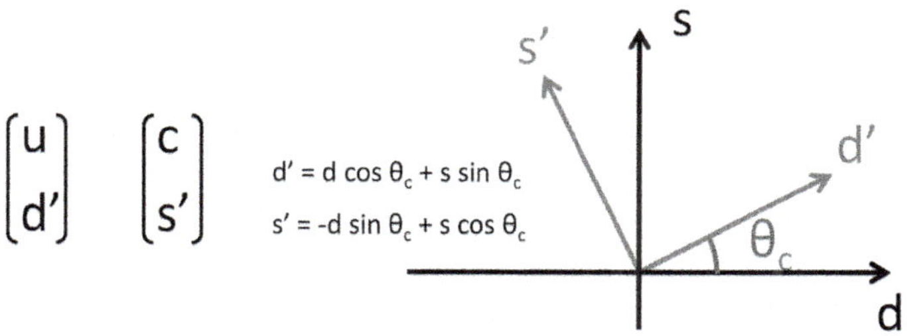

Fig. 6.42 The two orthogonal combinations of the quarks s and d in the d' and s' states

fourth quark, the charm c, to symmetrize the weak currents, organizing the quarks into two SU(2) doublets. Such scheme, known as the GIM mechanism, solves the FCNC puzzle and was spectacularly confirmed with the discovery of the J/ψ meson (see Sect. 5.7.2). FCNC are now suppressed by the cancelation of the two lowest diagrams in Fig. 6.41. In fact in the limit of equal masses the cancelation would be perfect. As the c mass is much higher than u mass the sum of the diagrams will lead to terms proportional to $m_c^2/m_{Z,W}^2$.

There are now two orthogonal combinations of the quarks s and d (Fig. 6.42):

$$d' = d\,\cos\theta_c + s\sin\theta_c$$

$$s' = -d\sin\theta_c + s\cos\theta_c$$

which couple, via the $W^{\pm}$, respectively to the u and c quarks.

The GIM mechanism can be translated in a matrix form as

$$\begin{pmatrix} d' \\ s' \end{pmatrix} = V_C \begin{pmatrix} d \\ s \end{pmatrix} = \begin{pmatrix} \cos\theta_c & \sin\theta_c \\ -\sin\theta_c & \cos\theta_c \end{pmatrix} \begin{pmatrix} d \\ s \end{pmatrix} .$$

where V_C is a 2×2 rotation matrix.

6.3.7 Extension to Three Quark Families: The CKM Matrix

A generic mixing matrix for three families can be written

$$V_{CKM} = \begin{pmatrix} V_{ud} & V_{us} & V_{ub} \\ V_{cd} & V_{cs} & V_{cb} \\ V_{td} & V_{ts} & V_{tb} \end{pmatrix}$$

in such a way that, for example, the square of the coupling of the b quark to the u quark in the weak transition (which is in turn proportional to the probability of the transition) can be written as

$$|g_{ub}|^2 = |V_{ub}|^2 g_W^2 .$$

The Japanese physicists Makoto Kobayashi and Toshihide Maskawa proposed this form of the quark mixing matrix in 1973. Their work was built on that of Cabibbo to extend the concept of quark mixing from two to three generations of quarks; at that time, the third generation had not been observed, yet, but, as we shall see, the extension to three families allows explaining the violation of the CP symmetry, i.e., of the product of the operations of charge conjugation and parity. In 2008, Kobayashi and Maskawa shared one half of the Nobel Prize in Physics "for the discovery of the origin of the broken symmetry which predicts the existence of at least three families of quarks in nature".

A-priori, being the V_{ij} complex numbers, the CKM matrix might have $2N^2$ degrees of freedom; however, the physical constraints reduce the free elements to $(N-1)^2$. The physical constraints are

- Unitarity. If there are only three quark families, one must have

$$V^{\dagger}V = I \tag{6.86}$$

  where I is the identity matrix. This will guarantee that in an effective transition each u-type quark will transform into one of the three d-type quarks (i.e., that the current is conserved and no fourth generation is present). This constraint reduces the number of degrees of freedom to N^2; the six equations underneath can be written explicitly as

$$\sum_k |V_{ik}|^2 = 1 \ (i = 1, 2, 3) \tag{6.87}$$

(the above is called weak invariance) and

$$\sum_k V^*_{jk} V_{ik} = 0 \ (i > j) \,. \tag{6.88}$$

This last equation is a constraint on three sets of three complex numbers, telling that these numbers form the sides of a triangle in the complex plane. There are three independent choices of i and j, and hence three independent triangles; they are called unitarity triangles, and we shall discuss them later in larger detail.

- Phase invariance. $2N - 1$ of these parameters leave physics invariant, since one phase can be absorbed into each quark field, and an overall common phase is unobservable. Hence, the total number of free variables is $N^2 - (2N - 1) = (N - 1)^2$.

Four independent parameters are thus required to fully define the CKM matrix ($N = 3$). This implies that the most general 3×3 unitary matrix cannot be constructed using real numbers only: Eq. (6.86) implies that a real matrix has only three degrees of freedom, and thus at least one imaginary parameter is required.

Many parameterizations have been proposed in the literature. An exact parametrization derived from the original work by Kobayashi and Maskawa (KM) extends the concept of Cabibbo angle; it uses three angles θ_{12}, θ_{13}, θ_{23}, and a phase δ:

$$V_{KM} = \begin{pmatrix} c_{12}c_{13} & s_{12}c_{13} & s_{13}e^{-i\delta} \\ -s_{12}c_{23} - c_{12}s_{23}s13e^{i\delta} & c_{12}c_{23} - s_{12}s_{23}s13e^{i\delta} & s_{23}c_{13} \\ s_{12}s_{23} - c_{12}c_{23}s13e^{i\delta} & -c_{12}s_{23} - s_{12}c_{23}s13 & c_{23}c_{13} \end{pmatrix} , \tag{6.89}$$

with the standard notations $s_{ij} = \sin\theta_{ij}$ (θ_{12} is the Cabibbo angle) and $c_{ij} = \cos\theta_{ij}$.

Another frequently used parametrization of the CKM matrix is the so-called Wolfenstein parametrization. It refers to four free parameters λ, A, ρ, and η, defined as

$$\lambda = s_{12} = \frac{|V_{us}|}{\sqrt{|V_{us}|^2 + |V_{ud}|^2}}$$
$$A = s_{23}/\lambda^2$$
$$s_{13}e^{i\delta} = A\lambda^3(\rho + i\eta)$$

(λ is the sine of the Cabibbo angle). We can use the experimental fact that $s_{13} \ll s_{23} \ll s_{12} \ll 1$ and expand the matrix in powers of λ. We obtain at order λ^4:

$$V_W \simeq \begin{pmatrix} 1-\frac{1}{2}\lambda^2 & \lambda & A\lambda^3(\rho - i\eta) \\ -\lambda & 1-\frac{1}{2}\lambda^2 & A\lambda^2 \\ A\lambda^3(1-\rho-i\eta) & -A\lambda^2 & 1 \end{pmatrix}. \quad (6.90)$$

As we shall see in the following, the combination of parameters $\bar{\rho} = \rho(1 - \lambda^2/2)$ and $\bar{\eta} = \eta(1 - \lambda^2/2)$ can be very useful.

The experimental knowledge of the terms of the CKM matrix comes essentially for the comparative study of probability of transitions between quarks. It is anyway challenging and difficult, since quarks are embedded in hadrons, and form factors for which only numerical QCD calculations are possible play a relevant role. In any case, the present experimental knowledge of the CKM matrix can be summarized in terms of the Wolfenstein parameters as

$$\begin{aligned} \lambda &= 0.22535 \pm 0.00065 \\ A &= 0.817 \pm 0.015 \\ \bar{\rho} = \rho(1-\lambda^2/2) &= 0.136 \pm 0.018 \\ \bar{\eta} = \eta(1-\lambda^2/2) &= 0.348 \pm 0.014\,. \end{aligned}$$

6.3.8 CP Violation

Weak interactions violate the parity and the charge conjugation symmetries. But, for a while, it was thought that the combined action of charge and parity transformation (CP) would restore the harmony physicists like so much. Indeed a left-handed neutrino transforms under CP into a right-handed anti-neutrino and the conjugate CP world still obeys to the V-A theory. However, surprisingly, the study of the $K^0 - \overline{K}^0$ system revealed in 1964 a small violation of the CP symmetry. In the turn of the century, CP violation was observed in many channels in the B sector. Since then, an intense theoretical and experimental work has been developed for the interpretation of these effects in the framework of the Standard Model, in particular by the precise determination of the parameters of the CKM matrix and by testing its self-consistency.

6.3.8.1 $K^0 - \overline{K}^0$ Mixing

Already in 1955, Gell-Mann and Pais had observed that the $K^0(d\bar{s})$ and the $\overline{K}^0(s\bar{d})$, which are eigenstates of the strong interaction, could mix through weak box diagrams as those represented in Fig. 6.43.

A pure K^0 ($\overline{K}^0$) beam will thus develop a $\overline{K}^0$ (K^0) component and at each time a linear combination of the K^0 and of the $\overline{K}^0$ may be observed. Since CP is conserved

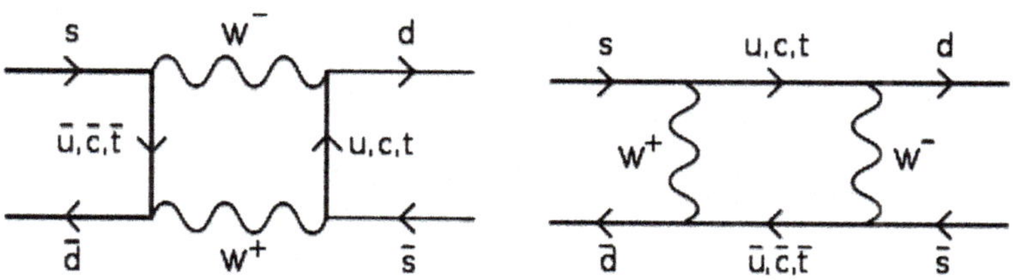

Fig. 6.43 Leading box diagrams for the $K^0 - \overline{K}^0$ mixing

in hadronic decays, the combinations which are eigenstates of CP are of particular relevance.

K^0 or $\overline{K}^0$ are not CP eigenstates: in fact they are the anti-particle of each other and the action of the CP operator may be, choosing an appropriated phase convention, written as

$$\mathrm{CP}\left|K^0\right\rangle = +\left|\overline{K}^0\right\rangle$$
$$\mathrm{CP}\left|\overline{K}^0\right\rangle = +\left|K^0\right\rangle .$$

Then the linear combinations

$$|K_1\rangle = \frac{1}{\sqrt{2}}\left(\left|K^0\right\rangle + \left|\overline{K}^0\right\rangle\right) \tag{6.91}$$
$$|K_2\rangle = \frac{1}{\sqrt{2}}\left(\left|K^0\right\rangle - \left|\overline{K}^0\right\rangle\right) \tag{6.92}$$

are CP eigenstates with eigenvalues $+1$ and -1, respectively.

The K_1 can thus decay into a two-pion system (which has CP $= +1$ eigenvalue), while the K_2 can, if CP is conserved, only decay into a three-pion system (which has CP $= -1$ eigenvalue).

The phase spaces associated to these decay modes are however quite different: $(m_K - 2m_\pi) \sim 220$ MeV; $(m_K - 3m_\pi) \sim 80$ MeV. Thus the corresponding lifetimes are also quite different:

$$\tau\,(K_1 \to \pi\pi) \sim 0.1 \text{ ns} \; ; \; \tau\,(K_2 \to \pi\pi\pi) \sim 52 \text{ ns}.$$

The short and the long lifetime states are usually designated by *K-short* (K_S) and *K-long* (K_L), respectively. These states are eigenstates of the free-particle Hamiltonian, which includes weak mixing terms, and if CP were a perfect symmetry, they would coincide with $|K_1\rangle$ and $|K_2\rangle$, respectively. The K_S and K_L wavefunctions evolve with time, respectively, as

$$|K_S(t)\rangle = |K_S(t=0)\rangle e^{-(im_S+\Gamma_S/2)t} \tag{6.93}$$
$$|K_L(t)\rangle = |K_L(t=0)\rangle e^{-(im_L+\Gamma_L/2)t} \tag{6.94}$$

where m_S (m_L) and Γ_S (Γ_L) are, respectively, the mass and the width of the K_S (K_L) mesons (see Sect. 2.6).

K^0 and $\bar{K}^0$, being a combination of K_S and K_L, will also evolve in time. Indeed, considering initially a pure beam of pure K^0 with an energy of a few GeV, just after a few tens of cm, the large majority of the K_S mesons will decay and the beam will become a pure K_L beam. The probability to find a K^0 in this beam after a time t can be expressed as:

$$P_{K^0 \longrightarrow K^0}(t) = \left| \frac{1}{\sqrt{2}} \left(\langle K^0 | K_S(t) \rangle + \langle K^0 | K_L(t) \rangle \right) \right|^2$$
$$= \frac{1}{4} \left(e^{-\Gamma_S t} + e^{-\Gamma_L t} + 2e^{-(\Gamma_S + \Gamma_L)t/2} \cos\left(\Delta m \, t\right) \right) \quad (6.95)$$

where $\Gamma_S = 1/\tau_s$, $\Gamma_L = 1/\tau_L$, and Δm is the difference between the masses of the two eigenstates. The last term, coming from the interference of the two amplitudes, provides a direct measurement of Δm.

In the limit $\Gamma_S \longrightarrow 0$, $\Gamma_L \longrightarrow 0$, a pure flavor oscillation between K^0 or $\overline{K}^0$ would occur:

$$P_{K^0 \longrightarrow K^0}(t) = \frac{1}{2}\left(1 + \cos\left(\Delta m \, t\right)\right) = \cos^2\left(\Delta m \, t\right) \, .$$

In the real case, however, the oscillation is damped and the survival probability converges quickly to a value close to 1/4. Measuring the initial oscillation through the study of semileptonic decays, which will be discussed later on in this section, Δm was determined to be

$$\Delta m \sim (3.483 \pm 0.006) \times 10^{-15} \text{ GeV} \, .$$

K_S and K_L have quite different lifetimes but almost the same mass.

6.3.8.2 CP Violation in 2π Modes

In 1964 Christenson, Cronin, Fitch and Turlay[7] performed the historical experience (Fig. 6.44) that revealed by the first time the existence of a small fraction of two-pion decays in a K_L beam:

---

[7]The Nobel Prize in Physics 1980 was awarded to James ("Jim") Cronin and Val Fitch "for the discovery of violations of fundamental symmetry principles in the decay of neutral K-mesons". Cronin (Chicago 1931) received his Ph.D. from the University of Chicago in 1955. He then worked at Brookhaven National Laboratory, in 1958 became a professor at Princeton University, and finally in Chicago. Later he moved to astroparticle physics, being with Alan Watson the founder of the Auger cosmic ray observatory. Fitch (Merriman, Nebraska, 1923—Princeton 2015) was interested in chemistry and he switched to physics in the mid-1940s when he participated to the Manhattan Project. Ph.D. in physics by Columbia University in 1954, he later moved to Princeton.

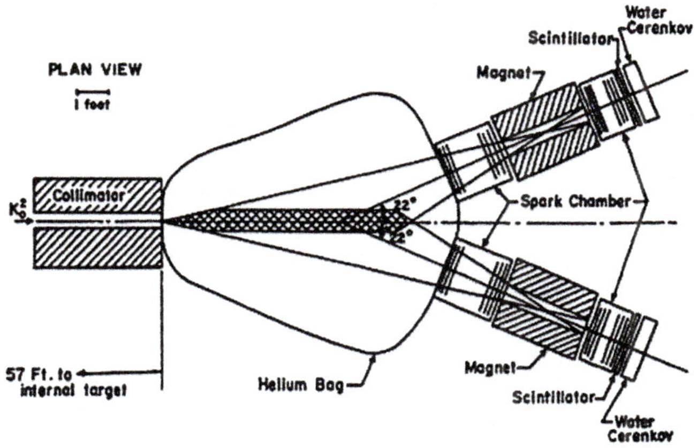

Fig. 6.44 Layout of the Christenson, Cronin, Fitch and Turlay experiment that demonstrated the existence of the decay $K_L \to \pi^+\pi^-$. ©The Nobel Foundation

$$R = \frac{\Gamma\left(K_L \to \pi^+\pi^-\right)}{\Gamma\left(K_L \to \text{all charged modes}\right)} = (2.0 \pm 0.4) \times 10^{-3}\,.$$

The K_L beam was produced in a primary target placed 17.5 m downstream the experiment and the observed decays occurred in a volume of He gas to minimize interactions. Two spectrometers each composed by two spark chambers separated by a magnet and terminated by a scintillator and a water Cherenkov measured and identified the charged decay products.

The presence of two-pion decay modes implied that the long-lived K_L was not a pure eigenstate of CP. The K_L should then have a small component of K_1 and so it could be expressed as:

$$|K_L\rangle = \frac{1}{\sqrt{1+|\varepsilon|^2}}\left(|K_2\rangle + \varepsilon\,|K_1\rangle\right) \tag{6.96}$$

where ε is a small complex parameter whose phase ϕ_ε depends on the phase convention chosen for the CP action on K^0 and $\overline{K}^0$. With the present phase convention

$$\phi_\varepsilon \simeq \tan^{-1}\frac{2\Delta m}{\Delta\Gamma}$$

where Δm and $\Delta\Gamma$ are, respectively, the differences between the masses and the decay widths of the two eigenstates.

Alternatively, K_L and K_s can be expressed as a function of the flavor eigenstates K^0 and $\overline{K}^0$ as

$$|K_L\rangle = \frac{1}{\sqrt{2\left(1+|\varepsilon|^2\right)}}\left((1+\varepsilon)\left|K^0\right\rangle - (1-\varepsilon)\left|\overline{K}^0\right\rangle\right) \tag{6.97}$$

$$|K_s\rangle = \frac{1}{\sqrt{2\left(1+|\varepsilon|^2\right)}}\left((1+\varepsilon)\left|K^0\right\rangle + (1-\varepsilon)\left|\overline{K}^0\right\rangle\right) \tag{6.98}$$

or, inverting the last two equations,

$$\left|K^0\right\rangle = \frac{1}{1+\varepsilon}\sqrt{\frac{1+|\varepsilon|^2}{2}}\,\left(|K_s\rangle + |K_L\rangle\right) \tag{6.99}$$

$$\left|\overline{K}^0\right\rangle = \frac{1}{1+\varepsilon}\sqrt{\frac{1+|\varepsilon|^2}{2}}\,\left(|K_s\rangle - |K_L\rangle\right)\,. \tag{6.100}$$

The probability that a state initially produced as a pure K^0 or $\bar{K}^0$ will decay into a 2π system will evolve in time. A "2π asymmetry" can thus be define as

$$A_{\pm}(t) = \frac{\Gamma\left(\overline{K}^0 \longrightarrow \pi^+\pi^-\right) - \Gamma\left(K^0 \longrightarrow \pi^+\pi^-\right)}{\Gamma\left(\overline{K}^0 \longrightarrow \pi^+\pi^-\right) + \Gamma\left(K^0 \longrightarrow \pi^+\pi^-\right)}\,. \tag{6.101}$$

This asymmetry, measured for instance by the CPLEAR experiment at CERN (Fig. 6.45), evolves in time and depends on ε and Δm. Fixing Δm to the world average, it was obtained:

$$|\varepsilon| = (2.264 \pm 0.023)\,10^{-3},\; \phi_\varepsilon = (43.19 \pm 0.53)^\circ\,.$$

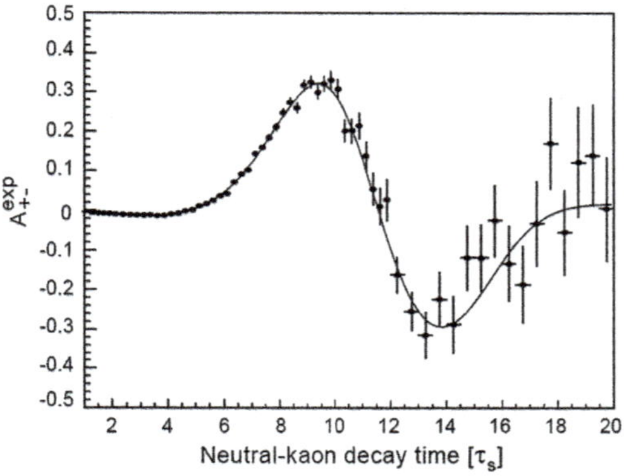

Fig. 6.45 Asymmetry in 2π decays between K^0 and $\overline{K}^0$ tagged events. Time is measured in K_s lifetimes. From A. Angelopoulos et al., Physics Reports 374 (2003) 165

6.3.8.3 CP Violation in Semileptonic $K^0, \overline{K}^0$ Decays

K^0 and $\overline{K}^0$ may also decay semileptonically through the channels:

$$K^0 \to \pi^- e^+ \nu_e \; ; \; \overline{K}^0 \to \pi^+ e^- \overline{\nu}_e$$

and thus CP violation can also be tested measuring the charge asymmetry A_L

$$A_L = \frac{K_L \to \pi^- l^+ \nu - K_L \to \pi^+ l^- \overline{\nu}}{K_L \to \pi^- l^+ \nu + K_L \to \pi^+ l^- \overline{\nu}}.$$

This asymmetry can be related to the CP violating parameter ε:

$$A_L = \frac{(1+\varepsilon)^2 - (1-\varepsilon)^2}{(1+\varepsilon)^2 + (1-\varepsilon)^2} \approx 2\, Re\,(\varepsilon)\,.$$

The measured value A_L is positive and it is in good agreement with the measurement of ε obtained in the 2π decay modes. The number of K_L having in their decay products an electron is slighter smaller (0.66 %) than the number of K_L having in their decay products a positron. There is thus an unambiguous way to define what is matter and what is antimatter.

6.3.8.4 Direct CP Violation

CP violation was so far discussed, in the mixing system $K^0 - \overline{K}^0$, in terms of a not perfect identification between the free-particle Hamiltonian eigenstates (K_S, K_L) and the CP eigenstates (K_1, K_2):

$$|K_S\rangle = \frac{1}{\sqrt{1+|\varepsilon|^2}} \left(|K_1\rangle + \varepsilon\, |K_2\rangle\right)$$

$$|K_L\rangle = \frac{1}{\sqrt{1+|\varepsilon|^2}} \left(|K_2\rangle + \varepsilon\, |K_1\rangle\right).$$

In this context, the decays of K_s and K_L into 2π modes are only due to the presence in both states of a K_1 component. It is then expected that the ratio of the decay amplitudes of the K_L and of the K_s into 2π modes should be equal to ε and independent of the charges of the two pions:

$$\eta = \frac{A\,(K_L \to \pi\pi)}{A\,(K_s \to \pi\pi)} = \varepsilon\,.$$

However, it was experimentally established that

$$\eta^{+-} = \frac{A\left(K_L \to \pi^+\pi^-\right)}{A\left(K_S \to \pi^+\pi^-\right)}$$

and

$$\eta^{00} = \frac{A\left(K_L \to \pi^0\pi^0\right)}{A\left(K_S \to \pi^0\pi^0\right)}$$

(both about 2×10^{-3}) are slightly, but significantly, different. In fact the present experimental ratio is

$$\left|\frac{\eta^{00}}{\eta^{+-}}\right| = 0.9950 \pm 0.0008\,.$$

This difference is interpreted as the existence of a *direct* CP violation in the K_1 decays. In other words, the decay rate of a meson to a given final state is not equal to the decay rate of its antimeson to the corresponding CP-conjugated final state:

$$\Gamma\left(M \to f\right) \neq \Gamma\left(\overline{M} \to \overline{f}\right)\,.$$

The CP violation discussed previously in the mixing of the system $K_0 - \overline{K}_0$ is now denominated *indirect* CP violation. This CP violation is related to the observation that the oscillation of a given meson to its antimeson may be different from the inverse oscillation of the antimeson to the meson:

$$\Gamma\left(M \to \overline{M}\right) \neq \Gamma\left(\overline{M} \to M\right).$$

Finally, CP violation may also occur whenever both the meson and its antimeson can decay to a common final state with or without $M - \overline{M}$ mixing:

$$\Gamma\left(M \to f\right) \neq \Gamma\left(\overline{M} \to f\right).$$

The *direct* CP violation is usually quantified by a parameter ε'. Assuming that this direct CP violation occurs in the K decays into 2π modes due to the fact that the 2π system may be formed in different isospin states ($I = 0,\ 2$) and the corresponding decay amplitudes may interfere, it can be shown that η^{+-} and η^{00} can be written as

$$\eta^{+-} = \varepsilon + \varepsilon'$$

$$\eta^{00} = \varepsilon - 2\,\varepsilon'\,.$$

The ratio between the CP violating parameters can also be related to the double ratio of the decay probabilities K_L and K_s into specific 2π modes:

$$Re\left(\frac{\varepsilon'}{\varepsilon}\right) = \frac{1}{6}\left(1 - \frac{|\eta^{00}|^2}{|\eta^{\pm}|^2}\right) = \frac{1}{6}\left(1 - \frac{\Gamma\left(K_L \to \pi^0\pi^0\right)\Gamma\left(K_s \to \pi^+\pi^-\right)}{\Gamma\left(K_L \to \pi^+\pi^-\right)\Gamma\left(K_s \to \pi^0\pi^0\right)}\right).$$

The present experimental value for this ratio is

$$Re\left(\frac{\varepsilon'}{\varepsilon}\right) \approx \frac{\varepsilon'}{\varepsilon} = (1.66 \pm 0.23) \times 10^{-3}\,.$$

6.3.8.5 CP Violation in the *B* Sector

Around 40 years after the discovery of the CP violation in the $K^0 - \overline{K}^0$ system, a large CP violation in the $B^0 - \overline{B}^0$ system was observed. The B^0 ($\overline{B}^0$) differs at the quark level from the K^0 ($\overline{K}^0$) just by the replacement of the s ($\overline{s}$) quark by a b ($\overline{b}$) quark. Thus B^0 and $\overline{B}^0$ should mix through similar weak box diagrams (Fig. 6.46) and the CP eigenstates should be also a combination of both.

However, these CP eigenstates have similar lifetimes since the b quark has a much larger mass than the s quark and thus the decay phase space is large for both CP eigenstates. These eigenstates are called *B-Light* (B_L) and *B-Heavy* (B_H) according to their masses, although their mass difference, $\Delta m_{B^0} \sim (3.337 \pm 0.033) \times 10^{-13}$ GeV, is small. The B_L and B_H meson cannot, therefore, be disentangled just by allowing one of them to decay and thus there are no pure B_L or B_H beams. Another strategy has to be followed.

In fact, the observation of the CP violation in the B sector was first found studying the time evolution of the decay rates of the B^0 and the $\overline{B}^0$ mesons to a common final state ($\Gamma\left(M \to f\right) \neq \Gamma\left(\overline{M} \to f\right)$), namely to $J/\psi\ K_S$.

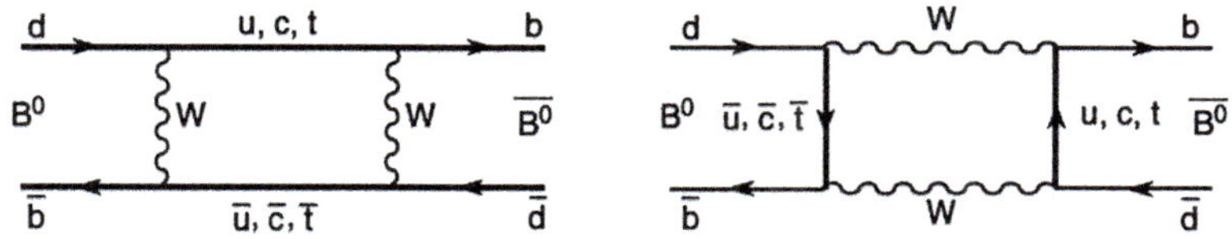

Fig. 6.46 Leading box diagrams for the $B^0 - \overline{B}^0$ mixing. From S. Braibant, G. Giacomelli, and M. Spurio, "Particles and fundamental interactions", Springer 2012

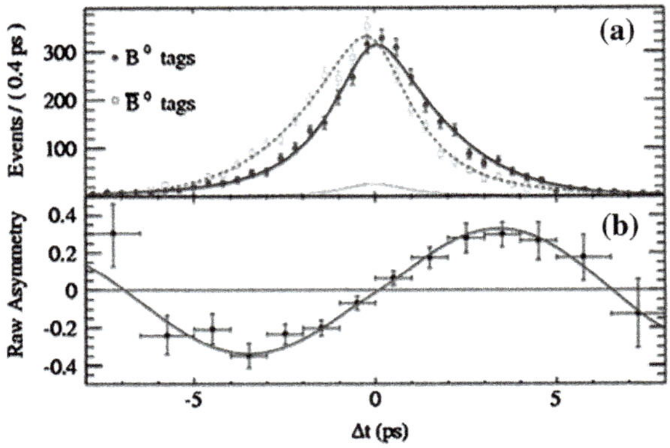

Fig. 6.47 Decay rate to $J/\psi\ K_S$ as a function of time of each of the B flavor states (*top*) and the derived time asymmetry (*bottom*). From C. Chen (BaBar), Contribution to the 34th International Conference on High-Energy Physics (July 2008)

At the experiment BaBar,[8] B mesons pairs were produced in the reaction

$$e^+e^- \rightarrow \Upsilon(4S) \rightarrow B^0\overline{B}^0 .$$

The $B^0\overline{B}^0$ states evolved entangled and therefore if one of the mesons was observed ("tagged") at a given time, the other had to be its anti-particle. The "tag" of the flavor of the B mesons could be done through the determination of the charge of the lepton in B semileptonic decays:

$$B^0 \rightarrow D^- l^+ \nu_l\ (\overline{b} \rightarrow \overline{c}\, l^+ \nu_l)$$

$$\overline{B}^0 \rightarrow D^+ l^- \overline{\nu}_l (b \rightarrow c\, l^- \overline{\nu}_l) .$$

It was thus possible to determine the decay rate of the untagged B meson to $J/\psi\ K_S$ as a function of its decay time. This rate is shown, both for "tagged" $B^0$and $\overline{B}^0$ in Fig. 6.47. The observed asymmetry:

$$A_{CP}(t) = \frac{\Gamma\left(\overline{B}^0 \rightarrow J/\psi\ K_S\right) - \Gamma(B^0 \rightarrow J/\psi\ K_S)}{\Gamma\left(\overline{B}^0 \rightarrow J/\psi\ K_S\right) + \Gamma(B^0 \rightarrow J/\psi\ K_S)}$$

[8]The BaBar detector was a cylindrical detector located at the Stanford Linear Accelerator Center in California. Electrons at an energy of 9 GeV collided with 3.1 GeV antielectrons to produce a center-of-mass collision energy of 10.58 GeV, corresponding to the Υ(4S) resonance. The Υ(4S) decays into a pair of B mesons, charged or neutral. The detector had the classical "onion-like" structure, starting from a Silicon Vertex Tracker (SVT) detecting the decay vertex, passing through a Cherenkov detector for particle identification, and ending with an electromagnetic calorimeter. A magnet produced a 1.5 T field allowing momentum measurement. BaBar analyzed some 100 million $B\bar{B}$ events, being a kind of "B factory".

is a clear proof of the CP violation in this channel. This asymmetry can be explained by the fact that the decays can occur with or without mixing. The decay amplitudes for these channels may interfere—in the case of the B^0, the relevant amplitudes are $A_1\left(B^0 \rightarrow J/\psi\, K_S\right)$ and $A_2\left(B^0 \rightarrow \overline{B}^0 \rightarrow J/\psi\, K_S\right)$.

Nowadays, after the experiments Belle and BaBar at the B factories at KEK and SLAC, and after the first years of the LHCB experiment at LHC, there is already a rich spectrum of B channels where CP violation was observed at a level above 5σ. These results allowed a precise determination of most of the parameters of the CKM matrix and intensive tests of its unitarity as it will be briefly discussed.

6.3.8.6 CP Violation in the Standard Model

CP violation in weak interactions can be linked to the existence of the complex phase of the CKM matrix which is expressed by the parameters δ and η, respectively, in the KM and in the Wolfenstein parametrizations (see the previous section). As a consequence, a necessary condition for the appearance of the complex phase, and thus for CP violation, is the presence of at least three generations of quarks (this clarifies the power of the intuition by Kobayashi and Maskawa). The reason why a complex phase in the CKM matrix causes CP violation can be seen as follows. Consider a process $A \rightarrow B$ and the CP-conjugated $\overline{A} \rightarrow \overline{B}$ between their anti-particles, with the appropriate helicity reversal. If there is no CP violation, the amplitudes, let us call them M and $\tilde{M}$, respectively, must be the same complex number—except that the CKM terms get conjugated. We can separate the magnitude and phase by writing

$$M = |M_1|\, e^{i\phi_1} e^{i\delta_1}$$

$$\tilde{M} = |M_1|\, e^{i\phi_1} e^{-i\delta_1}$$

where δ_1 is the phase term introduced from the CKM matrix (called often "weak phase") and ϕ_1 is the phase term generated by CP-invariants interactions in the decay (called often "strong phase"). The exact values of these phases depend on the convention but the differences between the weak phases and between the strong phases in any two different terms of the decay amplitude is independent of the convention.

Since physically measurable reaction rates are proportional to $|M|^2$, so far nothing is different. However, consider a process for which there are different routes (say for simplicity two routes). Now we have:

$$M = |M_1|\, e^{i\phi_1} e^{i\delta_1} + |M_2|\, e^{i\phi_2} e^{i\delta_2}$$

$$\tilde{M} = |M_1|\, e^{i\phi_1} e^{-i\delta_1} + |M_2|\, e^{i\phi_2} e^{i\delta_2}$$

and in general $|M|^2 \neq \left|\tilde{M}\right|^2$. Thus, a complex phase may give rise to processes that proceed at different rates for particles and antiparticles, and the CP symmetry may be violated. For example, the decay $B^0 \to K^+\pi^-$ is 13 % more common than its CP conjugate $\overline{B}^0 \to K^-\pi^+$.

The unitarity of the CKM matrix imposes, as we have discussed in the previous section, three independent orthogonality conditions:

$$\sum_k V_{jk}^* V_{ik} = 0 \; (i > j).$$

These conditions are sums of three complex numbers and thus can be represented in a complex plane as triangles, usually called the unitarity triangles.

In the triangles obtained by taking scalar products of neighboring rows or columns, the modulus of one of the sides is much smaller than the other two. The equation for which the moduli of the triangle are most comparable is

$$V_{ud} V_{ub}^* + V_{cd} V_{cb}^* + V_{td} V_{tb}^* = 0 \, .$$

The corresponding triangle is shown in Fig. 6.48.

The triangle is represented in the $(\overline{\rho}, \overline{\eta})$ phase space (see the discussion on the Wolfenstein parametrization in the previous section); its sides were divided by $\left|V_{cd} V_{cb}^*\right|$, which is the best known element in the sum; and it is rotated in order that the side with unit length is aligned along the real $(\overline{\rho})$ axis. The apex of the triangle is by construction located at $(\overline{\rho}, \overline{\eta})$ and the angles can be defined by:

$$\alpha \equiv \arg\left(-\frac{V_{td} V_{tb}^*}{V_{ud} V_{ub}^*}\right) \; ; \; \beta \equiv \arg\left(-\frac{V_{cd} V_{cb}^*}{V_{td} V_{tb}^*}\right) \; ; \; \gamma \equiv \arg\left(-\frac{V_{ud} V_{ub}^*}{V_{cd} V_{cb}^*}\right) .$$

It can also be demonstrated that the areas of all unitarity triangles are the same, and they equal half of the so-called Jarlskog invariant (from the Swedish physicist Cecilia Jarlskog), which can be expressed as $J \simeq A^2\lambda^6\eta$ in the Wolfenstein parametrization.

The fact that the Jarlskog invariant is proportional to η shows that the unitarity triangle is a measure of CP violation: if there is no CP violation, the triangle degenerates into a line. If the three sides do not close to a triangle, this might indicate that

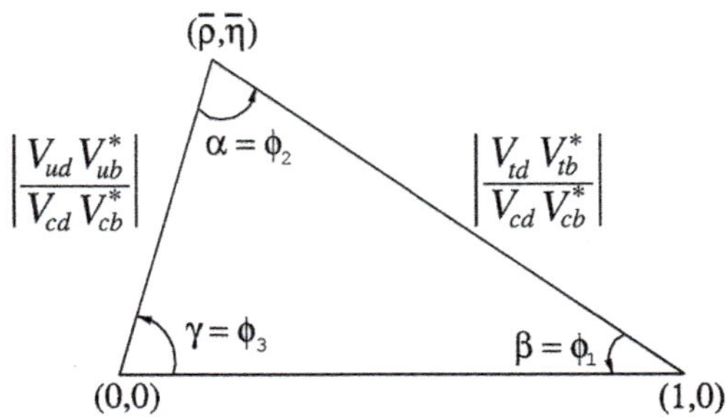

Fig. 6.48 One of the six unitarity triangles. The description of the sides in terms of the parameters in the Wolfenstein parametrization is shown

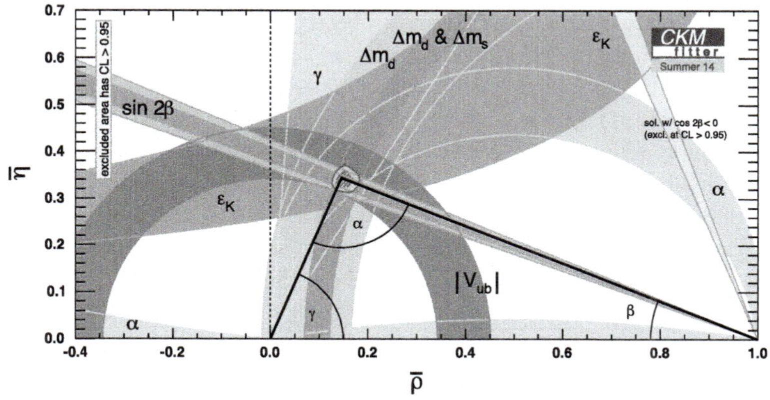

Fig. 6.49 Unitarity triangle and global CKM fit in the plane $(\overline{\rho}, \overline{\eta})$. From CKMfitter Group (J. Charles et al.), Eur. Phys. J. C41 (2005) 1, updated results and plots available at http://ckmfitter.in2p3.fr

the CKM matrix is not unitary, which would imply the existence of new physics—in particular, the existence of a fourth quark family.

The present experimental constrains on the CKM unitarity triangle, as well as a global fit to all the existing measurements by the CKMfitter group,[9] are shown in Fig. 6.49.

All present results are thus consistent with the CKM matrix being the only source of CP violation in the Standard Model. Nevertheless, it is widely believed that the observed matter–antimatter asymmetry in the Universe (see next section), requires the existence of new sources of CP violation that might be revealed either in the quark sector as small inconsistencies at the CKM matrix, or elsewhere, like in precise measurements of the neutrino oscillations or of the neutron electric dipole moments. The real nature of CP violation is still to be understood.

6.3.9 Matter–Antimatter Asymmetry

The existence of antimatter predicted by Dirac in 1930 and discovered by Anderson (see Chap. 3) is still today the object of intense study and speculation: would the physics of an antimatter dominated Universe be identical to the physics of the matter dominated Universe we are leaving in? Is there any other CP violation process than the tiny ones observed so far? How, in the framework of the Big Bang model, did the Universe became matter dominated?

Antiparticles are currently produced in accelerators, and observed in cosmic rays interactions in a small amount level (for instance, $\overline{p}/p \sim 10^{-4}$) (see Chap. 10).

---

[9]The CKMfitter group provides once or twice per year an updated analysis of standard model measurements and average values for the CKM matrix parameters.

At CERN the study of antimatter atoms has been pursued in the last 20 years. Antihydrogen atoms have been formed and trapped for periods as long as 16 min and recently the first antihydrogen beams were produced. The way is open, for instance, to detailed studies of the antihydrogen hyperfine transitions and to the measurement of the gravitational interactions between matter and antimatter. The electric charge of one antihydrogen atom was found by the ALPHA experiment to be compatible with zero to eight decimal places ($Q_{\overline{H}} \simeq (-1.3 \pm 1.1 \pm 0.4)\ 10^{-8} e$).

No primordial antimatter was observed so far, while the relative abundance of baryons (n_B) to photons (n_γ) was found to be (see Sect. 8.1.3):

$$\eta = \frac{n_B}{n_\gamma} \sim 5 \times 10^{-10} \ .$$

Although apparently small, this number is many orders of magnitude higher than what would be expected if there would be in the early Universe a equal number of baryons and antibaryons. Indeed in such case the annihilation between baryons and anti baryons would have occurred until its interaction rate equals the expansion rate of the Universe (see Sect. 8.1.2) and then the expected ratios can be computed to be:

$$\frac{n_B}{n_\gamma} = \frac{n_{\overline{B}}}{n_\gamma} \sim 10^{-18} \, .$$

The excess of matter over antimatter should then be present before nucleons and antinucleons may be formed. On the other hand, inflation (see Sect. 8.3.2) would wipe out any excess of baryonic charge present in the beginning of the Big Bang. Thus, this excess had to be originated by some unknown mechanism (baryogenesis) after inflation and before or during the quark–gluon plasma stage.

In 1967, soon after the discovery of the CMB and of the violation of CP in the $K^0 - \overline{K^0}$ system (see previous section), Andrej Sakharov modeled the Universe evolution from a $B = 0$ initial state to the $B \neq 0$ present state (B indicates here the baryonic number). This evolution needed three conditions which are now known as the Sakharov conditions:

1. Baryonic number (B) should be violated.
2. Charge (C) and Charge and Parity (CP) symmetries should be violated.
3. Baryon-number violating interactions should have occurred in the early Universe out of thermal equilibrium.

The first condition is obvious. The second is necessary since if C and CP were conserved any baryonic charge excess produced in a given reaction would be compensated by the conjugated reaction. The third is more subtle: the particles and their corresponding antiparticles do not achieve thermal equilibrium due to rapid expansion decreasing the occurrence of pair annihilation.

Thermal equilibrium may have been broken when symmetry breaking processes had occurred. Whenever two phases are present, the boundary regions between these (for instance the surfaces of bubbles in boiling water) are out of thermal equilibrium. In the framework of the Standard Model (see Chap. 7) this fact could in principle had occurred at the electroweak phase transition. However, it was demonstrated analytically and numerically that, for a Higgs with a mass as the one observed recently ($m_H \sim 125\,\text{GeV}$), the electroweak phase transition does not provide the thermal instability required for the formation of the present baryon asymmetry in the Universe.

The exact mechanism responsible for the observed matter-antimatter asymmetry in the Universe is still to be discovered. The Standard Model is not clearly the end of physics.

6.4 Strong Interactions and QCD

The quark model simplifies the description of hadrons. We saw that deep inelastic scattering evidences a physical reality for quarks—although the interaction between these particles is very peculiar, since no free quarks have been observed up to now. A heuristic form of the potential between quarks with the characteristics needed has been shown.

Within the quark model, we needed to introduce a new quantum number—color—to explain how bound stated of three identical quarks can exist and not violate the Pauli exclusion principle. Invariance with respect to color can be described by a symmetry group $SU(3)_c$, where the subscript c indicates color.

The theory of quantum chromodynamics (QCD) enhances the concept of color from a role of label to the role of charge, and is the basis for the description of the interactions binding quarks in hadrons. The phenomenological description through an effective potential can be seen as a limit of this exact description, and the strong interactions binding nucleons can be explained as van der Waals forces between neutral objects.

QCD has been extensively tested, and is very successful. The American physicists David J. Gross, David Politzer and Frank Wilczek shared the 2004 the Nobel Prize for physics by devising an elegant mathematical framework to express the asymptotic (i.e., in the limit of very short distances, equivalent to the high momentum transfer limit) freedom of quarks in hadrons, leading to the development of QCD.

However, a caveat should be stressed. At very short distances, QCD is essentially a theory of free quarks and gluons—with relatively weak interactions, and observables can be perturbatively calculated. However at longer wavelengths, of the order the proton size $\sim 1\,\text{fm} = 10^{-15}\,\text{m}$, the coupling constant between partons becomes too large to compute observables (we remind that exact solutions are in general impossible, and perturbative calculations must be performed): the Lagrangian of QCD, that in principle contains all physics, becomes de facto of little help in this regime. Parts of QCD can thus be calculated in terms of the fundamental parameters using the full dynamical (lagrangian) representation, while for other sectors one should use models, guided by the characteristics of the theory, whose effective parameters cannot be calculated but which can be constrained by experimental data.

6.4.1 Yang–Mills Theories

Before formulating QCD as a gauge theory, we must extend the formalism shown for the description of electromagnetism (Sect. 6.2.6) to a symmetry group like SU(3). This extension is not trivial, and it was formulated by Yang and Mills in the 1950s.

U(1). Let us first summarize the ingredients of the U(1) gauge theory—which is the prototype of the *abelian* gauge theories, i.e., of the gauge theories defined by symmetry groups for which the generators commute. We have seen in Sect. 6.2.3 that the requirement that physics is invariant under local U(1) phase transformation implies the existence of the photon gauge field. QED can be derived by requiring the Lagrangian to be invariant under local U(1) transformations of the form $U = e^{iq\chi(x)I}$—note the identity operator I, which, in the case of U(1), is just unity. The recipe is:

- Find the gauge invariance of the theory—in the case of electromagnetism U(1):

$$\psi(x) \to \psi'(x) = U(x)\psi(x) = \psi(x)e^{iq\chi(x)} . \tag{6.102}$$

- Replace the derivative in the Lagrangian with a *covariant* derivative

$$\partial_\mu \to D_\mu = \partial_\mu + iqA_\mu(x) \tag{6.103}$$

  where A_μ transforms as

$$A_\mu \to A'_\mu = A_\mu + \partial_\mu \chi . \tag{6.104}$$

The Lagrangian

$$\mathcal{L}_{\text{QED}} = \bar{\psi}(i\gamma^\mu D_\mu - m)\psi - \frac{1}{4}F_{\mu\nu}F^{\mu\nu} \tag{6.105}$$

with

$$F_{\mu\nu} = \partial_\mu A_\nu - \partial_\nu A_\mu = \frac{1}{iq}[D_\mu, D_\nu] \tag{6.106}$$

is invariant for the local gauge transformation, and the field A_μ and its interactions with ψ are defined by the invariance itself. Note that the Lagrangian can be written as

$$\mathcal{L} = \mathcal{L}_{\text{loc}} + \mathcal{L}_{\text{gf}}$$

where $\mathcal{L}_{\text{loc}}$ is the locally invariant Lagrangian for the particle, $\mathcal{L}_{\text{gf}}$ is the field Lagrangian.

Abelian Symmetry Groups. What we have seen for U(1) can be trivially extended to symmetries with more than one generator, if the generators commute.

Non-Abelian Symmetry Groups and Yang–Mills Theories. When the symmetry group is non-abelian, i.e., generators do not commute, the above recipes must be generalized. If the generators of the symmetry are T^a, with a =1, ..., n, one can write the gauge invariance as

$$\psi(x) \to \psi'(x) = e^{i\, g_s \sum_a \epsilon_a(x)\, T^a} \,. \tag{6.107}$$

From now on, we shall not explicitly write the sum over a—the index varying within the set of the generators, or of the gauge bosons, which will be assumed implicitly when the index is repeated; generators are a group. We do not associate any particular meaning to the fact that A is subscript or superscript.

If the commutation relations hold

$$[T^a, T^b] = i f^{abc} T^c \tag{6.108}$$

One can define the covariant derivative as

$$D_\mu = \partial_\mu + i g T^a \mathcal{A}^a_\mu \tag{6.109}$$

where $\mathcal{A}^a_\mu$ are the vector potentials, and g is the coupling constant. In four dimensions, the coupling constant g is a pure number and for a SU(N) group one has $a, b, c = 1 \ldots N^2 - 1$.

The gauge field Lagrangian has the form

$$\mathcal{L}_{\text{gf}} = -\frac{1}{4} F^{a\mu\nu} F^a_{\mu\nu} \,. \tag{6.110}$$

The relation

$$F^a_{\mu\nu} = \partial_\mu \mathcal{A}^a_\nu - \partial_\nu \mathcal{A}^a_\mu + g f^{abc} \mathcal{A}^b_\mu \mathcal{A}^c_\nu \tag{6.111}$$

can be derived by the commutator

$$[D_\mu, D_\nu] = -i g T^a F^a_{\mu\nu} \,. \tag{6.112}$$

The field is self-interacting: from the given Lagrangian one can derive the equations

$$\partial^\mu F^a_{\mu\nu} + g f^{abc} \mathcal{A}^{\mu b} F^c_{\mu\nu} = 0 \,. \tag{6.113}$$

A source J^a_μ enters into the equations of motion as

$$\partial^\mu F^a_{\mu\nu} + g f^{abc} \mathcal{A}^{b\mu} F^c_{\mu\nu} = -J^a_\nu \,. \tag{6.114}$$

One can demonstrate that a Yang–Mills theory is not renormalizable for dimensions greater than four.

6.4.2 The Lagrangian of QCD

QCD is based on the gauge group SU(3), the Special Unitary group in 3 dimensions (each dimension is a color, conventionally *Red*, *Green*, *Blue*). This group is represented by the set of unitary 3×3 complex matrices with determinant one (see Sect. 5.3.5).

Since there are nine linearly independent unitary complex matrices, there are a total of eight independent directions in this matrix space, i.e., the carriers of color (called gluons) are eight. Another way of seeing that the number of gluons is eight is that SU(3) has eight generators; each generator represents a color exchange, and thus a gauge boson (a gluon) in color space.

These matrices can operate both on each other (combinations of successive gauge transformations, physically corresponding to successive gluon emissions and/or gluon self-interactions) and on a set of complex 3-vectors, representing quarks in color space.

Due to the presence of color, a generic particle wave function can be written as a three-vector $\psi = (\psi_{qR}, \psi_{qG}, \psi_{qB})$ which is a superposition of fields with a definite color index $i = Red,\ Green,\ Blue$. The SU(3) symmetry corresponds to the freedom of rotation in this three-dimensional space. As we did for the electromagnetic gauge invariance, we can express the local gauge invariance as the invariance of the Lagrangian with respect to the gauge transformation

$$\psi(x) \to \psi'(x) = e^{i\, g_s \epsilon_a(x)\, t^a} \tag{6.115}$$

where the t^a ($a = 1 \ldots 8$) are the eight generators of the SU(3) group, and the $\epsilon_a(x)$ are generic local transformations. g_s is the strong coupling, related to α_s by the relation $g_s^2 = 4\pi\alpha_s$; we shall return to the strong coupling in more detail later.

Usually, the generators of SU(3) are written as

$$t^a = \frac{1}{2}\lambda^a \tag{6.116}$$

where the λ are the so-called Gell–Mann matrices, defined as:

$$\lambda^1 = \begin{pmatrix} 0 & 1 & 0 \\ 1 & 0 & 0 \\ 0 & 0 & 0 \end{pmatrix} ;\ \lambda^2 = \begin{pmatrix} 0 & -i & 0 \\ i & 0 & 0 \\ 0 & 0 & 0 \end{pmatrix} ;\ \lambda^3 = \begin{pmatrix} 1 & 0 & 0 \\ 0 & -1 & 0 \\ 0 & 0 & 0 \end{pmatrix}$$

$$\lambda^4 = \begin{pmatrix} 0 & 0 & 1 \\ 0 & 0 & 0 \\ 1 & 0 & 0 \end{pmatrix} ;\ \lambda^5 = \begin{pmatrix} 0 & 0 & -i \\ 0 & 0 & 0 \\ i & 0 & 0 \end{pmatrix}$$

$$\lambda^6 = \begin{pmatrix} 0 & 0 & 0 \\ 0 & 0 & 1 \\ 0 & 1 & 0 \end{pmatrix} ;\ \lambda^7 = \begin{pmatrix} 0 & 0 & 0 \\ 0 & 0 & -i \\ 0 & i & 0 \end{pmatrix} ;\ \lambda^8 = \frac{1}{\sqrt{3}} \begin{pmatrix} 1 & 0 & 0 \\ 0 & 1 & 0 \\ 0 & 0 & -2 \end{pmatrix} .$$

As discussed in Sect. 5.3.5, these generators are just the SU(3) analogs of the Pauli matrices in SU(2) (one can see it by looking at λ^1, λ^2 and λ^3). Note that superscribing or subscribing an index for a matrix makes no difference in this case.

As a consequence of the local gauge symmetry, eight massless fields $\mathcal{A}^a_\mu$ will appear (one for each generator); these are the gluon fields. The covariant derivative can be written as

$$D_\mu = \partial_\mu + i g_s t^a \mathcal{A}^a_\mu \,. \tag{6.117}$$

Finally, the QCD Lagrangian can be written as

$$\mathcal{L} = \bar{\psi}_q (i\gamma^\mu)(D_\mu)\psi_q - m_q \bar{\psi}_q \psi_q - \frac{1}{4} G^a_{\mu\nu} G^{a\mu\nu} \,, \tag{6.118}$$

where m_q is the quark mass, and $G^a_{\mu\nu}$ is the gluon field strength tensor for a gluon with color index a, defined as

$$G^a_{\mu\nu} = \partial_\mu \mathcal{A}^a_\nu - \partial_\nu \mathcal{A}^a_\mu + g_s f^{abc} \mathcal{A}^b_\mu \mathcal{A}^c_\nu \,, \tag{6.119}$$

and the f^{abc} are defined by the commutation relation $[t^a, t^b] = i f^{abc} t^k$. This terms arises since the generators do not commute.

To guarantee the local invariance, the field $\mathcal{A}^c$ transforms as

$$\mathcal{A}^c_\mu \to \mathcal{A}'^c_\mu = \mathcal{A}^c_\mu - \partial_\mu \epsilon_c - g_s f^{abc} \epsilon_a \mathcal{A}^b_\mu \,.$$

6.4.3 Vertices in QCD; Color Factors

The only stable hadronic states are neutral in color. The simplest example is the combination of a quark and antiquark, which in color space corresponds to

$$3 \otimes \overline{3} = 8 \oplus 1 \,. \tag{6.120}$$

A random (color-uncorrelated) quark–antiquark pair has a $1/N^2 = 1/9$ chance to be in a singlet state, corresponding to the symmetric wave function $\frac{1}{\sqrt{3}}$ $\left(|R\bar{R}\rangle + |G\bar{G}\rangle + |B\bar{B}\rangle\right)$; otherwise it is in an overall octet state (Fig. 6.50).

Correlated production processes like $Z \to q\bar{q}$ or $g \to q\bar{q}$ will project out specific components (here the singlet and octet, respectively).

In final states, we average over all incoming colors and sum over all possible outgoing ones. *Color factors* are thus associated to QCD processes; such factors basically count the number of "paths through color space" that the process can take, and multiply the probability for a process to happen.

A simple example is given by the decay $Z \to q\bar{q}$ (see the next Chapter). This vertex contains a δ_{ij} in color space: the outgoing quark and antiquark must have

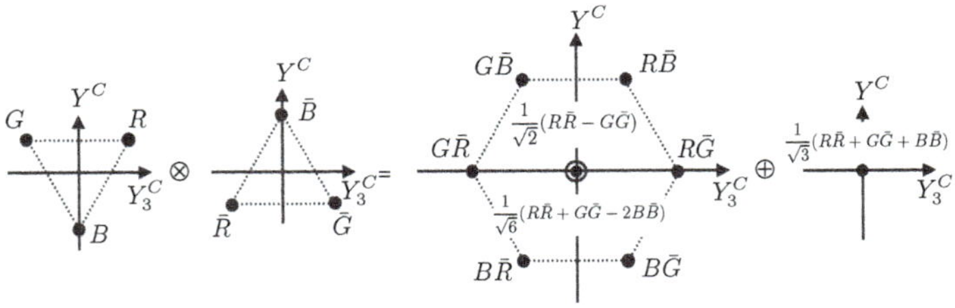

Fig. 6.50 Combinations of a quark and an antiquark in color space

identical (anti-)colors. Squaring the corresponding matrix element and summing over final state colors yields a color factor

$$e^+e^- \to Z \to q\bar{q} \quad : \quad \sum_{\text{colors}} |M|^2 \propto \delta_{ij}\delta^*_{ji} = \text{Tr}\{\delta\} = N_C = 3\,, \qquad (6.121)$$

since i and j are quark indices.

Another example is given by the so-called Drell-Yan process, $q\bar{q} \to \gamma^*/Z \to \ell^+\ell^-$ (Sect. 6.4.7) which is just the reverse of the previous one. The square of the matrix element must be the same as before, but since the quarks are here incoming, we must *average* rather than sum over their colors, leading to

$$q\bar{q} \to Z \to e^+e^- \quad : \quad \frac{1}{9}\sum_{\text{colors}} |M|^2 \propto \frac{1}{9}\delta_{ij}\delta^*_{ji} = \frac{1}{9}\text{Tr}\{\delta\} = \frac{1}{3}\,, \qquad (6.122)$$

and the color factor entails now a *suppression* due to the fact that only quarks of matching colors can produce a Z boson. The chance that a quark and an antiquark picked at random have a corresponding color-anticolor is $1/N_C$.

Color factors enter also in the calculation of probabilities for the vertices of QCD. In Fig. 6.51 one can see the definition of color factors for the three-body vertices $q \to qg$, $g \to gg$ (notice the difference from QED: being the gluon colored, the "triple gluon vertex" can exist, while the $\gamma \to \gamma\gamma$ vertex does not exist) and $g \to q\bar{q}$.

After tedious calculations, the color factors are

$$T_F = \frac{1}{2} \qquad\qquad C_F = \frac{4}{3} \qquad\qquad C_A = N_C = 3\,. \qquad (6.123)$$

6.4.4 The Strong Coupling

When we discussed QED, we analyzed the fact that renormalization can be absorbed in a running value for the charge, or a running value for the coupling constant.

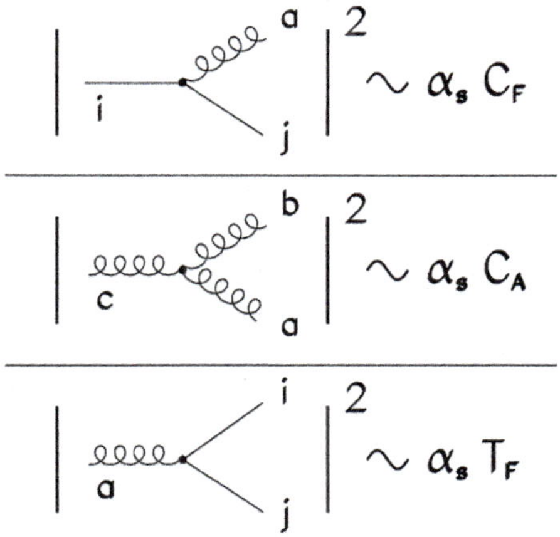

Fig. 6.51 Basic three-body vertices of QCD, and definition of the color factors

This can be interpreted physically as follows. A point-like charge polarizes the vacuum, creating electron-positron pairs which orient themselves as dipoles screening the charge itself. As q^2 increases (i.e., as the distance from the bare charge decreases), the effective charge perceived increases, because there is less screening. Mathematically, this is equivalent to the assumption that the coupling constant increases as q^2 increases.

Also in the case of QCD, the calculation based on the currents gives a logarithmic expression for the coupling constant, which is governed by the so-called *beta function*,

$$Q^2 \frac{\partial \alpha_s}{\partial Q^2} = \frac{\partial \alpha_s}{\partial \ln Q^2} = \beta(\alpha_s)\,, \tag{6.124}$$

where

$$\beta(\alpha_s) = -\alpha_s^2(b_0 + b_1\alpha_s + b_2\alpha_s^2 + \cdots), \tag{6.125}$$

with

$$b_0 = \frac{11C_A - 4T_R n_f}{12\pi}\,, \tag{6.126}$$

$$b_1 = \frac{17C_A^2 - 10T_R C_A n_f - 6T_R C_F n_f}{24\pi^2} = \frac{153 - 19\,n_f}{24\pi^2}. \tag{6.127}$$

In the expression for b_0, the first term is due to gluon loops and the second to the quark loops. In the same way, the first term in the b_1 coefficient comes from double gluon loops, and the others represent mixed quark–gluon loops.

At variance with the QED expression (6.66), the running constant increases with decreasing q^2.

$$\alpha_s(q^2) = \alpha_s(\mu^2)\frac{1}{1 + b_0\alpha_s(\mu^2)\ln\frac{q^2}{\mu^2} + \mathcal{O}(\alpha_s^2)} . \tag{6.128}$$

There is thus no possibility to define a limiting value for $q^2 \to \infty$, starting from which a perturbative expansion could be made (this was the case for QED). The value of the strong coupling must thus be specified at a given reference scale, typically $q^2 = M_Z^2$ (where most measurements have been performed thanks to LEP), from which we can obtain its value at any other scale by solving Eq. 6.124,

$$\alpha_s(q^2) \simeq \alpha_s(M_Z^2)\frac{1}{1 + b_0\alpha_s(M_Z^2)\ln\frac{Q^2}{M_Z^2}} . \tag{6.129}$$

The running coupling constant is shown as calculated from $\alpha_s(M_Z) = 0.1185$, in Fig. 6.52, and compared to the experimental data.

The dependence of b_0 on the number of flavors n_f entails a dependence of the slope of the energy evolution on the number of contributing flavors: the running changes slope across quark flavor thresholds. However, from $q \sim 1$ GeV to present accelerator energies an effective $n_f = 3$ approximation is reasonable, being the production of heavier quarks strongly suppressed.

Notice that in QCD quark–antiquark pairs screen the color charge, like e^+e^- pairs in QED. Antiscreening (which leads to increase the charge at larger distances) comes from gluon loops; getting closer to a quark the antiscreening effect of the virtual gluons is reduced. Since the contribution from virtual quarks and virtual gluons to screening is opposite, the winner is decided by the number of different flavors. For standard QCD with three colors, antiscreening prevails for $n_f < 16$.

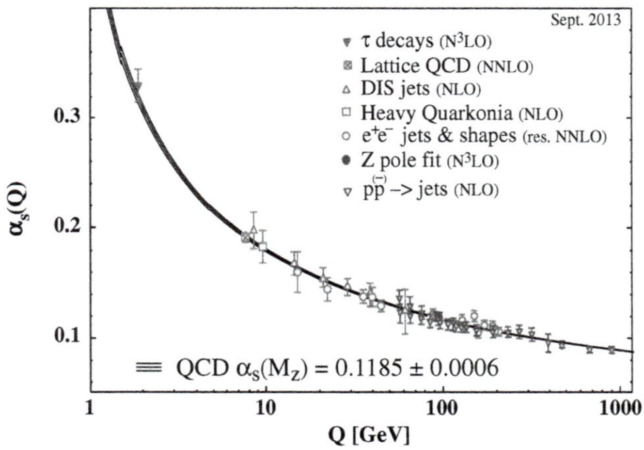

Fig. 6.52 Dependence of α_s on the energy scale Q; a fit to QCD is superimposed. From K.A. Olive et al. (Particle Data Group), Chin. Phys. C 38 (2014) 090001

6.4.5 Asymptotic Freedom and Confinement

When quarks are very close to each other, they behave almost as free particles. This is the famous "asymptotic freedom" of QCD. As a consequence, perturbation theory becomes accurate at higher energies (Eq. 6.128). Conversely, the potential grows at large distances.

In addition, the evolution of α_s with energy must make it comparable to the electromagnetic and weak couplings at some (large) energy, which, looking to our present extrapolations, may lie at some 10^{15}–10^{17} GeV—but such "unification" might happen at lower energies if new, yet undiscovered, particles generate large corrections to the evolution. After this point, we do not know how the further evolution could behave.

At a scale

$$\Lambda \sim 200\,\text{MeV} \tag{6.130}$$

the perturbative coupling (6.128) starts diverging; this is called the Landau pole. Note however that Eq. 6.128 is perturbative, and more terms are needed near the Landau pole: strong interactions indeed do not exhibit a divergence for $Q \to \Lambda$.

6.4.5.1 Quark–Gluon Plasma

Asymptotic freedom entails the fact that at extremely high temperature and/or density a new phase of matter should appear due to QCD. In this phase, called *quark–gluon plasma* (QGP), quarks and gluons become free: the color charges of partons are screened. It is believed during the first few ms after the big bang the Universe was in a QGP state, and flavors were equiprobable.

QGP should be formed when temperatures are close to 200 MeV and density is large enough. This makes the ion–ion colliders the ideal place to reproduce this state.

One characteristic of QGP should be that jets are "quenched": the high density of particles in the "fireball" which is formed after the collision absorbs jets in such a way that in the end no jet or just one jet appears.

Many experiments at hadron colliders tried to create this new state of matter in the 1980s and 1990s, and CERN announced indirect evidence for QGP in 2000. Current experiments at the Relativistic Heavy Ion Collider (RHIC) at BNL and at CERN's LHC are continuing this effort, by colliding relativistically accelerated gold (at RHIC) or lead (at LHC) ions. Also RHIC experiments have claimed to have created a QGP with a temperature 4 $T \sim 4 \times 10^{12}$ K (about 350 MeV).

6.4.6 Hadronization; Final States from Hadronic Interactions

Hadronization is the process by which a set of colored partons becomes a set of color-singlet hadrons.

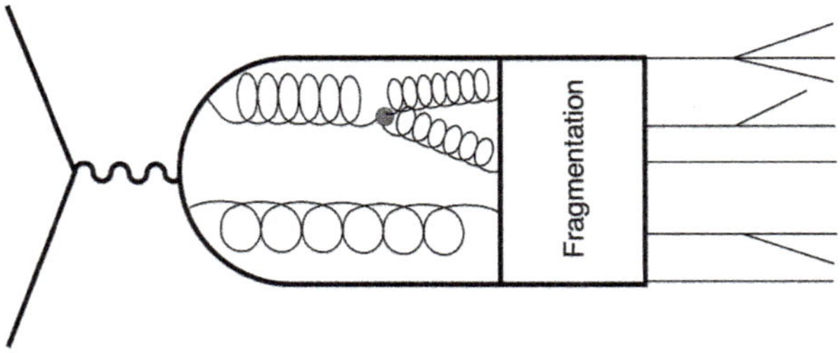

Fig. 6.53 The creation of a multihadronic final state from the decay of a Z boson or from a virtual photon state generated in an e^+e^- collision

At large energies, QCD processes can be described directly from the QCD Lagrangian. Quarks will radiate gluons, which branch into gluons or generate $q\bar{q}$ pairs, and so on. This is a parton shower, quite similar in concept to the electromagnetic showers described by QED.

However, at a certain *hadronization scale* Q_{had} we are not able anymore to perform perturbative calculations. We must turn to QCD-inspired phenomenological models to describe a transition of colored partons into colorless states, and the further branchings.

The problem of hadron generation from a high-energy collision is thus modeled through four steps (Fig. 6.53):

1. Evolution of partons through a parton shower.
2a. Grouping of the partons onto high-mass color-neutral states. Depending on the model these states are called "strings" or "clusters"—the difference is not relevant for the purpose of this book; we shall describe in larger detail the "string" model in the following.
2b. Map of strings/clusters onto a set of primary hadrons (via string break or cluster splitting).
3. Sequential decays of the unstable hadrons into secondaries (e.g., $\rho \to \pi\pi$, $\Lambda \to n\pi$, $\pi^0 \to \gamma\gamma$, ...).

The physics governing steps 2a and 2b is non-perturbative, and pertains to hadronization; some properties are anyway bound by the QCD Lagrangian.

An important result in lattice QCD,[10] confirmed by quarkonium spectroscopy, is that the potential of the color-dipole field between a charge and an anticharge at distances $r \gg 1$ fm, can be approximated as $V \sim kr$ (Fig. 6.54). This is called "linear confinement", and it justifies the *string model of hadronization*, discussed below in Sect. 6.4.6.1.

[10]This formulation of QCD in discrete rather than continuous spacetime allows pushing momentum cut-offs for calculations to the lowest values, below the hadronization scale; however, lattice QCD is computationally very expensive, requiring the use of the largest available supercomputers.

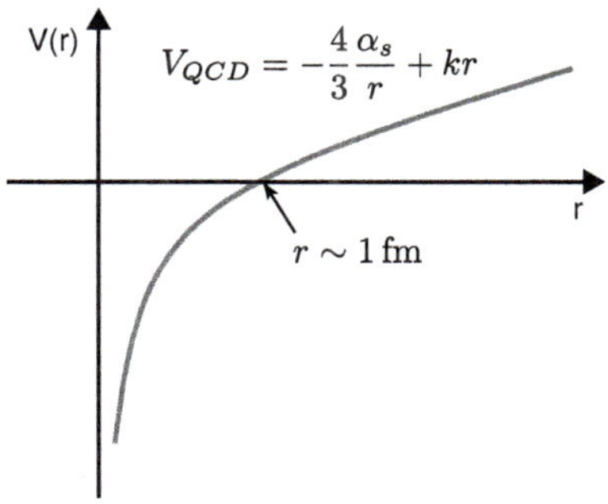

Fig. 6.54 The QCD effective potential

6.4.6.1 String Model

The Lund string model, implemented in the Pythia [F6.1] simulation software, is nowadays commonly used to model hadronic interactions. We shall shortly describe now the main characteristics of this model; many of the basic concepts are shared by any string-inspired method. A more complete discussion can be found in the book by Andersson [F6.2].

Consider the production of a $q\bar{q}$ pair, for instance in the process $e^+e^- \to \gamma^*/Z \to q\bar{q} \to$ hadrons. As the quarks move apart, a potential

$$V(r) = \kappa r \tag{6.131}$$

is stretched among them (at short distances, a Coulomb term proportional to $1/r$ should be added). Such a potential describes a string with energy per unit length κ, which has been determined from hadron spectroscopy and from fits to simulations to have the value $\kappa \sim 1$ GeV/fm ~ 0.2 GeV^2 (Fig. 6.54). The color flow in a string stores energy (Fig. 6.55).

A soft gluon possibly emitted does not affect very much the string evolution (string fragmentation is “infrared safe” with respect to the emission of soft and collinear gluons). A hard gluon, instead, can store enough energy that the qg and the $g\bar{q}$ elements operate as two different strings (Fig. 6.56). The quark fragmentation is different from the gluon fragmentation since quarks are only connected to a single string, while gluons have one on either side; the energy transferred to to strings by gluons is thus roughly double compared to quarks.

As the string endpoints move apart, their kinetic energy is converted into potential energy stored in the stringitself (Eq. 6.131). This process continues until by quantum

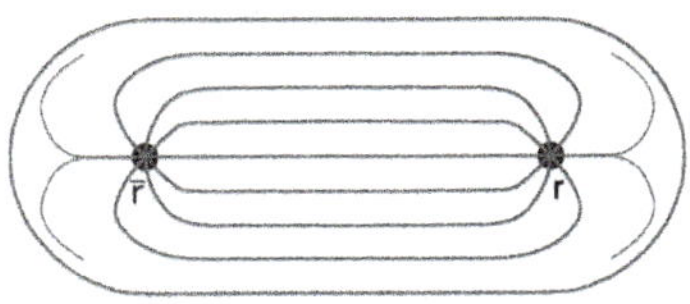

Fig. 6.55 The color flow in a string stores energy in a tube. Adapted from a lecture by T. Sjöstrand

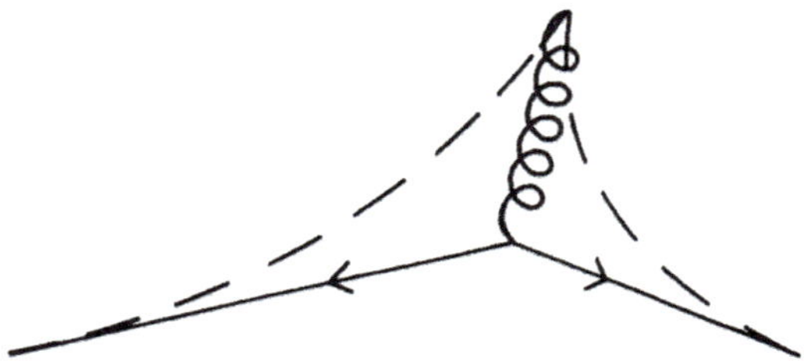

Fig. 6.56 Illustration of a $qg\bar{q}$ system. Color conservation entails the fact that the color string goes from quarks to gluons and vice-versa rather than from quark to antiquark

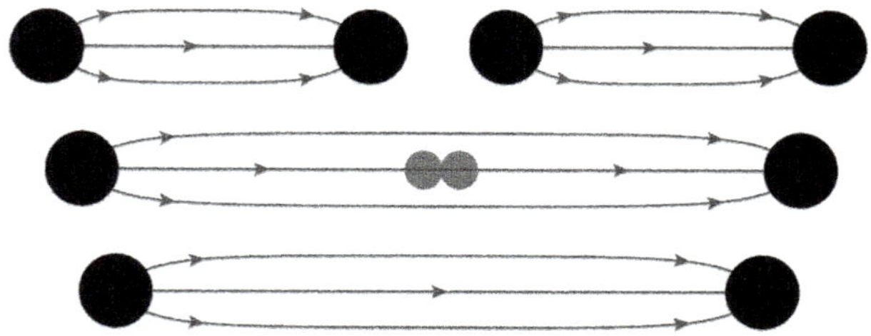

Fig. 6.57 String breaking by quark pair creation in the string field; time evolution goes from *bottom* to *top*

fluctuation a quark–antiquark pair emerges transforming energy from the string into mass. The original endpoint partons are now screened from each other, and the string is broken in two separate color-singlet pieces, $(q\bar{q}) \rightarrow (q\bar{q}') + (q'\bar{q})$, as shown in Fig. 6.57. This process then continues until only final state hadrons remain, as described in the following.

The individual string breaks are modeled from quantum mechanical tunneling, which leads to a suppression of transverse energies and masses:

$$\mathrm{Prob}(m_q^2, p_{\perp q}^2) \propto \exp\left(\frac{-\pi m_q^2}{\kappa}\right)\exp\left(\frac{-\pi p_{\perp q}^2}{\kappa}\right), \tag{6.132}$$

where m_q is the mass of the produced quark and $p_\perp$ is the transverse momentum with respect to the string. The $p_\perp$ spectrum of the quarks is thus independent of the quark flavor, and

$$\left\langle p_{\perp q}^2 \right\rangle = \sigma^2 = \kappa/\pi \sim (250\,\mathrm{MeV})^2 \,. \tag{6.133}$$

The mass suppression implied by Eq. 6.132 is such that strangeness suppression with respect to the creation of u or d, $s/u \sim s/d \sim$, is 0.2–0.3. This suppression is consistent with experimental measurements, e.g., of the K/π ratio in the final states from Z decays.

By inserting the charm quark mass in Eq. 6.132 one obtains a relative suppression of charm of the order of 10^{-11}. Heavy quarks can therefore be produced only in the perturbative stage and not during fragmentation.

Baryon production can be incorporated in the same picture if string breaks occur also by the production of pairs of *diquarks*, bound states of two quarks in a $\bar{3}$ representation (e.g., "red + blue = antigreen"). The relative probability of diquark–antidiquark to quark–antiquark production is extracted from experimental measurements, e.g., of the p/π ratio.

The creation of excited states (for example, hadrons with nonzero orbital momentum between quarks) is modeled by a probability that such events occur; this probability is again tuned on the final multiplicities measured for particles in hard collisions.

With $p_\perp^2$ and m^2 in the simulation of the fragmentation fixed from the extraction of random numbers distributed as in Eq. 6.132, the final step is to model the fraction, z, of the initial quark's longitudinal momentum that is carried by the final hadron; in first approximation, this should scale with energy for large enough energies. The form of the probability density for z used in the Lund model, the so-called fragmentation function $f(z)$, is

$$f(z) \propto \frac{1}{z}(1-z)^a \exp\left(-\frac{b\,(m_h^2 + p_{\perp h}^2)}{z}\right), \qquad (6.134)$$

which is known as the *Lund symmetric fragmentation function* (normalized to unit integral). These functions can be flavor-dependent and they are tuned from the experimental data. The mass dependence in $f(z)$ suggests a harder fragmentation function for heavier quarks (Fig. 6.58): this means that charm and beauty primary hadrons take most of the energy.

The process of iterative selection of flavors, transverse momenta, and z values for pairs breaking a string is illustrated in Fig. 6.59. A quark u produced in a hard process at high energy emerges from the parton shower, and lies at one extreme of

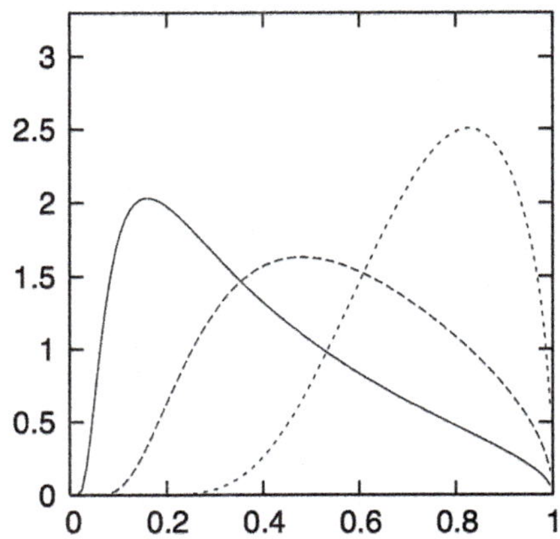

Fig. 6.58 Fragmentation function in the Lund parametrisation for quark–antiquark strings. *Curves* from *left* to *right* correspond to higher masses. Adapted from a lecture by T. Sjöstrand

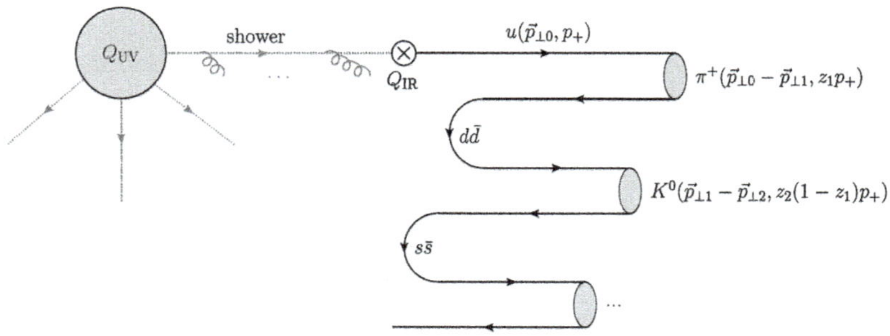

Fig. 6.59 Iterative selection of flavors and momenta in the Lund string fragmentation model. From P. Skands, http://arxiv.org/abs/1207.2389

a string. A $d\bar{d}$ pair is created from the vacuum; the $\bar{d}$ combines with the from the u and forms a π^+, which carries off a fraction z_1 of the total momentum p_+. The next hadron takes a fraction z_2 of the remaining momentum, etc. The z_i are random numbers generated according to a probability density function corresponding to the Lund fragmentation function.

6.4.6.2 Multiplicity in Hard Fragmentation

Average multiplicity is one of the basic observables characterizing hadronic final states. It is extensively studied both theoretically and experimentally at several center-of-mass energies. Experimentally, since the detection of charged particles is simpler than the detection of neutrals, one studies the average charged particle multiplicity. In the limit of large energies most of the particles in the final state are pions, and one can assume, by isospin symmetry, that the number of neutral pions is half the number of charged pions (pions are an isospin triplet).

In order to define the number of particles, one has to define what a stable hadron is. Typically, multiplicity is computed at a time $\Delta t = 10^{-12}$ s after the collision (this interval is larger than the typical lifetime of particles hadronically decaying, 10^{-23} s, but shorter than the typical weak decay lifetimes.

The problem of the energy dependence of the multiplicity was already studied by Fermi and Landau in the 30s. With simple thermodynamical arguments they concluded that the multiplicity from a hard interaction should be proportional to the square root of the center-of-mass energy:

$$\langle n \rangle (E_{CM}) = a\sqrt{E_{CM}} \tag{6.135}$$

A more precise expression has been obtained from QCD. The expression including leading and next-to-leading order calculation is:

$$\langle n \rangle (E_{CM}) = a[\alpha_s(E_{CM})]^b e^{c/\sqrt{\alpha_s(E_{CM})}} \left(1 + O(\sqrt{\alpha_s(E_{CM})}\right) , \tag{6.136}$$

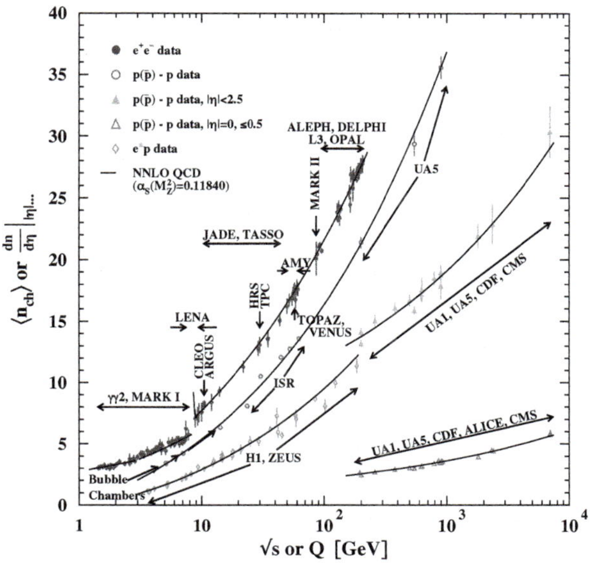

Fig. 6.60 Charged particle multiplicity in e^+e^- and $p\bar{p}$ collisions, pp and ep collisions versus the center-of-mass energy. From K.A. Olive et al. (Particle Data Group), Chin. Phys. C 38 (2014) 090001

where a is a parameter (not calculable from perturbation theory) whose value should be fitted from the data. The constants $b = 0.49$ and $c = 2.27$ are calculated from the theory.

The summary of the experimental data is shown in Fig. 6.60; a plot comparing the charge multiplicity in e^+e^- annihilations with expression 6.136 in a wide range of energies will be discussed in larger detail in the next chapter (Fig. 7.18). The charged particle multiplicity at the Z pole, 91.2 GeV, is about 21 (the total multiplicity including π^0 before their decays is about 30).

6.4.6.3 Jets in Electron-Positron Annihilation

In the quark–antiquark fragmentation into hadrons at low energies, the dominant feature is the production of resonances.

When energy increases, however, primary quarks and antiquarks start carrying a relevant momentum, large enough to allow string breakings. The fragmentation, as seen in the previous section, is essentially a soft process for what is related to the generation of transverse momenta. The phenomenological consequence is the materialization of jets of particles along the direction of the primary quark and antiquark (Fig. 6.61, left).

Since transverse momenta are almost independent of the collision energy while longitudinal momenta are of the order of half the center-of-mass energy, the collimation of jets increases as energy increases.

The angular distribution of jet axes in a blob of energy generated by e^+e^- annihilation follows the dependence

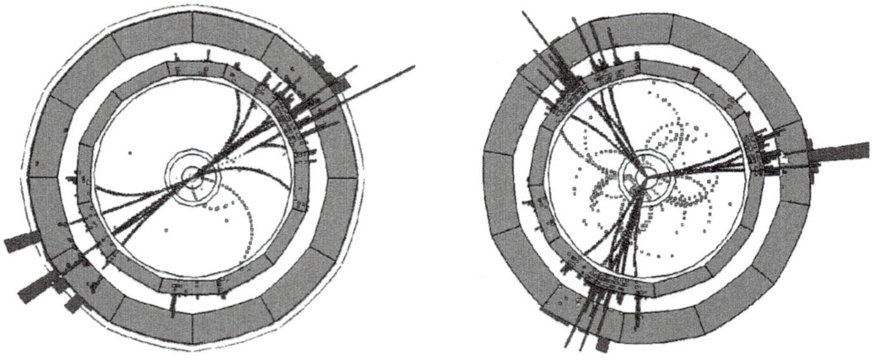

Fig. 6.61 A two-jet event (*left*) and a 3-jet event (*right*) observed by the ALEPH experiment at LEP. *Source* CERN

$$\frac{d\sigma}{d\cos\theta} \propto (1 + \cos^2\theta)$$

expected for spin 1/2 objects.

Some characteristics of quarks can be seen also by the ratio of the cross section into hadrons to the cross section into $\mu^+\mu^-$ pairs, as discussed in Sect. 5.4.2. QED predicts that this ratio should be equal to the sum of squared charges of the charged hadronic particles produced; due to the nature of QCD, the sum has to be extended over quarks and over colors. For $2m_t \gg \sqrt{s} \gg 2m_b$,

$$R = 3\left(\frac{1}{9} + \frac{4}{9} + \frac{1}{9} + \frac{4}{9} + \frac{1}{9}\right) = \frac{11}{3}\,.$$

The O(α_S) process $qg\bar{q}$ (Fig. 6.56) can give events with three jets (Fig. 6.61, right). Notice that, as one can see from Fig. 6.56, one expects an excess of particles in the direction of the gluon jet, with respect of the opposite direction, since this is where most of the color field is. This effect is called the string effect, and has been observed by the LEP experiments at CERN in the 1990s; we shall discuss it in the next chapter. This is evident also from the comparison of the color factors—as well as from considerations based on color conservation.

Jet production was first observed at e^+e^- colliders only in 1975. It was not an easy observation, and the reason is that the question "how many jets are there in an event", which at first sight seems to be trivial, is in itself meaningless, because there is arbitrariness in the definition of jets. A jet is a bunch of particles flying into similar directions in space; the number of jets in a final state of a collision depends on the clustering criteria which define two particles as belonging to the same bunch.

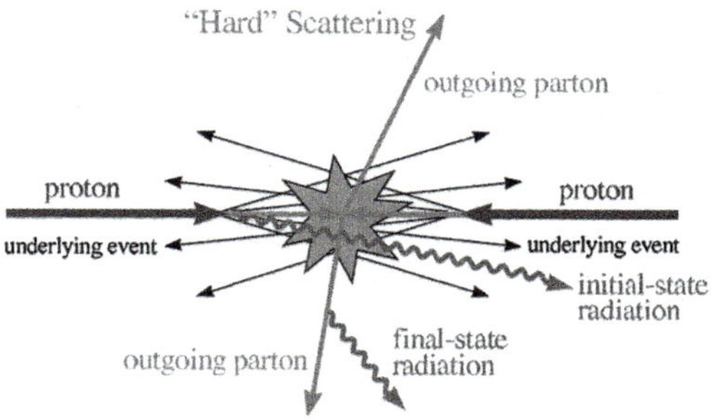

Fig. 6.62 Pictorial representation of a hadron-hadron interaction. From J.M. Campbell et al., Rept. Prog. Phys. 70 (2007) 89

6.4.6.4 Jets in Hadron-Hadron Collisions

The situation is more complicated when final state hadrons come from a hadron-hadron interaction. On top of the interaction between the two partons responsible for a hard scattering, there are in general additional interactions between the beam remnant partons; the results of such interaction are called the "underlying event" (Fig. 6.62).

Usually, the underlying event comes from a soft interaction involving low momentum transfer; therefore perturbative QCD cannot be applied and it has to be described by models. Contributions to the final energy may come from additional gluon radiation from either the initial state or final state partons; typically, the products have small transverse momentum with respect to the direction of the collision (in the center-of-mass system). In particular, in a collision at accelerators, many final products of the collision will be lost in the beam pipe.

To characterize QCD interactions, a useful quantity is the so-called "rapidity" y of a particle:

$$y = \frac{1}{2} \ln \frac{E + p_z}{E - p_z}, \tag{6.137}$$

where z is the common direction of the colliding hadrons in the center-of-mass[11] (the "beam" axis).

Under a boost in the z direction, rapidity transforms by the addition of a fixed quantity. This means that rapidity differences between pairs of particles are invariant with respect to Lorentz boosts along z.

In most collisions in high-energy hadronic scattering, the distribution of final state hadrons is approximately uniform in rapidity, within kinematic limits. That is, the distribution of final state hadrons is approximately invariant under boosts in the

[11] Rapidity is also a useful variable also for the study of electron–positron collisions. However there is nothing special in that case about the beam direction, apart from the $(1 + \cos\theta^2)$ dependence of the jet axis; rapidity in e^+e^- is thus usually defined with respect to the $q\bar{q}$ axis, and it still has, for 2-jet events, the property that the distribution of final state hadrons is approximately uniform in rapidity.

z direction. Thus detector elements should be approximately uniformly spaced in rapidity—indeed they are.

For a non-relativistic particle, rapidity is the same as velocity along the z-axis:

$$y = \frac{1}{2} \ln \frac{E + p_z}{E - p_z} \simeq \frac{1}{2} \ln \frac{m + mv_z}{m - mv_z} \simeq v_z. \tag{6.138}$$

Note that nonrelativistic velocities transform as well additively under boosts (as guaranteed by the Galilei transformation).

The rapidity of a particle is not easy to measure, since one should know its mass. We thus define a variable easier to measure: the pseudorapidity η

$$\eta = -\ln \tan \frac{\theta}{2}, \tag{6.139}$$

where θ is the angle of the momentum of the particle relative to the $+z$ axis. One can derive an expression for rapidity in terms of pseudorapidity and transverse momentum:

$$y = \ln \frac{\sqrt{m^2 + p_T^2 \cosh^2 \eta} + p_T \sinh \eta}{\sqrt{m^2 + p_T^2}}. \tag{6.140}$$

In the limit $m \ll p_T$, $y \to \eta$. This explains the name "pseudorapidity." Angles, and hence pseudorapidity, are easy to measure—but it is really the rapidity that is of physical significance.

To make the distinction between rapidity and pseudorapidity clear, let us examine the limit on the rapidities of the produced particles of a given mass at a given center-of-mass energy. There is clearly a limit on rapidity, but there is no limit on the pseudorapidity, since a particle can be physically produced at zero angle (or at 180°), where its pseudorapidity is infinite. The particles for which the distinction is very significant are those for which the transverse momentum is substantially less than the mass. Note that $y < \eta$ always.

6.4.7 Hadronic Cross Section

The two extreme limits of QCD, asymptotic freedom (perturbative) and confinement (non-perturbative), translate in two radical different strategies in the computation of the cross sections of the hadronic processes. At large momentum transfer (hard processes) cross sections can be computed as the convolution of the partonic (quarks and gluons) elementary cross sections over the parton distribution functions (PDFs). At low transfer momentum (soft interactions) cross sections must be computed using phenomenological models that describe the distribution of matter inside hadrons and

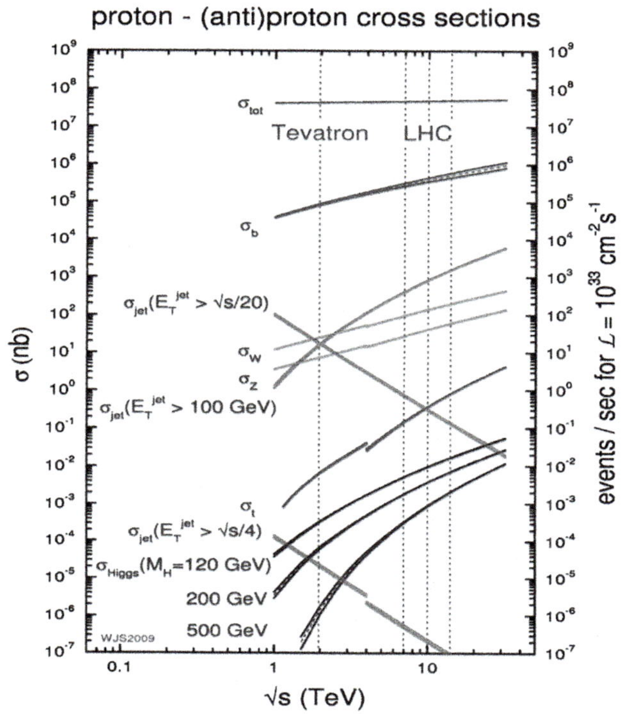

Fig. 6.63 Proton–(anti)proton cross sections at high energies. Cross section values for several important processes are given. The right vertical axis reports the number of events for a luminosity value $L = 10^{33}$ cm^{-2}s^{-1}. From N. Cartiglia, arXiv:1305.6131 [hep-ex]

whose parameters must be determined from the data. The soft processes are dominant. At the LHC for instance (Fig. 6.63), the total proton–proton cross section is of the order of 100 milibarn while the Higgs production cross section is of the order of tens of picobarn (a difference of 10 orders of magnitude!).

At high momentum transfer, the number of partons, mostly gluons, at small x, increases very fast as shown in Fig. 5.25. This fast rise, responsible for the increase of the total cross sections, can be explained by the possibility, at these energies, of gluons radiated by the valence quarks radiate themselves new gluons forming gluonic cascades. However, at higher energies, the gluons in the cascades interact with each other suppressing the emission of new soft gluons and a saturation state often described as the *color Glass Condensate* (*CGC*) is reached. In high energy, heavy-ion collisions, high densities may be accessible over extended regions and the QGP may be formed.

6.4.7.1 Hard Processes

In hadronic hard processes the factorisation assumption, tested first in the deep inelastic scattering, holds. The time scale of the elementary interaction between partons (or as in case of deep inelastic scattering between the virtual photon and the quarks), is basically given by the inverse of the transferred momentum Q

$$\tau_{int} \sim Q^{-1} \tag{6.141}$$

while the hadron time scale is given by the inverse of the QCD non-perturbative scale

$$\Lambda_{\text{QCD}} \sim 200\ \text{MeV} \tag{6.142}$$

$$\tau_{\text{had}} \sim \Lambda_{\text{QCD}}^{-1}\ . \tag{6.143}$$

Hence, whenever $\tau_{\text{int}} \ll \tau_{\text{had}}$ the processes at each timescale can be considered independent. Thus the inclusive cross section of the production of the final state X (for instance one $\mu^+\mu^-$ dilepton, or one multijet system, or one Higgs boson, ...) in the collision of two hadrons h_1 and h_2 with, respectively, four-momentum p_1 and p_2:

$$h_1\,(p_1)\ h_2\,(p_2) \rightarrow X + \cdots \tag{6.144}$$

can be given in leading order (LO) by (see Fig. 6.64):

$$\sigma_{h_1 h_2 \rightarrow X} = \sum_{ij} \int_0^1 dx_1 \int_0^1 dx_2\ f^i_{h_1}(x_1, Q)\ \ f^j_{h_2}(x_2, Q)\ \ \widehat{\sigma}_{ij\rightarrow X}\left(\hat{s}\right) \tag{6.145}$$

where $f^i_{h_1}$ and $f^j_{h_2}$ are the parton distribution functions evaluated at the scale Q, x_1 and x_2 are the fraction of momentum carried, respectively, by the partons i and j and $\widehat{\sigma}_{ij\rightarrow X}$ the partonic cross section evaluated at an effective square of the center-of-mass energy

$$\hat{s} = x_1 x_2 s \tag{6.146}$$

being s the square of the center-of-mass energy of the hadronic collision.

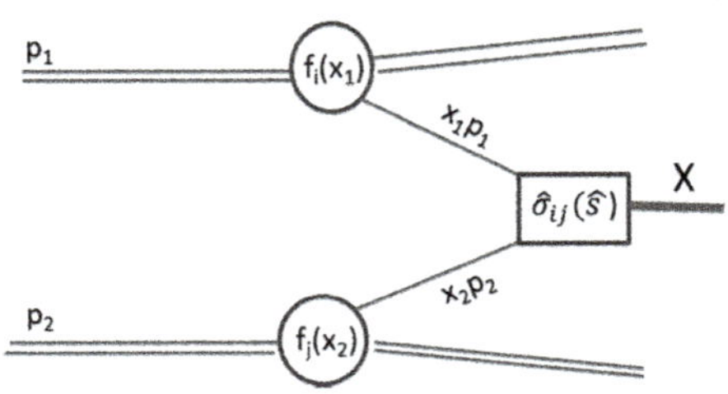

Fig. 6.64 First-order representation of a hadronic hard interaction producing a final state X

The scale Q is usually set to the effective center-of-mass energy $\sqrt{\hat{s}}$ (if X is a resonance, its mass) or to half of the jet transverse energy for high $p_\perp$ processes. The exact value of this scale is somehow arbitrary. If one would be able to compute all order diagrams involved in a given process then the final result would not depend on this particular choice. However, in practice, it is important to set the right scale in order that the corrections of higher order diagrams would have a smaller contribution.

Lower order diagrams give the right order of magnitude but to match the present experimental accuracy (in particular in the LHC) higher order diagrams are needed. Next-to-leading order (NLO) one-loop calculations were computed for many processes since many years and nowadays several predictions at two-loop level, next-to- next leading order (NNLO), are already available for several processes, as for instance the Higgs boson production at the LHC.

The partons not involved in this hard scattering (spectator partons) carry a non-negligible fraction of the total energy and may have small momentum transfer interactions. This underlying event needs to be subtracted from the hard process and may be estimated from minimum bias events with only one vertex (to avoid pile-up effects).

Drell–Yan Processes. The production of dileptons in the collision of two hadrons (known nowadays as the Drell-Yan process) was first interpreted in terms of quark–antiquark annihilation by Sydney Drell and Tung-Mow Yan in 1970. Its leading order diagram (Fig. 6.65) follows the factorisation scheme discussed above where the annihilation cross section $\widehat{\sigma}_{q\overline{q}\to l\overline{l}}$ is a pure QED process given by:

$$\sigma_{q\overline{q}\to l\overline{l}} = \frac{1}{N_c} {Q_q}^2 \frac{4\pi\alpha^2}{3M^2} \,. \tag{6.147}$$

Q_q is the quark charge and M^2 is the square of the center-of-mass energy of the system of the two colliding quark–antiquark (i.e., the square of the invariant mass of the dilepton system). M^2 is thus given by

$$M^2 = \hat{s} = x_1 x_2 s \,. \tag{6.148}$$

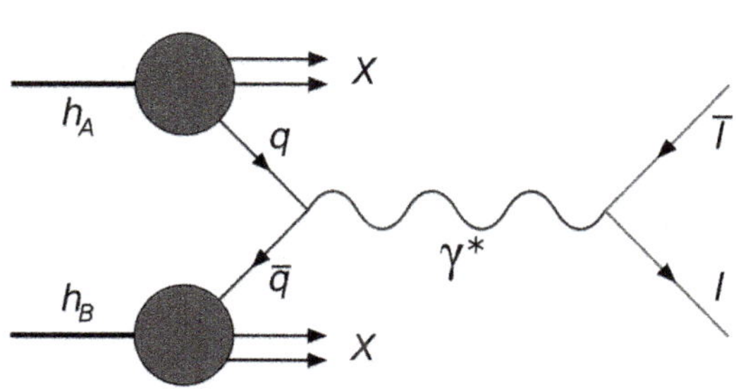

Fig. 6.65 Leading order diagram of the Drell-Yan process. By user:E2m [public domain], via Wikimedia Commons

Finally note that, as it was already discussed in Sect. 5.4.2, the color factor N_c appears in the denominator (average over the incoming colors) in contrast with what happens in the reverse process $l\bar{l} \rightarrow q\bar{q}$ (sum over outgoing colors) whose cross section is given by

$$\sigma_{l\bar{l}\rightarrow q\bar{q}} = N_c\, Q_q{}^2 \frac{4\pi\alpha^2}{3s}\,. \tag{6.149}$$

There is a net topological difference between the final states of these two processes. While in the e^+e^- interactions, the scattering into two leptons or two jets implies a back-to-back topology, in the Drell–Yan there is a back-to back topology in the plane transverse to the beam axis but, since each quark or antiquark carries an arbitrary fraction of the momentum of the parent hadron, the system has in general a momentum along the beam axis.

It is then important to observe that the rapidity of the dilepton system is by energy-momentum conservation equal to the rapidity of the quark–antiquark system,

$$y = y_{l\bar{l}} = y_{q\bar{q}}\,. \tag{6.150}$$

Neglecting the transverse momentum the rapidity is given by

$$y \equiv \frac{1}{2}\ln\frac{E_{l\bar{l}} + P_{Zl\bar{l}}}{E_{l\bar{l}} - P_{Zl\bar{l}}} = \frac{1}{2}\ln\frac{E_{q\bar{q}} + P_{Zq\bar{q}}}{E_{q\bar{q}} - P_{Zq\bar{q}}} = \frac{1}{2}\ln\frac{x_1}{x_2}\,. \tag{6.151}$$

Then, if the mass M and the rapidity y of the dilepton are measured, the momentum fractions of the quark and antiquark can, in this particular case, be directly accessed. In fact, inverting the equations which relates M and y with x_1, x_2 one obtains:

$$x_1 = \frac{M}{\sqrt{s}}\, e^{y} \tag{6.152}$$

$$x_2 = \frac{M}{\sqrt{s}}\, e^{-y}. \tag{6.153}$$

The Drell–Yan differential cross section can now be written in terms of M and y. Computing the Jacobian of the change of the variables from (x_1, x_2) to (M, y),

$$\frac{d\,(x_1 x_2)}{d\,(y, M)} = \frac{2M}{s}. \tag{6.154}$$

It can be easily shown that the differential Drell–Yan cross section for the collision of two hadrons is just:

$$\frac{d\sigma}{dMdy} = \frac{8\pi\alpha^2}{9Ms} f(x_1; x_2) \tag{6.155}$$

where $f(x_1; x_2)$ is the combined PDF for the fractions of momentum carried by the colliding quark and antiquark weighted by the square of the quark charge. For instance, in the case of proton–antiproton scattering one has, assuming that the quark pdfs in the proton are identical to the antiquark pdfs in the antiproton and neglecting the contributions of the antiquark (quark) of the proton (antiproton) and of other quarks than the u and d:

$$f(x_1; x_2) = \left(\frac{4}{9} \, u(x_1) \; u(x_2) + \frac{1}{9} \, d(x_1) \; d(x_2) \right) \tag{6.156}$$

where

$$u(x) \; = \; u^p(x) = \overline{u}^{\overline{p}}(x) \tag{6.157}$$

$$d(x) \; = \; d^p(x) = \overline{d}^{\overline{p}}(x) \, . \tag{6.158}$$

At the LHC in proton–proton collisions the antiquark must come from the sea. Anyhow, to have a good description of the dilepton data (see Fig. 6.66) it is not enough to consider the leading order diagram discussed above. In fact the peak observed around $M \sim 91$ GeV corresponds to the Z resonance, not accounted in the naïve Drell-Yan, and next-to-next leading order (NNLO) diagrams are needed to have such a perfect agreement between data and theory.

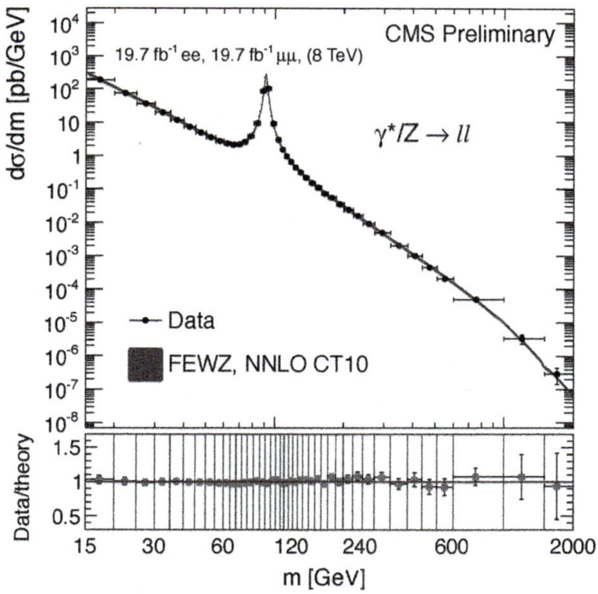

Fig. 6.66 Dilepton cross section measured by CMS. From V. Khachatryan et al. (CMS Collaboration), The European Physical Journal C75 (2015) 147

Multijet Production. Multijet events in hadronic interactions at high energies are abundant and provide both a direct test of QCD and an important background for all the physics channels with final hadronic states, in particular for the searches for new physics. At high transferred momentum interactions, their cross section may be computed following the factorisation scheme discussed above but involving at LO already a large number of elementary two-parton diagrams ($qq \rightarrow qq, qg \rightarrow qg, gg \rightarrow qg, q\overline{q} \rightarrow gg, \ldots$).

The transverse momentum (P_T) of the jets is, in these processes, a key final state variable and together with the jets rapidities (y_i) has to be related to the partonic variables in order that a comparison data/theory may be possible. For instance, in the production of two jets from the t-channel gluon exchange the elementary LO cross section is given by

$$\frac{d\sigma}{dQ^2dx_1dx_2} = \frac{4\pi{\alpha_s}^2}{9Q^2}\left[1+\left(1-\frac{Q^2}{\hat{s}}\right)\right] \tag{6.159}$$

and the following relations can be established between the partonic and the final state variables:

$$x_1 = \frac{P_T}{\sqrt{s}}\left(e^{y_1}+e^{y_2}\right) \tag{6.160}$$

$$x_2 = \frac{P_T}{\sqrt{s}}\left(e^{-y_1}+e^{-y_2}\right) \tag{6.161}$$

$$Q^2 = {P_T}^2\left(1+e^{y_1-y_2}\right) . \tag{6.162}$$

In practice, such calculations are performed numerically using sophisticated computer programs. However, the comparison of the prediction from this calculation

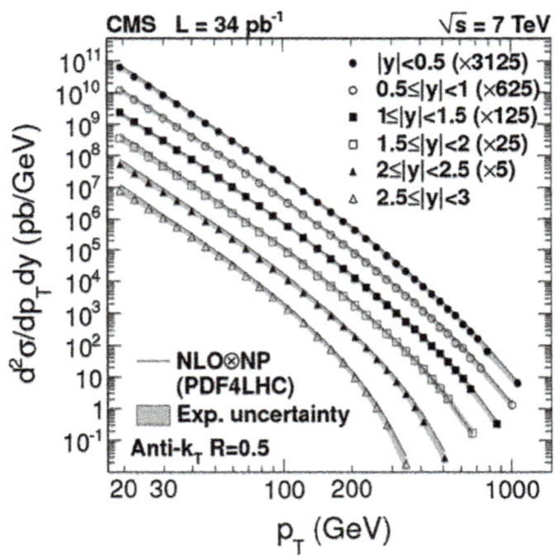

Fig. 6.67 Inclusive jet cross section measured by CMS. From S. Chatrchyan et al. (CMS Collaboration), Phys. Rev. Lett. 107 (2011) 132001

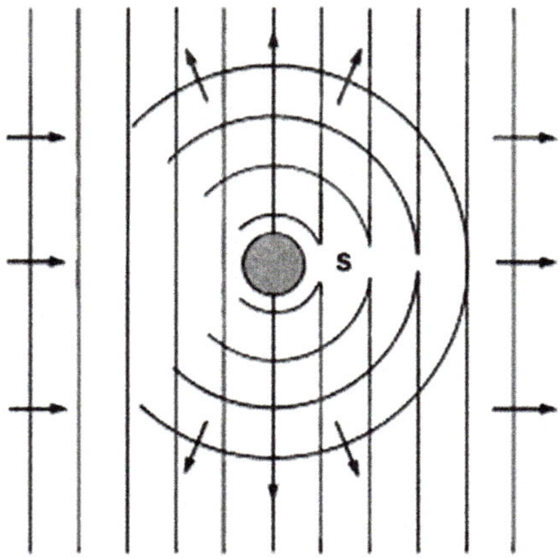

Fig. 6.68 Plane wave scattering by a diffusion center having as result an outcoming spherical wave

with the LHC data provides a powerful test of QCD which spans by many orders of magnitude (see Fig. 6.67).

6.4.7.2 Soft Processes

At low momentum transfer the factorization assumption breaks down. Therefore, it is no longer possible to compute the cross sections adding up perturbative interactions between partons, being the non-perturbative aspects of the hadrons "frozen" in the Parton Distribution Functions. The interaction between hadrons is thus described by phenomenological models.

One popular strategy is to recover optical models and their application in quantum mechanics (for an extended treatment see Ref. [6.4] in the "Further readings"). There, the interaction of a particle with a well-defined momentum, $\vec{p} = \hbar \vec{k}$, over a target may be seen as the scattering of a plane wave by a diffusion center (see Fig. 6.68).

The final state at large radius can then be described by the superposition of the incoming plane wave with one outcoming spherical scattered wave:

$$\psi\left(\vec{r}\right) \sim e^{ikz} + F\left(E, \theta\right) \frac{e^{i\vec{k}\cdot\vec{r}}}{r} \tag{6.163}$$

where z is the coordinate along the beam axis, θ is the scattering angle, E the energy and $F\left(E, \theta\right)$ is denominated as the elastic scattering amplitude.

The elastic differential cross section can be shown to be

$$\frac{d\sigma}{d\Omega} = \left|F\left(E, \theta\right)\right|^2 . \tag{6.164}$$

In the forward region ($\theta = 0$), the interference between the incident and the scattered waves is non-negligible. In fact, this term has a net effect on the reduction of the incident flux that can be seen as a kind of "shadow" created by the diffusion center. An important theorem, the *Optical Theorem*, connects the total cross section with the imaginary part of the forward elastic scattering amplitude:

$$\sigma_{\text{tot}}\ (E) = \frac{4\pi}{k}\ \text{Im}F\ (E, 0)\,. \tag{6.165}$$

The elastic cross section is just the integral of the elastic differential cross section,

$$\sigma_{el}\ (E) = \int |F\ (E, \theta)|^2\ d\Omega \tag{6.166}$$

and the inelastic cross section just the difference of the two cross sections

$$\sigma_{\text{inel}}\ (E) = \sigma_{\text{tot}}\ (E) - \sigma_{el}\ (E)\ . \tag{6.167}$$

It is often useful to decompose the elastic scattering amplitude in terms of angular quantum number l (for spinless particles scattering the angular momentum $\overrightarrow{L}$ is conserved; in the case of particles with spin the good quantity will be the total angular momentum $\overrightarrow{J} = \overrightarrow{L} + \overrightarrow{S}$):

$$F\ (E, \theta) = \frac{1}{k} \sum_{l=0}^{l=\infty} (2l + 1)\ f_l\ (E)\ \ P_l\ (\cos\theta) \tag{6.168}$$

where the functions $f_l\ (E)$ are the *partial wave amplitudes* and P_l are the Legendre polynomials which form an orthonormal basis.

Cross sections can be also written as a function of the *partial wave amplitudes:*

$$\sigma_{el}\ (E) = \frac{4\pi}{K^2} \sum_{l=0}^{l=\infty} (2l + 1) |f_l\ (E)|^2 \tag{6.169}$$

$$\sigma_{\text{tot}}\ (E) = \frac{4\pi}{K^2} \sum_{l=0}^{l=\infty} (2l + 1)\ \text{Im} f_l\ (E) \tag{6.170}$$

and again σ_{inel} is simply the difference between σ_{tot} and σ_{el}.

The optical theorem applied now at each partial wave imposes the following relation (unitarity condition):

$$\text{Im} f_l\ (E) \geq |f_l\ (E)|^2\,. \tag{6.171}$$

Noting that

$$\left| f_l - \frac{i}{2} \right|^2 = |f_l|^2 - \text{Im} f_l + \frac{1}{4} \tag{6.172}$$

this condition can be expressed as

$$\left| f_l - \frac{i}{2} \right|^2 \leq \frac{1}{4} . \tag{6.173}$$

This relation is automatically satisfied if the partial wave amplitude is written as

$$f_l = \frac{i}{2} \left(1 - e^{2i\delta_l}\right) \tag{6.174}$$

being δ_l a complex number.

Whenever δ_l is a pure real number

$$\text{Im} f_l(E) = |f_l(E)|^2 \tag{6.175}$$

and the scattering is totally elastic (the inelastic cross section is zero).

On the other hand, if the wavelength associated to the beam particle is much smaller than the target region,

$$\lambda \sim \frac{1}{k} \ll R \tag{6.176}$$

a description in terms of the classical impact parameter b (Fig. 6.69) may make sense.

Defining

$$b \equiv \frac{1}{k} \left(l + \frac{1}{2}\right) \tag{6.177}$$

the elastic scattering amplitude can then be expressed as

$$F(E, \theta) = 2k \sum_{b=1/k}^{l=\infty} b \, \Delta b \; f_{bk-1/2}(E) \; P_{bk-1/2}(\cos\theta) \tag{6.178}$$

with $\Delta b = 1/k$ which is the granularity of the sum.

In the limit $k \to \infty$, $\Delta b \to 0$, and the sum becomes an integral

$$F(E, \theta) = 2k \int_0^\infty b \, db \, a(b, E) \; P_{bk-1/2}(\cos\theta) \tag{6.179}$$

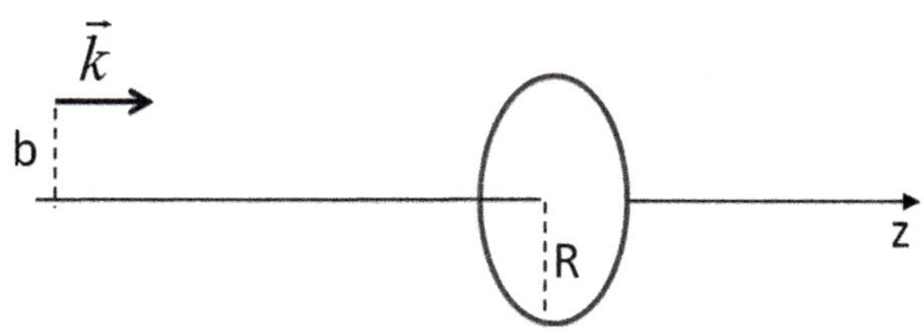

Fig. 6.69 Impact parameter definition in the scattering of particle with momentum $\vec{k}$ over a target region with radius R

where the Legendre polynomials $P_l(\cos\theta)$ were replaced by the Legendre functions $P_\nu(\cos\theta)$ being ν a real positive number, and the partial wave amplitudes f_l were interpolated giving rise to the scattering amplitude $a(b, E)$.

For small scattering angles, the Legendre functions may be approximated by a zeroth order Bessel Function $J_0(b, \theta)$ and finally one can write

$$F(E, \theta) \cong 2k \int_0^\infty b\, db\, a(b, E)\;\, J_0(b, \theta)\,. \qquad (6.180)$$

The scattering amplitude $a(b, E)$ is thus related to the elastic wave amplitude discussed above basically by a Bessel–Fourier transform.

Following a similar strategy to ensure automatically unitarity, $a(b, s)$ may be parametrized as

$$a(s, b) = \frac{i}{2}\left(1 - e^{i\chi(b,s)}\right) \qquad (6.181)$$

where

$$\chi(b, s) = \chi_R(b, s) + i\,\chi_I(b, s) \qquad (6.182)$$

is called the eikonal function.

It can be shown that the cross sections are related with the eikonal by the following expressions:

$$\sigma_{el}(s) = \int d^2b \left|1 - e^{i\chi(b,s)}\right|^2 \qquad (6.183)$$

$$\sigma_{tot}(s) = 2\int d^2b \left(1 - \cos(\chi_R(b, s))\, e^{-\chi_I(b,s)}\right) \qquad (6.184)$$

$$\sigma_{inel}(s) = \int d^2b\, \left(1 - e^{-2\chi_I(b,s)}\right) \qquad (6.185)$$

(the integrations run over the target region with a radius R).

Note that:

- if $\chi_I = 0$ then $\sigma_{inel} = 0$ and all the interactions are elastic.
- if $\chi_R = 0$ and $\chi_I \to \infty$ for $b \le R$, then $\sigma_{inel} = \sigma_{el}$ and $\sigma_{tot} = 2\pi R^2$. This is the so-called black disk limit.

In a first approximation, hadrons may be described by gray disks with mean radius R and $\chi(b, s) = i\Omega(s)$ for $b \le R$ and 0 otherwise. The opacity Ω is a real number $(0 < \Omega < \infty)$. In fact, the main features of proton–proton cross sections can be reproduced in such a simple model (Fig. 6.70). At the high energy limit, the gray disk tends asymptotically to a black disk and thus thereafter the increase of the cross section, limited by the *Froissart Bound* to $\ln^2(s)$, is just determined by the increase of the mean radius.

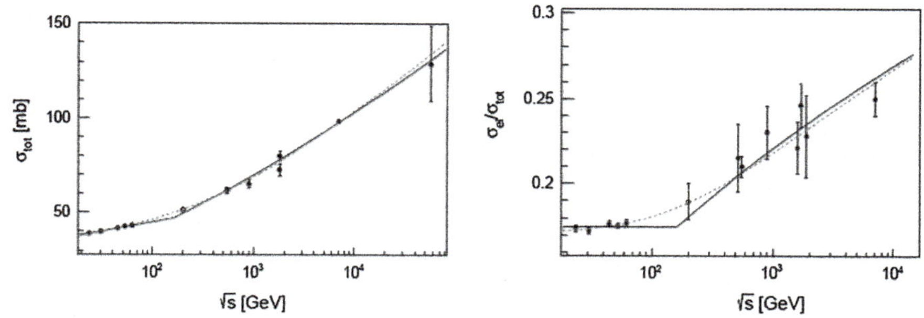

Fig. 6.70 The total cross section (*left*) and the ratio of the elastic and total cross sections in proton–proton interactions as a function of the center-of-mass energy. Points are experimental data and the lines are coming from a fit using one gray disk model. From R. Conceiąo et al., Nuclear Physics A 888 (2012) 58

The eikonal has no dimensions; it is just a complex number as a function of the impact parameter. The imaginary part can be, using a semiclassical argument, associated to the mean number of parton-parton collisions $\overline{n}\,(b, s)$. In fact, if such collisions were independent (no correlations means no diffraction), the probability to have n collisions at an impact parameter b would follow a Poisson distribution:

$$P\,(n, \overline{n}) = \frac{(\overline{n})^n e^{-\overline{n}}}{n!} \tag{6.186}$$

The probability to have at least one collision is given by

$$\sigma_{\text{inel}}\,(s, b) = 1 - e^{-\overline{n}} \tag{6.187}$$

so

$$\sigma_{\text{inel}}\,(s) = \int d^2b \left(1 - e^{-\overline{n}}\right). \tag{6.188}$$

Hence in this approximation

$$\chi_I\,(b, s) = \frac{1}{2}\overline{n}\,(b, s)\ . \tag{6.189}$$

Following this reasoning, $\chi_I\,(b, s)$ is then often computed as the sum of the different kind of parton-parton interactions factorizing each term into a transverse density function and the corresponding cross section:

$$\chi_I\,(b, s) = \sum G_i\,(b, s)\sigma_i\ . \tag{6.190}$$

For instance,

$$\chi_I(b,s) = G_{qq}(b,s)\,\sigma_{qq} + G_{qg}(b,s)\,\sigma_{qg} + G_{gg}(b,s)\,\sigma_{gg} \tag{6.191}$$

where qq, qg, gg stays for, respectively, the quark–quark, quark–gluon, and gluon–gluon interactions.

On the other hand, there are models where χ_I is divided in perturbative (hard) and nonperturbative (soft) terms:

$$\chi_I(b,s) = G_{\text{soft}}(b,s)\,\sigma_{\text{soft}} + G_{\text{hard}}(b,s)\,\sigma_{\text{hard}}. \tag{6.192}$$

The transverse density functions $G_i(b,s)$ must take into account the overlap of the two hadrons and can be computed as the convolution of the Fourier transform of the form factors of the two hadrons.

This kind of strategy can be extended to nuclei–nuclei interactions which are then seen as an independent sum of nucleon–nucleon interactions. This approximation known as the *Glauber*[12] *model*, can be written as:

$$\sigma_{NN}(s) = \int d^2b \left(1 - e^{-G(b,s)\,\sigma_{nn}(s))}\right). \tag{6.193}$$

The function $G(b,s)$ takes now into account the geometrical overlap of the two nuclei and indicates the probability per unit of area of finding simultaneously one nucleon in each nuclei at a given impact parameter.

6.4.7.3 High Density, High Energy

At high density and/or high energy new phenomena may appear.

At high density, whenever one is able to pack densely hadronic matter, as for instance in the core of dense neutron stars, in the first seconds of the Universe (the Big Bang) or in heavy-ion collisions at high energy (the little bangs), naively it is expected that some kind of color screening will occur and partons will become asymptotically free. In fact, the confinement scale is basically set by the size of the hadrons with an energy density ε of the order of 1 GeV/fm^3 and thus, if in larger space regions this kind of energy density is attained, a free gas of quarks and gluons may be formed. That order of magnitude, which corresponds to a transition temperature of around 170–190 MeV, is confirmed by QCD non-perturbative calculations using lattices (see Fig. 6.71). At this temperature, following a simplified Stefan–Boltzmann law for a relativistic free gas, there should be a fast increase of the energy density

[12]Roy Jay Glauber (New York, 1925) is an American physicist, recipient of the 2005 Nobel Prize "for his contribution to the quantum theory of optical coherence," a fundamental contribution to the field of quantum optics. For many years before, Glauber participated to the organization of the Ig Nobel Prize: he had the role of "keeper of the broom," sweeping the paper airplanes thrown during the event out from the stage.

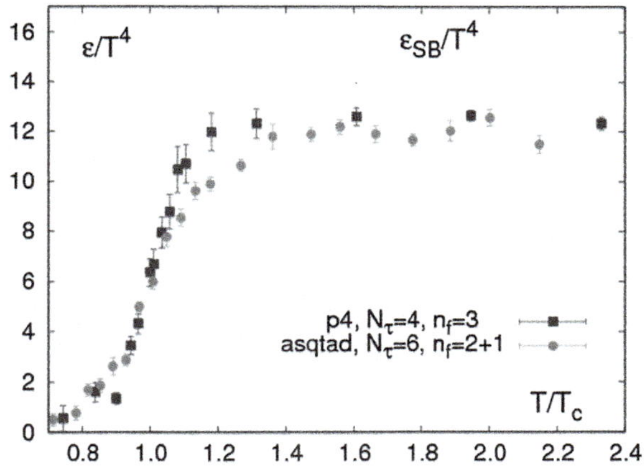

Fig. 6.71 Energy density of hadronic state of matter, with baryonic number zero, according to one lattice calculation. A sharp rise is observed near the critical temperature $T_c \sim 170$–$190\,\text{MeV}$. From C. Bernard et al., hep-lat/0610017

corresponding to the increase of the effective internal number degrees of freedom g_* from a free gas of pions ($g_* = 3$) to a new state of matter where quarks and gluons are asymptotically free ($g_* = 37$, considering two quark flavors). This new matter state is usually denominated as the Quark–Gluon Plasma (QGP).

The phase transition between hadronic and QGP states depends also strongly on the net baryon contents of the system. At the core of dense neutron stars QGP may occur at very low temperatures. The precise QCD phase diagram is therefore complex and still controversial. A simplified sketch is presented in Fig. 6.72 where the existence of a possible critical point is represented.

In Pb–Pb collisions at the LHC, center-of-mass energies per nucleon of 2.76 TeV, corresponding to an energy density for central events (head-on collisions, low-impact parameters) above 10 GeV/fm^3, have been attained. The multiplicity of such events

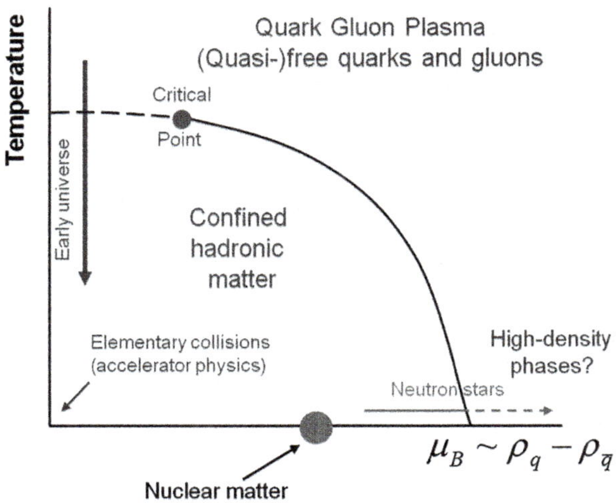

Fig. 6.72 Schematic representation of the QCD phase diagram as a function of the temperature and of the baryonic potential (measure the difference in the quark and antiquark contents of the system). From http://www.phys.uu.n/~leeuw179/bachelor_research/Bachelor_QCDQGP_2012.ppt

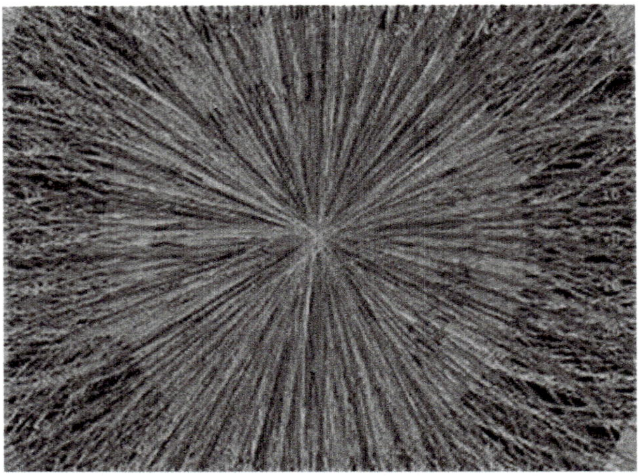

Fig. 6.73 First lead-lead event recorded by ALICE detector at LHC at center-of-mass energy per nucleon of 2.76 TeV. Thousands of charged particles were recorded by the time-projection chamber. *Source* CERN

is huge with thousand of particles detected (Fig. 6.73). Such events are an ideal laboratory to study the formation and the characteristics of the QGP. Both global observables, as the asymmetry of the flow of the final states particles, and hard probes like high transverse momentum particles, di-jets events, and specific heavy hadrons are under intense scrutiny.

The asymmetry of the flow of the final states particles can be predicted as a consequence of the anisotropies in the pressure gradients due to the shape and structure of the nuclei–nuclei interaction region (Fig. 6.74). In fact, more and faster particles are expected and seen in the region of the interaction plane (defined by the directions of

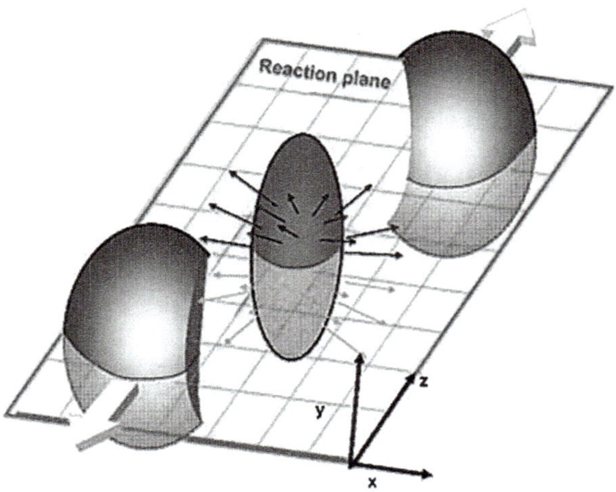

Fig. 6.74 Artistic representation of one heavy-ion collision. The reaction plane is defined by the momentum vectors of the two ions and the shape of the interaction region is due to the sharp pression gradients. From https://www.bnl.gov/rhic/news/061907/story2.asp

the two nuclei in the center-of-mass reference frame) where compression is higher. Although the in-out modulation (elliptic flow) is qualitatively in agreement with the predictions, quantitatively the effect is smaller than the expected with the assumption of one QGP formed by a free gas of quarks and gluons. Some kind of collective phenomena should exist which could be related to a nonzero QGP fluid viscosity. QGP behaves rather like a low viscosity liquid than an ideal gas. Such behavior was first discovered at lower energies at the RHIC collider at Brookhaven.

Partons resulting from elementary hard processes inside the QGP have to cross a high dense medium and thus may suffer significant energy losses or even be absorbed in what is generically called "quenching". The most spectacular observation of such phenomena is in di-jet events, where one of the high P_T jets loose a large fraction of its energy (Fig. 6.75). This "extinction" of jets is usually quantified in terms of the *nuclear suppression factor* R_{AA} defined as the ratio between differential P_T distributions in nuclei–nuclei and in proton–proton collisions:

$$R_{AA} = \frac{d^2 N_{AA}/dy dP_T}{N_{\rm coll} d^2 N_{pp}/dy dP_T} \,. \tag{6.194}$$

Where $N_{\rm coll}$ is the average number of nucleon–nucleon collisions at each specific rapidity bin.

In the absence of "medium effects" R_{AA} may reflect a possible modification of the parton distributions functions (pdfs) in nuclei as compared to the ones in free nucleons but should not be far from the unity. The measurement at the LHC (Fig. 6.76) showed however a clear suppression demonstrating clearly significant energy losses in the medium and in this way it can provide information of the dynamical properties of the medium, such as the density.

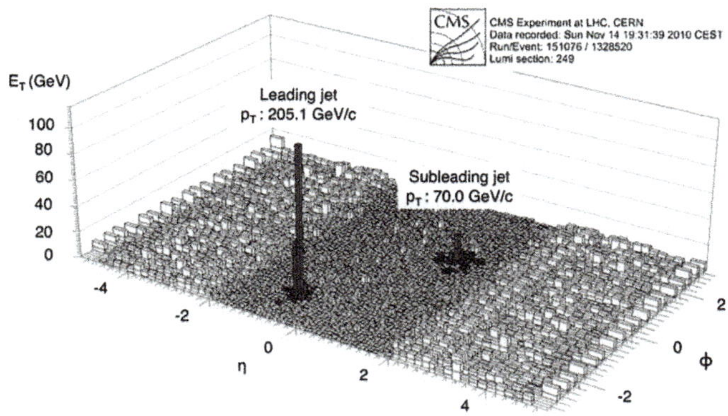

Fig. 6.75 Display of an unbalanced di-jet event recorded by the CMS experiment at the LHC in lead–lead collisions at a center-of-mass energy of 2.76 TeV per nucleon. The plot shows the sum of the electromagnetic and hadronic transverse energies as a function of the pseudorapidity and the azimuthal angle. The two identified jets are highlighted. From S. Chatrchyan et al. (CMS Collaboration), Phys. Rev. C84 (2011) 024906

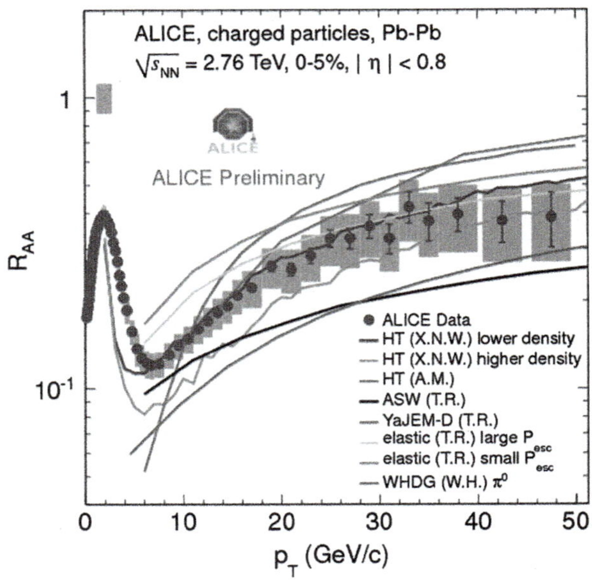

Fig. 6.76 The nuclear modification factor R_{AA} as a function of P_T, measured by ALICE experiment at LHC at center-of-mass energy per nucleon of 2.76 TeV, compared to different models. From J. Otwinowski, "High-$p_\perp$ processes measured with ALICE at the LHC", arXiv:1301.5285

But not only loss processes may occur in the presence of a hot and dense medium (QGP). The production of high-energy quarkonia (bound states of heavy quark–antiquark pairs) may also be suppressed whenever QGP is formed as initial proposed on a seminal paper in 1986 by Matsui and Satz in the case of the J/ψ ($c\overline{c}$ pair) production in high-energy heavy-ion collisions. The underlined proposed mechanism was a color analog of Debye screening which describes the screening of electrical charges in the plasma. Evidence of such suppression was soon reported at CERN in fixed target oxygen–uranium collisions at 200 GeV per nucleon by the NA38 collaboration. Many other results were published in the following years and a long discussion was held on whether the observed suppression was due to the absorption of these fragile $c\overline{c}$ states by the surrounding nuclear matter or to the possible existence of the QGP. In 2007 the NA60 reported, in indium–indium fixed target collisions at 158 GeV per nucleon, the existence of an anomalous J/ψ suppression not compatible with the nuclear absorption effects. However, this anomalous suppression did not increase at higher center-of-mass energies at the RHICH heavy-ion collider, and even recently showed a clear decrease at the LHC (Fig. 6.77). Meanwhile, the possible (re)combination of charm and anti-charm quarks at the boundaries of the QGP region was proposed as an enhancement production mechanism and such mechanism seems to be able to describe the present data.

The study of the J/ψ production, as well as of other quarkonia states, is extremely important to study QGP as it allows for a thermal spectroscopy of the QGP evolution. The dissociation/association of these $q\overline{q}$ pairs is intrinsically related to the QGP temperature; as such, as this medium expands and cools down, these pairs may recombine and each flavor has a different recombination temperature. However, the

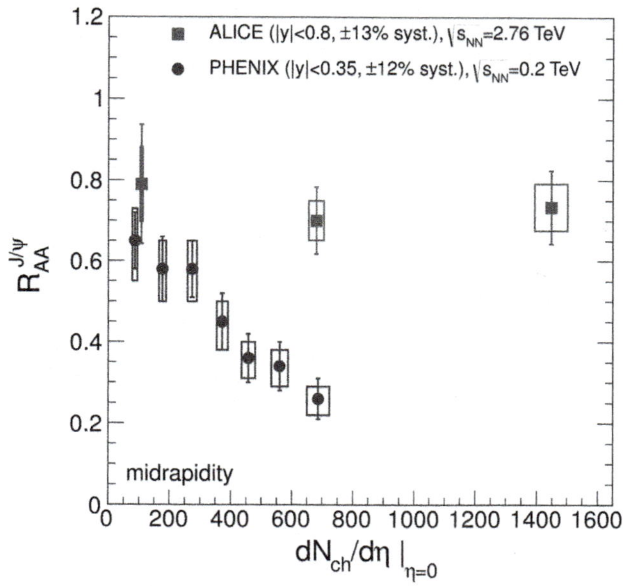

Fig. 6.77 The nuclear modification factor R_{AA} for inclusive J/ψ production at mid rapidity as reported by PHENIX (RHIC) and ALICE (LHC) experiments at center-of-mass energy per nucleon of 0.2 and 2.76 TeV, respectively. From http://cerncourier.com/cws/article/cern/48619

competition between the dissociation and association effects is not trivial and so far it was not yet experimentally assessed.

The process of formation of the QGP in high-energy heavy-ions collisions is theoretically challenging. It is generally accepted that in the first moments of the collisions the two nuclei had already reached the saturation state described by the color glass condensate (CGC) referred at the beginning of the Sect. 6.4.7. Then a fast thermalisation process occur ending in the formation of a QGP state described by relativistic hydrodynamic models. The intermediated stage, not experimentally accessible and not theoretically well established, is designated as *glasma*. Finally, the QGP "freezes-out" into a gas of hadrons. Such scheme is pictured out in an artistic representation in Fig. 6.78.

In ultra-high-energy cosmic ray experiments (see Chap. 10), events with center-of-mass energies well above those presently attainable in human-made accelerators are detected. Higher Q^2 and thus smaller scales ranges can then be explored opening a new possible window to test hadronic interactions.

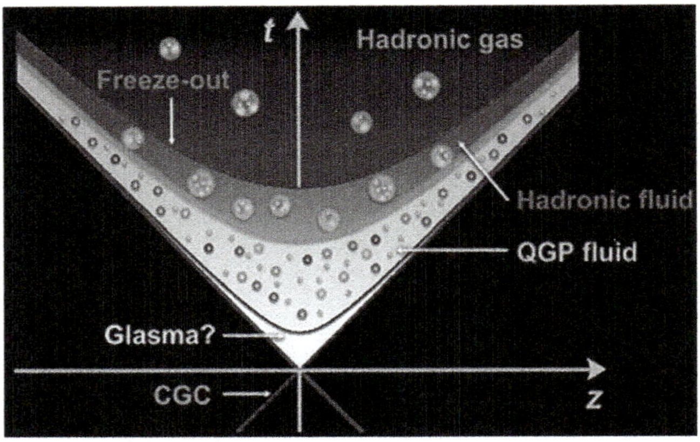

Fig. 6.78 An artistic representation of the time–space diagram of the evolution of the states created in Heavy Ion collisions. From "Relativistic Dissipative Hydrodynamic description of the Quark-Gluon Plasma" A. Monai 2014 (http://www.springer.com/978-4-431-54797-6)

Further Reading

[F6.1] M. Thomson, "Modern Particle Physics," Cambridge University Press 2013. A recent, pedagogical and rigorous book covering the main aspects of particle physics at advanced undergraduate and early graduate level.

[F6.2] A. Bettini, "Introduction to Elementary Particle Physics" (second edition), Cambridge University Press 2014. A good introduction to Particle Physics at the undergraduate level starting from the experimental aspects and deeply discussing relevant experiments.

[F6.3] D. Griffiths, "Introduction to Elementary Particles" (second edition), Wiley-VCH 2008. A reference book at the undergraduate level with many proposed problems at the end of each chapter; rather oriented on the theoretical aspects.

[F6.4] S. Gasiorowicz, "Quantum Physics" (third edition), Wiley 2003. Provides a concise and solid introduction to quantum mechanics. It is very useful for students that had already been exposed to the subject.

[F6.5] I.J.R. Aitchison, A.J.G. Hey, "Gauge Theories in Particle Physics: A Practical Introduction" (fourth edition—2 volumes), CRC Press, 2012. Provides a pedagogical and complete discussion on gauge field theories in the Standard Model of Particle Physics from QED (vol. 1) to electroweak theory and QCD (vol. 2).

Exercises

1. *Spinless particles interaction.* Determine, in the high-energy limit, the electromagnetic differential cross section between two spinless charged nonidentical particles.
2. *Dirac equation invariance.* Show that the Dirac equation written using the covariant derivative is gauge-invariant.
3. *Bhabha scattering.* Draw the QED Feynman diagrams at low order for the elastic e^+e^- scattering and discuss why the Bhabha scattering measurements at LEP are done at very small polar angle.
4. *Bilinear covariants.* Show that

   (a) $\overline{\psi}\psi$ is a scalar;
   (b) $\overline{\psi}\gamma^5\psi$ is a pseudoscalar;
   (c) $\overline{\psi}\gamma^\mu\psi$ is a four-vector;
   (d) $\overline{\psi}\gamma^\mu\gamma^5\psi$ is a pseudo four-vector.

5. *Chirality and helicity.* Show that the right helicity eigenstate $u_\uparrow$ can be decomposed in the right (u_R) and left (u_L) chiral states as follows:

$$u_\uparrow = \frac{1}{2}\left(1 + \frac{p}{E+m}\right) u_R + \frac{1}{2}\left(1 - \frac{p}{E+m}\right) u_L \, .$$

6. ***ν_μ** neutrino beams.* Consider a beam of ν_μ produced through the decay of a primary beam containing pions (90 %) and kaons (10 %). The primary beam has a momentum of 10 GeV and an intensity of $10^{10}\ \mathrm{s}^{-1}$.

   (a) Determine the number of pions and kaons that will decay in a tunnel 100 m long.
   (b) Determine the energy spectrum of the decay products.
   (c) Calculate the contamination of the ν_μ beam, i.e., the fraction of ν_e present in that beam.

7. ***ν_μ** semileptonic interaction.* Considering the process $\nu_\mu p \longrightarrow \mu^- X$:

   (a) Discuss what X could be (start by computing the available energy in the center-of-mass).
   (b) Write the amplitude at lower order for the process for the interaction of the ν_μ with the valence quark d ($\nu_\mu d \longrightarrow \mu^- u$).
   (c) Compute the effective energy in the center-of-mass for this process supposing that the energy of the ν_μ is 10 GeV and the produced muon takes 5 GeV and is detected at an angle of 10° with the ν_μ beam.
   (d) Write the cross section of the process $\nu_\mu p \longrightarrow \mu^- X$ as a function of the elementary cross section $\nu_\mu d \longrightarrow \mu^- u$.

8. *Neutrino and anti-neutrino deep inelastic scattering.* Determine, in the framework of the quark parton model, the ratio:

$$\frac{\sigma\left(\overline{\nu}_{\mu} N \longrightarrow \mu^{+} X\right)}{\sigma\left(\nu_{\mu} N \longrightarrow \mu^{-} X\right)}$$

where N stands for an isoscalar (same number of protons and neutrons) nucleus. Consider that the involved energies are much higher than the particle masses. Take into account only diagrams with valence quarks.

9. *Top pair production.* Consider the pair production of the top/anti-top quarks in a proton-antiproton collider. Draw the dominant first-order Feynman diagram for this reaction and estimate what should be the minimal beam energy of a collider to make the process happen. Discuss which channels have a clear experimental signature.
10. *c-quark decay.* Consider the decay of the c quark. Draw the dominant first-order Feynman diagrams of this decay and express the corresponding decay rates as a function of the muon decay rate and of the Cabibbo angle. Make an estimation of the c quark lifetime knowing that the muon lifetime is about 2.2 μs.
11. *Gray disk model in proton–proton interactions.* Determine, in the framework of the gray disk model, the mean radius and the opacity of the proton as a function of the center-of-mass energy (you can use Fig. 6.70 to extract the total and the elastic proton–proton cross sections).

Chapter 7
The Higgs Mechanism and the Standard Model of Particle Physics

The basic interactions affecting matter at the particle physics level are electromagnetism, strong interaction, and weak interaction. They can be unified by a Lagrangian displaying gauge invariance with respect to the SU(3) ⊗ SU(2) ⊗ U(1) *local symmetry group; this unification is called the standard model of particle physics. Within the standard model, an elegant mechanism, called the Higgs mechanism, accounts for the appearance of masses of particles and of some of the gauge bosons. The standard model is very successful, since it brilliantly passed extremely accurate precision tests and several predictions have been confirmed—in particular, the Higgs particle has been recently discovered in the predicted mass range. However, it can hardly be thought as the final theory of nature: some physics beyond the standard model must be discovered to account for gravitation and to explain the energy budget of the Universe.*

In the previous chapter, we have characterized three of the four known interactions: the electromagnetic, strong interaction, and weak interaction.

We have presented an elegant mechanism for deriving the existence of gauge bosons from a local symmetry group. We have carried out in detail the calculations related to the electromagnetic theory, showing that the electromagnetic field naturally appears from imposing a local U(1) gauge invariance. However, a constraint imposed by this procedure is that the carriers of the interactions are massless. If we would like to give mass to the photons, we would violate the gauge symmetry:

$$\frac{1}{2}M_A^2 A_\mu A^\mu \rightarrow \frac{1}{2}M_A^2 \left(A_\mu - \frac{1}{e}\partial_\mu \alpha\right)\left(A^\mu - \frac{1}{e}\partial^\mu \alpha\right) \neq \frac{1}{2}M_A^2 A_\mu A^\mu \qquad (7.1)$$

if $M_A \neq 0$. This is acceptable in this particular case, being the carriers of electromagnetic interaction identified with the massless photons, but not in general.

A similar situation applies to the theory of strong interactions, QCD. The symmetry with respect to rotation in color space, SU(3), entails the appearance of eight massless quanta of the field, the gluons, which successfully model the experimental observations.

A. De Angelis and M.J.M. Pimenta, *Introduction to Particle and Astroparticle Physics*, Undergraduate Lecture Notes in Physics,
DOI 10.1007/978-88-470-2688-9_7

The representation of the weak interaction has been less satisfactory. We have a SU(2) symmetry there, a kind of isospin, but the carriers of the force must be massive to explain the weakness and the short range of the interaction—indeed they are identified with the known $W^{\pm}$ and Z particles, with a mass of the order of 100 GeV. But we do not have a mechanism for explaining the existence of massive gauge particles, yet. Another problem is that, as we shall see, incorporating the fermion masses in the Lagrangian by brute force via a Dirac mass term $m\bar{\psi}\psi$ would violate the symmetry related to the weak interaction.

Is there a way to generate the gauge boson and the fermion masses without violating gauge invariance? The answer is given by the so-called Higgs mechanism, proposed in the 1960s. This mechanism is one of the biggest successes of fundamental physics, and requires the presence of a new particle; the Higgs boson, responsible for the masses of particles. This particle has been found experimentally in 2012 after 50 years of searches—consistent with the standard model parameters measured with high accuracy at LEP.

The Higgs mechanism allowed to formulate a quantum field theory—relativistically covariant—that explains all currently observed phenomena at the scale of elementary particles: the standard model of particle physics (in short "standard model," also abbreviated as SM). The SM includes all the known elementary particles and the three interactions relevant at the particle scale: the electromagnetic interaction, the strong interaction, and the weak interaction. It does not include gravitation, which, for now, cannot be described as a quantum theory. It is a $\mathrm{SU}(3) \otimes \mathrm{SU}(2) \otimes \mathrm{U}(1)$ symmetrical model.

The SM is built from two distinct interactions affecting twelve fundamental particles (quarks and leptons) and their antiparticles: the electroweak interaction, coming from the unification of the weak force and electromagnetism (QED), and the strong interaction explained by QCD. These interactions are explained by the exchange of gauge bosons (the vectors of these interactions) between elementary fermions.

All our knowledge about fundamental particles and interactions, which we have described in the previous chapters, can be summarized in the following table (taken from Iliopulos, 2012 CERN School on Particle Physics).

TABLE OF ELEMENTARY PARTICLES		
QUANTA OF RADIATION		
Strong Interactions		Eight gluons
Electromagnetic Interactions		Photon (γ)
Weak Interactions		Bosons W^+ , W^- , Z
Gravitational Interactions		Graviton (?)
MATTER PARTICLES		
	Leptons	Quarks
1st Family	(ν_e , e^-)	(u , d)
2nd Family	(ν_μ , μ^-)	(c , s)
3rd Family	(ν_τ , τ^-)	(t , b)
HIGGS BOSON		

Some remarks about the Table:

- Elementary particles are found to be all spin one-half particles. They are divided into quarks, which are sensitive to the strong interaction, and "leptons" which have no strong interactions. No reason is known neither for their number, nor for their properties, such as their quantum numbers.
- Quarks and leptons can be organized into three distinct groups, or "families." No deep explanation is known.
- Each quark species, called "flavor," appears under three charges, called "colors."
- Quarks and gluons do not appear as free particles. They are confined in bound states, the hadrons.

Let us see which mechanism can explain the mass of gauge bosons.

7.1 The Higgs Mechanism and the Origin of Mass

The principle of local gauge invariance works beautifully for electromagnetic interactions. Veltman and 't Hooft[1] proved in the early 1970s that gauge theories are renormalisable. But gauge fields appear to predict the existence of massless gauge bosons, while we know in nature that weak interaction is mediated by heavy vectors $W^{\pm}$ and Z.

How to introduce mass in a gauge theory? We have seen that a quadratic term $\mu^2 A^2$ in the gauge boson field spoils the gauge symmetry; gauge theories seem, at face value, to account only for massless gauge particles.

The idea to solve this problem came from fields different from particle physics, and it is related to spontaneous symmetry breaking.

7.1.1 Spontaneous Symmetry Breaking

Spontaneous symmetry breaking (SSB) was introduced into particle physics in 1964 by Englert and Brout, and independently by Higgs.[2] Higgs was the first to mention explicitly the appearance of a massive scalar particle associated with the curvature of

---

[1]Martinus Veltman (1931) is a Dutch physicist. He supervised the Ph.D. thesis of Gerardus't Hooft (1946), and during the thesis work, in 1971, they demonstrated that gauge theories were renormalizable. For this achievement, they shared the Nobel Prize for Physics in 1999.

[2]Peter Higgs (Newcastle, UK, 1929) has been taught at home having missed some early schooling. He moved to city of London School, and then to King's College also in London, at the age of 17 years, where he graduated in molecular physics in 1954. In 1980, he was assigned the chair of Theoretical Physics at Edinburgh. He shared the 2013 Nobel prize in physics with François Englert "for the theoretical discovery of a mechanism that contributes to our understanding of the origin of mass of subatomic particles, and which recently was confirmed through the discovery of the predicted fundamental particle." François Englert (1932) is a Belgian physicist; he is a Holocaust

Fig. 7.1 The potential described by the shape of the *bottom* of an empty bottle. Such a potential is frequently called "mexican hat" or "cul-de-bouteille," depending on the cultural background of the author

the effective potential that determines the SSB; the mechanism is commonly called the Higgs mechanism, and the particle is called the Higgs boson.

Let us see how SSB can create in the Lagrangian a mass term quadratic in the field. We shall concentrate on a scalar theory, but the extension to a vector theory does not add conceptually.

The idea is that the system has at least two phases:

- The unbroken phase: the physical states are invariant with respect to all symmetry groups with respect to which the Lagrangian displays invariance. In a local gauge theory, massless vector gauge bosons appear.
- The spontaneously broken phase: below a certain energy, a phase transition might occur. The system reaches a state of minimum energy (a "vacuum") in which part of the symmetry is hidden from the spectrum. For a gauge theory, we shall see that some of the gauge bosons become massive and appear as physical states.

Infinitesimal fluctuations of a system which is crossing a critical point can decide on the system's fate, by determining which branch among the possible ones is taken. Such fluctuations arise naturally in quantum physics, where the vacuum is just the point of minimal energy and not a point of zero energy.

It is this kind of phase transition that we want to study now.

7.1.2 An Example from Classical Mechanics

Consider the bottom of an empty wine bottle (Fig. 7.1). If a ball is put at the peak of the dome, the system is symmetrical with respect to rotating the bottle (the potential is rotationally symmetrical with respect to the vertical axis). But below a certain energy (height), the ball will spontaneously break this symmetry and move into a point of lowest energy. The bottle continues to have symmetry, but the system no longer does.

(Footnote 2 continued)
survivor. After graduating in 1959 at Université Libre de Bruxelles, he was nominated full professor at the same University in 1980, where he worked with Brout. Brout had died in 2011, and could not be awarded the Nobel prize.

In this case, what happens to the original symmetry of the equations? It still exists in the sense that, if a symmetry transformation is applied (in this case a rotation around the vertical axis) to the asymmetric solution, another asymmetric solution which is degenerate with the first one is obtained. The symmetry has been "spontaneously broken." A spontaneously broken symmetry has the characteristics, evident in the previous example, that a critical point, i.e., a critical value of some external quantity which can vary (in this case, the height from the bottom, which corresponds to the energy), exists, which determines whether symmetry breaking will occur. Beyond this critical point, frequently called a "false vacuum," the symmetric solution becomes unstable, and the ground state becomes asymmetric—and degenerate.

Spontaneous symmetry breaking appears in many phenomena, for example, in the orientation of domains in a ferromagnet, or in the bending of a rod pushed at its extremes (beyond a certain pressure, the rod must bend in a direction, but all directions are equivalent).

We move to applications to field theory, now.

7.1.3 *Application to Field Theory: Massless Fields Acquire Mass*

We have seen that a Lagrangian density

$$\mathcal{L}_0 = (\partial^\mu \phi^*)(\partial_\mu \phi) - M^2 \phi^* \phi \tag{7.2}$$

with M^2 real and positive and ϕ a complex scalar field

$$\phi(x) = \frac{1}{\sqrt{2}} (\phi_1(x) + i\phi_2(x))$$

describes a free scalar particle of mass M.

Let us now consider a complex scalar field whose dynamics is described by the Lagrangian density:

$$\mathcal{L}_1 = (\partial^\mu \phi^*)(\partial_\mu \phi) - \mu^2 \phi^* \phi - \lambda(\phi^* \phi)^2 \tag{7.3}$$

with $\lambda > 0$; let μ^2 be just a (real) parameter now.

The Lagrangians (7.2) and (7.3) are invariant under the group U(1) describing the global symmetry of rotation:

$$\phi(x) \rightarrow e^{i\theta} \phi(x) . \tag{7.4}$$

Let us try now to find the points of stability for the system (7.3). The potential associated to the Lagrangian is

$$V(\phi) = \mu^2 \phi^* \phi + \lambda(\phi^* \phi)^2 \tag{7.5}$$

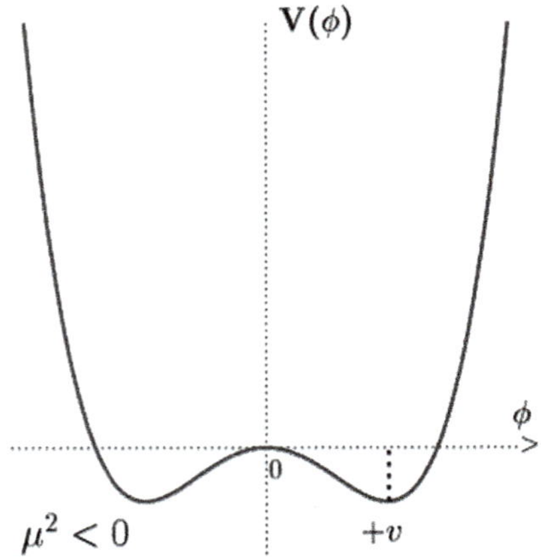

Fig. 7.2 The potential $V(\phi)$ with $\mu^2 < 0$ (cut on a plane containing the V axis)

and, as a function of the component fields,

$$V(\phi) = \frac{1}{2}\mu^2(\phi_1^2 + \phi_2^2) + \frac{1}{4}\lambda(\phi_1^2 + \phi_2^2)^2 . \tag{7.6}$$

The position of the minimum depends on the sign of μ^2:

- for $\mu^2 > 0$, the minimum is at $\phi = 0$;
- for $\mu^2 < 0$, there is a circle of minima at the complex ϕ-plane with radius v

$$v = (-\mu^2/\lambda)^{1/2}$$

(Fig. 7.2). Any point on the circle corresponds to a spontaneous breaking of the symmetry of (7.4). Spontaneous symmetry breaking occurs, if the kinetic energy is smaller than the potential corresponding to the height of the dome. We call v the vacuum expectation value: $|\phi| = v$ is the new vacuum for the system, and the argument, i.e., the angle in the complex plane, can be whatever. The actual minimum is not symmetrical, although the Lagrangian is.

Let us assume, for simplicity, that the actual minimum chosen by the system is at argument 0 (ϕ is real); this assumption does not affect generality. We now define a new coordinate system in which a coordinate σ goes along ϕ_1 and a coordinate ξ is perpendicular to it (Fig. 7.3). Notice that the coordinate ξ does not have influence on the potential, since the latter is constant along the circumference. We examine the

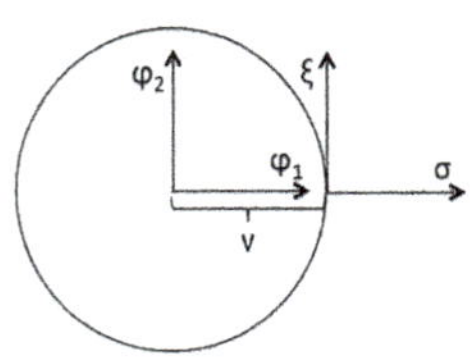

Fig. 7.3 Definition of the new fields σ and ξ

Lagrangian in the vicinity of the minimum. Choosing an appropriate scaling for ξ and σ, one can write

$$\phi = \frac{1}{\sqrt{2}}\left[(v+\sigma)+i\xi\right] \simeq \frac{1}{\sqrt{2}}(v+\sigma)e^{i\xi/v}$$

and thus

$$\partial_\mu \phi = \frac{i}{v}\partial_\mu \xi \phi + \frac{1}{\sqrt{2}} e^{i\xi/v}\partial_\mu \sigma \, .$$

Hence, taking as zero the point of minimum,

$$\mathcal{L}_1 = \frac{1}{2}\partial_\mu \sigma \, \partial^\mu \sigma + \frac{1}{2}\partial_\mu \xi \, \partial^\mu \xi - \frac{1}{2}(-2\mu^2)\sigma^2 + \text{const.} + \mathcal{O}(3)\, . \qquad (7.7)$$

The σ^2 term is a mass term, and thus Lagrangian (7.7) describes a scalar field of mass $m_\sigma^2 = -2\mu^2 = 2\lambda v^2$.

Since there are now nonzero cubic terms in σ, the reflexion symmetry is broken by the ground state: after choosing the actual vacuum, the ground state does not show all the symmetry of the initial Lagrangian (7.3). But a U(1) symmetry operator still exists, which turns one vacuum state into another one along the circumference.

Note that the initial field ϕ had two degrees of freedom. One cannot create or cancel degrees of freedom; in the new system, one degree of freedom is taken by the field σ, while the second is now absorbed by the massless field ξ, which moves the potential along the "cul de bouteille." The appearance of massless particles is an aspect of the *Goldstone theorem,* which we shall not demonstrate here. The Goldstone theorem states that, if a Lagrangian is invariant under a group of transformations $\mathcal{G}$ with n generators, and if there is a spontaneous symmetry breaking such that the new vacuum is invariant only under a group of transformations $\mathcal{G}' \subset \mathcal{G}$ with $m < n$ generators, then a number $(n-m)$ of massless scalar fields appear. These are called Goldstone bosons.

In the previous example, the Lagrangian had a U(1) symmetry (one generator). After the SSB, the system had no symmetry. One Golstone field ξ appeared.

7.1.4 *From SSB to the Higgs Mechanism: Gauge Symmetries and the Mass of Gauge Bosons*

We have seen that spontaneous symmetry breaking can give a mass to a field otherwise massless, and as a consequence some additional massless fields appear—the Goldstone fields.

In this section, we want to study the consequences of spontaneous symmetry breaking in the presence of a local gauge symmetry, as seen from the case $\mu^2 < 0$ in the potential (7.5). We shall see that (some of the) gauge bosons will become massive, and one or more additional massive scalar field(s) will appear—the Higgs field(s). The Goldstone bosons will disappear as an effect of the gauge invariance: this is called the Higgs mechanism.

We consider the case of a local U(1) symmetry: a complex scalar field coupled to itself and to an electromagnetic field A_μ

$$\mathcal{L} = -\frac{1}{4} F_{\mu\nu} F^{\mu\nu} + (D_\mu \phi)^* (D^\mu \phi) - \mu^2 \phi^* \phi - \lambda \, (\phi^* \phi)^2 \tag{7.8}$$

with the covariant derivative $D_\mu = \partial_\mu + ieA_\mu$. If $\mu^2 > 0$, this Lagrangian is associated to electrodynamics between scalar particles; and it is invariant with respect to local gauge transformations

$$\phi(x) \to \phi'(x) = e^{i\epsilon(x)} \, \phi(x)$$
$$A_\mu(x) \to A'_\mu(x) = A_\mu(x) - \frac{1}{e} \, \partial_\mu \epsilon(x) \, .$$

If $\mu^2 < 0$, we shall have spontaneous symmetry breaking. The ground state will be

$$\langle \phi \rangle = v = \sqrt{-\frac{\mu^2}{\lambda}} > 0 \, ; \tag{7.9}$$

and as in the previous section, we parametrize the field ϕ starting from the vacuum as

$$\phi(x) = \frac{1}{\sqrt{2}} (v + \sigma(x)) e^{i\xi(x)/v} \, . \tag{7.10}$$

We have seen in the previous section that the field $\xi(x)$ was massless and associated to the displacement between the degenerate ground states. But here the ground states are equivalent because of the gauge symmetry. Let us examine the consequences of this. We can rewrite Eq. (7.8) as

$$\begin{aligned} \mathcal{L} = -\frac{1}{4} \, F_{\mu\nu} F^{\mu\nu} + \frac{1}{2} \, \partial_\mu \sigma \, \partial^\mu \sigma + \frac{1}{2} \, \partial_\mu \xi \, \partial^\mu \xi \, + \frac{1}{2} e^2 v^2 A_\mu A^\mu \\ - veA_\mu \partial^\mu \xi + \mu^2 \sigma^2 + \mathcal{O}(3) \, . \end{aligned} \tag{7.11}$$

Thus the σ field acquires a mass $m_\sigma^2 = -2\mu^2$; there are in addition mixed terms between A_μ and ξ. Let us make a gauge transformation

$$\epsilon(x) = -\frac{\xi(x)}{v} \, ; \tag{7.12}$$

thus

$$\phi(x) \to \phi'(x) = e^{-i\xi(x)/v} \phi(x) = \frac{1}{\sqrt{2}} (v + \sigma(x)) \tag{7.13}$$

$$A_\mu(x) \to A'_\mu(x) + \frac{1}{ev}\partial_\mu \xi \tag{7.14}$$

and since the Lagrangian is invariant for this transformation, we must have

$$\begin{aligned} \mathcal{L}(\phi, A_\mu) &= \mathcal{L}(\phi', A'_\mu) \\ &= \frac{1}{2}\left[(\partial_\mu - ieA'_\mu)(v+\sigma)\right]\left[(\partial^\mu + ieA'^\mu)(v+\sigma)\right] \\ &\quad - \frac{1}{2}\mu^2(v+\sigma)^2 - \frac{1}{4}\lambda(v+\sigma)^4 - \frac{1}{4}F'_{\mu\nu}F'^{\mu\nu} . \end{aligned} \tag{7.15}$$

The Lagrangian in Eq. (7.15) can be also written as

$$\mathcal{L} = -\frac{1}{4}F'_{\mu\nu}F'^{\mu\nu} + \frac{1}{2}\partial_\mu\sigma\,\partial^\mu\sigma + \frac{1}{2}e^2v^2A'_\mu A'^\mu - \lambda v^2\sigma^2 + \mathcal{O}(3) , \tag{7.16}$$

and now it is clear that both σ and A_μ have acquired mass:

$$m_\sigma = \sqrt{2\lambda v^2} = \sqrt{-2\mu^2} \tag{7.17}$$

$$m_A = ev . \tag{7.18}$$

Notice that the field ξ is disappeared. This is called the *Higgs mechanism*; the massive (scalar) σ field is called the Higgs field. In a gauge theory, the Higgs field "eats" the Goldstone field.

Notice that the number of degrees of freedom of the theory did not change: now the gauge field A_μ is massive (three degrees of freedom), and the field σ has one degree of freedom (a total of four degrees of freedom). Before the SSB, the field had 2 degrees of freedom, and the massless gauge field an additional two—again, a total of four.

The exercise we just did is not appropriate to model electromagnetism—after all, the photon A_μ is massless to the best of our knowledge. However, it shows completely the technique associated to the Higgs mechanism.

We shall now apply this mechanism to explain the masses of the vectors of the weak interaction, the Z, and the $W^\pm$; but first, let us find the most appropriate description for the weak interaction, which is naturally linked to the electromagnetic one.

7.2 Electroweak Unification

The weak and electromagnetic interactions, although different, have some common properties which can be exploited for a more satisfactory—and "economical" description.

Let us start from an example, taken from experimental data. The $\Sigma^+(1189)$ baryon, a uus state, decays into $p\pi^0$ via a strangeness-changing weak decay (the basic transition at the quark level being $s \to ud\bar{u}$), and it has a lifetime of about 10^{-10} s, while the $\Sigma^0(1192)$, a uds state decaying electromagnetically into $\Lambda\gamma$, has a lifetime of the order of 10^{-19} s, the basic transition being $u \to u\gamma$. The phase space for both decays is quite similar, and thus the difference in lifetime must be due to the difference of the couplings for the two interaction, being the amplitude (and thus the inverse of the lifetime) proportional to the square of the coupling. The comparison shows that the weak coupling is smaller by a factor of order of $\sim 10^{-4}$ with respect to the electromagnetic coupling. Although weak interactions take place between all quarks and leptons, the weak interaction is typically hidden by the much greater strong and electromagnetic interactions, unless these are forbidden by some conservation rule. Observable weak interactions involve either neutrinos or quarks with a flavor change—flavor change being forbidden in strong and electromagnetic interactions, since photons and gluons do not carry flavor.

The factor 10^{-4} is very interesting, and suggests that the weak interactions might be weak because they are mediated by gauge fields, $W^\pm$ and Z, which are very massive and hence give rise to interactions of very short range. The strength of the interaction can be written as

$$f(q^2) = \frac{g_W^2}{q^2 + M^2},$$

where M is the mass of the W or Z boson.

In the low-q^2 limit, the interaction is point-like, and the strength is given by the Fermi coupling $G_F \simeq 10^{-5}$ GeV^{-2}. The picture sketched above looks indeed consistent with the hypothesis that $g_W \sim e$ (we shall obtain a quantitative relation at the end of this Section). In fact

$$G_F \simeq \frac{e^2}{M_Z^2}.$$

Glashow, Weinberg, and Salam[3] proposed in the 1960's—twenty years before the experimental discovery of the W and Z bosons—that the coupling of the W and Z to leptons and quarks is closely related to that of the photon; the weak and electromagnetic interactions are thus unified into an electroweak interaction. Mathematically, this unification is accomplished under a SU(2) $\otimes$ U(1) gauge group.

[3] Sheldon Lee Glashow (New York City 1932) shared with Steven Weinberg (New York City 1933) and Abdus Salam (Jhang, Pakistan, 1926 - Oxford 1996) the Nobel Prize for Physics in 1979 "for their complementary efforts in formulating the electroweak theory. The unity of electromagnetism and the weak force can be explained with this theory." Glashow was the son of Jewish immigrants from Russia. He and Weinberg were members of the same classes at the Bronx High School of Science, New York City (1950), and Cornell University (1954); then Glashow became full professor in Princeton, and Weinberg in Harvard. Salam graduated in Cambridge, where he became full professor of mathematics in 1554, moving then to Trieste.

A problem is apparently given by the mass of the vector bosons: the photon is massless, while the W and Z bosons are highly massive. Indeed an appropriate spontaneous symmetry breaking of the electroweak Lagrangian explains the masses of the $W^{\pm}$ and of the Z keeping the photon massless, and predicts the existence of a Higgs boson, which is called the standard model Higgs boson. The same Higgs boson can account for the masses of fermions. We shall see now how this unification is possible.

7.2.1 The Formalism of the Electroweak Theory

We used the symmetry group SU(2) to model weak interactions, while U(1) is the symmetry of QED. The natural space for a unified electroweak interaction appears thus to be SU(2) ⊗ U(1)—this is what the Glashow–Weinberg–Salam electroweak theory assumed at the end of the 1960s.

Let us call W^1, W^2, and W^0 the three gauge fields of SU(2). We call $W^a_{\mu\nu}$ ($a = 1, \ldots, 3$) the field tensors of SU(2) and $B_{\mu\nu}$ the field tensor of U(1). Notice that $B_{\mu\nu}$ is not equal to $F_{\mu\nu}$, as W^0 is not the Z field: since we use a tensor product of the two spaces, in general, the neutral field B can mix to the neutral field W^0, and the photon and Z states are a linear combination of the two.

The Lagrangian of the electroweak interaction needs to accommodate some experimental facts, which we have discussed in Chap. 6:

- Only the left-handed (right-handed) (anti)fermion chiralities participate in weak transitions—therefore, the interaction violates parity P and charge conjugation C; however, the combined transformation CP is still a good symmetry.
- The $W^{\pm}$ bosons couple to the left-handed fermionic doublets, where the electric charges of the two fermion partners differ in one unit. This leads to the following decay channels for the W^-:

$$W^- \rightarrow e^-\bar{\nu}_e\,,\ \mu^-\bar{\nu}_\mu\,,\ \tau^-\bar{\nu}_\tau\,,\ d'\bar{u}\,,\ s'\bar{c}\,,\ b'\bar{t} \tag{7.19}$$

  the latest being possible only as a virtual decay, since $m_t > m_W$.
  The doublet partners of up, charm, and top are mixtures of the three charge $-\frac{1}{3}$ quarks:

$$\begin{pmatrix} d' \\ s' \\ b' \end{pmatrix} = V_{CKM} \begin{pmatrix} d \\ s \\ b \end{pmatrix} . \tag{7.20}$$

  Thus, the weak eigenstates $d'\,, s'\,, b'$ are different from the mass eigenstates $d\,,\ s\,,\ b$. They are related through the 3×3 unitary matrix V_{CKM}, which characterizes flavor-mixing phenomena.
- The neutral carriers of the electroweak interactions have fermionic couplings with the following properties:

- All interacting vertices conserve flavor. Both the γ and the Z couple to a fermion and its own antifermion, i.e., $\gamma f \bar{f}$ and $Z f \bar{f}$.
- The interactions depend on the fermion electric charge Q_f. Neutrinos do not have electromagnetic interactions ($Q_\nu = 0$), but they have a nonzero coupling to the Z boson.
- Photons have the same interaction for both fermion chiralities.

• The strength of the interaction is universal, and lepton number is conserved.

We are ready now to draft the electroweak theory.

To describe weak interactions, the left-handed fermions should appear in doublets, and the right-handed fermions in singlets, and we would like to have massive gauge bosons $W^\pm$ and Z in addition to the photon. The simplest group with doublet representations having three generators is SU(2). The inclusion of the electromagnetic interactions implies an additional U(1) group. Hence, the symmetry group to consider is then

$$G \equiv \mathrm{SU}(2)_L \otimes \mathrm{U}(1)_Y \,, \tag{7.21}$$

where L refers to left-handed fields (this will represent the weak sector). We shall specify later the meaning of the subscript Y.

Let us first analyze the SU(2) part of the Lagrangian.

The $\mathrm{SU}(2)_L$ **part**. We have seen that the $W^\pm$ couple to the left chirality of fermionic doublets—what was called a $(V - A)$ coupling in the "old" scheme (Sect. 6.3.3). Let us start for simplicity our "modern" description from a leptonic doublet

$$\chi_L = \begin{pmatrix} \nu \\ e \end{pmatrix}_L \,.$$

Charged currents exist, coupling the members of the doublet:

$$j_\mu^+ = \bar{\nu}\gamma_\mu \left(\frac{1}{2}(1-\gamma_5)\right) e = \bar{\nu}_L \gamma_\mu e_L \tag{7.22}$$

$$j_\mu^- = \bar{e}\gamma_\mu \left(\frac{1}{2}(1-\gamma_5)\right) \nu = \bar{e}_L \gamma_\mu \nu_L \,. \tag{7.23}$$

These two currents are associated, for example, with weak decays of muons and neutrons. Notice that

$$\left(\frac{1}{2}(1-\gamma_5)\right)\left(\frac{1}{2}(1-\gamma_5)\right) = \left(\frac{1}{2}(1-\gamma_5)\right) ; \tag{7.24}$$

$$\left(\frac{1}{2}(1-\gamma_5)\right)\left(\frac{1}{2}(1+\gamma_5)\right) = 0 \,. \tag{7.25}$$

This should be evident by the physical meaning of these projectors; we leave for the Exercises a formal demonstration of these properties.

In analogy to the case of hadronic isospin, where the proton and neutron are considered as the two isospin eigenstates of the nucleon, we define a *weak isospin* doublet structure ($T = 1/2$)

$$\chi_L = \begin{pmatrix} \nu \\ e \end{pmatrix}_L \begin{array}{l} T_3 = +1/2 \\ T_3 = -1/2 \end{array}, \tag{7.26}$$

with raising and lowering operators between the two components of the doublet

$$\tau_\pm = \frac{1}{2}(\tau_1 \pm i\tau_2) \tag{7.27}$$

where the τ_i are the Pauli matrices.

The same formalism applies to a generic quark doublet, for example

$$\chi_L = \begin{pmatrix} u \\ d' \end{pmatrix}_L \begin{array}{l} T_3 = +1/2 \\ T_3 = -1/2 \end{array}. \tag{7.28}$$

With this formalism, we can write the charged currents as

$$j_\mu^+ = \bar{\chi}_L \gamma_\mu \tau_+ \chi_L \tag{7.29}$$

$$j_\mu^- = \bar{\chi}_L \gamma_\mu \tau_- \chi_L \,. \tag{7.30}$$

When imposing the SU(2) symmetry, one has two vector fields W^1 and W^2 corresponding to the Pauli matrices τ_1 and τ_2. Notice that they do not correspond necessarily to "good" particles, since, for example, they are not necessarily eigenstates of the electric charge operator. However, we have seen that they can be combined to physical states corresponding to the charged currents $W^\pm$ (Eq. 7.27):

$$W^\pm = \sqrt{\frac{1}{2}}(W^1 \pm iW^2)\,. \tag{7.31}$$

A third vector field W^0 associated to the third generator τ_3 corresponds to a neutral transition (analogous to the π^0 in the case of the isospin studied for strong interactions):

$$j_\mu^3 = \bar{\chi}_L \gamma_\mu \left(\frac{1}{2}\tau_3\right) \chi_L \,. \tag{7.32}$$

We have finally a triplet of currents

$$j_\mu^i = \bar{\chi}_L \gamma_\mu \left(\frac{1}{2}\tau_i\right) \chi_L \tag{7.33}$$

with algebra

$$[\tau_i, \tau_j] = i\epsilon_{ijk}\tau_k ; \tag{7.34}$$

from this, we can construct the Lagrangian according to the recipes in Sect. 6.4.1. Before doing so, let us examine the U(1) part of the Lagrangian.

The $\mathrm{U}(1)_Y$ **part**. The electromagnetic current

$$j^{em}_\mu = \bar{e}\gamma_\mu e = \bar{e}_L\gamma_\mu e_L + \bar{e}_R\gamma_\mu e_R$$

is invariant under $\mathrm{U}(1)_Q$, the gauge group of QED associated to the electromagnetic charge. It is, however, not invariant under $\mathrm{SU}(2)_L$: it contains e_L instead of χ_L.

The neutral isospin

$$j^3_\mu = \bar{\chi}_L\gamma_\mu\left(\frac{1}{2}\tau_3\right)\chi_L = \bar{\nu}_L\left(\frac{1}{2}\gamma_\mu\right)\nu_L - \bar{e}_L\left(\frac{1}{2}\gamma_\mu\right)e_L \tag{7.35}$$

couples only left-handed particles, while we know that neutral current involves both chiralities.

To have a consistent picture, we must construct a $\mathrm{SU}(2)_L$-invariant U(1) current. We define a *hypercharge*

$$Y = 2(Q - T_3) . \tag{7.36}$$

We can thus write

$$j^Y_\mu = 2j^{em}_\mu - 2j^3_\mu = -2\bar{e}_R\gamma_\mu e_R - \bar{\chi}_L\gamma_\mu\chi_L . \tag{7.37}$$

The last expression is invariant with respect to $\mathrm{SU}(2)_L$ (e_R is a weak hypercharge singlet).

Construction of the Electroweak Lagrangian. The part of the Lagrangian related to the interaction between gauge fields and fermion fields can now be written as

$$\mathcal{L}_{int} = -igj^a_\mu W^{a\mu} - i\frac{g'}{2}j^Y_\mu B^\mu , \tag{7.38}$$

while the part related to the gauge field is

$$\mathcal{L}_g = -\frac{1}{4}W^{\mu\nu}_a W^a_{\mu\nu} - \frac{1}{4}B^{\mu\nu}B_{\mu\nu} , \tag{7.39}$$

where $W^{a\mu\nu}(a = 1, 2, 3)$ and $B^{\mu\nu}$ are the field strength tensors for the weak isospin and weak hypercharge fields. In the above, we have called g the strength of the SU(2) coupling, and g' the strength of the hypercharge coupling. The field tensors above can be explicitly written as

$$W^a_{\mu\nu} = \partial_\mu W^a_\nu - \partial_\nu W^a_\mu - g\, f^{bca} W^b_\mu W^c_\nu \tag{7.40}$$

$$B_{\mu\nu} = \partial_\mu B_\nu - \partial_\nu B_\mu \,. \tag{7.41}$$

Finally, including the kinetic part, the Lagrangian can be written as

$$\mathcal{L}_0 = \sum_{\text{families}} \overline{\chi}_f (i\gamma^\mu D_\mu)\chi_f - \frac{1}{4} W^a_{\mu\nu} W^{a\mu\nu} - \frac{1}{4} B_{\mu\nu} B^{\mu\nu} \,, \tag{7.42}$$

with

$$D_\mu = \partial_\mu + igW^a_\mu \frac{\tau^a}{2} + ig'YB_\mu \,. \tag{7.43}$$

At this point, the four gauge bosons W^a and B are massless. But we know that the Higgs mechanism can solve this problem.

In addition, we did not introduce fermion masses, yet. When discussing electromagnetism and QCD as gauge theories, we put fermion masses "by hand" in the Lagrangian. This is not possible here, since an explicit mass term would break the SU(2) symmetry. A mass term $-m_f \overline{\chi}_f \chi_f$ for each fermion f in the Lagrangian would give, for the electron for instance,

$$- m_e \bar{e} e = -m_e \bar{e} \left(\frac{1}{2}(1-\gamma_5) + \frac{1}{2}(1+\gamma_5) \right) e = -m_e(\bar{e}_R e_L + \bar{e}_L e_R) \tag{7.44}$$

which is non-invariant under the isospin symmetry transformations, since e_L is a member of an $SU(2)_L$ doublet while e_R is a singlet.

We shall see that fermion masses can come "for free" from the Higgs mechanism.

7.2.2 *The Higgs Mechanism in the Electroweak Theory, and the Mass of the Electroweak Bosons*

In Sect. 7.1.4, the Higgs mechanism was used to generate a mass for the gauge boson corresponding to a U(1) local gauge symmetry. In this case, three Goldstone bosons will be required (we need them to give mass to W^+, W^-, and Z). In addition, after symmetry breaking, there will be (at least) one massive scalar particle corresponding to the field excitations in the direction picked out by the choice of the physical vacuum.

The simplest Higgs field, which has the necessary four degrees of freedom, consists of two complex scalar fields, placed in a weak isospin doublet. One of the scalar fields will be chosen to be charged with charge +1 and the other to be neutral. The hypercharge of the doublet components will thus be $Y = 2(Q - T_3) = 1$. The Higgs doublet is then written as

$$\phi = \begin{pmatrix} \phi^+ \\ \phi^0 \end{pmatrix} = \frac{1}{\sqrt{2}} \begin{pmatrix} \phi_1 + i\phi_2 \\ \phi_3 + i\phi_4 \end{pmatrix} . \tag{7.45}$$

We choose as a vacuum the point $\phi_1 = \phi_2 = \phi_4 = 0$ and $\phi_3 = v$, and we expand

$$\phi(x) = \frac{1}{\sqrt{2}} \begin{pmatrix} 0 \\ v + h(x) \end{pmatrix} .$$

To the SM Lagrangian discussed in the previous subsection (Eq. 7.42), we need to add the Higgs potential $V(\phi)$. We obtain, in a compact form, for the free Higgs Lagrangian:

$$\begin{aligned} \mathcal{L}_{Higgs} &= (D^{\mu}\phi)^{\dagger}(D_{\mu}\phi) - V(\phi^{\dagger}\phi) \qquad (7.46) \\ &= (D^{\mu}\phi)^{\dagger}(D_{\mu}\phi) - \mu^2\phi^{\dagger}\phi - \lambda(\phi^{\dagger}\phi)^2 . \end{aligned}$$

with the covariant derivative D_{μ} given by Eq. 7.43.

After SSB, this Lagrangian can be approximated around the minimum as

$$\begin{aligned} \mathcal{L}_{free} &= \text{constant} + \text{kinetic terms} + \\ &+ \frac{1}{2}\left(-2\mu^2\right) h^2 \qquad (7.47) \\ &+ \frac{1}{2}\left(\frac{1}{4}\, g^2 v^2\right) W^1_{\mu}\, W^{1\mu} + \frac{1}{2}\left(\frac{1}{4}\, g^2 v^2\right) W^2_{\mu}\, W^{2\mu} \qquad (7.48) \\ &+ \frac{1}{8}\, v^2 \begin{pmatrix} W^{3\mu} & B^{\mu} \end{pmatrix} \begin{pmatrix} g^2 & -gg'Y \\ -gg'Y & g'^2 \end{pmatrix} \begin{pmatrix} W^3_{\mu} \\ B_{\mu} \end{pmatrix} . \qquad (7.49) \\ &+ \mathcal{O}(3) , \end{aligned}$$

where as in the case of Sect. 7.1.4, we have introduced the new field around the point of minimum (we call it h instead of σ).

- As usual in the SSB, the h field acquires mass; we shall call the corresponding particle H. This is the famous standard model Higgs boson, and its mass is

$$m_H = \sqrt{-2\mu^2} = \sqrt{2\lambda}\, v . \tag{7.50}$$

- We now analyze the term (7.48). We have two massive charged bosons W^1 and W^2 with the same mass $gv/2$. We have seen, however, that physical states of integer charge ± 1 can be constructed by a linear combination of them (Eq. 7.31):

$$W^{\pm} = \sqrt{\frac{1}{2}}(W^1 \pm i W^2) .$$

The mass is as well

$$M_{W^{\pm}} = \frac{1}{2}\, gv\,, \tag{7.51}$$

and these states correspond naturally to the charged current vectors.

- Finally, let us analyze the term (7.49).

Here, the fields W^3 and B couple through a non-diagonal matrix; they thus are not mass eigenstates. The physical fields can be obtained by an appropriate rotation which diagonalizes the mass matrix

$$M = \begin{pmatrix} g^2 & -gg'Y \\ -gg'Y & g'^2 \end{pmatrix}.$$

For $Y = \pm 1$ (we recall our choice $Y = 1$), the determinant of the matrix is 0, and when we shall diagonalize, one of the two eigenstates will be massless. If we introduce the fields A_μ and Z_μ defined as

$$A_\mu = \sin\theta_W\, W^0_\mu + \cos\theta_W\, B_\mu \tag{7.52}$$

$$Z_\mu = \cos\theta_W\, W^0_\mu - \sin\theta_W\, B_\mu\,, \tag{7.53}$$

where the angle θ_W, called the Weinberg angle, parametrizes the electroweak mixing:

$$\tan\theta_W = \frac{g'}{g}\,, \tag{7.54}$$

the term (7.49) becomes

$$\frac{1}{8}\, v^2 \begin{pmatrix} A^\mu & Z^\mu \end{pmatrix} \begin{pmatrix} 0 & 0 \\ 0 & g^2 + g'^2 \end{pmatrix} \begin{pmatrix} A_\mu \\ Z_\mu \end{pmatrix}. \tag{7.55}$$

A_μ is then massless (we can identify it with the photon). Note that

$$M_Z = \frac{1}{2}\, v\sqrt{g^2 + g'^2}\,, \tag{7.56}$$

and thus

$$M_W = M_Z \cos\theta_W\,. \tag{7.57}$$

From the above expression, using the measured masses of the W and Z bosons, we can get an estimate of the Weinberg angle:

$$\sin^2\theta_W \simeq 1 - \left(\frac{M_W}{M_Z}\right)^2 \simeq 1 - \left(\frac{80.385\,\text{GeV}}{91.188\,\text{GeV}}\right)^2 \simeq 0.22\,. \tag{7.58}$$

Note the use of $\simeq$ symbols: this result has been obtained only at tree level of the electroweak theory, while in the determination of the actual masses of the W and Z, boson higher-order terms enter, related for example to QCD loops. Higher-order processes ("radiative corrections") should be taken into account to obtain a fully consistent picture, see later; the current "best fit" value of the Weinberg angle provides

$$\sin\theta_W = 0.2318 \pm 0.0006 \,. \tag{7.59}$$

Electric charge is obviously conserved, since it can be associated to a generator which is the linear combination of the isospin generator and of the generator of hypercharge. For eigenvectors of SU(2) ⊗ U(1), one has

$$Q = Y + \frac{T_3}{2} \,. \tag{7.60}$$

Thus the covariant derivative can be written as

$$\begin{aligned} D_\mu &= \left(\partial_\mu + igW_\mu^a \frac{\tau^a}{2} + ig'\frac{1}{2}B_\mu\right) \\ &= \partial_\mu + i\frac{g}{\sqrt{2}}\, W_\mu^+ \tau^+ + \frac{g}{\sqrt{2}}\, W^- \tau^- + ig\sin\theta_W Q A_\mu + i\frac{g}{\cos\theta_W}\left(\frac{\tau_3}{2} - \sin^2\theta_W Q\right) Z_\mu \end{aligned} \tag{7.61}$$

and thus

$$g\sin\theta_W = e \,. \tag{7.62}$$

The above relation holds also at tree level only: we shall see in Sect. 7.4 that radiative corrections can influence at a measurable level the actual values of the observables in the standard model.

7.2.3 The Fermion Masses

Up to now the fermion fields in the theory are massless, and we have seen (Eq. 7.44) that we cannot insert them by hand in the Lagrangian as we did in the case of QED and QCD. A simple way to explain the fermion masses consistent with the electroweak interaction is to ascribe such a mass to the Higgs mechanism, again: masses appear after the SSB.

The problem can be solved by imposing in the Lagrangian coupling of the Higgs doublet to the fermions by means of gauge invariant terms like (for the electron)

$$\mathcal{L}_{eeh} = -\frac{\lambda_e}{\sqrt{2}}\left[(\bar{\nu}_e, \bar{e})_L \begin{pmatrix} 0 \\ v+h \end{pmatrix} e_R + \bar{e}_R (0, v+h) \begin{pmatrix} \nu \\ e \end{pmatrix}_L\right] . \tag{7.63}$$

This generates fermion mass terms and Higgs-fermion interaction terms:

$$\mathcal{L}_{eeh} = -\frac{\lambda_e v}{\sqrt{2}}\bar{e}e \;-\; \frac{\lambda_e}{\sqrt{2}}\bar{e}e\,h\,. \tag{7.64}$$

Since the symmetry breaking term λ_f for each fermion field is unknown, the masses of fermions in the theory are free parameters, and must be determined by experiment. Setting $\lambda_f = \sqrt{2}m_f/v$, the part of the Lagrangian describing the fermion masses and the fermion-Higgs interaction for each fermion is

$$\mathcal{L}_{ffh} = -m_f\,\bar{f}f \;-\; \frac{m_f}{v}\,\bar{f}f\,h\,. \tag{7.65}$$

Notice that the coupling to the Higgs is proportional to the mass of the fermion: this is a strong prediction, to be verified by experiment.

We stress the fact that we need just one Higgs field to explain all massive particles of the standard model: the weak vector bosons $W^{\pm}$, Z, fermions, and the Higgs boson itself. The electromagnetic symmetry and the SU(3) color symmetry both remain unbroken—the former being an accidental symmetry of $\mathrm{SU}(2)_L \otimes \mathrm{U}(1)_Y$. This does not exclude, however, that additional Higgs fields exist—it is just a matter of economy of the theory or, if you prefer, it is just a consequence of the Occam's razor, to the best of our present knowledge.

7.2.4 Interactions Between Fermions and Gauge Bosons

Using Eq. (7.61), we can write the interaction terms between the gauge bosons and the fermions (we use a lepton doublet as an example, but the result is general) as

$$\begin{aligned}\mathcal{L}_{int} = &-\frac{g}{2\sqrt{2}}\,\overline{\nu}_e\gamma^{\mu}(1-\gamma_5)e\,W_{\mu}^{+} - \frac{g}{2\sqrt{2}}\,\overline{e}\gamma^{\mu}(1-\gamma_5)\nu_e\,W_{\mu}^{-}\\ &-\frac{g}{4\cos\theta_W}\left[\overline{\nu}_e\,\gamma^{\mu}(1-\gamma_5)\nu_e - \overline{e}\gamma^{\mu}(1-4\sin^2\theta_W - \gamma_5)e\right]Z_{\mu}\\ &-(-e)\,\overline{e}\gamma^{\mu}e\,A_{\mu}\,.\end{aligned} \tag{7.66}$$

The term proportional A_{μ} is the QED interaction.

Usually, the Z interaction in Eq. (7.66) is written (for a generic fermion f) as

$$\begin{aligned}\mathcal{L}_{int}^{Z} = &-\frac{g}{\cos\theta_W}\left[\overline{\nu}_e\gamma^{\mu}(g_V^{\nu} - g_A^{\nu}\gamma_5)\nu_e + \overline{e}\gamma^{\mu}(g_V^{e} - g_A^{e}\gamma_5)e\right]Z^{\mu}\\ &+\text{other fermions}\\ = &-\frac{g}{\cos\theta_W}\sum_f \overline{\chi}_f\gamma^{\mu}(g_V^f - g_A^f\gamma_5)\chi_f Z_{\mu}\end{aligned} \tag{7.67}$$

from which we define the couplings

$$g_V^f = \frac{1}{2}T_3^f - Q^f \sin^2\theta_W \, ; \qquad g_A^f = \frac{1}{2}T_3^f \, . \tag{7.68}$$

The Z interaction can also be written considering the left and right helicity states of the fermions. Indeed, for a generic fermions f,

$$\overline{\psi}_f \gamma^\mu (g_V^f - g_A^f \gamma_5)\psi_f = \overline{\psi}_f \gamma^\mu \left[\frac{1}{2}(g_V^f - g_A^f)(1-\gamma_5) + \frac{1}{2}(g_V^f - g_A^f)(1+\gamma_5)\right]\psi_f \tag{7.69}$$

$$= \overline{\psi}_{fL}\gamma^\mu g_L \psi_{fL} + \overline{\psi}_{fR}\gamma^\mu g_R \psi_{fR} \tag{7.70}$$

where the left and right couplings g_L and g_R are thus given by

$$g_L = \frac{1}{2}(g_V + g_A) \tag{7.71}$$

$$g_R = \frac{1}{2}(g_V - g_A)\, . \tag{7.72}$$

In the case of neutrinos, $Q = 0$, $g_V = g_A$ and thus $g_R = 0$. The right neutrino has then no interaction with the Z and as, by construction, it has also no interactions with the γ, $W^\pm$, and gluons. The right neutrino is therefore, if it exists, *sterile*.

On the contrary, for electrical charged fermions, $g_V \neq g_A$ and thus the Z boson couples both with left and right helicity states although with different strengths ($g_L \neq g_R \neq 0$).

Parity is also violated in the Z interactions.

These results are only valid in the one-family approximation. When extending to three families, there is a complication: in particular, the current eigenstates for quarks q' are not identical to the mass eigenstates q. If we start by u-type quarks being mass eigenstates, in the down-type quark sector, the two sets are connected by a unitary transformation

$$(d', s', b') = V_{\mathrm{CKM}}(d, s, b)\, . \tag{7.73}$$

Let us compute as an example the differential cross section for processes involving electroweak currents. We should not discuss the determination of the absolute value, but just the dependence on the flavor and on the angle.

Let us examine the fermion antifermion, $f\bar{f}$, production in e^+e^- annihilations.

At a center-of-mass energy smaller than the Z mass, the photon coupling will dominate the process. The branching fractions will be dominated by photon exchange, and thus proportional to Q_f^2 (being zero in particular for neutrinos).

Close to the Z mass, the process will be dominated by decays $Z \to f\bar{f}$ and the amplitude will be proportional to $(g_V^f + g_A^f)$ and $(g_V^f g_A^f)$, respectively, for left and right fermions. The width into $f\bar{f}$ will be then proportional to

$$\left[(g_V^f + g_A^f)^2 + (g_V^f - g_A^f)^2\right] = g_V^{f\,2} + g_A^{f\,2}. \tag{7.74}$$

Hence, one has

$$\Gamma_{f\bar{f}} \simeq \frac{M_Z}{12\pi}\left(\frac{g}{\cos\theta_W}\right)^2\left[g_V^{f\,2} + g_A^{f\,2}\right]. \tag{7.75}$$

Expressing the result in terms of the Fermi constant

$$\frac{G_F}{\sqrt{2}} = \frac{g^2}{8M_W^2} = \left(\frac{g}{\cos\theta_W}\right)^2\frac{1}{8M_Z^2} \tag{7.76}$$

one has

$$\Gamma = \frac{2G_F M_Z^3}{3\sqrt{2}\pi}\left[g_V^{f\,2} + g_A^{f\,2}\right]. \tag{7.77}$$

For example, for $\mu^+\mu^-$ pairs,

$$\Gamma(Z \to \mu^+\mu^-) \simeq 83.4\,\mathrm{MeV}. \tag{7.78}$$

As a consequence of parity violation in electroweak interactions, a forward-backward asymmetry will characterize the Z decays into $f\bar{f}$. The forward-backward asymmetry for the decay of a Z boson into a fermion pair is[4] in the core of unpolarized electron/positron beams,

$$A_{FB}^f \equiv \left[\int_0^{+1}\frac{d\sigma}{d\cos\theta} - \int_{-1}^{0}\frac{d\sigma}{d\cos\theta}\right]/\sigma_{tot} \overset{\sqrt{s}=M_Z}{\simeq} \frac{3}{4}A_e A_f \tag{7.79}$$

where the combinations A_f are given, in terms of the vector and axial vector couplings of the fermion f to the Z boson, by

$$A_f = \frac{2g_V^f g_A^f}{g_V^{f2} + g_A^{f2}}, \tag{7.80}$$

and A_e is the corresponding combination for the specific case of the electron.

The tree-level expressions discussed above give results which are correct at the percent level—in the case of b-quark final states, additional mass effects $\mathcal{O}(4m_b^2/M_Z^2)$, also ~0.01, have to be taken into account. For the production of e^+e^- final states, the t-channel gauge boson exchange contributions have to be included (this process allows to determine the absolute luminosity at e^+e^- colliders, making it particularly important), and it is dominant at low angles, the cross section being proportional to $\sin^3\theta$. However, one needs to include the one-loop radiative corrections so that the

[4]For a deduction, see for instance Chap. 16.2 of Reference [F7.2] in the "Further readings."

Z properties can be described accurately, and possibly some important higher-order effects, which will be shortly discussed later.

7.2.5 Self-interactions of Gauge Bosons

Self-couplings among the gauge bosons are present in the SM as a consequence of the non-abelian nature of the $\mathrm{SU(2)_L} \otimes \mathrm{U(1)_Y}$ symmetry. These couplings are dictated by the structure of the symmetry group as discussed before and, for instance, the triple self-couplings among the W and the $V = \gamma, Z$ bosons are given by

$$\mathcal{L}_{WWV} = i g_{WWV} \left[W^{\dagger}_{\mu\nu} W^{\mu} B^{\nu} - W^{\dagger}_{\mu} B_{\nu} W^{\mu\nu} + W^{\dagger}_{\mu} W_{\nu} B^{\mu\nu} \right] \tag{7.81}$$

with $g_{WW\gamma} = e$ and $g_{WWZ} = e / \tan\theta_W$.

7.2.6 Feynman Diagram Rules for the Electroweak Interaction

We have already shown how to compute the invariant amplitude $\mathcal{M}$ for a scalars fields in Sect. 6.2.7. We give here only the Feynman rules for the propagators (Fig. 7.4) and vertices (Fig. 7.5) of the standard model that we can use in our calculations—or our estimates, since the calculation of the complete amplitude including spin effects can be very lengthy and tedious. We follow here Ref. [F7.5] in the "Further readings"; a complete treatment of the calculation of amplitudes from the Feynman diagrams can be found in Ref. [F7.1]. Note that we do not provide QCD terms, since the few perturbative calculations practically feasible in QCD involve a very large number of graphs.

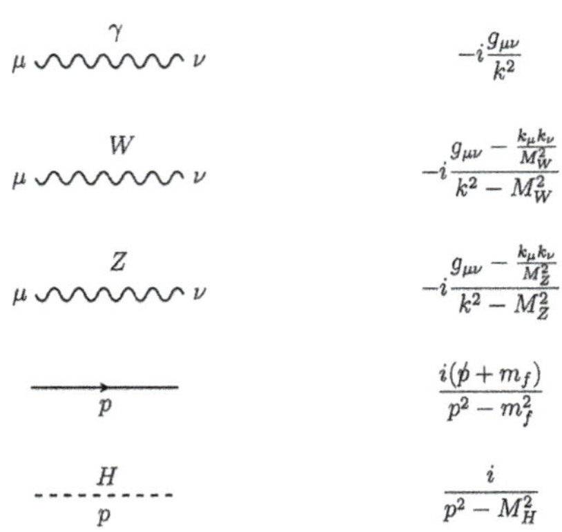

Fig. 7.4 Terms associated to propagators in the electroweak model. From [F7.5]

Charged Current

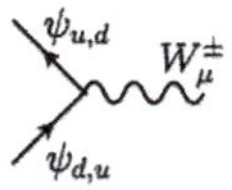

$$-i\frac{g}{\sqrt{2}}\gamma_\mu\frac{1-\gamma_5}{2}$$

Neutral Current

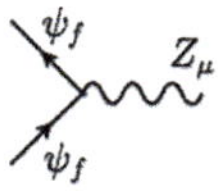

$$-i\frac{g}{\cos\theta_W}\gamma_\mu\left(g_V^f - g_A^f\gamma_5\right)$$

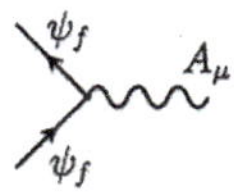

$$-ieQ_f\gamma_\mu$$

where

$$g_V^f = \frac{1}{2}T_f^3 - Q_f\sin^2\theta_W, \quad g_A^f = \frac{1}{2}T_f^3 .$$

Higgs Interactions

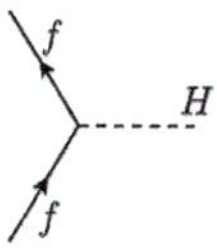

$$-i\frac{g}{2}\frac{m_f}{M_W} = -i\frac{m_f}{v}$$

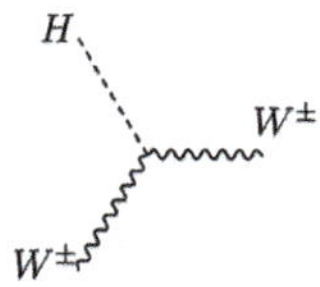

$$ig\,M_W = 2i\frac{M_W^2}{v}$$

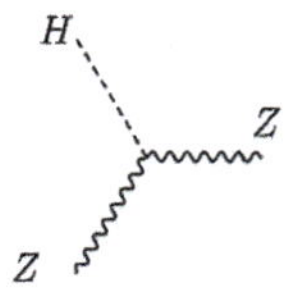

$$i\frac{g}{2\cos\theta_W}M_Z = i\frac{M_Z^2}{v}$$

Fig. 7.5 Terms associated to vertices in the electroweak model. Adapted from [F7.5]

7.3 The Lagrangian of the Standard Model

The Lagragian of the standard model is the sum of the electroweak Lagrangian (including the Higgs terms, which are responsible for the masses of the $W^{\pm}$ bosons and of the Z, and of the leptons) plus the QCD Lagrangian without the fermion mass terms.

7.3.1 The Higgs Particle in the Standard Model

The SM Higgs boson is thus the Higgs boson of the electroweak Lagrangian.

In accordance with relation (7.65), the interaction of the Higgs boson with a fermion is proportional to the mass of this fermion itself: $g_{Hff} = \frac{m_f}{v}$.

One finds that the Higgs boson couplings to the electroweak gauge bosons are instead proportional to the squares of their masses:

$$g_{HWW} = 2\frac{M_W^2}{v}, \quad g_{HHWW} = \frac{M_W^2}{v^2}, \quad \text{and} \quad g_{HZZ} = \frac{M_Z^2}{v}, \quad g_{HHZZ} = \frac{M_Z^2}{2v^2}. \tag{7.82}$$

Among the consequences, the prediction of the branching fractions for the decay of the Higgs boson is discussed later.

7.3.2 Standard Model Parameters

The standard model describes in detail particle physics at least at energies below or at the order of the electroweak scale (gravity is not considered here). Its power has been intensively and extensively demonstrated in the past thirty years by an impressive number of experiments (see later in this Chapter). However, it has a relatively large set of "magic numbers" not defined by the theory, which thus have to be obtained from measurements. The numerical values of these parameters were found to differ by more than 10 orders of magnitude (e.g., $m_\nu < 0.1$ eV, $m_t \sim 0.2$ TeV).

These free parameters may be listed in the hypothesis that neutrinos are standard massive particles (the hypothesis that they are "Majorana" particles, i.e., fermions coincident with their antiparticles, will be discussed in Chap. 9), as follows:

- In the gauge sector:
  - three gauge constants (respectively, $SU(2)_L$, $U(1)_Y$, $SU(3)$):

$$g, \; g', \; g_s \, .$$

- In the Higgs sector:
  - the two parameters of the standard Higgs potential:

$$\mu, \; \lambda \, .$$

- In the fermionic sector:
  - the twelve fermion masses (or alternatively the corresponding Higgs-fermion Yukawa couplings):

$$m_{\nu 1}, m_{\nu 2}, m_{\nu 3}, m_e, m_\mu, m_\tau, m_u, m_d, m_c, m_s, m_t, m_b\ ;$$

– the four quark CKM mixing parameters (in the Wolfenstein parametrization, see Sect. 6.3.7):

$$\lambda,\ A,\ \rho,\ \eta\ ;$$

– the four neutrino PMNS mixing parameters (see Sect. 9.1.1), which can be three real angles and one phase:

$$\theta_{12},\ \theta_{13},\ \theta_{23},\ \delta\ .$$

- In the strong CP sector:

  – A "CP violating phase" θ_{CP}. The U(1) symmetry cannot host CP violation (one can easily see it in this case since there is no room for the addition of nontrivial complex phases; a general demonstration holds for Abelian groups). We know instead that the electroweak sector contains a CP-violating phase. In principle, the QCD Lagrangian could also have a CP-violating term $i\,\theta_{CP}\epsilon_{\mu\nu\rho\sigma}F_a^{\mu\nu}F_a^{\rho\sigma}$; experiments tell, however, that the effective CP violation in strong interactions, if existing, is extremely small ($\theta_{CP,eff} < 10^{-10}$). It is common then to assume

$$\theta_{CP} = 0\ . \tag{7.83}$$

  It is difficult to imagine why the value of θ_{CP} should be accidentally so small—or zero: this is called the "strong CP problem". A viable solution would be to introduce an extra symmetry in the Lagrangian, and this was the solution proposed by Peccei and Quinn in the 1970s. An extra symmetry, however, involves a new gauge boson, which was called the axion; the axion should possibly be observed to confirm the theory.

However, the choice of the 26 "fundamental" parameters is somehow arbitrary because there are several quantities that are related to each other. In the Higgs sector, the vacuum expectation value v is often used; in the gauge sector, α, G_F, and α_s are the most common "experimental" choices to describe the couplings of the electromagnetic, weak interaction, and strong interaction; finally, $\sin^2\theta_W$ is for sure one of the most central quantities measured in the different decays or interaction channels. Indeed

$$v = \sqrt{-\frac{\mu^2}{\lambda}}\ ;\ \tan(\theta_W) = \frac{g'}{g}\ ;\ \alpha = \frac{e^2}{4\pi} = \frac{(g\sin\theta_W\,)^2}{4\pi}\ ;\ G_F = \frac{1}{\sqrt{2}\,v^2}\ ;\ \alpha_s = \frac{{g_s}^2}{4\pi}\ .$$

The bare masses of the electroweak gauge bosons, as well as their couplings, are derived directly from the standard model Lagrangian after the Higgs spontaneous symmetry breaking mechanism (see the previous sections):

- $m_\gamma = 0$, by construction
- $m_Z = \frac{1}{2}\sqrt{g^2 + g'^2}\,v$

- $m_W = \frac{1}{2} g v$

and $\sin^2\theta_W$ can also be expressed as

$$\sin^2\theta_W = 1 - \frac{{m_W}^2}{{m_Z}^2} = \frac{\pi\alpha}{\sqrt{2} G_F {m_W}^2} = \frac{\pi\alpha}{\sqrt{2} G_F {m_Z}^2 \cos^2\theta_W} . \tag{7.84}$$

The couplings of the electroweak gauge bosons to fermions were discussed in Sect. 7.2.4 and are proportional to

- $e = g \sin\theta_W$ for the photon γ;
- g, for the $W^\pm$ which only couples to left-handed fermions;
- $g/\cos\theta_W \; (g_V + g_A)$ and $g/\cos\theta_W \; (g_V - g_A)$ for the Z, respectively, for couplings to left and right fermions. $g_V = I_3 - 2\, Q_f \sin^2\theta_W$ and $g_A = I_3$ being Q_f and I_3, respectively, the electric charge and the third component of the weak isospin of the concerned fermion.

Finally, the mass of the Higgs boson is given, as seen Sect. 7.2.2, by

$$m_H = \sqrt{2\lambda} v . \tag{7.85}$$

7.3.3 Accidental Symmetries

The standard model exhibits additional global symmetries, collectively denoted accidental symmetries, which are continuous U(1) global symmetries which leave the Lagrangian invariant. By Noether's theorem, each symmetry has an associated conserved quantity; in particular, the conservation of baryon number (where each quark is assigned a baryon number of 1/3, while each antiquark is assigned a baryon number of − 1/3), electron number (each electron and its associated neutrino is assigned an electron number of + 1, while the anti-electron and the associated anti-neutrino carry a − 1 electron number), muon number, and tau number are accidental symmetries. Note that somehow these symmetries, although mathematically "accidental," exist by construction, since, when we designed the Lagrangian, we did not foresee gauge particles changing the lepton number or the baryon number as defined before.

In addition to the accidental symmetry, but nevertheless exact symmetries, described above, the standard model exhibits several approximate symmetries. Two of them are particularly important:

- The SU(3) quark flavor symmetry, which reminds us the symmetries in the "old" hadronic models. This obviously includes the SU(2) quark flavor symmetry—the strong isospin symmetry, which is less badly broken (only the two light quarks being involved).
- The SU(2) custodial symmetry, which keeps

$$r = \frac{m_W{}^2}{m_Z{}^2 \cos^2\theta_W} \simeq 1$$

limiting the size of the contributions from loops involving the Higgs particle (see the next Section). This symmetry is exact before the SSB.

7.4 Observables in the Standard Model

As it was discussed in the case of QED (see Sect. 6.2.9), measurable quantities are not directly bare quantities present in the Lagrangian, but correspond to effective quantities which "absorb" the infinities arising at each high-order diagrams due to the presence of loops for which integration over all possible momentum should be performed. These effective renormalized quantities depend on the energy scale of the measurement. This was the case for α and α_s as discussed in Sects. 6.2.10 and 6.4.4. The running of the electromagnetic coupling α

$$\alpha\left(m^2{}_e\right) \sim \frac{1}{137};\ \alpha\left(m^2{}_Z\right) \sim \frac{1}{129}$$

implies for instance a sizeable change on the values of m_Z and m_W from those that could be computed using the relations listed above ignoring this running and taking for G_F and $\sin^2\theta_W$ the values measured at low energy (muon decay for G_F and deep inelastic neutrino scattering for s_W).

In addition, QCD corrections to processes involving quarks can be important. For example, at the first perturbative order α_s (essentially keeping into account the emission of one gluon)

$$\frac{\Gamma(Z \to q\bar{q})}{\Gamma(Z \to q\bar{q})_{\text{leading}}} \simeq 1 + \frac{\alpha_s}{\pi}; \tag{7.86}$$

radiation of quarks and gluons increases the decay amplitude linearly in α_s. In fact these QCD corrections are known to $\mathcal{O}(\alpha_s^3)$ for $\Gamma_{q\bar{q}}$ (see later), and the measurement of the hadronic Z width provides the most precise estimate of α_s—see later a better approximation of Eq. (7.86).

High-order diagrams have thus to be carefully computed in order that the high-precision measurements that have been obtained in the last decades (see the next section) can be related to each other and, in that way, determine the level of consistency (or violation) of the standard model. These corrections are introduced often as a modification of previous lowest order formulas as, for example,

$$\sin^2\theta_W \cos^2\theta_W = \frac{\pi\alpha}{\sqrt{2} G_F m_Z{}^2 \,(1 - \Delta r)} \tag{7.87}$$

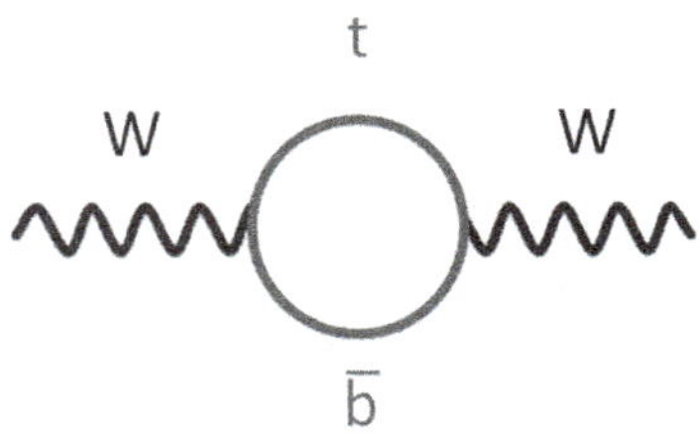

Fig. 7.6 $t\,\overline{b}$ loop contributing to self-energy corrections to the W propagator

and detailed calculations provide

$$\Delta r \sim -\frac{3\alpha}{16\pi\,\sin^4\theta_W}\frac{m_t{}^2}{m_Z{}^2}+\frac{11\alpha}{24\pi\,\sin^2\theta_W}\ln\left(\frac{m_H}{m_Z}\right)+\cdots. \tag{7.88}$$

The determination of Δr or of any electroweak correction is far beyond the scope of the present book. We just stress that Δr, and most of the radiative corrections, are in the largest part an effect of loops involving top quarks and Higgs particles. These enter in the calculations as $m_t{}^2$ and $\ln(m_H)$, respectively, and the total effect is at a some percent level. The quadratic dependence on m_t may appear as a surprise since in QED the contributions of loops involving heavy fermions are suppressed by inverse powers of the fermion mass. $SU(2)_L$ is however a chiral broken symmetry (the masses of the fermions are not degenerated) and for instance the self-energy corrections to the W propagator involving $t\,\overline{b}$ (or $\overline{t}\,b$) loops (Fig. 7.6) are proportional to $\left(m_t{}^2 - m_b{}^2\right)$. Both the quadratic dependence on m_t as well as the logarithmic dependence on m_H are a consequence of the way how the Higgs sector and the symmetry breaking mechanism are built in the standard model, leaving a remnant approximate symmetry (the "custodial" symmetry we examined in the previous section).

In addition, Δr can be sensitive to "new physics": the presence of additional virtual loops involving yet undiscovered particles affects the radiative corrections.

In a similar way, an electroweak form factor ρ_Z^f can be introduced to account for higher-order corrections to the Z couplings:

$$g_V = \sqrt{\rho_Z^f}\,\left(I_3 - 2\,Q_f\sin^2\theta_W\right) \tag{7.89}$$

$$g_A = \sqrt{\rho_Z^f}\;I_3\,. \tag{7.90}$$

The departure of ρ_Z^f from the unity ($\Delta\rho_Z^f$) is again a function of $m_t{}^2$ and $\ln(m_H)$.

Another way to incorporate radiative corrections in the calculations, frequently used in the literature, is to absorb them in the Weinberg angle, which then becomes an "effective" quantity. The form of the calculations then stays the same as that at leading order, but with an "effective" angle instead of the "bare" angle.

This approach is the simplest—and probably the most common in the literature—but one has to take into account that, at higher orders, the "effective" value of the angle will be different for different processes.

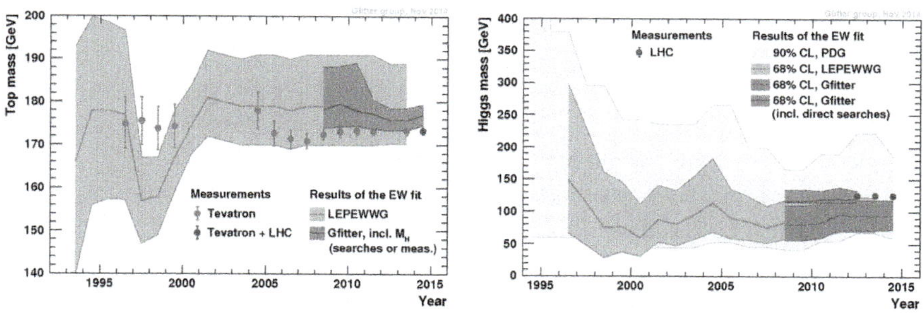

Fig. 7.7 Calculated mass of the top quark (*left*) and of the Higgs boson (*right*) from fits of experimental data to the standard model, as function of the year. The *lighter bands* represent the theoretical predictions at 95 % confidence level, while the *dark bands* at 68 % confidence level. The points with error bars represent the experimental values after the discovery of these particles. From http://project-gfitter.web.cern.ch/project-gfitter/

Global fits to electroweak precision data (see for instance the Gfitter project at CERN), taking properly into account correlations between standard model observables, have so far impressively shown the consistency of the standard model and its predictive power.

Both the mass of the top quark and mass of the Higgs boson were predicted before their discovery (Fig. 7.7), and the masses measured by experiment confirmed the prediction.

Many other consistency tests at accelerators confirmed the validity of the standard model; we shall discuss them in the next section.

7.5 Experimental Tests of the SM at Accelerators

The SM has been widely tested, also in the cosmological regime, and with high-precision table-top experiments at sub-GeV energies. The bulk of the tests, however, has been performed at particle accelerators, which span a wide range of high energies.

In particular, from 1989 to 1995, the Large Electron–Positron collider (LEP) at CERN provided collisions at center-of-mass energies near the Z mass; four large state-of-the-art detectors (ALEPH, DELPHI, L3 and OPAL) recorded about 17 million Z decays. Almost at the same time, the SLD experiment at the SLAC laboratory near Stanford, California, collected 600,000 Z events at the SLAC Linear Collider (SLC), with the added advantage of a longitudinally polarized electron beam—polarization provided additional opportunities to test the SM. LEP was upgraded later to higher energies starting from 1996, and eventually topped at a center-of-mass energy of about 210 GeV at the end of 2000. In this second phase, LEP could produce and study at a good rate of all SM particles—except the top quark and the Higgs boson; it produced in particular a huge statistics of pairs of W and Z bosons.

The Tevatron circular accelerator at the Fermilab near Chicago collided protons and antiprotons in a 7-km ring to energies of up to 1 TeV. It was completed in 1983, and its main achievement was the discovery of the top quark in 1995 by the scientists of the CDF and D0 detectors. The Tevatron ceased operations in 2011 because of the completion of the LHC, which had started stable operations in early 2010.

Finally, the Large Hadron Collider (LHC) was built in the 27-km-long LEP tunnel; it collides pairs of protons (and sometimes of heavy ions). It started stable operation in 2010, and increased its center-of-mass energy up the design energy of 14 TeV for proton–proton collisions, reached in 2015. Its main result has been the discovery of the Higgs boson.

All these accelerators provided extensive tests of the standard model; we shall review them in this section.

Before summarizing the main SM results at LEP/SLC, at the Tevatron and at the LHC, let us shortly remind some of the earlier results at accelerators which gave the scientific community confidence in the electroweak part of the standard model, and were already presented in Chap. 6:

- The discovery of the weak neutral currents by Gargamelle in 1972.
  A key prediction of the electroweak model was the existence of neutral currents mediated by the Z. These currents are normally difficult to reveal, since they are hidden by the most probable photon interactions. However, the reactions

$$\bar{\nu}_{\mu} + e^- \rightarrow \bar{\nu}_{\mu} + e^- \quad ; \quad \nu_{\mu} + N \rightarrow \nu_{\mu} + X$$

  cannot happen via photon exchange, nor via W exchange. The experimental discovery of these reactions happened in bubble-chamber events, thanks to Gargamelle, a giant bubble chamber. With a length of 4.8 m and a diameter of nearly 2 m, Gargamelle held nearly 12 m^3 of liquid freon, and operated from 1970 to 1978 with a muon neutrino beam produced by the CERN Proton Synchrotron. The first neutral current event was observed in December 1972, and the detection was published with larger statistics in 1973; in the end, approximately 83,000 neutrino interactions were analyzed, and 102 neutral current events observed. Gargamelle is now on exhibition in the CERN garden.
- The discovery of a particle made of the charm quark (the J/ψ) in 1974. Charm was essential to explain the absence of strangeness-changing neutral currents (by the so-called GIM mechanism, discussed in the previous Chapter).
- The discovery of the W and Z bosons at the CERN $Sp\bar{p}S$ collider in 1983, in the mass range predicted, and consistent with the relation $m_Z \simeq m_W / \cos\theta_W$.

7.5.1 Data Versus Experiments: LEP (and the Tevatron)

LEP has studied all the SM particles, except the top and the Higgs, which could not be produced since the c.m. energy was not large enough. Most of the results on the SM parameters are thus due to LEP. We shall see, however, that Tevatron and the LHC are also crucial for the test of the SM.

7.5.1.1 Electroweak Precision Measurements

In the context of the Minimal Standard Model (MSM) neglecting the neutrino masses which are anyway very small, electroweak processes can be computed at tree level from the electromagnetic coupling α, the weak coupling G_F, the Z mass M_Z, and from the elements of the CKM mixing matrix.

When higher-order corrections and phase space effect are included, one has to add to the above α_s, m_H, and the masses of the particles. The calculations show that the loops affecting the observables depend on the top mass through terms (m_t^2/M_Z^2), and on the Higgs mass through terms showing a logarithmic dependence $\ln(m_H^2/M_Z^2)$—plus, of course, on any kind of "heavy new physics" (see Sect. 7.4).

The set of the three SM variables which characterize the interaction is normally taken as $M_Z = 91.876 \pm 0.0021$ GeV (derived from the Z line shape, see later), $G_F = 1.1663787(6) \times 10^5$ GeV^{-2} (derived from the muon lifetime), and the fine structure constant in the low-energy limit $\alpha = 1/137.035999074(44)$, taken from several electromagnetic observables; these quantities have the smallest experimental errors.

One can measure the SM parameters through thousands of observables, with partially correlated statistical and systematic uncertainties; redundancy can display possible contradictions, pointing to new physics. This large set of results has been reduced to a more manageable set of 17 precision results, called electroweak observables. This was achieved by a model-independent procedure, developed by the LEP and Tevatron Electroweak Working Groups (a group of physicist from all around the world charged of producing "official" fits to precision observables in the SM).

About three-fourth of all observables arise from measurements performed in electron–positron collisions at the Z resonance, by the LEP experiments ALEPH, DELPHI, L3, and OPAL, and the SLD experiment. The Z-pole observables are five observables describing the Z lineshape and leptonic forward-backward asymmetries, two observables describing polarized leptonic asymmetries measured by SLD with polarized beams and at LEP through the tau polarization, six observables describing $b-$ and $c-$quark production at the Z pole, and finally the inclusive hadronic charge asymmetry. The remaining observables are the mass and total width of the W boson measured at LEP and at hadron accelerators, the top quark mass measured at hadron accelerators. Recently, also the Higgs mass has been added to the list; the fact that the Higgs mass has been found in the mass range predicted by the electroweak observables is another success of the theory.

Figure 7.8 shows the comparison of the electroweak observables with the best fit to the SM. One can appreciate the fact that the deviations from the fitted values are consistent with statistical fluctuations.

Figure 7.9 shows the evolution of the hadronic cross section $\sigma(e^+e^- \rightarrow hadrons)$ with energy, compared with the predictions of the SM. This is an incredible success of the SM, which quantitatively accounts for experimental data over a wide range of energies:

- starting from a region (above the Υ threshold and below some 50 GeV) where the production is basically due to photon exchange, and $\sigma \propto 1/s$,
- to a region in which the contributions from Z and γ are important and the Z/γ interference has to be taken into account,
- to a region of Z dominance (Eq. 7.93), and
- to a region in which the WW channel opens and triple boson vertices become relevant.

We describe in larger detail three of the most significant electroweak tests at LEP in phase I: the partial widths of the Z, the forward-backward asymmetries, and the study of the Z line shape, which has important cosmological implications. Finally, in this section, we examine the characteristics of vertices involving three gauge bosons.

Partial Widths of the Z. The partial widths of the Z, possibly normalized to the total width, are nontrivial parameters of the SM. Indeed, the evolution of the branching fractions with energy due to the varying relative weights of the Z and γ couplings are a probe into the theory.

Final states of the Z into $\mu^+\mu^-$ and $\tau^+\tau^-$ pairs can easily be identified. The e^+e^- final state is also easy to recognize, but in this case, the theoretical interpretation is less trivial, being the process dominated by t−channel exchange at low angles. Among

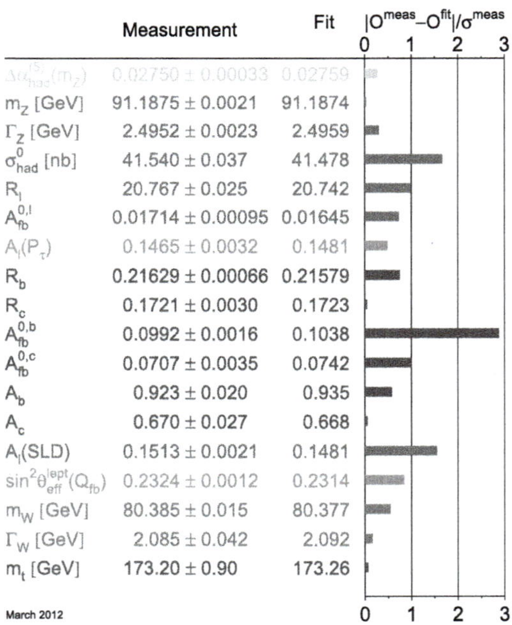

Fig. 7.8 Pull comparison of the fit results with the direct measurements in units of the experimental uncertainty. The absolute value of the pull (i.e., of the difference between the measured value and the fitted value divided by the uncertainty) of the Higgs mass is 0.0 (its value is completely consistent with the theoretical fit)

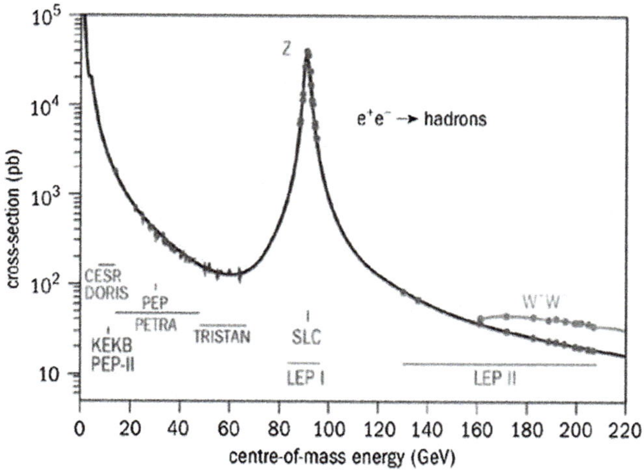

Fig. 7.9 Evolution of the hadronic cross section $\sigma(e^+e^- \rightarrow hadrons)$ with energy, compared with the predictions of the SM

hadronic final states, the $b\bar{b}$ and $c\bar{c}$ can be tagged using the finite lifetimes of the primary hadrons (the typical lifetime of particles containing c quarks and weakly decaying is of the order of 0.1 ps, while the typical lifetime of particles containing c quarks and weakly decaying is of the order of 1 ps). The tagging of $s\bar{s}$ final states is more difficult and affected by larger uncertainties.

All these measurements gave results consistent with the predictions from the SM (Table 7.1). By considering that the decay rates include the square of these factors, and all possible diagrams, the relative strengths of each coupling can be estimated (e.g., sum over quark families, and left and right contributions). The relative strengths of each coupling can be estimated by considering that the decay rates include the square of these factors, and all possible diagrams (e.g., sum over quark families, and left and right contributions). As we are considering only tree-level diagrams in the electroweak theory, this is naturally only an estimate.

Also, the energy evolution of the partial widths from lower energies and near the Z resonance is in agreement with the Z/γ mixing in the SM.

Z Asymmetries and $\sin^2\theta_{eff}$. Like the cross section $Z \rightarrow f\bar{f}$, the forward-backward asymmetry

$$A_{FB}^{f} \equiv \frac{\sigma_F - \sigma_B}{\sigma_F + \sigma_B} \simeq \frac{3}{4} A_e A_f \,, \tag{7.91}$$

where F (forward) means along the e^- direction, where the combinations A_f are given, in terms of the vector and axial vector couplings of the fermion f to the Z boson, by

$$A_f = \frac{2 g_V^f g_A^f}{g_V^{f2} + g_A^{f2}} \,, \tag{7.92}$$

Table 7.1 Relative branching fractions of the Z into $f\bar{f}$ pairs: predictions at leading order from the SM (for $\sin^2\theta_W = 0.23$) are compared to experimental results

Particle	g_V	g_A	Predicted (%)	Experimental (%)
Neutrinos (all)	1/4	1/4	20.5	(20.00 ± 0.06)
Charged leptons (all)			10.2	(10.097 ± 0.003)
Electron	$-1/4 + \sin^2\theta_W$	$-1/4$	3.4	(3.363 ± 0.004)
Muon	$-1/4 + \sin^2\theta_W$	$-1/4$	3.4	(3.366 ± 0.007)
Tau	$-1/4 + \sin^2\theta_W$	$-1/4$	3.4	(3.367 ± 0.008)
Hadrons (all)			69.2	(69.91±0.06)
Down-type quarks d, s, b	$-1/4 + 1/3 \sin^2\theta_W$	$-1/4$	15.2	(15.6±0.4)
Up-type quarks u, c	$1/4 - 2/3 \sin^2\theta_W$	1/4	11.8	(11.6±0.6)

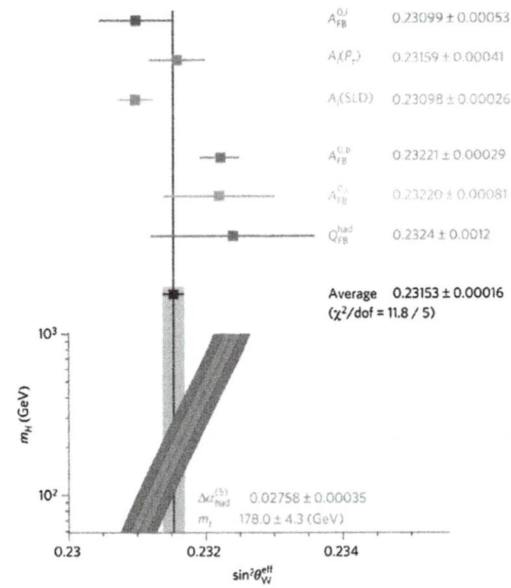

Fig. 7.10 Comparison of the effective electroweak mixing angle $\sin^2\theta_{\text{eff}}^{\text{lept}}$ derived from measurements depending on lepton couplings only (*top*) and on quark couplings as well (*bottom*). Also shown is the standard model prediction as a function of the Higgs mass, m_H. From M. Grunewald, CERN Courier, November 2005

can be measured for all charged lepton flavors, for heavy quarks, with smaller accuracy for $s\bar{s}$ pairs, and for all five quark flavors inclusively (overall hadronic asymmetry). It thus allows a powerful test of the SM.

One thus expects at the Z asymmetry values of about 7 % for up-type quarks, about 10 % for down-type quarks, and about 2 % for leptons. Figure 7.8 shows that results are consistent with the SM predictions, and this observable is powerful in constraining the value of $\sin\theta_W$ (Fig. 7.10).

Since the e^+e^- annihilation as a function of energy scans the γ/Z mixing, the study of the forward-backward asymmetry as a function of energy is also very impor-

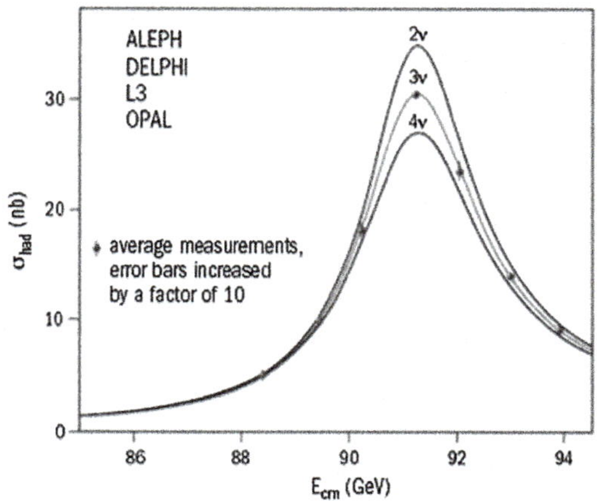

Fig. 7.11 Measurements of the hadron production cross section around the Z. The curves indicate the predicted cross section for two, three, and four neutrino species with standard model couplings and negligible mass. From M. Grunewald, CERN Courier, November 2005

tant. The energy evolution of the asymmetries from lower energies and near the Z resonance is in agreement with the Z/γ mixing in the SM.

The Z Lineshape and the Number of Light Neutrino Species. One of the most important measurements at LEP concerns the mass and width of the Z boson. While the Z mass is normally taken as an input to the standard model, its width depends on the number of kinematically available decay channels and the number of light neutrino species (Fig. 7.11). As we shall see, this is both a precision measurement confirming the SM and the measurement of a fundamental parameter for the evolution of the Universe.

Why is the number of quark and lepton families equal to three? Many families including heavy charged quarks and leptons could exist, without these heavy leptons being ever produced in accessible experiments, because of a lack of energy. It might be, however, that these yet undiscovered families include "light" neutrinos, kinematically accessible in Z decays—and we might see a proof of their existence in Z decays. The Z lineshape indeed obviously depends on the number of *kinematically accessible* neutrinos. Let us call them "light" neutrinos.

Around the Z pole, the $e^+e^- \to Z \to f\bar{f}$ annihilation cross section (s-channel) can be written as

$$\sigma_{s,Z} \simeq \frac{12\pi(\hbar c)^2}{M_Z^2} \frac{s\Gamma_e\Gamma_f}{(s - M_Z^2)^2 + s^2\Gamma_Z^2/M_Z^2} + \text{corrections.} \qquad (7.93)$$

The term explicitly written is the generic cross section for the production of a spin one particle in an e^+e^- annihilation, decaying into visible fermionic channels—just a particular case of the Breit–Wigner shape. The peak sits around the Z mass, and has a width Γ_Z. $B_f\Gamma_Z = \Gamma_f$ is the partial width of the Z into $f\bar{f}$. As we have seen, the branching fraction of the Z into hadrons is about 70 %, each of the leptons represents 3 %, while three neutrinos would contribute for approximately a 20 %. The term "corrections" includes radiative corrections and the effects of the presence of the photon. We remind that the branching fractions of the photon are proportional to Q_f^2, where Q_f is the electric charge of the final state. However, at the peak, the total electromagnetic cross section is less than 1 % of the cross section at the Z resonance. Radiative corrections, instead, are as large as 30 %; due to the availability of calculations up to second order in perturbation theory, this effect can be corrected for with a relative precision at the level of 10^{-4}. The effect of a number of neutrinos larger than three on the formula (7.93) would be to increase the width, and to decrease the cross section at the resonance.

The technique for the precision measurement of the Z cross section near the peak is not trivial; we shall just sketch it here. The energy of the beam is varied, and all visible channels are classified according to four categories: hadrons, electron pairs, muon pairs, and tau pairs. The extraction of the cross section from the number of events implies the knowledge of the luminosity of the accelerator. This is done by measuring at the same time another process with a calculable cross section, the elastic scattering $e^+e^- \to e^+e^-$ in the t−channel (Bhabha scattering), which results in an electron–positron pair at small angle. Of course one has to separate this process from the s-channel, and a different dependence on the polar angle is used for this purpose: the Bhabha cross section depends on the polar angle as $1/\sin^3\theta$, and quickly goes to 0 as θ grows. Another tool can be leptons universality: in the limit in which the lepton masses are negligible compared to the Z mass, the branching fractions of all leptons are equal.

In the end, one has a measurement of the total hadronic cross section from the LEP experiments (with the SLAC experiments contributing to a smaller extent due to the lower statistics) which is plotted in Fig. 7.11. The best fit to Eq. (7.93), assuming that the coupling of neutrinos is universal, provides

$$N_\nu = 2.9840 \pm 0.0082\,. \tag{7.94}$$

Notice that the number of neutrinos could be fractional—in case a fourth generation is relatively heavy—and universality is apparently violated due to limited the phase space.

The best-fit value of the Z width is

$$\Gamma_Z = 2.4952 \pm 0.0023\,\mathrm{GeV}\,. \tag{7.95}$$

The existence of additional light neutrinos would have considerable cosmological consequences: the evolution of the Universe immediately after the Big Bang would be affected. The creation of neutrons and protons is controlled by reactions involving the electron neutrino, such as $\nu_e\, n \to p\, e$ and is consequently sensitive to the number of light neutrino families N_ν which compete with electron neutrinos. Primordial nucleosynthesis (Sect. 8.1.3) is sensitive to this number.

A Fundamental Test: the WW and ZZ Cross Sections. One of the main scientific goals of the LEP II (i.e., at energies above the Z) program has been the measurement

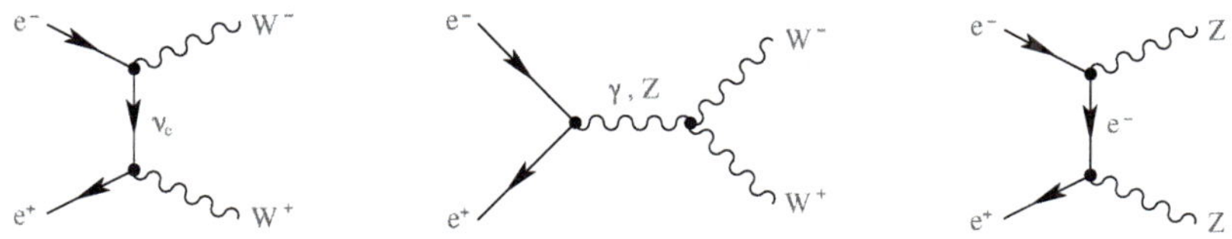

Fig. 7.12 Feynman diagrams contributing to $e^+e^- \to W^+W^-$ and $e^+e^- \to ZZ$

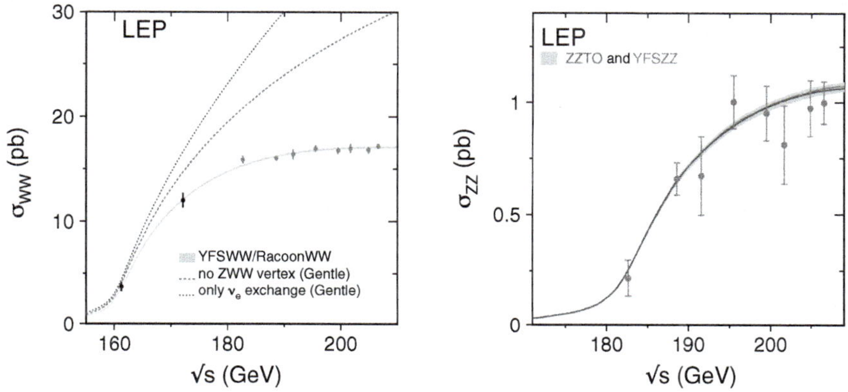

Fig. 7.13 Measured energy dependence of $\sigma(e^+e^- \to W^+W^-)$ (*left*) and $\sigma(e^+e^- \to ZZ)$ (*right*). The curves shown for the W-pair production cross section correspond to only the ν_e-exchange contribution (*upmost curve*), ν_e exchange plus photon exchange (*middle curve*), and all contributions including also the ZWW vertex (*lowest curve*). Only the e-exchange mechanism contributes to Z-pair production. From ALEPH, DELPHI, L3, OPAL Collaborations and the LEP Electroweak Working Group, Phys. Rept. 532 (2013)119

of the triple gauge vertices, through the experimental channels $e^+e^- \to WW$ and $e^+e^- \to ZZ$.

At tree level, the W-pair production process $e^+e^- \to W^+W^-$ involves three different contributions (Fig. 7.12), corresponding to the exchange of ν_e, γ, and Z. If the ZWW vertex would not exist, and the WW production occurred only via the neutrino exchange diagram, the WW production cross section would diverge for large values of $\sqrt{s}$. As shown in Fig. 7.13, the existence of the ZWW vertex is crucial in order to explain the data. At very high energies, the vertex HW^+W^- is also needed to prevent the divergence of the cross section.

Since the Z does not interact with the photon as it is electrically neutral, the SM does not include any local γZZ vertex. This leads to a $e^+e^- \to ZZ$ cross section that involves only the contribution from e exchange.

The agreement of the SM predictions with the experimental measurements in both production channels, W^+W^- and ZZ, is an important test for the gauge self-interactions. There is a clear signal of the presence of a ZWW vertex, with the predicted strength. Moreover, there is no evidence for any γZZ or ZZZ interactions. The gauge structure of the $\mathrm{SU}(2)_L \otimes \mathrm{U}(1)_Y$ theory is nicely confirmed by the data.

The experimental data at LEP II and at hadronic accelerators (mostly the Tevatron) have allowed the determination of the W mass and width with high accuracy:

$$M_W = 80.385 \pm 0.015\,\text{GeV} \tag{7.96}$$

$$\Gamma_W = 2.085 \pm 0.042\,\text{GeV}\,. \tag{7.97}$$

7.5.1.2 QCD Tests at LEP

After our considerations on the electroweak observables, let us now summarize the tests of the remaining building block of the SM: QCD.

LEP is an ideal laboratory for QCD studies since the center-of-mass energy is high with respect to the masses of the accessible quarks and (apart from radiative corrections which are important above the Z) well defined. As a consequence of the large center-of-mass energy, jets are collimated and their environment is clean: the hadron level is not so far from the parton level. The large statistics collected allows investigating rare topologies.

In particular, LEP confirmed the predictions of QCD in the following sectors—among others:

- QCD is not Abelian: the jet topology is inconsistent with an Abelian theory and evidences the existence of the three-gluon vertex.
  Angular correlations within 4-jet events are sensitive to the existence of the gluon self-coupling (Fig. 7.14), and were extensively studied at LEP.
  As a consequence of the different couplings in the $gq\bar{q}$ and ggg vertices, the distribution of the Bengtson–Zerwas angle (Fig. 7.15, left) between the cross product of the direction of the two most energetic jets in 4-jet events and the cross product of the direction of the two least energetic jets is substantially different in the predictions of QCD and of an Abelian theory where the gluon self-coupling does not exist.

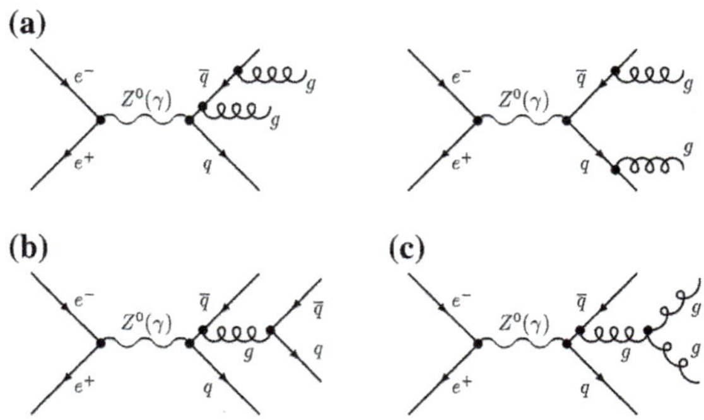

Fig. 7.14 Diagrams yielding four-parton final states **a** double gluon bremsstrahlung; **b** secondary $q\bar{q}$ pair production; **c** triple-gluon vertex

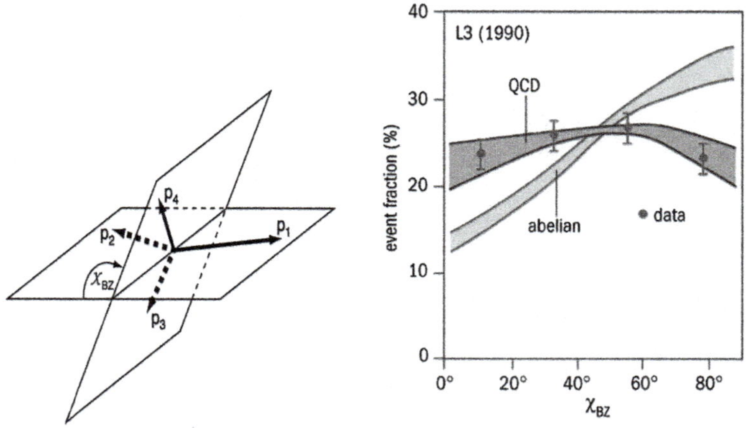

Fig. 7.15 *Left* Definition of the Bengtsson–Zerwas angle in 4-jet events. From P.N. Burrows, SLAC-PUB-7434, March 1997. *Right* Distribution for the data, compared with the predictions for QCD and for an Abelian theory. The experimental distribution is compatible with QCD, but it cannot be reproduced by an Abelian field theory of the strong interactions without gauge boson self-coupling. From CERN Courier, May 2004

The LEP result is summarized in Fig. 7.15, right; they are in excellent agreement with the gauge structure of QCD and proved to be inconsistent with an Abelian theory, i.e., the three-gluon vertex is needed to explain the data.

- Structure of QCD: measurement of the color factors.
  QCD predicts that quarks and gluons fragment differently due to their different color charges. Gluon jets are expected to be broader than quark jets; the multiplicity of hadrons in gluon jets should be larger than in quark jets of the same energy, and particles in gluon jets are expected to be less energetic. All these properties have been verified in the study of symmetric 3-jet events, in which the angle between pairs of consecutive jets is close to 120°—the so-called "Mercedes" events, like the event in Fig. 6.61, right. In these 3-jet events, samples of almost pure gluon jets could be selected by requiring the presence of a particle containing the b quark in both the other jets—this can be done due to the relatively long lifetime associated to the decay ($\tau_b \simeq 1$ ps, which corresponds to an average decay length of 300 μm for $\gamma = 1$, well measurable for example with Silicon vertex detectors).
  Many observables at hadron level can be computed in QCD using the so-called "local parton-hadron duality" (LPHD) hypothesis, i.e., computing quantities at parton level with a cutoff corresponding to a mass just above the pion mass, and then rescaling to hadrons with a normalization factor.
  Gluon jets in hadronic 3-jet events at LEP have indeed been found experimentally to have larger hadron multiplicities than quark jets. Once correcting for hadronization effects, one obtains

$$\frac{C_A}{C_F} = 2.29 \pm 0.09\,(\text{stat.}) \pm 0.15(\text{theory})\,,$$

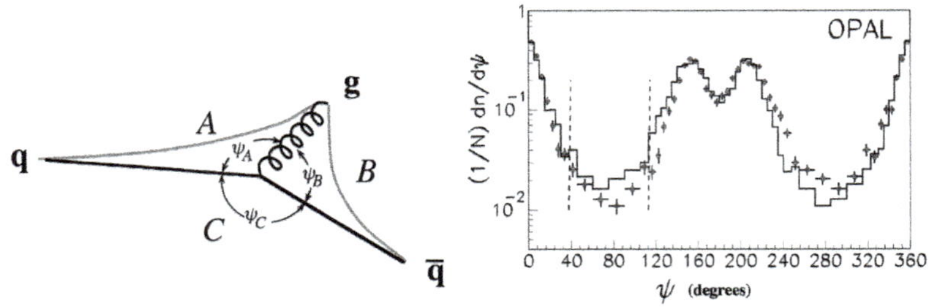

Fig. 7.16 *Left* Sketch of a 3-jet event in e^+e^- annihilations. In the Lund string model for fragmentation, string segments span the region between the quark q and the gluon g and between the antiquark $\bar{q}$ and the gluon. *Right* Experimental measurement of the particle flow $(1/N)dn/d\psi$, for events with $\psi_A = 150° \pm 10°$ and $\psi_C = 150° \pm 10°$. The points with errors show the flow from the higher energy quark jet to the low-energy quark jet and then to the gluon jet; in the histogram, it is shown the measured particle flow for the same events, starting at the high energy quark jet but proceeding in the opposite sense. The *dashed lines* show the regions, almost free of the fragmentation uncertainties, where the effect is visible. From OPAL Collaboration, Phys. Lett. B261 (1991) 334

consistent with the ratio of the color factors $C_A/C_F = 9/4$ that one can derive from the theory at leading order assuming LPHD (see Eq. 6.124).

- String effect.
  As anticipated in Sect. 6.4.6 and as one can see from Fig. 7.16, left, one expects in a Mercedes event an excess of particles in the direction of the gluon jet with respect to the opposite direction, since this is where most of the color field is. This effect is called the string effect and has been observed by the LEP experiments at CERN in the 1990s. This is evident also from the comparison of the color factors, as well as from considerations based on color conservation.
  A direct measurement of the string effect in Mercedes events is shown in Fig. 7.16, right.
- Measurement of α_s and check of its evolution with energy.
  One of the theoretically best known variables depending on α_s is the ratio of the Z partial decay widths R^0_{lept}, which is known to $\mathcal{O}(\alpha_s^3)$:

$$R^0_{\text{lept}} = \frac{\Gamma_{\text{hadrons}}}{\Gamma_{\text{leptons}}} = 19.934\left[1 + 1.045\left(\frac{\alpha_s}{\pi}\right) + 0.94\left(\frac{\alpha_s}{\pi}\right)^2 - 15\left(\frac{\alpha_s}{\pi}\right)^3\right] + \mathcal{O}\left(\frac{\alpha_s}{\pi}\right)^4 . \tag{7.98}$$

From the best-fit value $R^0_{lept} = 20.767 \pm 0.025$ (derived by assuming lepton universality), one obtains

$$\alpha_s(m_Z) = 0.124 \pm 0.004(\text{exp.})\,^{+0.003}_{-0.002}(\text{theory}). \tag{7.99}$$

The advantage of evaluating α_s from Eq. (7.98) is that non-perturbative corrections are suppressed since this quantity does not depend on hadronization, and the dependence on the renormalization scale μ is small. This renormalization scale is often responsible for the dominant uncertainty of α_s measurements. A fit to all electroweak Z pole data from LEP, SLD and to the direct measurements of m_t and m_W leads to

$$\alpha_s(m_Z) = 0.1183 \pm 0.0027. \tag{7.100}$$

One could ask if these are the most reliable evaluations of $\alpha_s(m_Z)$ using the LEP data. The problem is that the quoted results depend on the validity of the electroweak sector of the SM and thus small deviations can lead to large changes. One can also measure α_s from infrared safe hadronic event shape variables like jet rates, etc., not depending on the electroweak theory. A fit to the combined data results in a value

$$\alpha_s(m_Z) = 0.1195 \pm 0.0047,$$

where the error is almost entirely due to theoretical uncertainties (renormalization scale effects). The consistency of this value with the result in Eq. (7.100) is in itself a confirmation of QCD. Measurements of α_s at LEP energies using a multitude of analysis methods are collected in Fig. 7.17; one can appreciate their consistency.

- Non-perturbative QCD: evolution of average charge multiplicity with center-of-mass energy.
  As already seen in Chap. 6, average multiplicity is one of the basic observables characterizing hadronic final states; experimentally, since the detection of charged particles is simpler than the detection of neutrals, one studies the average charged particle multiplicity. In the limit of large energies, most of the particles in the final state are pions, and one can assume, by isospin symmetry, that the number of neutral pions is half the number of charged pions.
  LEP in particular has studied multiplicity of charged particles with lifetimes larger than 1 ps in a wide range of energies; using radiative events (i.e., events $e^+e^- \to Z'\gamma$ where the Z' is off-shell with respect to the Z), one could obtain, thanks to

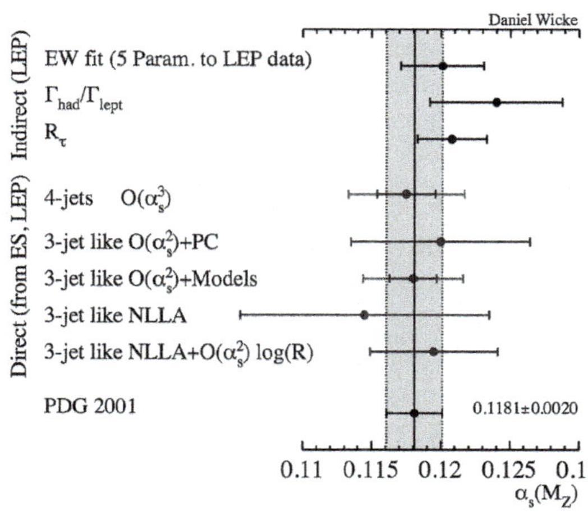

Fig. 7.17 Summary of α_s measurements at LEP compared to the world average. The theoretical uncertainty for all five measurements from event shapes (ES) is evaluated by changing the renormalization scale μ by a factor of 2

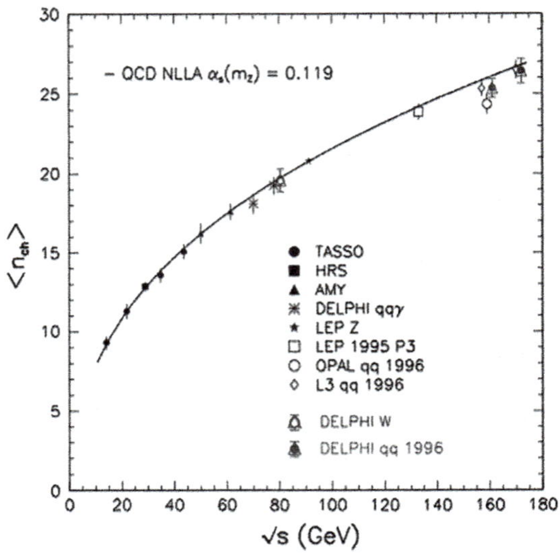

Fig. 7.18 Measured average charged particle multiplicity in $e^+e^- \to q\bar{q}$ events as a function of center-of-mass energy $\sqrt{s}$. DELPHI high energy results are compared with other experiments and with a fit to the prediction from QCD in next-to-leading order. The average charged particle multiplicity in W decays is also shown at an energy corresponding to the W mass. The measurements have been corrected for the different proportions of $b\bar{b}$ and $c\bar{c}$ events at the various energies

the large statistics collected at LEP in phase I, information also on the behavior of this observable at center-of-mass energies below the Z peak.

The QCD prediction including leading and next-to-leading order calculation is

$$\langle n\rangle(E_{CM}) = a[\alpha_s(E_{CM})]^b e^{c/\sqrt{\alpha_s(E_{CM})}}\left[1 + O(\sqrt{\alpha_s(E_{CM})})\right], \qquad (7.101)$$

where a is the LPHD scaling parameter (not calculable from perturbation theory) whose value should be fitted from the data; the constants $b = 0.49$ and $c = 2.27$ are instead calculated from the theory. The summary of the experimental data available is shown in Fig. 7.18 with the best fit to the QCD prediction.

Energy distribution of hadrons can be computed form LPHD; the coherence between radiated gluons causes a suppression at low energies. Experimental evidence for this phenomenon comes from the "hump-backed" plateau of the distribution of the variable $\xi = -\ln(2E_h/E_{CM})$ shown in Fig. 7.19, left.

The increase with energy of the maximum value, ξ^*, of these spectra is strongly reduced compared to expectations based on phase space (Fig. 7.19, right).

7.5.1.3 The Discovery of the Top Quark at the "Right" Mass at the Tevatron

LEP could not discover the top quark but was able to indirectly estimate its mass, since the top quark mass enters into calculations of characteristics of various electroweak observables, as seen before.

In 1994, the best (indirect) estimate for the top quark mass by LEP was $m_t = 178 \pm 20$ GeV.

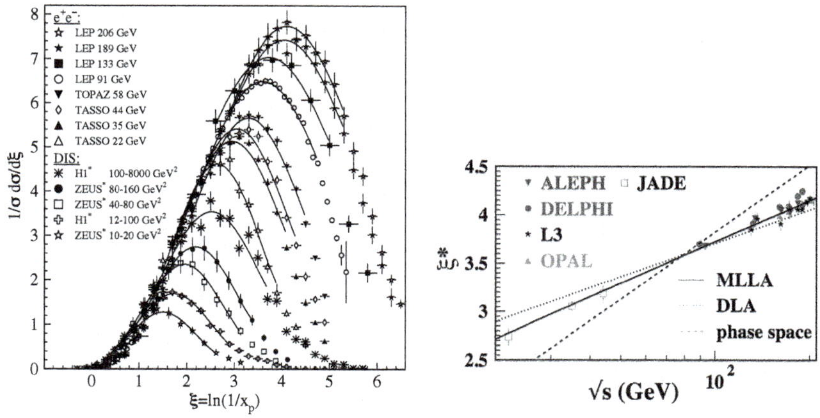

Fig. 7.19 *Left* Center-of-mass energy dependence of the spectra of charged hadrons as a function of $\xi = -\ln x$; $x = 2E_h/E_{CM}$. From K.A. Olive et al. (Particle Data Group), Chin. Phys. C 38 (2014) 090001. *Right* Energy dependence of the maximum of the ξ distribution, ξ^*

In March 1995, the two experiments CDF and D0 running at Fermilab at a center-of-mass energy of 1.8 TeV jointly reported the discovery of the top at a mass of 176 ± 18 GeV. The cross section was consistent with what predicted by the standard model. Figure 7.7, left, compares the indirect measurements of the top mass with the direct measurements.

7.5.2 LHC and the Discovery of the Higgs Boson

Despite the incredible success on the precision measurements related to standard model properties, LEP missed one of its most important targets: the discovery of the Higgs boson. The Higgs particle was found more than a decade after the LEP shutdown, thanks to the LHC proton–proton collider.

7.5.2.1 The Legacy of Indirect Measurements and Previous Unsuccessful Searches

The most important processes which could possibly produce a Higgs boson at LEP were, besides the unobserved decays $Z \to H + \gamma$ or $Z \to H + Z^*(Z^* \to f\bar{f})$, (a) the so-called Higgs-strahlung $e^+e^- \to Z + H$; and (b) the vector boson (W^+W^- or ZZ) fusion into a H boson and a lepton–antilepton pair (Fig. 7.20). The direct process $e^+e^- \to H$ as a negligible probability because of the small H coupling to e^+e^-, given the mass difference.

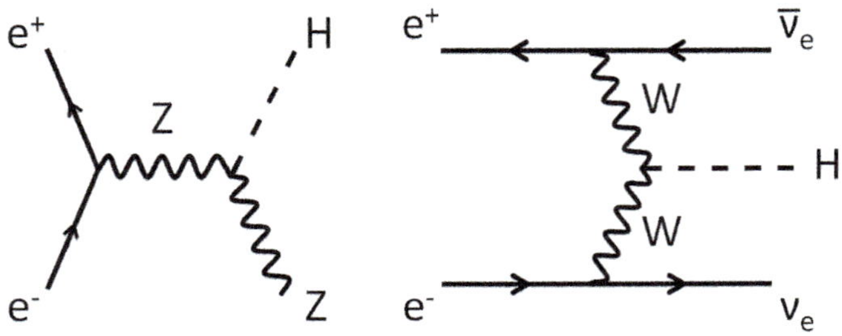

Fig. 7.20 Main Higgs production mechanisms at LEP: Higgs-strahlung (*left*) and vector boson fusion (*right*)

A first limit on the Higgs mass was obtained shortly after switching on the accelerator: the fact that no decays of the Z into H were observed immediately implies that the Higgs boson must be heavier than the Z. Then the center-of-mass energy of LEP was increased up to 210 GeV, still without finding evidence for the Higgs. Indirect experimental bounds on the SM Higgs boson mass (in the hypothesis of a minimal SM) were obtained from a global fit of precision measurements of electroweak observables at LEP described in the previous subsection; the uncertainty on radiative corrections was dominated by the uncertainty on the yet undiscovered Higgs boson—and, to a smaller extent, by the error on the measurement of the top mass: solid bounds could thus be derived.

LEP shut down in the year 2000. The global fit to the LEP data, with the constraints given by the top mass measurements at the Tevatron, suggested for the Higgs a mass of 94^{+29}_{-24} GeV (the likelihood distribution was peaked toward lower Higgs mass values, as shown in Fig. 7.21). On the other hand, direct searches for the Higgs boson conducted by the experiments at the LEP yielded a lower limit at 95 % C.L. (confidence limit)

$$m_H > 114.4\,\mathrm{GeV}/c^2\,. \tag{7.102}$$

Higgs masses above 171 GeV were also excluded at 95 % C.L. by the global electroweak fit. The negative result of searches at the Tevatron and at LHC conducted before 2011 excluded the range between 156 and 177 GeV; thus one could conclude, still at 95 % C.L.,

$$m_H < 156\,\mathrm{GeV}/c^2\,. \tag{7.103}$$

Scientists were finally closing on the most wanted particle in the history of high energy physics.

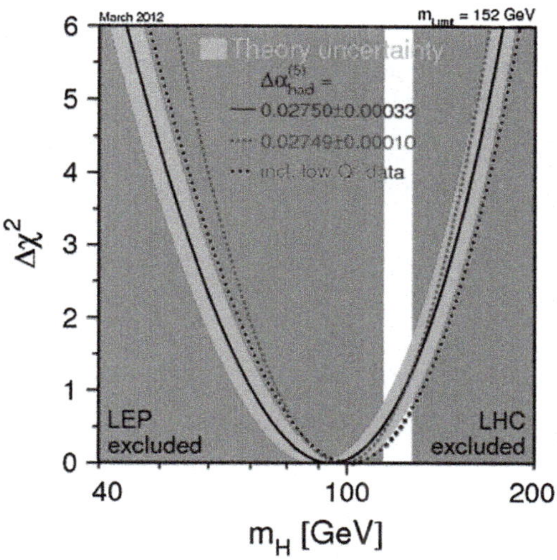

Fig. 7.21 Probability distribution for the mass of the Higgs boson before its direct discovery: fit to the standard model. Fit from the Electroweak Working Group

7.5.2.2 LHC and the Higgs

LHC started operation in September 2008 for a test run at center-of-mass energy smaller than 1 TeV, and then in a stable conditions in November 2009 after a serious accident in the test run damaged the vacuum tubes and part of the magnets. Starting from March 2010, LHC reached an energy of 3.5 TeV per beam, and thus an excellent discovery potential for the Higgs; the energy was further increased to 4 TeV per beam in mid 2012. Within the strict bound defined by (7.102) and (7.103), coming from the constraints of the indirect measurements, a mass interval between 120 and 130 GeV was expected as the most likely for the Higgs.

A Higgs particle around that mass range is mostly produced via gluon–gluon fusion. Gluon–gluon fusion can generate a virtual top quark loop, and since the Higgs couples to mass, this process is very effective in producing Higgs bosons—the order of magnitude of the cross section being of 10 pb. The second largest rate pertains to the so-called weak-boson fusion (WBF) process $qq \rightarrow qqH$ via the radiation of one W by each quark and the subsequent fusion of the two W bosons into a Higgs particle. The cross section is one order of magnitude smaller, i.e., about 1 pb.

A Higgs particle between 120 and 130 GeV is difficult to detect at LHC, because the W^+W^- decay channel is kinematically forbidden (one of the Ws has to be highly virtual). The total decay width is about 4 MeV; the branching fractions predicted by the SM are depicted in Fig. 7.22.

The dominant decay modes are $H \rightarrow b\bar{b}$ and $H \rightarrow WW^*$. The first one involves the production of jets, which are difficult to separate experimentally in the event; the second one involves jets and/or missing energy (in the semileptonic W, decay part of the energy is carried by an undetected neutrino, which makes the event reconstruction difficult). The decay $H \rightarrow ZZ$ might have some nice experimental features: the final

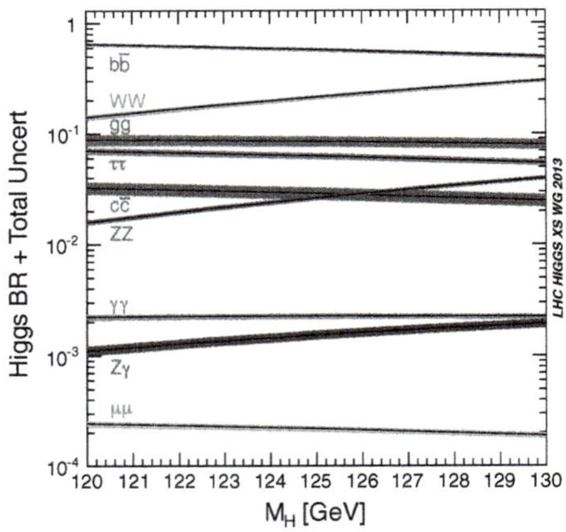

Fig. 7.22 Branching ratios expected for the decay of a Higgs boson of mass between 120 and 130 GeV. From K.A. Olive et al. (Particle Data Group), Chin. Phys. C 38 (2014) 090001

state into four light charged leptons is relatively easy to separate from the background. The decay $H \rightarrow \gamma\gamma$, although suppressed at the per mil level, has a clear signature; since it happens via a loop, it provides indirect information on the Higgs couplings to WW, ZZ, and $t\bar{t}$.

On July 4, 2012, a press conference at CERN followed worldwide finally announced the observation at the LHC detectors ATLAS and CMS of a narrow resonance with a mass of about 125 GeV, consistent with the SM Higgs boson. The evidence was statistically significant, above five standard deviations in either experiment; decays to $\gamma\gamma$ and to $ZZ \rightarrow 4$ leptons were detected, with rates consistent with those predicted for the SM Higgs.

Two candidate events are shown in Fig. 7.23, and we stress the word "candidate." Detection involves a statistically significant excess of such events, but the background from accidental $\gamma\gamma$ or four-lepton events is important (Fig. 7.24).

Later, the statistics increased, and the statistical significance as well; a compilation of the experimental data by the PDG is shown in Fig. 7.25. The present fitted value for the mass is

$$m_H = 125.09 \pm 0.24 \,\mathrm{GeV}/\mathrm{c}^2\,, \tag{7.104}$$

consistent with the bounds (7.102) and (7.103).

The 2013 Nobel prize in physics was awarded to François Englert and Peter Higgs "for the theoretical discovery of a mechanism that contributes to our understanding of the origin of mass of subatomic particles, and which recently was confirmed through the discovery of the predicted fundamental particle, by the ATLAS and CMS experiments at CERN's Large Hadron Collider."

The discovery of the Higgs was of enormous resonance. First of all, it concluded a 50-year long search based on a theoretical prediction. Then, it was the solution of a puzzle: the Higgs particle is the "last particle" in the minimal standard model. It should be emphasized, however, that this discovery does not conclude the research in fundamental physics. Some phenomena are not explained by the standard model: neither gravity, nor the presence of dark matter. In addition, as seen in Sect. 7.3.2, the

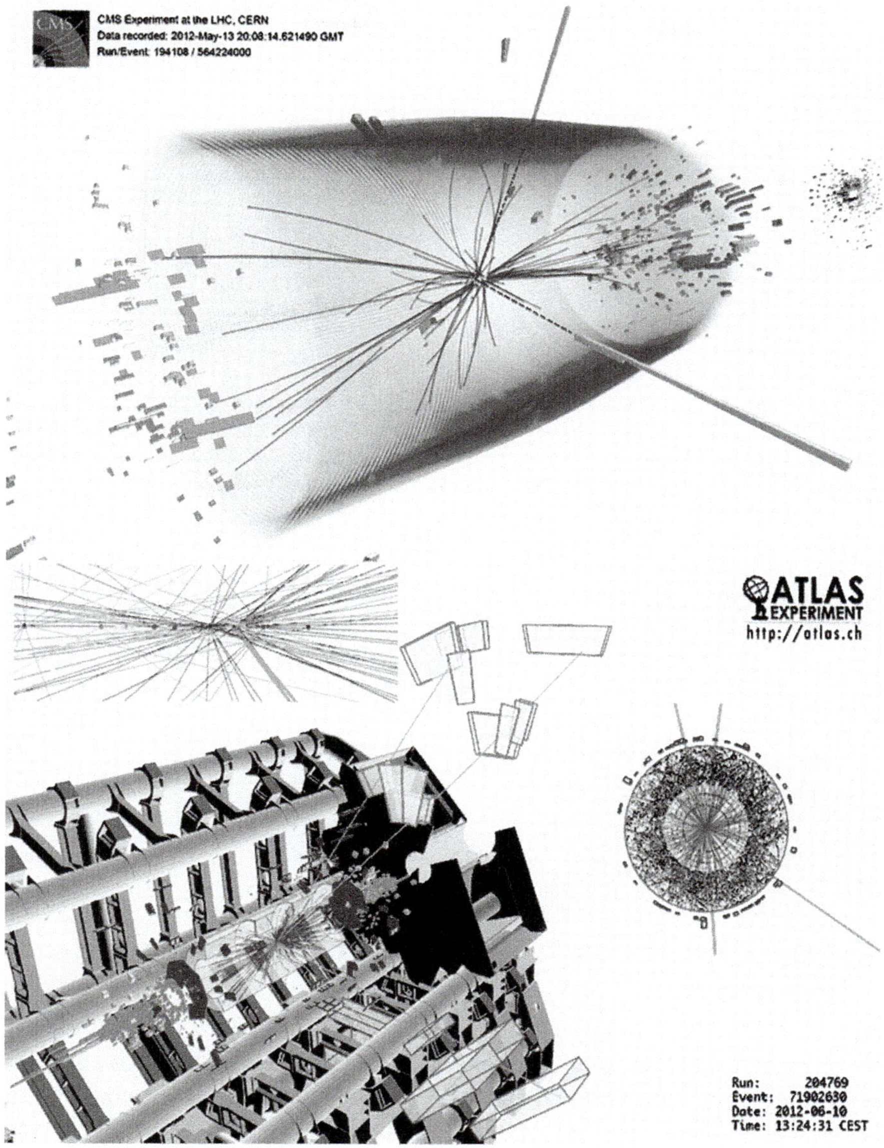

Fig. 7.23 Candidate Higgs boson events at the LHC. The *upper* is a Higgs decay into two photons (*dashed lines* and *towers*) recorded by CMS. The *lower* is a decay into four muons (*thick solid tracks*) recorded by ATLAS. *Source* CERN

SM has many free parameters: is this a minimal set, or some of them are calculable? It is very likely that the SM emerges from a more general (and maybe conceptually simpler) theory.

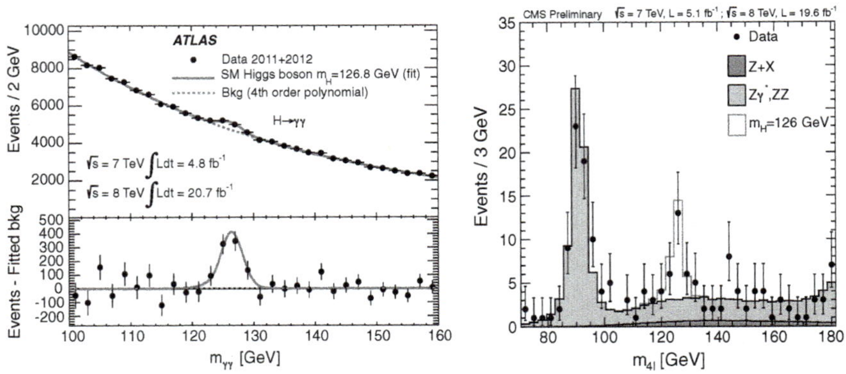

Fig. 7.24 Invariant mass of the $\gamma\gamma$ candidates in the ATLAS experiment (*left*; in the *lower* part of the plot the residuals from the fit to the background are shown) and of the 4-lepton events in CMS (*right*; the expected background is indicated by the *dark area*, including the peak close to the Z mass and coming from $Z\gamma^*$ events). From K.A. Olive et al. (Particle Data Group), Chin. Phys. C 38 (2014) 090001

Fig. 7.25 Decay channels currently measured for the Higgs boson. From K.A. Olive et al. (Particle Data Group), Chin. Phys. C 38 (2014) 090001

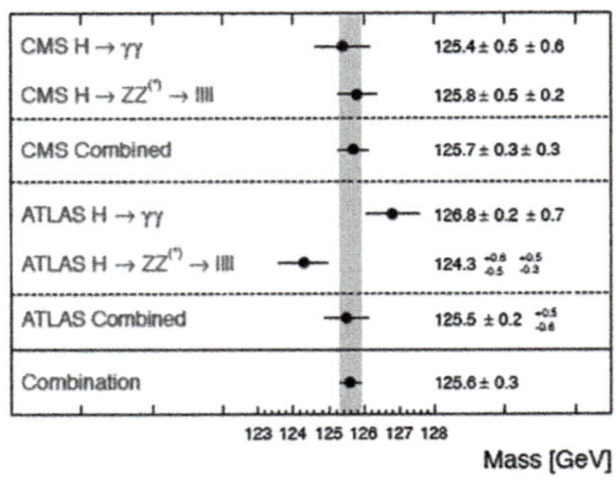

7.6 Beyond the Minimal SM of Particle Physics; Unification of Forces

We have studied the standard model of particle physics, and we have seen that this model is very successful in describing the behavior of matter at the subatomic level.

Can it be the final theory? This looks very unlikely: the SM seems rather an ad hoc model, and the $\mathrm{SU}(3) \otimes \mathrm{SU}(2) \otimes \mathrm{U}(1)$ looks like a low-energy symmetry which must be part of a bigger picture.

First of all, the standard model looks a bit too complicated to be thought as the fundamental theory. There are many particles, suggesting some higher symmetries (between families, between quarks and leptons, between fermions and bosons) grouping them in supermultiplets. There are many free parameters, as we have seen in Sect. 7.3.2.

Then, it does not describe gravity, which is the interaction driving the evolution of the Universe at large scale.

It does not describe all particles: as we said in Chap. 1, and as we shall discuss in larger detail in the next chapters, we have good reasons to believe that matter in the Universe is dominated by a yet undiscovered kind of particles difficult to accommodate in the standard model, the so-called dark matter.

Last but not least, one of the most intriguing questions is discussed as follows. The fundamental constants have values consistent with conditions for life as we know; sometimes this requires a fine tuning. Take for example the difference between the mass of the neutron and the mass of the proton, and the value of the Fermi constant: they have just the values needed for a Sun-like star to develop its life cycle in a few billions of years, which is the time needed for life as we know to develop and evolve. Is this just a coincidence or we miss a global view of the Universe? A minimal explanation is an "anthropic coincidence," which leads to the so-called "anthropic principle". The anthropic principle in its weak form can be expressed as "conditions that are observed in the universe must allow the observer to exist" (which sounds more or less like a tautology; note that the conditions are verified "here" and "now"), while one of the strongest forms states that "the Universe must have those properties which allow life to develop within it at some stage in its history." It is clear that on this question the borderline between physics and philosophy is very narrow, but we shall meet concrete predictions about the anthropic principle when shortly introducing the superstring theory. Just to conclude this argument which should deserve a deeper treatment, we cannot avoid the observation that discussions about existence are relevant only for civilizations evolute enough to think of the question—and we are one.

To summarize, many different clues indicate that the standard model is a work in progress and will have to be extended to describe physics at higher energies. Certainly, a new framework will be required close to the Planck scale $\sim 10^{18}$ GeV, where quantum gravitational effects become important. Probably, there is a simplified description of Nature at higher energies, with a prospect for the unification of forces.

As we have seen, renormalization entails the idea of coupling constants "running" with energy. At the energies explored up to now, the "strong" coupling constant is larger than the electromagnetic constant, which in turn is larger than the weak constant. The strong constant, however, decreases with increasing energy, while the weak and electromagnetic constants increase with energy. It is thus very tempting to conjecture that there will be an energy at which these constant meet—we know already that the weak and electromagnetic constants meet at a large energy scale. The evolution of the couplings with energy could be, qualitatively, shown in Fig. 7.26.

However, if we evolve the coupling constants on the basis of the known physics, i.e., of the standard model of particle physics, they will fail to meet at a single point (Fig. 7.29, left). The plot suggests the possibility of a grand unification scale at about 10^{16} eV, but, if we want that unification of the relevant forces happens, we must assume that there is new physics beyond the standard model. If also gravity will enter this grand unification scheme, there is no clue of how and at what energy level such unification will happen—but we shall see later that a hint can be formulated.

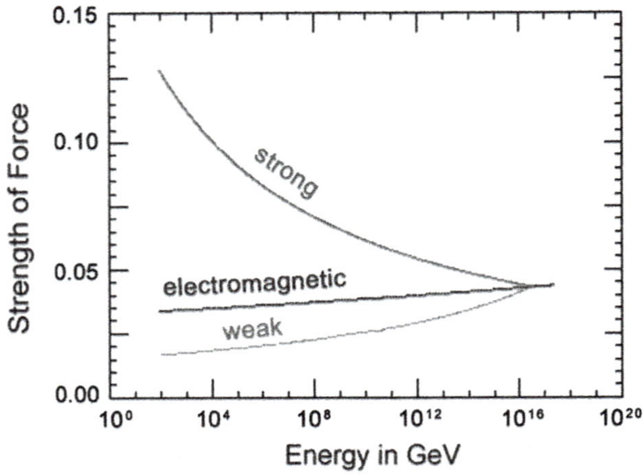

Fig. 7.26 Artistic scheme (qualitative) of the unification of the interaction forces

A unification mechanism requires symmetry groups which include the SM group; it should symmetrise the known particles. Some "grand unification" mechanisms have been proposed; in the following, we shall review the most popular.

7.6.1 GUT

The gauge theory of electroweak interactions is unified at high energy under the group $SU_L(2) \otimes U_Y(1)$. This symmetry is spontaneously broken at low energy splitting the electromagnetic and weak interactions. Given the structure of gauge theories in the standard model, it is tempting to explore the possibility that $SU_c(3) \otimes SU_L(2) \otimes U_Y(1)$ is unified by a larger group G at very large scales of energy such that

$$G \rightarrow SU_c(3) \otimes SU_L(2) \otimes U_Y(1) .$$

The smallest group including $SU_c(3) \otimes SU_L(2) \otimes U_Y(1)$ is SU(5), proposed by Georgi and Glashow in 1974; this approach, being the first proposed, is called by default the GUT (grand unified theory)—but of course any group including SU(5) can play the game. We shall describe in some detail in this section this "minimal" SU(5) GUT since it is the lowest rank (it has the smallest number of generators) GUT model and provides a good reference point for non-minimal GUTs. However, we should take into account the fact that this simple model has been shown experimentally inadequate, as discussed later.

The symmetry group has 24 generators, which include the extension to rank 5 of the generators of the standard model. A five-dimensional state vector allows to include quarks and leptons in the same vector. As an example, right states will be described all together in a spinor

$$\psi = (d_R, d_G, d_B, e^+, \bar{\nu}_e)\,, \tag{7.105}$$

where the subscript appended to quarks indicates the color.

In addition to the usual transitions involving the exchange of color and the W exchange, gauge bosons corresponding to some of the generators can mediate transitions between quarks and leptons (Fig. 7.27) via the exchange of two new gauge bosons X and Y with electric charges $-4/3$ and $-1/3$, respectively. When extrapolating at masses of order $M_U \sim 10^{15}$ GeV, where M_U is the unification mass, all the processes are characterized by a single "grand unified coupling constant" g_U. At energies $E \ll M_U$, processes involving the exchange of the X and Y bosons are heavily suppressed because of the large masses of these gauge fields, in the same way, as the $W^\pm$ exchange processes are suppressed relative to electromagnetic ones at energies $E \ll M_W$ in the unified electroweak theory.

The Georgi–Glashow GUT is elegant, in the sense that it allows an energy evolution of constants toward a possible unification of forces. In addition, it explains why the quark charges are fractional. We remind that generators of a special unitary group are traceless: the charge operator comes out to be one of the generators of SU(5).

Moreover, with one free parameter only (for example, assuming that there is actually unification of the interaction strengths, the unification scale, $M_U \simeq 10^{15}$ GeV), the theory predicts a value of $\sin^2\theta_W$ close to the one which has been experimentally determined, and the evolution of the $\mathrm{SU}(3)_C$, $\mathrm{SU}(2)_L$, and $\mathrm{U}(1)_Y$ coupling constants.

Unfortunately, the theory predicts twelve new gauge bosons which are color triplets and flavor doublets as well—they are thus called lepto-quarks. These gauge particles should acquire mass near the unification scale M_U and give rise to the new physics beyond the standard model; among the consequent new phenomena, proton decay has been the object of an intensive and so far unsuccessful experimental search. The $quark \rightarrow lepton$ transition can make the proton unstable via the diagrams of Fig. 7.28. Note that the decay channel

$$p \rightarrow e^+\pi^0$$

has a clear experimental signature. From the unification mass M_U, one can compute

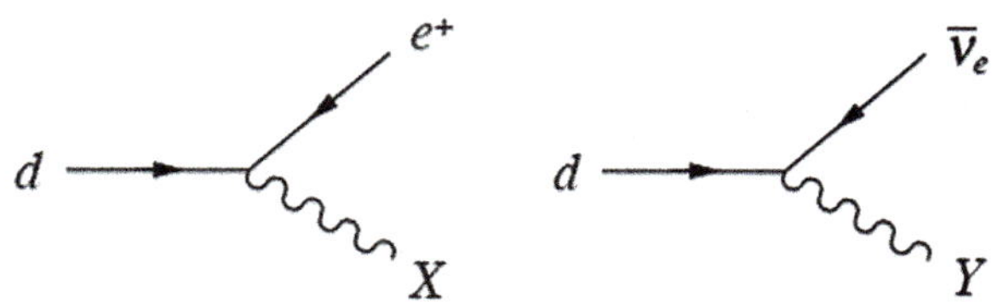

Fig. 7.27 Transitions between quarks and leptons are possible in GUTs

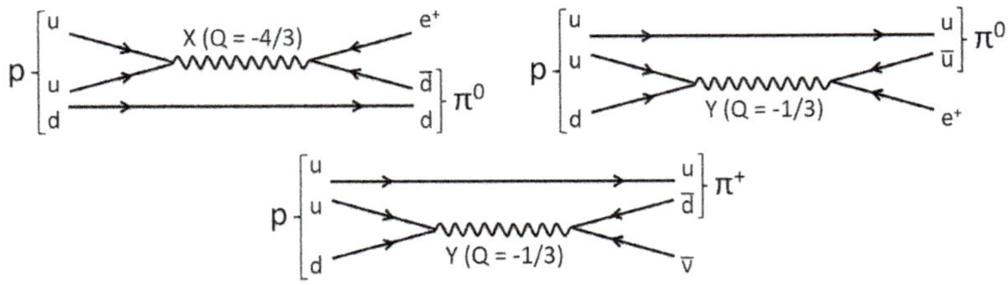

Fig. 7.28 Some mechanisms for proton decay in the SU(5) GUT

$$\tau_p \sim 10^{29} \text{ years}.$$

This is a strong prediction: baryonic mass should be unstable.

The experimental lower limit on the proton lifetime,

$$\tau_p > 5.9 \times 10^{33} \text{ years} \tag{7.106}$$

assuming the branching fractions computed by means of the minimal GUT, rules out the theory. In addition, the LEP precision data indicate that the coupling constants fail to meet exactly in one point for the current value of $\sin^2\theta_W$, as accurately measured by LEP. Of course one can save GUT by going to "non-minimal" versions, in particular with larger groups and a larger number of Higgs particles; in this way, one loses however simplicity and part of the elegance of the idea—apart possibly for the unification provided by supersymmetry, which we shall examine in next Section.

In his book, "The trouble with physics" (2007), Lee Smolin writes: *"After some twenty-five years, we are still waiting. No protons have decayed. We have been waiting long enough to know that* SU(5) *grand unification is wrong. It's a beautiful idea, but one that nature seems not to have adopted. [...] Indeed, it would be hard to underestimate the implications of this negative result.* SU(5) *is the most elegant way imaginable of unifying quarks with leptons, and it leads to a codification of the properties of the standard model in simple terms. Even after twenty-five years, I still find it stunning that* SU(5) *doesn't work."*

7.6.2 Supersymmetry

The most popular among non-minimal GUTs in particle physics is supersymmetry. Supersymmetry (SUSY) involves a symmetry between fermions and bosons: a SUSY transformation changes a boson into a fermion and vice versa. A supersymmetric theory is invariant under such a transformation. As a result, in a supersymmetric theory, each fermion has a superpartner which is a boson. In the same way, each boson possesses a superpartner which is a fermion. Supersymmetry interconnects different spin particles. This implies an equal number of fermionic and bosonic degrees of freedom.

By convention, the superpartners are denoted by a tilde. Scalar superpartners of fermions are identified by adding an "s" to the name of normal fermions (example: the selectron is the partner of the electron), while fermionic superpartners of bosons are identified by adding a "ino" at the end of the name (the photino is the superpartner of the photon). In the Minimal Supersymmetric Standard Model (MSSM), the Higgs sector is enlarged with respect to the SM, having at least two Higgs doublets. The spectrum of the minimal supersymmetric standard model therefore reads as in Table 7.2.

SUSY is clearly an approximate symmetry, otherwise the superpartners of each particle of the standard model would have been found, since they would have the same mass as the normal articles. But as of today, no supersymmetric partner has been observed. For example, the selectron would be relatively easy to produce in accelerators.

Superpartners are distinguished by a new quantum number called R-parity: the particles of the standard model have parity R = 1, and we assign a parity R = −1 to their superpartners. R-parity is a multiplicative number; if it is conserved, when attempting to produce supersymmetric particles from normal particles, they must be produced in pairs. In addition, a supersymmetric particle may disintegrate into lighter particles but one will have always at least a supersymmetric particle among the products of disintegration.

Always in the hypothesis of R-parity conservation (or small violation), a stable (or stable over cosmological times) lightest supersymmetric particle must exist, which can no longer disintegrate. The nature of the lightest supersymmetric particle (LSP) is a mystery. If it is the residue of all the decays of supersymmetric particles from the beginning of the Universe, one would expect that LSPs are abundant. Since we did not find it, yet, it must be neutral and it does not interact strongly.

The LSP candidates are then the lightest sneutrino and the lightest neutralino χ_0 (four neutralinos are the mass eigenstates coming from the mixtures of the zino and

Table 7.2 Fundamental particles in the minimal supersymmetric standard model: particles with $R = 1$ (left) and $R = -1$ (right)

Symbol	Spin	Name	Symbol	Spin	Name
e, μ, τ	1/2	Leptons	$\tilde{e}, \tilde{\mu}, \tilde{\tau}$	0	Sleptons
ν_e, ν_μ, ν_τ	1/2	Neutrinos	$\tilde{\nu}_e, \tilde{\nu}_\mu, \tilde{\nu}_\tau$	0	Sneutrinos
d, u, s, c, b, t	1/2	Quarks	$\tilde{d}, \tilde{u}, \tilde{s}, \tilde{c}, \tilde{b}, \tilde{t}$	0	Squarks
g	1	Gluon	$\tilde{g}$	1/2	Gluino
γ	1	Photon	$\tilde{\gamma}$	1/2	Photino
$W^\pm, Z$	1	EW gauge bosons	$\tilde{W}^\pm, \tilde{Z}$	1/2	Wino, zino
H_1, H_2	0	Higgs	$\tilde{H}_1, \tilde{H}_2$	1/2	Higgsinos

the photino, and the neutral higgsinos; in the same way, the mass eigenstates coming from the mixture of the winos and the charged higgsinos are called *charginos*). The LSP is stable or almost stable, and difficult to observe because neutral and weakly interacting.

The characteristic signature of the production of a SUSY LSPs would be missing energy in the reaction. For example, if the LSP is a neutralino (which has a "photino" component), the production of a selectron–antiselectron pair in e^+e^- collisions at LEP could be followed by the decay of the two selectrons in final states involving an LSP, the LSPs being invisible to the detection. Since no such events have been observed, a firm limit

$$M_{\mathrm{LSP}} > M_Z/2$$

can be set.

An attractive feature of SUSY is that it naturally provides the unification of forces. SUSY affects the evolution of the coupling constants, and SUSY particles can effectively contribute to the running of the coupling constants only for energies above the typical SUSY mass scale (the mass of the LSP). It turns out that within the Minimal Supersymmetric Standard Model (MSSM), i.e., the SUSY model requiring the minimal amount of particles beyond the standard model ones, a perfect unification of interactions can be obtained as shown in Fig. 7.29, right. From the fit requiring unification, one finds preferred values for the break point M_{LSP} and the unification

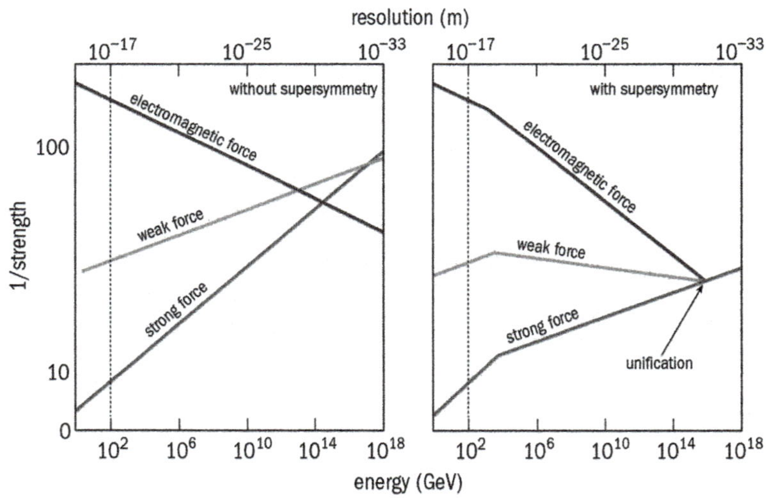

Fig. 7.29 The interaction couplings $\alpha_i = g_i^2/4\pi\hbar c$ fail to meet at a single point when they are extrapolated to high energies in the standard model, as well as in SU(5) GUTs. Minimal SUSY SU(5) model (*right*) does cause the couplings to meet in a point. While there are other ways to accommodate the data, this straightforward, unforced fit is encouraging for the idea of supersymmetric grand unification (Adapted from S. James Gates, Jr., http://live.iop-pp01.agh.sleek.net/2014/09/25/sticking-with-susy/; adapted from Ugo Amaldi, CERN)

point M_{GUT}:

$$M_{\rm LSP} = 10^{3.4\pm1.0}\ \text{GeV}, \qquad (7.107)$$
$$M_{\rm GUT} = 10^{15.8\pm0.4}\ \text{GeV}.$$

The observation in Fig. 7.29, right, was considered as the first "evidence" for supersymmetry, especially since $M_{\rm LSP}$ and $M_{\rm GUT}$ have "good" values with respect to a number of open problems.

In addition, the LSP provides a natural candidate for the yet unobserved competent of matter, the so-called dark matter that we introduced in Chap. 1 and we shall further discuss in the next chapter, and which is believed to be the main component of the matter in the Universe. Sneutrino-dominated dark matter is, however, ruled out in the MSSM due to the current limits on the interaction cross section of dark matter particles with ordinary matter. These limits have been provided by direct detection experiments—the sneutrino interacts via Z boson exchange and would have been detected by now if it makes up the dark matter.

Neutralino dark matter is thus the favored possibility. Neutralinos come out in SUSY to be Majorana fermions, i.e., each of them is identical with its antiparticle. Since these particles only interact with the weak vector bosons, they are not directly produced at hadron colliders in copious numbers. A neutralino in a mass consistent with Eq. 7.107 would provide, as we shall see, the required amount of "dark matter" to comply with the standard model of cosmology.

Gravitino dark matter is a possibility in non-minimal supersymmetric models incorporating gravity in which the scale of supersymmetry breaking is low, around 100 TeV. In such models, the gravitino can be very light, of the order of one eV. The gravitino is sometimes called a super-WIMP, as happens with dark matter, because its interaction strength is much weaker than that of other supersymmetric dark matter candidates.

7.6.3 Strings and Extra Dimensions; Superstrings

Gravity could not be turned into a renormalisable field theory up to now. One big problem is that classical gravitational waves carry spin $j = 2$, and present gauge theories in four dimensions are not renormalisable—the quantum loop integrals related to the graviton go to infinity for large momenta, i.e., as distance scales go to zero. Gravity could be, however, renormalisable in a large number of dimensions.

The starting point for string theory is the idea that the point-like elementary particles are just our view of one-dimensional objects called strings (the string scale being smaller than what is measurable by us, i.e., the extra dimension is compactified at our scales).

The analog of a Feynman diagram in string theory is a two-dimensional smooth surface (Fig. 7.30). The loop integrals over such a smooth surface do not meet the zero

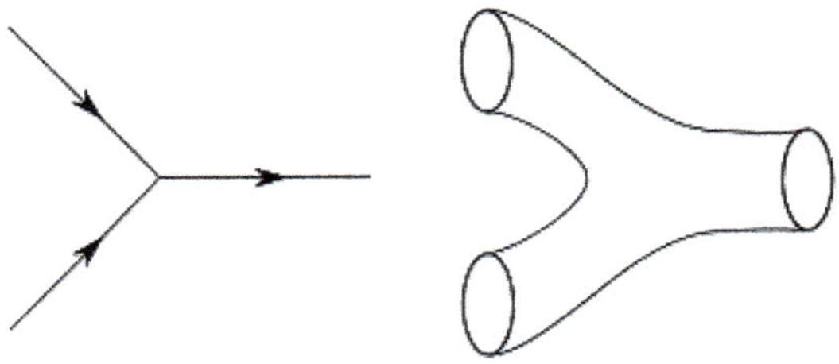

Fig. 7.30 *Left* In the Feynman representation, interactions can occur at zero distance—but gravity cannot be renormalized at zero distance. *Right* Adding an extra dimension and treating particles as strings solves the problem

distance, infinite momentum problems of the integrals over particle loops. In string theory, infinite momentum does not even mean zero distance. Instead, for strings, the relationship between distance and momentum is, roughly,

$$\Delta x \sim \frac{1}{p} + \frac{p}{T_s}$$

where T_s is called the string tension, the fundamental parameter of string theory. The above relation implies a minimum observable length for a quantum string theory of

$$L_{\text{min}} \sim \frac{1}{\sqrt{T_s}} .$$

The zero-distance behavior which is so problematic in quantum field theory becomes irrelevant in string theories, and this makes string theory very attractive as a theory of quantum gravity. The string theory should be the theory for quantum gravity, then this minimum length scale should be at least the size of the Planck length, which is the length scale made by the combination of Newton's constant, the speed of light, and Planck's constant:

$$L_{\text{min}} \sim \sqrt{G} \sim 10^{-35}\,\text{m} . \tag{7.108}$$

One can further generalize the concept on string adding more than one dimension: in this case, we speak more properly of *branes*. In dimension p, these are called $p-$branes.

String theories that include fermionic vibrations, incorporating supersymmetry, are known as superstring theories; several kinds have been described, based on symmetry groups as large as $SO(32)$, but all are now thought to be different limits of a general theory called M-theory. In string theories, spacetime is at least ten-dimensional—it is eleven-dimensional in M-theory.

7.6.3.1 Extra Dimensions Can Reduce the Number of Elementary Particles

An infinite number of N-dimensional particles arises naturally in a $N + 1-$ dimensional particle theory. To a N-dimensional observer, the velocity and momentum of a given particle in the hidden extra dimension, which is too small to observe, are invisible. But a particle moving in the $(N + 1)$th dimension has a nonzero energy, and the $N-$dimensional observer attributes this energy to the particle's mass. Therefore, for a given particle species living in $N + 1$ dimensions, each allowed energy level gives rise to a new elementary particle from the N-dimensional perspective.

A different way of expressing the same concept is that at distance scales larger than the string radius, each oscillation mode of the string gives rise to a different species of particle, with its mass, charge, and other properties determined by the string's dynamics. Particle emission and absorption correspond to the splitting and recombination of string, giving rise to the interactions between particles.

7.6.3.2 Criticism

Although successful, elegant, and theoretically fascinating, string theory is subject to many criticisms as the candidate to the theory of everything (ToE). In particular, it hardly meets the criterion of being falsifiable, since the energies needed to test it can be pushed to values so large that they cannot be reached experimentally. In addition, several versions of the theory are acceptable, and, once one is chosen, it lacks uniqueness of predictions. The vacuum structure of the theory contains an infinite number of distinct meta-stable vacua—some believe that this is a good thing, because it allows a natural anthropic explanation of the observed values of the physical constants.

7.6.4 Compositeness

As we observed in the beginning of this section, one of the characteristics of the standard model which makes it unlikely the final theory is the presence of three families of quarks and three families of leptons, the second and the third family looking more or less like replicas of the first one.

We were assuming up to now that quarks and leptons are fundamental; we should not forget that in the past, just a century ago, scientists thought that atoms were fundamental—and they could describe approximately their interactions. It was then discovered that atoms are composed of protons, neutrons, and electrons, and protons and neutrons are, in turn, composed of quarks, and some more fundamental interactions are regulating the components. Are these interactions, that the SM of particle physics describes successfully, really fundamental?

We should consider the possibility that quarks and leptons (and maybe also the vector bosons) are composite particles, made of even more elementary constituents.

This would change completely our view of nature, as it happened twice in the last century—thanks to relativity and quantum physics. The 12 known elementary particles have their own repeating patterns, suggesting they might not be truly fundamental, in the same way as the patterns on the atomic structure evidenced by Mendeleev suggested that atoms are not fundamental.

The presence of fundamental components of quarks and leptons could reduce the number of elementary particles and the number of free parameters in the SM. A number of physicists have attempted to develop a theory of "pre-quarks," which are called, in general, *preons*.

A minimal number of two different preons inside a quark or a lepton could explain the lightest family of quarks and leptons; the other families could be explained as excitations of the fundamental states. For example, in the so-called *rishon* model, there are two fundamental particles called rishons (which means "primary" in Hebrew); they are spin 1/2 fermions called T ("Third" since it has an electric charge of $e/3$) and V ("Vanishing," since it is electrically neutral). All leptons and all flavors of quarks are ordered triplets of rishons; such triplets have spin 1/2. They are built as follows: TTT = anti-electron; VVV = electron neutrino; TTV, TVT and VTT = three colors of up quarks; TVV, VTV and VVT = three colors of down antiquarks.

In the rishon model, the baryon number and the lepton number are not individually conserved, while $B - L$ is (demonstrate it); more elaborated models use three preons.

At the mass scale at which preons manifest themselves, interaction cross sections should rise, since new channels are open; we can thus set a lower limit of some 10 TeV to the possible mass of preons, since they have not been found at LHC. The interaction of UHE cosmic rays with the atmosphere reaches some 100 TeV in the center-of-mass, and thus cosmic rays are the ideal laboratory to observe such pre-constituents, if they exist and if their mass is not too large.

Further Reading

[F7.1] F. Halzen and Martin, "Quarks and Leptons: An Introductory Course in Modern Particle Physics," Wiley 1984. A book at early graduate level providing in a clear way the theories of modern physics in a how-to approach which teaches people how to do calculations.

[F7.2] M. Thomson, "Modern Particle Physics," Cambridge University Press 2013. A recent, pedagogical, and rigorous book covering the main aspects of Particle Physics at advanced undergraduate and early graduate level.

[F7.3] B.R. Martin and G.P. Shaw, "Particle Physics," Wiley 2009. A book at undergraduate level teaching the main concepts with very little calculations.

[F7.4] M. Merk, W. Hulsbergen, I. van Vulpen,"Particle Physics 1," Nikhef 2014. Lecture notes for one semester master course covering in a clear way the basics of electrodynamics, weak interactions, and electroweak unification and in particular symmetry breaking.

[F7.5] J. Romão, 'Particle Physics," 2014, http://porthos.ist.utl.pt/Public/textos/fp.pdf. Lecture notes for one semester master course on theoretical particle physics which is also a very good introduction to quantum field theory.

Exercises

1. *Symmetry breaking introducing a real scalar field.* Consider a simple model where a real scalar field Φ is introduced being the Lagrangian of such field:

$$L = \frac{1}{2}\left(\partial_{\mu}\Phi\right)^{2} - \frac{1}{2}\mu^{2}\Phi^{2} - \frac{1}{4}\lambda\Phi^{4}$$

Discuss the particle spectrum originated by small quantum perturbation around the minimum of the potential (vacuum) for the cases $\mu^2 > 0$ and $\mu^2 < 0$. Can such model accommodate a Goldstone boson?

2. *Handling left and right projection operators.* Demonstrate that

$$\left(\frac{1}{2}(1-\gamma_5)\right)\left(\frac{1}{2}(1-\gamma_5)\right) = \left(\frac{1}{2}(1-\gamma_5)\right);$$

$$\left(\frac{1}{2}(1-\gamma_5)\right)\left(\frac{1}{2}(1+\gamma_5)\right) = 0.$$

3. *Fermion mass terms.* Show that the fermion mass term $\mathcal{L}_f = -m_f\bar{\psi}_f\psi_f$ is gauge invariant in QED but not in $\mathrm{SU}(2)_L \otimes \mathrm{U}(1)_Y$.
4. *Mass of the photon.* Show that in the standard model the diagonalization of the $(W3_{\mu}, B_{\mu})$ mass matrix ensures the existence of a massless photon.
5. *Fermion couplings.* Verify that choosing the weak hypercharge according to the Gell-Mann Nishijima formula ($Y = 2Q - I_3$) ensures the right couplings of all fermions with the electroweak neutral bosons (Z, Y).
6. $\sin^2\theta_W$. Determine the value of $\sin^2\theta_W$ from the experimental measurements of

   (a) G_F and M_W;
   (b) M_W and M_Z.

7. *W decays.* Compute at leading order the ratio of the probabilities that a $W^{\pm}$ boson decays into leptons to the probability that it decays into hadrons.
8. *GIM mechanism.* Justify the GIM mechanism, discussed in Sect. 6.3.6.
9. *Fourth Family exclusion limit at LEP.* One of the first results of LEP was the exclusion at a level of 5σ of a fourth family of light neutrinos. Estimate the number of hadronic events that had to be detected to establish such a limit taking into account only statistical errors.

10. *Higgs decays into* ZZ. The Higgs boson was first observed in the $H \to \gamma\gamma$ and $H \to ZZ \to 4\ leptons$ decay channels. Compute the branching fraction of $ZZ \to \mu^+\mu^-\mu^+\mu^-$ normalized to $ZZ \to anything$.
11. *Higgs decays into* $\gamma\gamma$. Design the lowest order Feynman diagrams for the decay of the Higgs boson in $\gamma\gamma$ and discuss why this channel was a golden channel in the discovery of the Higgs boson.

Chapter 8
The Standard Model of Cosmology and the Dark Universe

This chapter introduces the observational data on the structure, composition, and evolution of the Universe, within the framework of the theory of general relativity, and describes the model currently providing the best quantitative description. In particular, we will illustrate the experimental evidence suggesting the existence of new forms of matter of energy, and describe the expansion, the chemical evolution, and the formation of structures, from the beginning of time—that, we believe, started with a phase transition from a singularity: the "big bang".

The origin and fate of the Universe is, for many researchers, the fundamental question. Many answers were provided over the ages, a few of them built over scientific observations and reasoning. During the last century important scientific theoretical and experimental breakthroughs occurred after Einstein's proposal of the General Theory of Relativity in 1915, with precise and systematic measurements establishing the expansion of the Universe, the existence the Cosmic Microwave Background, and the abundances of light elements in the Universe. The fate of the Universe can be predicted from its energy content—but, although the chemical composition of the Universe and the physical nature of its constituent matter have occupied scientists for centuries, we do not know yet this energy content well enough.

We are made of protons, neutrons, and electrons, combined into atoms in which most of the energy is concentrated in the nuclei (baryonic matter); and we know a few more particles (photons, neutrinos, ...) accounting for a limited fraction of the total energy of atoms. However, the motion of stars in galaxies, as well as results about background photons in the Universe (both will be discussed in the rest of this chapter) are inconsistent with the presently known laws of physics, unless we assume that a new form of matter exists. This matter is not visible, showing little or no interaction with photons—we call it "dark matter". It is, however, important in the composition of the Universe, because its energy is a factor of five larger than the energy of baryonic matter.

A. De Angelis and M.J.M. Pimenta, *Introduction to Particle and Astroparticle Physics*, Undergraduate Lecture Notes in Physics,
DOI 10.1007/978-88-470-2688-9_8

Recently, the composition of the Universe has become even more puzzling, as observations imply an accelerated expansion. Such an acceleration can be explained by a new, unknown, form of energy—we call it "dark energy"—generating a repulsive gravitational force. Something is ripping the Universe apart.

The current view on the distribution of the total budget between these forms of energy is shown in Fig. 1.7. Note that we are facing a new Copernican revolution: we are not made of the same matter that most of the Universe is made of. Moreover, the Universe displays a global behavior difficult to explain, as we shall see in Sect. 8.1.1.

Today, at the beginning of the twenty-first century, the Big Bang model with a large fraction of dark matter (DM) and dark energy is widely accepted as "the standard model of cosmology", but no one knows what the "dark" part really is, and thus the Universe and its ultimate fate remain basically unknown.

8.1 Experimental Cosmology

About one century ago, we believed that the Milky Way was the only galaxy; today, we have a more-refined view of the Universe, and the field of experimental cosmology probably grows at a faster rate than any other field in physics. In the last century, we obtained unexpected results about the composition of the Universe, and its global structure.

8.1.1 The Universe Is Expanding

As introduced in Chap. 1, striking evidence that the Universe is expanding comes from the observation that most galaxies are receding in all directions with radial velocities v proportional to their distance d from us. This is the famous Hubble law

$$v = H_0\, d, \tag{8.1}$$

where $H_0 \simeq 68$ km s^{-1}Mpc^{-1} is the so-called Hubble constant (we shall see that it is not at all constant, and can change during the history of the Universe) which is often expressed as a function of a dimensionless parameter h defined as

$$h = \frac{H_0}{100 \text{ km s}^{-1} \text{ Mpc}^{-1}}\,. \tag{8.2}$$

However, velocity and distance are not directly measured. The main observables are the redshift z—i.e., the fractional wavelength shift observed in specific absorption lines (hydrogen, sodium, magnesium, ...) of the measured spectra of objects (Fig. 8.1)

$$z = \frac{\lambda_{\text{observed}} - \lambda_{\text{emitted}}}{\lambda_{\text{emitted}}} = \frac{\Delta\lambda}{\lambda_{\text{emitted}}}, \tag{8.3}$$

and the apparent luminosity of the celestial objects (stars, galaxies, supernovae, ...), for which we assume we know the intrinsic luminosity.

A redshift occurs whenever $\Delta\lambda > 0$ which is the case for the large majority of galaxies. There are notable exceptions ($\Delta\lambda < 0$, a blueshift) as the one of M31, the nearby Andromeda galaxy, explained by a large intrinsic velocity (peculiar velocity) oriented toward us.

The wavelength shifts were first observed by the US astronomer James Keeler at the end of the nineteenth century in the spectrum of the light reflected by the rings of Saturn, and later on, at the beginning of the twentieth century, by the US astronomer Vesto Slipher, in the spectral lines of several galaxies. In 1925 there were already around 40 galaxies with measured spectral lines.

These wavelength shifts were (and still often are) incorrectly identified as simple special relativistic Doppler shifts due to the movement of the sources. In this case z would be given by

$$z = \sqrt{\frac{1+\beta}{1-\beta}} - 1, \tag{8.4}$$

which in the limit of small β becomes

$$z \simeq \beta; \tag{8.5}$$

in terms of z the Hubble law can then be written as:

$$z \simeq \frac{H_0}{c} d\,. \tag{8.6}$$

However, the limit of small β is not valid for high redshift objects with z as high as 11 that have been observed in the last years—the list of the most distant object comprises more than 100 objects with $z > 7$ among galaxies or protogalaxies (the most abundant category), quasars, black holes, and even star or protostars. On the other hand, high redshift supernovae (typically $z \sim 0.1$ to 1) have been extensively studied. From these studies an interpretation of the expansion based on special relativity is clearly excluded: one has to invoke general relativity.

In terms of general relativity (see Sect. 8.2) the observed redshift is not due to any movement of the cosmic objects but due to the expansion of the proper space between them. This expansion has no center: an observer at any point of the Universe will see the expansion in the same way with all objects in all directions receding with radial velocities given by the same Hubble law and not limited by the speed of light (in fact for $z > \sim 1.5$ radial velocities are, in a large range of cosmological models, higher than c): it is the distance scale in the Universe that is changing.

Let us now write the distance between two objects as

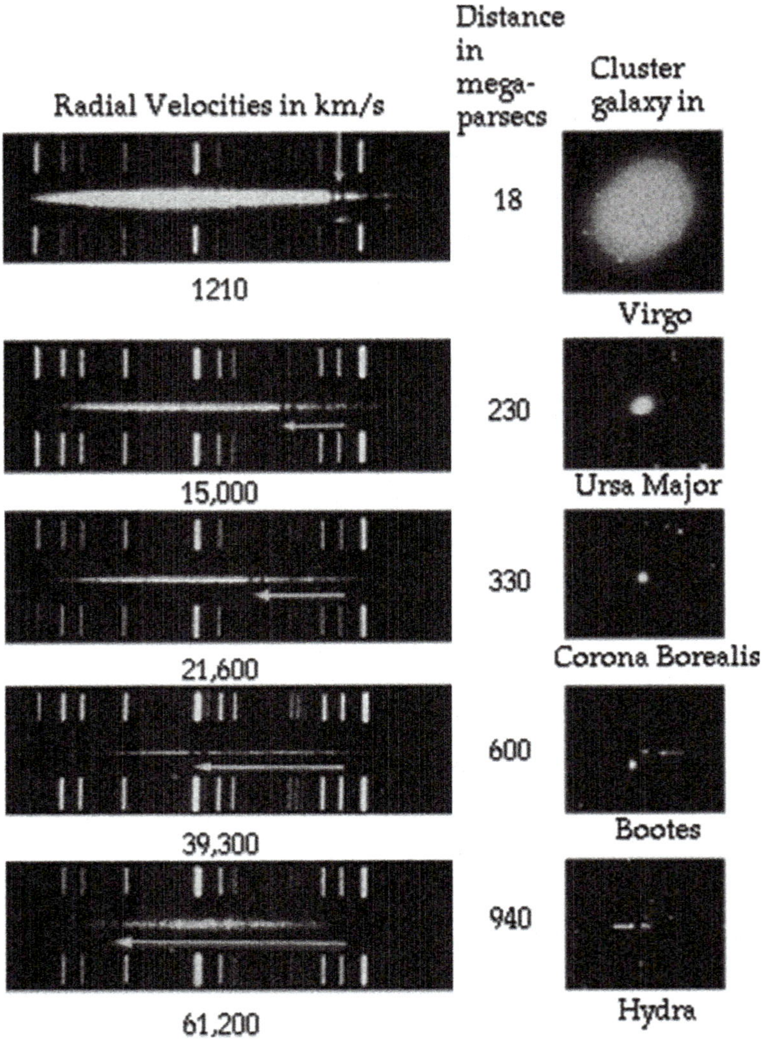

Fig. 8.1 Wavelength shifts observed in spectra of galaxies depending on their distance. From J. Silk, "The big bang", Times Books 2000

$$d = a(t)x, \tag{8.7}$$

where $a\,(t)$ is a scale that may change with time and x by definition is the present $(t = t_0)$ distance between the objects $(a\,(t_0) = 1)$ that does not change with time (comoving distance). Then

$$\dot{d} = \dot{a}\,x$$

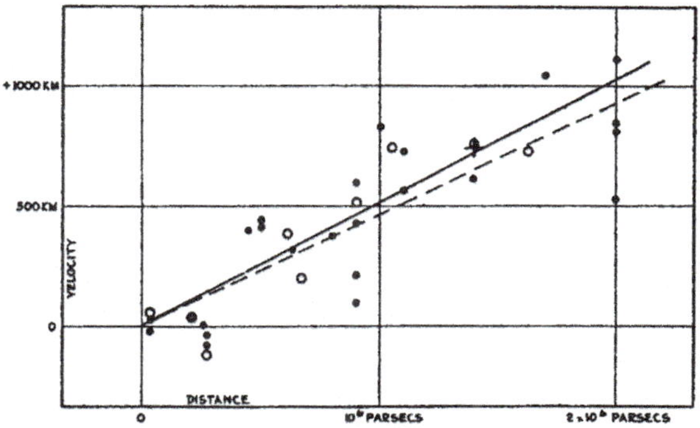

Fig. 8.2 The velocity–distance relation measured by Hubble (the original "Hubble plot"). From E. Hubble, Proceedings of the National Academy of Sciences 15 (1929) 168

$$v = H_0\, d$$

with

$$H_0 = \frac{\dot{a}}{a}\,. \tag{8.8}$$

The Hubble constant, in this simple model, is just the expansion rate of the distance scale in the Universe.

Let us come back to the problem of the measurement of distances. The usual method to measure distances is to use reference objects (standard candles), for which the absolute luminosity L is known. Then, assuming isotropic light emission in an Euclidean Universe (see Sect. 8.2) and measuring the corresponding light flux f on Earth, the distance d can be estimated as

$$d = \sqrt{\frac{L}{4\pi f}}\,. \tag{8.9}$$

Hubble in his original plot shown in Fig. 8.2 used as standard candles Cepheid[1] stars, as well as the brightest stars in the Galaxy, and even entire galaxies (assuming the absolute luminosity of the brightest stars and of the Galaxies to be approximately constant).

The original Hubble result showed a linear correlation between v and d but the slope (the Hubble constant) was wrong by a factor of 7 due to an overall calibration error caused mainly by a systematic underestimate of the absorption of light by dust.

A constant slope would mean that the scale distance $a(t)$ discussed above would increase linearly with time:

---

[1]Cepheids are variable red supergiant stars with pulsing periods strongly correlated with their absolute luminosity. This extremely useful propriety was discovered by the US astronomer Henrietta Leavitt at the beginning of twentieth century and has been used by Hubble to demonstrate in 1924 that the Andromeda Nebula M31 was too far to be part of our own galaxy, the Milky Way.

$$a(t) = a(t_0) + \dot{a}(t - t_0) ,$$

i.e.,

$$\frac{a(t)}{a(t_0)} = 1 + H_0 (t - t_0) . \tag{8.10}$$

Hubble in his original article suggested, under the influence of a model by de Sitter, that this linear behavior could be just a first-order approximation. In fact until recently (1998) most people were convinced that at some point the expansion should be slowed down under the influence of gravity which should be the dominant (attractive) force at large scale.

This is why the next term added to the expansion is usually written by introducing a deceleration parameter q_0 (if $q_0 > 0$ the expansion slows down) defined as

$$q_0 = -\frac{\ddot{a}\, a}{\dot{a}^2} = -\frac{\ddot{a}}{{H_0}^2 a}, \tag{8.11}$$

and then

$$\frac{a(t)}{a(t_0)} = 1 + H_0 (t - t_0) - \frac{1}{2} q_0 {H_0}^2 (t - t_0)^2 . \tag{8.12}$$

The relation between z and d must now be modified to include this new term.

However, in an expanding Universe the computation of the distance is much more subtle. Two distances are usually defined between two objects: the proper distance d_p and the luminosity distance d_L.

d_p is defined as the length measured on the spatial geodesic connecting the two objects at a fixed time (a geodesic is defined to be a curve whose tangent vectors remain parallel if they are transported along it. Geodesics are (locally) the shortest path between points in space, and describe locally the infinitesimal path of a test inertial particle). It can be shown (see Ref. [F8.2]) that

$$d_p \simeq \frac{c}{H_0} z \left(1 - \frac{1 + q_0}{2} z\right) ; \tag{8.13}$$

for small z the usual linear Hubble law is recovered.

d_L is defined as the distance that is experimentally determined using a standard candle assuming a static and Euclidean Universe as noted above:

$$d_L = \sqrt{\frac{L}{4\pi f}} . \tag{8.14}$$

The relation between d_p and d_L depends on the curvature of the Universe (see Sect. 8.2.1). Even in a flat (Euclidean) Universe (see Sect. 8.2.1 for a formal definition;

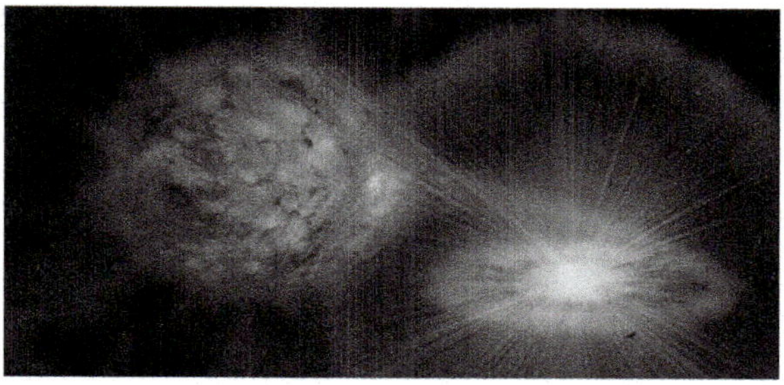

Fig. 8.3 Artistic representation of the formation and explosion of a supernova Ia (Image from A. Hardy, David A. Hardy/www.astroart.org)

for the moment, we rely on an intuitive one, and think of flat space as a space in which the sum of the internal angles of a triangle is always π) the flux of light emitted by an object with a redshift z and received at Earth is attenuated by a factor $(1+z)^2$ due to the dilation of time ($\gamma \simeq (1+z)$) and the increase of the photon's wavelength ($a^{-1} = (1+z)$). Then if the Universe was basically flat

$$d_L = d_p\,(1+z) \simeq \frac{c}{H_0} z \left[1 + \frac{1-q_0}{2} z \right] . \tag{8.15}$$

To experimentally determine q_0 one needs to extend the range of distances in the Hubble plot by a large amount. New and brighter standard candles are needed.

The type Ia supernovae have been revealed themselves as an optimal option in recent years. Supernovae Ia occur whenever, in a binary system formed by a small white dwarf and another star (for instance a red giant), the white dwarf accretes matter from its companion reaching a total critical mass of about 1.4 solar mass. At this point a nuclear fusion reaction starts, leading to a gigantic explosion (with a luminosity about 10^5 times larger than the brightest Cepheids; see Fig. 8.3 for an artistic representation).

The results obtained by the "Supernova Cosmology Project" and by the "High-z Supernova Search Team" resulted in extended Hubble plots (Fig. 8.4) that were a surprise and triggered a revolution in our understanding of the content and evolution of the Universe.[2] The striking point is that the fit to the experimental supernova (z, d) data leads to negative values of q_0 meaning that, contrary to what was expected, the expansion of the Universe is nowadays accelerating.

[2]The Supernova Cosmology Project is a collaboration, led by Saul Perlmutter, dedicated to the study of distant supernovae of type Ia, that started collecting data in 1988. Another collaboration also searching for distant supernovae of type Ia was formed by Brian Schmidt Adam Riess in 1994, the High-z Supernova Search Team. These teams found over 50 distant supernovae of type Ia for which the light received was weaker than expected—which implied that the rate of expansion of the Universe was increasing. Saul Perlmutter, born in 1959 in Champaign–Urbana, IL, USA, Ph.D. from University of California, Berkeley; Brian P. Schmidt, U.S. and Australian citizen, born in 1967 in USA, Ph.D. from Harvard; Adam G. Riess, born in 1969 in Washington, DC, USA. Ph.D. from Harvard, all professors in the US, were awarded the 2011 Nobel Prize in Physics "for the discovery of the accelerating expansion of the Universe through observations of distant supernovae."

An alternative method to the use of standard candles to determine extragalactic distances is the use of "standard rulers". Let us suppose that we know the absolute length l of an object (the standard ruler) that is placed at some distance transversally to the observation line. Then the distance of the object can be obtained from its angular size $\delta\theta$ by the simple formula:

$$d_A = \frac{l}{\delta\theta}, \tag{8.16}$$

where d_A is known as the angular diameter distance. In a curved and/or expanding Universe d_A does not coincide with the proper (d_p) and the luminosity (d_L) distances defined above but it can be shown (see Ref. [8.2]) that:

$$d_A = \frac{d_L}{(1+z)^2}. \tag{8.17}$$

Several candidates for standard rulers have been discussed in the last years and, in particular, the observation of Baryon Acoustic Oscillations (BAO) opened a new and promising path. BAO use the Fourier transform of the distance correlation function between specific astrophysics objects (for instance luminous red galaxies, blue galaxies, etc.) to discover, as function of the redshift z, possible clustering scales of the baryonic matter. These scales may be related to the evolution of initial density perturbations in the early Universe (see Sect. 8.3). The correlation function ξ between pairs of galaxies is just the probability that the two galaxies are at the distance r and thus a sharp peak in $\xi(r)$ will correspond in its Fourier transform to an oscillation spectrum with a well-defined frequency.

8.1.1.1 Dark Energy

There is no classical explanation for the accelerated expansion of the Universe. A new form of energy is invoked, permeating the space and exerting a negative pressure. This kind of energy can be described in the general theory of relativity (see later) and associated, e.g., to a "cosmological" term Λ; it is called "dark" energy—and to the present knowledge has the largest energy share in the Universe.

In Sect. 8.3.3 the current overall best picture able to accommodate all present experimental results (the so-called ΛCDM model) will be discussed.

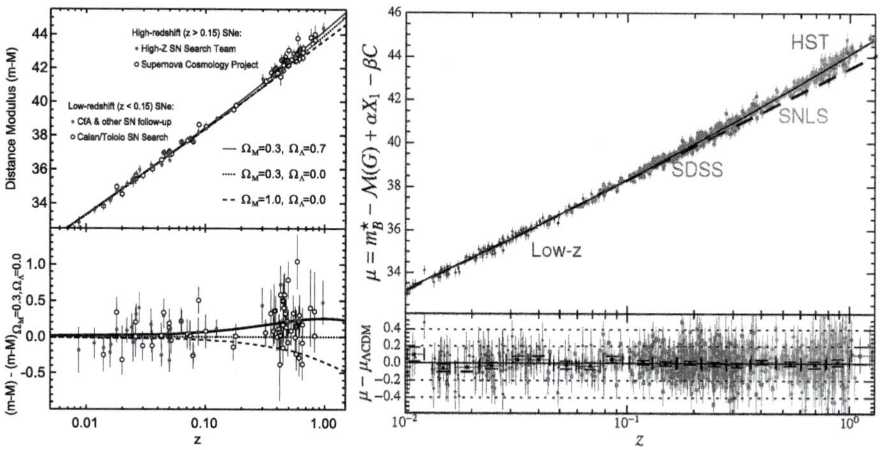

Fig. 8.4 *Left* The "Hubble plot" obtained by the "High-*z* Supernova Search Team" and by the "Supernova Cosmology Project". The *lines* represent the prediction of several models with different energy contents of the Universe (see Sect. 8.3.3). The best fit corresponds to an accelerating expansion scenario. From "Measuring Cosmology with Supernovae", by Saul Perlmutter and Brian P. Schmidt; Lecture Notes in Physics 2003, Springer. *Right* an updated version by the "Supernova Legacy Survey" and the "Sloan Digital Sky Survey" projects, M. Betoule et al. arXiv:1401.4064

8.1.2 Cosmic Microwave Background

In 1965 Penzias and Wilson,[3] two radio astronomers working at Bell Laboratories in New Jersey, discovered by accident that the Universe is filled with a mysterious isotropic and constant microwave radiation corresponding to a blackbody temperature around 3 K.

Penzias and Wilson were just measuring a small fraction of the black body spectrum. Indeed they were measuring the region in the tail around wavelength $\lambda \sim 7.5$ cm while the spectrum peaks around $\lambda \sim 2$ mm. To fully measure the density spectrum it is necessary to go above the Earth's atmosphere, which absorbs wavelengths lower than $\lambda \sim 3$ cm. These measurements were eventually performed in several balloon and satellite experiments. In particular, the Cosmic Background Explorer (COBE), launched in 1989, was the first to show that in the 0.1 to 5 mm range the spectrum is well described by the Planck blackbody formula

$$\varepsilon_\gamma(\nu)\,d\nu = \frac{8\pi h}{c^3}\,\frac{\nu^3 d\nu}{e^{\left(\frac{h\nu}{k_B T}\right)}-1}\,, \tag{8.18}$$

[3]Arno Penzias (1933–) was born in Munich, Germany. In 1939 his family was rounded up for deportation, but they managed to escape to the US, where he could graduate in Physics at Columbia University. Robert Wilson (1936–) grew up in Huston, Texas, and studied at Caltech. They shared the 1978 Nobel prize in Physics "for their discovery of the cosmic microwave background radiation."

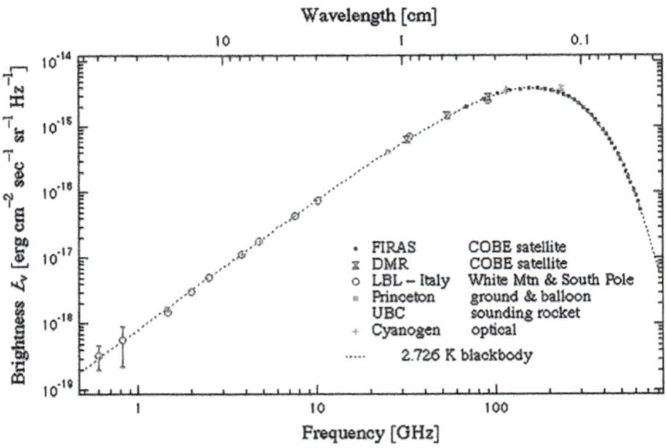

Fig. 8.5 The CMB intensity plot as measured by COBE and other experiments (from http://aether.lbl.gov/www/projects/cobe/CMB_intensity.gif)

where k_B is the Boltzmann constant. Other measurements at longer wavelengths confirmed that the cosmic microwave background (CMB) spectrum is well described by the spectrum of a single temperature blackbody (Fig. 8.5) with a mean temperature of

$$T = (2.726 \pm 0.001)\,\mathrm{K}\,.$$

The total photon energy density is then obtained by integrating the Planck formula over the entire frequency range, resulting in the Stefan–Boltzmann law

$$\varepsilon_\gamma = \frac{\pi^2}{15}\frac{(k_B\,T)^4}{(\hbar\,c)^3} \simeq 0.26\,\mathrm{eV\,cm^{-3}}\,; \tag{8.19}$$

moreover, the number density of photons is given by

$$n_\gamma \simeq \frac{2.4}{\pi^2}\left(\frac{k_B\,T}{\hbar c}\right)^3 \simeq 410\,\mathrm{cm^{-3}}\,. \tag{8.20}$$

The existence of CMB had been predicted in the 1940s by George Gamow, Robert Dicke, Ralph Alpher, and Robert Herman in the framework of the Big Bang model.

8.1.2.1 Recombination and Decoupling

In the Big Bang model the expanding Universe cools down going through successive stages of lower energy density (temperature) and more complex structures. Radiation materializes into pairs of particles and antiparticles, which, in turn, give origin to the existing objects and structures in the Universe (nuclei, atoms, planets, stars, galaxies, …). In this context, the CMB is the electromagnetic radiation left over when electrons and protons combine to form neutral atoms (the, so-called, *recombination* phase). After this stage, the absence of charged matter allows photons to be basically free of interactions, and evolve independently in the expanding Universe (*photon decoupling*).

In a simple, but reasonable, approximation (neglecting heavier elements, in particular helium) recombination occurs as the result of the balance between the formation and the photodisintegration of hydrogen atoms:

$$p + e^- \rightarrow \mathrm{H} + \gamma \,;\; \mathrm{H} + \gamma \rightarrow p\ e^- .$$

If these reactions are in equilibrium at a given temperature T (high enough to allow the photodisintegration and low enough to consider $e,\ p, H$ as nonrelativistic particles) the number density of electrons, protons, and hydrogen atoms may be approximated by the Maxwell–Boltzmann distribution (see Sect. 8.3.1)

$$n_x = g_x \left(\frac{m_x\, k_B\, T}{2\pi\, \hbar^2}\right)^{\frac{3}{2}} e^{-\left(\frac{m_x\, c^2}{k_B\, T}\right)} , \tag{8.21}$$

where g_x is a statistical factor accounting for the spin.

The ratio $n_\mathrm{H}/\left(n_p n_e\right)$ can then be approximately modeled by the Saha equation

$$\frac{n_H}{n_p n_e} \simeq \left(\frac{m_e k_B\, T}{2\pi\, \hbar^2}\right)^{-\frac{3}{2}} e^{\left(\frac{Q}{k_B\, T}\right)} , \tag{8.22}$$

where

$$Q = \left(m_p + m_e - m_H\right) c^2 \simeq 13.6\ \mathrm{eV} \tag{8.23}$$

is the hydrogen binding energy.

Defining X as the fractional ionization ($X = 1$ for complete ionization, whereas, $X = 0$ when all protons are inside neutral atoms),

$$X = \frac{n_p}{n_p + n_\mathrm{H}} , \tag{8.24}$$

and assuming that there is no total net charge, ($n_p = n_e$), the Saha equation can be rewritten as

$$\frac{1-X}{X} \simeq n_p \left(\frac{m_e\, k_B\, T}{2\pi\, \hbar^2}\right)^{-\frac{3}{2}} e^{\left(\frac{Q}{k_B\, T}\right)} . \tag{8.25}$$

On the other hand at thermal equilibrium, the energy density of photons as a function of the frequency ν follows the usual blackbody distribution corresponding, as we have seen before, to a photon density number of:

$$n_\gamma \simeq \frac{2.4}{\pi^2}\left(\frac{k_B\,T}{\hbar\,c}\right)^3 . \tag{8.26}$$

For the typical photodisintegration temperatures ($k_B\,T \sim 13.6$ eV)

$$n_\gamma \gg n_B ,$$

where n_B is the total number of baryons, which in this simple approximation is defined as

$$n_B = n_p + n_{\mathrm{H}} = \frac{n_p}{X} ; \tag{8.27}$$

we have

$$\eta = \frac{n_B}{n_\gamma} = \frac{n_p}{X\,n_\gamma} \ll 1 \tag{8.28}$$

in terms of the baryon to photon ratio.

After decoupling, n_B and n_γ evolve independently from each other, and both evolve as $a\,(t)^{-3}$, where $a\,(t)$ is the scale factor of the Universe, see Sect. 8.1.1. Thus, η is basically a constant, which can be measured at the present time through the measurement of the content of light elements in the Universe (see Sect. 8.1.3):

$$\eta \sim 5 - 6\; 10^{-10} . \tag{8.29}$$

The Saha equation can then be written as a function of η and T, and used to determine the recombination temperature (assuming $X \sim 0.5$):

$$\frac{1-X}{X^2} = 2 \simeq 3.84\,\eta\left(\frac{k_B\,T}{m_e\,c^2}\right)^{\frac{3}{2}} e^{\left(\frac{Q}{k_B\,T}\right)} . \tag{8.30}$$

The solution of this equation gives a remarkably stable value of temperature for a wide range of η (see Fig. 8.6). For instance $\eta \sim 5.5\; 10^{-10}$ results into

$$k_B\,T_{\mathrm{rec}} \simeq 0.323\,\mathrm{eV} \Longrightarrow T_{\mathrm{rec}} \simeq 3740\ \mathrm{K} . \tag{8.31}$$

This temperature is much higher than the measured CMB temperature reported above. The difference is attributed to the expansion of the Universe between the recombination epoch and the present. Indeed, as discussed in Sect. 8.1.1, the photon wavelength increases during the expansion of a flat Universe by a factor $(1+z)$. The entire CMB spectrum was expanded by this factor, and then it could be estimated that recombination had occurred at

$$z_{\mathrm{rec}} \sim 1300 - 1400 . \tag{8.32}$$

After recombination the Universe becomes substantially transparent to photons. The photon decoupling time is defined as the moment when the interaction rate of photons Γ_{γ_scat} equals the expansion rate of the Universe (which is given by the Hubble parameter)

$$\Gamma_{\gamma\_\text{scat}} \sim H\,. \qquad (8.33)$$

The dominant interaction process is the photon–electron Thomson scattering. Then

$$\Gamma_{\gamma\_\text{scat}} \simeq n_e \sigma_T c,$$

where n_e and σ_T are, respectively, the free electron density number and the Thomson cross section.

Finally as n_e can be related to the fractional ionization $X(z)$ and the baryon density number, ($n_e = X(z)\, n_B$) the redshift at which the photon decoupling occurs (z_{dec}) is given by

$$X(z_{\text{dec}})\, n_B \sigma_T c \sim H\,. \qquad (8.34)$$

However, the precise computation of z_{dec} is subtle. Both n_B and H evolve during the expansion (for instance in a matter-dominated flat Universe, as it will be discussed in Sect. 8.2, $n_B\,(z) \propto n_{B,0}(1+z)^3$ and $H\,(z) \propto H_0(1+z)^{\frac{3}{2}}$). Furthermore, the Saha equation is not valid after recombination once electrons and photons are no longer in thermal equilibrium. The exact value of z_{dec} depends thus on the specific model for the evolution of the Universe and the final result is of the order of

$$z_{\text{dec}} \sim 1100. \qquad (8.35)$$

After decoupling the probability of a further scattering is extremely low except at the reionization epoch. In fact, after the formation of the first stars, there is a period ($6 < z < 20$) when the Universe is still small enough for the neutral hydrogen formed at recombination to be ionized by the radiation emitted by stars. Still, the

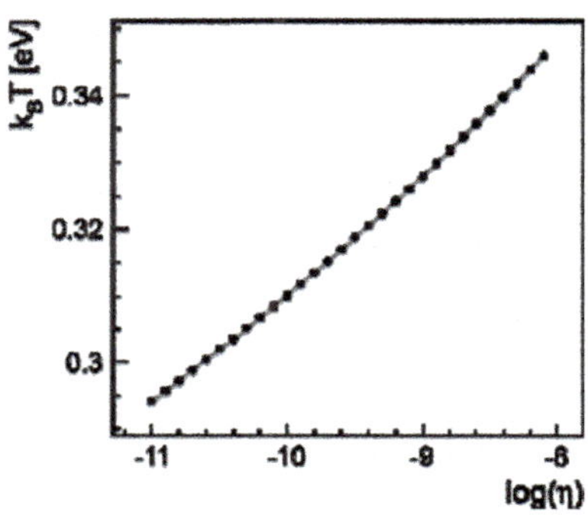

Fig. 8.6 Analytical solution for T as a function of η in the Saha equation

scattering probability of CMB photons during this epoch is small. To account for it, the reionization optical depth parameter τ is introduced, in terms of which the scattering probability is given by

$$P \sim 1 - e^{-\tau}.$$

The CMB photons follow then spacetime geodesics until they reach us. These geodesics are slightly distorted by the gravitational effects of the mass fluctuations close to the path giving rise to microlensing effects, which are responsible for a typical total deflection of ~2 arcminutes.

The spacetime points to where the last scattering occurred thereby define a region called the *last scattering surface* situated at a redshift, z_{lss}, very close to z_{dec}

$$z_{lss} \sim z_{dec} \sim 1100. \tag{8.36}$$

Beyond z_{lss} the Universe is opaque to photons and to be able to observe it other messengers, e.g., gravitational waves, will have to be studied. On the other hand, the measurement of the primordial nucleosynthesis (Sect. 8.1.3) allows to indirectly test the Big Bang model at times well before the recombination epoch.

8.1.2.2 Temperature Fluctuations

The COBE satellite[4] measured the temperature fluctuations in sky regions centered at different points with galactic coordinates (θ, Φ)

$$\frac{\delta T(\theta, \Phi)}{\langle T \rangle} = \frac{T(\theta, \Phi) - \langle T \rangle}{\langle T \rangle} \tag{8.37}$$

and found that, apart from a dipole anisotropy of the order of 10^{-3}, the temperature fluctuations are of the order of 10^{-5}: the observed CMB spectrum is remarkably isotropic.

[4]Three satellite missions have been launched so far to study the cosmic background radiation. The first was COBE in 1989, followed by WMAP (Wilkinson Microwave Anisotropy Probe) in 2001, both of which were NASA missions. The latest (with the best angular resolution and sensitivity), called Planck, has been launched by the European Space Agency (ESA) with a contribution from NASA in 2009, and is still in orbit. In terms of sensitivity and angular resolution, WMAP improved COBE by a factor of 40, and Planck gave a further improvement by a factor of 4; in addition Planck measures polarization. The instruments onboard Planck are a low frequency (solid state) instrument from 30 GHz, and a bolometer—a device for measuring the power of incident electromagnetic radiation via the heating of a material with a temperature-dependent electrical resistance—for higher frequencies (up to 900 GHz). The total weight of the payload is 2 tons (it is thus classified as a large mission); it needs to be kept at cryostatic temperatures. John Mather, from the Goddard Space Flight Center, and George Smoot, at the University of California, Berkeley, shared the 2006 Nobel Prize in Physics "for their discovery of the blackbody form and anisotropy of the cosmic microwave background radiation".

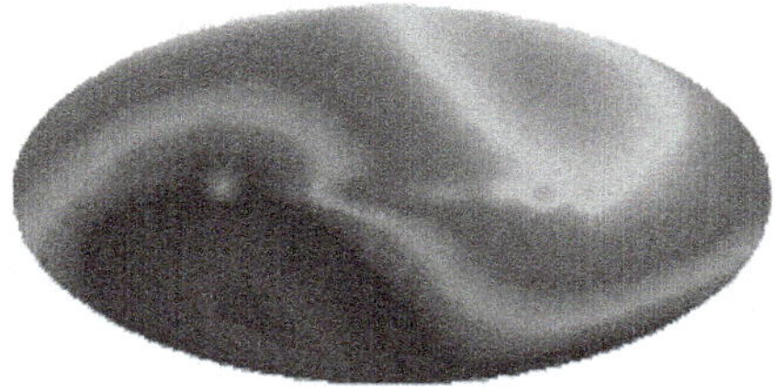

Fig. 8.7 Sky map (in galactic coordinates) of CMB temperatures measured by COBE after the subtraction of the emission from our Galaxy. A dipole component is clearly visible. (from http://apod.nasa.gov/apod/ap010128.html)

The dipole distortion (a slight blueshift in one direction of the sky and a redshift in the opposite direction—Fig. 8.7) observed in the measured average temperature can be attributed to a global Doppler shift due to the peculiar motion (COBE, Earth, Solar system, Milky way, local Group, Virgo cluster, ...) with respect to an hypothetical CMB isotropic reference frame characterized by a temperature T. Indeed

$$T^* = T\left(1 + \frac{v}{c}\cos\theta\right) \tag{8.38}$$

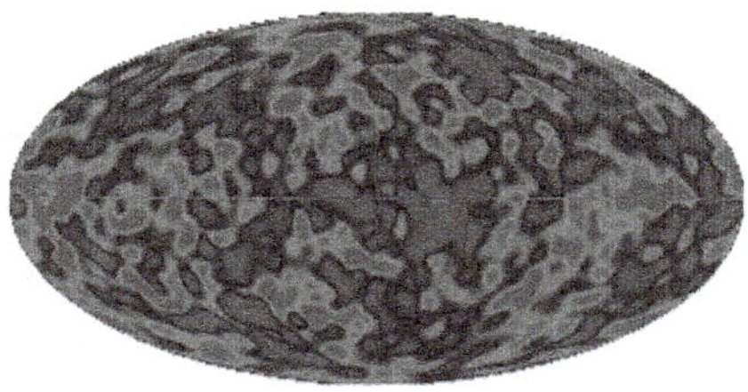

Fig. 8.8 CMB temperature fluctuations sky map as measured by COBE after the subtraction of the dipole component and of the emission from our Galaxy. http://lambda.gsfc.nasa.gov/product/cobe/

Fig. 8.9 CMB temperature fluctuations sky map as measured by the Planck mission after the subtraction of the dipole component and of the emission from our Galaxy. http://www.esa.int/spaceinimages/

with

$$v = (371 \pm 1)\,\mathrm{km/s}.$$

After removing this effect, the remaining fluctuations reveal a pattern of tiny inhomogeneities at the level of the last scattering surface. The original picture from COBE (Fig. 8.8) with an angular resolution of 7° was confirmed and greatly improved by the Wilkinson Microwave Anisotropy Probe WMAP, which obtained full sky maps with a 0.2° angular resolution. Very recently, the Planck satellite delivered sky maps with three times higher resolution and ten times higher sensitivity (Fig. 8.9), also covering a larger frequency range.

Once these maps are obtained it is possible to establish two-point correlations between any two spatial directions.

Technically, the temperature fluctuations are expanded using spherical harmonics

$$\frac{\delta T}{\langle T \rangle}(\theta, \Phi) = \sum_{l=0}^{\infty} \sum_{m=-l}^{l} a_{\mathrm{lm}} Y_{\mathrm{lm}}^{*}(\theta, \Phi)\,, \tag{8.39}$$

with

$$a_{\mathrm{lm}} = \int_{\theta=-\pi}^{\pi} \int_{\Phi=0}^{2\pi} \frac{\delta T}{\langle T \rangle}(\theta, \Phi)\; Y_{\mathrm{lm}}^{*}(\theta, \Phi)\; d\Omega. \tag{8.40}$$

Then the correlation between two directions $\hat{n}$ and $\hat{n}^{*}$ separated by an angle α is defined as

$$C(\alpha) = \left\langle \frac{\delta T}{\langle T \rangle}(\hat{n})\; \frac{\delta T}{\langle T \rangle}(\hat{n}^{*}) \right\rangle_{\hat{n}\,\cdot\,\hat{n}^{*} = \cos\alpha}$$

and can be expressed as

$$C(\alpha) = \frac{1}{4\pi} \sum_{l=0}^{\infty} (2l+1)\, C_l\, P_l(\cos\alpha)\,,$$

where P_l are the Legendre polynomials and the C_l, the multipole moments, are given by the variance of the harmonic coefficients a_{lm}:

$$C_l = \frac{1}{2l+1} \sum_{m=-l}^{l} \left\langle |\, a_{\mathrm{lm}}|^2 \right\rangle. \tag{8.41}$$

Each multipole moment corresponds to a sort of angular frequency l, whose conjugate variable is an angular scale α such that

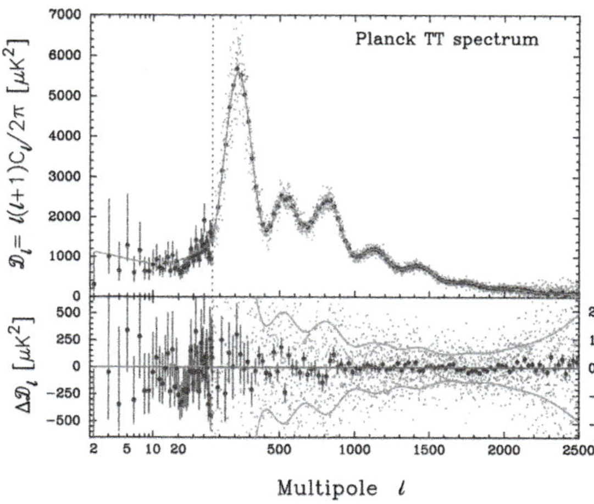

Fig. 8.10 Temperature power spectrum from the Planck experiment. In the lower panel the residuals with respect to the best ΛCDM fit (see later) are shown. Credits: Planck collaboration, P.A.R. Ade et al., Astr. Astrophys. 571 (2014) A16

$$\alpha = \frac{180^\circ}{l}\,. \tag{8.42}$$

The total temperature fluctuations (temperature powers spectrum) can be then expressed as a function of the multipole moment l (Fig. 8.10, top)

$$\langle \Delta T^2 \rangle = \left(\frac{l(l+1)}{2\pi} C_l \right) \langle T \rangle^2 \,. \tag{8.43}$$

Such a function shows a characteristic pattern with a first peak around $l \sim 200$ followed by several smaller peaks.

The first peak at an angular scale of 1° defines the size of the horizon at the time of last scattering (see Sect. 8.1.3) and the other peaks (acoustic peaks) are extremely sensitive to the specific contents and evolution model of the Universe at that time. The observation of very tiny fluctuations at large scales (much greater than the horizon, $l \ll 200$) leads to the hypothesis that the Universe, to be casually connected, went a through a very early stage of exponential expansion, called *inflation*.

Anisotropies can also be found studying the polarization of the CMB photons. Indeed at the recombination and reionization epochs the CMB may be partially polarized by Thomson scattering with electrons. It can be shown that linear polarization may also be originated by quadrupole temperature anisotropies. In general there are two polarization orthogonal modes, respectively, called B-mode (curl-like) and E-mode (gradient-like), but Thomson scattering only originates the E-mode while primordial gravitational waves are expected to display both polarization modes. Gravitational lensing of the CMB E-modes may also be a source of B-modes. E-modes were first measured in 2002 by the DASI telescope in Antarctica and recently the Planck collaboration published high resolution maps of the CMB polarization over the full sky. The detection and the interpretation of B-modes are very challenging since the signals are tiny and foreground contaminations, as the emission by galactic dust, are not always easy to estimate. The arrival angles of the CMB photons are smeared, due to the microlensing effects, by dispersions that are function of the integrated mass distribution along the photon paths. It is possible, however, to deduce these dispersions statistically from the observed temperature angular power spectra and/or from polarized E- and B-mode fields. The precise measurement of these dispersions will give valuable information for the determination of the cosmological parameters. It will also help constraining parameters, such as the sum of the neutrino masses or the dark energy content, that are relevant for the growth of structures in the Universe; and evaluating contaminations in the B-mode patterns from possible primordial gravity waves. In the last years, the detection of gravitational lensing was reported by several experiments such as the Atacama Cosmology Telescope, the South Pole Telescope, and the POLARBEAR experiment, and recently at a 40σ and 5σ significance by the Planck collaboration using, respectively, temperature and polarization data or only polarization data. Some of these aspects will be discussed briefly in Sect. 8.3 but a detailed discussion of the theoretical and experimental aspects of this fast-moving field is far beyond the scope of this book.

8.1.3 Primordial Nucleosynthesis

The measurement of the abundances of the light elements in the Universe (H, D, ^{3}He, ^{4}He,^{6}Li,^{7}Li) is the third observational "pillar" of the Big Bang model, after the Hubble expansion and the CMB. As it was proposed, and first computed, by the Russian American physicists, Ralph Alpher and George Gamow in 1948, the expanding Universe cools down and when it reaches temperatures of the order of the nuclei biding energies per nucleon ($\sim 1-10$ MeV) nucleosynthesis occurs if there are enough protons and neutrons available. The main nuclear fusion reactions are

- proton–neutron fusion:

$$p{+}n \;\rightarrow \mathrm{D}\,\gamma$$

- Deuterium–Deuterium fusion:

$$\mathrm{D{+}D} \;\rightarrow^3 \mathrm{He}\; n$$

$$\mathrm{D{+}D} \;\rightarrow^3 \mathrm{H}\; p$$

$$\mathrm{D{+}D} \;\rightarrow\; ^4\mathrm{He}\; \gamma$$

- other ^{4}He formation reactions:

$$^3\mathrm{He{+}D} \;\rightarrow\; ^4\mathrm{He}\; p$$

$$^3\mathrm{H{+}D} \;\rightarrow\; ^4\mathrm{He}\; n$$

$$^3\text{He}+n \rightarrow {}^4\text{He}$$

$$^3\text{H}+p \rightarrow {}^4\text{He}\,\gamma$$

- and finally the lithium and beryllium formation reactions (there are no stable nuclei with $A = 5$):

$$^4\text{He}+\text{D} \rightarrow {}^6\text{Li}\,\gamma$$

$$^4\text{He}+{}^3\text{H} \rightarrow {}^7\text{Li}\,\gamma$$

$$^4\text{He}+{}^3\text{He} \rightarrow {}^7\text{Be}\,\gamma$$

$$^7\text{Be}+\gamma \rightarrow {}^7\text{Li}\,p.$$

The absence of stable nuclei with $A = 8$ basically stops the primordial big bang nucleosynthesis chain. Heavier nuclei are produced in stellar (up to Fe) or supernova nucleosynthesis.[5]

The relative abundance of neutrons and protons, in case of thermal equilibrium at a temperature T, is fixed by the ratio of the usual Maxwell–Boltzmann distributions (similarly to what was discussed for the recombination—Sect. 8.1.2):

$$\frac{n_n}{n_p} = \left(\frac{m_n}{m_p}\right)^{\frac{3}{2}} \exp\left(-\frac{(m_n - m_p)\,c^2}{k\,T}\right). \qquad (8.44)$$

If $k_B\,T \gg (m_n - m_p)\,c^2 \longrightarrow n_n/n_p \sim 1$; if $k_B\,T \ll (m_n - m_p)\,c^2 \longrightarrow n_n/n_p \sim 0$.

Thermal equilibrium is established through the weak processes connecting protons and neutrons:

$$n+\nu_e \rightleftharpoons p+e^-$$

$$n+e^+ \rightleftharpoons p+\overline{\nu}_e$$

as long as the interaction rate of these reactions $\Gamma_{n,p}$ is greater than the expansion rate of the Universe,

$$\Gamma_{n,p} \geq H.$$

Γ and H evolve during the expansion diminishing the former much faster than the latter. Indeed in a flat Universe dominated by radiation (Sect. 8.2)

---

[5]Iron (^{56}Fe) is the stable element for which the binding energy per nucleon is largest (about 8.8 MeV); it is thus the natural endpoint of fusion processes of lighter elements, and of fission of heavier elements.

$$\Gamma_{n,p} \sim G_F\, T^5, \tag{8.45}$$

$$H \sim \sqrt{g^*}\, T^2, \tag{8.46}$$

where G_F is the Fermi weak interaction constant and g^* the number of degrees of freedom that depends on the relativistic particles content of the Universe (namely, on the number of generations of light neutrinos N_μ, which, in turn, allows to set a limit on N_μ).

The exact calculation of the freeze-out temperature T_f at which

$$\Gamma_{n,p} \sim H$$

is out of the scope of this book. The values obtained for T_f are a little below the MeV scale,

$$k_B\, T_f \sim 0.8\,\text{MeV}\,. \tag{8.47}$$

At this temperature

$$\frac{n_n}{n_p} \sim 0.2.$$

After the freeze-out this ratio would remain constant if neutrons were stable. However, as we know, neutrons decay via beta decay,

$$n \to\ p\, e^- \overline{\nu}_e.$$

Therefore, the $\frac{n_n}{n_p}$ ratio will decrease slowly while all the neutrons will not be bound inside nuclei, so that

$$\frac{n_n}{n_p} \sim 0.2\, e^{-t/\tau_n} \tag{8.48}$$

with $\tau_n \simeq 885.7$ s.

The first step of the primordial nucleosynthesis is, as we have seen, the formation of Deuterium via proton–neutron fusion

$$p{+}n\ \rightleftharpoons \mathrm{D}\,\gamma$$

Although the deuterium binding energy, 2.22 MeV, is higher than the freeze-out temperature, the fact that the baryons to photons ratio η is quite small ($\eta \sim 5-6\ 10^{-10}$) makes photodissociation of the deuterium nuclei possible at temperatures lower than the blackbody peak temperature T_f (the Planck function has a long tail). The relative number of free protons, free neutrons, and deuterium nuclei can be expressed, using a Saha-like equation (Sect. 8.1.2), as follows:

$$\frac{n_{\rm D}}{n_p n_n} \simeq \frac{g_{\rm D}}{g_p g_n}\left(\frac{m_{\rm D}}{m_p\, m_n}\right)^{\frac{3}{2}}\left(\frac{k_B\,T}{2\pi\,\hbar^2}\right)^{-\frac{3}{2}} e^{\left(\frac{Q}{k_B\,T}\right)}, \tag{8.49}$$

where Q is now given by

$$Q = \left(m_p + m_n - m_{\rm D}\right)\, c^2 \sim 2.22\ \text{MeV}.$$

Expressing n_p as a function of η and n_γ and performing an order of magnitude estimation, we obtain

$$\frac{n_{\rm D}}{n_n} \propto \eta\, n_\gamma \left(\frac{m_p\ c^2\, k_B\, T}{\pi\,\hbar^2}\right)^{-\frac{3}{2}} e^{\left(\frac{Q}{k_B\,T}\right)}. \tag{8.50}$$

Replacing now n_γ by the Planck distribution

$$\frac{n_{\rm D}}{n_n} \propto \eta \left(\frac{k_B\,T}{m_p\ c^2}\right)^{\frac{3}{2}} e^{\left(\frac{Q}{k_B\,T}\right)}. \tag{8.51}$$

This is analogous to the formulation of the Saha equation used to determine the recombination temperature (Sect. 8.1.2). As we have shown its solution (for instance for $(n_{\rm D}/n_n) \sim 1$) gives a remarkably stable value of temperature. In fact there is a sharp transition around $k_B\ T_{\rm D} \sim 0.1$ MeV: above this value neutrons and protons are basically free; below this value all neutrons are substantially bound first inside D nuclei and finally inside ^{4}He nuclei, provided that there is enough time before the fusion rate of nuclei becomes smaller than the expansion rate of the Universe. Indeed, since the ^{4}He binding energy per nucleon is much higher than those of the D, ^3H, and ^{3}He, and since there is no stable nuclei with $A = 5$, then ^{4}He is the most favorite final state.

The primordial abundance of ^{4}He, Y_p, is defined usually as the fraction of mass density of ^{4}He nuclei, $\rho(^4\text{He})$, over the total baryonic mass density, $\rho\,(\text{Baryons})$

$$Y_p = \frac{\rho\,(^4\text{He})}{\rho\,(\text{Baryons})}. \tag{8.52}$$

In a crude way let us assume that after nucleosynthesis all baryons are H or ^{4}He, i.e., that

$$\rho\,(\text{H}) + \rho\left(^4\text{He}\right) = 1.$$

Thus

$$Y_p = 1 - \frac{\rho\,(\text{H})}{\rho\,(\text{Baryons})}$$

$$Y_p = 1 - \frac{n_p - n_n}{n_p + n_n} = \frac{2\frac{n_n}{n_p}}{1 + \frac{n_n}{n_p}}. \tag{8.53}$$

For $(n_n/n_p) \sim 0.2$, $Y_p = 0.33$.

In fact due to the decay of neutrons between $k_B\ T_f \sim 0.8$ MeV and $k_B\ T_\mathrm{D} \sim 0.1$ MeV

$$\frac{n_n}{n_p} \sim 0.13 - 0.15$$

and the best estimation for Y_p is in the range

$$Y_p \sim 0.23 - 0.26\,. \tag{8.54}$$

Around one quarter of the baryonic mass of the Universe is due to ^{4}He and around three quarters is made of hydrogen.

There are however small fractions of D, ^{3}He, and ^{3}H that did not turn into ^{4}He, and there are, thus, tiny fractions of ^{7}Li and ^{7}Be that could be formed after the production of ^{4}He and before the dilution of the nuclei due to the expansion of the Universe. Although their abundances are quantitatively quite small, the comparison

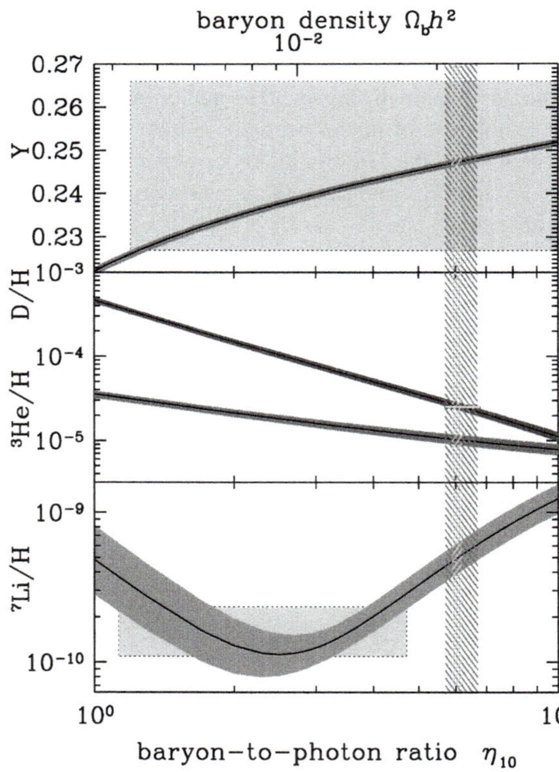

Fig. 8.11 The observed and predicted abundances of ^{4}He, D, ^{3}He, and ^{7}Li. Small (large) boxes represent the measurements within 2 sigma statistical (statistical and systematic) errors. The narrow *vertical* band represents the constraints on η (expressed in units of 10^{10} from the CMB power spectrum analysis while the wider is the best fit to experimental data). From K.A. Olive et al. (Particle Data Group), Chin. Phys. C 38 (2014) 090001

of the expected and measured ratios are important because they are rather sensitive to the ratio of baryons to photons, η.

In Fig. 8.11 the predicted abundances of ^{4}He, D, ^{3}He, and ^{7}Li computed in the framework of the standard model of Big Bang nucleosynthesis as a function of η are compared with recent measurements (for details see the Particle Data Book). An increase in η will increase slightly the Deuterium formation temperature T_{D} (there are less γ per baryon available for the photodissociation of the Deuterium), and therefore, there is more time for the development of the chain fusion processes ending at the formation of the ^{4}He. Therefore, the fraction of ^{4}He will increase slightly, in relative terms, and the fraction of D and ^{3}He will decrease much more significantly, again in relative terms. The evolution of the fraction of ^{7}Li is, on the contrary, not monotonous; it shows a minimum due to the fact that it is built up from two processes that have a different behavior (the fusion of ^{4}He and ^{3}H is a decreasing function of η; the production via ^{7}Be is an increasing function of η).

Apart from the measured value for the fraction of ^{7}Li all the other measurements converge to a common value of η that is, within the uncertainties, compatible with the value indirectly determined by the study of the acoustic peaks in the CMB power spectrum (see Sect. 8.3.3).

8.1.4 Astrophysical Evidence for Dark Matter

Evidence that the Newtonian physics applied to visible matter does not describe the dynamics of stars, galaxies, and galaxy clusters was well established in the twentieth century.

As a first approximation, one can estimate the mass of a galaxy based on its brightness: the brighter galaxies contain more stars. However, there are other ways to assess the total mass of a galaxy. In spiral galaxies, for example, stars rotate in quasicircular orbits around the center. The rotational speed of peripheral stars depends, according to Newton's law, on the total mass of the galaxy, and one has thus an independent measurement of this mass. Do these two methods give consistent results?

In 1933 the Swiss astronomer, Fritz Zwicky studied the motion of distant clusters of galaxies, in particular the Coma and the Virgin clusters. The masses of the galaxies in the clusters were estimated by Zwicky based on their brightnesses. Zwicky obtained the masses of the clusters from the sums of the galactic masses (based on the luminosity of galaxies) and compared them to the value of the total mass obtained independently from the measurement of the distributions of velocities of individual galaxies in the cluster (the mass can be derived from the average speed by the virial[6] theorem, stating that the average over time of the total kinetic energy

[6] The word *virial* comes from the latin *vis*, i.e., strength, or force; the term was coined by German physicist and mathematician Rudolf Clausius, one of the founders of the science of thermodynamics, around 1870.

of a stable system bound by a force proportional to $1/r^2$ is one-half of the average of the absolute value of its potential energy, when we set to zero value of potential energy at infinity). The second mass value was found to be 400 times larger than the first estimate, indicating the presence of nonluminous matter.

Despite this early evidence, it was only in the 1970s that scientists began to explore this discrepancy in a systematic way and that the existence of dark matter started to be quantified. It was realized that the discovery of dark matter would not only have solved the problem of the lack of mass in clusters of galaxies, but would also have had much more far-reaching consequences on our prediction of the evolution and fate of the Universe.

An important observational evidence of the need for dark matter was provided by the rotation curves of spiral galaxies—the Milky Way is one of them. Spiral galaxies contain a large population of stars placed on nearly circular orbits around the galactic center. Astronomers have conducted observations of the orbital velocities of stars in the peripheral regions of a large number of spiral galaxies, and found that the orbital speeds remain constant, contrary to the expected prediction of reduction at larger radii. The mass enclosed within the orbits radius must therefore gradually increase in regions even beyond the edge of the visible galaxy.

Later, another independent confirmation of Zwicky's findings came from gravitational lensing. Lensing is the effect predicted by general relativity that bends the light coming from distant objects, due to large massive bodies in the path to the observer. As such it constitutes another method of measuring the total content of gravitational matter. The obtained mass-to-light ratios in distant clusters match the dynamical estimates of dark matter in those clusters.

8.1.4.1 How Is DM Distributed in Galaxies?

If an amount of matter $M(r)$ is contained in a galaxy within a radius r (let us imagine, for simplicity, a spherically symmetric distribution), one can derive the velocity for a star orbiting at a fixed radius r from the fundamental laws of dynamics and gravitation (the centripetal acceleration being provided by the gravitational force). If μ is the mass of the star,

$$\frac{\mu v^2}{r} = G\frac{\mu M(r)}{r^2} \Longrightarrow v^2 = G\frac{M(r)}{r} .$$

As a consequence, in the halo of a Galaxy (where one expects from the distribution of luminous matter that $M(r) \simeq M_{\rm tot}$, the total mass of the Galaxy),

$$v \simeq \sqrt{\frac{GM_{\rm tot}}{r}} . \tag{8.55}$$

The above result holds also for a generic bounded orbit, as a consequence of the virial theorem.

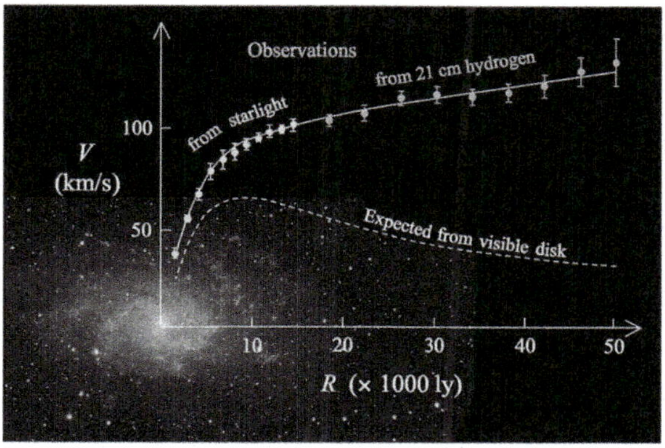

Fig. 8.12 Rotation curve of the Galaxy M33 (from http://en.wikipedia.org/wiki/Galaxy_rotation_curve#/media/File:M33_rotation_curve_HI.gif, public domain)

Experimental observations do not agree with this conclusion: the speed of stars as a function of the distance from the center (the so-called rotation curve of the galaxy) does not decrease as $1/\sqrt{r}$ in the halo (see for example, Fig. 8.12, referred to the Galaxy M33, the so-called Triangulum galaxy).

Stars at the outskirts of M33 have observed orbital velocities of about 150 kilometers per second, and would escape in reasonably short timescales if the galaxy were composed only of visible matter, since their orbital speeds are three times larger than the escape velocity from the galaxy.

The Milky Way itself displays a roughly saturating speed in the halo (Fig. 8.13).

To explain within Newtonian gravity an asymptotically constant or nearly constant speed in the halo, as observed in many spiral galaxies, an amount of matter $M(r)$ roughly proportional to the radius has to be considered—and thus a density ρ of dark matter decreasing as r^{-2} should exist, which is not observed in the luminous mass. Thus one can conclude that dark matter exists, having a distribution which does not fall off with radius as for luminous matter. Of course this matter is likely to be more

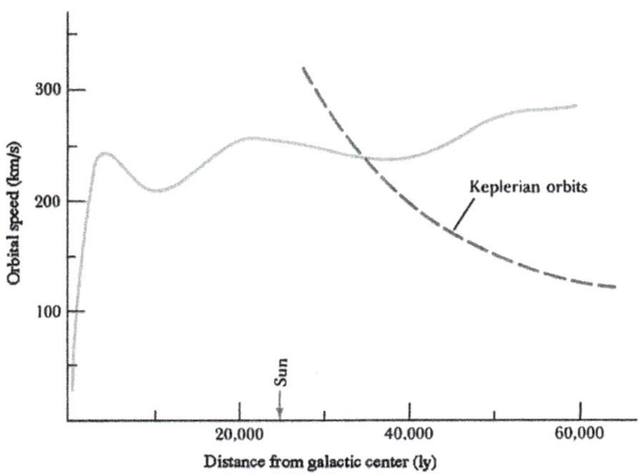

Fig. 8.13 Rotation curve of the Milky Way (from http://abyss.uoregon.edu)

densely packed in the center than luminous matter, but its effect is more visible in the halo, since there it is dominating.

Asymptotically for at large r, a density function ρ proportional to r^{-2} can be expressed as:

$$\text{Isothermal}: \ \rho(r) = \rho_{\text{iso}}(r) = \frac{\rho_s}{1 + (r/r_s)^2} \,. \tag{8.56}$$

This is the so-called isothermal spherical halo, since it can be derived also from simulations of a gas of particles in hydrostatic equilibrium and at uniform temperature. This model is described by two parameters: the central density, and the radius of the core r_s (alternately, the parameters can be the bulk radius and the limiting speed). The isothermal halo is therefore characterized as a rotation curve growing linearly starting from the center, approaching a flat rotation curve at large radii.

Profiles obtained in numerical simulations of dark matter including baryons are steeper in the center than those obtained from simulations with nonbaryonic dark matter only. The Navarro, Frenk, and White (NFW) profile, often used as a benchmark, follows a r^{-1} distribution at the center. On the contrary, the Einasto profile does not follow a power law at the GC, is smoother at kpc scales, and seems to fit better more recent numerical simulations. A value of about 0.17 for the shape parameter α in Eq. 8.57 is consistent with present data and simulations. Moore and collaborators have suggested profiles steeper than NFW.

The analytical expression of these profiles are

$$\begin{aligned}
\text{NFW}: \ & \rho_{\text{NFW}}(r) = \rho_s \frac{r_s}{r}\left(1 + \frac{r}{r_s}\right)^{-2} \\
\text{Einasto}: \ & \rho_{\text{Einasto}}(r) = \rho_s \exp\left\{-\frac{2}{\alpha}\left[\left(\frac{r}{r_s}\right)^{\alpha} - 1\right]\right\} \\
\text{Moore}: \ & \rho_{\text{Moore}}(r) = \rho_s \left(\frac{r_s}{r}\right)^{1.16}\left(1 + \frac{r}{r_s}\right)^{-1.84} \,.
\end{aligned} \tag{8.57}$$

Presently, there are no good observational measurements of the shape of the Milky Way; this is why one usually assumes a spherically symmetrical distribution. Figure 8.14 compares these different profiles with the constraint to fit the velocities in the halo of our galaxy.[7]

To distinguish between the functional forms for the halos is not easy. They vary among each other only in the central region, where the luminous matter is dominant. Needless to say, the high-density, central region is the most crucial for detection—and uncertainties there span three orders of magnitude.

---

[7] In the neighborood of the Solar System one has a DM density

$$\rho_{\text{DM,local}} \simeq 0.4\,\text{GeV}/\text{cm}^3 \,,$$

i.e., five orders of magnitude larger than the total energy density of the Universe.

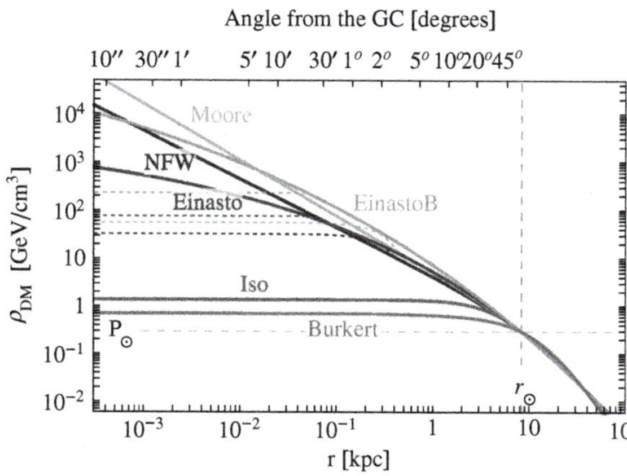

Fig. 8.14 Comparison of the densities as a function of the radius for DM profiles used in the literature, with values adequate to fit the radial distribution of velocities in the halo of the Milky Way. The curve EinastoB indicates an Einasto curve with a different α parameter. From M. Cirelli et al., "PPPC 4 DM ID: A Poor Particle Physicist Cookbook for Dark Matter Indirect Detection", arXiv:1012.4515, JCAP 1103 (2011) 051

Also because of this, one of the preferred targets for astrophysical searches for DM are small satellite galaxies of the Milky Way, the so-called dwarf spheroidals (dSph). For these galaxies the ratio between the estimate of the total mass M inferred from the velocity dispersion (velocities of single stars are measured with an accuracy of a few kilometers per second thanks to optical measurements) and the luminous mass L, inferred from the count of the number of stars, can be very large. The dwarf spheroidal satellites of the Milky Way could become tidally disrupted if they did not have enough dark matter. In addition these objects are not far from us: a possible DM signal should not be attenuated by distance dimming.

Table 8.1 A list of dSph satellites of the Milky Way that may represent the best candidates for DM searches according to their distance from the Sun, luminosity, and inferred M/L ratio

dSph	$D_\odot$ (kpc)	L ($10^3\ L_\odot$)	M/L ratio
Segue 1	23	0.3	>1000
UMa II	32	2.8	1100
Willman 1	38	0.9	700
Coma Berenices	44	2.6	450
UMi	66	290	580
Sculptor	79	2200	7
Draco	82	260	320
Sextans	86	500	90
Carina	101	430	40
Fornax	138	15500	10

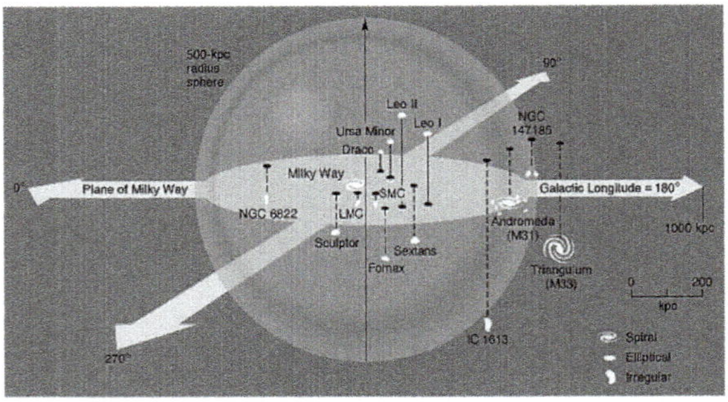

Fig. 8.15 The Local Group of galaxies around the Milky Way (from http://abyss.uoregon.edu/~js/ast123/lectures/lec11.html). The largest galaxies are the Milky Way, Andromeda, and M33, and have a spiral form. Most of the other galaxies are rather small and with a spheroidal form. These orbit closely the large galaxies, as is also the case of the irregular Magellanic Clouds, best visible in the southern hemisphere, which are located at a distance of about 120,000 ly to be compared with the Milky Way radius of about 50,000 ly

Table 8.1 shows some characteristics of dSph in the Milky Way; their position is shown in Fig. 8.15.

The observations of the dynamics of galaxies and clusters of galaxies, however, are not the only astrophysical evidence of the presence of DM. Cosmological models for the formation of galaxies and clusters of galaxies indicate that these structures fail to form without DM.

8.1.4.2 An Alternative Explanation: Modified Gravity

The dependence of v^2 on the mass $M(r)$ on which the evidence for DM is based relies on the virial theorem, stating that the absolute value of the kinetic energy is in average equal to the potential energy for a bound state (after defining zero energy at infinite distance). The departure from this Newtonian prediction could also be related to a departure from Newtonian gravity.

Alternative theories do not necessarily require dark matter, and replace it with a modified Newtonian gravitational dynamics. Notice that, in an historical perspective, deviations from expected gravitational dynamics already led to the discovery of previously *unknown matter* sources. Indeed, the planet Neptune was discovered following the prediction by Le Verrier in the 1840s of its position based on the detailed observation of the orbit of Uranus and Newtonian dynamics. In the late nineteenth century, the disturbances to the expected orbit of Neptune led to the discovery of Pluto. On the other hand, the precession of the perihelion of Mercury, which could not be *quantitatively* explained by Newtonian gravity, confirmed the prediction of general relativity—and thus a modified dynamics.

The simplest model of modified Newtonian dynamics is called MOND; it was proposed in 1983 by Milgrom, suggesting that for extremely small accelerations the

Newton's gravitational law may not be valid—indeed Newton's law has been verified only at reasonably large values of the gravitational acceleration. MOND postulates that the acceleration a is not linearly dependent on the gradient of the gravitational field ϕ_N at small values of the acceleration, and proposes the following modification:

$$\mu\left(\frac{a}{a_0}\right) a = \left|-\vec{\nabla}\phi_N\right| . \tag{8.58}$$

The function μ is positive, smooth, and monotonically increasing; it is approximately equal to its argument when the argument takes small values compared to unity (deep MOND limit), but approaches unity when that argument is large. a_0 is a constant with dimensions, approximately equal to $10^{-10}\,\mathrm{m\,s^{-2}}$.

Let us now consider again stars orbiting a galaxy with speed $v(r)$ at radius r. For large r values, a will be smaller than a_0 and we can approximate $\mu(x) \simeq x$. One has then

$$\frac{v^4}{r^2} \simeq a_0 \frac{GM}{r^2} .$$

In this limit, the rotation curve flattens at a typical value v_f given by

$$v_f = (MGa_0)^{1/4} . \tag{8.59}$$

MOND explains well the shapes of rotation curves. But for the masses of clusters, one finds an improvement but the problem is not completely solved.

The likelyhood that MOND is the full explanation for the anomaly observed in the velocities of stars in the halo of galaxies is not strong. An explanation through MOND would require an *ad hoc* theory to account for cosmological evidence as well. In addition, a recent (2004) astrophysical observation, the merging galaxy cluster 1E0657-58 (the so-called bullet cluster), has further weakened the MOND hypothesis. The bullet cluster consists of two colliding clusters of galaxies, at a distance of about 3.7 Gly. In this case (Fig. 8.16), the distance of the center of mass to the center of baryonic mass cannot be explained by changes in the gravitational law, as indicated by data with a statistical significance of 8σ.

One could also consider the fact that galaxies may contain invisible matter of known nature, either baryons in a form which is hard to detect optically, or massive neutrinos—MOND, then reduces the amount of invisible matter needed to explain the observations.

8.1.5 Age of the Universe

The age of the Universe is an old question. Has the Universe a finite age? Or is the Universe eternal and always equal to itself (steady state Universe)?

Fig. 8.16 The matter in the "bullet cluster" is shown in this composite image (from http://apod.nasa.gov/apod/ap060824.html, credits: NASA/CXC/CfA/ M. Markevitch et al.). In this image depicting the collision of two clusters of galaxies, the bluish areas show the distributions of dark matter in the clusters, as obtained with gravitational lensing, and the red areas correspond to the hot x-ray emitting gases. The individual galaxies observed in the optical image data have a total mass much smaller than the mass in the gas, but the sum of these masses is far less than the mass of dark matter. The clear separation of dark matter and gas clouds is a direct evidence of the existence of dark matter

For sure the Universe must be older than the oldest object that it contains and the first question has been then: how old is the Earth? In the eleventh century, the Persian astronomer, Abu Rayhan al-Biruni had already realized that Earth should have a finite age, but he just stated that the origin of Earth was too far away to possibly measure it. In the nineteenth century, there were finally the first quantitative estimates. From considerations, both, on the formation of the geological layers, and on the thermodynamics of the formation and cooling of Earth, it was estimated that the age of the Earth should be of the order of tens of millions of years. These estimates were in contradiction with both, some religious beliefs, and Darwin's theory of evolution. Rev. James Ussher, an Irish Archbishop, published in 1650 a detailed calculation concluding that according to the Bible "God created Heaven and Earth" some six thousand years ago (more precisely "at the beginning of the night of October 23rd in the year 710 of the Julian period", which means 4004 B.C.). On the other hand, tens or even a few hundred million years seemed to be a too short time to allow for the slow evolution advocated by Darwin. Only the discovery of radioactivity at the end of nineteenth century provided precise clocks to date rocks and meteorite debris with, and thus to allow for reliable estimates of the age of the Earth. Surveys in the Hudson Bay in Canada found rocks with ages of over four billion ($\sim$4.3 $\times$ 10^9) years. On the other hand measurements on several meteorites, in particular on the Canyon Diablo meteorite found in Arizona, USA, established dates of the order of (4.5−4.6) $\times$ 10^9 years. Darwin had the time he needed!

The proportion of elements other than hydrogen and helium (defined as the metallicity) in a celestial object can be used as an indication of its age. After primordial nucleosynthesis (Sect. 8.1.3) the Universe is basically composed by hydrogen and helium. Thus the older (first) stars should have lower metallicity than the younger ones (for instance our Sun). The measurement of the age of low metallicity stars imposes, therefore, an important constraint on the age of the Universe. The oldest stars with a well-determined age found so far are HE 1523-0901, a red giant at around 7500 light years away from us, and HD 140283, denominated the Methuselah star, located around 190 light years away. The age of HE 1523-0901 was measured to be 13.2 billion years, using mainly the decay of uranium and thorium. The age of HD 140283 was determined to be (14.46 ± 0.8) billion years.

The "cosmological" age of the Universe is defined as the time since the Big Bang, which at zeroth order is just given by the inverse of the Hubble constant:

$$T \simeq \frac{1}{H_0} \simeq 14 \times 10^9 \text{ years} . \qquad (8.60)$$

The precise value is determined by solving Friedmann equations (see Sect. 8.2) for a given set of the cosmological parameters . Within the ΛCDM model (see Sect. 8.3.3) the best fit value, taking into account the present knowledge of such parameters, is $T = (13.798 \pm 0.037) \times 10^9$ years.

Within uncertainties both the cosmological age and the age of the first stars are compatible, but the first stars had to be formed quite early in the history of the Universe.

Finally, we stress that a Universe with a finite age and in expansion will escape the nineteenth century Olbers' Paradox: "How can the night be dark?" This paradox relies on the argument that in an infinite static Universe with uniform star density (as the Universe was believed to be by most scientists until the mid of last century) the night should be as bright as the day. In fact, the light coming from a star is inversely proportional to the square of its distance, but the number of stars in a shell at a distance between r and $r + dr$ is proportional to the square of the distance r. From this it seems that any shell in the Universe should contribute the same amount light. Apart from some too crude approximations (such as, not taking into account the finite life of stars), redshift and the finite size of the Universe solve the paradox.

8.2 General Relativity

Special relativity, introduced in Chap. 2, states that one cannot distinguish on the basis of the laws of physics between two inertial frames (moving at constant speed one with respect to the other). Experience tells that it is possible to distinguish between an inertial frame and an accelerated frame. Can the picture change if we include gravity?

In classical mechanics, gravity is a force and determines the movement of a body according to Newton's second law. The gravitational force is proportional to the body's gravity charge, which is the gravitational mass m_g; this, in turn, is proportional to the inertial mass m_I, that characterizes the body's inertia to be accelerated by a force. The net result is that the local acceleration of a body, g, due to a gravitational field created by a mass M at a distance r, is proportional to the ratio m_g/m_I

$$F_g = m_g\, G\frac{M}{r^2}$$

$$F_g = m_I\, g,$$

and

$$g = \frac{m_g}{m_I}\, G\frac{M}{r^2},$$

where G is the universal gravitational constant.

Thus if m_g were proportional to m_I the movement of a body in a gravitational field would be independent of its mass and composition. In fact the experiments of Galilei on inclined planes showed the universality of the movement of rolling balls of different compositions and weights. Such universality was also found by measuring the period of pendulums with different weights and compositions but identical lengths, first again by Galilei, and later on with a much higher precision (better than 0.1 %) by Newton. Nowadays, m_g/m_I is experimentally known to be constant for all bodies, independent of their nature, mass, and composition, up to a relative resolution of $5\ 10^{-14}$. We then *choose* G in such a way that $m_g/m_I \equiv 1$. Space-based experiments, allowing improved sensitivities up to 10^{-17} on m_g/m_I, are planned for the next years.

It is difficult to believe that such a precise equality is just a coincidence. This equality has been thus promoted to the level of a principle, named the *weak equivalence principle*, and it led Einstein to formulate the *strong equivalence principle* which is a fundamental postulate of General Relativity. Einstein stated that it is not possible to distinguish infinitesimal movements occurring in an inertial frame due to gravity (which are proportional to the gravitational mass), from movements occurring in an accelerated frame due to "fictitious" inertial forces (which are proportional to the inertial mass).

A ball dropped in a gravitational field has, during an infinitesimal time interval, the same behavior that a free ball has in an accelerated frame if the acceleration a of the accelerated frame is opposite to the local acceleration g of gravity (Fig. 8.17).

Einstein proposed a thought experiment, in which an observer is inside an elevator, when the cables were cut. In such a freely falling frame gravity is locally canceled out by the "fictitious" forces due to the acceleration of the frame. The free fall frames are equivalent to inertial frames.

A horizontal light beam in such free fall elevator then moves in a straight line for an observer inside the elevator but it curves down for an observer outside the elevator (Fig. 8.18). Light must therefore curve in a gravitational field.

The bending of light passing near the Sun was discussed by Newton himself and computed by Cavendish and Soldner to be of about 0.9 arcsecond for a light ray

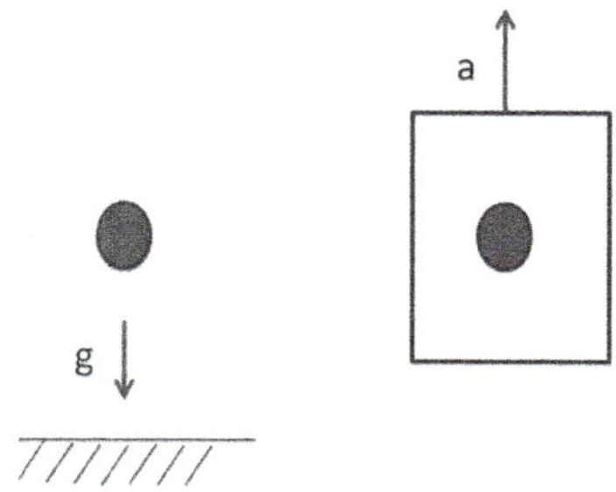

Fig. 8.17 Ball in a gravitational field (*left*) and inside an accelerated box (*right*)

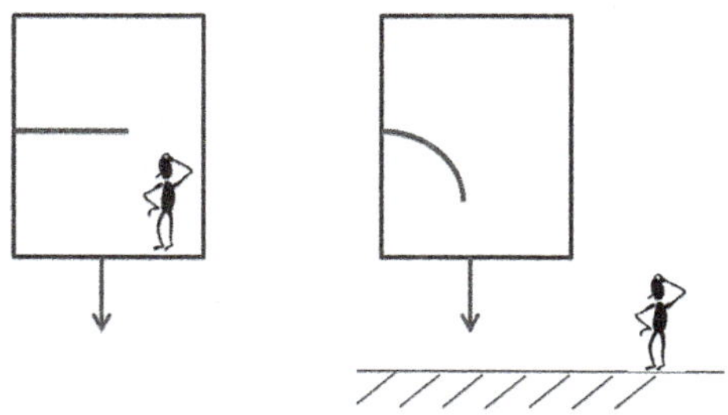

Fig. 8.18 Trajectory of a light beam in an elevator freely falling seen by an observer inside *left* and outside (*right*)

passing close to the Sun's limb; this result in its deduction assumes the Newton corpuscular theory of light. However, Einstein found, using the newborn equations of general relativity, a double value and then a clear test was in principle possible through the observation of the apparent position of stars during a total solar eclipse. In May 1919 Eddington and Dyson led two independent expeditions, respectively, to the equatorial islands of São Tomé e Príncipe and to Sobral, Brazil. The observations were perturbed by clouds (Príncipe) and by instrumental effects (Sobral) but nevertheless the announcement by Eddington that Einstein's predictions were confirmed had an enormous impact on public opinion and made general relativity widely known. Further and more robust observations were carried on in the following years and the predictions of general relativity on light deflection were firmly confirmed.

A light beam emitted by a source at rest in such an elevator (Fig. 8.19) is seen, by an outside observer at rest in the gravitational field, as emitted by a moving source and thus its frequency is Doppler shifted (redshifted if the elevator is running away from the outside observer):

$$z = \frac{\Delta\lambda}{\lambda} \sim \beta \sim \frac{g\,\Delta x}{c^2}\,. \tag{8.61}$$

For the outside observer a clock placed in a freely falling elevator runs slower (has a longer period) than a clock placed in another freely falling elevator situated *above* the first one having thus a lower velocity (Fig. 8.20).

This time dilation due to the gravitational field can also be seen as due to the differences in the energy losses of the photons emitted in both elevators by "climbing" out the gravitational field. In fact in a weak gravitational field the variation of the total

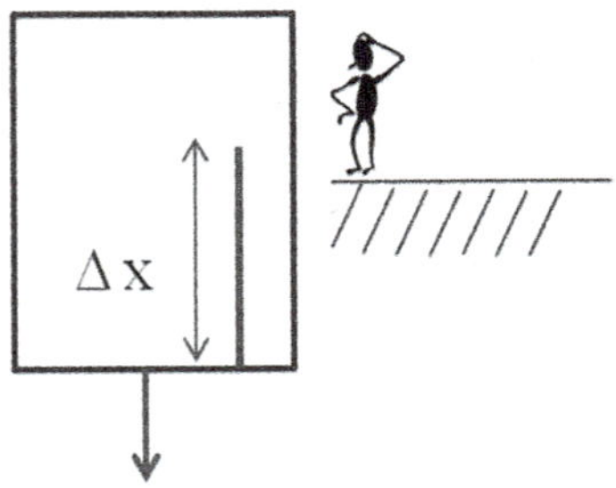

Fig. 8.19 Light beam emitted by a source in an the elevator seen by an outside observer at rest in the gravitational field

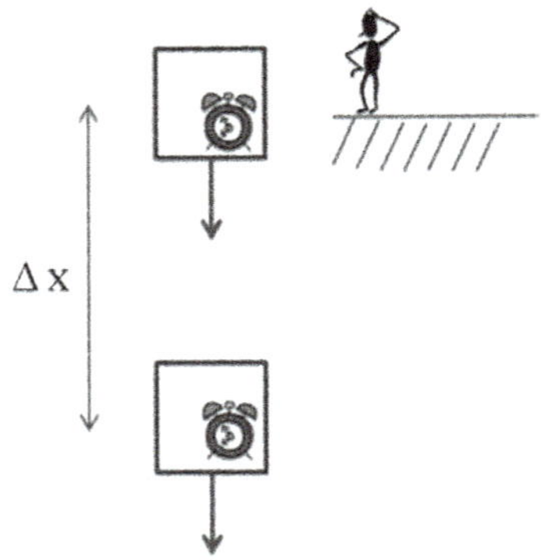

Fig. 8.20 Clocks at different heights in freely falling elevators seen by an observer at rest in the gravitational field

energy of a particle of mass m, assuming the equivalence principle, is independent of m:

$$\frac{\Delta E}{E} \simeq \frac{mg\Delta x}{mc^2} = \frac{g\Delta x}{c^2} \,.$$

Since, for a photon, energy and frequency are related by the Planck formula $E = h\nu$, the previous result is recovered:

$$\frac{\Delta E}{E} = \frac{\Delta \nu}{\nu} \sim \frac{\Delta \lambda}{\lambda} \sim \frac{g\Delta x}{c^2} \,.$$

Time runs slower in the presence of stronger gravitational fields.

Gravity in General Relativity is no longer a force (whose sources are the masses) acting in a flat spacetime Universe. Gravity (as it will be discussed in next sections) is embedded in the geometry of spacetime that is determined by the energy and momentum contents of the Universe.

8.2.1 *Flat and Curved Spaces*

Classical mechanics considers that we are living in a Euclidean three-dimensional space (flat, i.e., with vanishing curvature), where through each point outside a "straight" line (a geodesic) there is one, and only one, straight line parallel to the first one; the sum of the internal angles of a triangle is 180°; the circumference of a circle of radius R is $2\pi R$, and so on. However, it is interesting to consider what would happen if this was not the case.

To understand why a different approach could be interesting, let us consider that we are living on the surface of a sphere (the Earth is approximately a sphere). Such a surface has positive (ideally constant) curvature at all points (i.e., the spherical surface stays on just one side of the tangent planes to the surface at any given point): the small distance between two points is now the length of the arc of circle connecting the two points and whose center coincides with the center of the sphere (geodesic line in the sphere), and this is as close as we can get to a *straight line*; the sum of the angles of a triangle is greater than 180°; the circumference of a circle of radius R is less than $2\pi R$. Alternatively, let us imagine that we were living on a saddle, which has a negative curvature (the surface can curve away from the tangent plane in two different directions): then the sum of the angles of a triangle is less than 180°; the perimeter of a circumference is greater than $2\pi R$, and so on. The three cases are visualized in Fig. 8.21. The metrics of the sphere and of the saddle are not Euclidean.

8.2.1.1 2D Space

In a flat 2D surface (a plane) the square of the distance between two points is given in Cartesian coordinates by

$$ds^2 = dx^2 + dy^2 \tag{8.62}$$

or

$$ds^2 = g_{\mu\nu} dx^\mu dx^\nu \tag{8.63}$$

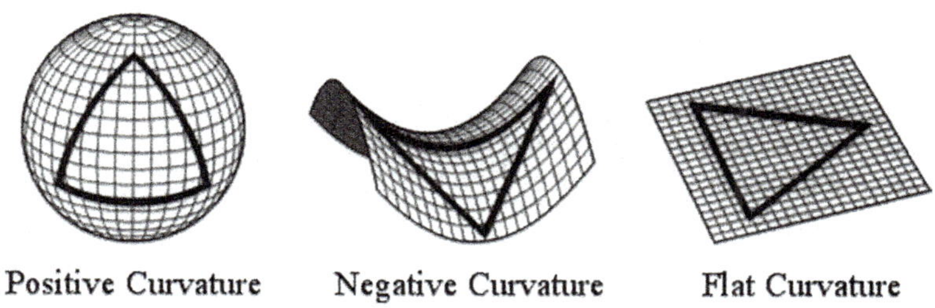

Fig. 8.21 2D surfaces with positive, negative, and null curvatures (from http://thesimplephysicist.com/?p=665, ©2014 Bill Halman/tdotwebcreations)

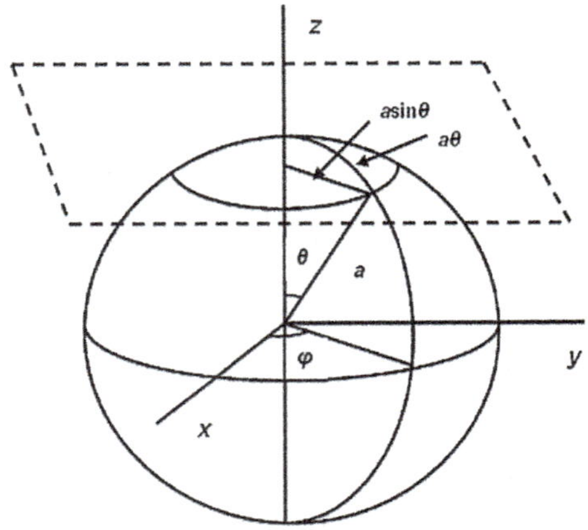

Fig. 8.22 Distances on a sphere of radius a. From A. Tan et al. DOI:10.5772/50508

with

$$g_{\mu\nu} = \begin{bmatrix} 1 & 0 \\ 0 & 1 \end{bmatrix} . \quad (8.64)$$

The metric $g_{\mu\nu}$ of the 2D flat surface is constant and the geodesics are straight lines.

The metric of a 2D spherical surface is a little more complex. The square of the distance between two neighboring close points situated on the surface of a sphere with radius a embedded in our usual 3D Euclidean space (Fig. 8.22) is given in spherical coordinates by

$$ds^2 = a^2 d\theta^2 + a^2 \sin^2\theta \, d\varphi^2 . \quad (8.65)$$

The maximum distance between two points on the sphere is bounded by $d = \sqrt{s^2} = \pi \, a$: the two points are the extrema of a half great circle.

Now the matrix representing the metrics in spherical coordinates,

$$g_{\mu\nu} = \begin{bmatrix} a^2 & 0 \\ 0 & a^2 \sin^2\theta \end{bmatrix} , \quad (8.66)$$

is no longer constant, because of the presence of the $\sin^2\theta$ term. It is not possible to cover the entire sphere with one unique plane without somewhat distorting the plane, although it is always possible to define locally at each point one tangent plane. The geodesics are not straight lines; they are indeed part of great circles, as it can be deduced directly from the metrics and its derivatives.

This metric can now be written introducing a new variable $r = \sin\theta$ as

$$ds^2 = a^2 \left(\frac{dr^2}{1 - Kr^2} + r^2 \, d\varphi^2 \right) \quad (8.67)$$

with

$$K = 1$$

for the case of the sphere.[8] Indeed, the sphere has a positive ($K = 1$) curvature at any point of its surface. However, the above expressions are valid both for the case of negative ($K = -1$) and null ($K = 0$) curvature. In the case of a flat surface, indeed, the usual expression in polar coordinates is recovered:

$$ds^2 = a^2 \left(dr^2 + r^2 \, d\varphi^2\right) . \tag{8.68}$$

The distance between two points with the same φ and, respectively, $r_1 = 0$ and $r_2 = \frac{R}{a}$, is given by:

$$s = \int_0^{\frac{R}{a}} a \, \frac{dr}{\sqrt{1 - Kr^2}} = a \, S_k \tag{8.69}$$

with

$$S_k = \begin{cases} \arcsin\left(\frac{R}{a}\right), & \text{if } K = 1 \\ R/a, & \text{if } K = 0 \\ \operatorname{arcsinh}\left(\frac{R}{a}\right), & \text{if } K = -1 \end{cases} . \tag{8.70}$$

The area of the sphere is now given by

$$A = 4 \, \pi \, a^2 {S_k}^2 . \tag{8.71}$$

The relation between the proper distance and the luminosity distance (Sect. 8.1.1) is now

$$d_L = d_p \, \frac{a}{R} \, S_k \, (1{+}z) \, , \tag{8.72}$$

and the metric can also be written in a more compact form using the function S_k:

$$ds^2 = a^2 \left(dr^2 + {S_k}^2 d\varphi^2\right) . \tag{8.73}$$

[8] r is dimensionless, with range $[0, 1]$; K is the curvature, which, in general, can be -1, 0, or $+1$. The more general change of coordinates $r' = a \sin\theta$ does not result in anything new, and can be recast in the form used above after setting $r = r'/a$. Of course with the r' coordinate, the curvature is not normalized, and can be, generically, negative, zero, or positive.

8.2.1.2 3D Space

For a homogeneous and isotropic 3D space the previous formula can be generalized (now r and θ are independent variables) leading to:

$$ds^2 = a^2 \left[\frac{dr^2}{1 - Kr^2} + r^2 \left(d\theta^2 + \sin^2\theta \, d\varphi^2 \right) \right] .$$

8.2.1.3 4D Spacetime

For a spatially homogeneous and isotropic 4D spacetime the generalization leads to the so-called Robertson–Walker metric ($c = 1$):

$$ds^2 = dt^2 - a^2(t) \left[\frac{dr^2}{1 - Kr^2} + r^2 \left(d\theta^2 + \sin^2\theta \, d\varphi^2 \right) \right] \tag{8.74}$$

where $a\,(t)$ is a radial scale factor which may depend on t (allowing for the expansion/contraction of the Universe).

Introducing the solid angle, $d\Omega^2 = d\theta^2 + \sin^2\theta \, d\varphi^2$, the Robertson–Walker metric can be written as

$$ds^2 = dt^2 - a^2\,(t) \left(\frac{dr^2}{1 - Kr^2} + r^2 \, d\Omega^2 \right) . \tag{8.75}$$

Finally, the Robertson–Walker metric can also be written using the functions S_k introduced above as

$$ds^2 = dt^2 - a^2\,(t) \left(dr^2 + S_k{}^2 \, d\Omega^2 \right) . \tag{8.76}$$

The special relativity Minkowski metric is a particular case ($K = 0, \quad a\,(t) =$ constant) of the Robertson–Walker metric.

The geodesics in a 4D spacetime correspond to the extremal (maximum or minimum depending on the metric definition) world lines joining two events in spacetime and not to the 3D space paths between the two points. The geodesics are determined, as before, just from the metric and its derivatives.

8.2.2 The Einstein Equations

In General Relativity the world lines of freely falling test particles are just the geodesics of the 4D spacetime of the Universe we are living in, whose geometry is locally determined by its energy and momentum contents as expressed by Einstein equations (which, below, are in the form where we neglect a cosmological constant term)

$$G_{\mu\nu} = R_{\mu\nu} - \frac{1}{2} g_{\mu\nu} \mathcal{R} = \frac{8\pi}{c^4} T_{\mu\nu}.$$

In the equations above $G_{\mu\nu}$ and $R_{\mu\nu}$ are, respectively, the Einstein and the Ricci tensors, which are built from the metric and its derivatives; $\mathcal{R}$ is the Ricci scalar $\left(\mathcal{R} = g^{\mu\nu} R_{\mu\nu}\right)$ and $T_{\mu\nu}$ is the energy–momentum tensor.

The energy and the momentum of the particles determine the geometry of the Universe which then determines the trajectories of the particles. Gravity is embedded in the geometry of spacetime. Time runs slower in the presence of gravitational fields.

Einstein equations are tensor equations and thus independent on the reference frame (the covariance of the physics laws is automatically ensured). They involve 4D symmetric tensors and represent in fact 10 independent nonlinear partial differential equations whose solutions, the metrics of spacetime, are in general very difficult to sort out. However, in particular and relevant cases, exact or approximate solutions can be found. Examples are the Minkowski metric (empty Universe); the Schwarzschild metric (spacetime metric outside a noncharged spherically symmetric nonrotating massive object—see Sect. 8.2.4); the Kerr metric (a cylindrically symmetric vacuum solution); the Robertson–Walker metric (homogeneous and isotropic Universe—see Sect. 8.2.3).

Einstein introduced at some point a new metric proportional term in his equations (Λ, the so-called "cosmological constant")

$$G_{\mu\nu} + g_{\mu\nu}\Lambda = \frac{8\pi}{c^4} T_{\mu\nu}. \tag{8.77}$$

His motivation was to allow for static cosmological solutions, as this term can balance gravitational attraction. Although later on Einstein discarded this term (the static Universe would be unstable), the recent discovery of the accelerated expansion of the Universe might give it again an essential role (see Sects. 8.2.3 and 8.3.3).

The energy–momentum tensor $T^{\mu\nu}$ in a Universe of free noninteracting particles with four-momentum p_i^{μ} moving along trajectories $r_i\,(t)$ is defined as

$$T^{\mu 0} = \sum_i p_i^{\mu}\,(t)\;\delta^3\,(r - r_i\,(t)) \tag{8.78}$$

$$T^{\mu k} = \sum_i p_i^{\mu}\,(t)\,\frac{dx_i^k}{dt}\,\delta^3\,(r - r_i\,(t))\;. \tag{8.79}$$

The $T^{\mu 0}$ terms can be seen as "charges" and the $T^{\mu k}$ terms as "currents", which then obey a continuity equation which ensures energy–momentum conservation. In general relativity local energy–momentum conservation generalizes the corresponding results in special relativity,

$$\frac{\partial}{\partial x^0} T^{\mu 0} + \nabla_i T^{\mu i} = 0 \tag{8.80}$$

or

$$\frac{\partial}{\partial x^\nu} T^{\mu\nu} = 0\,. \tag{8.81}$$

To get an intuitive grasp of the physical meaning of the energy–momentum tensor, let us consider the case of a special relativistic perfect fluid (no viscosity). In the rest frame of a fluid with energy density ρ and pressure $\mathcal{P}$

$$T^{00} = c^2 \rho \tag{8.82}$$

$$T^{0i} = 0 \tag{8.83}$$

$$T^{ij} = \mathcal{P}\,\delta_{ij} \tag{8.84}$$

The pressure has indeed the dimension of an energy density ($\delta W = F \cdot dx = \mathcal{P}\,dV$) and accounts for the "kinetic energy" of the fluid.

To appreciate a fundamental difference from the Newtonian case, we quote that for a perfect fluid with energy density ρ and pressure $\mathcal{P}$ the weak gravity field predicted by Newton is given by

$$\nabla^2 \phi = 4\,\pi G\,\rho, \tag{8.85}$$

from which we readily see that pressure does not contribute. On the contrary, the weak field limit of Einstein equations is

$$\nabla^2 \phi = 4\,\pi G\,\left(\rho + \frac{3\,\mathcal{P}}{c^2}\right). \tag{8.86}$$

Remembering that, in the case of a relativistic fluid

$$\mathcal{P} \sim \frac{1}{3}\,\rho\,c^2 \tag{8.87}$$

the weak gravitational field is then determined by

$$\nabla^2 \phi = 8\,\pi G\,\rho, \tag{8.88}$$

which shows that the gravitational field predicted by general relativity is twice the one predicted by Newtonian gravity. Indeed, the observed light deflection by Eddington in 1919 at S. Tomé and Príncipe islands in a solar eclipse was twice the one expected according to classical Newtonian mechanics.

Once the metric is known, the free fall trajectories of test particles are obtained "just" by solving the geodesic equations

$$\frac{d^2x^\sigma}{d\tau^2} + \Gamma^\sigma_{\mu\nu}\frac{dx^\mu}{d\tau}\frac{dx^\nu}{d\tau} = 0, \tag{8.89}$$

where $\Gamma^\sigma_{\mu\nu}$ are the Christoffel symbols given by

$$\Gamma^\sigma_{\mu\nu} = \frac{g^{\rho\sigma}}{2}\left(\frac{\partial g_{\nu\rho}}{\partial x^\mu} + \frac{\partial g_{\mu\rho}}{\partial x^\nu} - \frac{\partial g_{\mu\nu}}{\partial x^\rho}\right). \tag{8.90}$$

In the particular case of flat space in Cartesian coordinates the metric tensor is everywhere constant, $\Gamma^\sigma_{\mu\nu} = 0$, and then

$$\frac{d^2x^\mu}{d\tau^2} = 0\,.$$

The free particles classical straight world lines are then recovered.

8.2.3 The Friedmann–Lemaitre–Robertson–Walker Model

The present standard model of cosmology assumes the so-called cosmological principle, which assumes a homogeneous and isotropic Universe at large scales. Homogeneity means that, in Einstein's words, "all places in the Universe are alike" and isotropic just means that all directions are equivalent.

The Robertson–Walker metric discussed before (Sect. 8.2.1) embodies these symmetries leaving two independent functions, $a(t)$ and $K(t)$, which represent, respectively, the evolution of the scale and of the curvature of the Universe. The Russian physicist, Alexander Friedmann in 1922, and independently the Belgian Georges Lemaitre in 1927, solved the Einstein Equations for such a metric leading to the famous Friedmann equations, which are still the starting point for the standard cosmological model, also known as the Friedmann–Lemaitre–Robertson–Walker (FLRW) model.

8.2.3.1 Friedmann Equations

Friedmann equations can be written as ($c = 1$)

$$\left(\frac{\dot{a}}{a}\right)^2 + \frac{K}{a^2} = \frac{8\pi G}{3}\rho + \frac{\Lambda}{3} \tag{8.91}$$

$$\left(\frac{\ddot{a}}{a}\right) = -\frac{4\pi G}{3}(\rho + 3\mathcal{P}) + \frac{\Lambda}{3}. \tag{8.92}$$

These equations can be combined into a thermodynamics like equation

$$\frac{d}{dt}\left(\rho\, a^3\right) = -\mathcal{P}\frac{d}{dt}\left(a^3\right), \tag{8.93}$$

where by identifying a^3 with the volume V we can recognize adiabatic energy conservation

$$dE = -\mathcal{P}\, dV .$$

Moreover, remembering that the Hubble parameter is given by (Eq. 8.8):

$$H = \frac{\dot{a}}{a},$$

the first Friedmann equation is also often written as

$$H^2 + \frac{K}{a^2} = \frac{8\pi G}{3}\rho + \frac{\Lambda}{3}, \tag{8.94}$$

which shows that the Hubble constant is not a constant but a parameter that evolves with the evolution of the Universe.

8.2.3.2 Classical Newtonian Mechanics

"Friedmann-like" equations can also be formally deduced in the framework of classical Newtonian mechanics, as follows:

1. From Newton law of gravitation and from the Newton second law of motion we can write

$$m\ddot{R} = -\frac{GMm}{R^2} \Longrightarrow \left(\frac{\ddot{a}}{a}\right) = -\frac{4\pi G}{3}\rho. \tag{8.95}$$

2. From energy conservation

$$\frac{1}{2}m\,\dot{R}^2 - \frac{GMm}{R} = \text{constant} \Longrightarrow \left(\frac{\dot{a}}{a}\right)^2 - \frac{\text{constant}}{a^2} = \frac{8\pi G}{3}\rho. \tag{8.96}$$

The two Friedmann equations are "almost" recovered. The striking differences are that in classical mechanics the pressure does not contribute to the "gravitational mass", and that the Λ term must be introduced by hand as a form of repulsive potential.

The curvature of spacetime is, in this "classical" version, associated to (minus) the total energy of the system, which can somehow be interpreted as a "binding energy".

8.2.3.3 Single Component Universes

The two Friedmann equations determine, once the energy density ρ and the pressure $\mathcal{P}$ are known, the evolution of the scale $(a(t))$ and of the curvature $(K(t))$ of the Universe. However, ρ and $\mathcal{P}$ are nontrivial quantities. They depend critically on the amount of the different forms of energy and matter that exist in the Universe at each evolution stage.

In the simplest case of a Universe with just nonrelativistic particles (ordinary baryonic matter or "cold"—i.e., nonrelativistic—dark matter) the pressure is negligible with respect to the energy density ($\mathcal{P} \ll \rho_m c^2$) and the Friedmann equations can be approximated as

$$\frac{d}{dt}\left(\rho_m \; a^3\right) = 0 \tag{8.97}$$

$$\left(\frac{\ddot{a}}{a}\right) = -\frac{4\pi G}{3} \; (\rho_m) \, . \tag{8.98}$$

Solving these equations one finds

$$\rho_m \propto \frac{1}{a^3} \; ; \; a\,(t) \propto t^{\frac{2}{3}} \, . \tag{8.99}$$

In general for a Universe with just one kind of matter characterized by an equation of state relating ρ and $\mathcal{P}$ of the type

$$\mathcal{P} = \alpha \, \rho \tag{8.100}$$

the solutions are

$$\rho \propto a^{-3(\alpha+1)} \; ; \; a\,(t) \propto t^{\,\frac{2}{3(\alpha+1)}} \, . \tag{8.101}$$

For instance, in the case of a Universe dominated by relativistic particles (radiation or hot matter) $\alpha = 1/3$ and we obtain

$$a\,(t) \propto t^{\frac{1}{2}} \; ; \; \rho_{\mathrm{rad}} \propto \frac{1}{a^4} \, . \tag{8.102}$$

This last relation can be interpreted by taking, as an example, a photon-dominated Universe, where the decrease in the density number of photons ($n_\gamma \propto a^{-3}$) combines with a decrease in the mean photon energy ($E_\gamma \propto a^{-1}$) corresponding to wavelength dilation.

8.2.3.4 Static Universe and Vacuum Energy Density

To model a static Universe ($\dot{a} = 0$, $\ddot{a} = 0$) one should have:

$$\frac{K}{a^2} = \frac{8\pi G}{3}\rho\,;\;\rho + 3\mathcal{P} = 0\,. \tag{8.103}$$

K should then be positive ($K = 1$) and $\mathcal{P} = -\frac{1}{3}\rho$. This requires a "new form of energy" with negative α, which can be related to the "cosmological" constant term. By reading this term on the right-hand side of Einstein equations, we can formally include it in the energy–momentum tensor, thus defining a "vacuum" tensor as

$$T^{\Lambda}_{\mu\nu} = g_{\mu\nu}\Lambda = \begin{pmatrix} \rho_\Lambda & 0 & 0 & 0 \\ 0 & -\rho_\Lambda & 0 & 0 \\ 0 & 0 & -\rho_\Lambda & 0 \\ 0 & 0 & 0 & -\rho_\Lambda \end{pmatrix} \tag{8.104}$$

with

$$\rho_\Lambda = \frac{\Lambda}{8\pi G}. \tag{8.105}$$

This implies an equation of state of the form ($\alpha = -1$):

$$\mathcal{P}_\Lambda = -\rho_\Lambda. \tag{8.106}$$

Therefore, in a static Universe we would have

$$\rho = \rho_m + \rho_\Lambda \tag{8.107}$$

and

$$\rho_m = 2\,\rho_\Lambda\,. \tag{8.108}$$

8.2.3.5 De Sitter Universe

In a Universe dominated by the cosmological constant ($\rho \equiv \rho_\Lambda$), as first discussed by de Sitter

$$\frac{d}{dt}\left(\rho_\Lambda\, a^3\right) = \rho_\Lambda\,\frac{d}{dt}\left(a^3\right) \tag{8.109}$$

and

$$H^2 + \frac{K}{a^2} = \frac{\Lambda}{3} \tag{8.110}$$

implying

$$\rho_\Lambda = \text{constant};\; a(t) \sim e^{Ht} \tag{8.111}$$

with

$$H = \sqrt{\frac{\Lambda}{3}}. \tag{8.112}$$

Thus the de Sitter Universe has an exponential expansion while its energy density remains constant.

8.2.3.6 Critical Density

The curvature of the Universe depends, according to Friedmann equations, on the energy density of the Universe according to

$$\frac{K}{a^2} = \frac{8\pi G}{3}\rho - H^2 \tag{8.113}$$

Therefore, if

$$\rho = \rho_{\text{crit}} = \frac{3\,H^2}{8\pi G} \tag{8.114}$$

one obtains

$$K = 0$$

and the Universe is, in this case, spatially flat.

For the present value of H_0 this corresponds to

$$\rho_{\text{crit}} = (5.4 \pm 0.1)\,\text{GeV/m}^3\,, \tag{8.115}$$

i.e., less than 6 hydrogen atoms per cubic meter. The number of baryons per cubic meter one obtains from galaxy counts is, however, twenty times smaller—consistently with the result of the fit to CMB data.

8.2.3.7 Normalized Densities Ω_i, H, and q_0

The energy densities of each type of matter, radiation, and vacuum are often normalized to the critical density as follows:

$$\Omega_i = \frac{\rho_i}{\rho_{\text{crit}}} = \frac{8\pi G}{3\,H^2}\,\rho_i\,. \tag{8.116}$$

By defining also a normalized "curvature" energy density as

$$\Omega_K = -\frac{K}{H^2 a^2} = -\frac{K}{\dot{a}^2}, \tag{8.117}$$

the first Friedmann equation

$$\frac{8\pi G}{3H^2}\rho - \frac{K}{H^2 a^2} = 1 \tag{8.118}$$

takes then a very simple form

$$\Omega_m + \Omega_{\rm rad} + \Omega_\Lambda + \Omega_K = 1\,. \tag{8.119}$$

On the other hand, it can be shown that taking into account the specific evolution of each type of density with the scale parameter a, the evolution equation for the Hubble parameter can be written as

$$H^2 = H_0^2\left(\Omega_{0\Lambda} + \Omega_{0K}a^{-2} + \Omega_{0m}a^{-3} + \Omega_{0{\rm rad}}a^{-4}\right), \tag{8.120}$$

where the subscripts 0 indicate the values at present time ($t = t_0$, $a_0 = 1$).

Since the scale factor a is related to the redshift z, as discussed in Sect. 8.1.1, by

$$(1+z) = a^{-1},$$

the Hubble evolution equation can then be written as

$$H^2(z) = H_0^2\left(\Omega_{0\Lambda} + \Omega_{0K}(1+z)^2 + \Omega_{0m}(1+z)^3 + \Omega_{0{\rm rad}}(1+z)^4\right). \tag{8.121}$$

Finally, the deceleration parameter q_0 can also be expressed as a function of the normalized densities Ω_i. In fact q_0 was defined as (Sect. 8.1.1)

$$q_0 = -\frac{\ddot{a}}{{H_0}^2 a}\,. \tag{8.122}$$

Now, using the second Friedmann equation

$$\left(\frac{\ddot{a}}{a}\right) = -\frac{4\pi G}{3}\left(\rho + 3\mathcal{P}\right)$$

and the equations of state

$$\mathcal{P}_i = \alpha_i\,\rho_i,$$

one obtains

$$q_0 = -\frac{\ddot{a}}{{H_0}^2 a} = \frac{1}{2}\frac{8\pi G}{3{H_0}^2}\sum_i \rho_i\,(1+3\alpha_i)$$

$$q_0 = \frac{1}{2}\sum_i \Omega_i\,(1+3\alpha_i)$$

$$q_0 = \frac{1}{2}\Omega_{0m} + \Omega_{0\text{rad}} - \Omega_{0\Lambda} \tag{8.123}$$

These equations in H and q_0 are of the utmost importance, since they connect directly the experimentally measured quantities H_0 and q_0 to the densities of the various energy species in the Universe.

8.2.3.8 Experimental Determination of the Normalized Densities

The total density of baryons, visible or invisible, as inferred from nucleosynthesis, is about 0.26 baryons per cubic meter, i.e.,

$$\Omega_B = 0.042 \pm 0.004\,. \tag{8.124}$$

A small fraction of this is luminous—i.e., visible energy.

The currently most accurate determination of the overall densities comes from global fits of cosmological parameters to recent observations (see later). Using measurements of the anisotropy of the CMB and of the spatial distribution of galaxies, as well as the measured acceleration, the data indicate a fraction of nonbaryonic DM over the total energy content of the Universe given by

$$\Omega_c = \Omega_{\text{DM,nonbaryonic}} = 0.265 \pm 0.011\,. \tag{8.125}$$

According to the same fits, the total baryonic matter density is

$$\Omega_B = 0.050 \pm 0.002\,. \tag{8.126}$$

Part of the baryonic matter may contribute to DM in form of nonluminous Dark Matter, e.g., massive compact objects or cold molecular gas clouds (see later).

In summary, a remarkable agreement of independent astrophysical observations with cosmology indicates that the energy content of DM in the Universe could be about 25 % of the total energy of the Universe, compared to some 5 % due to ordinary matter.

The dark energy density can be measured from the observed curvature in the Hubble plot and from the position of the "acoustic peaks" in the angular power spectrum of the temperature fluctuations in the CMB; this gives the total density

$$\Omega_m + \Omega_{\rm rad} + \Omega_\Lambda = 1.000 \pm 0.004\,. \tag{8.127}$$

These results indicate a value for the dark energy density of

$$\Omega_\Lambda = 0.70 \pm 0.01\,. \tag{8.128}$$

Dark energy dominates the energy content of a remarkably flat Universe.

Friedmann equation (8.91) can also be rewritten as

$$\Omega = \frac{\rho}{\rho_{\rm crit}} = 1 + \frac{K}{H^2 a^2}\,, \tag{8.129}$$

where the *closure parameter* Ω is the sum of Ω_m, $\Omega_{\rm rad}$ and Ω_Λ, with $\Omega_{\rm rad} \simeq 5 \times 10^{-5}$ being negligible. This means that, in general, Ω is a function of time, unless $\Omega = 1$ and thus $K = 0$ (flat Universe).

Present experimental data indicate a value

$$\Omega = 1.000 \pm 0.004\, : \tag{8.130}$$

it would look very strange if this was a coincidence, unless Ω is identically one. For this reason this fact is at the heart of the standard model of cosmology, the ΛCDM model, which *postulates* $\Omega = 1$.

8.2.3.9 Estimation of the Age of the Universe from the Friedmann Equations

The evolution of the Hubble parameter can be used to estimate the age of the Universe for different composition of the total energy density. Indeed

$$H = \frac{\dot a}{a} = \frac{1}{a}\frac{da}{dt} = -\left(\frac{dz/dt}{1+z}\right)$$

$$dt = -\frac{dz}{(1+z)\;H}$$

$$(t_0 - t) = \frac{1}{H_0}\int_0^Z \frac{dz}{(1+z)\left(\Omega_{0\Lambda} + \Omega_{0K}(1+z)^2 + \Omega_{0m}(1+z)^3 + \Omega_{0\rm rad}(1+z)^4\right)^{1/2}}\,. \tag{8.131}$$

The solution to this equation has to be obtained numerically in most realistic situations. However, in some simplified scenarios, an analytical solution can be found. In particular for matter ($\Omega_m = 1$) and radiation ($\Omega_{0\rm rad} = 1$) dominated Universes the solutions are, respectively,

$$(t_0 - t) = \frac{2}{3\,H_0} \tag{8.132}$$

and

$$(t_0 - t) = \frac{1}{2\,H_0}\,. \tag{8.133}$$

In a flat Universe with matter and vacuum energy parameters close to the ones presently measured ($\Omega_m = 0.3,\ \ \Omega_{0\Lambda} = 0.7$) we obtain

$$(t_0 - t) \sim \frac{0.96}{H_0}\,. \tag{8.134}$$

8.2.3.10 Evolution Scenarios

Friedmann equations have four independent parameters which can be chosen as:

- the present value of the Hubble parameter, H_0;
- the present value of the energy density of radiation, $\Omega_{0\text{rad}}$;
- the present value of the energy density of matter, Ω_{0m};
- the present value of the energy density of vacuum, $\Omega_{0\Lambda}$.

If we know these parameters, the geometry and the past and future evolutions of the Universe are determined provided the dynamics of the interactions, annihilations and creations of the different particle components (see Sect. 8.3.1) are neglected. The solutions to these equations, in the general multicomponent scenarios, cannot be expressed in closed, analytical form, and require numerical approaches.

However, as we have discussed above, the evolution of the energy density of the different components scales with different powers of the scale parameter of the Universe a. Therefore, there are "eras" where a single component dominates. It is then reasonable to suppose that, initially, the Universe was radiation dominated (apart for a very short period where it is believed that inflation occurred—see Sect. 8.3.2), then that it was matter dominated and finally, at the present time, that it is the vacuum energy (mostly "dark" energy, i.e., not coming from quantum fluctuations of the vacuum of the known interactions) that is starting to dominate (Fig. 8.23).

The crossing point ($a = a_{\text{cross}}$) between the matter and radiation ages can be obtained, at first approximation, by just equalizing the corresponding densities:

$$\Omega_{\text{rad}}\left(a_{\text{cross}}\right) = \Omega_m\left(a_{\text{cross}}\right)$$

$$\Omega_{\text{rad}}\left(a_0\right)\left(\frac{a_{\text{cross}}}{a_0}\right)^{-4} = \Omega_m\left(a_0\right)\left(\frac{a_{\text{cross}}}{a_0}\right)^{-3}$$

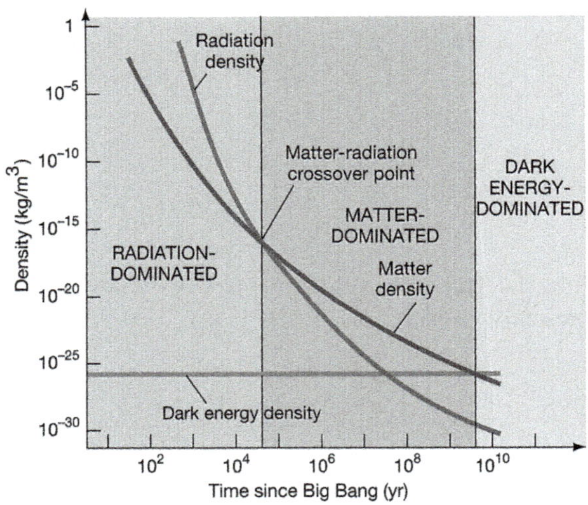

Fig. 8.23 The different ages the Universe passed through since the Big Bang (from http://scienceblogs.com/startswithabang/files/2013/06/AT_7e_Figure_27_01.jpeg, ©2011 Pearson education, Pearson Addison Wesley)

$$\left(\frac{a_{\rm cross}}{a_0}\right)^{-1} = 1 + z_{\rm cross} = \frac{\Omega_m\,(a_0)}{\Omega_{\rm rad}\,(a_0)}. \tag{8.135}$$

The time after the Big Bang when this crossing point occurs, can, approximately, be obtained from the evolution of the scale factor in a radiation dominated Universe

$$a_{\rm cross} \sim \left(2\ H_0\ \sqrt{\Omega_{\rm rad}\,(a_0)}\ \ t_{\rm cross}\right)^{\frac{1}{2}}, \tag{8.136}$$

or

$$t_{\rm cross} \sim {a_{\rm cross}}^2 \left(2\ H_0\ \sqrt{\Omega_{\rm rad}\,(a_0)}\ \right)^{-1}. \tag{8.137}$$

Using the current best fit values for the parameters (see Sect. 8.3.3), we obtain

$$z_{\rm cross}\ \sim 3\,200 \Longrightarrow t_{\rm cross} \sim 7\ 10^4 \text{ years}. \tag{8.138}$$

After this time (i.e., during the large majority of the Universe evolution) a two-component (matter and vacuum) description should be able to give a reasonable, approximate description. In this case the geometry and the evolution of the Universe are determined only by Ω_m and Ω_Λ. Although this is a restricted parameter phase space, there are several different possible evolution scenarios as shown in Fig. 8.24:

1. If $\Omega_m + \Omega_\Lambda\ = 1$ the Universe is flat but it can expand forever ($\Omega_\Lambda\ > 0$) or eventually recollapses ($\Omega_\Lambda < 0$).
2. If $\Omega_m + \Omega_\Lambda > 1$ the Universe is closed (positive curvature).
3. If $\Omega_m + \Omega_\Lambda < 1$ the Universe is open (negative curvature).

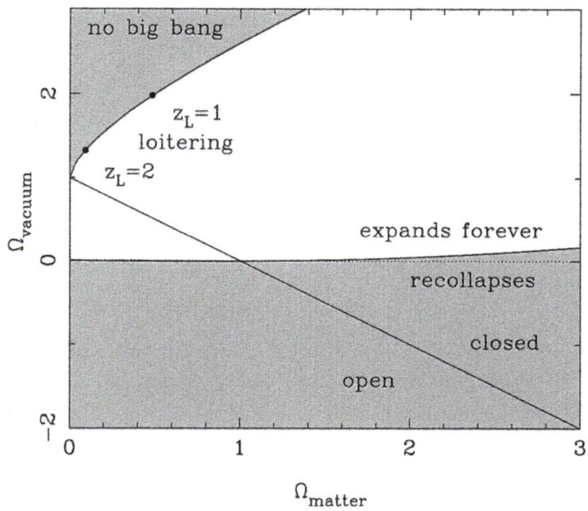

Fig. 8.24 Different scenarios for the expansion of the Universe. The Hubble constant was fixed to $H_0 = 70$ (km/s)/Mpc. From J.A. Peacock, "Cosmological Physics", Cambridge University press 1998

4. In a small phase space region with $\Omega_m + \Omega_\Lambda > 1$ and $\Omega_\Lambda > 0$ there is a solution with no Big Bang—the Universe bounces between a minimum and a maximum scale factor.

Some of these evolution scenarios are represented as functions of time in Fig. 8.25 for selected points in the parameter space discussed above. The green curve represents a flat, matter-dominated, critical density Universe (the expansion rate is continually slowing down). The blue curve shows an open, low density, matter-dominated Universe (the expansion is slowing down, but not as much). The orange curve shows a closed, high-density Universe (the expansion reverts to a "big crunch"). The red

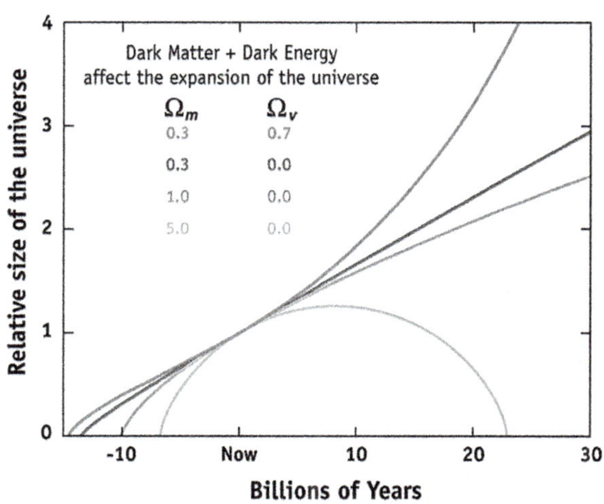

Fig. 8.25 Evolution of the Universe in a two-component model (matter and vacuum) for different $(\Omega_m, \Omega_\Lambda)$ values. (from http://map.gsfc.nasa.gov/universe/uni_fate.html)

curve shows a Universe with a large fraction of "dark energy" (the expansion of the Universe accelerates).

The present experimental evidence (see Sect. 8.3.3) highly favors the "dark energy" scenario, leading to a cold thermal death of the Universe.

8.2.4 Black Holes

The first exact analytical solution of Einstein equations was found in 1915, just a month after the publication of Einstein's original paper, by Karl Schwarzschild, a German physicist who died one year later from a disease contracted on the First World War battle field.

Schwarzschild's solution describes the gravitational field in the vacuum surrounding a single, spherical, nonrotating massive object. In this case the spacetime metric (the Schwarzschild metric) can be expressed as

$$ds^2 = \left(1 - \frac{r_S}{r}\right) c^2 dt^2 - \left(1 - \frac{r_S}{r}\right)^{-1} dr^2 - r^2 (d\theta^2 + \sin^2\theta d\phi^2), \qquad (8.139)$$

with

$$r_S = \frac{2GM}{c^2} \simeq 2.7\,\mathrm{km}\,\frac{M}{M_{\mathrm{Sun}}}\,. \qquad (8.140)$$

In the weak field limit, $r \to \infty$, we recover flat spacetime. According to this solution, a clock with period τ^* placed at a point r is seen by an observer placed at $r = \infty$ with a period τ given by:

$$\tau = \left(1 - \frac{r_S}{r}\right)^{-1} \tau^*\,. \qquad (8.141)$$

In the limit $r \to r_S$ (the Schwarzschild radius) the metric shows a coordinate singularity: the time component goes to zero and the radial component goes to infinity. From the point of view of an asymptotic observer, the period τ^* is seen now as infinitely large. No light emitted at $r = r_S$ is able to reach the $r > r_S$ world. This is what is usually called, following John Wheeler, a "black hole".

The existence of objects so massive that light would not be able to escape from them, was already predicted in the end of the eighteenth century by Michell in England and independently by Laplace in France. They just realized that, if the escape velocity from a massive object would have been greater than the speed of light, then the light could not escape from the object:

$$v_{\mathrm{esc}} = \sqrt{\frac{2\,G\,M}{r}} > c\,. \qquad (8.142)$$

Thus an object with radius R and a mass M would be a "black hole" if:

$$M > \frac{c^2}{G}\frac{R}{2}: \tag{8.143}$$

the"classical" radius and the Schwarzschild radius coincide.

The singularity observed in the Schwarzschild metric is not in fact a real physics singularity; it depends on the reference frame chosen (see [8.3] for a discussion). An observer in free fall frame will cross the Schwarzschild surface without feeling any discontinuity; (s)he will go on receiving signals from the outside world but (s)he will not be able to escape from the unavoidable, i.e., from crunching, at last, at the center of the black hole (the real physical singularity).

Schwarzschild black holes are however just a specific case. In 1963, New Zealand mathematician Roy Kerr found an exact solution to Einstein equations for the case of a rotating noncharged black hole and two years later the US Ezra Newman extended it to the more general case of rotating charged black holes. In fact it can be proved that an electrovacuum black hole can be completely described by three parameters: mass, angular momentum, and electric charge (the so-called no-hair theorem).

Black holes are not just exotic solutions of the General Theory of Relativity. They may be formed either by gravitational collapse or particle high-energy collisions. While so far there is no evidence of their formation in human-made accelerators, there is striking indirect evidence that they are part of several binary systems and that they are present in the center of most galaxies, including our own (the Milky Way hosts in its center a black hole of roughly 4 million solar masses, as determined from the orbit of nearby stars). Extreme high-energy phenomena in the Universe, generating the most energetic cosmic rays, may also be caused by supermassive black holes inside AGNs (Active Galactic Nuclei—see Chap. 10).

8.3 Past, Present, and Future of the Universe

8.3.1 Early Universe

In its "first" moments the Universe, according to the Big Bang model, is filled with a high-density, hot (high-energy) gas of relativistic particles at thermal equilibrium.

The assumption of thermal equilibrium is justified since the interaction rate per particle Γ ($\Gamma = n\sigma v : n$, number density; σ, cross section; v, mean relative velocity) and the Hubble parameter H ($H^2 \sim \frac{8\pi G}{3}\rho$) evolve with the energy density ρ as

$$\Gamma \propto n \propto \rho\,;\; H \propto \rho^{\frac{1}{2}}\,. \tag{8.144}$$

Thus at some point, going back in time, we should have had

$$\frac{\Gamma}{H} \gg 1 \ . \tag{8.145}$$

Since the early Universe is radiation dominated (Sect. 8.2)

$$\rho_{\text{rad}} \propto \frac{1}{a^4} \ ; \ a\,(t) \propto t^{\frac{1}{2}} \ . \tag{8.146}$$

The temperature is, by definition, proportional to the mean particle energy and thus, in the case of radiation, it increases proportionally to the inverse of the Universe scale:

$$T \propto a^{-1} \ . \tag{8.147}$$

On the other hand, at each temperature T the number density, the energy density, and the pressure of each particle type are given (neglecting chemical potentials) by standard quantum statistical mechanics

$$n_i = \frac{g_i}{(2\pi\hbar)^3} \int_0^{\infty} \frac{4\pi\, p^2}{e^{\left(\frac{E_i}{k_B\, T}\right)} \pm 1}\, dp \ , \tag{8.148}$$

$$\rho_i c^2 = \frac{g_i}{(2\pi\hbar)^3} \int_0^{\infty} \frac{4\pi\, p^2}{e^{\left(\frac{E_i}{k_B\, T}\right)} \pm 1}\, E_i\, dp \ , \tag{8.149}$$

$$P_i = \frac{g_i}{(2\pi\hbar)^3} \int_0^{\infty} \frac{4\pi\, p^2}{e^{\left(\frac{E_i}{k_B\, T}\right)} \pm 1}\, \frac{p_i c^2}{3\, E_i} dp \ , \tag{8.150}$$

where g_i are the internal degrees of freedom of the particles—the +/– signs are for bosons (Bose–Einstein statistics) and fermions (Fermi–Dirac statistics), respectively.

For $k_B\, T \gg m_i c^2$ (relativistic limit)

$$n_i = \begin{cases} g_i\, \frac{\zeta(3)}{\pi^2} \left(\frac{k_B\, T}{\hbar c}\right)^3, & \text{for bosons} \\ \frac{3}{4} \left[g_i\, \frac{\zeta(3)}{\pi^2} \left(\frac{k_B\, T}{\hbar c}\right)^3 \right], & \text{for fermions} \end{cases} \tag{8.151}$$

$$\rho_i c^2 = \begin{cases} g_i\, \frac{\pi^2}{30} k_B\, T \left(\frac{k_B\, T}{\hbar c}\right)^3, & \text{for bosons} \\ \frac{7}{8} \left[g_i\, \frac{\pi^2}{30} k_B\, T \left(\frac{k_B\, T}{\hbar c}\right)^3 \right], & \text{for fermions} \end{cases} \tag{8.152}$$

$$P_i = \frac{\rho_i c^2}{3},$$

where ζ is Riemann zeta function ($\zeta\,(3) \simeq 1.20206$).

For a nonrelativistic particle with $m_x\, c^2 \sim k_B\, T$ the classical Maxwell–Boltzmann distribution is recovered,

$$n_x = g_x \left(\frac{m_x\, k_B\, T}{2\pi\, \hbar^2}\right)^{\frac{3}{2}} e^{-\left(\frac{m_x\, c^2}{k_B\, T}\right)}.$$

The total energy density in the early Universe can be obtained summing over all possible relativistic particles and can be written as

$$\rho\, c^2 = g^*_{ef} \frac{\pi^2}{30} k_B\, T \left(\frac{k_B\, T}{\hbar c}\right)^3, \qquad (8.153)$$

where g^*_{ef} is defined as the total "effective" number of degrees of freedom, and is given by

$$g^*_{ef} = \sum_{\text{bosons}} g_i + \frac{7}{8} \sum_{\text{fermions}} g_j\,. \qquad (8.154)$$

However, the scattering interaction rate of some relativistic particles (like neutrinos, see below) may become at some point smaller than the expansion rate of the Universe and then they will be no more in thermal equilibrium with the other particles. It is said that they *decouple* and their temperature will evolve as a^{-1} independently of the temperature of the other particles. The individual temperatures T_i, T_j may be introduced in the definition of the "effective" number of degrees of freedom as

$$g_{ef} = \sum_{\text{bosons}} g_i \left(\frac{T_i}{T}\right)^4 + \frac{7}{8} \sum_{\text{fermions}} g_j \left(\frac{T_j}{T}\right)^4 \qquad (8.155)$$

(g_{ef} is of course a function of the age of the Universe). At a given time all the particles with $M_x c^2 \ll k_B\, T$ contribute.

The total energy density determines the evolution of the Hubble parameter

$$H^2 \sim \frac{8\pi G}{3} \frac{\pi^2}{30}\, g_{ef}\, k_B\, T \left(\frac{k_B\, T}{\hbar c}\right)^3 \qquad (8.156)$$

$$H \sim \left(\frac{4\pi^3 G}{45\,(\hbar c)^3}\right)^{1/2} g_{ef}{}^{1/2}\, (k_B\, T)^2, \qquad (8.157)$$

or, introducing the Planck mass (Sect. 2.11),

$$H \sim 1.66\, g_{ef}{}^{1/2} \frac{(k_B\, T)^2}{\hbar\, c^2\, m_P}. \qquad (8.158)$$

Remembering (Sect. 8.2) that in a radiation dominated Universe the Hubble parameter is related to time just by

$$H = \frac{1}{2\,t}\,, \tag{8.159}$$

time and temperature are related by

$$t = \left(\frac{45(\hbar c)^3}{16\,\pi^3 G}\right)^{1/2} \frac{1}{g_{ef}^{\;1/2}} \frac{1}{(k_B\,T)^2}, \tag{8.160}$$

or using standard units

$$t = \frac{2.4}{g_{ef}^{\;1/2}} \left(\frac{1\,\mathrm{MeV}}{k_B\,T}\right)^2 \mathrm{s}\,, \tag{8.161}$$

which is a kind of rule of thumb formula for the early Universe.

Finally, the expansion of the Universe is assumed to be adiabatic. In fact there is, by definition, no outside system and the total entropy is much higher than the small variations due to irreversible processes. The entropy of the early Universe can then be assumed to be constant.

Remembering that the entropy S can be defined as

$$S = \frac{\left(\rho c^2 + P\right)}{k_B T} V, \tag{8.162}$$

the entropy density s is then given as:

$$s \;=\; \frac{\rho c^2 + P}{k_B\,T}\,. \tag{8.163}$$

Summing over all possible particle types

$$s = g^s_{ef} \frac{2\pi^2}{45} \left(\frac{k_B\,T}{\hbar c}\right)^3, \tag{8.164}$$

where g^s_{ef} is defined similarly to g_{ef} as

$$g^s_{ef} = \sum_{\mathrm{bosons}} g_i \left(\frac{T_i}{T}\right)^3 + \frac{7}{8} \sum_{\mathrm{fermions}} g_j \left(\frac{T_j}{T}\right)^3. \tag{8.165}$$

At $k_B\,T \sim 1$ TeV ($T \sim 10^{16}$ K, $t \sim 10^{-12}$ s) all the standard model particles should contribute. In the SM there are 6 different types of bosons (γ, $W^{\pm}$, Z, g, H^0) and 24 types of fermions and antifermions (quarks and leptons): thus the total "effective" number of degrees of freedom is

$$g^*_{ef} = 106.75\,. \tag{8.166}$$

At early times the possibility of physics beyond the standard model (Grand Unified Theories (GUT) with new bosons and Higgs fields, SuperSymmetry with the association to each of the existing bosons or fermions of, respectively, a new fermion or boson, ...) may increases this number. The way up to the Planck time ($\sim 10^{-43}$ s), where general relativity meets quantum mechanics and all the interactions may become unified, remains basically unknown. Quantum gravity theories like string theory or loop quantum gravity have been extensively explored in the last years, but, for the moment, are still more elegant mathematical constructions than real physical theories. The review of such attempts is out of the scope of the present book. Only the decoupling of a possible stable heavy dark matter particle will be discussed in the following.

At later times the temperature decreases, and g_{ef} decreases as well. At $k_B\,T \sim$ 0.2 GeV hadronization occurs and quarks and gluons become confined into massive hadrons. At $k_B T \sim 1$ MeV ($t \sim 1$ s) the light elements are formed (primordial nucleosynthesis, see Sect. 8.1.3). Around the same temperature neutrinos also decouple as it will be discussed below. At $k_B T \sim 0.8$ eV the total energy density of nonrelativistic particles is higher than the total energy density of relativistic particles and then the Universe enters a matter-dominated era (see Sect. 8.2.3). Finally at $k_B T \sim 0.3$ eV recombination and decoupling occur (see Sect. 8.1.2). At that moment, the hot plasma of photons, baryons, and electrons, which was coherently oscillating under the combined action of gravity (attraction) and radiation pressure (repulsion), breaks apart: photons propagate away originating the CMB while the baryon oscillations stop (no more radiation pressure) leaving a density excess at a fixed radius (the sound horizon) which, convoluted with the initial density fluctuations, are the seeds for the subsequent structure formation. This entire evolution scenario is strongly constrained by the existence of dark matter which is gravitationally coupled to baryons and represented at that time 2/3 of the total energy of the Universe.

8.3.1.1 Dark Matter Freeze-Out

The present, most popular, cosmological model for the Universe (the, so-called, ΛCDM model—see Sect. 8.3.3) assumes that dark matter is formed by stable massive nonrelativistic particles. These particles must be weakly interacting—otherwise they would have been found (see later); the acronym WIMP (Weakly Interactive Massive Particle) is often used to name them, since for several reasons that will be discussed below, the favorite theoretical guess compatible with experiment is that they are heavier than 45 GeV. The lightest supersymmetric particle, possibly one of the neutralinos χ (see the previous Chapter), is for many the most likely candidate; we shall use often the symbol χ to indicate a generic WIMP. They must be neutral and we assume without loss of generality that they coincide with their antiparticle (as it is the case for the neutralino).

Nonrelativistic particles in thermal equilibrium (ensured by production and annihilation interactions like $f\,\bar{f} \rightarrow \chi\chi$ and $\chi\chi \rightarrow f\,\bar{f}$, respectively) follow, as it was discussed above, a Maxwell–Boltzmann distribution and therefore their equilibrium particle density is proportional to

$$n_{\chi} \propto e^{-\left(\frac{m_{\chi}\,c^2}{k_B\,T}\right)} . \tag{8.167}$$

Decoupling, once again, happens when

$$\Gamma_{\chi} = n\sigma v \sim H . \tag{8.168}$$

After decoupling annihilations cease and the density of dark matter particles will just decrease with a^{-3}. The value of the decoupling density is therefore an inverse function of $\langle \sigma v \rangle$, where the velocity v is small for a large mass particle. In Fig. 8.26 the number density of an hypothetical dark matter particle as a function of time (expressed in terms of the ratio $\frac{m_{\chi}\,c^2}{k_B\,T}$) for different assumed values of $\langle \sigma\ v \rangle$ is shown.

If this new particle χ has Fermi-type weak interactions (Chap. 6) its annihilation cross sections for low energies can be expressed as

$$\sigma \sim {G_{\chi}}^2\,{m_{\chi}}^2 \sim \frac{{g_{\chi}}^4}{{m_{\chi}}^2} \tag{8.169}$$

where G_{χ} and g_{χ} are the effective and the elementary coupling constants.

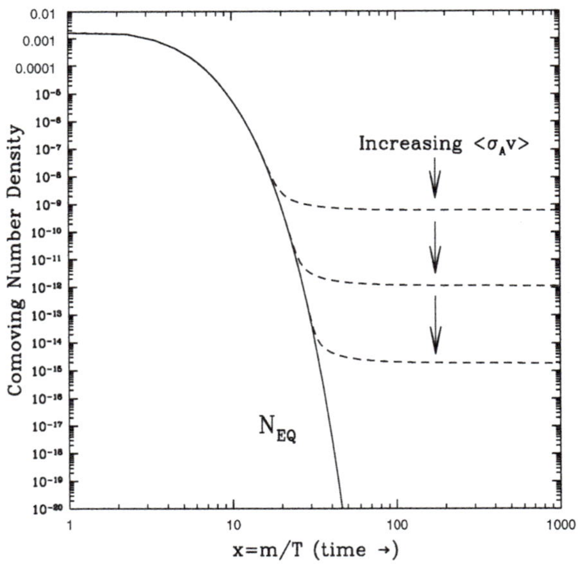

Fig. 8.26 The number density of a nonrelativistic massive particle as a function of time (expressed in terms of the ratio $\frac{m_{\chi}\,c^2}{k_B\,T}$) for different values of $\langle \sigma\ v \rangle$ (from D. Hooper, "TASI 2008 Lectures on Dark Matter", arXiv:0901.4090 [hep-ph])

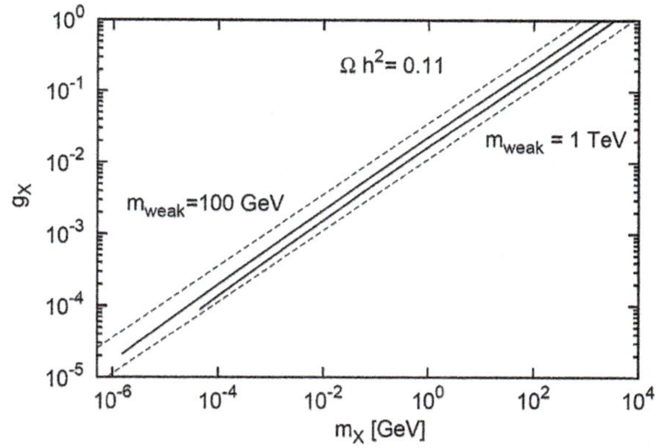

Fig. 8.27 g_χ as function of m_χ which satisfy the present relation (from http://physicslearning2.colorado.edu/tasi/tasi_2011/videos/Feng/lecture3/feng-lecture3.pdf)

A fixed relation between g_χ and m_χ, as shown in Fig. 8.27, can then ensure a given decoupling density of particles. In particular the expected values for a WIMP with $g_\chi \sim g_W \sim 0.6$ and $m_\chi \sim m_W \sim 80$ GeV give the right scale for the observed dark matter density ($\Omega_{DM} \sim 0.2$–0.3, see Sect. 8.3.3); this coincidence is called the *WIMP miracle*. A WIMP can indeed be the mysterious missing dark particle, but this is not the only available solution.

However, the Fermi interaction is a model valid only at low energies; its divergence at higher energies is regularized introducing the intermediate vector bosons $W^{\pm}$:

$$M_W{}^2 \rightarrow \left(q^2 - M_W{}^2\right)$$

which changes drastically the behavior of the dependence of cross section on energy from

$$\sigma \propto E^2$$

to

$$\sigma \propto E^{-2}.$$

A more careful treatment should be then carried on. In the present text, we just want to stress that this fact explains the usual break of the sensitivity of the present cross sections upper limits obtained on the WIMP searches for masses around M_W (Fig. 8.36) reported in Sect. 8.4.2.

8.3.1.2 Neutrino Decoupling and e^+ e^- Annihilations

Decoupling of neutrinos occurs, similarly to what was discussed in Sect. 8.1.3 in the case of the primordial nucleosynthesis, whenever the neutrinos interaction rate Γ_ν is of the order of the expansion rate of the Universe

$$\Gamma_\nu \sim H .$$

Neutrinos interact just via weak interactions (like $\nu\, e^- \rightarrow \nu\, e^-$) and thus their cross sections have a magnitude for $k_B\, T \sim \sqrt{s} \ll m_W$ of the order of

$$\sigma \sim {G_F}^2 s \sim {G_F}^2 (k_B\, T)^2 . \tag{8.170}$$

The neutrino interaction rate Γ_ν is proportional to T^5 ($\Gamma_\nu = n\sigma\, v$ and $n \propto T^3$, $v \sim c$) while H, as seen above (Eq. 8.161), is proportional to T^2. Therefore, there will be a crossing point, which, indeed, occurs for temperatures around a few MeV.

Before the decoupling photons and neutrinos have the same temperature. But from this point on, neutrinos will have basically no more interactions and, thus, their temperature decreases just with a^{-1}, while photons are still in thermal equilibrium with a plasma of electrons and positrons through production ($\gamma\gamma \rightarrow e^+e^-$) and annihilation ($e^+e^- \rightarrow \gamma\gamma$) reactions. For temperatures below 1 MeV this equilibrium breaks down as the production reaction is no more possible ($m_e\, c^2 \sim 0.5$ MeV). However, entropy should be conserved and therefore

$$g^s_{ef}\, T^3 = \text{constant}$$

$$g^{e\gamma}_{ef} T^3_{e\gamma} = g^\gamma_{ef} T^3_\gamma ;$$

before decoupling

$$g^{e\gamma}_{ef} = 2 \times 2 \times \frac{7}{8} + 2 = \frac{11}{2} \tag{8.171}$$

and after decoupling

$$g^\gamma_{ef} = 2 . \tag{8.172}$$

Therefore,

$$\frac{T_\gamma}{T_{e\gamma}} \sim \left(\frac{11}{4}\right)^{1/3} \simeq 1.4 . \tag{8.173}$$

The temperature of photons after the annihilation of the electrons and positrons is thus higher than the neutrino temperature at the same time (the so-called *reheating*).

The temperature of the neutrino cosmic background is therefore nowadays around 1.95 K, while the temperature of the CMB is around 2.73 K (see Sect. 8.1.2).

The ratio between the number density of cosmic background neutrinos (and antineutrinos) and photons can then be computed using Eq. 8.153 as:

$$\frac{N_\nu}{N_\gamma} = 3\,\frac{3}{11}\,,$$

where the factor 3 takes into account the existence of three relativistic neutrino families. Reminding that nowadays $N_\gamma \simeq 410/\text{cm}^3$, the number density of cosmological neutrinos should be:

$$N_\nu \simeq 340/\text{cm}^3\,.$$

The detection of such neutrinos (which are all around) remains an enormous challenge to experimental particle and astroparticle physicists.

8.3.2 *Inflation and Large-Scale Structures*

The early Universe should have been remarkably flat, isotropic, and homogeneous to be consistent with the present measurements of the total energy density of the Universe (equal or very near of the critical density) and with the extremely tiny temperature fluctuations ($\sim 10^{-5}$) observed in the CMB. On the contrary, at scales $\sim$50 Mpc, the observed Universe is filled with rather inhomogeneous structures, like galaxies, clusters, superclusters, and voids. The solution for this apparent paradox was given introducing in the very early Universe an exponential superluminal (receding velocities much greater than the speed of light) expansion. It is the, so-called, *inflation*.

8.3.2.1 The Inflaton Field

The possibility of a kind of exponential expansion was already discussed above in the framework of a Universe dominated by the cosmological constant (Sect. 8.2). The novelty was to introduce a mechanism that could, for a while, provide a vacuum energy density and a state equation ($\mathcal{P} = \alpha\rho$, with $\alpha < -1/3$) ensuring thus the necessary negative pressure. A scalar field filling the entire Universe (the "inflaton field") can serve these purposes.

In fact the energy density and the pressure of a scalar field $\phi(t)$ with an associated potential energy $V(\phi)$ are given by (For a discussion see [F 8.5]):

$$\rho = \frac{1}{2}\frac{1}{\hbar c^3}\dot{\phi}^2 + V(\phi)\;;\;\mathcal{P} = \frac{1}{2}\frac{1}{\hbar c^3}\dot{\phi}^2 - V(\phi)\,. \qquad (8.174)$$

Thus whenever

$$\frac{1}{2\hbar c^3}\dot{\phi}^2 < V(\phi)$$

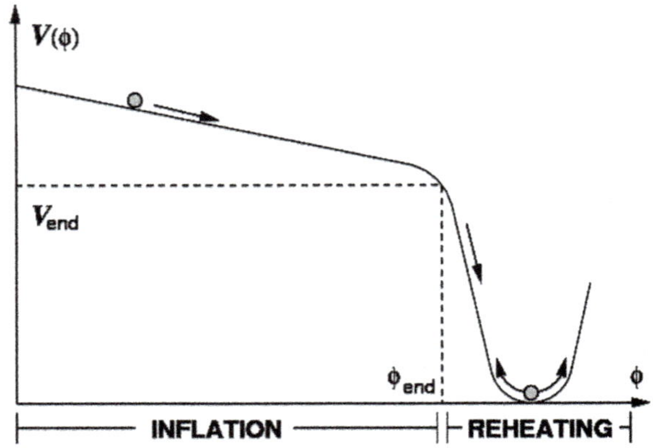

Fig. 8.28 An example of an inflaton potential. (from K. Dimopoulos, J. Phys.: Conf. Ser. 283 012010 (doi:10.1088/1742-6596/283/1/012010))

an exponential expansion will occur. This condition is satisfied by a reasonably flat potential, like the one sketched in Fig. 8.28.

In the first phase the inflaton field rolls down slowly starting from a state of false vacuum ($\phi = 0,\ V(\phi) \neq 0$) and inflation occurs. In this case the inflation period ends when the potential changes abruptly its shape going to a minimum (the true vacuum). The field then oscillates around this minimum dissipating its energy: this process will refill the empty Universe originated by the exponential expansion with radiation (*reheating*), which will then be the starting point of a "classical" hot big bang expansion.

During the inflation period a superluminal expansion will thus occur

$$a(t) \sim e^{Ht} \tag{8.175}$$

with (see Sect. 8.2.3)

$$H \sim \sqrt{\frac{8\pi G}{3c^2}\rho}\,.$$

In this period the scale factor will grow as

$$\frac{a\left(t_f\right)}{a\left(t_i\right)} \sim e^N, \tag{8.176}$$

with N (the number of e-foldings), typically, of the order of about 10^2.

8.3.2.2 Flatness, Horizon, and Monopole Problems

The energy density evolves with (see Sect. 8.2.3)

$$\Omega - 1 = \frac{Kc^2}{H^2a^2}.$$

Then, at the end of the inflationary period, the energy density will be very close to the critical density as it was predicted extrapolating back the present measured energy density values to the early moments of the Universe (the so-called flatness problem). For example, at the epoch of the primordial nucleosynthesis ($t \sim 1$ s) the deviation from the critical density should be $\leq 10^{-12}$–10^{-16}.

The exponential expansion will also give a solution to the puzzle that arises from the observations of the extreme uniformity of the CMB temperature measured all over the sky (the so-called horizon problem).

In the standard Big Bang model the horizon distance (the maximum distance light could have traveled since the origin of time) at last scattering ($t_{ls} \sim 3\ 10^5$ years, $z_{ls} \sim 1100$) is given by

$$d_H = a(t_{ls}) \int_0^{t_{ls}} \frac{c\,dt}{a(t)}. \tag{8.177}$$

If there was no expansion d_H would be as expected to be just $d_H = c\ t_{ls}$.

Basically, the horizon distance is just a consequence of the finite speed of light (which also solves the Olbers' paradox as referred in Sect. 8.1.5).

In a similar way the proper distance from last scattering to the present ($t_0 \sim 14\ 10^9$ years) is given by

$$D = a(t_0) \int_{t_{ls}}^{t_0} \frac{c\,dt}{a(t)}. \tag{8.178}$$

In the Big Bang model there was (see Sect. 8.2.3) first a radiation dominated expansion followed by a matter-dominated expansion with scale parameter evolution $a(t) \propto t^{\frac{1}{2}}$ and $a(t) \propto t^{\frac{2}{3}}$, respectively. The crossing point was computed to be around $t_{\rm cross} \sim 7\ 10^4$ years.

Then, assuming that during most of the time the Universe is matter dominated (the correction due to the radiation dominated period is small),

$$d_H \sim 3\ t_{ls},$$

$$D \sim 3\ t_0.$$

The regions causally connected at the time of last scattering (when the CMB photons were emitted) as seen by an observer on Earth have an angular size of

$$\delta\theta \sim \frac{d_H}{D}\ (1 + z_{ls})\ \frac{180^\circ}{\pi} \sim 1^\circ - 2^\circ, \tag{8.179}$$

where the $(1 + z_{ls})$ factor accounts for the expansion between the time of last scattering and the present.

Regions separated by more than this angular distance have in the standard Big Bang model no way to be in thermal equilibrium. Inflation, by postulating a superluminal expansion at a very early time, ensures that the entire Universe that we can now observe was causally connected in those first moments.

Finally, according to the Big Bang picture, at the very early moments of the Universe all the interactions should be unified. When later temperature decreases, successive phase transitions, due to spontaneous symmetry breaking, will originate the present world we live in, in which the different interactions are well individualized. The problem is that Grand Unified Theories (GUT) phase transition should give rise to a high density of magnetic monopoles. Although none of these monopoles were ever observed (the so-called "monopole problem"), if inflation had occurred just after the GUT phase transition the monopoles (or any other possible relics, e.g., the gravitino) would be extremely diluted and this problem would be solved.

It is then tempting to associate the inflaton field to some GUT breaking mechanism, but it was shown that potentials derived from GUTs do not work, and for this reason the inflaton potential is still, for the moment, an empirical choice.

8.3.2.3 Structure Formation

Nowadays, the most relevant and falsifiable aspect of inflation models is their predictions for the origin and evolution of the structures that are observed in the present Universe.

Quantum fluctuations of the inflaton field originate primeval density perturbations at all distance scales. During the inflationary period all scales that can be observed today went out of the horizon (the number of e-foldings is set accordingly) to reenter later (starting from the small scales and progressively moving to large scales) during the classical expansion (the horizon grows faster than the Universe scale). They evolve under the combined action of gravity, pressure, and dissipation, giving rise first to the observed acoustic peaks in the CMB power spectrum and, finally, to the observed structures in the Universe.

The spatial density fluctuations are usually decomposed into Fourier modes labeled by their wave number k or by their wavelength $\lambda = 2\pi/k$, and

$$\frac{\delta\rho}{\rho}\left(\vec{r}\right) = A \int_{-\infty}^{\infty} \delta_k \, e^{-i\vec{k}.\vec{r}} \, d^3k .$$

Each distance scale corresponds then to a density fluctuation wave characterized by amplitude and dispersion. Generic inflationary models predict density perturbations that are adiabatic (the perturbations in all particle species are similar if they are originated by one single field), Gaussian (the amplitudes follow a Gaussian probability distribution), and obeying a scalar power law spectrum of the type

$$\left\langle |\delta_k|^2 \right\rangle \sim A_s \, k^{n_s - 1} .$$

If $n_s = 1$ (Harrison–Zel'dovich spectrum) the amplitudes in the corresponding gravitational potential are equal at all the scales.

This power spectrum is distorted (in particular for high k, i.e., small scales) as each scale mode will reenter the horizon at a different moment and thus will evolve differently.

In the radiation-dominated phase baryonic matter is coupled to photons and thus the density perturbation modes that had reentered the horizon cannot grow due the existence of a strong radiation pressure which opposes gravity (the sound speed is high and therefore the Jeans scale is greater than the horizon). These perturbations on very small scales can be strongly (or even completely) suppressed, while on larger scales a pattern of acoustic oscillations is built up.

At recombination baryons and photons decouple, the sound speed decreases dramatically, and the Jean scale goes to zero (there is no more photon pressure to sustain gravitation). The baryonic density perturbations will then grow as a power law.

At recombination the regions with matter overdensities will originate cold spots in the CMB. In fact, it can be shown that due to the combined action of the gravitational potential ϕ and the Doppler shift, the temperature fluctuations at each point in space are proportional to the gravitational potential

$$\frac{\delta T}{\langle T \rangle} \cong \frac{1}{3} \phi .$$

Then the pattern of density acoustic oscillations at recombination remains imprinted in the CMB power spectrum, with the positions and amplitudes of the observed peaks strongly correlated with the Universe model parameters. For instance the position of the first peak is, as it was discussed in Sect. 8.1.2, a measurement of the size of the horizon at recombination, and thus strongly constrains the curvature of the Universe, while its amplitude depends on the baryon/photon ratio.

The pattern of the density oscillations at recombination should be also somehow imprinted in the matter distribution in the Universe, what starts to be revealed by the observation of the Baryon Acoustic Oscillations (see Sect. 8.1.1).

Dark matter is, by definition, not coupled to photons and therefore it is not subject to any dramatic change at recombination time. Whenever dark matter became cold (nonrelativistic) associated density perturbations could start to grow and build gravitational potential wells that have, then, been "filled" by baryons after recombination and boosted the formation of gravitational structures. The relative proportion of hot (for instance neutrinos) and cold (for instance WIMPs) dark matter may lead to different scenarios for the formation of large-scale structures. In presence of hot dark matter, a top to bottom formation scenario (from superclusters to galaxies) is favored, while in a cold dark matter (CDM) scenario, it is just the contrary: this second case is in agreement with observational evidence of the existence of supernovae almost as old as the Universe.

8.3.3 The ΛCDM Model

The ΛCDM model, also denominated as *the concordance model* or *the Standard Model of Cosmology*, is a Friedmann–Lemaitre–Robertson–Walker (FLRW) model (Sect. 8.2) with a reduced set of parameters. It postulates that we are living in a flat Universe ($\Omega_m + \Omega_{\text{rad}} + \Omega_\Lambda = 1$) dominated by dark energy in the form of a nonzero cosmological constant ($\mathcal{P} = -\rho$) and CDM; it also assumes a power law spectrum of primordial perturbations.

The assumption that ($\Omega_m + \Omega_{\text{rad}} + \Omega_\Lambda = 1$) is motivated by the fact that observations are consistent with this value with high accuracy. Indeed

$$\Omega_m + \Omega_{\text{rad}} + \Omega_\Lambda = 1.000 \pm 0.004 . \tag{8.180}$$

The key ΛCDM parameters are:

1. H_0, the Hubble parameter;
2. Ω_B, the baryonic matter density;
3. Ω_c, the cold dark matter density;
4. τ, the optical depth to reionization;
5. A_s, n_s, respectively, the amplitude and the spectral index of the primordial fluctuation spectrum.

All these parameters were introduced in previous sections. Since at present $\Omega_{\text{rad}} \simeq 0$, then $\Omega_\Lambda \simeq 1 - (\Omega_B + \Omega_c)$.

The first evidence for a nonzero cosmological constant came from the observations by the "Supernova Cosmology Project" and by the "High-z Supernova Search Team", which show that the Universe is in a state of accelerated expansion (see Sect. 8.1.1). In 2003 it was already possible to conclude that $\Omega_m \sim 0.3$ and $\Omega_\Lambda \sim 0.7$ (Fig. 8.29). The present best fit for observational data by the PDG provides for the main ΛCDM parameters the following values:

1. $H_0 = (67.3 \pm 1.2)$ km s^{-1} Mpc^{-1}
2. $\Omega_B = 0.0499 \pm 0.0022$
3. $\Omega_c = 0.265 \pm 0.011$
4. $\tau = 0.091 \pm 0.014$
5. $\ln\left(10^{10} A_s\right) = 3.103 \pm 0.072$; $n_s = 0.9616 \pm 0.0094$.

Relaxing some of the assumptions of the standard ΛCDM model, extra parameters like, for example, the total mass of the neutrinos, the number of neutrino families, the dark energy equation of state, the spatial curvature, can be added. No surprises have shown up.

As it is the case for particle physics, at the beginning of the twenty-first century we have a standard model also for cosmology that describes with remarkable precision the high quality data sets we were able to gather in the last years. Although we do not yet know how to deduce the parameters of this SM from first principles in a more complete theory, we do have, nevertheless, realized that a slight change in many of these parameters would jeopardize the chance of our existence in the Universe. Are we special?

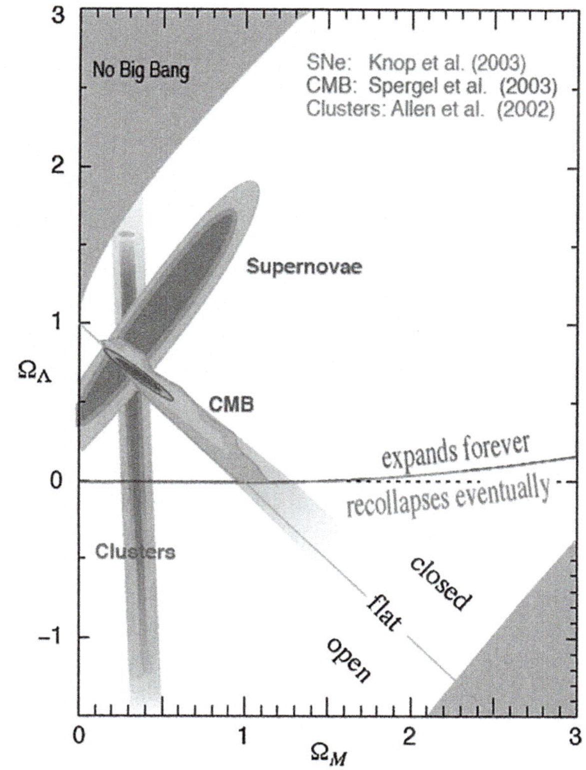

Fig. 8.29 Confidence regions in the plane (Ω_M, Ω_Λ). Credit: http://supernova.lbl.gov

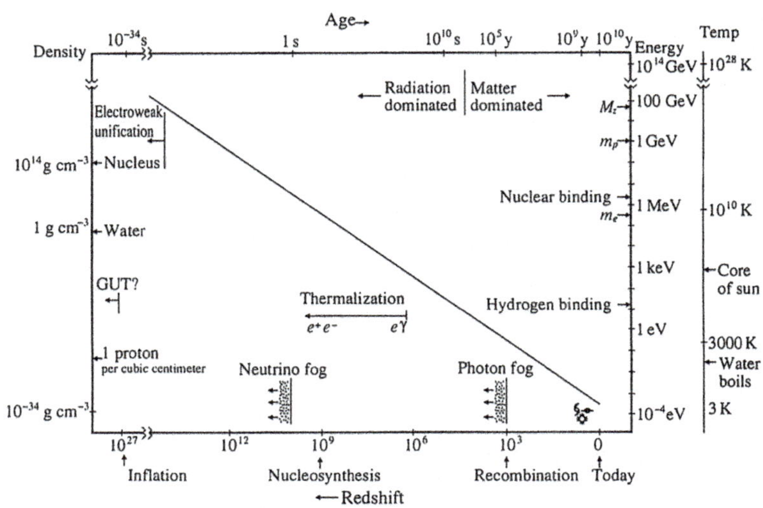

Fig. 8.30 The density, temperature, age, and redshift for the several Universe epochs. From Eric Linder, "First principles of Cosmology", Addison-Wesley 1997

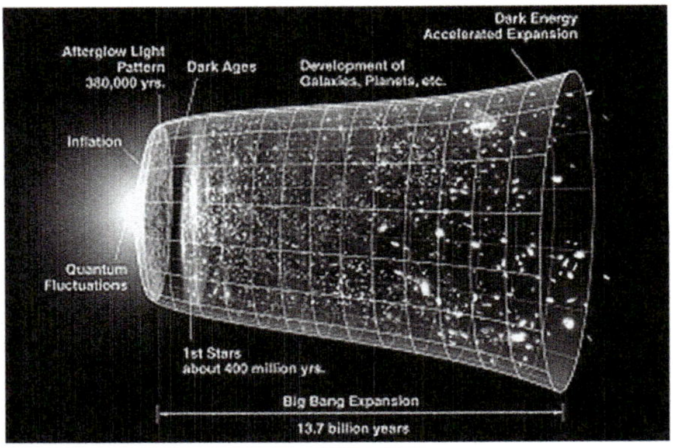

Fig. 8.31 Timeline of our Universe. Credits: NASA (http://map.gsfc.nasa.gov/)

At the same time, additional questions pop up. What is dark matter made of? And how about the dark energy? Why the "particle physics" vacuum expectation value originated from quantum fluctuations is 120 orders of magnitude higher than what is needed to account for dark energy?

Finally, the standard model of cosmology gives us a coherent picture of the evolution of the Universe (Figs. 8.30 and 8.31) starting from as back as Planck time, where even General Relativity is no longer valid. What happened before? Was there a single beginning, or our Universe is one of many? What will happen in the future? Is our Universe condemned to a thermal death? Questions for the twenty-first century. Questions for the present students and the future researchers.

8.4 What Is Dark Matter Made Of, and How Can It Be Found?

Observations indicate a large amount of dark matter or substantial modifications of the standard theory of gravitation (see Sect. 8.1.4).

Let us analyze now what DM can be made of, and what would be the related experimental signatures. We shall discuss in this chapter direct searches for dark matter, searches at accelerators, and shortly indirect searches; a more detailed discussion of indirect signatures in the context of multimessenger astrophysics will be presented in Sects. 10.5.3 and 10.5.5.

8.4.1 Baryonic Dark Matter

Dark matter is unlikely to consist of baryons.

- First, the ΛCDM model (Sect. 8.3.3) computes the total content of baryonic DM (i.e., nonluminous matter made by ordinary baryons) from the fit to the CMB spectrum, and the result obtained is only some 4 % of the total energy of the Universe; the structure of the Universe, computed from astrophysical simulations, is consistent with the fractions computed within the ΛCDM model.
- Second, the abundances of light elements depend on the baryon density, and the observed abundances are again consistent with those coming from the fit to Ω_b coming from the CMB data.

A direct search is however motivated by the fact that the hypotheses on which cosmological measurements are based might be wrong (as in the case of MOND, for example).

Baryonic DM should cluster into massive astrophysical compact objects (the so-called MACHOs), or into molecular clouds. Molecular clouds, when hot, are easily observable (for example in hot galaxy clusters); the result of observations is that the amount of DM due to molecular clouds is small. The main baryonic component should be thus concentrated in massive objects (MACHOs), including black holes. We can estimate the amount of this component using the gravitational field generated by it.

8.4.1.1 MACHOs and Gravitational Lensing

MACHO is a generic name for a compact structure composed of baryonic matter, which emits little or no radiation, and are thus very difficult to detect. MACHOs may be black holes or neutron stars, as well as brown or very faint dwarf stars, or large planets.

A MACHO may be detected when it passes in front of or very close to a star and the star light is bent by the MACHO's gravity. This causes more light to reach the observer and the star to look brighter. This effect is known as gravitational microlensing (Fig. 8.32).

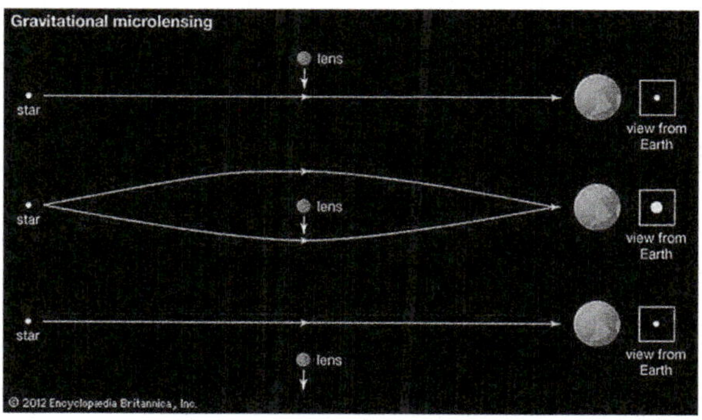

Fig. 8.32 The principle of gravitational microlensing (©Encyclopedia Britannica)

Several research groups have searched for MACHOs and found that only a few percent of the total DM can be attributed to them. Therefore, MACHOs do not solve the missing mass problem.

The microlensing technique is also one of the main sources of information in the search for extrasolar planets (see Chap. 11).

8.4.2 Weakly Interacting Particles

Candidates for nonbaryonic DM must interact very weakly with electromagnetic radiation (otherwise they would not be dark), and they must have the right density to explain about one-quarter of the energy content of the Universe. Candidates include generic weakly interacting particles. A new particle of mass above the eV and below some $M_Z/2$ (about 45 GeV) would have been already found by LEP: DM particles must be very heavy or very light if they exist. They must also be stable on cosmological time scales (otherwise they would have decayed by now).

We use the acronyms WIMP (weakly interacting massive particle) to indicate possible new "heavy" particles, and WISP (weakly interacting slim particle, or sub-eV particle) to indicate possible new light particles.

Among WISPs, neutrinos seem to be an obvious candidate. However, they have a free-streaming length larger than the size of a supercluster of galaxies (they thus enter in the category of the so-called "hot" dark matter). If neutrinos were the main constituent of dark matter, the first galaxies would have the sizes of superclusters; this is in contrast with the deep field observations from the Hubble Space Telescope (which looked in the past by sampling the Universe in depth). Observations from the Planck satellite allow to set an upper limit at 95 % CL

$$\Omega_\nu \le 0.004 .$$

After having excluded known matter as possible a DM candidate, we are only left with presently unknown—although sometimes theoretically hypothesized—matter.

8.4.2.1 New WISPs: Axions and ALPs

The axion is a hypothetical light pseudoscalar (spin-parity 0^+) particle originally postulated to explain the so-called strong CP problem. In principle, CP should not be a symmetry of the QCD Lagrangian; however, CP (and T) appear to be conserved, as opposed to what happens for weak interactions; this fact has been verified with very good accuracy. To fix this problem, Peccei and Quinn (1977) proposed a new global symmetry, spontaneously broken at a very-high-energy scale, and giving rise to an associated boson called the axion (see Sect. 7.3.2). Being pseudoscalar (like the π^0), the axion can decay into two photons, at a rate determined by the (small) coupling

$g_{a\gamma\gamma} = 1/M$—all quantities here are expressed in NU. The axion mass m is given by the formula

$$\frac{m}{1\,\text{eV}} \simeq \frac{1}{M/6 \times 10^6\,\text{GeV}} . \quad (8.181)$$

The axion lifetime would then be proportional to $1/M^5$, which is larger than the age of the Universe for $m < 10$ eV. An axion below this mass would thus be stable.

Since the axion couples to two photons, in a magnetic or electric field it could convert to a photon; vice versa, a photon in an external magnetic or electric field could convert into an axion; the amplitude of the process would be proportional to $g_{a\gamma\gamma}$.

Axion-like particles (ALPs) are a generalization of the axion: while the axion is characterized by a strict relationship between its mass m and $g_{a\gamma\gamma} = 1/M$, these two parameters are unrelated for ALPs. Depending on the actual values of their mass and coupling constant, ALPs can play an important role in cosmology, either as cold dark matter particles or as quintessential dark energy.

In order to account for dark matter, that is, to reach an energy density of the order of the critical density, axion masses should be at least 0.1 meV. Light, axions, and ALPs could still be DM candidates, since they are produced nonthermally, and thus they can be "cold".

Axion Searches. Attempts are being made to directly detect axions mostly by:

1. using the *light-shining-through-a-wall* technique: a laser beam travels through a region of high magnetic field, allowing the possible conversion of photons into axions. These axions can then pass through a wall, and on the other side they can be converted back into photons in a magnetic field;
2. trying to detect *solar axions*: the CAST (CERN Axion Solar Telescope) experiment looks for the X-rays that would result from the conversion of solar axions (produced in the Sun when X-rays scatter off electrons and protons) back into photons, using a 9-tons superconducting magnet.

Indirect searches are also possible, and are prevalently made by investigating three effects.

4. The birefringence of vacuum in high magnetic fields due to photon–axion mixing. It is common that different polarizations experience a different refractive index in matter—a common example is a uniaxial crystal. The vacuum is also expected to become birefringent in presence of an external magnetic field perpendicular to the propagation direction, due to the orientation of the virtual e^+e^- loops. The magnitude of this birefringence could be enhanced by the presence of an axion field, which provides further magnetic-dependent mixing of light to a virtual field (experiment PVLAS by E. Zavattini et al. 2006).
5. Possible anomalies in the cooling times of cataclysmic stars. Stars produce vast quantities of weakly interacting particles, like neutrinos and possibly hypothetical gravitons, axions, and other unknown particles. Although this flux of particles

cannot be measured directly, the properties of the stars would be different from expectations if they could loose too much energy in a new way. The results on the cooling times and the photon fluxes (since photons are coupled to axions) constrain the characteristics of the nonvisible axions.

6. ALPs can also directly affect the propagation of photons coming from astrophysical sources, by mixing to them. This possibility has been proposed in 2007 by De Angelis, Roncadelli, and Mansutti (DARMa), and by Simet, Hooper, and Serpico. The conversion of photons into axions (in the random extragalactic magnetic fields, or at the source and in the Milky Way), could give rise to a sort of cosmic light-shining-through-a-wall effect. This might enhance the yield of very-high-energy photons from distant blazars, which would be otherwise suppressed by the interaction of these photons with the background photons in the Universe (see Chap. 10).

Techniques (1) to (5), with negative results, have limited the region of mass and coupling allowed for ALPs. Failure to detect axions from the Sun with the CAST experiment sets the robust bound $M > 1.14 \cdot 10^{10}$ GeV for $m < 0.02$ eV. Furthermore, the stronger bound $M > 3 \cdot 10^{11}$ GeV holds for ALPs with $m < 10^{-10}$ eV, based on astrophysical observations (technique 5; in particular the photon flux and the cooling time of SN1987A).

A possible evidence for ALPs comes from anomalies in the propagation of very-high-energy photons from astrophysical sources (technique 6).

A summary of the present exclusion limits, and the possible observational window indicated by the propagation of VHE photons, are summarized in Fig. 8.33. The topic is very hot and many new experimental results are expected in the next years.

8.4.2.2 Searches for WIMPs

If dark matter particles are massive (above some $M_Z/2$ and below some TeV) and they are weakly interacting (WIMPs), the "WIMP miracle" guarantees that they can saturate the energy budget for dark matter and the weak interaction characterizing them can be the well-known electroweak interaction. Let us look in more detail to this economical solution.

WIMPs should be neutral, and should have a lifetime large enough in order to have survived from the early Universe until the present time. They "freeze out" once their interaction rate equals the Hubble expansion rate of the Universe—or alternatively when the mean free path equals the size of the Universe: $c/H_0 \sim 1/\Omega_\chi \sigma_{\text{ann}}$, where σ_{ann} is the cross section for WIMP pair annihilation. Inserting the appropriate constants, their relic density can be calculated as

$$\Omega_\chi \simeq \frac{0.6\,\text{pb}}{\sigma_{\text{ann}}} . \tag{8.182}$$

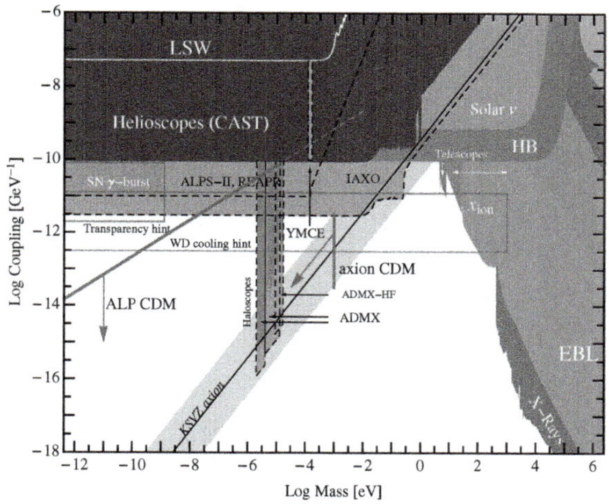

Fig. 8.33 Axion and ALP coupling to photons versus the ALP mass. Colored regions are: generic prediction for the QCD axion, which relate its mass with its coupling to photons (*yellow*), experimentally excluded regions (*dark green*), constraints from astronomical observations (gray) or from astrophysical or cosmological arguments (*blue*), and sensitivity of planned experiments (*light green*). Shown in *red* are boundaries where axions and ALPs can account for all the cold dark matter produced either thermally or nonthermally by the vacuum-realignment mechanism. Adapted from A. Ringwald, J. Phys. Conf. Ser. 485 (2014) 012013

If we assume that dark matter can be explained by just one particle χ, we have $\Omega_\chi \sim 0.22$, and

$$\sigma_{\rm ann} \sim 3\,\text{pb}\,. \tag{8.183}$$

An alternative formulation of Eq. 8.182 is:

$$\Omega_\chi h^2 \simeq \frac{3\,\text{cm}^3\text{s}^{-1}}{\langle\sigma_{\rm ann}|v_\chi|\rangle}\,, \tag{8.184}$$

where v_χ is the relative velocity between the two WIMPs; this leads, since $\Omega_\chi h^2 \sim 0.1$, to

$$\langle\sigma_{\rm ann}|v_\chi|\rangle \simeq 3\times 10^{-26}\,\text{cm}^3\text{s}^{-1}\,. \tag{8.185}$$

The results in Eqs. 8.183 and 8.185 are a natural benchmark for the behavior of DM particles. A little miracle is that if you take the electroweak cross section at an energy scale around 100 GeV (since we are in unification regime we can take for the order of magnitude the electromagnetic expression $\sigma(e^+e^- \to \mu^+\mu^-) \simeq \alpha^2/s$, see Eq. 5.57). The result is

$$\sigma(100\,\mathrm{GeV}) \sim \alpha^2/s \simeq \frac{100\mathrm{nb}}{s[\mathrm{GeV}^2]} \simeq 10\,\mathrm{pb}\,, \tag{8.186}$$

remarkably close to what is needed for a single particle in the 100 GeV range subject to an interaction strength typical of electroweak interactions.

Several extensions to the SM have proposed WIMP candidates, most notably supersymmetric models (SUSY) with $R-$parity conservation, in which the lightest supersymmetric particle, the putative neutralino χ, is stable and thus a serious candidate (Sect. 7.6.1). If its mass is between a few GeV and a few TeV, the parameter space of SUSY can accommodate weak auto-annihilation cross sections (the neutralino is a Majorana particle) . For this reason the neutralino is usually thought to be a "natural" DM candidate. However, more general models are also allowed.

WIMPs (we shall often identify them generically by the χ letter, irrespectively of their possible neutralino-like nature) could be detected directly, via elastic scattering with targets on Earth, or indirectly, by their self-annihilation products in high-density DM environments, or by their decay products.

WIMP velocities in the Earth's surroundings are expected to be of one order of magnitude smaller than the galactic escape velocity, i.e., they are nonrelativistic: thermalized WIMPs have typical speeds

$$\sqrt{\langle v_\chi^2 \rangle} \simeq \sqrt{\frac{2k_B T}{m_\chi}} \simeq 27 \left(\frac{100\,\mathrm{GeV}}{m_\chi}\right)^{1/2} \mathrm{m/s}\,.$$

These are smaller than the velocity $v_\odot$ of the Solar System with respect to the center of the Galaxy, which is of the order of $10^{-3}\,c$.

If the Milky Way's dark halo is composed of WIMPs, then, given the DM density in the vicinity of the Solar System and the speed of the Solar System with respect to the center of the Galaxy, the χ flux on the Earth should be about

$$\Phi_\chi \simeq v_\odot n_{\mathrm{DM,local}} \simeq 10^5 \frac{100\,\mathrm{GeV}}{m_\chi} \mathrm{cm}^{-2}\mathrm{s}^{-1}$$

(a local dark matter density of 0.4 GeV/cm^3 has been used to compute the number density of DM particles). This flux is rather large and a potentially measurable fraction might scatter off nuclei.

Direct detection of dark matter relies on observation of the scattering or other interaction of the WIMPs inside low-background Earth-based detectors.

Indirect detection relies on the observation of:

- The annihilation products of pairs of WIMPs—for example, in the halo of the galaxy, or as a result of their accumulation in the core of the Sun or of the Earth. Such annihilation can happen if the WIMP is a boson, or, if a fermion, a Majorana particle; indeed the SUSY neutralino, a likely WIMP candidate, is a Majorana particle.

- The decay products of the WIMP—WIMPs can be unstable, provided their lifetime is larger than the Hubble time.

Direct Detection of WIMPs at Accelerators. WIMPs can be created at colliders, but not observed, since they are neutral and weakly interacting. However, it is possible to infer their existence. Their signature would be missing energy when one tries to reconstruct the dynamics of the collision. There has been a huge effort to search for the appearance of these new particles.

The searches are complementary to the direct searches that will be described later; however, to compare with noncollider searches, dark matter, the limits need to be translated via an effective field theory into upper limits on WIMP-nucleon scattering or on WIMP annihilation cross sections. They are thus model dependent. However, for many channels foreseen in SUSY, they are complementary to the limits from direct searches (Fig. 8.36) and in particular they can exclude the region below 10 GeV and a cross section per nucleon of the order of 10^{-44} cm^2, where direct searches are not very sensitive.

Direct Detection of WIMPs in Underground Detectors. If the dark matter conjecture is correct, we live in a sea of WIMPs. For a WIMP mass of 50 GeV, there might be in our surroundings some 10^5 particles per cubic meter, moving at a speed smaller than the revolution velocity of the Earth around the Sun.[9] Experimental detection is based on the nuclear recoil that would be caused by WIMP elastic scattering.

The kinematics of the scattering is such that the transferred energy is in the keV range. The recoil energy E_K of a particle of mass M initially at rest after a nonrelativistic collision with a particle of mass m_χ traveling at a speed $10^{-3}c$ is approximately

$$E_K \simeq 50\,\text{keV}\left[\frac{M}{100\,\text{GeV}}\left(\frac{2}{1+M/m_\chi}\right)^2\right]. \tag{8.187}$$

The expected number of collisions is some 10^{-3} per day in a kilogram of material.

Detectors sensitive to the direct detection of a WIMP interaction should have a low energy threshold, a low-background noise, and a large mass. The energy of a nucleus after a scattering from a WIMP is converted into a signal corresponding to (1) ionization, (2) scintillation light; (3) vibration quanta (phonons). The main experimental problem is to distinguish the genuine nuclear recoil induced by a WIMP from the huge background due to environmental radioactivity. It would be useful to do experiments which can measure the nuclear recoil energy and if possible the direction. The intrinsic rejection power of these detectors can be enhanced by the simultaneous detection of different observables (for example, heat and ionization or heat and scintillation).

[9]From astrophysical observations the local WIMP density about 0.4 GeV/cm^3; the velocity distribution is maxwellian, truncated by the galactic escape velocity of 650 km/s. For a mass of 50 GeV, the RMS velocity is comparable to the speed of the solar system in the galaxy, $\sim$ 230 km/s.

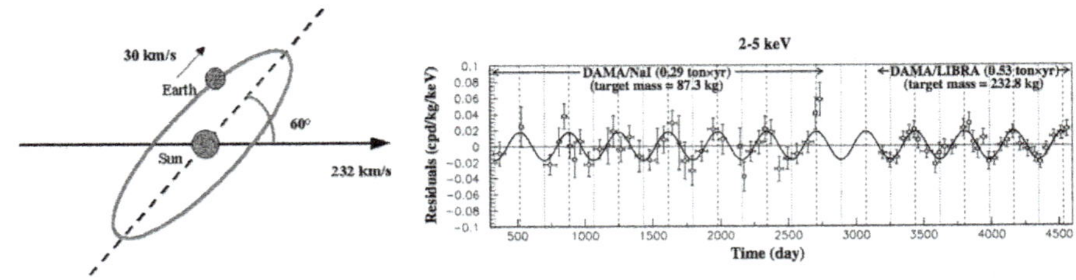

Fig. 8.34 *Left* the directions of the Sun's and the Earth's motions during a year. Assuming the WIMPs to be on average at rest in the Galaxy, the average speed of the WIMPs relative to the Earth is modulated with a period of 1 year. *Right* annual modulation of the total counting rate (background plus possible dark matter signal) in 7 years of data with the DAMA detector. A constant counting rate has been subtracted. From R. Bernabei et al., Riv. Nuovo Cim. 26 (2003) 1

The WIMP rate may be expected to exhibit some angular and time dependence. For example, there might be a daily modulation because of the shadowing effects of the Earth when turned away from the galactic center. An annual modulation in the event rate would also be expected as the Earth's orbital velocity around the Sun (about 30 km/s) adds to or subtracts from the velocity of the Solar System with respect to the galactic center (about 230 km/s), so that the number of WIMPs intercepted per unit time varies (Fig. 8.34, left).

The detectors have then to be well isolated from the environment, possibly shielded with active and passive materials, and constructed with very low activity materials. In particular, it is essential to operate in an appropriate underground laboratory to limit the background from cosmic rays and from natural radioactivity. There are many underground laboratories in the world, mostly located in mines or in underground halls close to tunnels, and the choice of the appropriate laboratory for running a low-noise experiment is of primary importance. Just to summarize some of the main characteristics,

- The thickness of the rock (to isolate from muons and from the secondary products of their interaction).
- The geology (radioactive materials produce neutrons that should be shielded) and the presence of Radon.
- The volume available (none of the present installations could host a megaton detector).
- The logistics.

Some of the largest underground detectors in the world are shown in Fig. 8.35.

As an example, the INFN Gran Sasso National Laboratory (LNGS), which is the largest underground European laboratory, hosts some 900 researchers from 30 different countries. LNGS is located near the town of L'Aquila, about 120 kilometers from Rome. The underground facilities are located on one side of the highway tunnel crossing the Gran Sasso mountain; there are three large experimental halls, each about 100 m long, 20 m wide, and 18 m high. An average 1400 m rock coverage gives a reduction factor of one million in the cosmic ray flux; the neutron flux is thousand

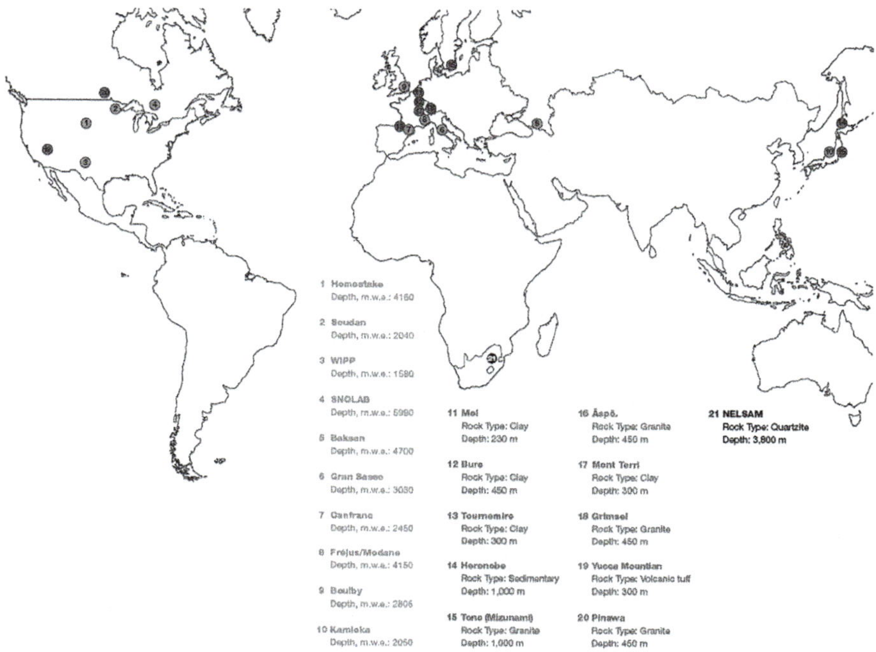

Fig. 8.35 Underground laboratories for research in particle physics (1–10) listed with their depth in meters water equivalent. Laboratories for research in the million-year scale isolation of nuclear waste are also shown (11–20). The NELSAM laboratory (21) is for earthquake research. From www.deepscience.org

times less than the one at the surface. One of the halls points to CERN, allowing long-baseline accelerator neutrino experiments.

Essentially three types of detectors operate searching directly for dark matter in underground facilities all around the world.

- Semiconductor detectors. The recoil nucleus or an energetic charged particle or radiation ionizes the traversed material and produces a small electric signal proportional to the deposited energy. Germanium crystals, which have a very small value of the gap energy (3 eV) and thus have a good resolution of 1 ‰ at 1 MeV, are commonly used as very good detectors since some years. The leading detectors are the CDMS, CoGeNT, CRESST, and EDELWEISS experiments. The bolometric technique (bolometers are ionization-sensitive detectors kept cold in a Wheatstone bridge; the effects measured are: the change in electric resistance consequent to the heating, i.e., the deposited energy; and ionization) increases the power of background rejection, and allows a direct estimate of the mass of the scattering particle.
- Scintillating crystals. Although their resolution is worse than Germanium detectors, no cooling is required. The scintillation technique is simple and well known, and large volumes can be attained because the cost per mass is low. However, these detectors are not performant enough to allow an event-by-event analysis.

For this reason, some experiments are looking for a time-dependent modulation of a WIMP signal in their data. As the Earth moves around the Sun, the WIMP flux should be maximum in June (when the revolution velocity of the Earth adds to the velocity of the solar system in the Galaxy) and minimum in December, with an expected amplitude variation of a few percent.
DAMA (now called in its upgrade DAMA/LIBRA) is the first experiment using this detection strategy. The apparatus is made of highly radio-pure NaI(Tl) crystals, each with a mass of about 10 kg, with two PMTs at the two opposing faces.
- Noble liquid detectors. Certainly the best technique, in particular in a low-background environment, is to use as detectors noble elements (this implies low background from the source itself) such as argon (A = 40) and xenon (A = 131). Liquid xenon (LXe) and liquid argon (LAr) are good scintillators and ionizers in response to the passage of radiation. Using pulse-shape discrimination of the signal, events induced by a WIMP can be distinguished from background electron recoil.
The main technique is to the present knowledge the "double phase" technique. A vessel is partially filled with noble liquid, with the rest of the vessel containing the same element in a gaseous state. Electric fields of about 1 kV/cm and 10 kV/cm are established across the liquid and gas volumes, respectively. An interaction in the liquid produces excitation and ionization processes. Photomultiplier tubes are present in the gas volume and in the liquid. The double phase allows reconstruction of the topology of the interaction (the gas allowing a TPC reconstruction), thus helping background removal.

Whatever the detector is, the energy threshold is a limiting factor on the sensitivity at low WIMP masses; but for high values of m_χ the flux decreases as $1/m_\chi$ and the sensitivity for fixed mass densities also drops. The best sensitivity is attained for WIMP masses close to the mass of the recoiling nucleus (Fig. 8.36).

The experimental situation is presently unclear. Possible WIMP detection signals were claimed by the experiment DAMA, based on a large scintillator (NaI (Tl)) volume, and the CRESST and CoGeNT data show some stress with respect to experiments finding no signal. The data analyzed by DAMA corresponded to 6 years of exposure with a detector mass of 250 kg, to be added to 6 years of exposure done earlier with a detector mass of 100 kg. Based on the observation of a signal at 8.9 σ (Fig. 8.34, right) modulated with the expected period of 1 year and the correct phase (with a maximum near June 2), DAMA collaborators propose two possible scenarios: a WIMP with $m_\chi \simeq 50$ GeV and $\sigma_A \simeq 7 \times 10^{-6}$ pb and a WIMP with $m_\chi \simeq 8$ GeV and $\sigma_A \simeq 10^{-3}$ pb. The DAMA signal is controversial, as it has not presently been reproduced by other experiments with comparable sensitivity but wiith different types of detectors.

The current best limits come from:

- The XENON100 detector, a 165 kg liquid xenon detector located in LGNS with 62 kg in the target region and the remaining xenon in an active veto together with high purity Germanium detectors.

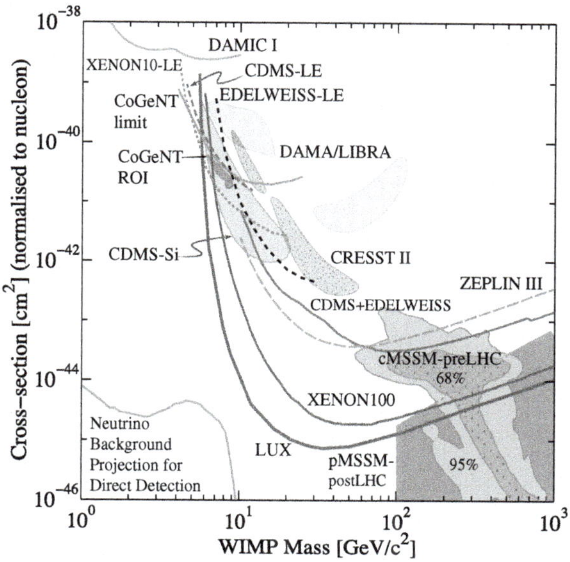

Fig. 8.36 Compilation of experimental results on cross sections of WIMPs versus masses. The areas labeled as DAMA/LIBRA, CREST II, CDMS-Si, and CoGeNT indicate regions of possible signals from those experiments. For context, some supersymmetry implications are also given. From K.A. Olive et al. (Particle Data Group), Chin. Phys. C 38 (2014) 090001, in which experiments are also described in better detail

- The LUX detector, a 370 kg xenon detector installed in the Homestake laboratory (now called SURF) in the US. The LUX collaboration has recently announced results from an 85-days run and an average operational mass of 118 kg.

A new liquid xenon-based project, XENON1t, is planned in the LNGS, with 3.5 tons of liquid xenon.

Indirect Detection of WIMPs. WIMPs are likely to annihilate in pairs; it is also possible that they are unstable, with lifetimes comparable with the Hubble time, or larger. In these cases one can detect secondary products of WIMP decays. Let us concentrate now on the case of annihilation in pairs—most of the considerations apply to decays as well.

If the WIMP mass is below the W mass, the annihilation of a pair of WIMPs should proceed mostly through $f\bar{f}$ pairs. The state coming from the annihilation should be mostly a spin-0 state (in the case of small mutual velocity the s-wave state is favored in the annihilation; one can derive a more general demonstration using the Clebsch–Gordan coefficients). Helicity suppression entails that the decay in the heaviest accessible fermion pair is preferred, similar to what seen in Chap. 6 when studying the $\pi^\pm$ decay (Sect. 6.3.4): the probability decay into a fermion–antifermion pair is proportional to the square of the mass of the fermion. In the mass region between 10 GeV and 80 GeV, the decay into $b\bar{b}$ pairs is thus preferred (this consideration does not hold if the decay is radiative, and in this case a generic $f\bar{f}$ pair will be produced). The $f\bar{f}$ pair will then hadronize and produce a number of secondary particles.

In the case of the annihilation in the cores of stars, the only secondary products which could be detected would be neutrinos. However, no evidence for a significant

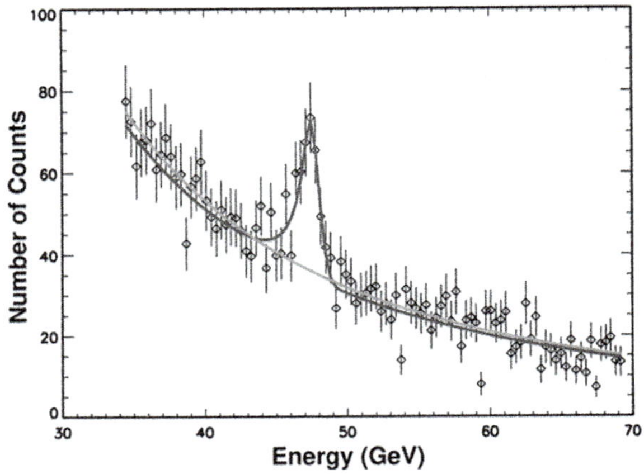

Fig. 8.37 γ-ray signature of neutralino self-annihilation or of neutralino decay. Simulation from the Fermi-LAT collaboration

extra flux of high-energy neutrinos from the direction of the Sun or from the Earth's core has ever been found.

One could have annihilations in the halos of galaxies or in accretion regions close to black holes or generic cusps of dark matter density. In this case one could have generation of secondary particles, including gamma rays, or antimatter which would appear in excess to the standard rate.

We shortly present here the possible scenarios for detections, which will be discussed in larger details in Chap. 10, in the context of multimessenger astrophysics.

Gamma Rays. The self-annihilation of a heavy WIMP χ can generate photons (Fig. 8.37) in three main ways.

(a) Directly, via annihilation into a photon pair ($\chi\chi \to \gamma\gamma$) or into a photon—$Z$ pair ($\chi\chi \to \gamma Z$) with $E_\gamma = m_\chi$ or $E_\gamma = (m_\chi - m_Z)^2/(4m_\chi)$, respectively; these processes give a clear signature at high energies, as the energy is monochromatic, but the process is suppressed at one loop, so the flux is expected to be very faint.
(b) Via annihilation into a quark pair which produces jets emitting in turn a large number of γ photons ($qq \to$ *jets* $\to$ *many photons*); this process produces a continuum of gamma rays with energies below the WIMP mass. The flux can be large but the signature might be difficult to detect, since it might be masked by astrophysical sources of photons.
(c) Via internal bremsstrahlung; also in this case one has an excess of low energy gamma rays with respect to a background which is not so well known. Besides the internal bremsstrahlung photons, one will still have the photons coming from the processes described at the two previous items.

The γ-ray flux from the annihilation of dark matter particles of mass m_χ can be expressed as the product of a particle physics component times an astrophysics component:

$$\frac{dN}{dE} = \underbrace{\frac{1}{4\pi}\frac{\langle\sigma_{\text{ann}}v\rangle}{2m_\chi^2}\frac{dN_\gamma}{dE}}_{\text{Particle Physics}} \times \underbrace{\int_{\Delta\Omega - l.o.s.} dl(\Omega)\rho_\chi^2}_{\text{Astrophysics}} \,. \tag{8.188}$$

The particle physics factor contains $\langle\sigma_{\text{ann}}v\rangle$, the velocity-weighted annihilation cross section (there is indeed a possible component from cosmology in v), and dN_γ/dE, the γ-ray energy spectrum for all final states convoluted with the respective branching rations. The part of the integral over line of sight ($l.o.s.$) in the observed solid angle of the squared density of the dark matter distribution constitutes the astrophysical contribution.

It is clear that the expected flux of photons from dark matter annihilations, and thus its detectability, depend crucially on the knowledge of the annihilation cross section σ_{ann} (which even within SUSY has uncertainties of one to two orders of magnitude for a given WIMP mass) and of ρ_χ, which is even more uncertain, and enters squared in the calculation. A cusps in the dark matter profile, or even the presence of local clumps, could make the detection easier by enhancing ρ_χ—and we saw that the density in the cusps is uncertain by several orders of magnitude within current models (Sect. 8.1.4.1). In the case of WIMP decays, the density term will be linear.

The targets for dark matter searches should be not extended, with the highest density, with no associated astrophysical sources, close to us, and possibly with some indication of small luminosity/mass ratio from the stellar dynamics.

The galactic center is at a distance of about 8 kpc from the Earth. A black hole of about 3.6×10^6 solar masses, Sgr A*, lies there. Because of its proximity, this region might be the best candidate for indirect searches of dark matter. Unfortunately, there are other astrophysical γ-ray sources in the field of view (e.g., the supernova remnant Sgr A East), and the halo core radius makes it an extended rather than a point-like source.

The best observational targets for dark matter detection outside the Galaxy are the Milky Way's dwarf spheroidal satellite galaxies (for example, Carina, Draco, Fornax, Sculptor, Sextans, Ursa Minor). For all of them (e.g., Draco), there is observational evidence of a mass excess with respect to what can be estimated from luminous objects, i.e., a high M/L ratio.

The experimental searches will be discussed in Sect. 10.5.3.

Neutrinos. Neutrino–antineutrino pairs can also be used for probing WIMP annihilation or decay; neutrinos are produced via the same mechanisms producing gamma rays, apart from the fact that neutrino radiation is negligible. Besides the smaller astrophysical background, the advantage of neutrinos is that they can be observed even if the annihilation happens in the corse of opaque astrophysical objects (the Sun or compact objects in particular); apart form these cases, discussed in the beginning of this section, the sensitivity of the gamma-ray channel is by far superior, due to the experimental difficulty of detecting neutrinos for the present and next generation of detectors.

Matter–Antimatter and Electron Signatures. Another indirect manifestation of the presence of new WIMPs would be given by their decay (or self-annihilation) producing

- Electron–positron pairs. The electron yield can also be studied in gamma-ray detectors (in particular Cherenkov telescopes). A smoking gun signature would be the presence of a peak in the energy of the collected electrons, indicating a two-body decay. A shoulder reaching $m_\chi/2$ could also be a signature, but, in this last case, one could hypothesize astrophysical sources as well.
  A peak in the yield of electrons (plus positrons) has been claimed by the ATIC balloon experiment but was not confirmed by Fermi and H.E.S.S. (see Chap. 10). The study of the electron and positron spectrum at high energy is of great interest anyway, and Cherenkov telescopes can contribute in the range of VHE γ-rays.
- Democratically antimatter and matter, resulting in an excess of antimatter with respect to the prediction of models in which antimatter is just coming from secondary processes.
  Recently, PAMELA observed a positron abundance in cosmic radiation higher than that predicted by current models (see Chap. 10). This has been confirmed by the AMS-02 mission, reaching unprecedented accuracy. AMS-02 has also found an excess of antiprotons with respect to models in which only secondary production is accounted.
  The study of the ratio e^+/e^- and $\bar{p}/p$ at higher energies would give useful information; a smoking gun signature for the origin of positrons from the decay of a χ or from a $\chi\chi$ annihilation would be a drop-off of the ratio at a given energy. A more detailed discussion of experimental data will be presented in Chap. 10.

8.4.3 Other Nonbaryonic Candidates

Additional candidates, more or less theoretically motivated, have been proposed in the literature. We list them shortly here; they are less economic than the ones discussed before (WIMPs in particular).

Sterile Neutrinos. A possible DM candidate is a "sterile" neutrino, i.e., a neutrino which does not interact via weak interactions. We know that such neutrino states exist: the right-handed component of neutrinos in the standard model are sterile. Constraints form cosmology make it, however, unlikely that light sterile neutrinos can be the main component of dark matter; sterile neutrinos with masses of the order of the keV and above could be however, with some difficulty, accommodated in the present theories.

Kaluza–Klein States. If particles propagate in extra spacetime dimensions, they will have an infinite spectroscopy of partner states with identical quantum numbers; these states could be a DM candidate.

Matter in Parallel Branes; Shadow or Mirror Matter. Some theories postulate the presence of matter in parallel branes, interacting with our world only via gravity or via a super-weak interaction. In theories popular in the 1960s, a "mirror matter" was postulated to form astronomical mirror objects; the cosmology in the mirror sector could be different from our cosmology, possibly explaining the formation of dark halos. This mirror-matter cosmology has been claimed to explain a wide range of phenomena.

Superheavy Particles (WIMPzillas). Superheavy particles above the GZK cutoff (WIMPzillas) could have been produced in the early Universe; their presence could be detected by an excess of cosmic rays at ultrahigh energies.

Further Reading

[F8.1] J. Silk, "The big bang", Times Books 2000.
[F8.2] B. Ryden, "Introduction to Cosmology", Addison Wesley, Pearson Education 2003. This book provides a clear introduction to cosmology for upper level undergraduates.
[F8.3] E.F. Taylor and J.A. Wheeler, "Exploring Black Holes, introduction to general relativity", Addison Wesley 2000. This book provides an enlightening introduction to the physics of black holes emphasizing how they are "seen" by observers in different reference frames.
[F8.4] M.V. Berry, "Principles of Cosmology and Gravitation", Adam Hilger 1989. This book presents the fundamentals of general relativity and cosmology with many worked examples and exercises without requiring the use of tensor calculus.
[F8.5] V. Mukhanov, "Physical Foundations of Cosmology", Cambridge 2005. This book provides a comprehensive introduction to inflationary cosmology at early graduate level.

Exercises

1. *Cosmological principle and Hubble law*. Show that the Hubble law does not contradict the cosmological principle (all points in space and time are equivalent).
2. *Olbers Paradox*. Why is the night dark? Does the existence of interstellar dust (explanation studied by Olbers himself) solve the paradox?
3. *Steady state Universe*. In a steady state Universe with Hubble law, matter has to be permanently created. Compute in that scenario the creation rate of matter.
4. *Blackbody form of the Cosmic Microwave Background*. In 1965 Penzias and Wilson discovered that nowadays the Universe is filled with a cosmic microwave background which follows an almost perfect Planck blackbody formula. Show

that the blackbody form of the energy density of the background photons was preserved during the expansion and the cooling that had occurred in the Universe after photon decoupling.

5. *Nucleosynthesis and neutron lifetime.* The value of the neutron lifetime, which is anormaly long for weak decay processes (why?), is determinant in the evolution of the Universe. Discuss what would have been the primordial fraction of He if the neutron lifetime would have been one-tenth of its real value.
6. *GPS time corrections.* Identical clocks situated in a GPS satellite and at the Earth surface have different periods due general relativity effects. Compute the time difference in one day between a clock situated in a satellite in a circular orbit around Earth with a period of 12 h and a clock situated on the Equator at the Earth surface. Consider that Earth has a spherical symmetry and use the Schwarzschild metric.
7. *Asymptotically Matter-dominated Universe.* Consider a Universe composed only by matter and radiation. Show that whatever would have been the initial proportion of the matter and the radiation energy densities this Universe will be asymptotically matter dominated.
8. *Flatness of the Early Universe.* The present experimental data indicate a value for a total energy density of the Universe compatible with one within a few per mil. Compute the maximum possible value of $|\Omega - 1|$ at the scale of the electroweak symmetry breaking consistent with the measurements at the present time.
9. *WIMP "miracle".* Show that a possible Weak Interacting Massive Particle (WIMP) with a mass of the order of $m_\chi \sim 100$ GeV would have the relic density needed to be the cosmic dark matter (this is the so-called WIMP "miracle").

Chapter 9
The Properties of Neutrinos

This chapter is dedicated to the physics of neutrinos, which are neutral particles, partners of the charged leptons in SU(2) *multiplets, subject to the weak interaction only—besides the negligible gravitational interaction. Due to their low probability of interaction, they are very difficult to detect and as a consequence the neutrino sector is the less known in the standard model of particle physics. In the late 1990s it has been discovered that neutrinos of different flavors (electron, muon, or tau) "oscillate": neutrinos created with well-defined leptonic flavor may be detected in another flavor eigenstate. This phenomenon implies that neutrinos have a nonzero—although tiny even for the standards of particle physics—mass.*

Neutrinos have been important for the developments of particle physics since they were conjectured in the twentieth century, and are still at present at the center of many theoretical and experimental efforts. Their detection is difficult, since they are subject only to weak interactions (besides the even weaker gravitational interaction).

The existence of neutrinos was predicted by Wolfgang Pauli in 1930 in order to assure the energy momentum conservation in the β decay as it was recalled in Sect. 2.3. Then in 1933 Enrico Fermi established the basis of the theory of weak interactions in an analogy with QED but later on it was discovered that parity is not conserved in weak interactions: neutrinos should be (with probability close to one) left-handed, antineutrinos should be right-handed (see Chap. 6). The theory needed a serious update, which was performed by the electroweak unification (Chap. 7).

Neutrinos were experimentally discovered only in the second-half of the twentieth century: first the electron antineutrino in 1956 by Reines[1] and Cowan (Sect. 2.8); then

[1] Frederick Reines (1918–1998) was an American physicist, professor at the University of California at Irvine and formerly employed in the Manhattan project. He won the Nobel Prize in Physics 1995 "for pioneering experimental contributions to lepton physics;" his conational and coworker Clyde Cowan Jr. (1919–1974) had already passed away at the time of the recognition.

A. De Angelis and M.J.M. Pimenta, *Introduction to Particle and Astroparticle Physics*, Undergraduate Lecture Notes in Physics,
DOI 10.1007/978-88-470-2688-9_9

in 1962 the muon-neutrino by Lederman, Schwartz and Steinberger[2]; and finally, the tau neutrino in 2000 by the DONUT experiment at Fermilab (Sect. 5.6.2). Meanwhile it was established in 1991 in the LEP experiments at CERN that indeed there are only three kinds of light neutrinos (see Sect. 7.5.1).

Neutrinos cannot be directly observed: they are only detected through their interactions. Different neutrino flavors are defined by the flavors of the charged lepton they produce in weak interactions. The electron neutrino ν_e, for example, is the neutrino produced together with a positron, and its interaction will produce an electron. Similarly, for the muon and the tau neutrinos.

For many years it was thought that neutrinos are massless, and for the standard model of particle physics three generations of massless left-handed neutrinos were enough—a nonzero mass was not forbidden, but it implied new mass terms in the Lagrangian discussed in Chap. 7. There was anyway a cloud: the so-called solar neutrino problem—in short, the number of solar electron neutrinos arriving to the Earth was measured to be much smaller (roughly between one-third and 60 %, depending on the experiment's threshold) of what it should be according to estimates based on the solar power. This problem was solved when it was demonstrated that neutrinos can change flavor dynamically: neutrino species "mix", and quantum mechanics implies that they cannot be massless.

9.1 Sources and Detectors; Evidence of the Transmutation of the Neutrino Flavor

Neutrinos are generated in several processes, and their energy spans a wide range (Fig. 9.1). Correspondingly, there are different kinds of detectors to comply with the different fluxes and cross sections expected.

Let us start analyzing some of the neutrino sources. In particular solar neutrinos, atmospheric neutrinos, reactor neutrinos, and accelerator neutrinos have been complementary in determining the neutrino oscillation parameters, and thus, constraining the masses and the mixing matrix. Other sources of neutrinos, more relevant for neutrino astrophysics, will be discussed in Chap. 10.

9.1.1 Solar Neutrinos, and the Solar Neutrino Problem

In the so-called "Standard Solar Model" (SSM), the Sun produces energy via thermonuclear reactions in its center, in a kernel much smaller than the Sun's radius. Most

[2]The Nobel Prize in Physics 1988 was awarded jointly to Leon Lederman (New York 1922), Melvin Schwartz (New York 1931—Ketchum, Idaho, 2006) and Jack Steinberger (Bad Kissingen 1921) "for the neutrino beam method and the demonstration of the doublet structure of the leptons through the discovery of the muon neutrino".

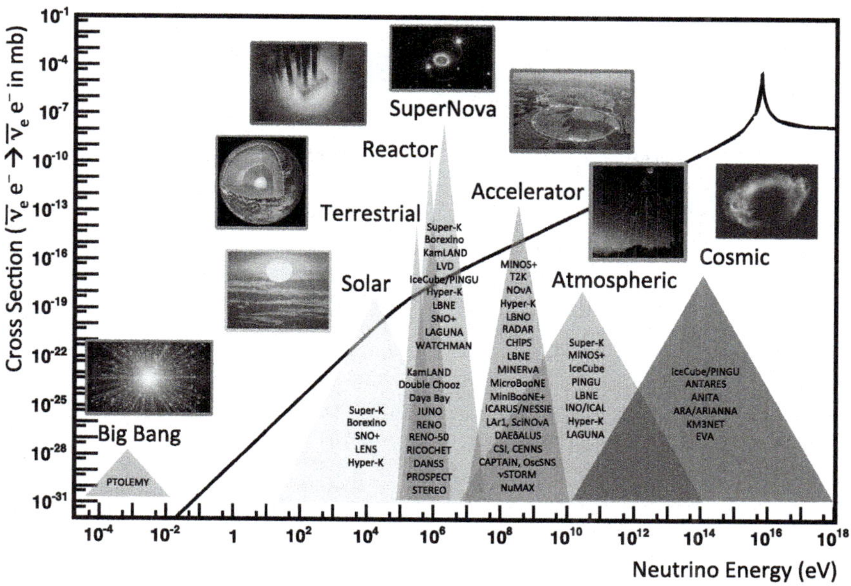

Fig. 9.1 Neutrino interaction cross section as a function of energy, showing typical energy regimes accessible by different neutrino sources and experiments. The *curve* shows the scattering cross section for an electron antineutrino on an electron. From A. de Gouvêa et al., arXiv:1310.4340v1

of the energy is released via MeV photons, which will originate the electromagnetic solar radiation through propagation and interaction processes that take a long time (~2 million years). The light emitted comes mostly from the thermal emission of the external region, the photosphere, which has a temperature of about 6000 K, and is heated by the moderation of these photons.

The fusion reactions in the Sun release 26.7 MeV per reaction and produce also a large flux of electron neutrinos that can be detected at Earth (the expected flux at Earth predicted by John Bahcall and collaborators in the Standard Solar Model (SSM) is $\sim 6 \times 10^{10}$ cm^{-2}s^{-1}). This flux is produced mainly by the nuclear reactions initiated by proton-proton (pp) fusions as sketched in Fig. 9.2. The contribution of the alternative CNO chain[3] is small.

The dominant pp reaction (>90 % of the total flux) produces ν_e which have a low energy endpoint (<0.42 MeV) as it is shown in Fig. 9.3. The ^{7}B line at 0.86 MeV is the second most relevant ν_e source (7–8 %) while the "pep" reaction producing ν_e with energy of 1.44 MeV contributes with just a 0.2 %.

The ^{8}B neutrinos are produced in the "$ppIII$" chain with energies <15 MeV and although their flux could appear marginal (~0.1 %) they have a major role in the

[3] The CNO cycle (for carbon-nitrogen-oxygen) is a set of alternative chains of conversion of hydrogen to helium. In the CNO cycle, four protons fuse, giving origin to one alpha particle, two positrons and two electron neutrinos; the cycle uses C, N, and O as catalysts. While the threshold of the pp-chain is around temperatures of 4 MK, the threshold of a self-sustained CNO chain is at approximately 15 MK. The CNO chain becomes dominant at 17 MK.

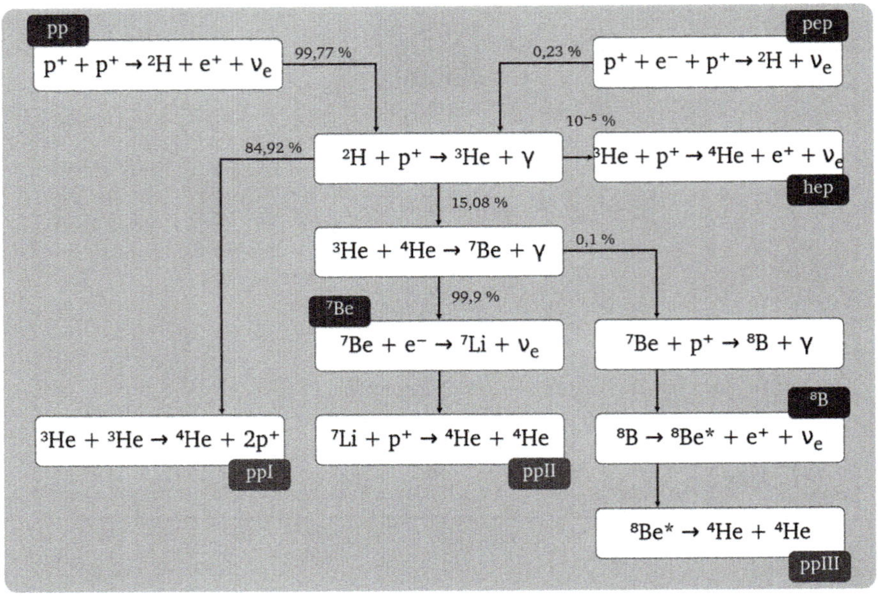

Fig. 9.2 Main nuclear fusion reactions that contribute to the solar neutrino flux. By Dorottya Szam [CC BY 2.5 http://creativecommons.org/licenses/by/2.5], via Wikimedia Commons

solar neutrino detection experiments. In fact, they were the dominant contribution in the historical Chlorine experiment and can be detected by Cherenkov experiments like Super-Kamiokande and SNO (Fig. 9.3).

The first solar neutrino experiment was done in the late 1960s by Ray Davis in the Homestake mine in South Dakota, USA, counting the number of ^{37}Ar atoms produced in 615 ton of C_2Cl_4 by the reaction involving chlorine:

$$\nu_e \, {}^{37}_{17}\mathrm{Cl} \rightarrow {}^{37}_{18}\mathrm{Ar} \, e^- \tag{9.1}$$

(Nobel prize for Davis, as we discussed in Chap. 4). The observed rate was just around one-third of the expected number of interactions. This unexpected result originated the so-called "solar neutrino problem" that for three decades lead to a systematic and careful work of a large community of physicists, chemists, and engineers which, at the end, confirmed both the predictions of the SSM and the experimental results of Ray Davis. The explanation should be in some fundamental properties of neutrinos. Indeed subsequent solar neutrino experiments based on different detection techniques found also a significant deficit in the observed ν_e fluxes; in particular, the GALLEX (at Gran Sasso in Italy) and the SAGE (at Baksan in Russia) experiments used also a radiochemical technique with a lower threshold, choosing Gallium as the detection medium, enabling thus the detection of *pp* neutrinos.

The Kamiokande and the Super-Kamiokande (described in Chap. 4; also called Super-K, or SK) experiments at Kamioka in Japan used water as the target material

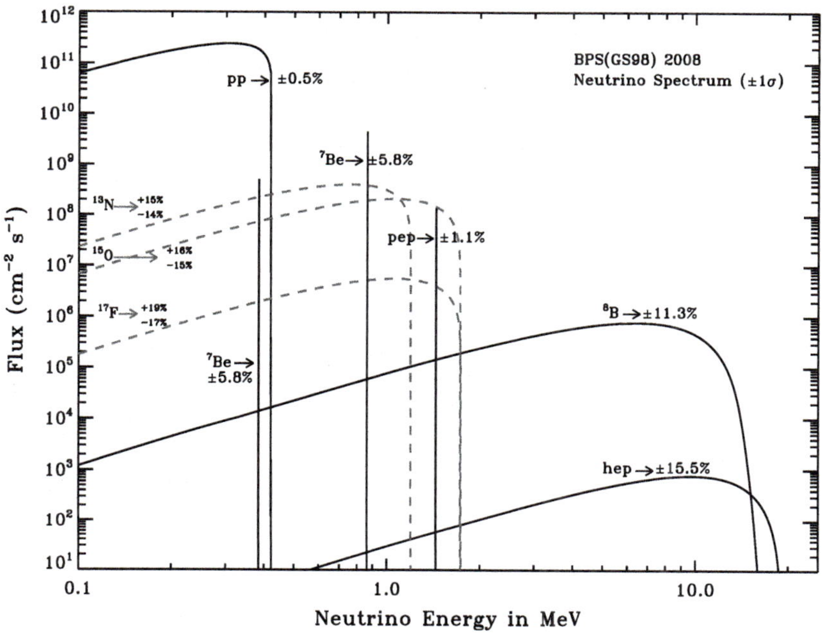

Fig. 9.3 The solar neutrino energy spectrum predicted by the SSM. For continuum sources, the fluxes are expressed in units of $cm^{-2}s^{-1}$ MeV^{-1} at the Earth's surface. For line sources, the units are number of neutrinos $cm^{-2}s^{-1}$. The total theoretical errors are quoted for each source. *Source* http://arxiv.org/abs/0811.2424

(50,000 tons in the case of Super-K) which allowed the detection, by Cherenkov radiation, of electrons produced in the interaction of MeV neutrinos on atomic electrons. The energy and the direction of the scattered electron could be measured determining respectively the number of photons and the orientation of the Cherenkov ring. In this way, as the electron keeps basically the direction of the incoming neutrino, it could be proved that indeed the neutrinos were coming from the Sun as it is shown by the obtained beautiful neutrino "photography" of the Sun (Fig. 9.4).

Also in this experiment the total observed flux, when interpreted as only ν_e interactions, is significantly lower than expected by the SSM.

Was the SSM wrong, or some electron neutrinos were disappearing on their way to the Earth?

The final answer was given by the Sudbury Neutrino Observatory (SNO) in Canada. SNO used 1000 tons of heavy water (D_2O) as the target material. Both charged- and neutral-current neutrino interactions with deuterium nuclei are then possible:

- $\nu_e\, d \to e^- p\; p$ (charged current, CC);
- $\nu_x\, d \to \nu_x\, n\; p$ (neutral current, NC).

Fig. 9.4 Image of the Sun obtained in a 500 days run of Super-Kamiokande. Credit: R. Svoboda and K. Gordan (Louisiana State University, USA)

While in the first reaction only the ν_e may interact (the neutrino energy is below the kinematic threshold for muon production), the νs of all flavors can contribute to the second one. The resulting e^- is detected by measuring the corresponding water Cherenkov ring. The neutron in the final state may be captured either with low efficiency in the deuterium nuclei or with higher efficiency in ^{35}Cl nuclei from 2 tons of salt (NaCl) that were added in the second phase of the experiment. In any case in those radiative captures γ photons are produced and these may produce, via Compton scattering, relativistic electrons which again originate Cherenkov radiation. In the third and final phase, an array of ^{3}He-filled proportional counters were deployed to provide an independent counting of the NC reaction. In addition to the two processes described above, the elastic scattering

$$\nu_x e^- \rightarrow \nu_x e^-$$

is also possible for all neutrino types—although with different cross sections, being the neutrino electron process favored with respect to the other neutrino types.

While ν_e, ν_μ, ν_τ can contribute to the NC, only ν_e contribute to the CC. Thus one has in SNO a clear way to separate the measurement of the ν_e flux from the measurement of the different active neutrino species (in a three-flavor model, $\nu_e + \nu_\mu + \nu_\tau$). SNO could determine that

$$\frac{\Phi(\nu_e)}{\Phi(\nu_x)} = 0.340 \pm 0.038\,(\text{stat.} + \text{syst.})$$

and thus indicate that electron neutrinos might transform themselves into different neutrino flavors during their travel from the Sun to the Earth. The result is compatible with a value of 1/3.

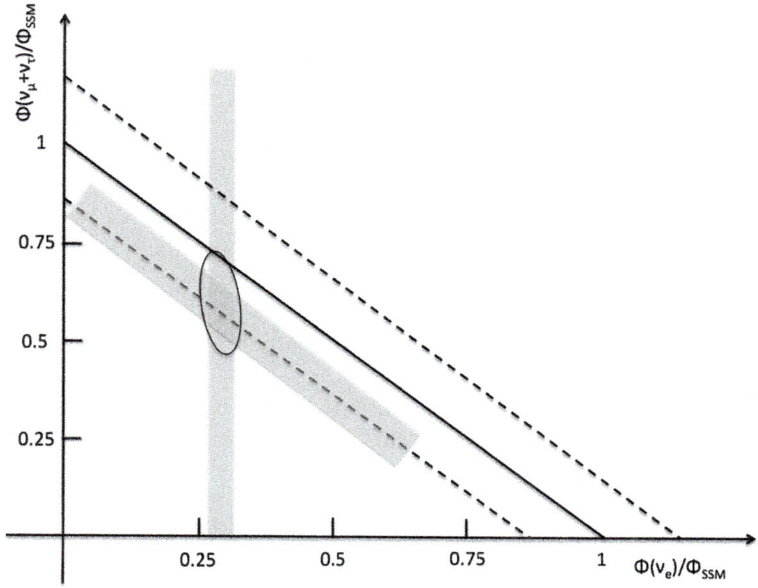

Fig. 9.5 The flux of muon plus tau neutrinos versus the flux of electron neutrinos as derived from the SNO data. The *vertical* band comes from the SNO charged-current analysis; the *diagonal* band from the SNO neutral-current analysis; the *ellipse* shows the 68 % confidence region from the best fit to the data. The predicted Standard Solar Model total neutrino flux is the *solid line* lying between the *dotted lines*

The results obtained by SNO are summarized in Fig. 9.5. The total measured neutrino flux is clearly compatible with the total flux expected from the SSM and the fraction of detected ν_e is consistent with being only one-third of the total number of the neutrinos.

The solar neutrino problem could be solved without relying on the SSM, and the solution was that solar neutrinos change their flavor during their way to the Earth; the oscillation appears to be complete, in the sense that electron neutrinos are only one third of the total.

Let us examine now the characteristics of the oscillation of neutrinos in the simplified hypothesis that there are only two flavors, and two eigenstates.

9.1.2 Neutrino Oscillation in a Two-Flavor System

The transmutation of neutrinos from one species to another implies in a quantum mechanical world an oscillation phenomenon, similar to what we have observed in the $K^0 - \bar{K}^0$ system. We examine now a simplified model of the neutrino oscillations, to see its implications.

In a world with two flavor (let us suppose for the moment they are ν_e, ν_μ) and two mass (ν_1, ν_2) eigenstates, the flavor eigenstates can be written as a function of a single real mixing angle θ as:

$$\nu_e = \nu_1 \cos\theta + \nu_2 \sin\theta \tag{9.2}$$

$$\nu_\mu = -\nu_1 \sin\theta + \nu_2 \cos\theta \tag{9.3}$$

or, using matrices,

$$\begin{pmatrix} \nu_e \\ \nu_\mu \end{pmatrix} = \begin{pmatrix} \cos\theta & \sin\theta \\ -\sin\theta & \cos\theta \end{pmatrix} \begin{pmatrix} \nu_1 \\ \nu_2 \end{pmatrix} . \tag{9.4}$$

Then, for instance, if a ν_e is produced at time $t = 0$ and position $\vec{x} = 0$, the space–time evolution of this quantum state ψ will be determined by the evolution of the corresponding mass eigenstates:

$$\psi = \nu_1 \cos\theta e^{-i(E_1 t - \vec{p}_1 . \vec{x})} + \nu_2 \sin\theta e^{-i(E_2 t - \vec{p}_2 . \vec{x})} \tag{9.5}$$

(note that, in order for the mixing to have an effect, the two masses must be different, i.e., at least one should be different from zero) or, expressing this quantum state again in terms of the weak eigenstates:

$$\psi = \left(\cos^2\theta e^{-i(E_1 t - \vec{p}_1 . \vec{x})} + \sin^2\theta e^{-i(E_2 t - \vec{p}_2 . \vec{x})}\right) \nu_e -$$
$$\left(\cos\theta \sin\theta \left(e^{-i(E_1 t - \vec{p}_1 . \vec{x})} - e^{-i(E_2 t - \vec{p}_2 . \vec{x})}\right)\right) \nu_\mu .$$

Note that at $(t = 0, \vec{x} = 0)$, $\psi = \nu_e$ but, at later times, there will be usually a mixture between the two-flavor states ν_e, ν_μ.

It can be seen from the equations above that the probability to find a state ν_μ at a distance L from the production point is given by:

$$P\left(\nu_e \to \nu_\mu\right) = \sin^2(2\theta) \sin^2\left(\frac{\Delta m^2 L}{4E_\nu}\right) \tag{9.6}$$

where

$$\Delta m^2 = \left(m_1^2 - m_2^2\right) . \tag{9.7}$$

The $\sin^2(2\theta)$ factor plays the role of the amplitude of the oscillation while the phase is given by $\Delta m^2 L/4E_\nu$. A phase too small or too large makes the measurement of the oscillation parameters quite difficult. Typically, an experiment is sensitive to:

$$\left|\Delta m^2\right| \sim \frac{E_\nu}{L} . \tag{9.8}$$

It is also usual to define an oscillation length L_ν as:

$$L_\nu = \frac{2\pi E_\nu}{\Delta m^2} \tag{9.9}$$

and then

$$P\left(\nu_e \to \nu_\mu\right) = \sin^2(2\theta)\sin^2\left(\frac{\pi}{2}\frac{L}{L_\nu}\right). \tag{9.10}$$

We stress the fact that, whenever $L \sim n\, L_\nu$ (with $n = 1, 3, \ldots$), the probability of oscillation is maximal.

The oscillation formula is often written using practical units:

$$P\left(\nu_e \to \nu_\mu\right) = \sin^2(2\theta)\sin^2\left(1.27\,\frac{\Delta m^2\left(\text{eV}^2\right)L\,(\text{km})}{E_\nu\,(\text{GeV})}\right) \tag{9.11}$$

and then the probability to find a state ν_e at the same distance L is by construction:

$$P\left(\nu_e \to \nu_e\right) = 1 - P\left(\nu_e \to \nu_\mu\right) \tag{9.12}$$

$$= 1 - \sin^2(2\theta)\sin^2\left(1.27\,\frac{\Delta m^2\left(\text{eV}^2\right)L\,(\text{km})}{E_\nu\,(\text{GeV})}\right). \tag{9.13}$$

The oscillation probabilities, in this two-flavor world, are just a function of two parameters: the mixing angle θ and the difference of the squares of the two masses $\Delta m^2 = \left(m_1^2 - m_2^2\right)$.

Experiments that measure the possible depletion of the initial neutrino beam are called *disappearance* experiments. Experiments that search for neutrinos with a flavor different from the flavor of the initial neutrino beam are called *appearance* experiments. An appearance experiment is basically sensitive to a given oscillation channel $\nu_i \to \nu_j$ with $i \neq j$ while a disappearance experiment is sensitive to transitions to all possible different neutrino species, or to pure disappearance—absorption, for example.

The determination of the parameters of neutrino oscillations has been one of the priorities of the research during the last years. If neutrinos oscillate, their masses, although small, cannot be zero. The direct measurement of such masses and of the mixing strengths has gained a renewed interest.

The theoretical origin of neutrino masses is not yet established: either it is the result of the Higgs mechanism as it is the case for all the other fermions (Dirac neutrino) or, as suggested by Majorana, the neutrino is its own antineutrino (Majorana neutrino). If the latter is the case, double beta decays—nuclear decays in which two neutrons become protons—could be neutrinoless (the simplest ways of viewing this fact is to think that the two neutrinos annihilate each other, or that the second neutron absorbs the neutrino emitted by the first one during its transition, and then undergoes the process $\nu n \rightarrow p$).

In addition, neutrinos travel a long way within the Sun, and it is also plausible that most of the neutrino oscillation happens in matter.

The neutrino oscillations may be enhanced (or suppressed) whenever neutrinos travel through matter. In fact, while all neutrino flavors interact equally with matter through neutral currents, charged current interactions with matter are flavor dependent (at solar neutrino energies, basically only electron neutrinos can interact). This is called the MSW effect, as it comes from works by Lincoln Wolfenstein, Stanislav Mikheyev, and Alexei Smirnov. Thus, the time evolution in matter of the electron neutrino and of the other neutrinos can be different. This effect is translated, in a two-flavor approximation, into a modified oscillation probability the oscillation probability $\nu_e \rightarrow \nu_x$ to:

$$P(\nu_e \rightarrow \nu_x) = \sin^2(2\theta_m)\sin^2\left(\frac{\pi}{2}\frac{L_e}{L_\nu}F\right) \tag{9.14}$$

where

$$\sin(2\theta_m) = \sin(2\theta)/F, \tag{9.15}$$

$$F = \sqrt{\left(\cos(2\theta) - \frac{L_\nu}{L_e}\right)^2 + \sin^2(2\theta)} \tag{9.16}$$

and

$$L_e = \pm 2\pi/\left(2\sqrt{2}G_F N_e\right). \tag{9.17}$$

L_e, the electron neutrino interaction length, is positive for neutrinos, negative for antineutrinos. G_F is the Fermi constant and N_e the electron density in matter.

L_ν, the neutrino oscillation length in vacuum, is, as defined before, a function of the neutrino energy and of the difference of the square of the masses:

$$L_\nu = \frac{2\pi E_\nu}{\Delta m^2}. \tag{9.18}$$

Note that the sign of L_ν is determined by the sign of Δm^2. In fact as it will be discussed in Sect. 9.2 there are two possibilities in the hierarchy of the neutrino mass and thus the sign of Δm^2 can be positive or negative.

The values of the mass eigenstates are also changed. The new eigenstates are given by:

$$M_{2,1}^2 = \frac{1}{2}\left[m_1^2 + m_2^2 + \Delta m^2\left(\frac{L_\nu}{L_e} \pm F\right)\right] \tag{9.19}$$

where the + sign is for M_2 and the − is for M_1.

Whenever

$$L_\nu = L_e \cos(2\theta) \tag{9.20}$$

the amplitude of the oscillation is maximal ($\sin^2(2\theta_m) = 1$). Thus, for a given set of (E, N_e) values, resonant oscillations are possible and the oscillation probability may be strongly enhanced independently of the value of θ in vacuum.

In the center of the Sun $N_e \sim 3\times10^{31}\,\text{m}^{-3}$ and then the value of L_e is $\sim 3\times10^5$ m which is a small number when compared with the Sun radius (10^8–10^9 m). In this way, the suppression of the electron neutrinos is a function of the neutrino energy for given values of Δm^2 and θ.

How to determine the oscillation parameters? More information comes from different neutrino sources.

9.1.3 *Long-Baseline Reactor Experiments and the Estimate of $\nu_e \to \nu_\mu$ Oscillation Parameters*

Nuclear reactors are abundant $\overline{\nu}_e$ sources via the β decays of several of the isotopes produced in the fission reactions. The $\overline{\nu}_e$ have an energy of a few MeV and can be detected by the inverse β decay reaction ($\overline{\nu}_e\, p \to e^+ n$). The results from reactors can be combined with the results obtained in the solar experiments supposing that ν_e and $\overline{\nu}_e$ have the same behavior (as guaranteed by CPT invariance). In reactor experiments the energies and the distances are much better determined than in solar experiments,

The KamLAND experiment (again in Kamioka in Japan), a 1000-ton liquid scintillator detector, is placed at distances of the order of 100 km of several nuclear reactors (the weighted average distance being of 180 km) and thus, as discussed in the previous section, is sensitive to long-scale oscillations.

Electron antineutrinos are detected through the reaction $\bar{\nu}_e p \to e^+ n$, which has a 1.8 MeV energy threshold. The prompt scintillation light from the positron allows to estimate the energy of the incident antineutrino. The neutron recoil energy is only a few tens of keV; the neutron is captured on hydrogen and a characteristic 2.2 MeV gamma ray is emitted after some 200 μs. This delayed-coincidence between the positron and the gamma ray signals provides a very powerful signature for distinguishing antineutrinos from backgrounds produced by other sources.

KamLAND detects a clear pattern of oscillation as shown in Fig. 9.6. In a two-flavor approximation (Eq. 9.13), the fit to this pattern provides for the Δ^2_m involved in the ν_e oscillations (we shall call it Δm^2_{Sun}) a value of

$$\Delta m^2_{\text{Sun}} \sim 7\times10^{-5}\text{eV}^2\,. \tag{9.21}$$

A value of

$$\tan^2(\theta_{\text{Sun}}) \sim 0.47 \tag{9.22}$$

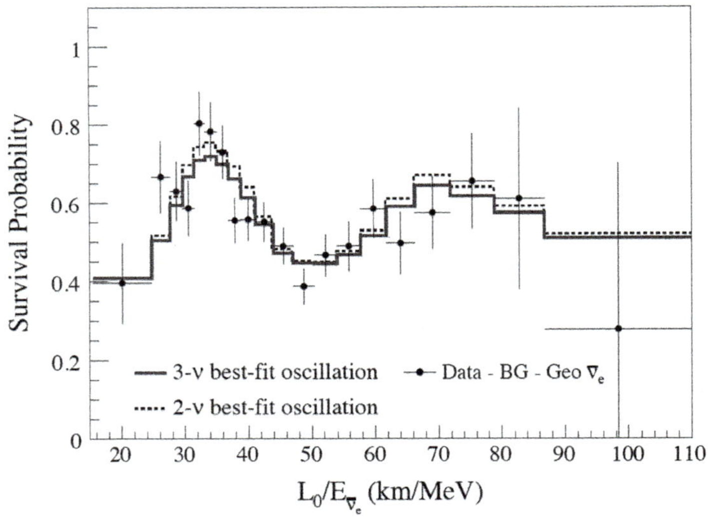

Fig. 9.6 The $\overline{\nu}_e$ survival probability as a function of L/E observed in the KamLAND experiment. Figure from A. Gando et al. (KamLAND Collab.), Phys. Rev. D83 (2011) 052002

is obtained by the combination of KamLAND data and of solar experiments, (Fig. 9.7).

9.1.4 Atmospheric Neutrinos, and the $\nu_\mu \to \nu_\tau$ Oscillation

Another solid evidence that neutrinos do oscillate came from the measurement at the Earth surface of the relative ratio of the ν_e and ν_μ produced in cosmic-ray showers (Fig. 9.8; see also Chap. 10) by the decays of the $\pi^\pm$ and in a smaller percentage of the $K^\pm$. The decay chains:

$$\pi^+ \to \mu^+\nu_\mu \; ; \; \mu^+ \to e^+\nu_e\overline{\nu}_\mu \tag{9.23}$$

$$\pi^- \to \mu^-\overline{\nu}_\mu \; ; \; \mu^- \to e^-\overline{\nu}_e\nu_\mu \tag{9.24}$$

imply that the ratio:

$$R = \frac{\nu_\mu + \overline{\nu}_\mu}{\nu_e + \overline{\nu}_e} \tag{9.25}$$

should be around 2. In fact the value of this ratio is slightly different from 2, because not all muons decay in their way to Earth and only around 63 % of the $K^\pm$ follow similar decay chains; this ratio is, thus, energy-dependent. Monte Carlo calculations allow the computation of these corrections.

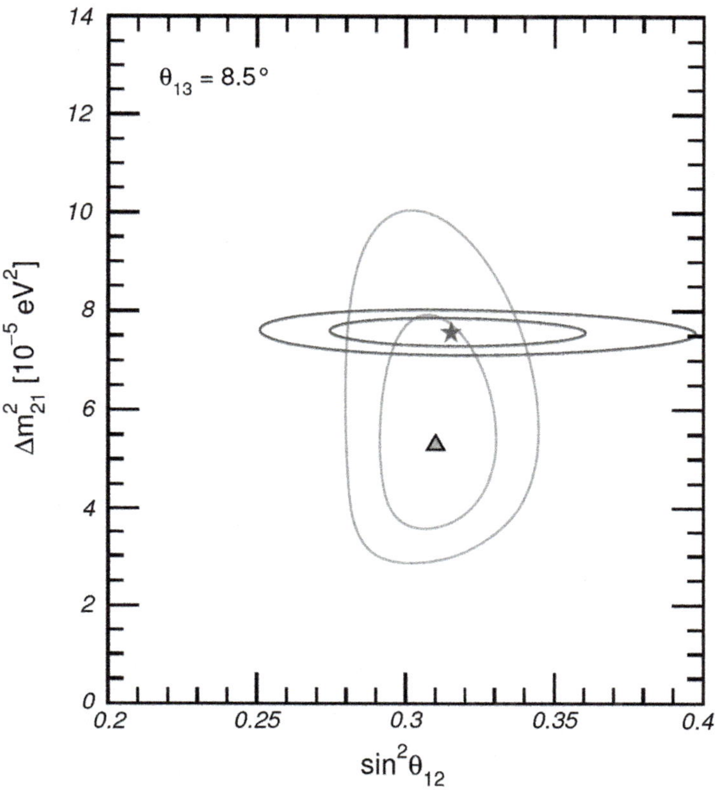

Fig. 9.7 Allowed parameter regions (at 1σ and 2σ) in the ($\sin^2(\theta_{12})$, Δm^2_{21}) space for the combined analysis of solar neutrino data and for the analysis of KamLAND data. The result for KamLAND is illustrated by the *ellipses* with *horizontal* major axis (with the best fit marked by a *green star*). Figure adapted from NuFIT2014: M.C. Gonzalez-Garcia, M. Maltoni and T. Schwetzd, "Updated fit to three neutrino mixing: status of leptonic CP violation", JHEP 11 (2014) 052 (Color in online)

The ratio measured by Kamiokande-II, Super-Kamiokande and, later on, by several other experiments, was however quite different from 2. There is, as it is shown in Fig. 9.9, left, a clear deficit of muon neutrinos coming mainly from below the detector. Indeed upward muon neutrinos ($\cos\theta < 0$, see Fig. 9.8) which traveled longer distances showed a higher probability to disappear. As the interaction cross section in the Earth is too small to explain such disappearance (and no deficit was observed for electron neutrinos), this phenomenon is due to muon neutrino oscillation in particular into tau neutrinos.

Since the number of electron neutrinos was found not to deviate from expectations, oscillations were interpreted as indeed mainly involving tau neutrinos (any undetected type of neutrino would anyway explain the observations). In fact the observed modulation pattern as a function of the zenith angle (Fig. 9.9, left) and as a function of L/E (Fig. 9.9, right) is very well reproduced considering the same

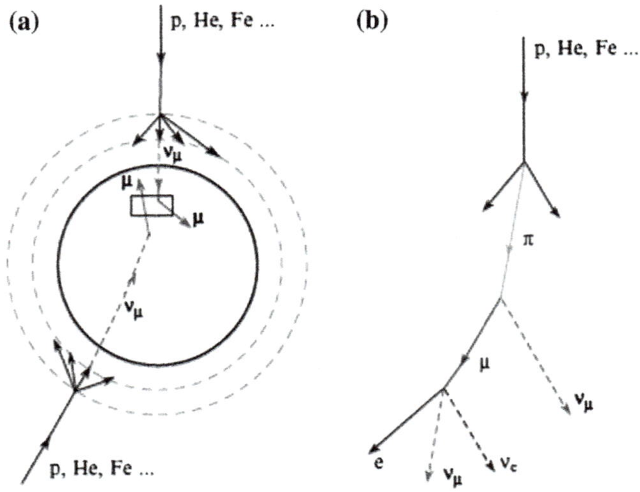

Fig. 9.8 The interaction of cosmic rays in the upper atmosphere give rise to particle showers comprising neutrinos (*right* picture), which originate from a 10–20 km thick atmospheric layer. A large volume detector placed underground, like Super-K, is used to detect them; downward-going neutrinos traveled only few tens of kilometers and had no "space" to oscillate, while upward-going neutrinos have traveled about 10,000 km and have likely oscillated. The detector (*left* picture) can distinguish between electron neutrinos and muon neutrinos: secondary muons are likely to escape the detector (non-contained or partially contained events), while secondary electrons formed by neutrino electrons interacting in the detector are likely to be absorbed (fully contained events). In the case of fully contained events the electron ring is "fuzzier" than the muon ring. From Braibant, Giacomelli and Spurio, "Particles and fundamental interactions,' Springer 2014

oscillation formula deduced in just a two-flavor scenario but now between the muon and the tau neutrinos.

The best fit to the data in a $\nu_\mu - \nu_\tau$ oscillation scenario provides (NuFIT2014)[4]

$$\Delta m^2_{\rm atm} \sim 0.0025\,{\rm eV}^2 \tag{9.26}$$

and a mixing consistent with being maximal

$$\sin^2(2\theta_{\rm atm}) \sim 1\,. \tag{9.27}$$

We now need to extend the phenomenology of flavor oscillation to three families to see the global picture.

[4]The NuFIT groups provides and regularly updates at the web site http://www.nu-fit.org/ a global analysis of neutrino oscillation measurements.

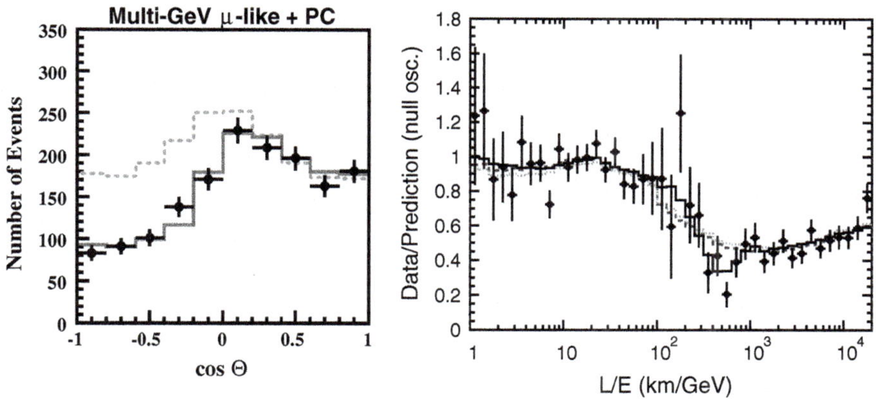

Fig. 9.9 *Left* Zenith angle distribution of the muon neutrinos in SK. The observed number of upward-going neutrinos was roughly half of the predictions. *Right* Survival probability of ν_μ in function of L/E. *Black dots* show the observations and the *lines* shows the prediction based on neutrino oscillation. Data show a dip around $L/E \simeq 500$ km/GeV. The prediction of two-flavor neutrino oscillations agrees well with the position of the dip. From http://www-sk.icrr.u-tokyo.ac.jp/sk/physics/atmnu-e.html and The Super-Kamiokande Collaboration, Y. Ashie et al., "Evidence for an Oscillatory Signature in Atmospheric Neutrino Oscillations," Phys. Rev. Lett. 93 (2004) 101801 (color in Online)

9.1.5 *Phenomenology of Neutrino Oscillations: Extension to Three Families*

Bruno Pontecorvo first suggested in 1957 that the neutrino may oscillate; in the 1960s it was suggested that the neutrino weak and mass eigenstates were not the same. Neutrinos would be produced in weak interactions in pure flavor states that would be a superposition of several mass states (preserving unitarity) which would determine their time-space evolution, giving rise to mixed flavor states.

We have shortly discussed in the beginning of this chapter a simplified model in which only two neutrinos and two mass eigenstates appear. Assuming three weak eigenstates (ν_e, ν_μ, ν_τ) and three mass eigenstates (ν_1, ν_2, ν_3), the mixing can be modeled, similarly to what seen for the CKM matrix, using a 3×3 unitary matrix, which we call today the Pontecorvo-Maki-Nakagawa-Sakata (PMNS) matrix

$$\begin{pmatrix} \nu_e \\ \nu_\mu \\ \nu_\tau \end{pmatrix} = \begin{pmatrix} U_{e1} & U_{e2} & U_{e3} \\ U_{\mu 1} & U_{\mu 2} & U_{\mu 3} \\ U_{\tau 1} & U_{\tau 2} & U_{\tau 1} \end{pmatrix} \begin{pmatrix} \nu_1 \\ \nu_2 \\ \nu_3 \end{pmatrix} . \tag{9.28}$$

Taking into account the relations imposed by unitarity and the fact that several phases can be absorbed in the definition of the fields (if the neutrinos are standard fermions) there are only three real parameters usually chosen as the mixing angles

$\theta_{12}, \theta_{13}, \theta_{23}$ and a single complex phase written in the form $e^{i\delta}$. If the mixing angles and δ are $\neq 0$, CP is violated.

The PMNS matrix can be written as the product of three 3×3 matrices:

$$\begin{pmatrix} U_{e1} & U_{e2} & U_{e3} \\ U_{\mu 1} & U_{\mu 2} & U_{\mu 3} \\ U_{\tau 1} & U_{\tau 2} & U_{\tau 1} \end{pmatrix} = \tag{9.29}$$

$$= \begin{pmatrix} 1 & 0 & 0 \\ 0 & \cos\theta_{23} & \sin\theta_{23} \\ 0 & -\sin\theta_{23} & \cos\theta_{23} \end{pmatrix} \begin{pmatrix} \cos\theta_{13} & 0 & \sin\theta_{13}e^{-i\delta} \\ 0 & 1 & 0 \\ -\sin\theta_{13}e^{i\delta} & 0 & \cos\theta_{13} \end{pmatrix} \begin{pmatrix} \cos\theta_{12} & \sin\theta_{12} & 0 \\ -\sin\theta_{12} & \cos\theta_{12} & 0 \\ 0 & 0 & 1 \end{pmatrix}. \tag{9.30}$$

This format puts in evidence what we observed: in the first approximation, both the oscillation $\nu_e \rightarrow \nu_\mu$ and the oscillation $\nu_\mu \rightarrow \nu_\tau$ can be described as oscillations between two weak eigenstates and two mass eigenstates. Thus, we can identify the two most important parameters for solar neutrinos, θ_{Sun} and Δm^2_{Sun}, with θ_{12} and Δm^2_{21}, respectively; while for atmospheric neutrinos we identify θ_{atm} and Δm^2_{atm}, with θ_{23} and $|\Delta m^2_{32}| \simeq |\Delta m^2_{31}|$, respectively (experimentally it was observed that $|\Delta m^2_{32}| \simeq |\Delta m^2_{31}| \gg |\Delta m^2_{21}|$).

The survival probability, for example, $\nu_e \rightarrow \nu_e$, in the case of three families is given by:

$$P(\nu_e \rightarrow \nu_e) = 1 - 4|U_{e1}|^2 |U_{e2}|^2 \sin^2\left(\frac{\Delta m^2_{21} L}{4E_\nu}\right) +$$
$$-4|U_{e1}|^2 |U_{e3}|^2 \sin^2\left(\frac{\Delta m^2_{31} L}{4E_\nu}\right) - 4|U_{e2}|^2 |U_{e3}|^2 \sin^2\left(\frac{\Delta m^2_{32} L}{4E_\nu}\right).$$

The fact that $|\Delta m^2_{32}| \simeq |\Delta m^2_{31}| \gg |\Delta m^2_{21}|$ leads to an oscillation characterized by two different length scales. Indeed assuming that $|\Delta m^2_{32}| = |\Delta m^2_{31}|$, imposing unitarity and expressing the matrix elements in terms of the PMNS parametrization reported above, one obtains:

$$P(\nu_e \rightarrow \nu_e) \simeq 1 - \cos^4(\theta_{13}) \sin^2(2\theta_{12}) \sin^2\left(\frac{\Delta m^2_{21} L}{4E_\nu}\right) - \sin^2(2\theta_{13}) \sin^2\left(\frac{\Delta m^2_{32} L}{4E_\nu}\right). \tag{9.31}$$

In reactor experiments the energy of the neutrino (in fact $\overline{\nu}_e$) beams are of the order of a few MeV. Thus, as $\Delta m^2_{21} \sim 10^{-5} - 10^{-4}$ eV^2 and $|\Delta m^2_{32}| \sim 10^{-3}$ eV^2, experiments placed at distances of the order of the km are sensitive to θ_{13} while experiments placed at distances of the order of the hundreds of km are sensitive to θ_{12}.

This two-length behavior is illustrated in Fig. 9.10 where the probability of ν_e survival is shown for fixed oscillation parameters.

In the first case ($L \sim$ km) the above formula can be simplified to:

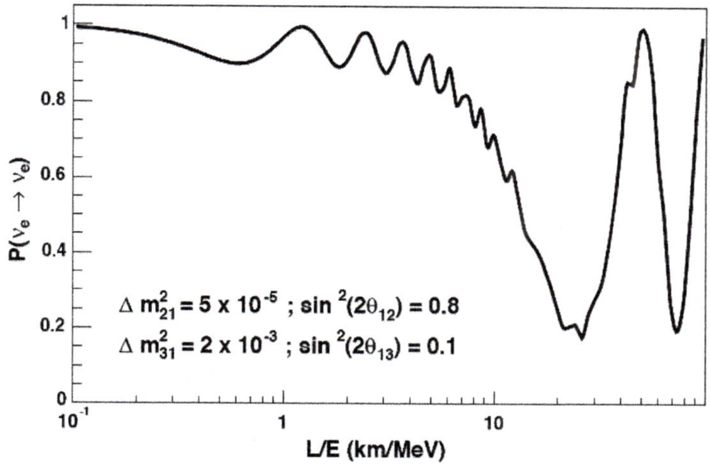

Fig. 9.10 The ν_e survival probability as a function of L/E for fixed oscillation parameters as indicated in the figure. From http://www.hep.anl.gov/minos/reactor13/reactor13.pdf

$$P\left(\overline{\nu}_e \to \overline{\nu}_e\right) \approx 1 - \sin^2\left(2\theta_{13}\right)\sin^2\left(\frac{\Delta m^2_{32}\, L}{4\, E_{\overline{\nu}}}\right) \tag{9.32}$$

while in the second case ($L \sim 100$ km) it can be simplified to:

$$P\left(\overline{\nu}_e \to \overline{\nu}_e\right) \approx 1 - \cos^4\left(\theta_{13}\right)\sin^2\left(2\theta_{12}\right)\sin^2\left(\frac{\Delta m^2_{21}\, L}{4\, E_{\overline{\nu}}}\right). \tag{9.33}$$

9.1.6 Short-Baseline Reactor Experiments, and the Determination of θ_{13}

Close to a fission reactor, where the long wavelength oscillation did not develop yet, the electron antineutrino survival probability can thus be approximated as computed in Eq. 9.33.

The Daya Bay experiment in China is a system of six 20-ton liquid scintillator detectors (antineutrino detectors, AD) arranged in three experimental halls (EH), placed near six nuclear reactors (the geometry is shown in Fig. 9.11, left); as a consequence of the distances and of the geometry it is sensitive to short oscillations which may occur in a 3×3 mixing matrix scenario (see Sect. 9.1.5). In fact Daya Bay reported in March 2012 the first evidence of such short-scale oscillations (Fig. 9.11, right), finding a value of

$$\sin^2 2\theta_{13} = 0.092 \pm 0.017\,(stat. + syst.)\,.$$

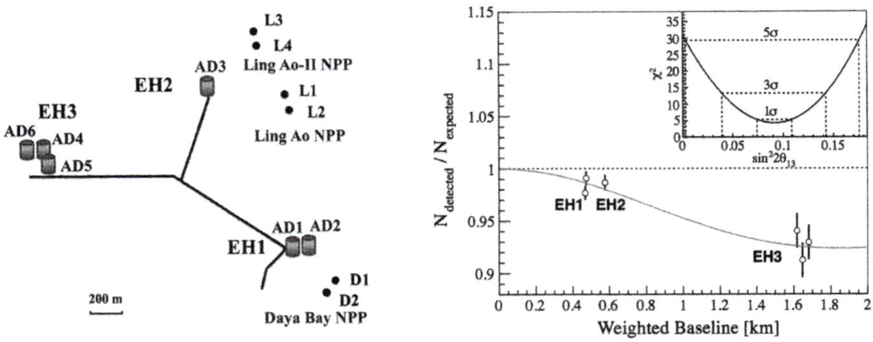

Fig. 9.11 *Left* Layout of the Daya Bay experiment. The *dots* represent reactors, labeled as D1, D2, L1, L2, L3, and L4; the locations of the detectors are labeled EH1, EH2 and EH3. *Right* The $\overline{\nu}_e$ disappearance as measured by the Daya Bay experiment. *Right* Ratio of the measured signal in each detector versus the signal expected assuming no oscillation. The oscillation survival probability at the best-fit $\sin^2 2\theta_{13}$ value is given by the smooth *curve*. The χ^2 versus $\sin^2 2\theta_{13}$ is shown in the *inset*. Figures from F.P. An et al., Phys. Rev. Lett. 108 (2012) 171803

Although small, a nonzero value of θ_{13} allows the phase $\delta \neq 0$ to produce CP violation in the neutrino sector. Later, the RENO experiment in South Korea confirmed the result.

9.1.7 Accelerator Neutrinos

The results from atmospheric neutrino experiments and reactor experiments can be tested in accelerator experiments, building intense and collimated ν_μ and $\overline{\nu}_\mu$ beams from the decay of secondary $\pi^\pm$ (and in a smaller percentage of $K^\pm$), and placing detectors both near (100–1000 m) and far (100–1000 km) from the primary target. The oscillation distance L is then fixed and the neutrino flux and the energy spectrum can be well predicted and precisely measured at the near detectors, constraining the elements of the neutrino mixing matrix.

The K2K (KEK to Kamioka) experiment, in Japan, was the first of such experiments (actually its construction started at the end of the 1990s before the discovery of the neutrino oscillations in Super-Kamiokande). The neutrino beam, with a mean energy of 1.3 GeV, was produced at KEK in Tsukuba and the interactions were measured in a nearby detector at 300 m and in the Super-Kamiokande detector at 250 km (Fig. 9.12). 112 events were detected while 158 ± 9 where expected without considering oscillations; a neutrino oscillation pattern compatible with the atmospheric neutrino results was observed.

The T2K (Tokai to Kamioka) experiment succeeded K2K sending muon neutrinos to the Super-Kamiokande detector. It is a second generation experiment situated at 295 km from Kamioka. The neutrino beam, produced in the the J-PARC facility

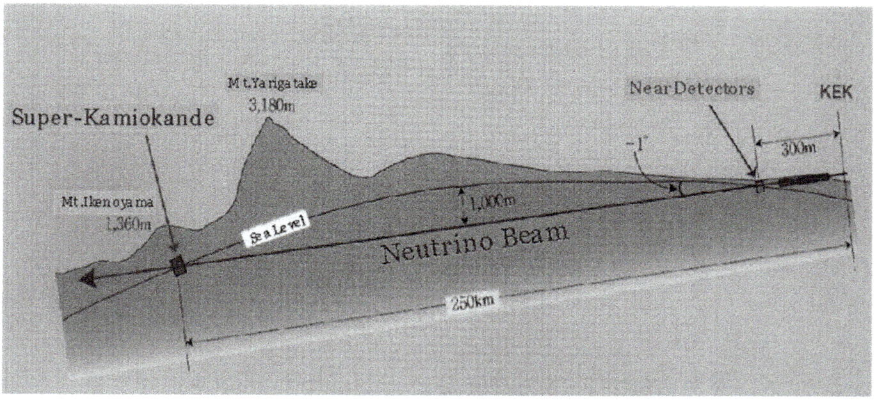

Fig. 9.12 Sketch of the neutrino path in the K2K long-baseline experiment. From http://neutrino.kek.jp

in Tokai, Eastern Japan, has a narrow range of energies around 600 MeV, selected in order to maximize the neutrino oscillation probability in their way to Super-Kamiokande. The intensity of the beam is two orders of magnitude larger than in K2K. The near detector (ND280), 280 m downstream the neutrino beam, is a segmented detector composed of neutrino targets inside a tracking system surrounded by a magnet. ND280 can measure the energy spectrum of the ν beam, its flux, flavor content, and interaction cross sections before the neutrino oscillation. We shall see later that, on top of precise measurements of the $\overline{\nu}_\mu$ disappearance (Fig. 9.13, left), T2K detected for the first time explicitly the appearance of $\overline{\nu}_e$ in a $\overline{\nu}_\mu$ beam.

In the USA, the MINOS experiment started taking data in 2005. The beam line at Fermilab is optimized to produce both ν_μ and $\overline{\nu}_\mu$ beams with mean energy of 3 GeV. The far detector is placed at a distance of 735 km in the Soudan mine. A distortion of the energy spectrum at the far detector compatible with the previous oscillation measurements was observed for ν_μ beams (Fig. 9.13, right).

These results can be once again interpreted in terms of oscillations in a two-flavor scenario (but now considering $\nu_\mu \to \nu_\tau$ as shown in Fig. 9.14). They confirm and improve the result from the atmospheric neutrinos. The mixing is almost maximal ($\sin^2(2\theta_{23}) > 0.90$) and the mass difference ($\left|\Delta m^2_{\text{atm}}\right| \sim 2.3 \times 10^{-3}$ eV^2) is again much smaller than the normal fermion masses but much higher than the values measured in the case of the electron neutrino beam, i.e., in the "solar" neutrinos ($\Delta m^2_{\text{Sun}} \sim 8 \times 10^{-5}$ eV^2) as discussed above. Accelerator and atmospheric experiments are complementary: in the former L is fixed and E known assuring a good resolution in the measurement of $\left|\Delta m^2_{23}\right|$ while in the latter the fluxes are high assuring a good resolution in the measurement of θ_{23} and being also sensitive to θ_{13}.

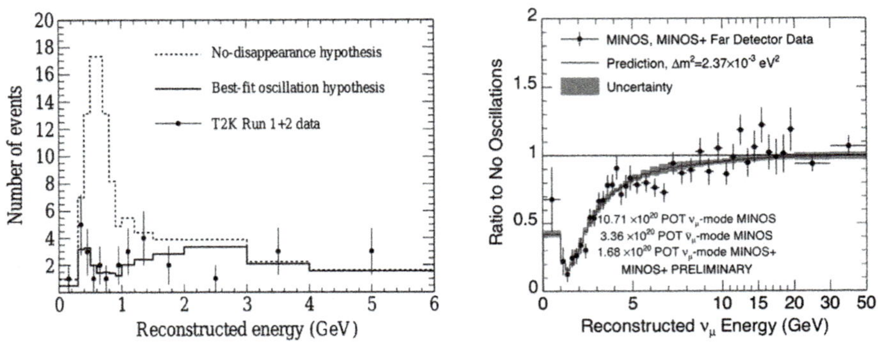

Fig. 9.13 *Left* The first T2K study on the disappearance of muon neutrinos: muon-antineutrino events with well-reconstructed energy recorded before 2011. The energy distribution is compared to the calculations with and without oscillations. From Phys. Rev. D 85, 031103 (2012). *Right* The ratio of the observed spectrum of muon neutrino interactions from MINOS to the predicted spectrum in the absence oscillations. The *dark band* represents the prediction assuming oscillations and its 1σ systematic uncertainty, using the best-fit oscillation parameters from MINOS. The observed data are well described by the oscillation model. From http://www-numi.fnal.gov/PublicInfo/forscientists.html

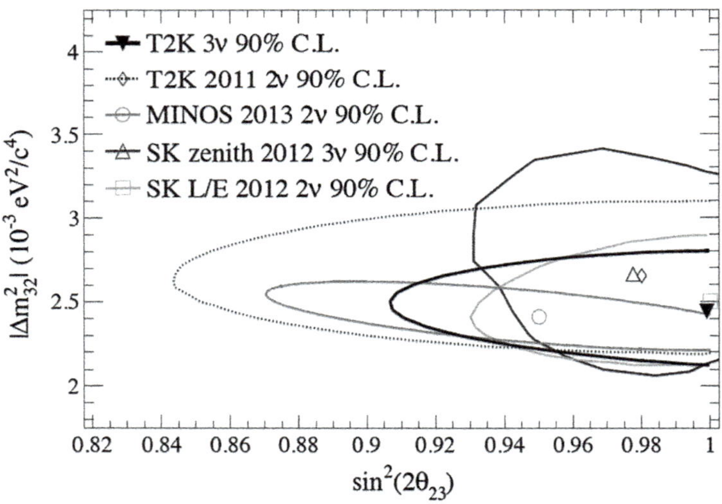

Fig. 9.14 The ($|\Delta m^2_{23}|$, $\sin^2(2\theta_{23})$) 90 % CL allowed regions. From K. Abe et al. (T2K Collab.), Phys. Rev. Lett. 111 (2013) 211803

9.1.8 Explicit Appearance Experiment

The SNO experiment was somehow an appearance experiment: the comparison of charged current events with neutral-current events was an indication that μ plus τ neutrinos were present in the flux of solar neutrinos. Recently, however, two exper-

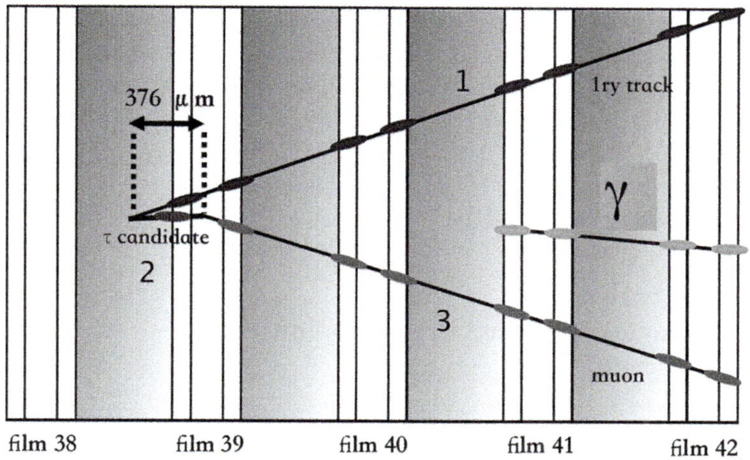

Fig. 9.15 One of the three tau antineutrino candidate events observed by OPERA. From http://operaweb.lngs.infn.it/

iments made an explicit detection of neutrinos of different flavor from the muon neutrinos an an accelerator beam.

The OPERA experiment situated at Gran Sasso, Italy, receives a 17 GeV muon neutrino beam produced at CERN located 730 km away. OPERA uses a sophisticated 1300 tons detector composed by a sandwich of photographic emulsion films and lead plates in order to be able to detect tau-leptons: it is thus an appearance experiment aiming to detect tau-neutrinos resulting from the oscillation of the initial muon neutrino beam. OPERA, which concluded data-taking, reported four tau-neutrino candidates corresponding to significance of about 4σ; one of them is shown in Fig. 9.15.

The first observation of the ν_e appearance from a high-purity ν_μ beam was recently reported by T2K: 28 electron neutrino events were detected while 4.92 ± 0.55 background events were expected in the case of no neutrino oscillation.

9.1.9 A Gift from Nature: Geo-Neutrinos

The interior of the Earth radiates heat at a rate of about 50 TW, which is about 0.1 % of the incoming solar energy. Part of this heat originates on the energy generated upon the decay of radioactive isotopes, while another part is due to the cooling of the Earth.

The Earth's radioactive elements (in particular ^{238}U, ^{232}Th, ^{40}K) are β^- emitters and thus natural sources of $\overline{\nu}_e$, in this case designated as geo-neutrinos. The fluxes are small (as an example, around 21 events/year in KamLAND) but their measurements may provide important geological information on Earth's composition and structure that can not be accessible by any other means. The main backgrounds are due to nuclear reactors, since the contribution of atmospheric neutrinos is negligible and the Sun emits exclusively ν_e. KamLAND and Borexino (a 280 ton liquid scintillator detector in Gran Sasso) reported in 2013 a total observed signal of 116^{+28}_{-27} and 14.3 ± 4.4 events, respectively. The current estimates are that, although with large errors, some 20 TW of energy from the Earth comes from nuclear processes.

Thanks to neutrino detectors, a new highly interdisciplinary field, neutrino geophysics, has just been born.

9.2 Neutrino Oscillations; Present Knowledge of the Oscillation Parameters

The simplified model in which neutrinos coming from two mass eigenstates oscillate between two flavors does not describe the full picture coming from the data. The large majority of the present experimental results are well-described assuming three weak eigenstates (ν_e, ν_μ, ν_τ) and three mass eigenstates (ν_1, ν_2, ν_3). Some researchers evidence a possible tension in the data, which for the first time was announced as the "LSND[5] anomaly. LSND claimed an oscillation with $|\Delta m| \sim 1$ eV, which would imply the existence of a neutrino with mass of at least one eV. The only way to accommodate this with the LEP results in the number of neutrino families is that this particle is a new kind of neutrino, which should be sterile. A recent review of short range ($L < 100$ m) reactor neutrino results, indicating a possible "reactor neutrino anomaly," can be found in G. Mention et al., Phys. Rev. D 83 (2011) 073006. These discrepancies are difficult to reconcile with cosmological bounds on the neutrino mass (see later), unless neutrinos have new properties; the discussion is beyond the scope of this book.

The mixing matrix between three states is the Pontecorvo-Maki-Nakagawa-Sakata (PMNS) matrix (see Sect. 9.1.5). However, it should be noted that a complete treatment of neutrino propagation requires subtle questions of field theory, and has close links to the foundation of quantum mechanics. Since different mass components travel at different speeds, the mixing spreads the neutrino wavefunction in space, with EPR-like[6] implications.

The parameters of the PMNS matrix are: two mass differences (Δm^2_{21} and Δm^2_{31}); three angles (θ_{12}, θ_{23} and θ_{13}); one single complex phase written in the form $e^{i\delta}$. Several groups provide regular fits imposing the unitarity of the mixing matrix, in particular the PDG group and the NuFIT group. Averages of the mass differences

[5]The Liquid Scintillator Neutrino Detector (LSND) was a 167-tons scintillation counter at Los Alamos National Laboratory that measured the flux of neutrinos produced by a near accelerator neutrino source.

[6]The Einstein-Podolski-Rosen (EPR) paradox originally involved two particles, A and B, which interact briefly and then move off in opposite directions. The two particles are then entangled, and any measurement on A (projection of A on an eigenstate) would have *immediately* implications on the state of B; this would violate locality. In the case of neutrinos, the neutrino wavefunction itself spreads during the travel, with possible nonlocal effects.

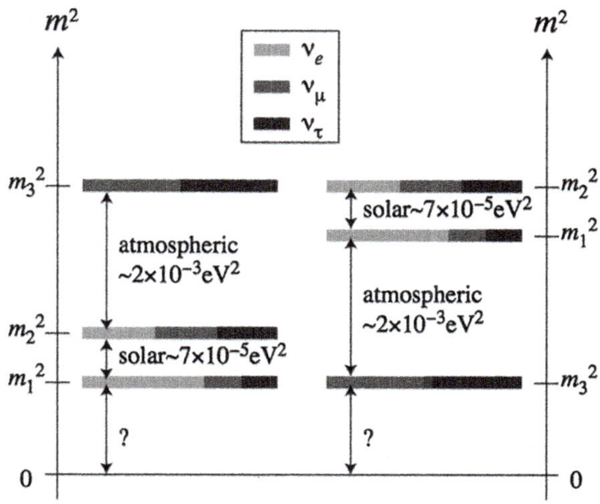

Fig. 9.16 Diagram of the relationship between the mass eigenstates (labeled 1, 2 and 3) for neutrinos and the flavor eigenstates (ν_e, ν_μ, ν_τ). Neutrinos from the Sun have been used to determine the relation between m_2 and m_1; m_3 may be greater or smaller than m_1 and m_2. The fractional contribution of each flavor to the mass eigenstates is indicted by the *colored bars*. (From S.F. King, arXiv:0712.1750)

and of the mixing angles (derived from PDG 2014 and PDG2013) are:

$$\Delta m_{21}^2 = (8.7 \pm 0.2\,\text{meV})^2 \tag{9.34}$$

$$|\Delta M|^2 \equiv \left| m_3^2 - \frac{m_2^2 + m_1^2}{2} \right| = (49.3 \pm 0.6\,\text{meV})^2 \simeq \Delta m_{31}^2 \simeq \Delta m_{21}^2 \tag{9.35}$$

$$\sin^2(2\theta_{12}) = 0.857 \pm 0.024 \tag{9.36}$$

$$\sin^2(2\theta_{23}) > 0.95 \tag{9.37}$$

$$\sin^2(2\theta_{13}) = 0.095 \pm 0.010\,. \tag{9.38}$$

Note the elaborate way in which we defined the parameter $|\Delta M|^2$ in Eq. 9.35. This comes from the fact that the sign of Δm_{31} (and thence of Δm_{32}) is not known. In fact only the sign of Δm_{21} is determined from the experimental measurements (solar neutrinos). There are two possibilities (Fig. 9.16):

- $m_1 < m_2 < m_3$ (the so-called "normal" hierarchy, NH—sometimes HI);
- $m_3 < m_1 < m_2$ (called the inverted hierarchy, IH).

The complex phase is, from NuFit 2014,

$$\delta/^\circ = 306^{+39}_{-70}\,(NH)\,;\;\; \delta/^\circ = 254^{+63}_{-62}\,(IH)\,. \tag{9.39}$$

It may thus be $\delta \neq 0$ (which implies the violation of CP in the neutrino sector) but there is not yet sensitivity to confirm firmly that hypothesis (the value is consistent with 360° in less than three standard deviations).

The PMNS matrix is then highly non-diagonal, which is very different from what it is observed in the quark sector (see Sect. 6.3.7). The best estimates of 3σ confidence intervals for its elements are:

$$\begin{pmatrix} \nu_e \\ \nu_\mu \\ \nu_\tau \end{pmatrix} = \begin{pmatrix} 0.801 \to 0.845 & 0.514 \to 0.580 & 0.137 \to 0.158 \\ 0.225 \to 0.517 & 0.441 \to 0.699 & 0.614 \to 0.793 \\ 0.246 \to 0.529 & 0.464 \to 0.713 & 0.590 \to 0.776 \end{pmatrix} \begin{pmatrix} \nu_1 \\ \nu_2 \\ \nu_3 \end{pmatrix} . \tag{9.40}$$

Future facilities are planned to improve our knowledge of the mixing matrix and possibly discover new physics; in particular, high-precision and high-luminosity long-baseline neutrino oscillation experiments have been proposed in the US, in Japan, and in Europe.

9.3 Neutrino Masses

The discovery of neutrino oscillations showed, as discussed above that the neutrino flavor eigenstates are not mass eigenstates and at least two of the mass eigenstates are different from zero.

Thanks to a huge experimental effort, we know quite well the neutrino mass differences. As of today we do not know, however, the absolute values of the neutrino masses. The values of Δm^2_{ij} (Eqs. 9.34 and 9.35) suggest masses of the order of 1–100 meV. However, the possibility that the mass of the lightest neutrino is much larger than this and that all three known neutrino masses are quasi-degenerate is not excluded.

Neutrino masses can only be directly determined via non-oscillation neutrino experiments. The most model-independent observable for the determination of the mass of the electron neutrino is the shape of the endpoint of the beta decay spectrum. Other probes of the absolute value of the neutrino masses include double beta decays, if neutrinos are of the Majorana type, discussed below, and maps of the large-scale structure of the Universe, which is sensitive to the masses of neutrinos—although this sensitivity depends on cosmological models.

9.3.1 The Constraints from Cosmological and Astrophysical Data

The neutrino mass is constrained by cosmological data. Indeed neutrinos contribute to the energy density of the Universe playing the role of "hot dark matter". The combined analyses of the CMB data and of the surveys of the large-scale structures in the Universe (see Chap. 8) sets a limit on the sum of the mass of the three neutrino species to

$$\sum m_{\nu_i} < 0.23 \text{ eV} . \tag{9.41}$$

at 95 % C.L.

A more conservative limit

$$\sum m_{\nu_i} < 0.6 \text{ eV} \tag{9.42}$$

can be extracted as follows, based on the density and sizes of structures in the Universe. Initial fluctuations seeded the present structures in the Universe, growing during its evolution. Neutrinos, due to their tiny masses, can escape from most structures being their speed larger than the escape speed. As a net result, neutrinos can erase the structures at scales smaller than a certain value D_F called the free streaming distance. The smaller the sum of the neutrino masses, the larger is D_F. The relevant observable is the mass spectrum, i.e., the probability of finding a structure of a given mass as a function of the mass itself. Cosmological simulations predict the shape of the mass spectrum in terms of a small number of parameters; the limit in Eq. 9.42 is the limit beyond which the predicted distribution of structures is inconsistent with the observed one.

Data from astrophysical neutrino propagation over large distances are less constraining. So far the only reported upper limit on the neutrino velocity was obtained comparing the energy and the arrival time of a few tens of neutrinos by three different experiments from the explosion of the supernova 1987A in the Large Magellanic Cloud at around 50 kpc from Earth. From these results a limit of about 6 eV was obtained on the masses of the neutrinos reaching the Earth. The present long-baseline accelerator experiments are not sensitive enough to set competitive limits. Recent claims of the observation of superluminal neutrinos by the OPERA experiment were proved to be experimentally wrong.

9.3.2 Direct Measurements of the Electron Neutrino Mass: Beta Decays

The study of the energy spectrum of the electrons produced in nuclear β decays is, one century after the first measurement, still the target of intense experimental efforts. In particular, the detailed measurement of the endpoint of this spectrum may allow the determination of the electron neutrino mass by direct energy conservation.

In fact it can be shown that whenever the parity of the initial and the final nuclei is the same, the spectrum of the outgoing electron is given by:

$$\frac{dN}{dE} = \frac{G_F^2 \cos^2\theta_c I^2}{2\pi^3} F(Z, R, E)\,|\vec{p}|\,E\,(E_0 - E)\sqrt{(E_0 - E)^2 - m_{\nu_e}^2} \tag{9.43}$$

where:

1. $\cos\theta_c$ is the cosine of the Cabibbo angle.
2. I an isospin factor that depends on the isospin of the initial and the final nucleus.
3. $F(Z, R, E)$ is the "Fermi function" accounting for the electrostatic interaction between the nuclei and the outgoing electron which depends on the nuclear charge Z, on the nuclear radius R and the electron energy.

4. $E_0 \simeq Q = M(Z, A) - M(Z+1, A) - m_e$ is the total energy available for the electron and the antineutrino.

The endpoint of this spectrum can then be graphically determined plotting the quantity $K(E)$ (Kurie plot, Fig. 9.17, left), where

$$K(E) = \frac{dN/dE}{F(Z, R, E)\,|\vec{p}|\,E}. \tag{9.44}$$

In the case of $m_{\nu_e} = 0$

$$K(E) \propto (E_0 - E) \tag{9.45}$$

and the plot is just a straight line. However, if $m_{\nu_e} \neq 0$, this line bends slightly near the end point (Fig. 9.17) and $K(E)$ becomes null now at:

$$E = E_0 - m_{\nu_e}. \tag{9.46}$$

Assuming a mixing scenario with three nondegenerate mass eigenvalues the spectrum at the endpoint would be the superposition of the Kurie plots corresponding to each of the mass eigenvalues (Fig. 9.17, right); indeed the measured mass will be a superposition m_β such that

$$m_\beta^2 = \sum |U_{ei}^2| m_i^2. \tag{9.47}$$

Two nuclides are of major importance to current β-decay experiments: tritium and ^{187}Re. The physics is the same in both cases, but the experimental technique differs. Tritium has a relatively high Q-value of 18.6 keV which makes the detection of the electron easier in the process $^3H \rightarrow ^3He + e^- + \bar{\nu}_e$; the detection of electrons from ^{187}Re (with a 2.5 keV Q-value) needs a micro calorimeter embedded in the radioactive material.

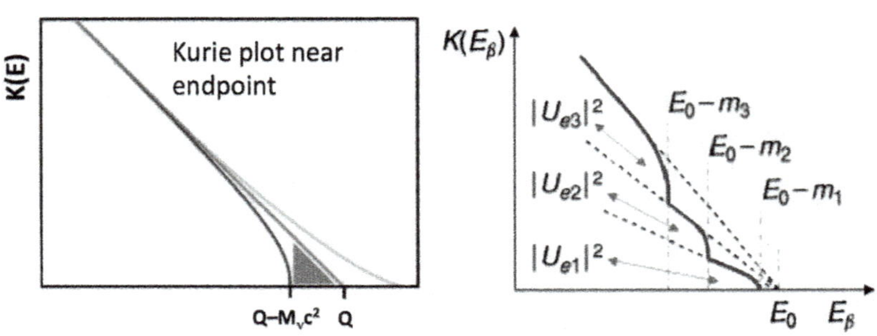

Fig. 9.17 *Left* Kurie plot. The *green line* represents the ideal case for $m_{\nu_e} = 0$, the *red line* the ideal case for $m_{\nu_e} \neq 0$ and the *blue line* the real case where a finite detector resolution introduces a smearing at the endpoint. *Right* Detail of the endpoint in case of a mixing scenario with three non-degenerated mass eigenvalues. Credit: Andrea Giuliani, *Review of Neutrino Mass Measurements*, Quark-Lepton conference Prague, 2005 (Color in online)

The best present results were obtained by experiments (Troitsk in Russia and Mainz in Germany) using as source the tritium ^3H. These experiments measure the electron energy using complex magnetic and electrostatic spectrometers. The current limit is

$$m_{\nu_e} < 2.0\,\text{eV} \tag{9.48}$$

at 95 % C.L.

Following this line an ambitious project (KATRIN in Karlsruhe, Germany) having a 200-ton spectrometer is presently in preparation. KATRIN aims either to improve this limit by an order of magnitude or to measure the mass, with a sensitivity of 0.2 eV. An alternative proposal (Project 8 in Yale, US) is to use uses the measurement of the cyclotron frequency of individual electrons to reach similar sensitivities.

9.3.3 Direct Measurements of the Muon- and Tau-Neutrino Masses

The muon and the tau neutrino masses were studied respectively in the decays of charged pions ($\pi^+ \to \mu^+\nu_\mu$ and $\pi^- \to \mu^-\overline{\nu}_\mu$), and in the three- and five-prongs decays of the tau lepton. Limits

$$m_{\nu_\mu} < 0.19\,\text{MeV} \tag{9.49}$$

and

$$m_{\nu_\tau} < 18.2\,\text{MeV} \tag{9.50}$$

were obtained at 95 % confidence level. They are not competitive with the cosmological limits.

9.3.4 Incorporating Neutrino Masses in the Theory

The first formulation of the standard model had to be extended to accommodate neutrino masses.

The most straightforward solution is to introduce in the SM Lagrangian mass terms for the neutrinos similar to the existing for the other fermions:

$$-\frac{g_\nu}{\sqrt{2}}v\left(\overline{\nu}_L\nu_R + \overline{\nu}_R\nu_L\right) \tag{9.51}$$

where g_ν is the Yukawa coupling, $v = 246$ GeV is the Higgs vacuum expectation value, and ν_L and ν_R are ,respectively, the left- and right-handed chiral[7] Dirac spinors. This mass term is built with the normal Dirac spinors and so these neutrinos are designated as Dirac neutrinos. The right-handed neutrino that was not necessary in the first formulation of the SM should then exist. In fact, in the case of massless neutrinos chirality is conserved and as the right-handed neutrino is a SU(2) singlet and has no weak interactions, as well as no strong and electromagnetic interactions: excluding gravitational effects, it is invisible.

However neutrinos and antineutrinos have, apart from the lepton number which is not generated by a fundamental interaction symmetry, the same quantum numbers and thus they can be a unique particle. This would not be possible in the case of the electrons/positrons, for example, as they have electric charge.

The neutrino and the antineutrino can, in this hypothesis first introduced by Ettore Majorana in 1937, be described only by two-component chiral spinors (instead of the four-component spinors in the case of the Dirac fermions). A left-handed neutrino is identical (but for a phase) to a right-handed antineutrino which may be described by the CP conjugate of the left-handed neutrino (ν_L^C). A mass term involving left-handed neutrinos and right-handed antineutrinos can then be written:

$$-\frac{1}{2}m\left(\overline{\nu}_L\nu_L^C + \overline{\nu_L^C}\nu_L\right) . \tag{9.52}$$

However $\overline{\nu_L^C}\nu_L$ has weak hypercharge $Y = -2$ and thus can not make a gauge invariant coupling with the Standard Model Higgs doublet which has $Y = +1$. In this way either the Higgs sector would be extended or again right-handed neutrinos (in this case of the Majorana type) are introduced. Indeed a left-handed antineutrino is identical (but for a phase) to a right-handed neutrino and may be described by the CP conjugate of the right-handed neutrino (ν_R^C). In this case mass terms involving right-handed neutrinos and left-handed antineutrinos like:

$$-\frac{1}{2}M\left(\overline{\nu_R^C}\nu_R + \overline{\nu}_R\nu_R^C\right) \tag{9.53}$$

are SU(2) singlets: they can be introduced directly in the Lagrangian without breaking the gauge invariance. No Higgs mechanism is therefore needed in this scenario in the neutrino sector. These Majorana neutrinos would not couple to the weak bosons.

Both Dirac and Majorana mass terms may be present. In the so-called "see-saw" mechanism, a Dirac term with mass m_D and a Majorana term with mass M as defined above are introduced. The physical states for which the mass matrix is diagonal are now a light neutrino with mass

$$m_\nu \sim \frac{m_D^2}{M} \tag{9.54}$$

[7] Hereafter in this section the designations "left"- and "right-handed" refer to chirality and not to helicity. Note that for massive neutrinos chirality and helicity are not equivalent (see chap. 6).

and a heavy neutrino with mass

$$m_N \sim M\,. \tag{9.55}$$

The interested reader can find the explanation in the additional material.

The light neutrino, in the limit of $M \gg m_D$, has basically the same couplings as the standard model neutrinos, while the heavy neutrino is basically right-handed and thus sterile and may have, if they are stable, a role on Dark Matter.

The extremely small values that experiments indicate for the neutrino masses are in this way generated thanks to the existence of a huge Majorana mass, while the scale of the Dirac mass is the same as for the other fermions.

9.3.5 Majorana Neutrinos and the Neutrinoless Double Beta Decay

If neutrinos are Majorana particles (i.e., if neutrinos and anti-neutrinos the same particle), neutrinoless double β decays ($0\nu\beta\beta$, also called $\beta\beta0\nu$) can occur, in particular for nuclei for which the normal β decays are forbidden by energy conservation. The lines corresponding to the emission of the $\overline{\nu}_e$ can be connected becoming an internal line (Fig. 9.18).

Considering that the nuclei before and after the decay are basically at rest, the sum of the energies of the two electrons in $0\nu\beta\beta$ decays is just the difference of the masses of the two nuclei ($Q = M(Z, A) - M(Z+2, A) - 2m_e$). Thus in these decays the energy spectrum of the emitted electrons should be a well-defined line while in the normal double β decay, with the emission of two $\overline{\nu}_e$, this spectrum accommodates a large phase space and the electron energy distribution is broad (Fig. 9.19).

The decay rate is proportional to the square of the sum of the several mass eigenstate amplitudes corresponding to the exchange of the electron (anti)neutrino (coherent sum of virtual channels). Then it is useful to define an effective Majorana mass as:

$$m_{\beta\beta} = \left| \sum |U_{ek}|^2 e^{i\alpha_k} m_k \right| \tag{9.56}$$

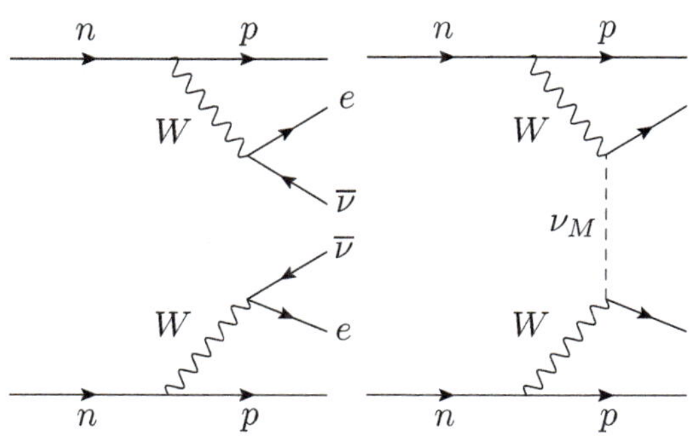

Fig. 9.18 Double β decay diagrams: on the *left* the case of Dirac neutrinos characterized by a final state with two $\overline{\nu}_e$; on the *right* the neutrinoless β decay allowed in the case of Majorana neutrinos

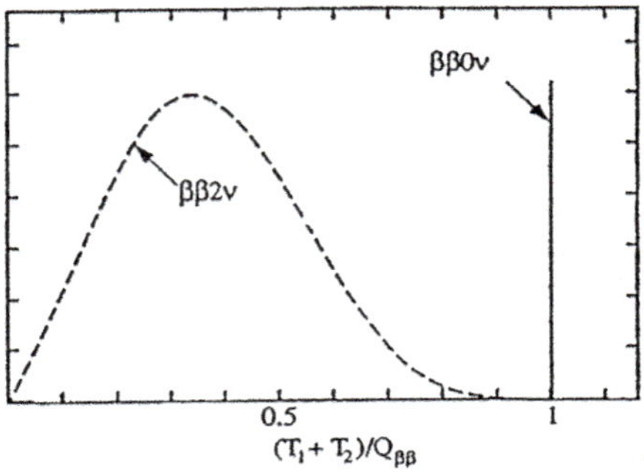

Fig. 9.19 Energy spectrum of the sum of the two electrons in the case of the double β decay of a nucleus with (broad distribution) or without (line) $\overline{\nu}_e$ emission

where α_i are the Majorana phases (one of them can be seen as a global phase and be absorbed by the neutrino wavefunctions but the two remaining ones cannot be absorbed, as it was the case for Dirac neutrinos).

Being a function both of the neutrino masses and of the mixing parameters, this effective mass depends on the neutrino mass hierarchy. In the case of the normal hierarchy total cancelation may occur for given range of masses of the lightest neutrino and $m_{\beta\beta}$ may be null.

The experimental measurement is extremely difficult due to the low decay rates and the backgrounds (the $2\nu\beta\beta$ background is difficult to reduce, unless the energy resolution is extremely good). An ideal experiment should then have a large source mass, an excellent energy resolution; a clean environment and techniques to suppress the background (such as particle identification, spatial resolution, and timing information) would help in general.

Several experimental strategies have been implemented in the last years, and eleven isotopes for which single beta decay is energetically forbidden have been experimentally observed undergoing double beta decay, for which half-lives (typically of the order of 10^{21} years) have been measured. Among the most interesting double beta decay emitters are:

- ^{136}Xe, with a high Q-value of about 2.5 MeV where background is small, which can can be dissolved in liquid scintillators or used as gas for a homogeneous detector providing both scintillation and ionization signals (this technique is exploited by the Enriched Xenon Observatory EXO, installed near Carlsbad, New Mexico, and by KamLAND-Zen in Kamioka).
- ^{76}Ge, which can be embedded in solid state detectors (GERDA at Gran Sasso and Majorana in the Sanford Underground Research Facility SURF, South Dakota).
- ^{130}Te, which has a large natural abundance and can be used to build a bolometric detector (experiment CUORE at LNGS). Bolometers measure the energy released by a particle using the change in the electric resistance induced by the heating of a crystal at very low (mK) temperatures. ^{130}Te can be also dissolved in liquid scintillators (experiment SNO+).

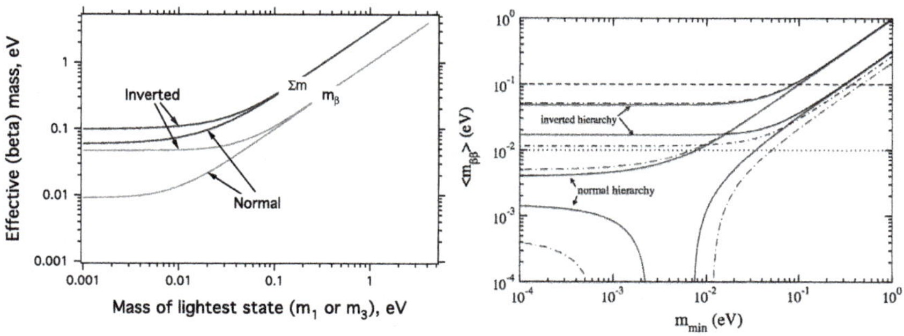

Fig. 9.20 *Left* m_β and the sum of the neutrino masses as a function of the lightest neutrino mass. In *green* (*upper band*) and in *red* (*lower band*) the allowed phase spaces, respectively, for the inverted and the normal hierarchy scenarios are plotted. *Right* $m_{\beta\beta}$ as a function of the lightest neutrino mass (three standard deviation intervals). In the *upper band* and in the *lower band* the allowed phase spaces, respectively, for the inverted and the normal hierarchy scenarios are plotted

No confirmed signal was so far established (limits for $m_{\beta\beta}$ of a few hundred meV were obtained from the limits on the neutrinoless half-lives). In the next years, the new generation of experiments may reach the inverted hierarchy mass region.

9.3.6 Present Mass Limits and Prospects

The measurements on the single β decay and, if Majorana on the double β decay, set limits, as discussed in the previous sections, respectively on the weighted m_{ν_e} and on the effective $m_{\beta\beta}$ electron neutrino mass. In the latter case case, we should note however that limits are directly measured on the lifetimes of unstable nucleons, and must be translated into $m_{\beta\beta}$.

The bound from cosmology can be similarly expressed in a graph of that form. In that case

$$\sum m_i = m_1 + \sqrt{m_1^2 + \delta m_{12}^2} + \sqrt{m_1^2 + \delta m_{13}^2} \,. \qquad (9.57)$$

These limits are plotted as a function of the lightest neutrino mass in Fig. 9.20.

Most of the phase space for the inverted mass hierarchy scenario may be fully probed by the next generation of beta and double beta experiments.

Further Reading

[F9.1] C. Giunti and C.W. Kim, "Fundamentals of Neutrino Physics and Astrophysics", Oxford 2007.

Exercises

1. *Neutrino interaction cross section.* Explain the peak in the cross section in Fig. 9.1.
2. *Neutrino oscillation probability.* Given a beam of a pure muon neutrino beam, with fixed energy E, derive the probability of observing another neutrino flavor at a distance L assuming two weak eigenstates related to two mass eigenstates by a simple rotation matrix.
3. *Tau antineutrinos appearance.* OPERA is looking for the appearance of tau antineutrinos in the CNGS (CERN neutrinos to Gran Sasso) muon-antineutrino beam. The average neutrino energy is 17 GeV and the baseline is about 730 km. Neglecting mass effects, calculate the oscillation probability

$$P(\bar{\nu}_\mu \to \bar{\nu}_\tau)$$

   and comment.
4. *Neutrino mass differences.* One neutrino experiment detects, at 200 m from the nuclear reactor that the flux of a 3 MeV antineutrino beam is $(90 \pm 10)\%$ of what was expected in case of no oscillation. Assuming a maximal mixing determine the value of Δm^2_ν at a confidence level of 95 %.
5. *Neutrino rotation angles.* Suppose there are three neutrino types (electron, muon, tau) and three mass values, related by the 3×3 PMNS matrix, usually factorized by three rotation matrices. Knowing that the three mass values are such that:

   – Δm^2 (solar) $= m_2^2 - m_1^2 \sim 10^{-5}\text{eV}^2$
   – Δm^2 (atmospheric) $= \left|m_3^2 - m_2^2\right| \sim 10^{-3}\text{eV}^2$

   discuss the optimization of reactor and accelerator experiments to measure each of the three rotation angles and to confirm such mass differences. Compare, for example, the pairs of experiments (KamLAND, DayaBay), (T2K,OPERA).
6. *Neutrino from Supernova 1987A.* In 1987, a Supernova explosion was observed in the Magellanic Cloud, and neutrinos were measured in three different detectors. The neutrinos, with energies between 10 and 50 MeV, arrived with a time span of 10s, after a travel distance of 5×10^{12} s, and 3 h before photons at any wavelength.

   (a) Can this information be used to determine a neutrino mass? Discuss the quantitative mass limits that could be derived from the SN1987A.
   (b) This was the only SN observed in neutrinos, up to now, but the same reasoning can be used in pulsed accelerator beams. Derive the needed time and position precision to measure ∼1 eV masses, given a beam energy E ∼1 GeV and distance L.

7. *Double β decay.* Double β decay is a rare process, possible only for a small fraction of the nuclear isotopes. The neutrinoless double β decay is only possible if lepton number is not conserved, and is one of the most promising channels to discover lepton number violation. Discuss the optimization (total mass, chosen isotope, backgrounds, energy resolution, ...) of the experiments looking for $0\nu\beta\beta$. List other possible experimental signals of lepton number violation you can think of.

Chapter 10
Messengers from the High-Energy Universe

By combining observations of a single phenomenon using different particles, it is possible to achieve a more complete understanding of the properties of the sources; this approach is known as multi-messenger astrophysics. Multi-messenger astrophysics uses mostly information coming from charged particles and from photons at different wavelengths; in the recent years it has been dominated by the discoveries due to high-energy gamma rays, and it might soon incorporate new information from neutrinos and gravitational radiation.

10.1 The Data

Cosmic rays were discovered at the beginning of the twentieth century (see Chap. 3). Since then an enormous number of experiments were performed on the Earth's surface, on balloons, or on airplanes, or even on satellites. We know today that cosmic rays of different nature, spanning many decades in energy, are of cosmic origin and travel through the interstellar space. Their origin and composition is a challenging question, and the combined study of cosmic rays of different nature and energies (multi-messenger astrophysics) can solve fundamental problems, in particular related to physics in extreme environments, and unveil the presence of new particles produced in high-energy phenomena and/or in earlier stages of the Universe.

As we have seen in the Introduction, we think that the ultimate engine of the acceleration of cosmic rays is gravity. In gigantic gravitational collapses, such as those occurred in supernovae and in the accretion of supermassive black holes at the expense of the surrounding matter, part of the potential gravitational energy is transformed into kinetic energy of particles. The mechanism is not fully understood, although we can model part of it; we shall give more details in this chapter. As usual in physics, experimental data are the key: we need to know as accurately as possible the origin, composition, and energy spectrum of cosmic rays.

The composition and energy spectrum of cosmic rays is not a well-defined problem: it depends on where experiments are performed. One could try a schematic

A. De Angelis and M.J.M. Pimenta, *Introduction to Particle and Astroparticle Physics*, Undergraduate Lecture Notes in Physics,
DOI 10.1007/978-88-470-2688-9_10

separation between "primary" cosmic rays—as produced by astrophysical sources—and "secondaries"—those produced in interactions of the primaries with interstellar gas or with nuclei in the Earth's atmosphere. Lithium, beryllium and boron, for example, are very rare products in stellar nucleosynthesis, and thus are secondary particles, as well as antiprotons and positrons—if some antimatter is primary is a question of primary interest.

The interaction with the atmosphere is particularly important since it changes drastically the composition of cosmic rays. In the cases in which the flux of cosmic rays has to be measured at ground (for example, high-energy cosmic rays, energies above tens/hundreds GeV, where the low flux makes the use of satellites ineffective) one needs nontrivial unfolding operations to understand the primary composition: experimental data are thus less clear.

Typically, the composition and energy spectra of nuclei are unfolded to inside the Galaxy; the lifetime of cosmic rays in the galaxy decreases with energy since the most energetic cosmic rays tend to escape (this fact is described by the "leaky-box" model, see later).

10.1.1 Charged Cosmic Rays

Charged cosmic rays arrive close to the Solar System after being deflected from the galactic magnetic fields (about 1 μG in intensity) and possibly by extragalactic magnetic fields, if they are of extragalactic origin; when getting closer to the Earth they start interacting with stronger magnetic fields—up to $O(1\text{G})$ at the Earth's surface, although for shorter distances. Fluxes of charged particles at lower energies, below 1 GeV, can thus be influenced, e.g., by the solar cycle which affects the magnetic field from the Sun.

The energy spectrum of the charged cosmic rays reaching the atmosphere spans over many decades in flux and energy (Fig. 10.1). At low energies, $E < 1$ GeV, the fluxes are high (thousands of particles per square meter per second) while at the highest energies, $E > 10^{11}$ GeV, they are extremely scarce (less than one particle per square kilometer per century).

The cosmic rays at the end of the known spectrum have thus energies well above the highest beam energies attained in any man-made accelerator and their interactions on the top of the Earth atmosphere have center-of-mass energies of a few hundred TeV (the design LHC beam energy is $E = 7 \times 10^3$ GeV).

Below a few GeV their flux is modulated by the solar wind, displaying an anti-correlation with it, and depends also on the local Earth geomagnetic field. Above a few GeV the intensity of the cosmic rays flux follows basically a power law,

$$I(E) \propto E^{-\gamma}$$

with the differential spectral index γ being typically between 2.7 and 3.3.

The small changes in the spectral index can be clearly visualized multiplying the flux by some power of the energy. Figure 10.2 shows a suggestive anthropomorphic

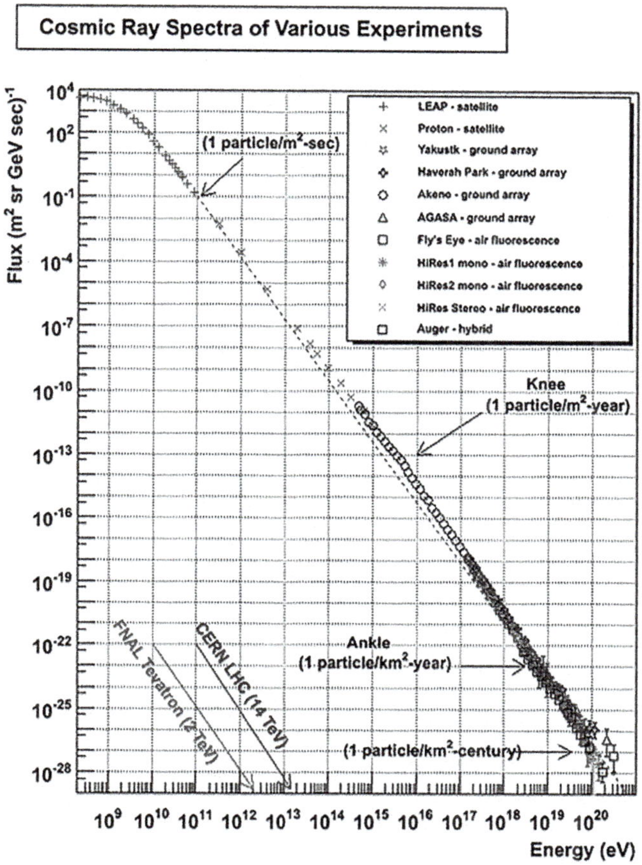

Fig. 10.1 Energy spectrum of charged cosmic rays. Credits: http://www.physics.utah.edu/~whanlon/spectrum.html

representation of the cosmic ray energy spectrum obtained multiplying the flux by $E^{-2.6}$.

Two clear break points corresponding to changes in the spectral index are observed. The first, called the knee, occurs around $E \simeq 5 \times 10^{15}$ eV, and it is commonly associated to the transition from galactic to extragalactic cosmic rays; it corresponds to a steep ending from a spectral index of about 2.7 to a spectral index of about 3.1. There is experimental evidence that the chemical composition of cosmic rays changes after the knee region with an increasing fraction of heavy nuclei at higher energy, at least up to about 10^{17} eV. At even higher energies the chemical composition remains matter of debate. The second clear feature, denominated the "ankle," occurs around $E \simeq 5 \times 10^{18}$ eV and its nature is still controversial. Another feature, called the second knee, marks a steepening to from about 3.1 to about 3.3, at an energy of about 400 PeV.

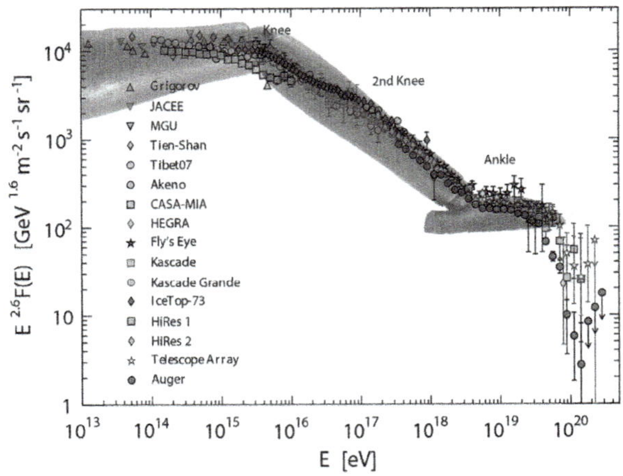

Fig. 10.2 Anthropomorphic representation of the cosmic ray energy spectrum coming from experimental measurements by different experiments; the spectrum has been multiplied by $E^{2.6}$. From Ref. [10.1] in the "Further reading"

The number of primary nucleons per GeV from some 10 GeV to beyond 100 TeV is approximately

$$N(E) \simeq 1.8 \times 10^4 E^{-2.7} \frac{\text{nucleons}}{\text{m}^2\,\text{s}\,\text{sr}\,\text{Gev}} \tag{10.1}$$

where E is the energy per nucleon in GeV.

A strong suppression at the highest energies, $E \simeq 5 \times 10^{19}$ eV, is nowadays clearly established (Fig. 10.2); it may result, as explained in Chaps. 1 and 2, from the so-called GZK mechanism due to the interaction of highly energetic protons with the Cosmic Microwave Background (CMB), but a scenario in which an important part of the effect is a change of composition (from protons to heavier nuclei, which undergo nuclear photodisintegration,[1] see later), and/or the exhaustion of the sources (see Sect. 10.3), is not excluded.

10.1.1.1 Composition

Cosmic rays are basically protons (~90 %) and heavier nuclei. The electron/positron flux at the top of the atmosphere is small (a few per mil of the total cosmic ray flux) but extremely interesting as it may be a signature of unknown astrophysical or Dark Matter sources (see Chap. 8). Antiprotons fluxes are even smaller (about four orders of magnitude) and so far compatible with secondary production by hadronic interactions of primary cosmic rays with the interstellar medium. Up to now there is no evidence for the existence of heavier anti-nuclei (in particular anti-deuterium and anti-helium) in cosmic rays.

[1]In the case of nuclei, the spallation due to the giant dipole resonance (a collective excitation of nucleons in nuclei due to the interaction with photons) has also to be considered. For all nuclei but iron the corresponding mean free paths are, at these energies, much smaller than the proton GZK mean free path.

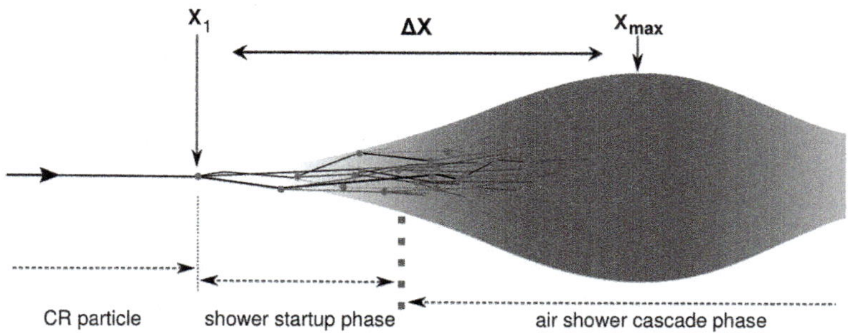

Fig. 10.3 Shower development scheme. Adapted from the Ph.D. thesis of R. Ulrich: "Measurement of the proton–air cross section using hybrid data of the Pierre Auger Observatory," http://bibliothek.fzk.de/zb/berichte/FZKA7389.pdf

Accessing the composition of cosmic rays can be done in the region below a few TeV, for example, by combining momentum measurement with the information from Cherenkov detectors, or transition radiation detectors. This is how the AMS-02 magnetic spectrometer performs the task.

For EAS detectors, being able to distinguish between a shower generated by a proton or by a heavier particle is a more difficult task. One of the most popular air shower variables which in principle allows disentangling between protons and heavier nuclei is the so-called X_{max}, the depth of the maximum number of particles in the shower, expressed in g/cm^2. X_{max} may be defined (Fig. 10.3) as the sum of the depth of the first interaction X_1 and a shower development length ΔX:

$$X_{max} = X_1 + \Delta X\,.$$

This discriminating variable uses the fact that the cross section for a nucleus grows with its atomic number.

The X_1 distribution is just a negative exponential, $\exp\left(-X_1/\Lambda_\eta\right)$, where Λ_η is the interaction length which is proportional to the inverse of the cosmic ray–air interaction cross section. The observed X_{max} distribution is thus the convolution of the X_1 distribution with the ΔX distribution (which has a shape similar to the X_{max} distribution) and a detector resolution function (Fig. 10.4). Note that the tail of the X_{max} distribution reflects the X_1 exponential distribution.

The absolute and relative fluxes of the main hadronic components of the cosmic rays arriving at Earth are shown, respectively, in Figs. 10.5 and 10.6 and compared, in this last figure, to the relative abundances existing in the solar system. To understand these figures, we should take into account the facts that:

- in nature nuclei with even number of nucleons are more stable, having higher binding energy because of pairing effects;
- features of charged particle acceleration must depend on *rigidity*

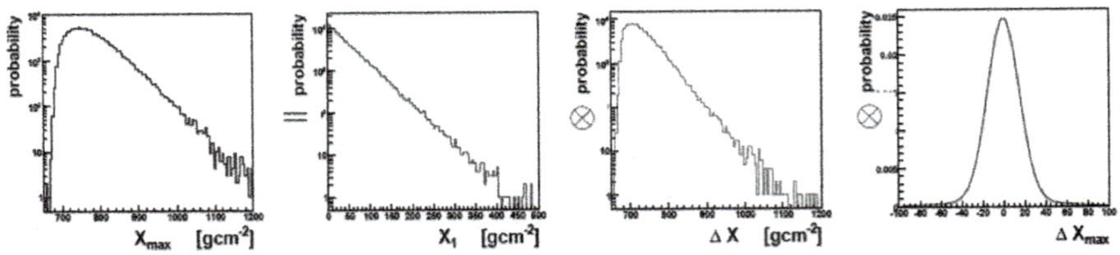

Fig. 10.4 Ingredients of the X_{max} distribution. Adapted from the Ph.D. thesis of R. Ulrich: "Measurement of the proton–air cross section using hybrid data of the Pierre Auger Observatory," http://bibliothek.fzk.de/zb/berichte/FZKA7389.pdf

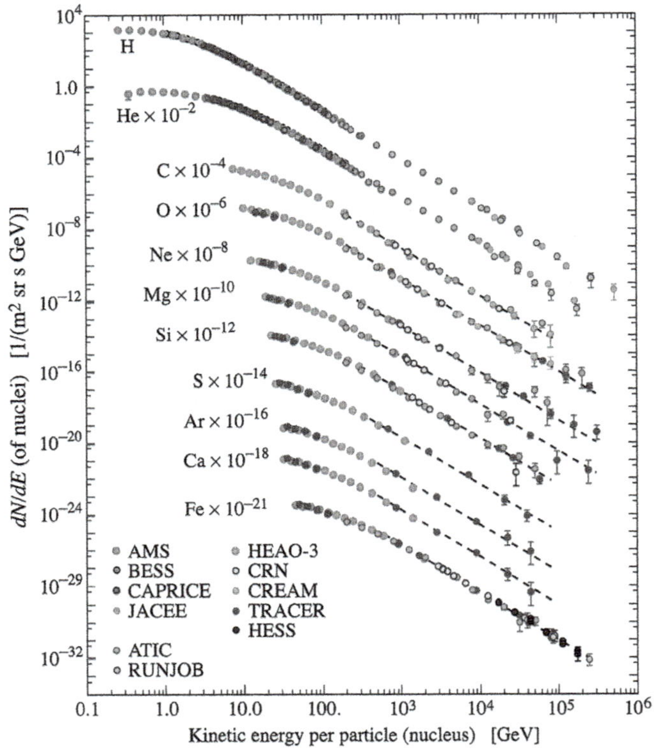

Fig. 10.5 Flux of the hadronic components of cosmic rays arriving at Earth for energies greater than 2 GeV. From K.A. Olive et al. (Particle Data Group), Chin. Phys. C 38 (2014) 090001

$$\mathcal{R} = r_L Bc = \frac{pc}{ze}$$

where r_L is the Larmor radius, and z the atomic number of the nucleus. Rigidity is measured in V.

Besides a clear deficit of hydrogen and helium in the cosmic rays compared to the composition of the Solar System, the main features from this comparison are

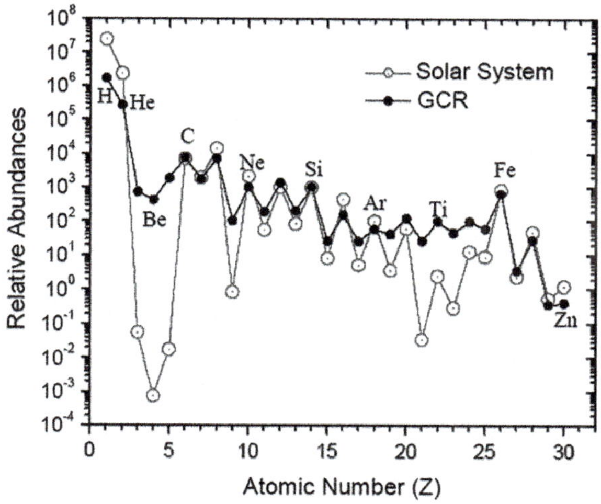

Fig. 10.6 Relative abundance of the main hadronic components present in the cosmic rays arriving at Earth and in the Solar System. Both are normalized to the abundance of $Si = 10^3$, and the relevant energy range is a few hundreds MeV/nucleon. Credit: ACE archives, http://www.srl.caltech.edu/ACE

the agreement on the "peaks" (more tightly bounded even-Z nuclei) and higher abundances for cosmic rays on the "valleys." These features can be explained within a scenario where primary cosmic rays are produced in stellar endproducts (see later), being the "valley" elements mainly secondaries produced in the interaction of the primaries cosmic rays with the interstellar medium ("spallation").

10.1.1.2 Anisotropy

The arrival direction of charged cosmic rays is basically isotropic—a fact which can find explanation in the effect of the galactic magnetic field. Milagro, IceCube, HAWC, ARGO and the Tibet air shower array have, however, observed small anisotropies (at the level of about one part per mil) in cosmic rays with energies above a few TeV, in space regions of size of about 10°; they might be due to nearby sources. The excesses, however, cannot be attributed to a common source—which means that deeper studies are needed to understand.

To accelerate particles up to the ultrahigh-energy (UHE) region above the EeV, 10^{18} eV, one needs conditions that are present in astrophysical objects such as the surroundings of supermassive black holes in active galactic nuclei (AGN), or transient high-energy events such as the ones generating gamma ray bursts (see later). Galactic objects are not likely to be acceleration sites for particles of such energy, also because UHECRs do not concentrate in the galactic plane; in addition, the galactic magnetic field cannot confine UHECRs above 10^{18} eV within our galaxy.

A search for anisotropies in the UHE region has been performed by the Pierre Auger and the Telescope Array collaborations. In the south galactic hemisphere the Pierre Auger Collaboration reports a correlation of events above 57 EeV with the

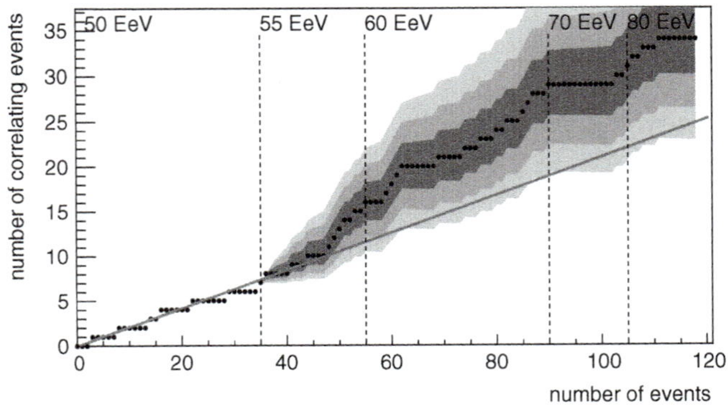

Fig. 10.7 Energy ordered correlation of events above 57 EeV with the Véron-Cetty Véron (VCV) AGN catalog as seen by the Pierre Auger Observatory up to June 2011. The *red line* represents the foreseen correlation from an isotropic sky. Image credit: Pierre Auger collaboration

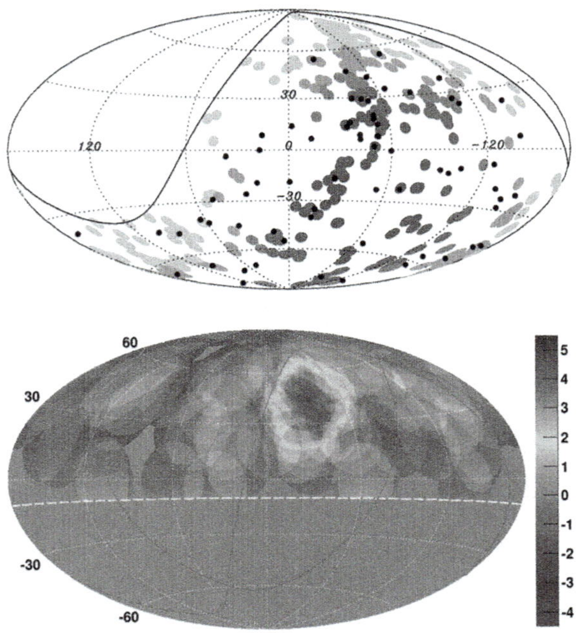

Fig. 10.8 *Top* Positions of the AGN within 71 Mpc used in the correlation search by Auger (*dots*) and *circles* centered at the arrival directions of the events with $E > 57$ EeV observed by Auger in galactic coordinates. *Bottom* Map in equatorial coordinates (note the difference of coordinates with respect to the plot on *top*) for the events above 57 EeV as seen by the Telescope Array collaboration up to May 2014. The *color* scale represents the estimated significance. For details see arXiv:1404.5890. Image credit: K Kawata/University of Tokyo Institute for Cosmic Ray Research

Véron-Cetty Véron (VCV) AGN catalog of the order of 33 %, with 21 % expected under the isotropic hypothesis (Fig. 10.7). A possible significant correlation exists with the region of Centaurus A (Fig. 10.8, top), an AGN at distance of about 13 million

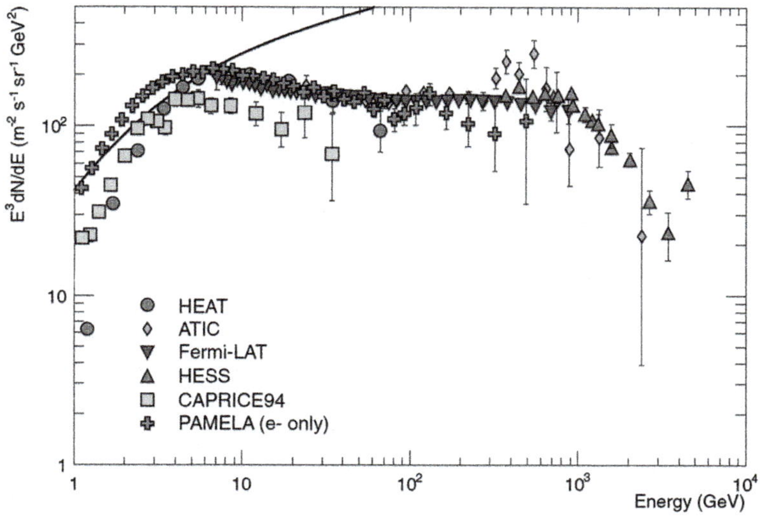

Fig. 10.9 Energy spectrum of e^+ plus e^- (except PAMELA data, for which electrons only are plotted) multiplied by E^3. The *line* shows 1 % of the proton spectrum. From K.A. Olive et al. (Particle Data Group), Chin. Phys. C 38 (2014) 090001

light-years—one of the closest AGN, and the fifth brightest galaxy in the sky, hosting a black hole of about 55 million solar masses and known to be a VHE gamma ray emitter. In the northern hemisphere the Telescope Array collaboration observed for events above 57 EeV a hotspot in the direction of the Ursa Major constellation with a diameter of 30°–40° (Fig. 10.8, bottom). No statistically significant correlation with AGNs from the VCV catalog was observed.

10.1.1.3 Electrons and Positrons

High-energy electrons and positrons have short propagation distances as they lose energy through synchrotron and inverse Compton processes as they propagate through the Galaxy. Their spectrum is therefore expected to be dominated by local electron accelerators or by the decay/interactions of heavier particles nearby. Positrons in particular could be the signature of the decay of dark matter particles.

Given these considerations, the experimental data on the flux of electrons plus positrons, shown in Fig. 10.9, are quite puzzling. First of all, the yield in the 100 GeV region and above indicates the presence of nearby sources. Second, the ATIC balloon experiment measured a bump-like structure at energies between 250 and 700 GeV. A bump-like structure might indicate the presence of a heavy particle decaying into electrons and positrons of a definite mass. However, this finding was not confirmed by later and more accurate instruments like the Fermi satellite Large Area Tracker (Fermi-LAT) and AMS-02. The H.E.S.S. Cherenkov array also measured the electron

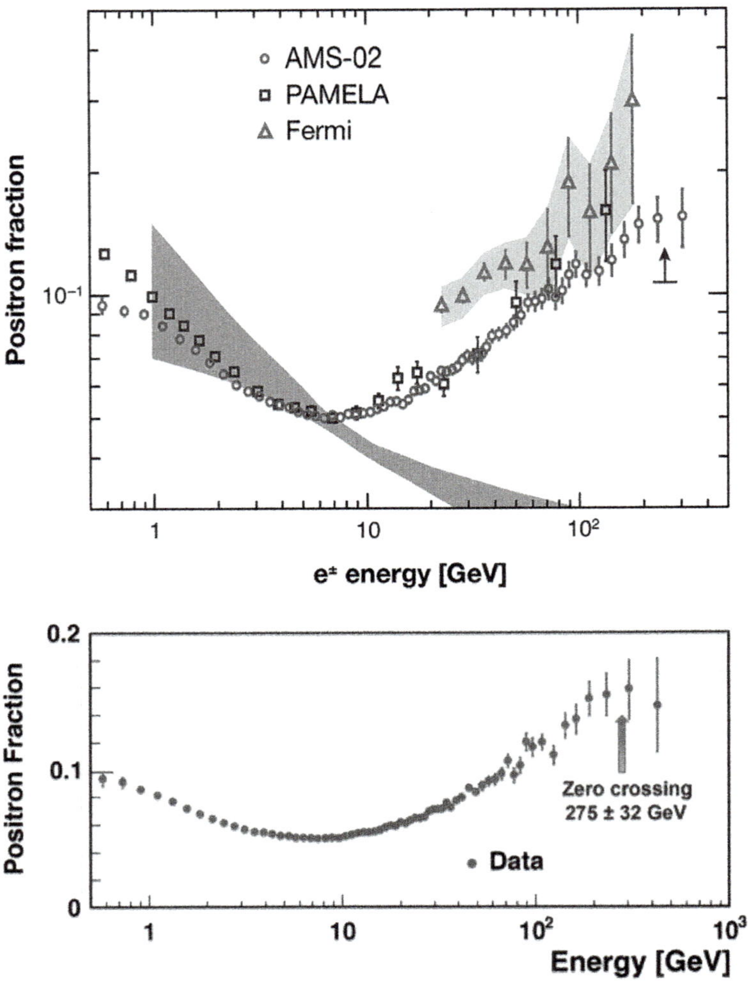

Fig. 10.10 *Top* The positron fraction in high-energy cosmic rays: measurement from the AMS-02, PAMELA, and Fermi-LAT satellites. The AMS measurement confirms an excess in the high-energy positron fraction, above what is expected from positrons produced in cosmic ray interactions, first observed by PAMELA (the *gray* band indicates the expected range in the positron fraction, which is based on standard cosmic ray models). *Bottom* Recently updated results by AMS-02. No sharp structures are observed, besides a hint of a maximum at (275 ± 32) GeV

flux at high energies, finding indications of a cutoff above 1 TeV, but no evidence for a peak.

The ratio as a function of energy between the flux of positrons and the total flux of electrons plus positrons is shown in Fig. 10.10. This is even more intriguing: in a matter-dominated Universe, one would expect this ratio to decrease with energy, unless specific sources of positrons are present nearby; but data show an anomaly called the "PAMELA anomaly." If these sources are heavy particles decaying into

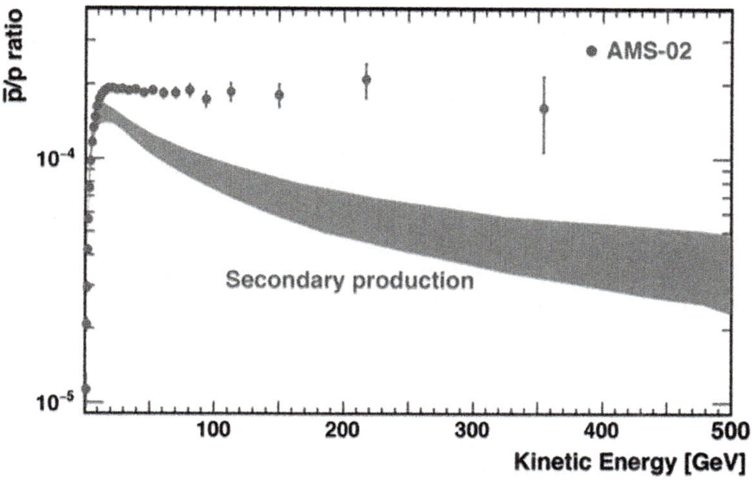

Fig. 10.11 Antiproton to proton ratio measured by AMS-02. Within the existing models of secondary production, the ratio lacks explanation

final states involving positrons, one could expect the ratio to increase, and then steeply drop after reaching half of the mass of the decaying particle. If an astrophysical source of high-energy positrons is present, a smooth spectrum is expected.

Data cannot give a definite answer yet. A possible decrease of the ratio at an energy of about 275 GeV might have been observed (Fig. 10.10, bottom). The most recent data on the abundance of high-energy pulsars nearby (see later) might justify an astrophysical explanation. However, the origin from a new massive particle of mass of the order of 1 TeV is not excluded yet. New data will tell more. Recent results by AMS-02 on the antiproton yield are particularly intriguing.

10.1.1.4 Antiprotons

Data are shown in Fig. 10.11. The antiproton to proton ratio stays constant from 20 to 400 GeV. This behavior cannot be explained by secondary production of antiprotons from ordinary cosmic ray collisions. At variance with the excess of positrons, the excess of antiprotons cannot be easily explained from pulsar origin. More study is needed, and this is certainly one of the next frontiers.

10.1.1.5 Cosmic Rays at the Earth's Surface: Muons

Most charged particles on the top of the atmosphere are electrons and protons; however, the interaction with the atoms of the atmosphere itself dramatically changes the composition. Secondary muons, photons, electrons/positrons and neutrinos are produced by the interaction of charged cosmic rays in air, in addition to less stable particles. Note that the nucleon composition changes dramatically in such a way that

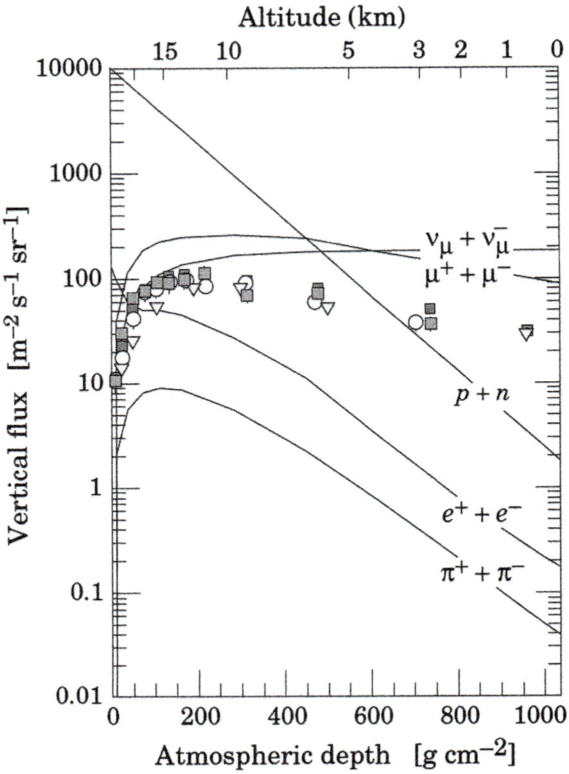

Fig. 10.12 Fluxes of cosmic rays with $E > 1$ GeV in the atmosphere as a function of height (simulation). The points show experimental measurements of negative muons. From K.A. Olive et al. (Particle Data Group), Chin. Phys. C 38 (2014) 090001

neutrons, which are 10 % of the total at the atmosphere's surface, become roughly 1/3 at the Earth's surface.

Primary muons can hardly reach the Earth's atmosphere due to their lifetime ($\tau \sim 2\,\mu$s); this lifetime is however large enough, that secondary muons produced in the atmosphere can reach the Earth's surface, offering a wonderful example of time dilation: the space crossed in average by such particles is $L \simeq c\gamma\tau$, and already for $\gamma \sim 50$ (i.e., an energy of about 5 GeV) they can travel 30 km, which roughly corresponds to the atmospheric depth. Muons lose some 3 GeV by ionization when crossing the atmosphere.

Charged particles at sea level are mostly muons (see Fig. 10.12), with a mean energy of about 4 GeV.

The flux of muons from above beyond an energy of 1 GeV at sea level is about 60 $m^{-2}\,s^{-1}\,sr^{-1}$. A horizontal detectors sees roughly one muon per square centimeter per minute. The zenith angular distribution for muons of $E \sim 3$ GeV is $\propto \cos^2\theta$, being steeper at lower energies and flatter at higher energies: low energy muons at large angles decay before reaching the surface. The ratio between μ^+ and μ^- is due to the fact that there are more π^+ than π^- in the proton-initiated showers; there are about 30 % more μ^+ than μ^- at momenta above 1 GeV/c.

A fortiori, among known particles only muons and neutrinos reach significant depths underground. The muon flux reaches $10^{-2}\,m^{-2}\,s^{-1}\,sr^{-1}$ under 1 km of water equivalent (corresponding to about 400 m of average rock) and becomes about $10^{-8}\,m^{-2}\,s^{-1}\,sr^{-1}$ at 10 km of water equivalent.

10.1.2 Photons

High-energy astrophysical processes generate photon radiation over a large range of wavelengths. Such photon radiation can be associated to the emitters, which is an advantage with respect to charged cosmic rays. In addition, photon radiation, besides being interesting per se, can give insights on the acceleration of charged particles, being photons secondary products of accelerated charged particles. In addition, they are likely to be present in the decay chain of unstable massive particles, or in the annihilation of pairs of particles like dark matter particles.

The experimental data on the cosmic photon radiation span some 30 energy decades; a compilation of the data is shown in Fig. 10.13. A bump is visible corresponding to the CMB, while the general behavior of the yield of gamma rays at high energies can be approximated by an energy dependence as a power law $E^{-2.4}$

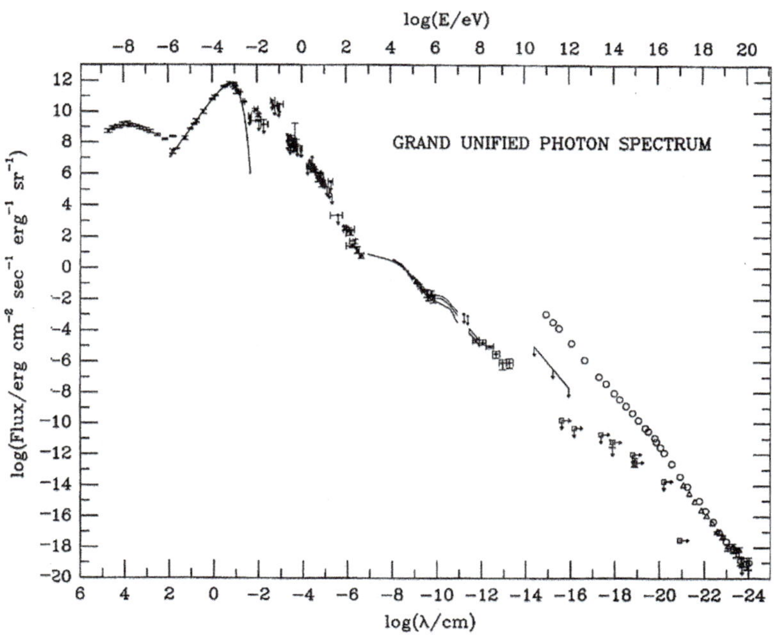

Fig. 10.13 Spectrum of photons was experimentally detected at different energies. From M.T. Ressell and M.S. Turner 1989, "The grand unified photon spectrum: a coherent view of the diffuse extragalactic background radiation," [Batavia, Ill.]: Fermi National Accelerator Laboratory. http://purl.fdlp.gov/GPO/gpo48234

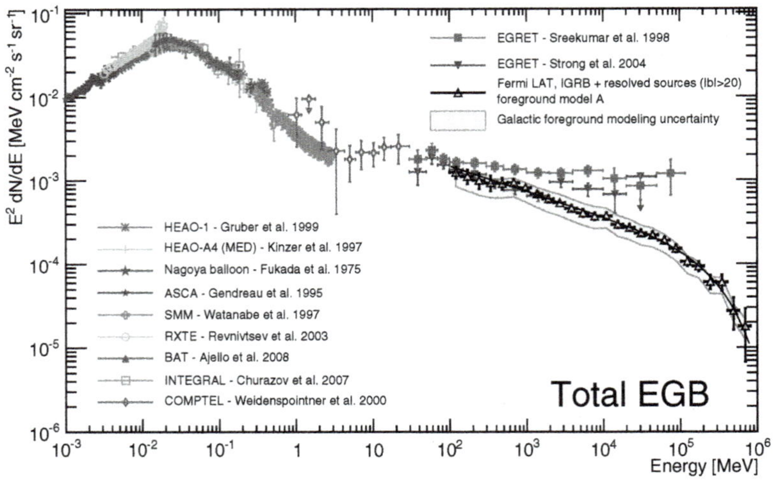

Fig. 10.14 Spectrum of the total extragalactic gamma ray emission measured by the Fermi-LAT. From M. Ackermann et al., The Astrophysical Journal 799 (2015) 86

(Fig. 10.14). A cutoff at energies close to 1 TeV might be explained by the absorption of higher energy photons by background photons near the visible populating the intergalactic medium—through creation of e^+e^- pairs, see later.

There is little doubt on the existence of the so-called ultra- and extremely-high-energy photons (respectively in the PeV-EeV and in the EeV-ZeV range), but so far cosmic gamma rays have been unambiguously detected only in the low (MeV), high (GeV) and very (TeV) high-energy domains, and the behavior above some 30 TeV is extrapolated by data at lower energies and constrained by experimental upper limits. The data related to diffuse (i.e., unassociated with sources) flux do not go beyond 1 TeV (Fig. 10.14).

In Chap. 4 we have defined as high energy (HE) the photons above 30 MeV—i.e., the threshold for the production of e^+e^- pairs plus some phase space; as very high energy (VHE) the photons above 30 GeV. The HE—and VHE in particular—regions are especially important related to the physics of cosmic rays and to fundamental physics. One of the possible sources of HE gamma rays is indeed the generation as a secondary product in conventional scenarios of acceleration of charged particles; in this case cosmic gamma rays are a probe into cosmic accelerators. The VHE domain is sensitive to energy scales important for particle physics. One is the 100 GeV scale expected for cold dark matter. A second energy scale is the TeV scale, where supersymmetric particles might appear. Finally, it might be possible to access the unification scale and the Planck scale: an energy $\sim 10^{19}$ GeV, corresponding to a mass $\sqrt{\hbar c/G}$—which is, apart from factors of order ~ 1, the mass of a black hole whose Schwarzschild radius equals its Compton wavelength.

Gamma rays provide at present the best window into the nonthermal Universe, being the "hottest" thermalized processes observed up to now in the accretion region of supermassive black holes at a temperature scale of 10^5 K, which corresponds to

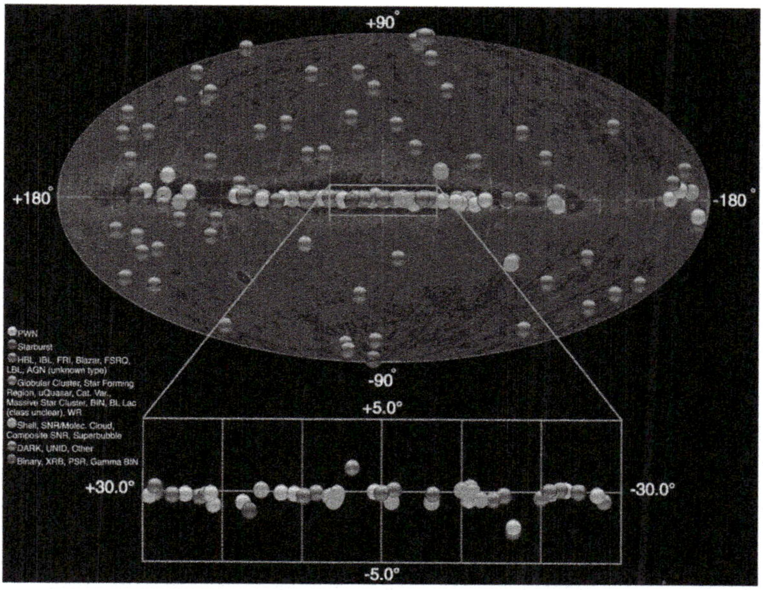

Fig. 10.15 Sources of VHE emission plotted in galactic coordinates. The background represents the high-energy gamma rays detected by Fermi-LAT. The region near the Galactic Center is enlarged. From http://tevcat.uchicago.edu/, May 2015

a few tens of eV—the X-ray region. Tests of fundamental physics with gamma rays are much beyond the reach of terrestrial accelerators.

Besides the interest for fundamental physics, the astrophysical interest of HE and VHE photons is evident: for some sources such as the AGN—supermassive black holes in the center of galaxies, powered by infalling matter—the total power emitted above 100 MeV dominates the electromagnetic dissipation.

A look to the sources of cosmic gamma rays in the HE region shows a diffuse background, plus a set of localized emitters. Roughly, 3000 HE emitters have been identified up to now, mostly by the Fermi-LAT, and some 200 of them are VHE emitters as well (Fig. 10.15).

About half of the gamma ray emitters are objects in our Galaxy; at VHE most of them can be associated to supernova remnants (SNR), while at MeV to GeV energies they are mostly pulsars; the remaining half are extragalactic, and the space resolution of present detectors (slightly better than 0.1°) is not good enough to associate them with particular points in the host galaxies; we believe, however, that they are produced in the vicinity of supermassive black holes in the galaxies—see later.

The strongest steady emitters are galactic objects; this can be explained by the fact that, being closer, they suffer a smaller attenuation. The observed strongest steady emitter at VHE is the Crab Nebula. The energy distribution of the photons from Crab Nebula is typical for gamma sources (see the explanation of the "double-hump" structure in Sect. 10.2.2.1), and it is shown in Fig. 10.16.

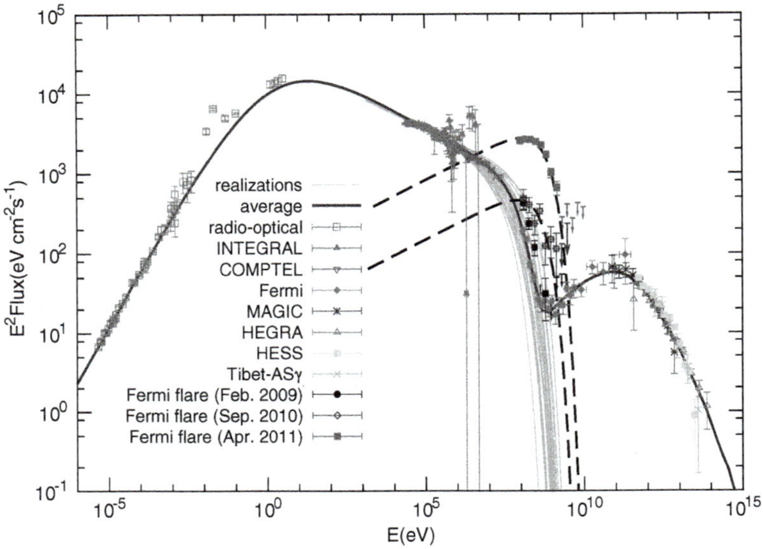

Fig. 10.16 Spectral energy distribution of the Crab Nebula (data from radio-optical up to 100 TeV). From Yuan, Yin et al., arXiv:1109.0075

10.1.2.1 Transient Phenomena and Gamma Ray Bursts; Quasiperiodical Emissions

Among cosmic rays, gamma rays are important not only because they point to the sources, but also because the sensitivity of present instruments is such that transient events (in jargon, "transients") can be recorded. Sources of HE and VHE gamma rays (some of which might likely be also sources of charged cosmic rays, neutrinos and other radiation) were indeed discovered to exhibit transient phenomena, with timescales from few seconds to few days.

The sky exhibits in particular transient events from steady emitters ("flares") and burst of gamma rays from previously dark regions ("gamma ray bursts"). The phenomenology of such events is described in the rest of this section.

Short timescale variability has been observed in the gamma emission at high energies for several astrophysical objects, both galactic and extragalactic, in particular binary systems, and AGN. For binary systems the variability is quasiperiodical and can be related to the orbital motion, while for AGN it must be related to some cataclysmic events; this is the phenomenon of flares. Flares observed from Crab Nebula have, as today, no universally accepted interpretation.

Flares. Flares are characteristic mostly of extragalactic emitters (AGN). Among galactic emitters, the Crab Nebula, which was for longtime used as a "standard candle" in gamma astrophysics, has been recently discovered to be subject to dramatic flares on timescales of $\sim$10 h, with transient emission briefly dominating the flux

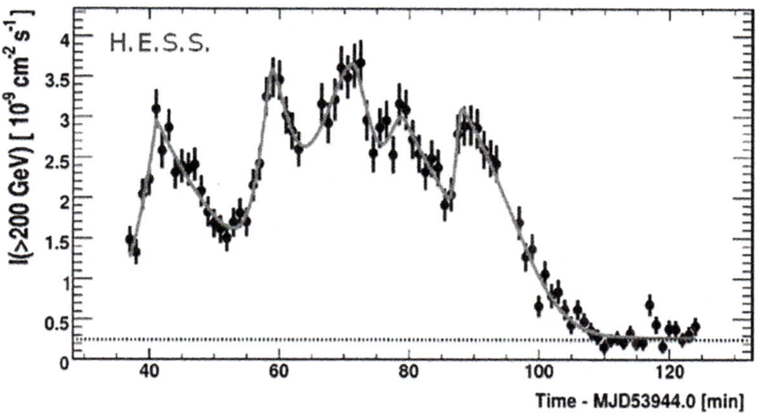

Fig. 10.17 Variability in the very-high-energy emission of the blazar PKS 2155-304. The *dotted horizontal line* indicates the flux from the Crab Nebula (from the H.E.S.S. experiment, http://www.mpi-hd.mpg.de/hfm/HESS)

from this object with a diameter of 10 light-years—which is the diameter of the shell including the pulsar remnant of the imploded star, and correspond to roughly 0.1° as seen from Earth.

Very short timescale emission from blazars have also been observed in the TeV band, the most prominent being at present the flare from the AGN PKS 2155-304 shown in Fig. 10.17: a factor >100 flux increase with respect to the quiescent state, with variability on timescales close to 1 min. Note that the radius of the black hole powering PKS2155 is about 10^4 light seconds (corresponding to 10^9 solar masses), which has implications on the mechanisms of emission of gamma rays (see later).

Indeed the gamma ray sky looks like a movie rather than a picture, the most astonishing phenomenon being the explosion of gamma ray bursts.

Gamma Ray Bursts. Gamma Ray Bursts (GRBs) are extremely intense and fast shots of gamma radiation. They last from fractions of a second to a few seconds and sometimes up to a thousand seconds, often followed by "afterglows" orders of magnitude less energetic than the primary emission after minutes, hours, or even days. GRBs are detected once per day in average, typically in X-rays and soft gamma rays. They are named GRByymmdd after the date on which they were detected: the first two numbers after "GRB" correspond to the last two digits of the year, the second two numbers to the month, and the last two numbers to the day. A progressive letter ("A," "B," ...) might be added—it is mandatory if more than one GRB was discovered in the same day, and it became customary after 2010.

Their position appears random in the sky (Fig. 10.18), which suggests that they are of extragalactic origin. A few of them per year have energy fluxes and energies large enough that the Fermi-LAT can detect them (photons of the order of few tens of GeV have been detected in a few of them). Also in this case the sources appear to be isotropic.

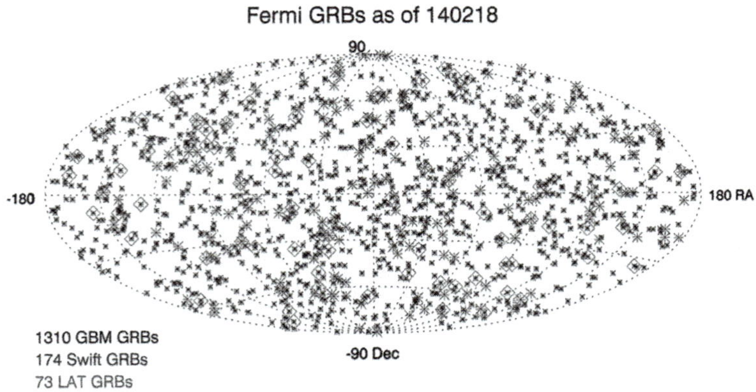

Fig. 10.18 Skymap of the GRBs located by the GRB monitor of Fermi and by the Fermi-LAT. Some events also seen by the Swift satellite are also shown. Credit: NASA

The energy spectrum is nonthermal and varies from event to event, peaking at around a few hundred keV and extending up to several GeV. It can be roughly fitted by phenomenological function (a smoothly broken power law) called "Band spectrum" (from the name of David Band who proposed it). The change of spectral slope from a typical slope of -1 to a typical slope of -2 occurs at a break energy E_b which, for the majority of observed bursts, is in the range between 0.1 and 1 MeV. Sometimes HE photons are emitted in the afterglows.

During fractions of seconds, their energy emission in the gamma ray band exceeds in some cases the energy flux of the rest of the Universe in the same band. The time integrated fluxes range from about 10^{-7} to about 10^{-4} erg/cm^2. If the emission were isotropic, the energy output would on average amount to a solar rest-mass energy, about 10^{54} erg; however, if the mechanism is similar to the one in AGN the emission should be beamed,[2] with a typical jet opening angle of a few degrees. Thus the actual average energy yield in γ rays should be $\sim 10^{51}$ erg. This value can be larger than the energy content of a typical supernova explosion, of which only 1 % emerges as visible photons (over a time span of thousands of years).

The distribution of their duration is bimodal (Fig. 10.19), and allows a first phenomenological classification between "short" GRBs (lasting typically 0.3 s; duration is usually defined as the time T90 during which 90 % of the photons are detected) and "long" GRBs (lasting more than 2 s, and typically 40 s). Short GRBs are on average harder than long GRBs.

[2] Accretion disks, in particular in AGN, produce sometimes two opposite collimated jets, with a fast outflow of matter and energy from close to the disc. The direction of the jet is determined by the rotational axis of the accretion disk and/or of the black hole. The resolution of astronomical instruments is in general too poor, especially at high energies, to resolve jet morphology in gamma rays, and as a consequence observations cannot provide explanations for the mechanism yet. The limited experimental information available comes from the radio waveband, where very-long-baseline interferometry can image at sub-parsec scales the emission of synchrotron radiation near the black hole—but radiation should be present from the radio through to the gamma ray range.

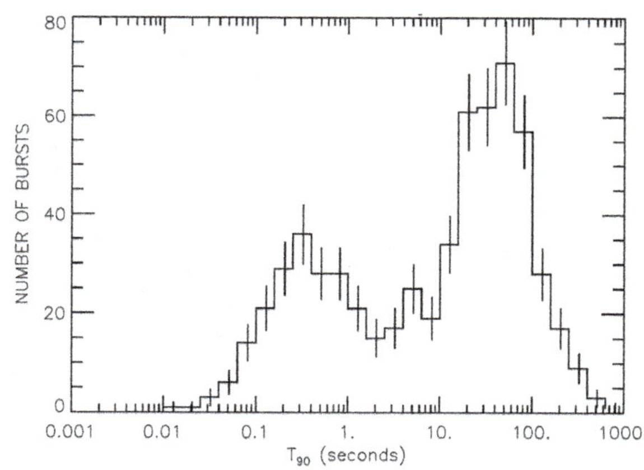

Fig. 10.19 Time distribution of GRBs detected by the BATSE satellite as a function of the time T90 during which 90 % of the photons are detected. It is easy to see "short" and "long" GRBs. Credit: NASA

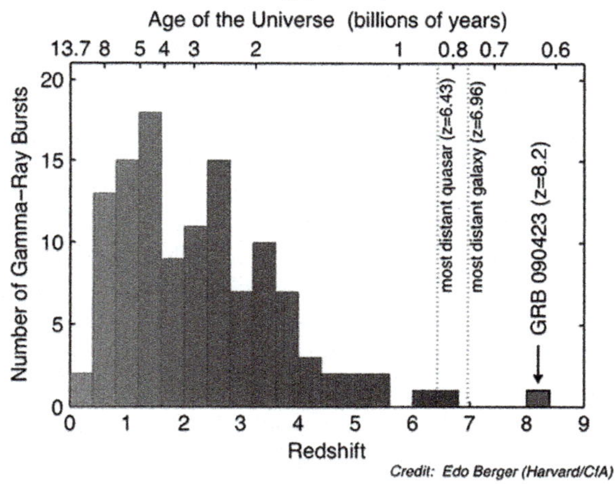

Fig. 10.20 Distribution of redshifts and corresponding age of the Universe for gamma ray bursts detected by NASA's Swift satellite. Credit: Edo Berger (Harvard), 2009

GRBs are generally far away, typically at $z \sim 1$ and beyond (Fig. 10.20). The farthest event ever detected is a 10-s long GRB at $z \simeq 8.2$, called GRB090423, observed by the Swift satellite (the burst alert monitor of Swift being sensitive to energies up to 0.35 MeV).

Short GRBs are very difficult to associate to known objects. For long GRBs in several cases the emission has been associated with a formation of a supernova, presumably of very high mass (a "hypernova"). Possible mechanisms for GRBs will be discussed later.

A historical curiosity: the first GRB was discovered in 1967 by one of the US satellites of the Vela series, but the discovery has been kept secret for six years. The Vela satellites had been launched to verify if Soviet Union was respecting the nuclear test ban treaty imposing non-testing of nuclear devices in space. After the observation of the GRB, it took some time to be sure that the event was of astrophysical origin. Unfortunately, we do not know anything about possible similar discoveries by the Soviet Union.

Binary Systems. Binary stars (i.e., pairs of stars bound by gravitational interaction) are frequent in the Universe: most solar-size and larger stars reside in binaries. Binary systems in which one object is compact (a pulsar, a neutron star, or a black hole) have been observed to be periodical emitters of gamma radiation.

A particular class of binary systems is the microquasars, binary systems comprising a black hole, which exhibit relativistic jets (they are morphologically similar to the AGN). In quasars, the accretion object is a supermassive (millions to billions of solar masses) black hole; in microquasars, the mass of the compact object is only a few solar masses.

10.1.2.2 Large Regions of Photon Emission; The Fermi Bubbles

As the space resolution of the Fermi-LAT and of the Cherenkov telescopes are of the order of 0.1°, we can image diffuse structure only in the Milky Way: the other galaxies will mostly appear like a point. Morphology studies at VHE are basically limited to structures within our galaxy.

Morphology of SNR is in particular one of the keys to understand physics in the vicinity of matter at high density—and one of the tools to understand the mechanism of acceleration of cosmic rays. Sometimes SNRs and the surrounding regions are too large to be imaged by Cherenkov telescopes, which typically have fields of view of 3°–4°. A large field of view is also essential to understand the nature of primary accelerators in pulsar wind nebulae (PWN), as discussed in Sect. 10.3.1.2: it would be important to estimate the energy spectrum as a function of the angular distance to the center of the pulsar to separate the hadronic acceleration from the leptonic acceleration. The highest energy electrons lose energy quickly as they propagate away from the source; this is not true for protons.

Intermediate emission structures, a few degrees in radius, have been observed by MILAGRO and ARGO, which can be attributed to diffusion of protons within the interstellar medium.

A surprising discovery by Fermi-LAT was the existence of a giant structure emitting photons in our Galaxy, with size comparable to the size of the Galaxy itself: the so-called Fermi bubbles. These two structures, about 50,000-light-years across (Fig. 10.21), have quite sharp boundaries end emit rather uniformly in space with an energy spectrum peaking at a few GeV but yielding sizable amount of energy still up to 20 GeV.

Although the parts of the bubbles closest to the galactic plane shine in microwaves as well as gamma rays, about two-thirds of the way out the microwaves fade and only gamma rays are detectable.

Possible explanations of such a large structure are related to the past activity of the black hole in the center of our Galaxy. A large-scale structure of the magnetic field in the bubble region might indicate an origin from the center of the Galaxy, where magnetic fields are of the order of 100 μG, and might also explain the mechanism of emission as synchrotron radiation from trapped electrons. However, this explanation is highly speculative, and as of today the reason for the emission is unknown.

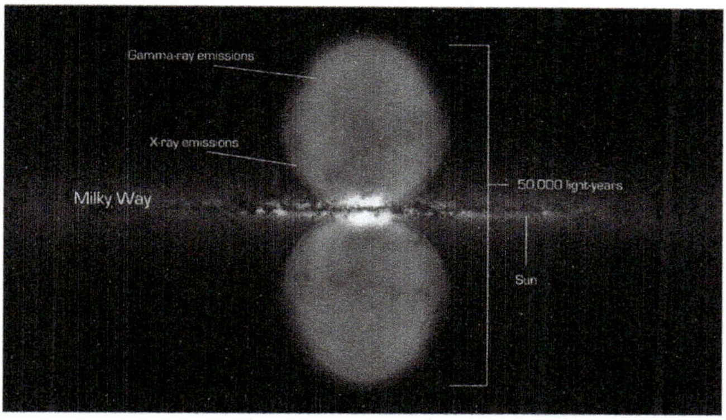

Fig. 10.21 The Fermi bubbles (smoothed). Credit: NASA

10.1.3 Neutrinos

Neutrinos play a special role in particle astrophysics. Their cross section is very small and they can leave production sites without interacting. Differently from photons, neutrinos can carry information about the core of the astrophysical objects that produce them. Different from photons, they practically do not suffer absorption during their cosmic propagation.

1. Neutrinos can be produced in the nuclear reactions generating energy in stars. For example, the Sun emits about 2×10^{38} neutrinos/s (see Chap. 9).
2. They should be also produced in the most violent phenomena, including the Big Bang, supernovae, and the accretion of supermassive black holes.
3. They are the main output of the cooling of astrophysical objects, including neutron stars and red giants.
4. They are produced as secondary by-products of cosmic ray collisions:
   - with photons or nuclei near the acceleration regions (these are "astrophysical" neutrinos, like the ones at items 2. and 3.);
   - with the CMB in the case of ultrahigh-energy cosmic rays suffering the GZK effect (these are called cosmogenic neutrinos, or also "GZK neutrinos," although the mechanism was proposed by Berezinsky and Zatsepin in 1969);
   - and also in the atmosphere (they are called atmospheric neutrinos).
5. Finally, they are likely to be present in the decay chain of unstable massive particles, or in the annihilation of pairs of particles like dark matter particles.

Sources 2., 4. and 5. in the list above are also common to photons; however, detection of astrophysical neutrinos could help constraining properties of the primary cosmic ray spectrum more effectively than high-energy photons. Neutrinos produced by reactions of ultrahigh-energy cosmic rays can provide information on otherwise inaccessible cosmic accelerators.

Experimental data on neutrinos are however scarce: their small cross section makes the detection difficult, and if a detector can be built with a sensitivity large

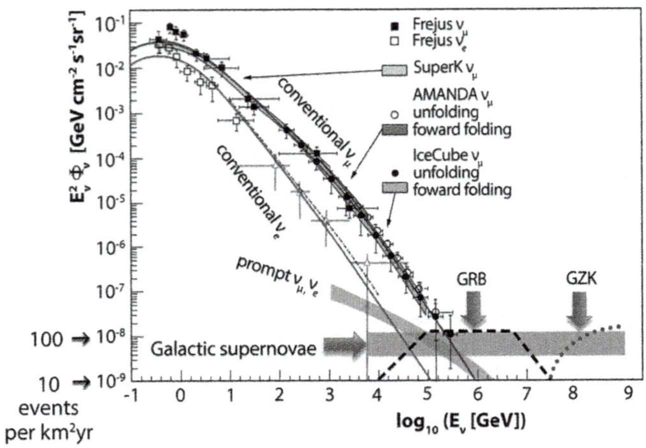

Fig. 10.22 The flux of neutrinos: experimental data. Credit: F. Halzen

enough to obtain useful information on astrophysical neutrinos is still an open question. We discussed in Chap. 4 the problems of such detectors.

The spectrum of neutrinos experimentally detected is shown in Fig. 10.22. Most of them are expected to be atmospheric neutrinos, i.e., coming from the interaction of cosmic rays with the atmosphere.

We recall, however, the results on the detection of VHE neutrinos from a supernova collapse.

10.1.3.1 Neutrinos from SN1987A

On February 24, 1987, a supernova was observed in the Large Magellanic Cloud (LMC), a small galaxy satellite of the Milky Way (about 10^{10} solar masses, i.e., 1 % of the Milky Way) at a distance of about 50 kpc from the Earth. As it was the first supernova observed in 1987, it was called SN1987A; it was also the first supernova since 1604 visible with the naked eye. The event was associated with the collapse of the star Sanduleak-69202, a main sequence star of mass about 20 solar masses.

No very high-energy gamma emission was detected (in 1987 gamma detectors were not operating), but gamma rays at the intermediate energies characteristic of gamma transitions could be recorded.

Three hours before the optical detection, a bunch of neutrinos was observed on Earth. SN1987A was the first (and unique, up to now) neutrino source other than the Sun: three water Cherenkov detectors, Kamiokande-II, the Irvine–Michigan–Brookhaven (IMB) experiment, and the Baksan detector observed 12, 8, and 5 neutrino interaction events, respectively, over a 13 s interval. This time interval is consistent with the estimated duration of a core collapse. The energy of neutrinos can be inferred from the energy of the recoil electrons to be in the tens of MeV range, consistent with the origin from a collapse.

SN1987A allowed also investigations on particle physics properties of neutrinos. The neutrino arrival time distribution sets an upper limit of 10 eV on the neutrino mass; the fact that they did not spread allows setting an upper limit on the magnetic moment $< 10^{-12} \mu_B$, where μ_B is the Bohr magneton. A determination of the neutrino velocity can also be derived, being consistent with the speed of light within two parts in 10^{-9}.

10.1.3.2 Very-High-Energy Neutrinos

IceCube reported during the last years the observation of a few neutrinos with deposited energy above 1 PeV and a population of several tens of events above 30 TeV; this significantly exceeds the expected atmospheric backgrounds (above 100 TeV the expected atmospheric neutrino background falls to the level of one event per year; the background above 30 TeV is expected to be about 40 %): the excess must be of astrophysical or cosmogenic origin.

To identify any bright neutrino sources in the data, IceCube passed the data on the detected neutrinos through a clustering algorithm (Fig. 10.23) as well as searching for directional correlations with the TeV gamma ray sources in Fig. 10.15. No significant evidence of clustering or correlations was found when comparing the significance with random datasets with the same statistics.

The overall high-energy spectrum is shown in (Fig. 10.24).

10.1.4 Gravitational Radiation

The graviton, a massless spin 2 particle (this condition is required by the fact that gravity is attractive only), is the proposed mediator of any field theory of gravity,

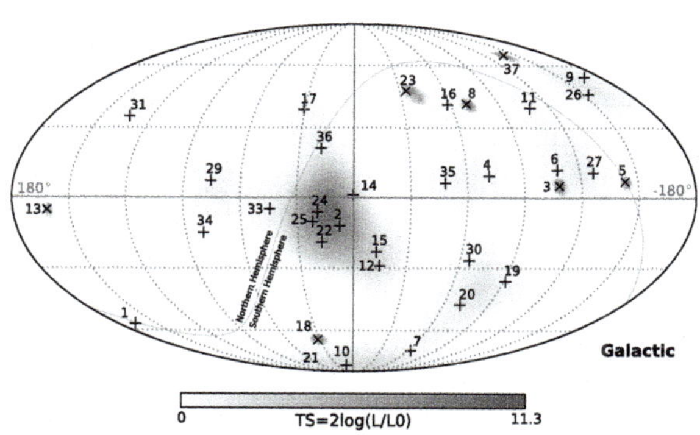

Fig. 10.23 Arrival directions of the IceCube events above 30 TeV in galactic coordinates (numbers correspond to the numbering in IceCube's catalog). Shower-like events (angular resolution ~15°) are marked with "+" and those containing muon tracks (angular resolution ~1°) with "x." The *gray line* denotes the equatorial plane. *Colors* show the test statistic (TS) for the point source clustering test at each location. No significant clustering was observed. Credit: IceCube collaboration

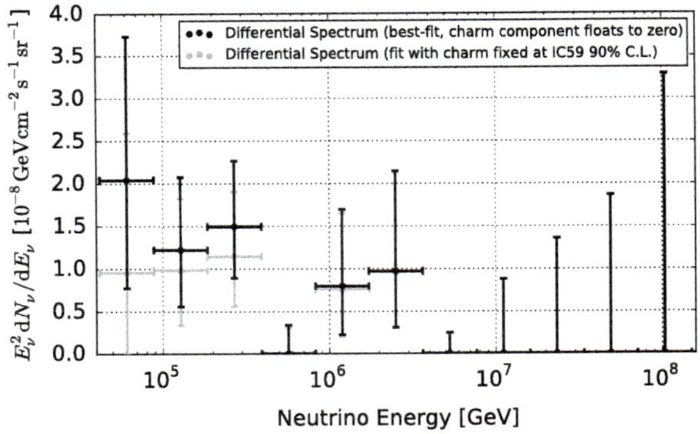

Fig. 10.24 Extraterrestrial neutrino flux ($\nu + \bar{\nu}$) as a function of energy, after subtraction of the background form atmospheric neutrinos (*gray lines* indicate an extreme model of background). Credit: IceCube collaboration

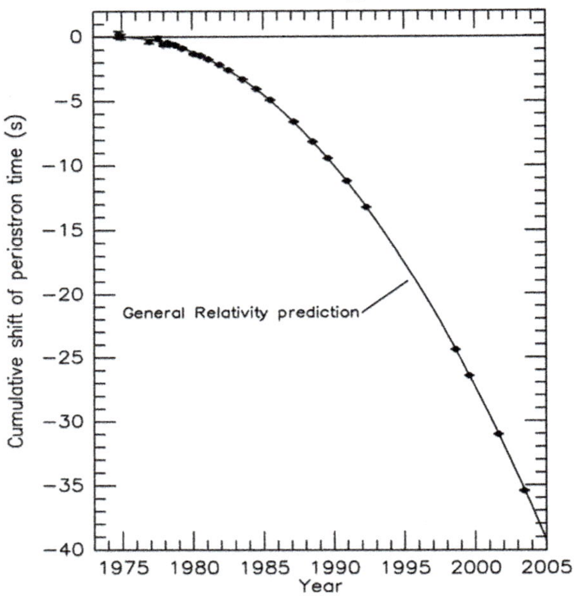

Fig. 10.25 Observed accumulated shifts of the times of periastron in the PSR 1913+16 compared with the general relativity prediction from gravitational radiation (from Joseph H. Taylor Jr.—©The Nobel Foundation 1993)

but has been not detected yet. Indeed the coupling of the graviton with matter is predicted to be extremely weak and thus its direct detection is extremely difficult.

However, indirect evidence of gravitational radiation was clearly demonstrated by Hulse and Taylor studying in 1974 the binary pulsar PSR 1913+16. In such system, composed by two neutron stars, it was possible to deduce, from the precise time of arrivals of the recorded pulses, all the binary orbital parameters and thus verify that the orbital period is decreasing in agreement with the prediction of Einstein general theory of relativity (about 40 s in 30 years, see Fig. 10.25). The total energy of the system is thus decreasing.

Gravitational waves must then exist, which propagate the lost energy; their detection will open a new and unique channel to observe the Universe—complementary to the lines described in the previous sections.

10.2 How Are High-Energy Cosmic Rays Produced?

At a scientific conference in 1933, Zwicky and Baade advanced a revolutionary conjecture: massive stars end their lives in explosions which blow them apart; such explosions produce cosmic rays, and leave behind a collapsed star made of densely packed neutrons. It took 16 years before Fermi could devise a model in which this conjecture could be explained; indeed we are convinced nowadays that most of the accelerators of cosmic rays in our galaxy are, indeed, SNR. But how can a supernova remnant (or whatever remnant of a gravitational collapse) accelerate particles?

10.2.1 Acceleration of Charged Cosmic Rays

Charged cosmic rays produced by particle ejection in the several possible astrophysical sources may be accelerated in the regions of space with strong turbulent magnetic fields. Permanent magnetic fields are not a good candidate since they cannot accelerate particles, static electric fields would be quickly neutralized, while variable magnetic fields may induce variable electrical fields and thus accelerate, provided the particles are submitted to many acceleration cycles.

In 1949 Fermi proposed a mechanism in which particles could be accelerated in stochastic collisions; this mechanism could model acceleration in shock waves which can be associated to the remnant of a gravitational collapse (for example, a stellar collapse, but also, as we know today, the surrounding of a black hole accreted in the center of a galaxy). Let us suppose (see Fig. 10.26) a charged particle with energy E_1 (velocity v) in the "laboratory" frame scattering inside a shock wave, i.e., a moving boundary between regions of different density.

The cloud has a velocity $\beta = V/c$, and θ_1 and θ_2 are the angles between, respectively, the initial and final particle momentum and the cloud velocity.

The energy of the particle E_1^* in the cloud reference frame is given by (neglecting the particle mass with respect to its kinetic energy):

$$E_1^* = \gamma E_1 (1 - \beta \cos\theta_1) .$$

Fig. 10.26 Scattering of a cosmic ray in an interstellar cloud

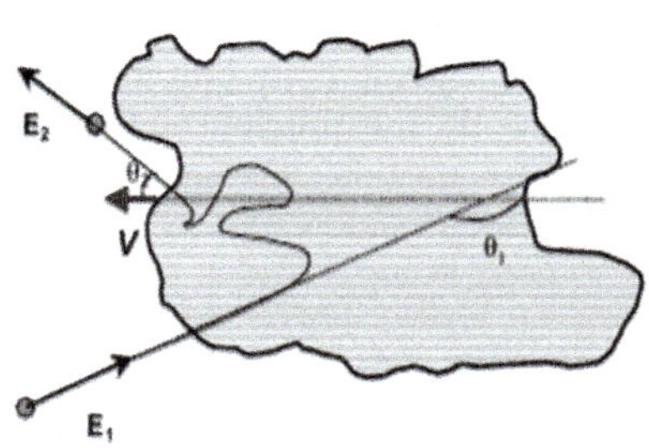

In the cloud reference frame $E_2^* = E_1^*$ (collisions to a wall!) and in the laboratory frame the energy of the particle after the collision is:

$$E_2 = \gamma E_2^*(1 + \beta \cos\theta_2) \,.$$

Thus the relative energy change is given by:

$$\frac{\Delta E}{E} = \frac{1 - \beta\cos\theta_1 + \beta\cos\theta_2 - \beta^2\cos\theta_1\cos\theta_2}{1-\beta^2} - 1 \,. \tag{10.2}$$

The particle suffers a great number of collisions inside the cloud so its output angle is basically random, it does not conserve the memory of the input direction. Then

$$\langle\cos\theta_2\rangle = 0$$

and

$$\left\langle\frac{\Delta E}{E}\right\rangle = \frac{1 - \beta\langle\cos\theta_1\rangle}{1-\beta^2} - 1 \,.$$

The probability P of a collision between the cosmic ray and the cloud is not constant as a function of the relative angle θ_1; it is rather proportional to their relative velocity (it is more probable that a particle hits a cloud that is coming against it, that it hits a cloud that it is running away from it).

$$P \propto (v - V\cos\theta_1) \simeq (1 - \beta\cos\theta_1)$$

and thus

$$\langle\cos\theta_1\rangle = \frac{\int_{-1}^{1}\cos\theta_1(1-\beta\cos\theta_1)d\cos\theta_1}{\int_{-1}^{1}(1-\beta\cos\theta_1)d\cos\theta_1} = -\frac{\beta}{3} \,.$$

The average energy increases then by a factor

$$\left\langle\frac{\Delta E}{E}\right\rangle \simeq \frac{4}{3}\beta^2 \,. \tag{10.3}$$

This mechanism is known as the second-order Fermi acceleration mechanism. Its energy gain is quadratic in the cloud velocity and, as this velocity is usually small ($\beta \simeq 10^{-4}$), it is not able to explain the cosmic ray energy spectrum.

In the presence of shock waves a Diffusive Shock Acceleration (DSA) occur turning the energy gain linear in β (1st order Fermi mechanism). What changes here is that, rather than assuming the directions for the shock wave to be randomly distributed, we assume that they are strongly correlated—in the limit, a "plane wave."

Let us suppose a shock wave originated, for instance, in a supernova explosion (see Fig. 10.27) with supersonic speed, creating a high-density region (compression factors ~4) propagating with a locally plane wave front.

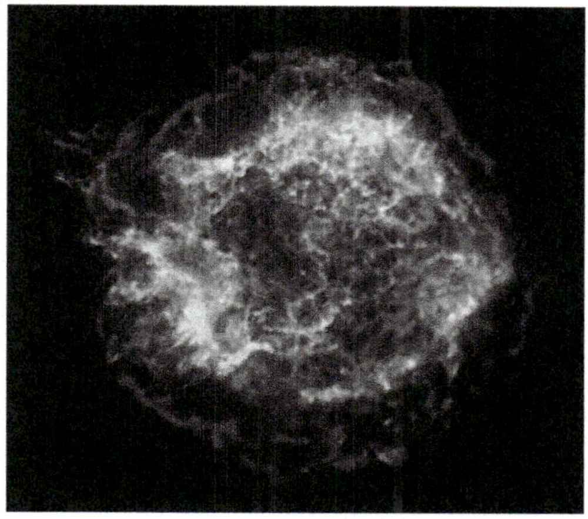

Fig. 10.27 X-ray image of the Cassiopeia A Supernova remnant. Credits NASA

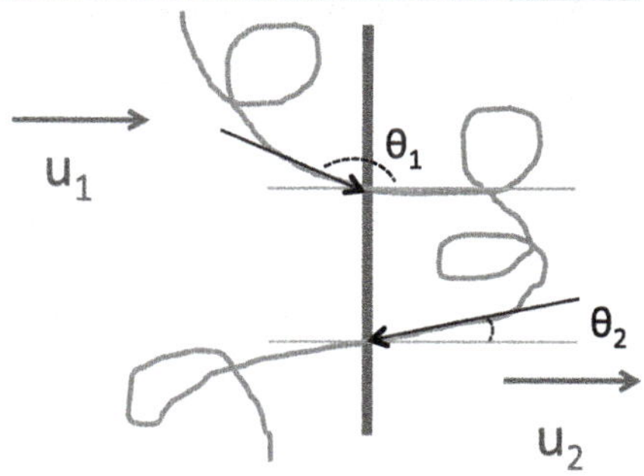

Fig. 10.28 Cosmic ray acceleration in the shock wave rest frame

In the shock wave rest frame the medium ahead of the shock (upstream) runs into the shock front with a velocity $\overrightarrow{u_1}$, while the shocked gas (downstream) moves away with a velocity $\overrightarrow{u_2}$ (Fig. 10.28). Thus in the laboratory frame a particle coming from upstream to downstream will meet a high-density magnetized gas travelling with $V = \overrightarrow{u_1} - \overrightarrow{u_2}$ in a head-on collision. The particle may suffer a great number of elastic scatterings inside this medium (due to the turbulent magnetic field) and invert the direction of its initial velocity crossing multiple times the shock front.

The angle $\theta_1(\theta_2)$ between the particle initial (final) velocity and the shock velocity (see Fig. 10.26) is now constrained to this specific geometry: $-1 \leq \cos\theta_1 \leq 0$ $(0 \leq \cos\theta_2 \leq 1)$; on the other hand the probability of crossing the wave front is proportional to $\cos\theta_1$ $(\cos\theta_2)$.

The mean values are

$$\langle\cos\theta_1\rangle = \frac{\int_{-1}^{0}\cos^2\theta_1 d\cos\theta_1}{\int_{-1}^{0}\cos\theta_1 d\cos\theta_1} = -\frac{2}{3}$$

$$\langle\cos\theta_2\rangle = \frac{\int_{0}^{1}\cos^2\theta_2 d\cos\theta_2}{\int_{0}^{1}\cos\theta_2 d\cos\theta_2} = \frac{2}{3}\,.$$

Thus reminding that

$$\frac{\Delta E}{E} = \frac{1 - \beta \cos\theta_1 + \beta \cos\theta_2 - \beta^2 \cos\theta_1 \cos\theta_2}{1 - \beta^2} - 1$$

the energy gain is given by:

$$\frac{\Delta E}{E} = \epsilon \simeq \frac{4}{3}\beta \,. \tag{10.4}$$

The turbulent magnetic fields in each side of the wave front will make the particle cross the wave front back and forth many times.

In each cycle the particle gains then a small fraction of energy ϵ. After n cycles the energy of the particle is:

$$E_n = E_0(1+\epsilon)^n$$

or the number of cycles needed to a particle attains a given energy E is:

$$n = \ln\left(\frac{E}{E_0}\right) / \ln(1+\epsilon)\,. \tag{10.5}$$

On the other hand a particle may escape from the shock region with some probability P_i (which is proportional to the velocity V) and then the probability P_{E_n} that a particle escapes from the shock region with an energy greater or equal to E_n is:

$$P_{E_n} = P_i \sum_{j=n}^{\infty} (1 - P_j)^n = (1 - P_i)^n\,. \tag{10.6}$$

Replacing n by the formula deduced above one has:

$$P_{E_n} = (1 - P_i)^{\ln\left(\frac{E}{E_0}\right)/\ln(1+\epsilon)}$$

$$\ln P_{E_n} = \frac{\ln\left(\frac{E}{E_0}\right)}{\ln(1+\epsilon)} \ln(1 - P_i)$$

$$\ln P_{E_n} = \frac{\ln(1 - P_i)}{\ln(1+\epsilon)} \ln(\frac{E}{E_0})\,.$$

Then

$$\frac{N}{N_0} = P_{E_n} = (\frac{E}{E_0})^{-\alpha}$$

and

$$\frac{dN}{dE} \propto \left(\frac{E}{E_0}\right)^{-\gamma} \tag{10.7}$$

with

$$\alpha = -\frac{\ln(1-P_i)}{\ln(1+\epsilon)} \cong \frac{P_i}{\epsilon} \tag{10.8}$$

and

$$\gamma = \alpha + 1 \,. \tag{10.9}$$

The Fermi mechanism predicts then that the energy spectrum is a power law with an almost constant index (both ϵ and P_i are proportional to $\langle\beta\rangle$).

In the case of supersonic shock α is predicted to be around 1 ($\gamma \sim 2$). However, the detected spectrum at Earth should be steeper! In its long journey from the Galactic sources to the Earth the probability that the particles escapes from the Galaxy is proportional to its energy (see Sect. 10.4.1):

$$\left|\frac{dN}{dE}\right|_{\text{Earth}} \propto \left(\frac{dN}{dE}\right)_{\text{sources}} \times E^{-\delta} \propto \left(\frac{E}{E_0}\right)^{-\gamma-\delta} .$$

Indeed the experimental values for the observed spectral index, as it was discussed in Sect. 10.1.1, are between 2.7 and 3.3. From this, γ is measured to be between 2.0 and 2.3.

SNR through Fermi first-order acceleration mechanisms are commonly recognized nowadays as responsible for most of the high-energy cosmic rays. However, the proof that this mechanism can accelerate cosmic rays all the way up to the knee region is still missing.

10.2.1.1 The Knee

The maximum energy that a charged particle could achieve in the supernova remnant is then simply the rate of energy gain, times the time T_S spent in the shock. In the Fermi first-order model,

$$\frac{dE}{dt} \simeq \beta \frac{E}{T_{\text{cycle}}} ,$$

where $T_{\text{cycle}} \simeq \lambda_{\text{cycle}}/(\beta c)$ is the time between two crossings. Since $\lambda_{\text{cycle}} \simeq r_L \simeq E/(ZeB)$ (r_L is the Larmor radius),

$$T_{\text{cycle}} \simeq \frac{E}{ZeB\beta c} \Longrightarrow \frac{dE}{dt} \simeq (\beta^2 c)ZeB \,.$$

Finally

$$E_{\text{max}} \simeq T_S \frac{dE}{dt} \simeq ZeBR_S\beta \,.$$

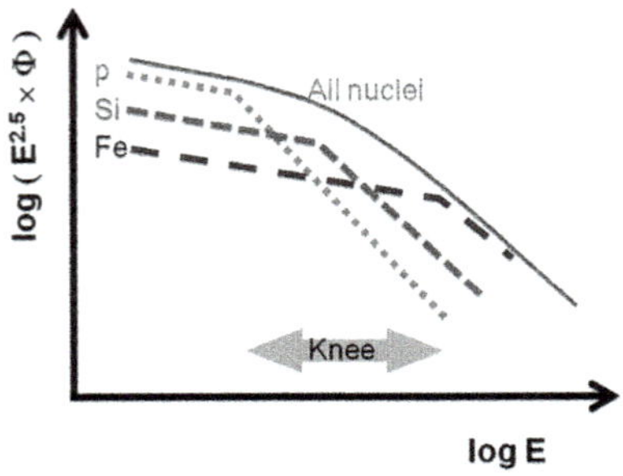

Fig. 10.29 The interpretation of the knee as due to the dependence of the maximum energy on the nuclear charge Z. The flux of each nuclear species sharply decreases after a given cutoff. The behavior of hydrogen, silicon ($Z = 14$) and iron ($Z = 26$) nuclei are depicted in figure. From [F10.1]

Inserting 4 μG as a typical value of the galactic magnetic field B, and assuming $T_S \simeq R_S/(\beta c)$, where R_S is the radius of the supernova remnant, we obtain:

$$E_{\max} \simeq \beta\, Ze\, B\, R_S \simeq 300\, Z\, \text{TeV}\,. \tag{10.10}$$

The shock acceleration of interstellar particles in SNR explains the spectrum of cosmic ray protons up to few hundreds of TeV, close to the region where the knee begins.

An important consequence of (10.10) is that the maximum energy is proportional to the charge Z of the ion, and it is thus higher for multiply ionized nuclei with respect to a single-charged proton. Thus, in this model, the knee is explained as a structure due to the different maximum energy reached by nuclei with different charge Z (Fig. 10.29).

10.2.2 Production of High-Energy Gamma Rays

High-energy photons can be produced by the interaction of high-energy charged particles (for example, electrons, protons, ions accelerated by the shock waves of remnants of gravitational collapses) with nuclear targets such as molecular clouds or radiation fields (magnetic fields, photon fields). An alternative could be the production via the decay of heavy particles. We distinguish between purely leptonic mechanisms of production and models in which photons are secondary products of hadronic interactions; the latter provide a direct link between high-energy photon production and the acceleration of charged cosmic rays.

10.2.2.1 Leptonic Models

Being neutral, photons cannot be directly accelerated; however, mechanisms exist such that photons of rather large energies are radiated. We examine in this subsection radiation processes just involving leptons (they are called "leptonic" mechanisms of photoproduction). In particular, we shall sketch the simplest self-sustaining acceleration mechanism, the synchrotron self-compton in one region.

Synchrotron Radiation. High-energy photon emission in the beginning is generally due to synchrotron radiation. The dynamics of charged particles is strongly influenced by the magnetic fields present is astrophysical environments through the Lorentz force. Accelerated relativistic electrons radiate synchrotron photons; the power loss can be expressed as

$$-\frac{dE}{dt} = \frac{e^4 B_\perp^2}{6\pi\epsilon_0 \mathrm{cm}^2}\beta^2\gamma^2 \tag{10.11}$$

(since in the rest frame of the particle the acceleration is small, we can use the nonrelativistic expression for the radiation rate).

It is immediately evident from Eq. 10.11 that synchrotron energy loss is by far more important for electrons than for protons.

Compton scattering and "inverse Compton." The Compton scattering of a photon by an electron is a relativistic effect, by which the frequency of a photon changes due to a scattering. In the scattering of a photon by an electron at rest, the wavelength shift of the photon can be expressed as

$$\frac{\lambda' - \lambda}{\lambda} = \frac{\hbar\omega}{m_e c^2}(1 - \cos\alpha)\,,$$

where α is the angle of the photon after the collision with respect to its line of flight. As evident from the equation and from the physics of the problem, the energy of the scattered photon always decreases.

However, when low-energy photons collide with high-energy electrons instead than with electrons at rest their energy can increase: such process is called inverse Compton (IC) scattering. This mechanism is very effective for increasing the photon energy (for this reason it is called "inverse"), and is important in regions of high soft-photon energy density and energetic electron number density.

Synchrotron Self-Compton. The simplest purely leptonic mechanism we can draw for photon "accelerate"—a mechanism we have seen at work in astrophysical objects—is the so-called self-synchrotron Compton (SSC) mechanism. The SSC is a purely leptonic mechanism, in which ultrarelativistic electrons accelerated in a magnetic field—such as the field present in the accretion region of AGN, or in the surrounding of SNR—generate synchrotron photons. The typical values of the fields involved are such that the synchrotron photons have an energy spectrum peaked in the infrared/X-ray range. Such photons in turn interact via Compton scattering with

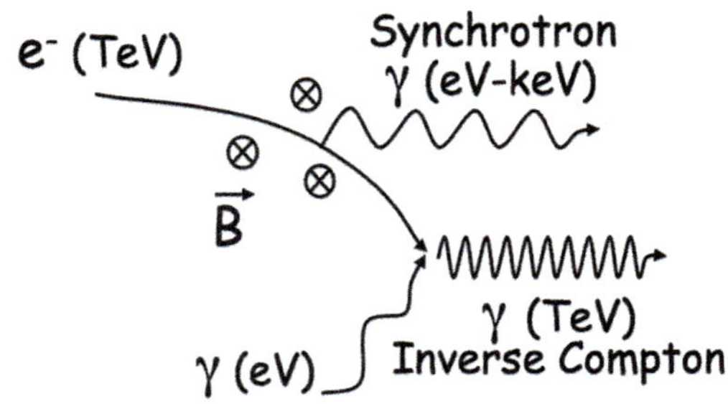

Fig. 10.30 Scheme of the SSC mechanism

their own parent electron population (Fig. 10.30); since electrons are ultrarelativistic (with a Lorentz factor $\gamma_e \sim 10^{4-5}$), the energy of the upscattered photon can be boosted by a large factor.

For a power law population of relativistic electrons with a differential spectral index q and a blackbody population of soft photons at a temperature T, mean photon energies and energy distributions can be calculated for electron energies in the Thomson regime and in the relativistic Klein-Nishina regime:

$$\langle E_\gamma \rangle \simeq \tfrac{4}{3}\gamma_e^2 \langle \eta \rangle \quad \text{for } \gamma_e \eta \ll m_e c^2 \text{ (Thomson limit)} \tag{10.12}$$

$$\simeq \tfrac{1}{2}\langle E_e \rangle \quad \text{for } \gamma_e \eta \gg m_e c^2 \text{ (Klein-Nishina limit)} \tag{10.13}$$

$$\frac{dN_\gamma}{dE_\gamma} \propto E_\gamma^{-\frac{q+1}{2}} \quad \text{for } \gamma_e \eta \ll m_e c^2 \text{ (Thomson limit)} \tag{10.14}$$

$$\propto E_\gamma^{-(q+1)} \ln(E_\gamma) \quad \text{for } \gamma_e \eta \gg m_e c^2 \text{ (Klein-Nishina limit)} \tag{10.15}$$

where E_γ denotes the scattered photon's energy, E_e denotes the energy of the parent electron, and η denotes the energy of the seed photon. A useful approximate relation linking the electron's energy and the Comptonised photon's energy is given by:

$$E_\gamma \simeq 6.5 \left(\frac{E_e}{\text{TeV}}\right)^2 \left(\frac{\eta}{\text{meV}}\right) \text{GeV} .$$

The Compton component can peak at GeV–TeV energies; the two characteristic synchrotron and Compton peaks are clearly visible on top of a general E_γ^{-2} dependence. Figure 10.31 shows the resulting energy spectrum. This behavior has been verified with high accuracy on the Crab Nebula, see Fig. 10.16. If in a given region the photons from synchrotron radiation can be described by a power law with spectral index p, in the first approximation the tails at the highest energies from both the synchrotron and the Compton mechanisms will have a spectral index p.

A cornerstone characteristics of the SSC model is a definite correlation between the yields from synchrotron radiation and from IC during a flare (it would be difficult to accommodate in the theory an "orphan flare," i.e., a flare in the IC region not accompanied by a flare in the synchrotron region). Although most of the flaring activities occur almost simultaneously with TeV gamma ray and X-ray fluxes, observations of

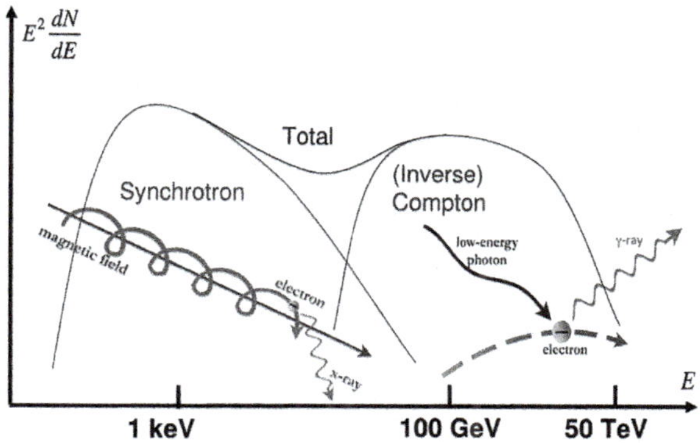

Fig. 10.31 Differential energy spectrum of photons in the SSC model

1ES 1959+650 and other AGN have exhibited VHE gamma ray flares without their counterparts in X-rays. Flares observed in VHE gamma rays with absence of high activity in X-rays are very difficult to reconcile with the standard SSC, although this model has been very successful in explaining the SED of blazars.

10.2.2.2 Hadronic Model

Alternative and complementary models of VHE emission involve two populations: one primary population of photons coming from the synchrotron radiation or the bremsstrahlung of electrons accelerated, and the other generated by cascades initiated by primary protons/nuclei that had been accelerated in the system. The beam of accelerated hadrons collides with a target of nucleons (for example, a molecular cloud) or by a sea of photons (hadronic photoproduction).

In both cases, the energy of the primary protons is expected by the physics of hadronic cascades to be one or two orders of magnitude larger than the energy of gamma rays, since the dominant mechanism for photon production is the decay of the secondary π^0 mesons into $\gamma\gamma$ pairs at the end of the hadronic cascade. The study of γ rays can thus provide insights on the production of charged cosmic rays.

Photons coming from π^0 decay have energies larger than photons from synchrotron radiation. In the rest frame of the neutral pion, both photons have energy $E_\gamma = m_\pi c^2/2 \simeq 70$ MeV and momentum opposite to each other. Once boosted for the energy of the emitting π^0, the probability to emit a photon of energy E_γ is constant over the range of kinematically allowed energies (see the Exercise 3).

When an energy spectrum of accelerated protons at the source decreasing as E^{-p} (with p typically about 2) is considered, due to the Lorentz boost the decay spectrum for an observer in the laboratory frame drops steeply below about 100–200 MeV and the energy dependence is similar to that of the parent protons at energies beyond GeV. A bump corresponding to pion decay, rising up to 70 MeV and then decreasing (possibly extending to energies of the GeV due to the Lorentz boost) identifies the presence of gamma rays of hadronic origin.

The existence of a hadronic component has been demonstrated from the experimental data (see later) and could explain the production of hadrons up to almost the knee.

Since neutrinos cannot be absorbed nor radiated, the neutrino would be a unique tracer of a hadronic cascade.

10.2.2.3 Top-Down Mechanisms

Finally, top-down mechanisms might be at the origin of photons of extremely high energies.

Many believe that at the highest energy cosmic rays are the decay products of remnant particles or topological structures created in the early universe. In these scenarios the highest energy cosmic rays are predominantly photons. A topological defect from a phase transition in grand unified theories with typical energy scale of 10^{24} eV will suffer a chain decay into GUT particles X and Y that subsequently decay to known particles; in the long term the number of neutral pions (decaying into photons) is two orders of magnitude larger than the number of protons. Therefore, if the decay of topological defects is the source of the highest energy cosmic rays, they must be photons—and neutrinos, which are more difficult to detect.

In the VHE region, photons might come from the decay of heavier particles (dark matter particles for example), or from blobs of energy coming from the annihilation of pairs of such particles. Experimental data collected up to now do not support the existence of such mechanisms—which are anyway searched for actively, since they might shed light on new physics.

10.2.3 Production of High-Energy Neutrinos

VHE neutrinos can come from the same sources as VHE gamma rays, but not from purely leptonic mechanisms: top-down mechanisms and hadronic cascades are thus neutrino sources. As a consequence, evidence of neutrinos in an accelerator would be a smoking gun for the presence of a hadronic component among accelerated particles—or of new physics.

10.3 Possible Acceleration Sites and Sources

In Sect. 10.2.1 we explained how a particle can be accelerated. In which astrophysical objects such acceleration process can take place?

In order to effectively accelerate a particle, this must be contained in a region of size R such that the magnetic field and the energy must be in a relation

Table 10.1 Typical values of radii and magnetic fields in acceleration sites, and the maximum attainable energy

Source	Magnetic field	Radius	Maximum energy (eV)
SNR	30 μG	1 pc	3×10^{16}
AGN	300 μG	10^4 pc	$>10^{21}$
GRB	10^9 G	10^{-3} AU	0.2×10^{21}

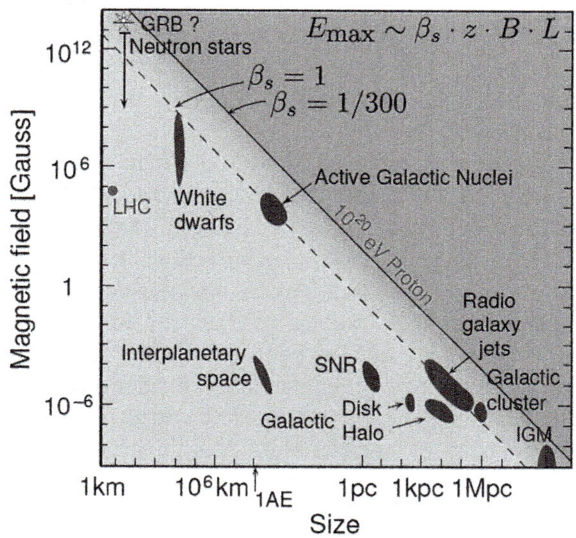

Fig. 10.32 The "Hillas plot" represents astrophysical objects which are potential cosmic ray accelerators on a two-dimensional diagram where on the *horizontal* direction the size linear extension R of the accelerator, and on the vertical direction the magnetic field strength B are plotted. The maximal acceleration energy E is proportional to $ZRBv$, where v is the shock velocity in units of the speed of light and Z is the absolute value of the particle charge in units of the electron charge. Particular values for the maximal energy correspond to *diagonal lines* in this diagram and can be realized either in a large, low field acceleration region or in a compact accelerator with high magnetic fields. Typical shock velocities go from $v \sim 1$ in extreme environments down to $v \sim 1/300$. From http://astro.uni-wuppertal.de/~kampert

$$\frac{E}{1\,\mathrm{PeV}} \simeq \frac{B}{1\,\mu\mathrm{G}} \times \frac{R}{1\,\mathrm{pc}} \tag{10.16}$$

$$\frac{E}{1\,\mathrm{PeV}} \simeq 0.2\frac{B}{1\,\mathrm{G}} \times \frac{R}{1\,\mathrm{AU}}\,. \tag{10.17}$$

This entails the so-called Hillas relation, which is illustrated by (Table 10.1) and Fig. 10.32. We remind that the energies in the Hillas plot are maximum attainable energies: besides the containment, one must have an acceleration mechanism.

In the following, known possible acceleration sites are described.

10.3.1 Stellar Endproducts

We have seen that most of the VHE gamma emission in the Galaxy can be associated to SNR. More than 90 % of the TeV galactic sources discovered up to now are, indeed, SNR in a large sense (we include here in the set of "SNR" at large also pulsar wind nebulae, see later).

The term "supernova" indicates a very energetic "stella nova," a term invented by Galileo Galilei to indicate objects that appeared to be new stars, that had not been observed before in the sky. The name is a bit ironic, since Galilei's diagnosis was wrong: supernovae are actually stars at the end of their life cycle, exploding with a collapse of their nuclei.

In the beginning, a massive star burns the hydrogen in its core. When the hydrogen is exhausted, the core contracts until the density and temperature conditions are reached such that the fusion $3\alpha \rightarrow {}^{12}\text{C}$ can take place, which continues until helium is exhausted. This pattern (fuel exhaustion, contraction, heating, and ignition of the ashes of the previous cycle) might repeat several times depending on the mass, leading finally to an explosive burning. A 25-solar mass star can go through a set of burning cycles ending up in the burning of Si to Fe in about 7 My (as discussed in Chap. 1, Fe is stable with respect to fusion), with the final explosion stage taking a few days.

A supernova remnant (SNR) is the structure left over after a supernova explosion: a high-density neutron star (or a black hole) lies at the center of the exploded star, whereas the ejecta appear as an expanding bubble of hot gas that shocks and sweeps up the interstellar medium. A star with mass larger than 1.4 times the mass of the Sun cannot die into a white dwarf and will collapse; it will become a neutron star or possibly, if its mass is larger than 5–10 times the mass of the Sun, into a black hole. Most collapsing stars are expected to produce a neutron star.

10.3.1.1 Neutron Stars; Pulsars

When a star collapses into a neutron star, its size shrinks to some 10–20 km, with a density of 5×10^{17} kg/m^3. Since angular momentum is conserved, the rotation can become very fast, with periods of the order of a few ms up to 1 s. Neutron stars in young SNRs are typically pulsars (short for pulsating stars), i.e., they emit a pulsed beam of electromagnetic radiation. Since the magnetic axis is in general not aligned to the rotation axis, two peaks corresponding to each of the magnetic poles can be seen for each period (Fig. 10.33).

The millisecond rotating period for young pulsars can be estimated using basic physics arguments. A star like our Sun has a radius $R \sim 7 \times 10^5$ km and a rotation period of $T \simeq 30$ days, so that the angular velocity is $\omega \sim 2.5 \times$ µrad/s. After the collapse, the neutron star has a radius $R_{NS} \sim 10$ km. From angular momentum conservation, one can write:

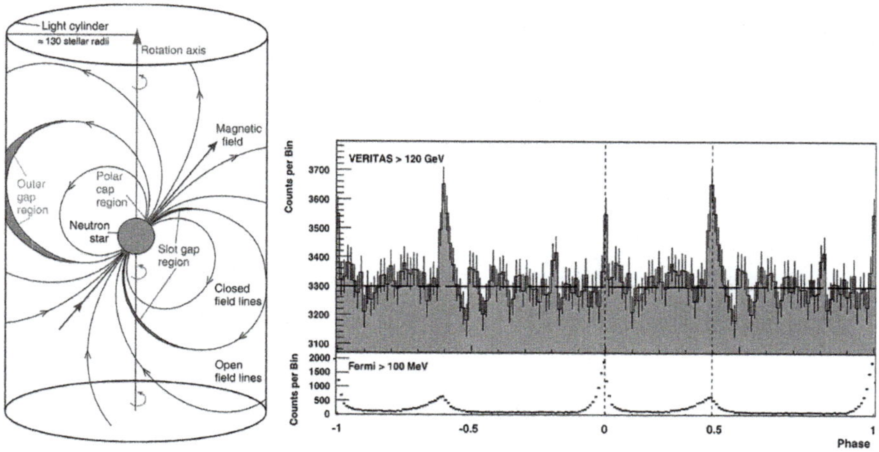

Fig. 10.33 *Left* Schematic of the Crab Pulsar. Electrons are trapped and accelerated along the magnetic field lines of the pulsar and can emit electromagnetic synchrotron radiation. Vacuum gaps or vacuum regions occur at the "polar cap" close to the neutron star surface and in the outer region; in these regions density varies and thus one can have acceleration. From MAGIC Collaboration, Science 322 (2008) 1221. *Right* Time-resolved emission from the Crab Pulsar at HE and VHE; the period is about 33 ms. From VERITAS Collaboration, Science 334 (2011) 69

$$R^2\omega \sim R_{NS}^2\omega_{NS} \Longrightarrow \omega_{NS} = \omega\frac{R^2}{R_{NS}^2} \Longrightarrow T_{NS} \simeq 0.5\,\text{ms}\,.$$

The gravitational collapse amplifies the stellar magnetic field. As a result, the magnetic field B_{NS} near the NS surface is extremely high. To obtain an estimate of its magnitude, let us use the conservation of the magnetic flux during the contraction. Assuming the magnetic field to be approximately constant over the surface,

$$B_{\text{star}}R^2 = B_{NS}R_{NS}^2 \Longrightarrow B_{NS} = B_{\text{star}}\frac{R^2}{R_{NS}^2}\,.$$

For a typical value of $B_{\text{star}} = 1$ kG, the magnetic fields on the surface of the neutron star is about 10^{12} G. This estimate has been experimentally confirmed by measuring quantised energy levels of free electrons in the pulsar strong magnetic fields.

Typical pulsars emitting high-energy radiation have cutoffs of the order of a few GeV. 117 HE pulsars emitting at energies above 100 MeV have been discovered by the Fermi-LAT until 2013. They are very close to the Solar System (Fig. 10.34, left), most of the ones for which the distance has been measured being less that a few kpc away. A typical spectral energy distribution is shown in Fig. 10.34, right.

The pulsar in Crab Nebula is not typical, being one of the two (together with the Vela pulsar) detected up to now in VHE (Fig. 10.35)—Crab and Vela were also the first HE pulsars discovered in the late 1970s.

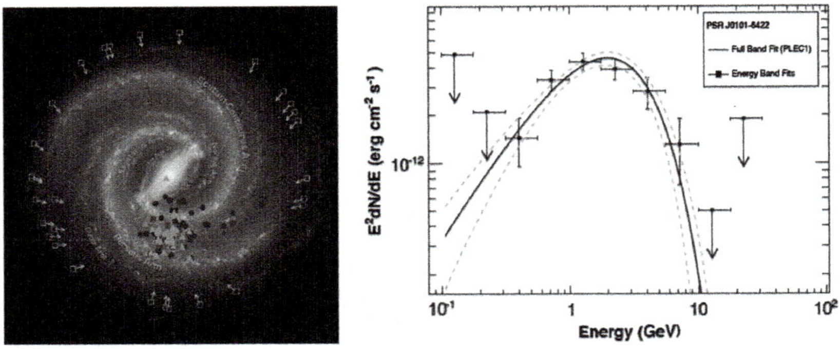

Fig. 10.34 *Left* Map of the pulsars detected by the Fermi-LAT (the Sun is roughly in the center of the distribution). The *open squares* with *arrows* indicate the *lines* of sight toward pulsars for which no distance estimates exist. Credit: NASA. *Right* Spectral energy distribution from a typical high-energy pulsar. Credit: NASA

10.3.1.2 Supernova Remnants

Typical velocities for the expulsion of the material out of the core of the explosion are of the order of $c/100$. The shock slows down over time as it sweeps up the ambient medium, but it can expand over tens of thousands of years and over tens of parsecs before its speed falls below the local sound speed.[3]

Based on their emission and morphology (which are actually related), SNRs are generally classified under three categories: shell-type, pulsar wind nebulae (PWN), and composite (a combination of the former, i.e., a shell-type SNR containing a PWN).

Shell Supernova Remnants. As the shockwave from the supernova explosion plows through space, it heats and stirs up any interstellar material it encounters, which produces a big shell of hot material in space. This process can continue up to 10^4–10^5 years before the energy release becomes negligible.

Magnetic field strengths are estimated to be $B \sim 10\,\mu$G to 1 mG.

Pulsar Wind Nebulae. Pulsar Wind Nebulae (PWNe) are SNR with a young pulsar slowing down in rotation: the typical rate of decrease of kinetic energy lies in the range $\dot{E} \sim 10^{32}$–10^{39} erg/s. In most cases only a negligible fraction of this energy goes into the pulsed electromagnetic radiation observed from the pulsar, whereas most of it is deposited into an outflowing relativistic magnetized particle wind: due to external pressure, this wind decelerates abruptly at a termination shock, and beyond this the pulsar wind thermalizes and radiates synchrotron photons which might possibly undergo IC scattering—so resulting in a PWN.

The best known case of a PWN is the Crab Nebula, powered by the central young ($\sim$1000 year) pulsar B0531+21.

---

[3] The speed of shock is the speed at which pressure waves propagate, and thus it determines the rate at which disturbances can propagate in the medium, i.e., the "cone of causality."

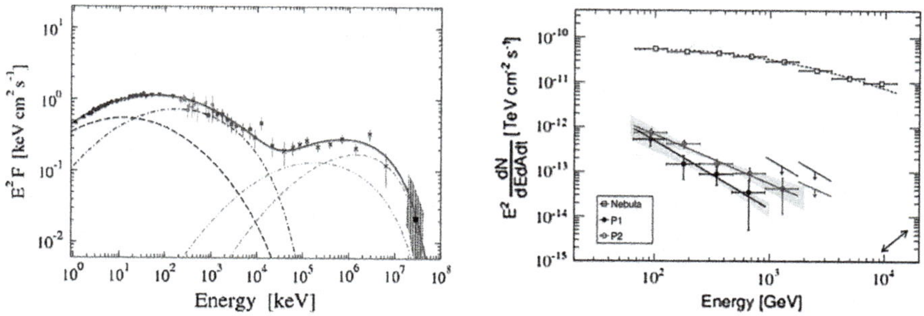

Fig. 10.35 *Left* Spectral energy distribution of the Crab Pulsar. *Right* The VHE energy emission, compared with the emission from the pulsar wind nebula powered by the pulsar itself (see later). The two periodical peaks are separated. Credits: MAGIC Collaboration

Crab Nebula emits radiation across a large part of the electromagnetic spectrum, as seen in Fig. 10.16.

One can separate the contribution of the pulsar itself to the photon radiation from the contribution of the PWN (Fig. 10.35).

In the early stages of a PWN's evolution, the pulsar spindown luminosity provides a steady source of energy to the PWN which then expands supersonically into the surrounding low-density medium—the PWN radius evolving as $\propto t^{6/5}$. Eventually, inside the PWN volume there will be two zones:

- an inner one dominated by the freely outgoing particles and fields that ends at a termination shock where the wind, whose pressure decreases outwards, is contained by external pressure and the particles are accelerated and get scrambled so that synchrotron radiation can be emitted,
- an outer zone which is the emitting region of the PWN.

The morphology of a PWN in this early phase is revealed by the X-ray image of the Crab Nebula, which is dominated by a bright torus, whose ring-shaped inner part is thought to correspond to the wind termination shock.

A PWN enters a new phase when its outer boundary collides with the reverse shock from the embedding SNR. For a 10^{51} erg supernova explosion ejecting 10 $M_\odot$ into an ambient interstellar medium of density 1 atom cm^{-3} (typical case in the Galaxy), this typically happens ~7000 year after the explosion. The encounter with the reverse shock eventually compresses the PWN, which leads to an increase in the magnetic field and hence to a brightening of the PWN.

10.3.2 Other Galactic Sources

Particle acceleration could be a more common phenomenon than indicated above, and be characteristic of many astrophysical objects. For example, TeV emission has been found from binary sources, like the binary system LS I +61 303, harboring a compact object and a massive star.

The likely mechanism is that leptonic sources of electromagnetic radiation (electrons) and probably also high-energy hadrons are produced by matter accreting onto the stellar companion. Due to the motions of ionized matter, very strong electromagnetic fields are produced in the vicinity of the compact object, and charged particles can be accelerated to high energies.

However, most of the Galactic emitters of VHE gamma rays are SNRs (at large: our definition includes PWNe). SNRs can, in principle, reach energies not larger than a few PeV, being limited by the product of radius time the magnetic field—see the Hillas plot. However, photons above a few tens of TeV have never been observed, and the question is if this is due to the limitation of present detectors.

Finally, the galactic zoo could be more varied: next generation detectors will tell if other emitters exist.

10.3.3 Extragalactic Sources

Among the extragalactic emitters that may be observed from Earth, AGN, and Gamma Ray burst could have the energetics to reach the highest energies.

10.3.3.1 Active Galactic Nuclei

Supermassive black holes of $\sim 10^6$–10^{10} solar masses ($M_\odot$) and beyond reside in the cores of most galaxies—for example, the center of our Galaxy, the Milky Way, hosts a black hole of roughly 4 million solar masses, its mass having been determined by the orbital motion of nearby stars. The mass of the black hole is correlated to the velocity dispersion of the stars in the galaxy bulge.[4]

In approximately 1 % of the cases such black hole is active, i.e., it displays strong emission and has signatures of accretion: we speak of an active galactic nucleus (AGN). Around 10 % of these AGN exhibit relativistic jets powered by accretion onto a supermassive BH. Despite the fact that AGN have been studied for several decades, the knowledge of the emission characteristics up to the highest photon energies is mandatory for an understanding of these extreme particle accelerators.

Infalling matter onto the black hole can produce a spectacular activity; when they are active, i.e., they are powered by accretion at the expenses of nearby matter, these black holes are called AGN.

[4]The so-called $M - \sigma$ relation is an empirical correlation between the stellar velocity dispersion σ of a galaxy bulge and the mass M of the SMBH at its center:

$$\frac{M}{10^8 M_\odot} \approx 1.9 \left(\frac{\sigma}{200\ \mathrm{km/s}}\right)^{5.1}$$

where $M_\odot$ is the solar mass. A relationship exists also between galaxy luminosity and BH mass, but with a larger scatter.

An AGN emits over a wide range of wavelengths from γ ray to radio: typical luminosities range from about 10^{37} to 10^{40} W. The energy spectrum of an AGN is radically different from an ordinary galaxy, whose emission is due to its constituent stars. Very large luminosities are possible (up to 10,000 times a typical galaxy). The maximum luminosity (in equilibrium conditions) is set by requirement that gravity (inward) is equal to radiation pressure (outward); this is called the Eddington luminosity. Approximately, the Eddington luminosity (in units of the solar luminosity) is 40,000 times the BH mass expressed in solar units. For short times, the luminosity can be larger than the Eddington luminosity.

Matter falling into the central black hole will conserve its angular momentum and will form a rotating accretion disk around the BH itself. In about 10 % of AGN, the infalling matter turns on powerful collimated jets that shoot out in opposite directions, likely perpendicular to the disk, at relativistic speeds (see Fig. 10.36). Jets have been observed close to the BH having a size of about 0.01 pc, orders of magnitude smaller than the radius of the black hole and a fraction 10^{-5} of the length of jets themselves.

Frictional effects within the disk raise the temperature to very high values, causing the emission of energetic radiation—the gravitational energy of infalling matter accounts for the power emitted. The typical values of the magnetic fields are of the order of 10^4 G close to the BH horizon, quickly decaying along the jet axis.

Many AGN vary substantially in brightness over very short timescales (days or even minutes). Since a source of light cannot vary in brightness on a timescale shorter than the time taken by light to cross it, the energy sources in AGN must be very compact, much smaller than their Schwarzschild radii—the Schwarzschild radius of the BH is $3\,\text{km} \times (M/M_\odot)$, i.e., 20 AU (about 10^4 light seconds) for a supermassive black hole mass of $10^9 M_\odot$.

The so-called "unified model" accounts for all kinds of active galaxies within the same basic model. The supermassive black hole and its inner accretion disk are surrounded by matter in a toroidal shape, and according to the "unified model" the type of active galaxy we see depends on the orientation of the torus and jets relative to our line of sight. The jet radiates mostly along its axis, also due to the Lorentz enhancement—the observed photon frequency in the laboratory frame is boosted by a Doppler factor which is obtained by the Lorentz transformation of a photon from the jet fluid frame into the laboratory frame; in addition, the jet is beamed by the Lorentz boost.

- An observer looking very close to the jet axis will observe essentially the emission from the jet, and thus will detect a (possibly variable) source with no spectral lines: this is called a blazar.
- As the angle of sight with respect to the jet grows, the observer will start seeing a compact source inside the torus; in this case we speak of a quasar.
- From a line of sight closer to the plane of the torus, the BH is hidden, and one observes essentially the jets (and thus, extended radio-emitting clouds); in this case, we speak of a radio galaxy (Fig. 10.36).

The class of jet dominated AGN corresponds mostly to radio-loud AGN. These can be blazars or nonaligned AGN depending on the orientation of their jets with

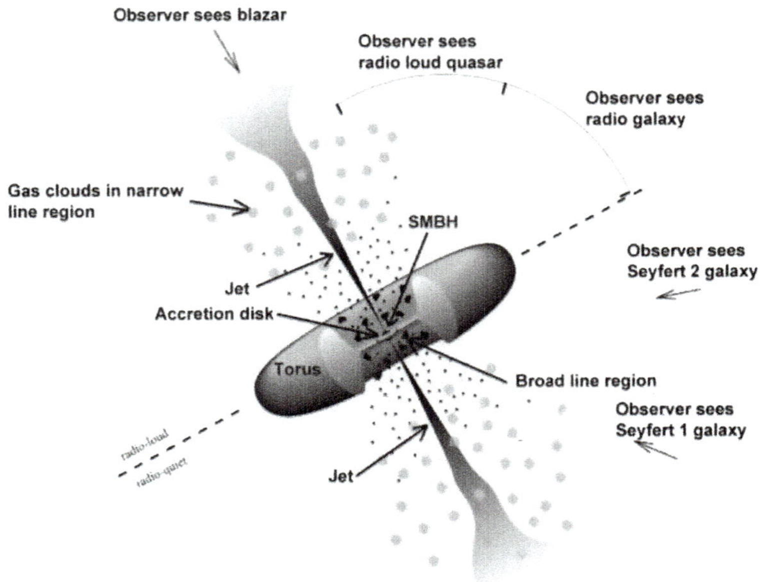

Fig. 10.36 Schematic diagram for the emission by an AGN. In the "unified model" of AGN, all share a common structure and only appear different to observers because of the angle at which they are viewed. From http://www.astro-photography.net, adapted

respect to the line of sight. In blazars, emission is modified by relativistic effects due to the Lorentz boost.

Blazars. Observationally, blazars are divided into two main subclasses depending on their spectral properties.

- FSRQs. They show broad emission lines in their optical spectrum.
- BL Lacertae objects (BL Lacs). They have no strong, broad lines in their optical spectrum. BL Lacs are moreover classified according to the energies of the peaks of their SED; they are called accordingly low-energy peaked BL Lacs (LBLs), intermediate-energy peaked BL Lacs (IBL) and high-energy peaked BL Lacs (HBL). Typically FSRQs have a synchrotron peak at lower energies than LBLs.

Blazar population studies at radio to X-ray frequencies indicate a redshift distribution for BL Lacs that seems to peak at $z \sim 0.3$, with only few sources beyond $z \sim 0.8$, while the FSRQ population is characterized by a rather broad maximum between $z \sim 0.6$–1.5.

Non-AGN Extra Galactic Gamma Ray Sources. At TeV energies, the extragalactic γ ray sky is completely dominated by blazars. At present, more than 50 objects have been discovered and are listed in the online TeV Catalog. The two most massive close by starburst galaxies NGC 253 and M82 are the only non-AGN sources detected at TeV energies. Only 3 radio galaxies have been detected at TeV energies (Centaurus A, M87 and NGC 1275).

At GeV energies, a significant number (about 1/3) of AGN of uncertain type has been detected by the Fermi-LAT (emitters that Fermi could not associate to any known object) and few non-AGN objects have been discovered. Among non-AGN objects, there are several local group galaxies (LMC, SMC, M31) as well as galaxies in the star formation phase galaxies (NGC 4945, NGC 1068, NGC 253, and M82)—CRs below the knee can be accelerated by SNRs or other objects that are related to star formation activity.

The Gamma Ray Yield From Extragalactic Objects. The observed VHE spectra at high energies are usually described by a power law $dN/dE \propto E^{-\Gamma}$. The spectral indices Γ need to be fitted from a distribution deconvoluted from absorption in the Universe, since the transparency of the Universe depends on energy; they typically range in the interval from 2 to 4, with some indications for spectral hardening with increasing activity. Emission beyond 10 TeV has been established, for close galaxies like Mrk 501 and Mrk 421. Some sources are usually detected during high states (flares) only, with low states falling below current sensitivities.

Observed VHE flux levels for extragalactic objects typically range from 1 % of the Crab Nebula steady flux (for the average/steady states) up to 10 times as much when the AGN are in high activity phases. Since TeV instruments are now able to detect sources at the level of 1 % of the Crab, the variability down to few minute scale of the near and bright TeV-emitting blazars (Mrk 421 and Mrk 501) can be studied in detail. Another consequence of the sensitivity of Cherenkov telescopes is that more than one extragalactic object could be visible in the same field of view.

The study and classification of AGN and their acceleration mechanisms require observations from different instruments. The spectral energy distributions (SEDs) of blazars can span almost 20 orders of magnitude in energy, making simultaneous multiwavelength observations a particularly important diagnostic tool to disentangle the underlying nonthermal processes. Often, SEDs of various objects are obtained using nonsimultaneous data—which limits the accuracy of our models.

In all cases, the overall shape of the SEDs exhibit the typical broad double-hump distribution, as shown in Fig. 10.37 for three AGN at different distances. The SEDs of all AGN considered show that there are considerable differences in the position of the peaks of the two components and in their relative intensities. According to current models, the low-energy hump is interpreted as due to synchrotron emission from highly relativistic electrons, and the high-energy bump is related to inverse Compton emission of various underlying radiation fields, or a π^0 decays, depending on the production mechanism in action (Sect. 10.2.2). Large variability is present, especially at optical/UV and X-ray frequencies.

Variability is also a way to distinguish between hadronic and leptonic acceleration modes. In a pure leptonic mode, one expects that in a flare the increase of the synchrotron hump is fully correlated to the increase of the IC hump; in a hadronic mode, vice versa, one can have a "orphan flare" of the peak corresponding to π^0 decay.

Studies on different blazar populations indicate a continuous spectral trend from FSRQ to LBL to IBL to HBL, called the "blazar sequence." The sequence is char-

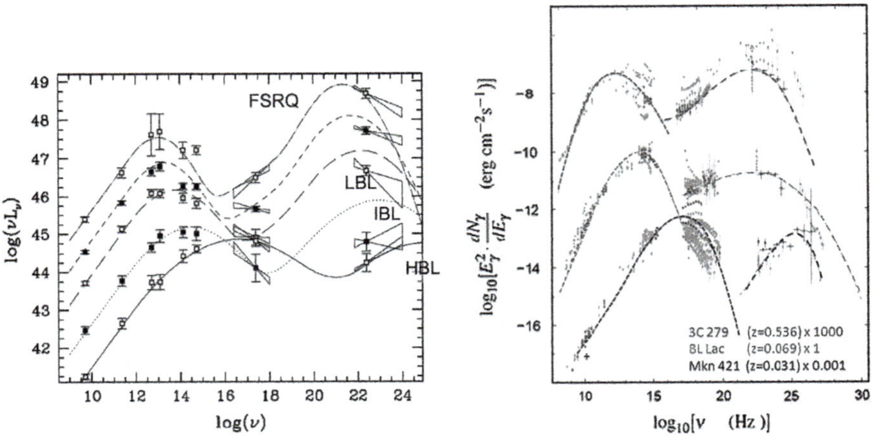

Fig. 10.37 *Left* The blazar sequence. From G. Fossati et al., Mon. Not. Roy. Astron. Soc. 299 (1998) 433. *Right* The SED of three different AGN at different distance from the Earth and belonging to different subclasses. To improve the visibility of the spectra, the contents of the farthest (3C 279) have been multiplied by a factor 1000, while that of the nearest (Mrk 421) by a factor 0.001. The *dashed lines* represent the best fit to the data assuming leptonic production. From D. Donato et al., Astron. Astrophys. 375 (2001) 739

acterized by a decreasing source luminosity, increasing synchrotron peak frequency, and a decreasing ratio of high- to low-energy component (Fig. 10.37).

10.3.3.2 Gamma Ray Bursts

Gamma ray bursts are another very important possible extragalactic acceleration site. As discussed before, GRBs are the most luminous events occurring in the gamma ray Universe (10^{51} erg on timescales of tens of seconds or less).

The connection between large mass supernovae (from the explosion of hyper-giants, stars with a mass of between 100 and 300 times that of the Sun) and long GRBs is proven by the observation of events coincident both in time and space, and the energetics would account for the emission—just by extrapolating the energetics from a supernova. During the abrupt compression of such a giant star the magnetic field could be squeezed to extremely large values, of the order of 10^{12}–10^{14} G, in a radius of some tens of kilometers.

Models of the origin of the short GRBs are, instead, more speculative. Observation of afterglows allowed the identification of host galaxies where short GRBs occurred. It was found that short GRBs are not associated with supernovae, and prefer galaxies that contain a considerable quantity of old stars. The current hypothesis attributes the origin of short GRBs to the merging of pairs of compact objects, possibly one neutron star and one black hole; the neutron star loses energy due to gravitational radiation, and thus spirals closer and closer until tidal forces disintegrate it providing an enormous quantity of energy before merging into the black hole. This process can last only a few seconds.

Although the two families of GRBs are likely to have different progenitors, the acceleration mechanism that gives rise to the γ rays themselves (and possibly to charged hadrons one order of magnitude more energetic, and thus also to neutrinos) can be the same.

The fireball model is the most widely used theoretical framework to describe the physics of the GRBs, either long or short.

In this model, first the black hole formed (or accreted) starts to pull in more stellar material; quickly an accretion disk forms, with the inner portion spinning around the BH at a relativistic speed. This creates a magnetic field which blasts outward two jets of electrons, positrons and protons at ultrarelativisic speed in a plane out of the accretion disk. Photons are formed in this pre-burst.

Step two is the fireball shock. Each jet behaves in its way as a shock wave, plowing into and sweeping out matter like a "fireball". Gamma rays are produced as a result of the collisions of blobs of matter; the fireball medium does not allow the light to escape until it has cooled enough to become transparent—at which point light can escape in the direction of motion of the jet, ahead of the shock front. From the point of view of the observer, the photons first detected are emitted by a particle moving at relativistic speed, resulting in a Doppler blueshift to the highest energies (i.e., gamma rays). This is the gamma ray burst.

An afterglow results when material escaped from the fireball collides with the interstellar medium and creates photons. The afterglow can persist for months as the energies of photons decrease.

Figure 10.38 shows a scheme of the fireball shock model.

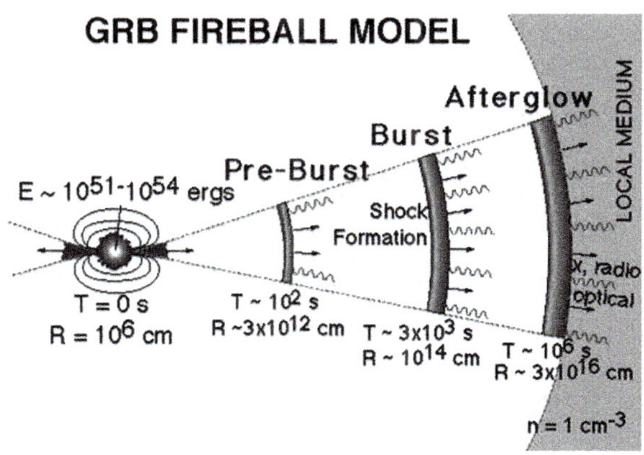

Fig. 10.38 The fireball model. Credits: http://www.swift.ac.uk/about/grb.php

10.3.4 Gamma Rays and the Origin of Cosmic Rays

10.3.4.1 Gamma Rays and the Origin of Cosmic Rays from SNRs

Among the categories of possible cosmic ray accelerators, several have been studied in an effort to infer the relation between gamma rays and charged particles. In the Milky Way in particular, SNRs are, since the hypothesis formulated by Baade and Zwicky in 1934, thought to be possible accelerators to energies of the order of 1 PeV and beyond. The particle acceleration in SNRs is accompanied by production of gamma rays due to interactions of accelerated protons and nuclei with the ambient medium.

The conjecture has a twofold justification. From one side, SNRs are natural places in which strong shocks develop and such shocks can accelerate particles. On the other side, supernovae can easily account for the required energetics. Nowadays, as a general remark, we can state that there is no doubt that SNR accelerate (part of the) galactic CR, the open questions being: which kind of SNR; in which phase of their evolution SNR really do accelerate particles; and if the maximum energy of these accelerated particles can go beyond $\sim$1 PeV.

A very important step forward in this field of research was achieved in the recent years with an impressive amount of experimental data at TeV energies, by Cherenkov telescopes (H.E.S.S., MAGIC, VERITAS), and at GeV energies, by the Fermi-LAT and AGILE satellites.

In SNRs with molecular clouds, in particular, a possible mechanism involves a source of cosmic rays illuminating clouds at different distances, and generating

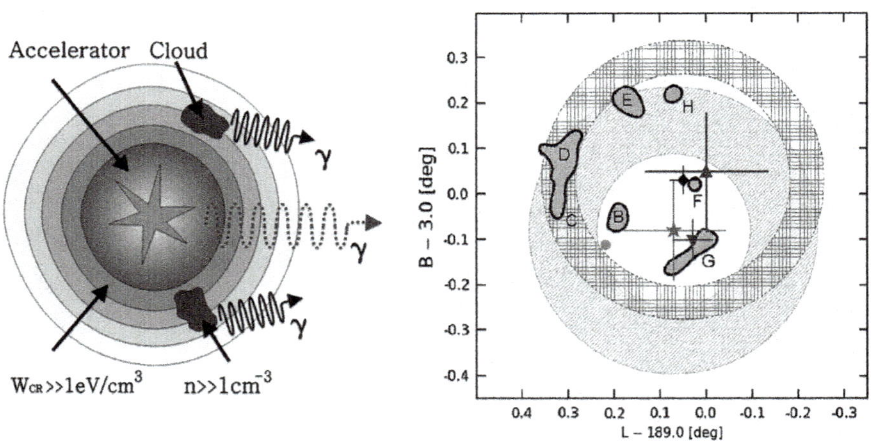

Fig. 10.39 *On the left* Scheme of the generation of a hadronic cascade in the dump of a proton with a molecular cloud. *On the right*, IC443: centroid of the emission from different gamma detectors. The position measured by Fermi-LAT is marked as a diamond that by MAGIC as a downwards oriented triangle; the latter is consistent with the molecular cloud G.

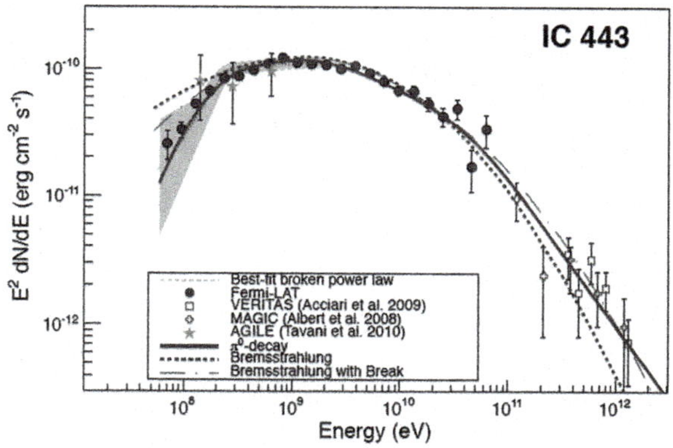

Fig. 10.40 Spectral energy distribution of photons emitted by the SNR IC443. The fit requires on top of photons coming from a leptonic acceleration mechanism also photons from π^0 decay. Credit: Fermi-LAT Collaboration (M. Ackermann et al.), Science 339 (2013) 807

hadronic showers by pp collisions. This allows to spot the generation of cosmic rays by the study of photons coming from π^0 decays in the hadronic showers.

Recent experimental results support the "beam dump" hypothesis. An example of such a mechanism at work is the SNR IC443. In Fig. 10.39, a region of acceleration at GeV energies is seen by the LAT space telescope onboard the Fermi-LAT significantly displaced from the centroid of emission detected at higher energies by the MAGIC gamma ray telescope—which, in turn, is positionally consistent with a molecular cloud. The spectral energy distribution of photons also supports a two-component emissions, with a rate of acceleration of primary electrons approximately equal to the rate of production of protons.

Such a 2-region displaced emission morphology has been also detected in several other SNRs (W44 and W82 for example).

Besides indications from the studies of the morphology, the detection of photons of energies of the order of 100 TeV and above could be a direct indication of production via π^0 decay, since the emission via leptonic mechanisms should be strongly suppressed at those energies where the inverse Compton scattering cross section enters the Klein–Nishina regime. Finally, in some sources the fit to the spectral energy distribution seems to indicate a component from π^0 decay (Fig. 10.40).

Let us shortly examine the relation between the spectra of secondary photons and the spectra of primary cosmic rays (we shall assume protons). We shall also examine the case of neutrinos, which could become, if large enough detectors are built, another powerful tool for the experimental investigation.

In beam dump processes, almost the same number of π^0, π^- and π^+ are produced. The π^0 decay immediately into two gamma rays; the charged pions decay into $\mu\,\nu_\mu$, with the μ decaying into $e\nu_e\nu_\mu$ (charge conjugates are implicitly included). Thus, there are three neutrinos for each pion and six neutrinos every two gamma rays.

When the collision happens between an accelerated proton and a nucleon at rest in a molecular cloud, for a c.m. energy of 1 TeV (which corresponds to an energy of the accelerated proton of 0.5 PeV in the frame of the cloud), around 60 charged and 30 neutral pions are produced. Taking into account the fact that in the final state of the charged pion decay there are three neutrinos and one electron/positron, the energy fraction transferred to each neutrino is about 1/4. Typically, thus, each neutrino carries (1/4) × (1/90) ≃ 0.3 % of the primary energy. This fraction decreases roughly logarithmically with energy.

In photoproduction processes ($\gamma p \to n\pi^+$, $\gamma p \to p\pi^0$), the neutrino energy from the π^+ decay is related to the parent proton energy through the relation

$$E_\nu \simeq E_p/20\,.$$

This arises because the average energy of the pion in photoproduction is some 20 % of the energy of the parent proton, and in the π^+ decay chain four leptons are produced, each of which roughly carries 1/4 of the pion energy.

The fraction of energy transferred to photons in π^+ decays is 1/4 (because the positron annihilates producing additional photons); for each π^+ one has in addition a π^0, which decays into photons. Following such simple arguments, the ratio of neutrino to photon flux in photoproduction processes is

$$L_\nu/L_\gamma \simeq 3/5\,.$$

Here, and in the following, under the symbol ν we always consider the sum of neutrinos and antineutrinos.

The existence of neutrino sources not observed in gamma rays is not excluded. If a source is occulted by the presence of thick clouds or material along the line of sight to the Earth, gamma rays are absorbed while neutrinos survive.

10.3.4.2 Testing if Cosmic Rays Originate from AGN

As the energetics of SNRs might explain the production of galactic CR, the energetics of AGN might explain the production of CR up to the highest energies. In the Hillas relation, the magnetic field and the typical sizes are such that acceleration is possible (Table 10.1).

Where molecular clouds are not a likely target, as, for example, in the vicinity of supermassive black holes, proton–photon interactions can start the hadronic shower.

Although the spatial resolution of gamma ray telescopes is not yet good enough to study the morphology of extragalactic emitters, a recent study of a flare from the nearby galaxy M87 (at a distance of about 50 Mly, i.e., a redshift of 0.0004) by the main gamma telescopes plus the VLBA radio array has shown, based on the VLBA imaging power, that this AGN accelerates particles to very high energies in the immediate vicinity (less than 60 Schwarzschild radii) of its central black hole. This galaxy is very active: its black hole, of a mass of approximately 7 billion solar masses, accretes by 2 or 3 solar masses per year. A jet of energetic plasma originates at the core and extends outward at least 5000 ly.

Also Centaurus A, the near AGN for which some hint of correlation with the Auger UHE data exists, has been shown to be a VHE gamma emitter.

10.3.5 Sources of Neutrinos

Up to now, the following sources of neutrinos have been identified.

1. Solar neutrinos. The first detection of neutrinos from the Sun happened in the 1960s (Chaps. 4 and 9). The deficit of solar neutrinos with respect to the flux expected from the energy released by the Sun paved the way to the discovery of neutrino oscillations and thus, ultimately, of a nonzero neutrino mass (Chap. 9).
2. Neutrinos from core-collapse supernovae. The burst of neutrinos produced in a galactic core-collapse supernova is detectable with detectors like Super-Kamiokande and SNO; however, this has been detected only once up to now. On February 23, 1987, a neutrino burst from a supernova in the LMC, some 0.2 Mly from Earth, was observed in the proton-decay detectors Kamiokande and IMB. Approximately 20 events were seen, spread over approximately 10 s. The optical counterpart reached an apparent magnitude[5] of about 3, and could be observed a few hours after the neutrino burst. Within the limited statistics achieved by these first-generation detectors, the number of events and the burst duration were consistent with standard estimates of the energy release and cooling time of a supernova. This is the only unambiguous detection of neutrinos that can be localized from a source other from the Sun up to now.

The sources discussed before typically emit neutrinos at energies ranging from the cosmic microwave temperature to some 10 GeV.

Neutrino sources associated with some of nature's most spectacular accelerators however exist similar to photon sources. The program of experiments to map out the high-energy neutrino spectrum is now very active, guided by existing data on cosmic ray protons, nuclei, and γ rays, which constrain possible neutrino fluxes. Recently the IceCube Collaboration observed a few neutrino events with estimated energies around 1 PeV. They are the highest energy neutrinos ever observed, and they must come, directly or indirectly, from extra solar sources; at the present status, however, these events do not appear to cluster to a common source.

On top of these astrophysical sources, as we have seen in Chap. 9, neutrinos can be produced as a result of interactions in the Earth's atmosphere. When primary cosmic protons and nuclei hit the atmosphere, the hadronic reactions with atmospheric nuclei can produce in particular secondary pions, kaons, muons. Atmospheric neutrinos are generated by the decay of these secondaries. The dominating processes are

---

[5]The apparent magnitude (m) of a celestial object measures its brightness as seen by an observer on Earth. The magnitude scale originates in the Hellenistic practice of dividing stars into six magnitudes. The brightest stars were said to be of first magnitude ($m = 1$), while the faintest were of sixth magnitude ($m = 6$), the limit of naked eye human visibility; each grade of magnitude was considered twice the brightness of the following grade. The system is today formalized by defining a first magnitude star as a star that is 100 times as bright as a sixth magnitude star; thus, a first magnitude star is $\sqrt[5]{100}$ (about 2.512) times as bright as a second magnitude star (obviously the brighter an object appears, the lower the value of its magnitude). The stars Arcturus and Vega have an apparent magnitude approximately equal to 0. The problem of the relation between apparent magnitude, intrinsic magnitude, and distance, is related also to cosmology, as discussed in Chap. 8.

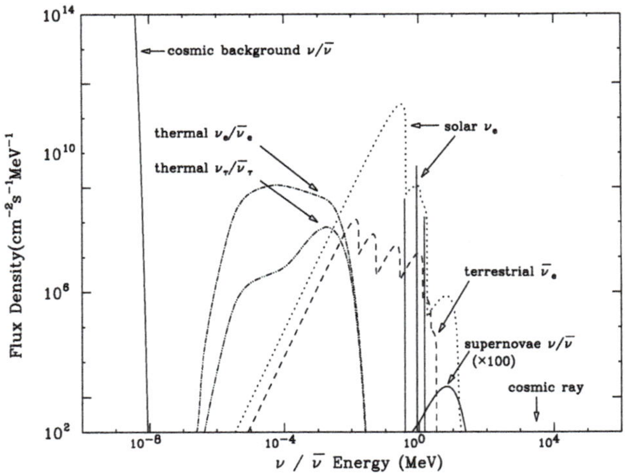

Fig. 10.41 Sources of neutrinos with energies below 1 TeV. From W.C. Haxton, arXiv:1209.3743, to appear in Wiley's Encyclopedia of Nuclear Physics

$$\pi^{\pm}(K^{\pm}) \rightarrow \mu^{\pm} + \nu_{\mu}(\overline{\nu}_{\mu}),$$
$$\mu^{\pm} \rightarrow e^{\pm} + \nu_{e}(\overline{\nu}_{e}) + \overline{\nu}_{\mu}(\nu_{\mu}) \,. \tag{10.18}$$

The neutrino sources just discussed are displayed in Fig. 10.41 according to their contributions to the terrestrial flux density. The figure includes low-energy sources, such as the thermal solar neutrinos of all flavors, and the terrestrial neutrinos—i.e., neutrinos coming from the Earth's natural radioactivity—not explicitly discussed here. Beyond the figure's high-energy limits there exist neutrino sources associated with some of nature's most energetic accelerators. Existing data on cosmic ray protons, nuclei, and γ rays constrain possible neutrino fluxes (see Fig. 10.42). The high-energy spectrum is one of the frontiers of neutrino astronomy.

Apart from the (low-energy) data from the Sun and from the Earth, and from the data from the flare of SN1987A, astrophysical neutrino fluxes detected up to now are consistent to be isotropic.

10.3.6 Sources of Gravitational Waves

The equations of Einstein's General Relativity (see Chap. 8) couple the metric of space–time with the energy and momentum of matter and radiation, thus providing the mechanism to generate gravitational waves at all scales. At the largest scales (extremely low frequencies 10^{-15}–10^{-18}Hz) the expected sources are the fluctuations of the primordial Universe. At most lower scales (frequencies 10^{-4}–10^{4}Hz) the expected sources are:

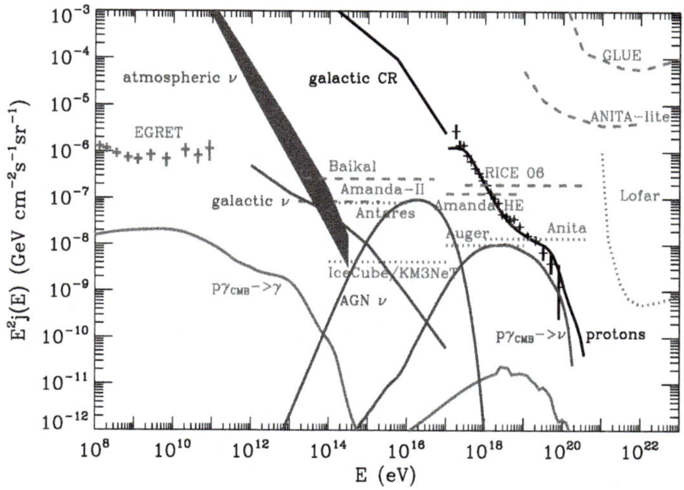

Fig. 10.42 A theoretical model of high-energy neutrino sources. The figure includes experimental data, limits, and projected sensitivities to existing and planned telescopes. Figure from G. Sigl, arXiv:0612240

- compact massive astrophysics objects and systems producing continuous gravitational waves with well-defined frequencies (e.g., spinning neutron stars or binaries in early coalescence phase);
- gravitational waves with increasing frequency patterns (e.g., the last coalescence stage of a binary system);
- gravitational wave bursts whenever there is a catastrophically disruption (for example, nonspherically symmetric supernovae explosions or gamma ray bursts).

It is commonly believed that the first detection of some of these sources will be at reach of the new generation of gravitational wave detectors now being installed. Figure 4.63 illustrates the sensitivity of some current and planned installations to possible sources of gravitational waves.

10.4 The Propagation Process

10.4.1 Propagation of Charged Cosmic Rays

The presence of magnetic fields in the Universe limits the possibility to investigate sources of emission of charged cosmic rays, as they are deflected by such fields.

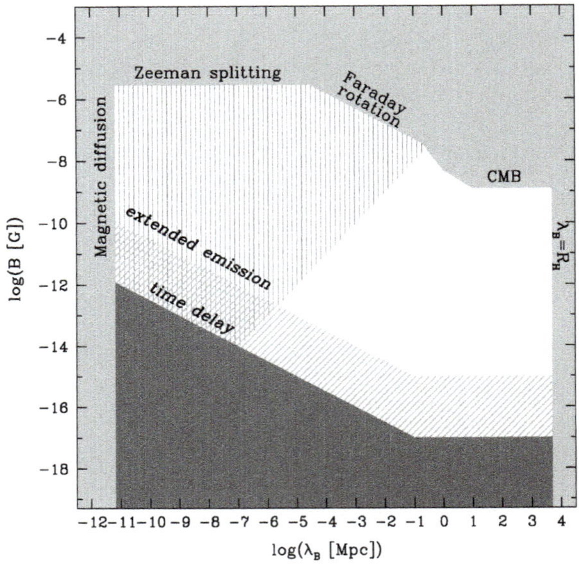

Fig. 10.43 Known observational bounds on the strength and correlation length of EGMF; the best fit from this analysis is indicated by the *cyan* region. Adapted from A. Neronov and I. Vovk, Science 328 (2010) 73

10.4.1.1 Magnetic Fields in the Universe

We know from studies of the Faraday rotation of polarization that the Galactic magnetic fields are of the order of a few μG; the structure is highly directional and maps exist.

In spite of intense efforts during the last few decades, the origin and structure of cosmic (i.e., extragalactic) magnetic fields remain instead elusive. Observations have detected the presence of nonzero magnetic fields in galaxies, clusters of galaxies, and in the bridges between clusters. Generally, the values of extragalactic magnetic fields (EGMF) are estimated to be between $<10^{-9}$ G and $<10^{-15}$ G. The determination of the strength and topology of large-scale magnetic fields is crucial because of their role in the propagation of ultrahigh-energy cosmic rays and, possibly, on structure formation.

Large-scale magnetic fields are believed to have a cellular structure. Namely, the magnetic field B is supposed to be have a correlation length λ, randomly changing its direction from one domain to another but keeping approximately the same strength. The present knowledge of EGMF is summarized in Fig. 10.43.

Since the Larmor radius of a particle of unit charge in a magnetic field can be written as

$$\frac{R_L}{1\text{pc}} \simeq \frac{\frac{E}{1\text{PeV}}}{\frac{B}{1\mu\text{G}}},$$

in order to "point" to the Galactic center, which is about 8 kpc from the Earth, for a Galacic field of 1μG one needs protons of energy of at least 2×10^{19} eV. The flux is very small at this energy; moreover, a black hole of 4 million solar masses like the

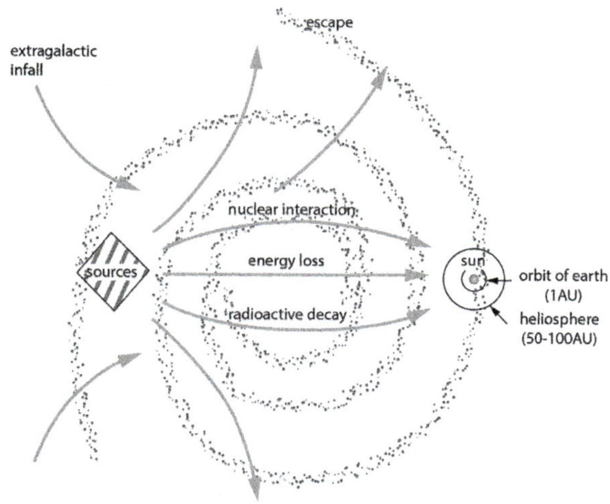

Fig. 10.44 Galactic cosmic ray propagation

black hole in the center of our Galaxy is not likely to accelerate particles up to this energy.

There is thus a need to use neutral messengers to study the emission of charged cosmic rays. Unfortunately, the yield of photons at an energy of 1 TeV is only 10^{-3} times the yield of protons, and the yield of neutrinos is expected to be of the same order. In addition, the detection of neutrinos is experimentally very challenging, as discussed in Chap. 4.

10.4.1.2 Propagation of Galactic Cosmic Rays

Cosmic rays produced in distant sources have to cross a long way before reaching Earth. Those produced in our Galaxy (Fig. 10.44) suffer diffusion in magnetic fields of the order of a μG, convection by Galactic winds, spallation in the interstellar medium, radioactive decays, as well as energy losses or gains (reacceleration). At some point they may arrive to Earth or just escape the Galaxy. Low-energy cosmic rays typically stay within the Galaxy for quite long times ($\sim 10^7$ years).

All these processes must be accounted in coupled transport equations which then determines the density N_i of each cosmic ray species. These differential equations, which are also function of the energy, can, for instance, be written as:

$$\begin{aligned}\frac{\partial N_i}{\partial t} &= Q_i + \vec{\nabla} \cdot \left(D \vec{\nabla} N_i - \vec{V} N_i \right) + \frac{\partial}{\partial E} \left(b\left(E\right) N_i \right) + \\ &-\left(\beta c \sigma_i^{\rm spall} + \frac{1}{\tau_i^{\rm decay}} + \frac{1}{\tau_i^{\rm escap}} + \right) N_i + \\ &+ \sum_{j>i} \left(\beta c \sigma_{ji}^{\rm spall} + \frac{1}{\tau_{ji}^{\rm decay}} \right) N_j \,. \end{aligned} \tag{10.19}$$

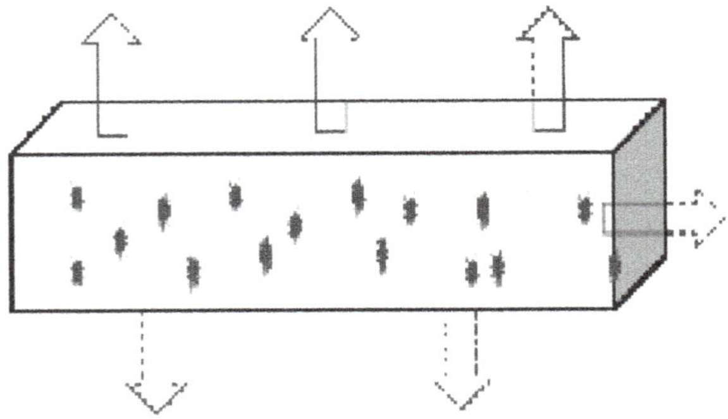

Fig. 10.45 The leaky-box model: a sketch

The first term on the right side accounts for the sources (injection spectrum), the second one for the diffusion and convection (D is the diffusion coefficient and $\vec{V}$ the convection velocity), the third for the changes in the energy spectrum due to energy losses or reacceleration, the fourth for the losses due to spallation, radioactive decays, and probability of escaping the Galactic, and the fifth the gains due to the spallation or decays of heavier elements.

These equations may thus include all the physics process and all spatial and energy dependences but the number of parameters is high and the constraints from experimental data (see below) are not enough to avoid strong correlations between them. The solutions can be obtained in a semi-analytical way or in a numerical using sophisticated codes (ex GALPROP), where three-dimensional distributions of sources (as traced by pulsars) and the interstellar medium can be included.

Simpler models, like Leaky-Box, are used to cope with the main features of the data. In the simplest version it consists only in a volume (box) where there are sources uniformly distributed and charged cosmic rays freely propagates having however some probability to escape by the walls (see Fig. 10.45).

The stationary equation of the Leaky-Box can be written as:

$$0 = Q_i - N_i \left\{ \rho\beta c\sigma_i^{\text{spall}} + \frac{1}{\gamma\tau_i^{\text{decay}}} + \frac{1}{\tau_i^{\text{esc}}} \right\} + \sum_{j>i} N_j \left\{ \rho\beta c\sigma_{ji}^{\text{spall}} + \frac{1}{\gamma\tau_{ji}^{\text{decay}}} \right\} . \tag{10.20}$$

Here once again the first term on the right side accounts for the sources and the second and the third, respectively, for the losses (due to spallation, radioactive decays, and probability of escape) and the gains (spallation or decays of heavier elements). The effects of diffusion geometry and so are all hidden in the escape lifetime.

All these models are adjusted to the experimental data and in particular to the energy dependence of the ratios of secondary elements (produced by spallation of heavier elements during the propagation) over primary elements (produced directly at the sources) as well as the ratios between unstable and stable isotopes of the same element. (See Fig. 10.46).

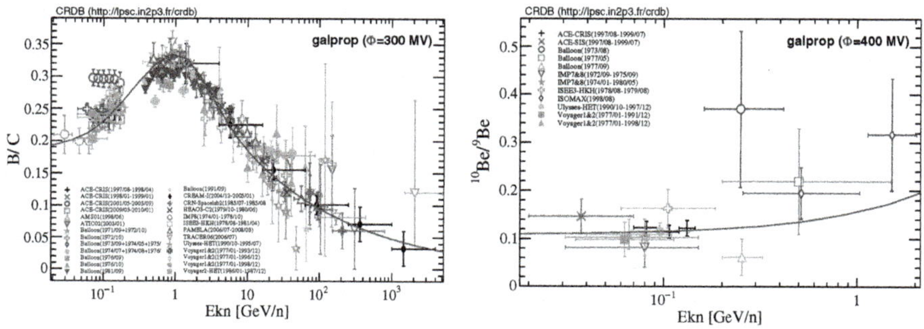

Fig. 10.46 *On the left* B/C (secondary over primary) as a function of the energy and *on the right* $^{10}Be/^{9}Be$ (unstable over stable). The data points were made using the Cosmic Ray database reference: Maurin, Melot, Taillet, A&A 569, A32 (2014) [arXiv:1302.5525]. The *full lines* are fits using Galprop model standard parameters. (reference: http://galprop.stanford.edu)

While primary/primary ratios (Fig. 10.47) basically do not depend on the energy, the secondary/primary ratio (Fig. 10.46, left) shows a strong dependence for high energies as a result of the increase of the probability of escape.

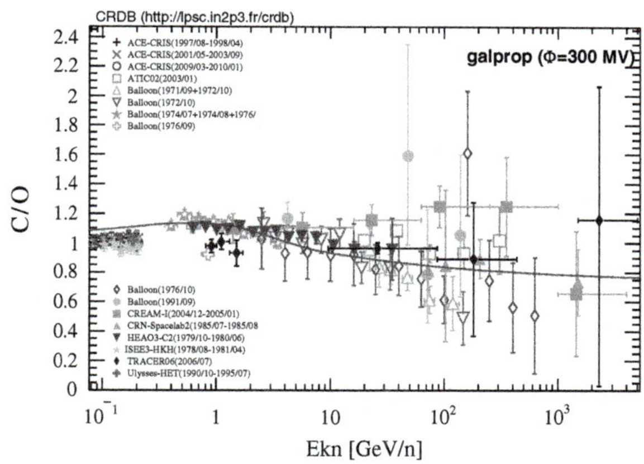

Fig. 10.47 The C/O ratio (primary/primary) as a function of the energy. The data points were made using the Cosmic Ray database reference: Maurin, Melot, Taillet, A&A 569, A32 (2014) [arXiv:1302.5525]. The *full lines* are fits using Galprop model standard parameters. (reference: http://galprop.stanford.edu)

One should note that in the propagation of electrons and positrons the energy losses are much higher (dominated by synchrotron radiation and inverse Compton scattering) and the escape probability much higher. Thus leaky-box like models do not apply to electrons and positrons.

10.4.1.3 Photon Background

The photon background in the Universe has the spectrum in Fig. 10.13.

The maximum photon density corresponds to the CMB; its total density is about 410 photons per cubic centimeter.

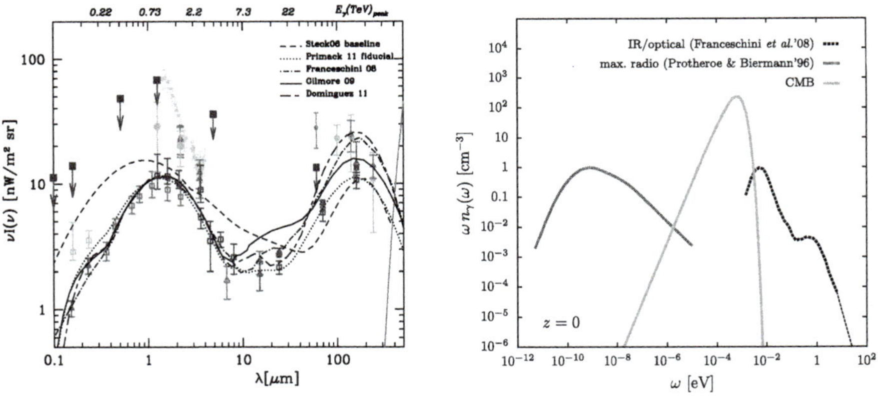

Fig. 10.48 *Left* Spectral energy distribution of the EBL as a function of the wavelength. Open symbols correspond to lower limits from galaxy counts while filled symbols correspond to direct estimates. The *curves* show a sample of different recent EBL models, as labeled. On the upper axis the TeV energy corresponding to the peak of the $\gamma\gamma$ cross section is plotted. From L. Costamante, IJMPD 22 (2013) 1330025. *Right* A summary of our knowledge about the density of background photons in intergalactic space, from the radio region to the CMB, to the infrared/optical/ultraviolet region. Reference: M. Ahlers et al., Astropart. Phys. 34 (2010) 106

A region of particular interest is the so-called extragalactic background light (EBL), i.e., the light in the visible and near infrared regions. It consists of the sum of starlight emitted by galaxies throughout their whole cosmic history, plus possible additional contributions, like, e.g., light from hypothetical first stars that formed before galaxies were assembled. Therefore, in principle the EBL contains important information about both the evolution of baryonic components of galaxies and the structure of the Universe in the pregalactic era. Note that the light is redshifted by a factor $(1 + z)$ due to the expansion of the Universe, and thus, the visible light from old sources is detected as infrared. The density of EBL photons in the region near the visible can be derived by direct deep field observations, and by constraints on the propagation of VHE photons (see later). A plot of the present knowledge on the density of photons in the EBL region is shown in Fig. 10.48, left.

Figure 10.48, right, shows a summary of the estimated photon number density of the background photons as composed by the radio background, the CMB, and the infrared/optical/ultraviolet background (EBL).

10.4.1.4 Extragalactic Cosmic Rays: The GZK Cutoff and the Photodisintegration of Nuclei

Extragalactic cosmic rays might cross large distances (tens or hundreds of Mpc) in the Universe. Indeed the Universe is full of CMB photons ($n_\gamma \sim 410\ \gamma/\mathrm{cm}^3$—see Chap. 8) with a temperature of $T \sim 2.73$ K ($\sim 10^{-3}$ eV). Greisen and Zatsepin and Kuzmin independently realized early in 1966 that for high-energy protons the inelastic channels

$$p\,\gamma_{CMB} \rightarrow \Delta^+ \rightarrow p\,\pi^0\ (n\,\pi^+)$$

are likely leading to a strong decrease of the proton interaction length (the so-called GZK cutoff). The threshold energy is just set by relativistic kinematics:

$$\left(p_p + p_\gamma\right)^2 = \left(m_p + m_\pi\right)^2$$

$$E_p = \frac{m_\pi^2 + 2m_p m_\pi}{4\ E_\gamma} \approx 6\ 10^{19}\ \text{eV} \tag{10.21}$$

reaching the pion photoproduction cross section, $\sigma_{\gamma p}$, values as high as $\sim$500 μb just above the threshold (with a plateau for higher energies slightly above $\sim$100 μb).

The mean free path of the protons above the threshold is thus:

$$\lambda_p \simeq \frac{1}{n_\gamma\ \sigma_{\gamma p}} \simeq 10\ \text{Mpc}\,. \tag{10.22}$$

In each GZK interaction the proton looses on average around 20 % of its initial energy.

A detailed computation of the effect of such cutoff on the energy spectrum of ultrahigh-energy cosmic ray at Earth would involve not only the convolution of the full CMB energy spectrum with the pion photoproduction cross section and the inelasticity distributions but also the knowledge of the sources, their location and energy spectrum as well as the exact model of expansion of the Universe (CMB photons are redshifted). An illustration of the energy losses of protons as a function of their propagation distance is shown in Fig. 10.49 without considering the expansion of the Universe. Typically, protons with energies above the GZK threshold energy after 50–100 Mpc loose the memory of their initial energy and end up with energies below the threshold.

The decay of the neutral and charged pions produced in these "GZK interactions" will originate, respectively, high-energy photons and neutrinos which would be a distinctive signature of such processes.

At a much lower energy ($E_p \sim 2\ 10^{18}$ eV) the conversion of a scattered CMB photon into an electron–positron pair may start to occur, what was associated by Hillas and later on by Berezinsky to the existence of the ankle (Sect. 10.1.1).

Heavier nuclei interacting with the CMB and Infrared Background (IRB) photons may disintegrate into lighter nuclei and typically one or two nucleons. The photo-disintegration cross section is high (up to $\sim$100 mb) and is dominated by the Giant Dipole resonance with a threshold which is a function of the nuclei binding energy per nucleon (for Fe the threshold of the photon energy in the nuclei rest frame is $\sim$10 MeV). Stable nuclei thus survive longer. The interaction length of Fe, the most stable nucleus, is at the GZK energy, similar to the proton GZK interaction length. Lighter nuclei have smaller interaction length and thus the probability of spallation during their way to Earth is higher.

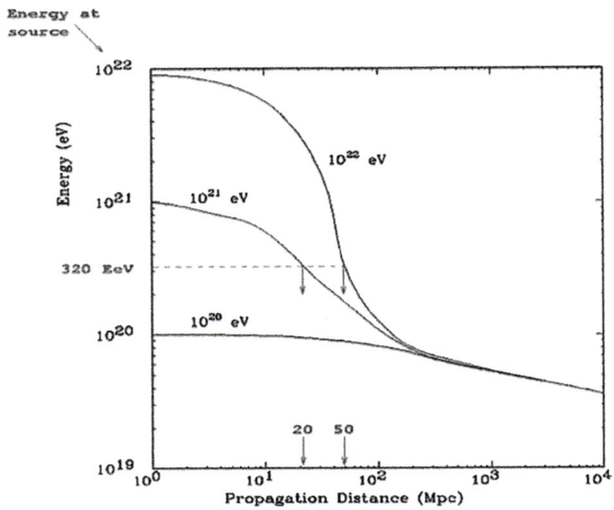

Fig. 10.49 Proton energy as a function of the propagation distance. From J.W. Cronin, Nucl. Phys. B Proc. Suppl. 28B (1992) 213

10.4.2 Propagation of Photons

Once produced, VHE photons must travel towards the observer. Electron–positron (e^-e^+) pair production in the interaction of VHE photons off extragalactic background photons is a source of opacity of the Universe to γ rays whenever the corresponding photon mean free path is of the order of the source distance or smaller.

The dominant process for the absorption is pair-creation

$$\gamma + \gamma_{\text{background}} \rightarrow e^+ + e^- ;$$

the process is kinematically allowed for

$$\epsilon > \epsilon_{\text{thr}}(E, \varphi) \equiv \frac{2\, m_e^2\, c^4}{E\,(1 - \cos\varphi)} , \tag{10.23}$$

where φ denotes the scattering angle, m_e is the electron mass, E is the energy of the incident photon and ϵ is the energy of the target (background) photon. Note that E and ϵ change along the line of sight in proportion of $(1 + z)$ because of the cosmic expansion. The corresponding cross section, computed by Breit and Wheeler in 1934, is

$$\sigma_{\gamma\gamma}(E, \epsilon, \varphi) = \frac{2\pi\alpha^2}{3m_e^2} W(\beta) \simeq 1.25 \cdot 10^{-25}\, W(\beta)\, \text{cm}^2 , \tag{10.24}$$

with

$$W(\beta) = \left(1 - \beta^2\right)\left[2\beta\left(\beta^2 - 2\right) + \left(3 - \beta^4\right)\ln\left(\frac{1+\beta}{1-\beta}\right)\right].$$

The cross section depends on E, ϵ and φ only through the speed β—in natural units—of the electron and of the positron in the center-of-mass

$$\beta(E, \epsilon, \varphi) \equiv \left[1 - \frac{2\, m_e^2\, c^4}{E\epsilon\,(1 - \cos\varphi)}\right]^{1/2}, \qquad (10.25)$$

and Eq. (10.23) implies that the process is kinematically allowed for $\beta^2 > 0$. The cross section $\sigma_{\gamma\gamma}(E, \epsilon, \varphi)$ reaches its maximum $\sigma_{\gamma\gamma}^{\max} \simeq 1.70 \cdot 10^{-25}\,\mathrm{cm}^2$ for $\beta \simeq 0.70$. Assuming head-on collisions ($\varphi = \pi$), it follows that $\sigma_{\gamma\gamma}(E, \epsilon, \pi)$ gets maximized for the background photon energy

$$\epsilon(E) \simeq \left(\frac{500\,\mathrm{GeV}}{E}\right)\,\mathrm{eV}\,, \qquad (10.26)$$

where E and ϵ correspond to the same redshift. For an isotropic background of photons, the cross section is maximized for background photons of energy:

$$\epsilon(E) \simeq \left(\frac{900\,\mathrm{GeV}}{E}\right)\,\mathrm{eV}\,. \qquad (10.27)$$

Explicitly, the situation can be summarized as follows:

- For $10\,\mathrm{GeV} \leq E < 10^5\,\mathrm{GeV}$ the EBL plays the leading role in the absorption. In particular, for $E \sim 10\,\mathrm{GeV}$ $\sigma_{\gamma\gamma}(E, \epsilon)$—integrated over an isotropic distribution of background photons—is maximal for $\epsilon \sim 90\,\mathrm{eV}$, corresponding to far-ultraviolet soft photons, whereas for $E \sim 10^5\,\mathrm{GeV}$ $\sigma_{\gamma\gamma}(E, \epsilon)$ is maximal for $\epsilon \sim 9 \cdot 10^{-3}\,\mathrm{eV}$, corresponding to soft photons in the far-infrared.
- For $10^5\,\mathrm{GeV} \leq E < 10^{10}\,\mathrm{GeV}$ the interaction with the CMB becomes dominant.
- For $E \geq 10^{10}\,\mathrm{GeV}$ the main source of opacity of the Universe, is the radio background.

The upper x-axis of Fig. 10.48, left, shows the energy of the incoming photon for which the cross section of interaction with a photon of the wavelength as in the lower x-axis is maximum.

From the cross section in Eq. (10.24), neglecting the expansion of the Universe one can compute a mean free path (Fig. 10.50); for energies smaller than some 10 GeV this is larger than the Hubble radius, but it becomes comparable with the distance of observed sources at energies above 100 GeV.

The attenuation suffered by observed VHE spectra can thus be used to derive constraints on the EBL density. Specifically, the probability P for a photon of observed energy E to survive absorption along its path from its source at redshift z to the observer plays the role of an attenuation factor for the radiation flux, and it is usually expressed in the form:

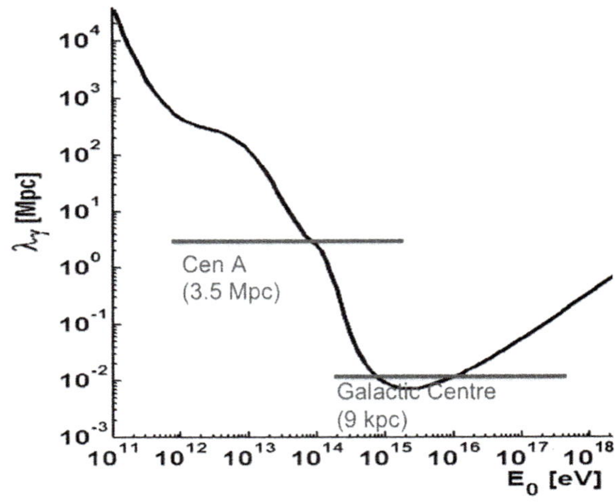

Fig. 10.50 Mean free path as a function of the photon energy, at $z = 0$. Adapted from A. de Angelis, G. Galanti, M. Roncadelli, MNRAS 432 (2013) 3245

$$P = e^{-\tau(E,z)} . \tag{10.28}$$

The coefficient $\tau(E, z)$ is called *optical depth.*

To compute the optical depth of a photon as a function of its observed energy E and the redshift z of its emission one has to take into account the fact that the energy E of a photon scales with the redshift z as $(1+z)$; thus when using Eq. 10.24 we must treat the energies as function of z and evolve $\sigma\big(E(z), \epsilon(z), \theta\big)$ for $E(z) = (1+z)E$ and $\epsilon(z) = (1+z)\epsilon$, where E and ϵ are the energies at redshift $z = 0$. The optical depth is then computed by convolving the photon number density of the background photon field with the cross section between the incident γ ray and the background target photons, and integrating the result over the distance, the scattering angle and the energy of the (redshifted) background photon:

$$\tau(E, z) = \int_0^z dl(z) \int_{-1}^1 d\cos\theta \frac{1-\cos\theta}{2} \times \\ \times \int_{\frac{2(m_e c^2)^2}{E(1-\cos\theta)}}^{\infty} d\epsilon(z)\, n_\epsilon\big(\epsilon(z), z\big)\, \sigma(E(z), \epsilon(z), \theta) \tag{10.29}$$

where θ is the scattering angle, $n_\epsilon\big(\epsilon(z), z\big)$ is the density for photons of energy $\epsilon(z)$ at the redshift z, and $l(z) = c\, dt(z)$ is the distance as a function of the redshift, defined by

$$\frac{dl}{dz} = \frac{c}{H_0} \frac{1}{(1+z)\left[(1+z)^2(\Omega_M\, z+1) - \Omega_\Lambda\, z(z+2)\right]^{\frac{1}{2}}} . \tag{10.30}$$

In the last formula (see Chap. 8) H_0 is the Hubble constant, Ω_M is the matter density (in units of the critical density, ρ_c) and Ω_Λ is the "dark energy" density (in units of ρ_c); therefore, since the optical depth depends also on the cosmological parame-

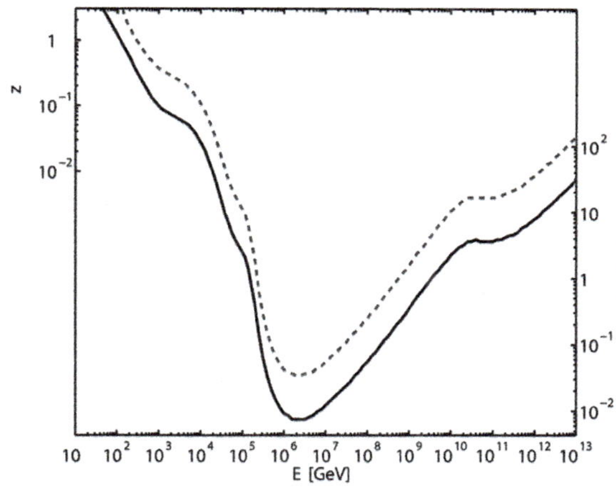

Fig. 10.51 *Curves* corresponding to the gamma ray horizon $\tau(E, z) = 1$ (*lower*) and to a survival probability of $e^{-\tau(E,z)} = 1\,\%$ (*upper*). Adapted from A. de Angelis, G. Galanti, M. Roncadelli, MNRAS 432 (2013) 3245

ters, its determination constrains the values of the cosmological parameters if the cosmological emission of galaxies is known.

The energy dependence of τ leads to appreciable modifications of the observed source spectrum (with respect to the spectrum at emission) even for small differences in τ, due to the exponential dependence described in Eq. (10.28). Since the optical depth (and consequently the absorption coefficient) increases with energy, the observed flux results steeper than the emitted one.

The *horizon* or *attenuation edge* for a photon of energy E is defined as the distance corresponding to the redshift z for which $\tau(E, z) = 1$, that gives an attenuation by a factor $1/e$ (see Fig. 10.51).

Other interactions than the one just described might change our picture of the attenuation of γ rays, and they are presently subject of thorough studies, since the present data on the absorption of photons are hardly compatible with the pure QED picture: from the observed luminosity of VHE photon sources, the Universe appears to be more transparent to γ rays than expected. One speculative explanation could be that γ rays might transform into sterile or quasi-sterile particles (like, for example, the axions which have been described in Chap. 8); this would increase the transparency by effectively decreasing the path length. A more detailed discussion will be given at the end of this chapter.

Mechanisms in which the absorption is changed through violation of the Lorentz invariance are also under scrutiny; such models are particularly appealing within scenarios inspired by quantum gravity (QG).

10.4.3 Propagation of Neutrinos

The neutrino cross section is the lowest among elementary particles. Neutrinos can thus travel with the smallest interaction probability and are the best possible astrophysical probe.

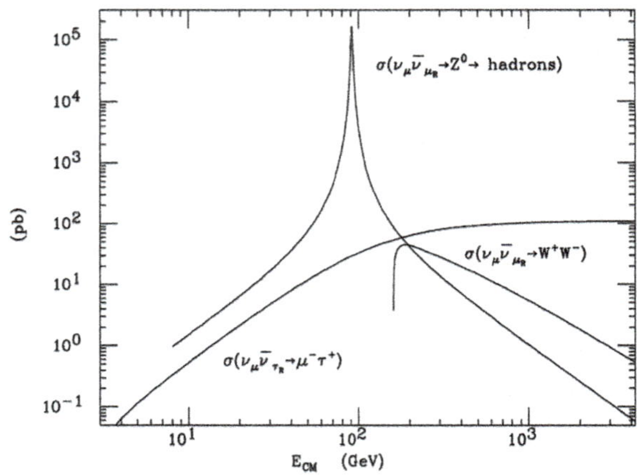

Fig. 10.52 $\nu\nu_c$ inelastic cross section as a function of the interaction center of mass. From D. Fargion and B. Mele, arXiv:astro-ph/9902024

Neutrinos of energies up to 10^{16} eV (which is the largest possible detectable energy, given the hypothesis of fluxes comparable with the photon fluxes, and the maximum size of neutrino detectors, of the order of a cubic kilometer) in practice travel undisturbed to the Earth.

On the other hand extremely high energetic neutrinos, if ever they exist in the Universe, will suffer a GZK-like interaction with the cosmological neutrinos ν_c. Indeed, the $\nu\nu_c$ cross section increases by several orders of magnitude whenever the center-of-mass energy of this interaction is large enough to open the inelastic channels as it is shown in Fig. 10.52. For instance, at $E_\nu \sim 10^{21}(4eV/m_\nu)$ the s-channel $\nu\nu_c \to Z$ is resonant. Thus, the Universe for these neutrinos of extreme energies becomes opaque.

10.4.4 Propagation of Gravitational Waves

Gravitational waves are oscillations of the space–time metrics which, accordingly to general relativity, propagates in the free space with the speed of light in the vacuum. Their coupling with matter and radiation is extremely weak and they propagate without significant attenuation, scattering, or dispersion in their way through the Universe. Their amplitude follows a $1/R$ dependence.

10.5 Frontier Physics and Open Questions

10.5.1 The Sources

The study of the galactic sources continues and their morphology and the SED of the emitted photons are telling us more and more, also in the context of multiwavelength

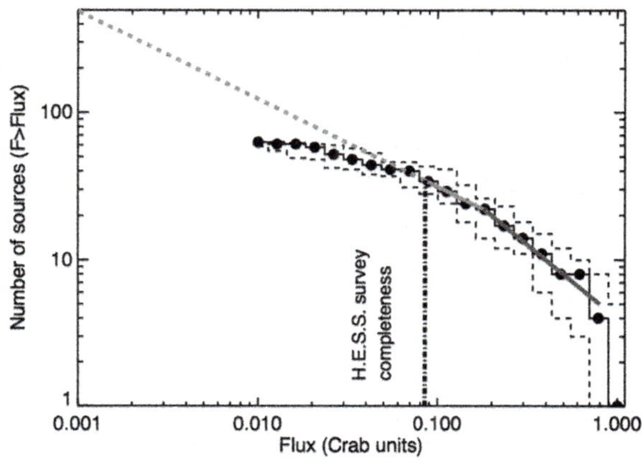

Fig. 10.53 "(log $N - \log S$)" diagram of the VHE Galactic sources. From M. Renaud, http://arxiv.org/abs/0905.1287

analyses; in the future, the planned Cherenkov Telescope Array (CTA) will give the possibility to explore the highest energies, and to contribute, together with high-energy CR detectors and possibly with neutrino detectors, to the final solution of the CR problem.

One of the main results from the next-generation detectors will probably be the discovery of new classes of CR sources. The key probably come from dedicating effort to surveys, which constitute an unbiased, systematic exploratory approach. Surveys of different extents and depths are amongst the scientific goals of all major planned facilities.

The key for such surveys are gamma detectors (and if possible, in the future, neutrino detectors).

More than half of the known VHE gamma sources (about 80) are located in the galactic plane. Galactic plane surveys are well suited to Cherenkov telescopes given the limited area to cover, as well as their low-energy thresholds and relatively good angular resolution (better than 0.1° to be compared to ∼1° for EAS detectors). CTA, investing 250 h (3 months) of observation, can achieve a 3 mCrab sensitivity (being the flux limit on a single pointing roughly proportional to $1/\sqrt{t_{obs}}$, where t_{obs} is the observation time) on the Galactic plane. More than 300 sources are expected at a sensitivity based on an extrapolation of the current "(log N − log S)" diagram[6] for VHE Galactic sources (Fig. 10.53).

All-sky VHE surveys are well suited to EAS arrays that observe the whole sky with high duty cycles and large field of view. MILAGRO and the Tibet air shower arrays have carried out a survey for sources in the Northern hemisphere down to an average sensitivity of 600 mCrab above 1 TeV; HAWC has a sensitivity of 50 mCrab in a year, at median energy around 1 TeV. EAS detectors like HAWC can then "guide" the CTA. A combination of CTA and the EAS can reach sensitivities better than 30

[6]The number of sources as a function of flux '(log $N - \log S$)" is an important tool for describing and investigating the properties of various types of source populations. It is defined as the cumulative distribution of the number of sources brighter than a given flux density S, and it base some regularity properties.

mCrab in large parts of the extragalactic sky. The survey could be correlated with maps obtained by UHE cosmic ray and high-energy neutrino experiments.

10.5.2 Ultrahigh-Energy Phenomena

Cosmic rays at ultrahigh energies are messengers from the extreme Universe and a unique opportunity to study particle physics at energies well above those reachable at the LHC. However, their limited flux and their indirect detection have not yet allowed answer to the basic and always present questions: Where are they coming from? What is their nature? How do they interact?

Under the commonly accepted assumptions of a finite horizon (due to a GZK-like interaction) and tiny extragalactic magnetic fields, it is expected that the possible number of sources is relatively small and thus some degree of anisotropy should be found studying the arrival directions of the highest energetic cosmic rays. Such searches have been performed extensively in the last years either by looking for correlations with catalogs of known astrophysical objects or by applying sophisticated self-correlation algorithms at all angular scales.

We already discussed the hot spot in the direction of the Ursa Major constellation claimed by the Telescope Array and the small excess in the region of Cen A by the Pierre Auger observatory. Both experiments have submitted proposals for the upgrade of their detectors and for the extension of their activity for a period of around 10 years. The quest for the origin of the extreme high-energy cosmic rays is still open.

The strong GZK-like suppression at the highest energies may be interpreted assuming different CR composition and sources scenarios. Indeed, both pure proton and mixed composition scenarios are able to describe the observed features. In the case of a pure proton scenario, the ankle would be described by the opening, at that energy, of the pair production channel in the interaction of the incoming protons with the CMB photons ($p\,\gamma_{CMB} \rightarrow p\,e^{+}e^{-}$) (this is called the dip model), while the suppression at the highest energies would be described in terms of the predicted GZK effect. In the case of mixed composition scenarios such features may be described playing with different sources distributions and injection spectra, assuming that the maximum energy that each nucleus may attain scales with its atomic number Z. An example of such fits is given in Fig. 10.54, where recent Telescope Array and Pierre Auger data are fitted respectively to a pure proton and a mixed composition scenario. The solution of such puzzle may only be found with the experimental determination of the cosmic ray composition from detailed studies on the observed characteristics of the extensive air showers.

The measured X_{max} distribution by the Pierre Auger collaboration in the energy bin 10^{18}–$10^{18.5}$eV for the 20 % of the most deeply penetrating showers is shown in Fig. 10.55. It follows the foreseen shape with a clear exponential tail. The selection of the most deeply penetrating showers strongly enhances the proton contents in

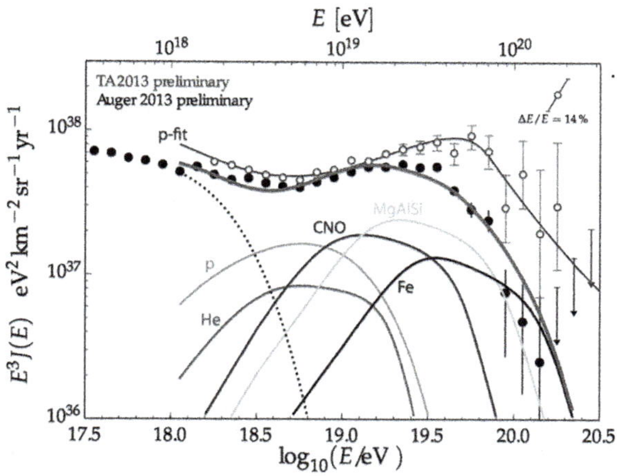

Fig. 10.54 Flux of UHE cosmic rays measured by the Telescope Array and by the Pierre Auger Observatory. Superposed are fits assuming pure proton (*light blue curve*) and mixed composition (*red curve*) scenarios. The *curves* below the result from the mixed mass composition fit. From K-H Kampert, P. Tinyakov, http://arxiv.org/abs/1405.0575 (Color in online)

Fig. 10.55 X_{max} distribution measured by the Pierre Auger Observatory in the energy interval 10^{18}–$10^{18.5}$eV. The *line* represents the likelihood fit performed to extract Λ_η. From P. Abreu et al., Phys. Rev. Lett. 109 (2012) 062002

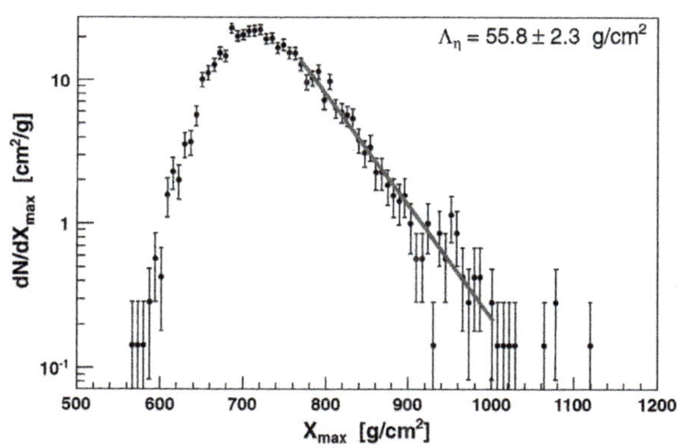

the data sample once the proton penetrate deeply in the atmosphere than any other nuclei.

The conversion of the exponential index of the distribution tail to a value of proton-air cross section is performed using detailed Monte Carlo simulations. The conversion to proton–proton total and inelastic cross section is then done using the Glauber model which takes into account the multi-scattering probability inside the nuclei (Sect. 6.4.7). The Auger result is shown in Fig. 10.56 together with accelerator data—namely with the recent LHC result as well as with the expected extrapolations of several phenomenological models. The several experimental results lies in a smooth extrapolation line which not only confirms the evolution as a function of the energy observed so far on the proton–proton cross section as it is a strong confirmation that at these energies there is an important fraction of protons in the cosmic ray "beam."

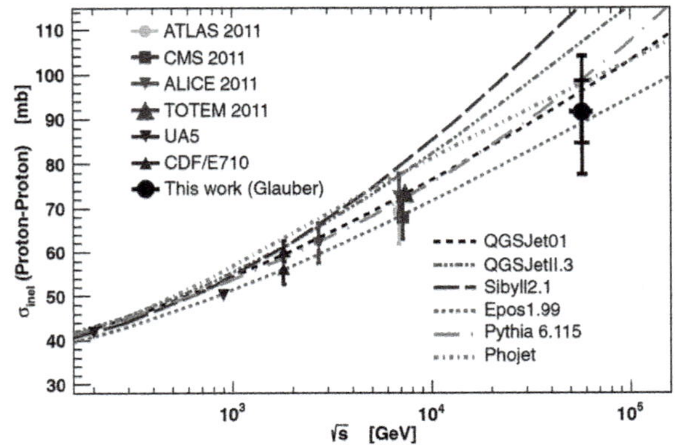

Fig. 10.56 Comparison of the inelastic proton–proton cross section derived by the Pierre Auger Observatory in the energy interval 10^{18}–$10^{18.5}$ eV to phenomenological model predictions and results from accelerator experiments at lower energies. From P. Abreu et al., Phys. Rev. Lett. 109 (2012) 062002

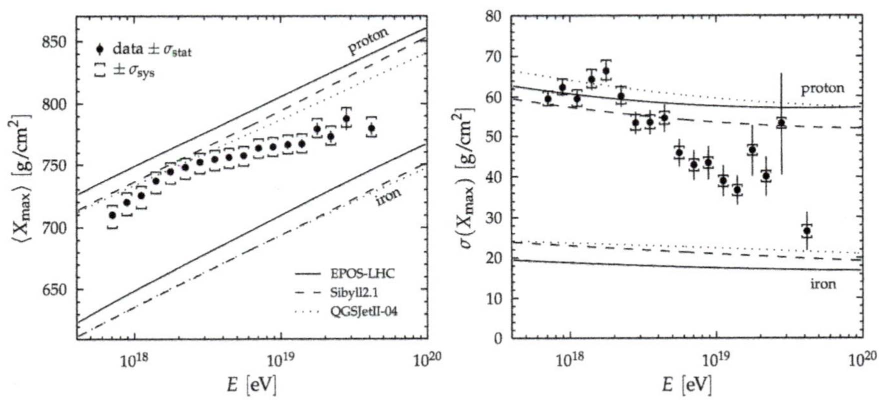

Fig. 10.57 Energy evolution of the mean (*left*) and the RMS (*right*) of the X_{max} distributions measured by the Pierre Auger Observatory. The *lines* represent the expectations for pure proton and iron composition for several simulation models which take into account the data obtained at the LHC. From A. Aab et al., Phys. Rev. D90 (2014) 122005

The first (mean) and second (RMS) momentum of the X_{max} distributions measured by the Pierre Auger collaboration as a function of the energy are shown in Fig. 10.57 where the lines represent the expected values for pure proton (red) or pure iron (blue) for several hadronic models tuned with recent LHC data.

These results may indicate some evidence of a change of the cosmic ray composition from light elements (with a large fraction of protons) at lower energies to heavy elements (with a large fraction of iron) at the highest energies, suggesting therefore a scenario of exhaustion of the sources. However, a simple bimodal proton–iron scenario does not fit the data of Fig. 10.57 well: as a matter of fact, none of the present simulations models is able to reproduce well all the observed ultrahigh-energy data.

A "new physics" solution providing a sudden increase of the proton-proton cross section (related to the access of a new scale of interaction below the parton scale, as discussed in Chap. 5) has also been proposed but, for the moment, it is still speculative.

Only qualitative and quantitative improvements in the understanding of the shower development, for example, accessing experimental information on the muon contents which is directly related with the development of the shower hadronic component, and/or improving the modeling of hadronic interactions in Monte Carlo simulations, may clarify this striking open question.

10.5.3 Top-Down Production Mechanisms for CR; WIMPs

Features in the spectra of known particles, in the GeV–TeV range, could show up if these particles originate in decays of exotic particles of very large mass possibly produced in the early Universe. Such long-lived heavy particles are predicted in many models, and the energy distribution of particles coming from their decay should be radically different from what predicted by the standard emission models from astrophysical sources.

Dark matter candidates (WIMPs in particular, as discussed in Chap. 8) are possible sources of, e.g., photons, electrons and positrons, and neutrinos via a top-down mechanism.

As discussed in Chap. 8, the relic density of DM can be expressed as

$$\Omega_\chi \simeq 0.24\,\frac{3\times 10^{-26}\mathrm{cm}^3\mathrm{s}^{-1}}{\langle\sigma_{ann} v\rangle}\,.$$

As the dark matter density is $\Omega_\chi \simeq 0.25$, the velocity-weighted annihilation cross section is about $3\times 10^{-26}\mathrm{cm}^3/\mathrm{s}$. The value for the interaction rate $\langle\sigma_{ann} v\rangle$ corresponds to a cross section of approximately 1 pb, typical for weak interactions. This is the so-called "WIMP miracle": a weakly interacting massive particle would be a good dark matter candidate. WIMP masses can be expected in the range between 10 GeV and a few TeV.

Given the expected amount of WIMP dark matter in the current Universe and the annihilation cross section, it is likely that dark matter is subject to self-annihilations. To be able to self-annihilate the dark matter particle much either coincide with its antiparticle, or be present in both the particle and antiparticle states. In the annihilation (or decay) of the dark matter particles all allowed standard model particles and antiparticles are produced, and gamma rays and/or charged particles are present in the final states (in the last case, with no preference between matter and antimatter, contrary to the standard sources of cosmic rays).

Where dark matter densities are large, the probability that WIMPs encounter each other and annihilate is enhanced. The problem is that we know the dark matter density in the halos, while the extrapolation to big density cores (like, for example, galactic centers are expected to be) relies on models.

If one can trust the extrapolations of dark matter density, one can predict the expected annihilation signal when assuming a certain interaction rate σv or put limits on the latter in the absence of a signal.

Matter and Antimatter. Dark matter particles annihilating or decaying in the halo of the Milky Way could potentially produce an observable flux of cosmic positrons and/or antiprotons, and thus be one possible explanation for the PAMELA anomaly, i.e., the excess of positron with respect to models just accounting for secondary production (Fig. 10.10). Most DM annihilation or decay models can naturally reproduce the observed rise of the positron fraction with energy, up to the mass of the DM candidate (or half the mass, depending if the self-annihilation or the decay hypothesis is chosen). This flux is expected not to be directional.

The measured antiproton flux also shows unexpected features with respect to the hypothesis of pure secondary production.

It is plausible that both the positron excess and the excess observed in the electron/positron yield with respect to current models (see Sect. 10.1.1) can be explained by the presence of nearby sources, in particular pulsars, which have indeed been copiously found by the Fermi-LAT (Sect. 10.3.1.1). AMS-02 is steadily increasing the energy range over which positrons and electrons are measured, as well as the statistics. If the positron excess is originated from a few nearby pulsars, it would probably give an anisotropy in the arrival direction of at the highest energies—there is a tradeoff here between distance and energy, since synchrotron losses are important; in addition, the energy spectrum should drop smoothly at the highest energies. A sharp cutoff in the positron fraction would instead be the signature of a DM origin of the positron excess, but the present data do not seem to support such a scenario.

The excess of antiprotons is difficult to accommodate in this explanation.

Search in Photon Channels. We shall refer in the following, unless explicitly specified, to a scenario in which secondary photons are produced in the annihilation of pairs of particles.

The expected flux of photons from dark matter annihilation can be expressed as

$$\frac{dN}{dE} = \underbrace{\frac{1}{4\pi}\frac{\langle\sigma_{ann}v\rangle}{2m_{DM}^2}\frac{dN_\gamma}{dE}}_{\text{Particle Physics}} \times \underbrace{\int_{\Delta\Omega-l.o.s.} dl(\Omega)\rho_{DM}^2}_{\text{Astrophysics}} \,. \tag{10.31}$$

DM-induced gamma rays can present sharp spectral signatures, like for instance $\gamma\gamma$ or $Z\gamma$ annihilation lines, with energies strictly related to the WIMP mass. However, since the WIMP is electrically neutral, these processes are loop suppressed and therefore should be rare. WIMP-induced gamma rays are thus expected to be dominated by a relatively featureless continuum of by-products of cascades and decays (mostly from π^0) following the annihilation in pairs of quarks or leptons. The number of resulting gamma rays depends quadratically on the DM density along the line of sight of the observer. This motivates search on targets, where one expects DM density enhancements. Among these targets are the Galactic center, galaxy clusters, and nearby dwarf spheroidal galaxies. Of course, an additional proof is given by proximity, to reduce the $1/d^2$ attenuation.

Unfortunately, as said before, dark matter densities are not known in the innermost regions of galaxies, where most of the signal should come from: data allow only the computation in the halos, and models helping in the extrapolation to the centers frequently disagree (Sect. 8.1.4). Observations of galaxy rotation curves favor constant density cores in the halos; unresolved "cusp" substructures can have a very large impact, but their existence is speculative—however, since they exist for baryonic matter, they are also likely to exist for DM. This uncertainty is typically expressed by the so-called "boost factor," defined as the ratio of the true, unknown, line-of-sight integral to the one obtained when assuming a smooth component without substructure.

As a consequence of all uncertainties described above, the choice of targets is somehow related to guesses, driven by the knowledge of locations where one expects large ratios of mass to luminous mass. Remembering Chap. 8, the main targets are:

- Dwarf Spheroidals.
  Dwarf spheroidal galaxies are a clean environment to search for dark matter annihilation: star formation is negligible, so astrophysical backgrounds that produce gamma rays are not a problem for the interpretation of data. The DM content can be determined from stellar dynamics and these objects have been found to be the ones with the largest mass-to-light ratios in the Universe. Some 25 dwarf satellite galaxies of the Milky Way are currently known and they are observed both by ground-based and by satellite-based gamma detectors. No signal has been found, and stringent limits have been calculated. In particular, a combined ("stacked") analysis of all known dwarf satellites with the Fermi-LAT satellite has allowed a limit to be set below the canonical thermal relic production cross section of $3 \times 10^{-26}\mathrm{cm}^3\mathrm{s}^{-1}$ for a range of WIMP masses (around 10 GeV) in the case of the annihilation into $b\bar{b}$ (the $b\bar{b}$ is used as a template due to the result obtained in Sect. 8.4.2).
- Galaxy clusters.
  Galaxy clusters are groups of hundreds to thousands galaxies bound by gravity. Galaxy clusters nearby (10–100 Mpc) include the Virgo, Fornax, Hercules, and Coma clusters. A very large aggregation known as the Great Attractor, dominated by the Norma Cluster, is massive enough to locally modify the trajectories in the expansion of the Universe.
  Galaxy clusters are much more distant than dwarf spheroidal galaxies or any of the other targets generally used for dark matter searches with gamma rays; however, like dwarf spheroidals, astrophysical dynamics shows that they are likely to be dark matter dominated—and if DM exists, one of the largest accumulators. The range of likely boost factors due to unresolved dark matter substructure can be large; however, when making conservative assumptions, the sensitivity to DM is several orders of magnitude away from the canonical thermal relic interaction rate.

- Galactic center.
  The GC is expected to be the brightest source of dark matter annihilation. However, the many astrophysical sources of gamma rays in that region complicate the identification of DM. In the GeV region the situation is further complicated by the presence of a highly structured and extremely bright diffuse gamma ray background arising from the interaction of the pool of cosmic rays with dense molecular material in the inner Galaxy. To limit problems, searches for dark matter annihilation/decay are usually performed in regions $0.3°$–$1°$ away form the central black hole.
  At TeV energies, Cherenkov telescopes detected a point source compatible with the position of the supermassive black hole in the center of our Galaxy and a diffuse emission coinciding with molecular material in the Galactic ridge. The galactic center source has a featureless power law spectrum at TeV energies with an exponential cutoff at $\sim$10 TeV not indicating a dark matter scenario; the signal is usually attributed to the supermassive black hole Sgr $A^\star$ or a to pulsar wind nebula in that region.
  Searches have been performed for a signal from the galactic dark matter halo close to the core; no signal has been found, and limits are comparable to those from dwarf spheroidals. However, these limits are more model dependent, since one must estimate the background.
  There have been however two hints of a signal in the galactic center region. An extended signal coinciding with the center of our Galaxy was reported above the galactic diffuse emission—however, the interaction of freshly produced cosmic rays with interstellar material is a likely explanation. The second claimed signal was the indication of a photon line at $\sim$130 GeV in regions of interest around the galactic center; we shall describe it better in the next item.
- Line Searches.
  The annihilation of WIMP pairs into γX would lead to monochromatic gamma rays with $E_\gamma = m_\chi(1 - m_X/4M_\chi^2)$. Such a signal would provide a smoking gun since astrophysical sources could very hardly produce it, in particular if such a signal is found in several locations. This process is expected to be loop suppressed being possible only at $\mathcal{O}(\alpha^2)$.
  The discovery of a hint of a signal at $\sim$130 GeV in the Fermi-LAT data, at a significance slightly larger than 4σ, was recently claimed as originating from regions near the GC. The search for the emission regions was driven by the data rather than by astrophysical considerations.
  The existence of this signal is strongly disputed, and some 5 more years of Fermi-LAT data will be needed to clarify the situation; also the Large Size Telescope of CTA (Chap. 4) will have the sensitivity to confirm or disprove the claim. The planned Russian/Italian space mission Gamma-400 for gamma detection in the MeV-GeV range can significantly improve the energy resolution in the 10 GeV region, which might be relevant for such line searches.

A summary of the present results is plotted in Fig. 10.58 together with extrapolation to next generation detectors.

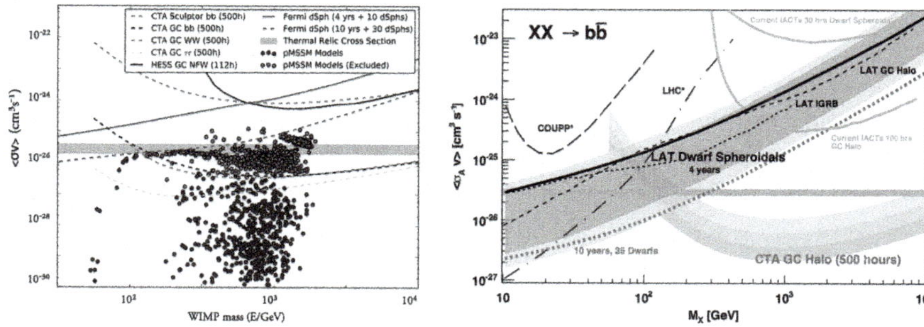

Fig. 10.58 Comparison of current (*solid lines*) and projected (*dashed lines*) limits on the DM annihilation interaction rate from different gamma ray searches as a function of WIMP mass. *Left* Expected limits for CTA are shown for WIMP annihilation compared to limits from Fermi-LAT and H.E.S.S. The *points* represent different choices of the parameters in SUSY. *Right* CTA is compared with the potential from direct detection (COUPP) and accelerator experiments (ATLAS and CMS at LHC). From S. Funk, arXiv:1310.2695

There is a unique region of the WIMP parameter space that CTA can best address in the near future: the high-mass ($\sim$1 TeV) scenario.

Neutrinos. Equation (10.31) holds for neutrinos as well, but the branching fractions into neutrinos are expected to be smaller, due to the fact that the radiative production of neutrinos is negligible. In addition, experimental detection is more difficult. However, the backgrounds are smaller with respect to the photon case.

Balancing the pros and the cons, gamma rays are the best investigation tool in case the emission comes from a region transparent to photons.

However, neutrinos are the best tool in case DM is concentrated in the center of massive objects, the Sun for example, which are opaque to gamma rays. Once gravitationally captured by such massive objects, DM particles lose energy in the interaction with nuclei and then settle into the core, where their densities and annihilation rates can be greatly enhanced; only neutrinos (and axions) can escape these dense objects. The centers of massive objects are among the places to look for a possible neutrino excess from DM annihilation using neutrino telescopes. No signal has been detected up to now (as in the case of axions from the Sun). A reliable prediction of the sensitivity is difficult, depending on many uncertain parameters like the annihilation cross section, the decay modes and the capture rate. The first two uncertainties are common to the photon channels.

10.5.4 Lorentz Symmetry Violation

The variability of the AGN in the VHE region can provide information about possible violations of the Lorentz invariance in the form of a dispersion relation for light expected, for example, in some QG models.

Lorentz invariance violation (LIV) at the n-th order in energy can be heuristically incorporated in a perturbation to the relativistic Hamiltonian:

$$E^2 \simeq m^2c^4 + p^2c^2 \left[1 - \xi_n \left(\frac{pc}{E_{\mathrm{LIV,n}}} \right)^n \right] , \tag{10.32}$$

which implies that the speed of light ($m = 0$) could have an energy dependence. From the expression $v = \partial E/\partial p$, the modified dispersion relation of photons can be expressed by the leading term of the Taylor series as an energy-dependent light speed

$$v(E) = \frac{\partial E}{\partial p} \simeq c \left[1 - \xi_n \frac{n+1}{2} \left(\frac{E}{E_{\mathrm{LIV,n}}} \right)^n \right] , \tag{10.33}$$

where n = 1 or n = 2 corresponds to linear or quadratic energy dependence, and $\xi_n = \pm 1$ is the sign of the LIV correction. If $\xi_n = +1$ ($\xi_n = -1$), high-energy photons travel in vacuum slower (faster) than low-energy photons.

The scale E_{LIV} at which the physics of space–time is expected to break down, requiring modifications or the creation of a new paradigm to avoid singularity problems, is referred to as the "QG energy scale", and is expected to be of the order of the Planck scale—an energy $E_P = M_P c^2 \simeq 1.2 \times 10^{19}$ GeV—or maybe lower, if new particles are discovered at an intermediate scale.

Because of the spectral dispersion, two GRB photons emitted simultaneously by the source would arrive on Earth with a time delay (Δt) if they have different energies. With the magnification of the cosmological distances of the GRBs and the high energies of these photons, the time delay (Δt) caused by the effect of Lorentz invariance violation could be measurable. Taking account of the cosmological expansion and using Eq. 10.33, we write the formula of the time delay as:

$$\Delta t = t_h - t_l = \xi_n \frac{1+n}{2H_0} \frac{E_h^n - E_l^n}{E_{LIV,n}^n} \int_0^z \frac{(1+z')^n dz'}{\sqrt{\Omega_m(1+z')^3 + \Omega_\Lambda}} . \tag{10.34}$$

Here, t_h is the arrival time of the high-energy photon, and t_l is the arrival time of the low-energy photon, with E_h and E_l being the photon energies measured at Earth.

For small z, and at first order,

$$t(E) \simeq d/c(E) \simeq \frac{zc_0}{H_0 c(E)} \simeq zT_H \left(1 - \xi_1 \frac{E}{E_P} \right)$$

where $T_H = 1/H_0 \simeq 5 \times 10^{17}$ s is Hubble's time.

AGN flares (Sect. 10.1.2.1) can be used as experimental tools: they are fast and photons arrive to us travel for long distances.

Mkn 501 (z = 0.034) had a spectacular flare between May and July 2005; it could be analyzed by the MAGIC telescope. The MAGIC data showed a negative correlation between the arrival time of photons and their energy (Fig. 10.59), yielding, if one assumes that the delay is due to linear QG effects, to an evaluation of $E_{\mathrm{LIV}} \sim 0.03\, E_P$. H.E.S.S. observations of the flare in PKS 2155 (Fig. 10.17), however, evidenced no effect, allowing to set a lower limit $E_{\mathrm{LIV}} > 0.04\, E_P$.

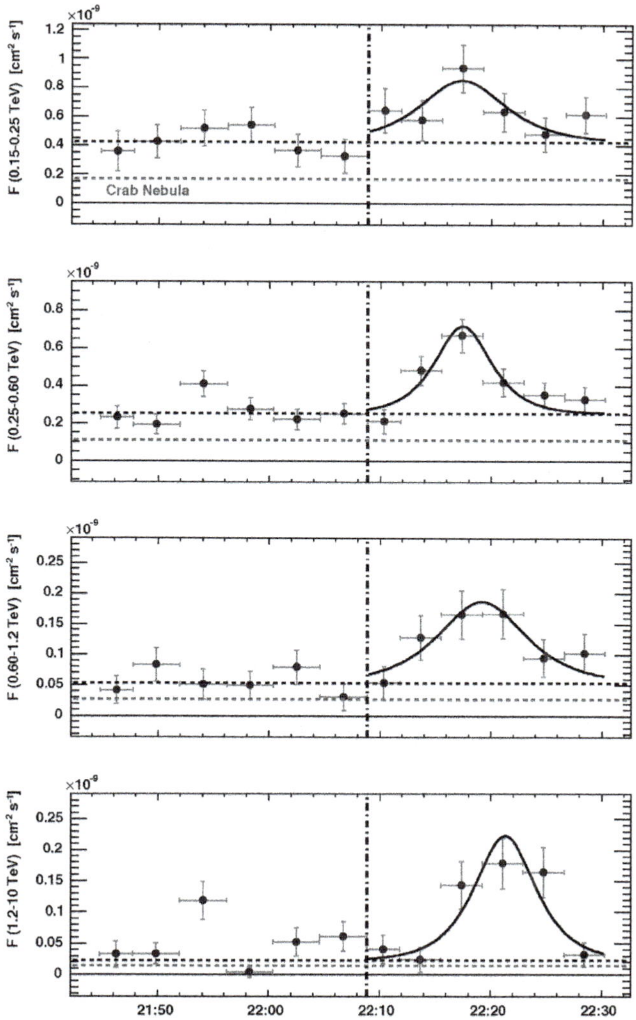

Fig. 10.59 Integral flux of Mkn 501 detected by MAGIC in four different energy ranges. From J. Albert et al., Phys. Lett. B668 (2008) 253

Lately, several GRBs observed by the Fermi satellite have been used to set more stringent limits. A problem, when setting limits, is that one does not know if photon emission at the source is ordered in energy; thus one has to make hypotheses—for example, that QG effects can only increase the intrinsic dispersion.

The Fermi satellite derived strong upper limits at 95 % C.L. from the total degree of dispersion, in the data of four GRBs:

$$E_{\mathrm{LIV},1} > 7.6 E_P \, .$$

In most QG scenarios violations to the universality of the speed of light happen at order larger than 1: $\Delta t \simeq (E/E_{\mathrm{LIV}})^{\nu}$ with $\nu > 1$. In this case the VHE detectors are even more sensitive with respect to other instruments like Fermi; for $\nu = 2$ the data from PKS 2155 give $E_{\mathrm{LIV}} > 10^{-9}\, E_P$.

10.5.5 Possible Anomalous Photon Propagation Effects

Some experimental indications exist, that the Universe might be more transparent to gamma rays than computed in Sect. 10.4.2.

As discussed before, the existence of a soft photon background in the Universe leads to a suppression of the observed flux of gamma rays from astrophysical sources through the $\gamma\gamma \rightarrow e^+e^-$ pair-production process. Several models have been proposed in the literature to estimate the spectral energy density (SED) of the soft background (EBL); since they are based on suitable experimental evidence (e.g., deep galaxy counts), all models yield consistent results, so that the SED of the EBL is fixed to a very good extent. Basically, the latter reproduces the SED of star-forming galaxies, which is characterized by a visibile/ultraviolet hump due to direct emission from stars and by an infrared hump due to the emission from the star-heated warm dust that typically hosts the sites of star formation.

However, the Universe looks more transparent than expected—this is called the "EBL crisis." Basically, two experimental evidences support this conjecture:

- When for each SED of high-z blazars, the data points observed in the optically thin regime ($\tau < 1$) are used to fit the VHE spectrum in optically thick regions, points at large attenuation are observed (Fig. 10.60). This violates the current EBL models, strongly based on observations, at some 5σ.
- The energy dependence of the gamma opacity τ leads to appreciable modifications of the observed source spectrum with respect to the spectrum at emission, due to the exponential decrease of τ on energy in the VHE gamma region. One would expect naively that the spectral index of blazars at VHE would increase with distance: due to absorption, the SED of blazars should become steeper at increasing distance. This phenomenon has not been observed (Fig. 10.61).

Among the possible explanations, a photon mixing with axion-like particles (ALPs), predicted by several extensions of the Standard Model (Sect. 8.4.2), can fix the EBL crisis, and obtain compatibility on the horizon calculation. Since ALPs are characterized by a coupling to two photons, in the presence of an external magnetic field B photon-ALP oscillations can show up. Photons are supposed to be emitted by a blazar in the usual way; some of them can turn into ALPs, either in the emission region, or during their travel. Later, some of the produced ALPs can convert back into photons (for example, in the Milky Way, which has a relatively large magnetic field) and ultimately be detected. In empty space this would obviously produce a flux dimming; remarkably enough, due to the EBL such a double conversion can make

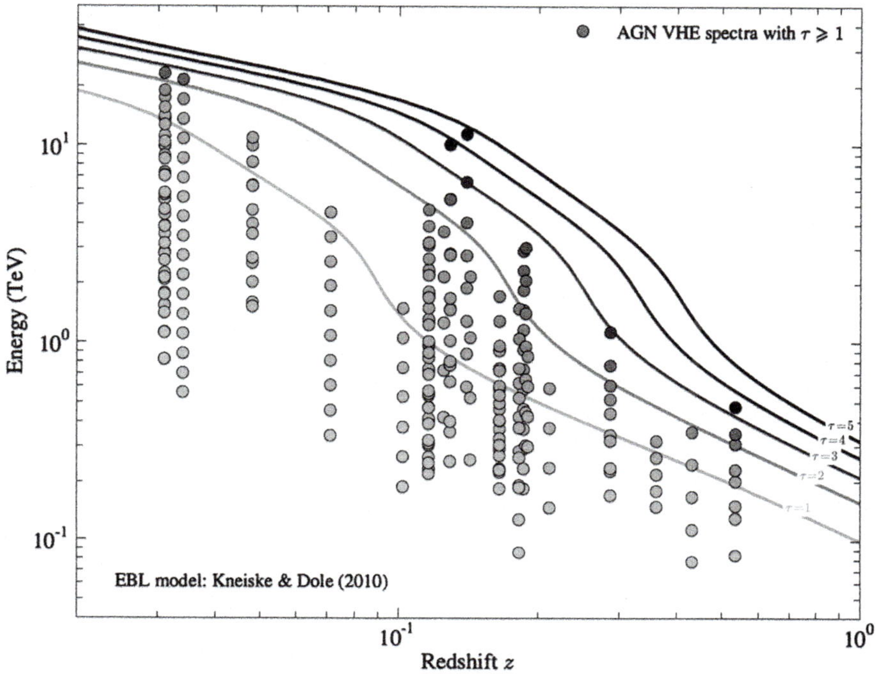

Fig. 10.60 For each individual spectral measurement including points at $\tau\rangle 1$, the corresponding value of z and E are marked in this diagram. The iso-contours for $\tau = 1,\ 2,\ 3,\ 4$ calculated using a minimum EBL model are overlaid. From D. Horns, M. Meyer, JCAP 1202 (2012) 033

the observed flux considerably larger than in the standard situation: in fact, ALPs do not undergo EBL absorption (Fig. 10.62).

We concentrate now on the photon transition to ALP in the intergalactic medium. The probability of photon-ALP mixing depends on the value and on the structure of the cosmic magnetic fields, largely unknown (see Sect. 10.4.1.1).

Both the strength and the correlation length of the cosmic magnetic fields do influence the calculation of the $\gamma \to a$ conversion probability. In the limit of low conversion probability, if s is the size of the typical region, the average probability $P_{\gamma\to a}$ of conversion in a region is

$$P_{\gamma\to a} \simeq 2\times 10^{-3}\left(\frac{B_T}{1\,\mathrm{nG}}\frac{\lambda_B}{1\,\mathrm{Mpc}}\frac{g_{a\gamma\gamma}}{10^{-10}\,\mathrm{GeV}^{-1}}\right)^2 , \tag{10.35}$$

where B_T is the transverse component of the magnetic field.

For a magnetic field of 0.1–1 nG, and a cellular size structure $\lambda_B \sim$ 1Mpc−10 Mpc (see Fig. 10.43), any ALP mass below 10^{-10} eV, with a coupling such that 10^{11} GeV $< M < 10^{13}$ GeV (well within the region experimentally allowed for mass and coupling) can explain the experimental results (Fig. 10.61).

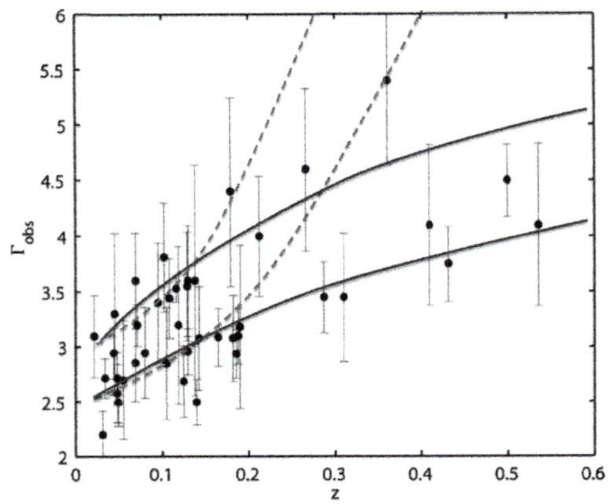

Fig. 10.61 The observed values of the spectral index for all blazars detected in VHE; superimposed is the predicted behavior of the observed spectral index from a source at constant intrinsic spectral index within two different scenarios. In the first one (area between the *two dotted lines*) Γ is computed from EBL absorption; in the second (area between the *two solid lines*) it is evaluated including also the photon-ALP oscillation. Original from A. de Angelis et al., Mon. Not. R. Astron. Soc. 394 (2009) L21; updated

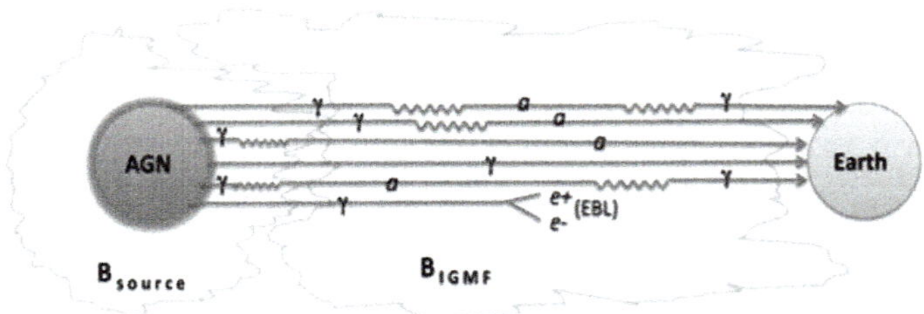

Fig. 10.62 Illustration of gamma ray propagation in the presence of oscillations between gamma rays and axion-like particles. From M.A. Sanchez-Conde et al., Phys. Rev. D79 (2009) 123511

Another possible explanation for the hard spectra of distant blazars, needing a more fine tuning, is that line-of-sight interactions of cosmic rays with CMB radiation and EBL generate secondary gamma rays relatively close to the observer.

10.5.5.1 LIV and Photon Propagation

A powerful tool to investigate Planck scale departures from Lorentz symmetry could be provided by a possible change in the energy threshold of the pair production process $\gamma_{VHE}\gamma_{EBL} \to e^+e^-$ of gamma rays from cosmological sources. This would affect the optical depth, and thus, photon propagation.

In a collision between a soft photon of energy ϵ and a high-energy photon of energy E, an electron–positron pair could be produced only if E is greater than the threshold energy E_{th}, which depends on ϵ and m_e^2.

Note that also the violation of the Lorentz invariance changes the optical depth. Using a dispersion relation as in Eq. (10.32), one obtains, for $n = 1$ and unmodified law of energy–momentum conservation, that for a given soft-photon energy ϵ, the process $\gamma\gamma \rightarrow e^+e^-$ is allowed only if E is greater than a certain threshold energy E_{th} which depends on ϵ and m_e^2. At first order:

$$E_{th}\epsilon + \xi(E_{th}^3/8E_p) \simeq m_e^2. \tag{10.36}$$

The $\xi \rightarrow 0$ limit corresponds to the special-relativistic result $E_{th} = m_e^2/\epsilon$. For $|\xi| \sim 1$ and sufficiently small values of ϵ (and correspondingly large values of E_{th}) the Planck scale correction cannot be ignored.

This provides an opportunity for tests based on dynamics. As an example, a 10 TeV photon and a 0.03 eV photon can produce an electron–positron pair according to ordinary special-relativistic kinematics, but they cannot produce a e^+e^- pair according to the dispersion relation in Eq. (10.32), with $n = 1$ and $\xi \sim -1$. The non-observation of EeV gamma rays has already excluded a good part of the parameter range of terms suppressed to first and second order in the Planck scale.

The situation for positive ξ is somewhat different, because a positive ξ decreases the energy requirement for electron–positron pair production.

10.5.5.2 A Win-Win Situation: Determination of Cosmological Parameters

If no indications of new physics (LIV, anomalous propagation) will be found after all, since the optical depth depends also on the cosmological parameters (Eq. 10.30), its determination constrains the values of the cosmological parameters if the cosmological emission of galaxies and EBL are known, and if only standard processes are at work.

A determination of Ω_M and Ω_Λ independent of the luminosity–distance relation currently used by the Supernovae 1 A observations can be obtained from the spectra of distant AGN.

Further Reading

[F10.1] M. Spurio, "Particles and Astrophysics (a Multi-Messenger Approach)," Springer 2015. Taking a systematic approach, this book comprehensively presents experimental aspects from the most advanced cosmic ray detectors, in particular detectors of photons at different wavelengths.

[F10.2] T. Stanev, "High-Energy Cosmic Rays," Springer 2010. A classic for experts in the discipline.

[F10.3] D.H. Perkins, "Particle Astrophysics," second edition, Oxford University Press 2008.

[F10.4] A. de Angelis, "Recent results from VHE gamma astrophysics related to fundamental physics and cosmic rays," arXiv:1104.5163.

Exercises

1. *Neutrinos from SN1987A.* Neutrinos from SN1987A, at an energy of about 50 MeV, arrived in a bunch lasting 13 s from a distance of 50 kpc, 3 h before the optical detection of the supernova. What can you say on the neutrino mass? What can you say about the neutrino speed (be careful...)?
2. *Time lag in light propagation.* Suppose that the speed of c of light depends on its energy E in such a way that

$$c(E) \simeq c_0 \left(1 + \xi \frac{E^2}{E_P^2}\right),$$

where E_P is the Planck energy (second-order Lorentz Invariance Violation). Compute the time lag between two VHE photons as a function of the energy difference and of the redshift z.

3. *Photon spectrum in hadronic cascades.* Demonstrate that in a decay $\pi^0 \to \gamma\gamma$, once boosted for the energy of the emitting π^0, the probability to emit a photon of energy E_γ is constant over the range of kinematically allowed energies.
4. *Extensive electromagnetic air showers.* The main characteristic of an electromagnetic shower (say, initiated by a photon) can be obtained using a simple Heitler model. Let E_0 be the energy of the primary particle and consider that the electrons, positrons and photons in the cascade always interact after traveling a certain atmospheric depth $d = X_0$, and that the energy is always equally shared between the two particles. With this assumptions, we can schematically represent the cascade as in Fig. 4.10.

   (a) Write the analytical expressions for the number of particles and for the energy of each particle at depth X as a function of d, n and E_0.
   (b) The multiplication of the cascade stop when the particles reach a critical energy, E_c (when the decay probability surpasses the interaction probability). Using the expressions obtained in the previous question, write as a function of E_0, E_c and $\lambda = d/\ln(2)$, the expressions, at the shower maximum, for:

      i. the average energy of the particles,
      ii. the number of particles, N_{max},
      iii. the atmospheric depth X_{max}.

5. *Extensive hadronic air showers.* Consider a shower initiated by a proton of energy E_0. We will describe it with a simple Heitler-like model: after each depth d an equal number of pions, n_π, and each of the 3 types is produced: π^0, π^+, π^-.

Neutral pions decay through $\pi^0 \to \gamma\gamma$ and their energy is transferred to the electromagnetic cascade. Only the charged pions will feed the hadronic cascade. We consider that the cascade ends when these particles decay as they reach a given decay energy E_{dec}, after n interactions, originating a muon (plus an undetected neutrino).

(a) How many generations are needed to have more that 90 % of the primary energy, E_0 in the electromagnetic component?
(b) Assuming the validity of the superposition principle, according to which a nucleus of mass number A and energy E_0 behaves like A nucleons of energy E_0/A, derive expressions for:

i. the depth where this maximum is reached, $X_{\max}$,
ii. the number of particles at the shower maximum,
iii. the number of muons produced in the shower, N_μ.

6. *Acceleration and propagation.* The transparency of the Universe to a given particle depends critically on its nature and energy. In fact, whenever it is possible to open an inelastic channel of the interaction between the *traveling* particle and the CMB, its mean free path diminishes drastically. Assuming that the only relevant phenomena that rules the mean free path of the *traveling* particle is the CMB (CνB), estimate the order of magnitude energies at which the transparency of the Universe changes significantly, for:

(a) Photons;
(b) Protons;
(c) Neutrinos.

Assume $\langle E_{\gamma CMB} \rangle \simeq 0.24$ meV; $\langle E_{\nu C\nu B} \rangle \simeq 0.17$ meV.

7. *Fermi acceleration mechanisms.* Calculate, for first- and second-order Fermi acceleration mechanism, how many times has the particle to cross the cloud (shock) to gain a factor 10 on its initial energy. Assume $\beta = 10^{-4}$ for the magnetic cloud and $\beta = 10^{-2}$ for the shock wave. Repeat the previous exercise assuming $\beta = 10^{-4}$ for both acceleration mechanisms.
8. *Imaging Array Cherenkov Telescopes.* In the isothermal approximation, the depth x of the atmosphere at a height h (i.e., the amount of atmosphere above h) can be approximated as

$$x \simeq X e^{-h/7\,\mathrm{km}} ,$$

with $X \simeq 1030$ g/cm^2. If a shower is generated by a gamma ray of $E = 1$ TeV penetrating the atmosphere vertically, considering that the radiation length X_0 of air is approximately 36.6 g/cm^2 (440 m) and its critical energy E_c is about 88 MeV and using Rossi approximation B (Table 4.1):

(a) Calculate the height h_M of the maximum of the shower in the Heitler model and in the Rossi approximation B.
(b) If 2000 useful Cherenkov photons per radiation length are emitted by charged particles in the visible and near UV, compute the total number N_γ of

Cherenkov photons generated by the shower (note: the critical energy is larger than the Cherenkov threshold).

(c) Supposing that the Cherenkov photons are all emitted at the center of gravity of the shower—that in the Heitler approximation is just the maximum of the shower minus one radiation length, compute how many photons per square meter arrive to a detector at a height h_d of 2000 m, supposing that the average attenuation length of photons in air is 3 km, and that the light pool can be derived by a opening of $\sim 1.3°$ from the shower maximum ($1.3°$ is the Cherenkov angle and $0.5°$, to be added in quadrature, comes from the intrinsic shower spread). Comment on the size of a Cherenkov telescope, considering an average reflectivity of the mirrors (including absorption in transmission) of 70 %, and a photodetection efficiency (including all the chains of acquisition) of 20 %.

(d) Redo the calculations for $E = 50$ GeV, and comment.

Chapter 11
Astrobiology and the Relation of Fundamental Physics to Life

> *How come that human beings are here on Earth today? How did the laws of physics made it possible that intelligent life evolved? Are we unique, or are we just one of many intelligent species populating the Universe? It is likely that in the vastness of the Universe we humans do not stand alone; in the near future we might be within reach of other civilizations, and we must understand how to identify them, and how to communicate with them. At the basis of all this is understanding what is life, and how it emerged on Earth and maybe elsewhere. The answer to these questions is written in the language of physics.*

The ultimate quest of astrophysics is probably to understand the role of the human beings in the Universe, and in this sense it converges with many different sciences. Astrobiology is the study of the origin, evolution, distribution, and future of life in the universe: both life on Earth and extraterrestrial life. This interdisciplinary field encompasses the study of the origin of the materials forming living beings on Earth, search for habitable environments outside Earth, and studies of the potential for terrestrial forms of life to adapt to challenges on Earth and in outer space. Astrobiology also addresses the question of how humans can detect extraterrestrial life if it exists, and how we can communicate with aliens. This relatively new field of science is a focus of a growing number of NASA and European Space Agency exploration missions in the solar system, as well as searches for extraterrestrial planets which might host life.

One of the main probes of astrobiology is to understand the question if we are unique, or just one of many intelligent species populating the Universe. The most important discovery of all in astrophysics would probably be to communicate with different beings: this would enrich us and probably change completely our vision of ourselves and of the Universe. But the question of life and of its meaning is central also in many other sciences, from biology to philosophy. In particular, biology wants to answer many questions, as the question of how life was born from non-living material (abiogenesis), a question that is central since Aristoteles. We are convinced that humans will soon be able to generate life from nonliving materials—and this will

A. De Angelis and M.J.M. Pimenta, *Introduction to Particle and Astroparticle Physics*, Undergraduate Lecture Notes in Physics,
DOI 10.1007/978-88-470-2688-9_11

probably be the most important discovery of all in biology, again changing radically our vision of ourselves. This would probably help also in understanding our origin as humans.

We shall see how astroparticle physics can help us in this research.

11.1 What is Life?

A proper definition of life, universally accepted, does not exist. We shall just try to clarify some of the conditions under which we might say that a system is living, i.e., to formulate a description.

Some of the characteristics most of us accept to define a living being are listed below.

- Presence of a body, distinguishing an individual. This definition is sometimes nontrivial (think for example of mushrooms, or of coral).
- Metabolism: Conversion of outside energy and materials into cellular components (anabolism) and decomposition of organic material (catabolism). Living bodies use energy to maintain internal organization (homeostasis) and the internal environment must be regulated to maintain characteristics different form the "external" environment.
- Growth: At least in a large part of life, anabolism is larger than catabolism, and growing organisms increase in size.
- Adaptation: Living beings change in response to the environment. This is fundamental to the process of evolution and is influenced by the organism's heredity, as well as by external factors.
- Response to stimuli (can go from the contraction of a unicellular organism to external chemicals, to complex reactions involving all the senses of multicellular organisms). Often the response generates motion—e.g., the leaves of a plant turning toward the sun (phototropism).
- Reproduction: the ability to produce new individual organisms. Clearly not everything that replicates is alive: in fact computers can replicate files; some machines can replicate themselves, but we cannot say that they are alive; on the other hand, some animals have no reproductive ability, such as most of the bees—reproduction has to be considered at the level of species rather than of individuals.

The above "physiological functions" have underlying physical and chemical bases. The living organisms we know have a body that is based on carbon: the molecules needed to form and operate the cells are made of carbon. But, why carbon? One reason is that carbon allows the lowest energy chemical bonds, and is a particularly versatile chemical element that can be bound to as many as four atoms at a time.

However, we can think of different elements. If we ask for a material which can allow the formation of complex structures, tetravalent elements (carbon, silicon, germanium, ...) are favored. The tetravalent elements heavier than silicon are heavier than iron, hence they can come only from supernova explosions, and are thus very rare; we are thus left only with silicon as a candidate for a life similar to our life other than carbon. Like carbon, silicon can create molecules large enough to carry biological information; it is however less abundant than carbon in the Universe.

Silicon has an additional drawback with respect to carbon: since silicon atoms are much bigger than carbon, having a larger mass and atomic radius, they have difficulty forming double bonds. This fact limits the chemical versatility required for metabolism.

A tranquilizing view on silicon-based aliens would be that in case of invasion they would rather eat our buildings than us...

11.1.1 Schrödinger's Definition of Life

In the pervious paragraphs we tried a descriptive definition of life. It would be useful to formulate a mathematical definition; attempts to do so, however, failed up to now.

Schrödinger tried to formulate a definition of life based on physics. In his view, everything was created from chaos but life tries to organize proteins, water atoms, etc.; so Schrödinger said life fights entropy, and gave the definition of negative entropy, as for living organization, or space-time structures. Schrödinger wrote: "When a system that is not alive is isolated or placed in a uniform environment, all motion usually comes to a standstill very soon as a result of various kinds of friction; differences of electric or chemical potential are equalized, substances which tend to form a chemical compound do so, temperature becomes uniform by heat conduction. After that the whole system fades away into a dead, inert lump of matter. A permanent state is reached, in which no observable events occur. The physicist calls this the state of thermodynamical equilibrium, or of 'maximum entropy' and, as he said,"It is by avoiding the rapid decay into the inert state of 'equilibrium' that an organism appears so enigmatic. What an organism feeds upon is negative entropy." An organism avoids decay "by eating, drinking, breathing and (in the case of plants) assimilating", and "everything that is going on in Nature, means an increase of the entropy, and so a organism continually increases its entropy, and thus tends to the state of maximum entropy, which means death; it can only try to stay alive by continually drawing its environment of negative entropy".

In summary, according to Schrödinger, life requires open systems able to decrease their internal entropy using substances or energy taken in from the environment, and subsequently reject material in a degraded form.

11.1.2 The Recipe of Life

In this section we will analyze what life needed and needs to develop on Earth, and what are the factors that influence it.

11.1.2.1 Water and Carbon

Liquid water is fundamental for life as we know it: it is very important because it is used like a solvent for many chemical reactions. On Earth, we have the perfect temperature to maintain water in liquid state, and one of the main reasons is the obliquity of Earth with respect to the ecliptic plane at about 23°, which allows seasonal changes.

Water can exchange organisms and substances with Earth, thanks to tides. The Moon is mostly responsible for the tides: the Moon's gravitational pull on the near side of the Earth is stronger than on the far side, and this difference causes tides (Fig. 11.1). The Moon orbits the Earth in the same direction as the Earth spins on its axis, so it takes about 24 h and 50 min for the Moon to return to the same location with respect to the Earth. In this time, it has passed overhead once and underfoot once, you have two tides. The Sun contributes to Earth's tides as well, but even if its gravitational force is much stronger than the Moon's, the solar tides are less than half that the one produced by the Moon: this is because the tides are caused by the difference in the gravity field, and the Earth's diameter is such a small fraction of the Sun–Earth distance that the gravitational force from the Sun changes only by a factor of 1.00017 across the Earth (see the first exercise). Tides are important because many biological organisms have biological cycles based on them, and if the Moon did not exist these types of cycles might not have arisen.

But, how did Earth come to possess water? Early Earth had probably oceans that are the result of several factors: first of all, volcanoes released gases and water vapor in the atmosphere, that condensed forming the oceans. Nevertheless, vapor from

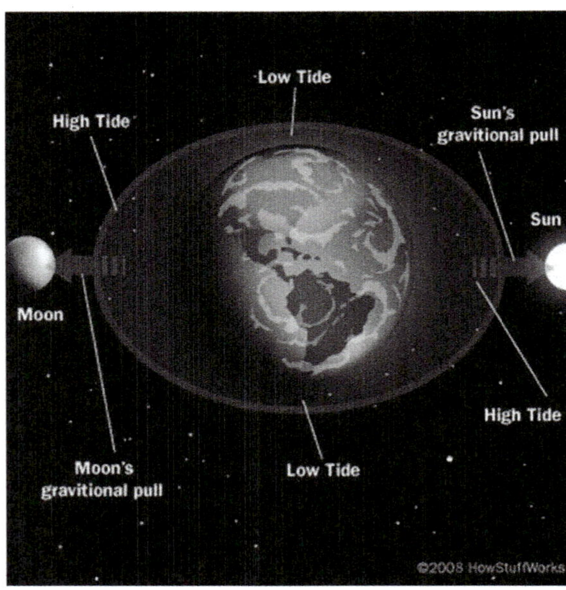

Fig. 11.1 The combination of the Moon's and Sun's gravitational pull causes the phenomenon of tides. From http://science.howstuffworks.com/environmental/earth/oceanography/ocean-current4.htm

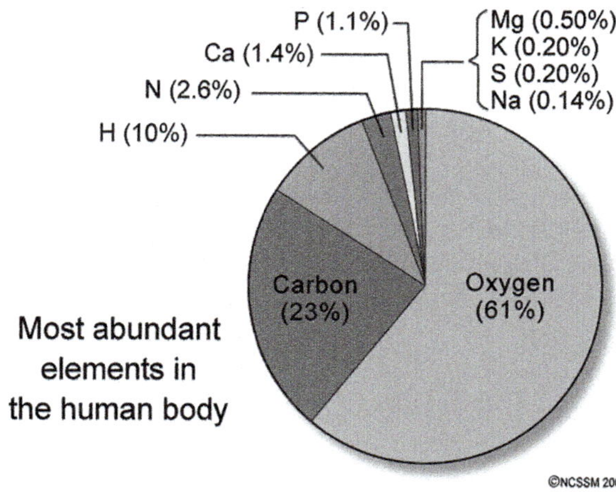

Fig. 11.2 Most abundant elements (in weight) that form the human body. From http://www.dlt.ncssm.edu/tiger/chem1.htm

the volcanoes is sterilized and no organisms can actually live in it: for this reason, many scientists think that some liquid water with seed of life may have been brought to Earth by comets and meteorites. The problem of how and where the water was generated on these bodies is not solved; it is, however, known that they carry water.

Life on Earth is based on more than 20 elements, but just 4 of them (i.e., oxygen, carbon, hydrogen, and nitrogen) make up 96 % of the mass of living cells (Fig. 11.2).

Non-water Solvents. An extraterrestrial life-form, however, might develop and use a solvent other than water, like ammonia, sulfuric acid, formamide, hydrocarbons, and (at temperatures lower than Earth's) liquid nitrogen.

Water has many properties important for life: in particular, it is liquid over a large range of temperatures, it has a high heat capacity—and thus it can help regulating temperature, it has a large vaporization heat, and it is a good solvent. Water is also amphoteric, i.e., it can donate and accept a H^+ ion, and act as an acid or as a base—this is important for facilitating many organic and biochemical reactions in water. In addition, water has the uncommon property of being less dense as a solid (ice) than as a liquid: thus masses of water freeze covering water itself by a layer of ice which isolates water from the external environment (fish in iced lakes swim at a temperature of 4 °C, the temperature of maximum density of water).

Ammonia (NH_3) is the best candidate to host life after water, being abundant in the Universe. Liquid ammonia is chemically similar to water, and numerous chemical reactions are possible in a solution of ammonia, which like water is a good solvent for most organic molecules, and in addition is capable of dissolving many elemental metals. Ammonia is amphoteric; is however flammable in oxygen, which could create problems for aerobic metabolism as we know it.

As the melting and boiling points of ammonia at normal pressure are between 195 and 240 K, a biosphere based on ammonia could exist at temperatures and air pressures extremely unusual in relation to life on Earth. The chemical being in general slower at low temperatures, ammonia-based life, if existing, would metabolize more slowly and evolve more slowly than life on Earth. On the other hand, lower temperatures might allow the development of living systems based on chemical species unstable at our temperatures. To be liquid at temperatures similar to the ones on Earth, ammonia needs high pressures: at 60 bar it melts at 196 K and boils at 371 K, more or less like water.

Since ammonia and ammonia–water mixtures remain liquid at temperatures far below the freezing point of water, they might be suitable for biochemical planets and moons that orbit outside of the "zone of habitability". This might apply, for example, to Titan, the largest moon of Saturn.

11.1.2.2 The Temperature and the Greenhouse Effects

A key ingredient affecting the development of life on our planet is temperature. One may think that the temperature on Earth is appropriate for liquid water because of the Earth's distance from the Sun; this is not true, because, for example, the Moon lies at the same distance from the Sun but its temperature, during the day, is 125 °C—and during night, −155 °C. The main reasons why the Earth has its current temperature are the interior heating, and the greenhouse effect.

The greenhouse effect slows down the infrared light's return to space: instrumental to this process are gases, like water vapor (H_2O), carbon dioxide (CO_2), and methane (CH_4), that are present in the atmosphere. They absorb infrared radiation and subsequently they release a new infrared photon. This latter photon can be absorbed by another greenhouse molecule, so the process may be repeated on and on: the result is that these gases tend to trap the infrared radiation in the lower atmosphere. Moreover, molecular motions contribute to heat the air, so both the low atmosphere and the ground get warmer (Fig. 11.3).

If the greenhouse effect did not take place, the average temperature on our planet would be −18 °C. A discriminating factor is the level of CO_2 in the atmosphere: on Earth most of the carbon dioxide is locked up in carbonate rocks, whereas only the 19 % is diffuse in the atmosphere. This prevents the temperature to get too hot, like it is on Venus where CO_2 is mostly distributed in the atmosphere and the temperature is hotter than on Mercury.

11.1.2.3 Shielding the Earth from Cosmic Rays

Life on our planet can develop because the atmosphere and the Earth's magnetic fields protects us from the high-energy particles and radiations coming from space. Cosmic rays are mostly degraded by the interaction with the atmosphere, which emerged in the first 500 million years of life from the vapor and gases expelled during the degassing of the planet's interior. Most of the gases of the atmosphere are thus the result of volcanic activity. In the early times, the Earth's atmosphere was composed of nitrogen and traces of CO_2 (<0.1 %), and very little molecular

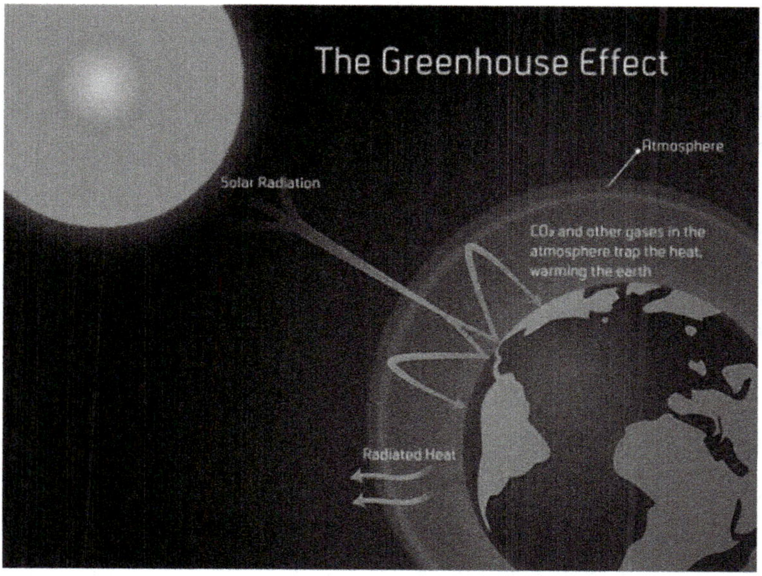

Fig. 11.3 Greenhouse gases trap and keep most of the infrared radiation in the low atmosphere. *Source* NASA

oxygen (O_2, which is now 21 %); the oxygen currently contained in the atmosphere increased as the result of photosynthesis by living organisms.

High-energy cosmic rays are not the only danger: also the charged particles coming from the Sun (the solar wind), and some of the Sun's radiation, can also be dangerous for life; for example, UV rays can damage proteins and DNA, destroying the molecules that product folic acid. Such a deficiency can cause (even in the 75 % of the cases) *spina bifida* in newborns. The ozone (O_3) layer in the upper atmosphere acts as a natural shield for UV rays, absorbing most of them.

The magnetic field of the Earth generates the magnetosphere that protects us from the lower energy cosmic rays that travel in the Galaxy (Fig. 11.4), in particular from the solar wind; the associated amount of energy would destroy life in our planet if there were no magnetosphere, that traps these particles and confines them.

Some of the cosmic rays are trapped in the Van Allen belts. The Van Allen belts were discovered in the late 1950s when Geiger counters were put on satellites. They are two main donut-shaped clouds.

- The outer belt is approximately toroidal, and it extends from an altitude of about three to ten Earth radii above the Earth's surface (most particles are around 4 to 5 Earth radii). It consists mainly of high-energy (0.1–10 MeV) electrons trapped by the Earth's magnetosphere.
- Electrons inhabit both belts; high-energy protons characterize the inner Van Allen belt, which goes typically from 0.2 to 2 Earth radii (1000–10000 km) above the Earth. When solar activity is particularly strong or in a region called the South

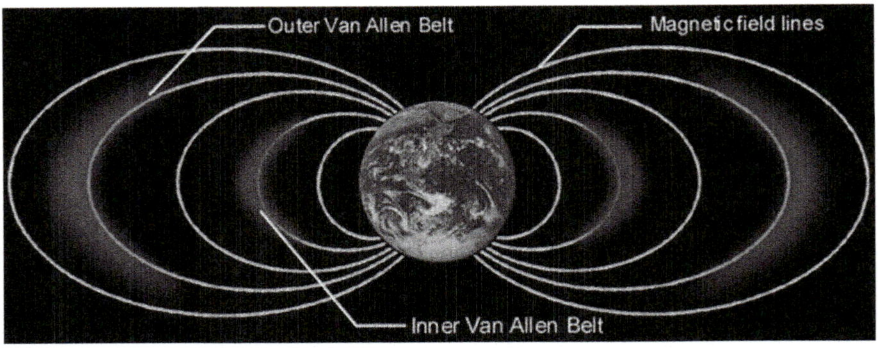

Fig. 11.4 The Earth's magnetic field and the Van Allen belts. From http://www.redorbit.com/

Atlantic Anomaly,[1] the inner boundary goes down to roughly 200 km above sea level. Energetic protons with energies up to 100 MeV and above are trapped by the strong magnetic fields in the region. The inner belt is a severe radiation hazard to astronauts working in Earth-orbit, and to some scientific instruments on satellite.

Close to the poles, charged particles trapped in the Earth's magnetic field can touch the atmosphere, and this reaction produces photons: this phenomenon is called Aurora Borealis in the North Pole, and Aurora Australis in the South Pole (Fig. 11.4).

11.1.3 Life in Extreme Environments

To understand what life is, and how to find life in the Universe, we can examine the most extreme living forms we know. We shall use this to define a *habitable region*—i.e., a region fulfilling a set of conditions under which we know life might occur. It is obviously not excluded that the actual conditions of life are wider.

Thanks to homeostasis, organisms on Earth called *extremophiles* exist, that can survive in extreme environments, such as:

- hot and cold places;
- salty and dry environments;
- acidic and basic places;
- environments of extreme pressure and radiations.

Let us analyze how such organisms can survive in these places.

- Hot and cold environments
  Examples of hot places are volcanoes in the deep oceans: there the temperature can goes from 50 to 80 °C and some organisms, called *hyperthermophiles*, evolved their proteins and membrane to resist at such high temperatures. An example of these

[1] The non-concentricity of the Earth and its magnetic dipole causes the magnetic field to be weakest in a region between South America and the south Atlantic; the solar wind can penetrate this region.

organisms is represented by the *Metharopyrus kandleri*, discovered on the wall of a black smokers in the Gulf of California at a depth of 2000 m, where the temperature range is 80–110 ℃; these organisms can survive and reproduce at 222 ℃.
Contrary to hyperthermophiles, there are organisms that can survive at very low temperatures. The Vostok Lake in the Arctic region is a good example of a cold place on Earth: *psychrophiles* evolved their membrane to survive at –15 ℃, as they create "antifreeze" proteins to keep their internal space liquid and protect their DNA.

- Salty and dry environments
  Halophiles can live in salty environments, with an external concentration of salt of 15–37% while keeping their own internal salts at a correct level; such organisms can be found in places like the Great Salt Lake (Utah, USA), Owens Lake (California, USA), the Dead Sea (Israel-Palestine-Jordan), and Makgadikgadi Pans (Botswana). Organisms called *xerophiles* can also live in very dry places with humidity lower than 1 %, like the Atacama desert in Chile.
- Acid and basic places
  Some organisms called *acidophiles* can live in acid places like sulfuric geysers, with pH ≤ 2.0, and some other organisms (called *alkaliphiles*) can live in basic places, with pH ≥ 8.5, like the Soda lakes in Africa, while still maintaining their own pH neutral.
- Extreme pressure and radiation
  On Earth we can find examples of organisms, called *piezophiles*, that survive at high pressures, like e.g., the Marianna Trench where pressure reaches 380 atmospheres, and organisms called *radio-resistant* that can survive high level of radiation that would ordinarily ionize and damage cells: the most radio resistant known organism is the *Thermococcus gammatolerans*, that can tolerate a radiation of gamma rays of 30000 Gy, and it was discovered in the Guaymas Basin.

In astrobiology, a specific class of extreme-resistant organisms is particularly important: the *polyextremophiles*, organisms that can simultaneously tolerate several extreme life conditions; an example is the *Deinococcus radiodurans*, that can live within high levels of radiation, at cold temperatures, and in dry environments.

11.1.4 The Kickoff

For thousands of years philosophers, scientists, and theologians have argued how life can come from non-life. Also in the interpretation of St. Augustine life came from non-living forms, although this biogenic process was mediated by God: "And God said, let the Earth bring forth the living creature after his kind, cattle, and creeping thing, and beast of the Earth after his kind: and it was so." Thus, God transferred to the Earth special life-giving powers, and using these powers the Earth generated plants and animals: "The Earth is said then to have produced grass and trees, that is, to have received the power of producing".

To avoid entering in controversial discussions, we shall assume here that at a certain time, somewhere in the Universe, life has emerged from non-life (abiogenesis), without invoking God and remaining within the realm of science boundaries.

Probably, if all the essential ingredients and appropriate conditions were present, life might have been generated in a long enough time—maybe having cosmic radiation as a catalyst. On the assumption that life originated spontaneously, many experiments showed that self-replicating molecules or their components could come into existence from their chemical components. However, there is no evidence to support the belief that life originated from non-life on Earth. We eagerly expect the day, maybe not far we think, when biologists on Earth will produce life from non-life.

An experiment by Miller and Urey in the 1950s used water, methane, ammonia, and hydrogen sealed inside a sterile glass flask connected to a flask half-full of liquid water. The liquid water in the smaller flask was heated to induce evaporation, and the water vapor was allowed to enter the larger flask. Continuous electrical sparks were fired between the electrodes to simulate lightning in the water vapor and gaseous mixture, and then the simulated atmosphere was cooled again so that the water condensed and trickled into a U-shaped trap at the bottom of the apparatus. The experiment yielded out of 20 aminoacids needed for life. Electric discharges might be present in some parts of the solar system, or the same catalytic effect could be provided by UV rays, or cosmic rays.

Note that the fact that the amount of oxygen in the beginning was scarce on Earth, is fundamental to the spontaneous development of life. Oxygen would destroy any organic material before abiogenesis—for example, when we die, we decay.

A popular hypothesis—called panspermia—is that life came to the Earth from other places, and that it can be transmitted to other places. According to this hypothesis, microscopic life—distributed by meteoroids, asteroids, or other small solar system bodies, or even pushed by radiation—may exist throughout the universe. The earliest clear evidence of life on Earth dates from 3.5 billion years ago, and is due to microbial mat fossils found in sandstone discovered in Australia. Cosmic dust permeating the universe contains complex organic substances. The panspermia hypothesis just pushes elsewhere and in some other time the problem of abiogenesis.

The problem of the very origin(s) of life, despite tremendous advances in biochemistry and in physics, remains a mystery. And also when, hopefully during the present century, the problem of the abiogenesis will be hopefully solved, it will certainly take longtime before understanding the transition from simple cells to complex organisms.

11.2 Life on the Solar System, Outside Earth

A first step to search for life in the solar system is to try to define a "habitable zone" that corresponds to the region where the temperature and the presence of water allow (or allowed) liquid water, there is an atmosphere, and appropriate conditions apply.

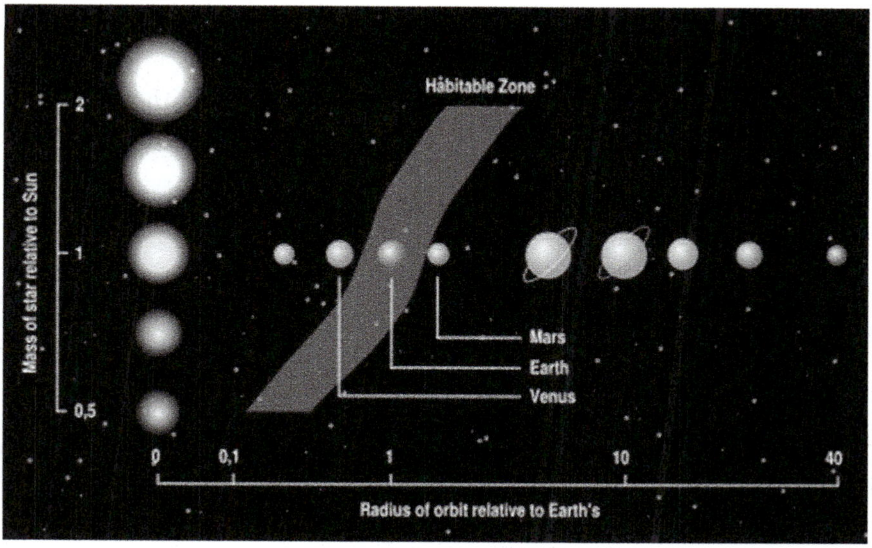

Fig. 11.5 The solar system's habitable zone. From http://www.universetoday.com/34731/habitable-planet/

Extremophiles suggest how life has a large range of definability, and that there is not a universal definition of habitability that suits every organism.

The habitable zone (Fig. 11.5) lies likely between Venus and Mars. This zone is not fixed because planets change their internal structure and conditions: they can get hotter or colder, and so they may not be forever habitable. Mercury, the first planet from the Sun—just 58 million km away—has a temperature ranging from 457 ℃ in the day to −173 ℃ in the night, not allowing the presence of liquid water or any chemical reaction; and it has no atmosphere, and its thus exposed to meteoric and cometary impacts.

The giant planets Jupiter and Saturn on the outer Solar System, having respectively a mass of 318 and 95 times the Earth's mass, seem also a very unlikely place for life. Jupiter, for example, is composed primarily of hydrogen and helium, plus small amounts of sulfur, ammonia, oxygen, and water. Temperatures and pressures are extreme. Jupiter does not have a solid surface, either—gravity can move a solid body to zones with high pressure. Saturn's atmospheric environment is also unfriendly due to strong gravity, high pressure, strong winds, and cold temperatures. Some of the moons of Jupiter and Saturn, however, can be thought as possible hosts of life.

Finally, the planets external to Saturn are too cold to be life-friendly.

11.2.1 Planets of the Solar System

In this section, we will discuss the possibility that conditions for life to develop may exist in other planets of the solar system, close to the habitability zone just defined.

11.2.1.1 Venus

Venus' structure and mass are very similar to the Earth's. However, although Venus, unlike Mercury, has an atmosphere, carbon-and water-based life cannot develop on Venus. The main problem is the high temperature of more than 400 ℃, due to the greenhouse effect. This effect is particularly strong on Venus because of volcano activity that fills the atmosphere with a large amount of gases. Pressure too is very high ($\sim$90 atmospheres), a condition that on Earth can be found only in the deepest oceans.

11.2.1.2 Mars

Mars orbits at approximately 228 million km from the Sun, and its mass is 11 % the Earth's (Fig. 11.6). Its atmosphere was originally similar to Venus' and Earth's (early) atmospheres, due to similar conditions during their formation.

Mars has always been one the best candidates for extraterrestrial life: a long time ago, Mars was probably warmer, probably it had liquid water (on its surface we recognize structures which can be attributed to past rivers, as shown in Fig. 11.6), and it must have had a deep atmosphere with gases produced by volcanic activity. But things have changed: volcanic activity stopped, and Mars has quickly lost its internal heat (due to its small mass) and most of its atmosphere. Mars was no longer protected by cosmic radiations and particles, and it began to cool down. This process lead to its current conditions: no liquid but frozen water, and temperatures impervious to life (summer: 27 to -130 ℃; winter: up to -143 ℃ at the poles). In 1976, two space probes (the Vikings) landed on the surface to find evidences of life, but found none.

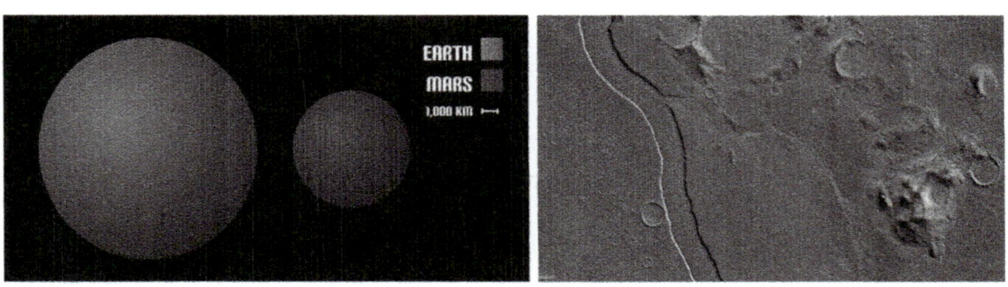

Fig. 11.6 *Left* Mars and Earth sizes. http://space-facts.com/mars-characteristics/. *Right* a structure on Mars' surface that can be related to the presence of ancient rivers. *Source* NASA

In July 2008, laboratory tests aboard NASA's Phoenix Mars Lander identified frozen water in a soil sample.

11.2.2 Satellites of Giant Planets

In this section, we will examine the particularities of three moons within the solar system: Europa (a satellite of Jupiter); and Titan and Enceladus (satellites of Saturn).

11.2.2.1 Europa

Jupiter's four main satellites are Io, Europa, Callisto, and Ganymede (the Galileian moons). Some them may have habitats capable of sustaining life: heated subsurface oceans of water may exist deep under the crusts of the three outer moons—Europa, Ganymede, and Callisto. The planned JUICE mission will study the habitability of these moons; Europa is seen as the main target.

It is the smallest of the four, having roughly the same size as our Moon. Its temperature reaches −160 °C. At such temperatures there is no liquid water, but what makes Europa so fascinating is hidden under its frozen surface: planetary geologists found out that only the oldest cracks appear to have drifted across the surface, which is rotating at a different rate respect to its interior, probably due to an underlying, 50 km thick, ocean layer of liquid water, methane, and ammonia. Figure 11.7 shows the hypothetical structure of Europa.

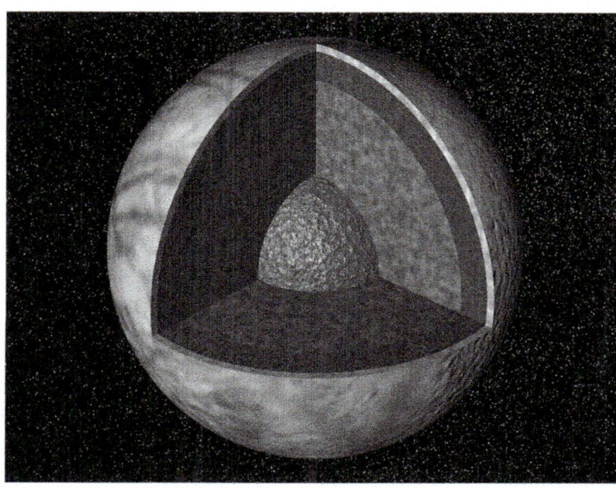

Fig. 11.7 Hypothetical structure of Europa: from outside in, we find the iced crust, the ocean, the rocky mantle, and the nuclear iron core. From NASA/Galileo Project and the University of Arizona

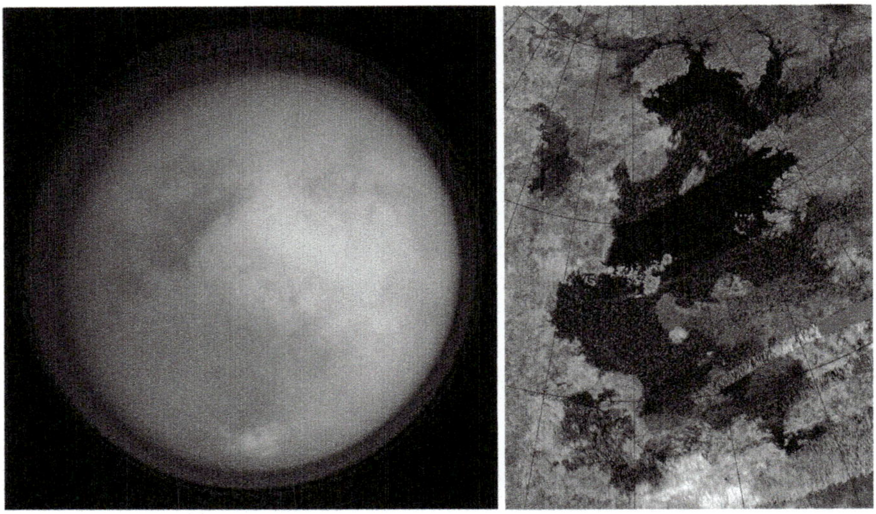

Fig. 11.8 *Left* Titan, a satellite of Saturn. *Right* Detail of Titan: note the presence of lakes on its surface. Credits: NASA (Cassini)

11.2.2.2 Titan

Titan (Fig. 11.8, left) is the largest moon of Saturn. Having a diameter of 5700 km it is bigger than Mercury, but has less than half its mass. Even if there is just a weak gasp of gas surrounding it, Titan has an atmosphere because it is situated in one of the coldest regions of the solar system. With a pressure of 1.5 atmospheres and a temperature of -170 °C, Titan can host solid, gas and liquid methane: in 1997 the Cassini space probe captured evidence of a giant methane lake, the Kraken sea (Fig. 11.8, right), that has a surface of about 400,000 km^2.

11.2.2.3 Enceladus

Discovered in 1789 by William Herschel, Enceladus is the sixth largest moon of Saturn; its diameter is about 500 km, roughly a tenth of that of Titan. It is mostly covered by ice and the surface temperature at noon only reaches -200 °C. In 2005, the Cassini spacecraft discovered that volcanoes near the south pole shoot geyser-like jets of water vapor, other volatiles, and solid material, including sodium chloride crystals and ice particles, into space; some of the water vapor falls back as snow. Cassini later discovered a large subsurface ocean of liquid water with a thickness of around 10 km. Enceladus is geologically active, and suffers tidal forces from another satellite (Dione). This moon could provide a habitable zone for microorganisms in the places where internal liquid from its interior is jetting out of its surface: some extremophiles living on Earth could live on Encelado's geysers. In view of the relatively accessible distance of Saturn's satellites, it is conceivable to think of a return space mission.

11.3 Search for Life Outside the Solar System

In the previous section, we saw how difficult is to find life on the other planets and moons of the solar system, because they hardly have the characteristics that life based on liquid water and carbon needs. But, what about the rest of the Galaxy? Are we alone?

Our Galaxy is 30 kpc large and it contains about 4×10^{11} stars, most with a planetary system: it seems impossible that we represent the only forms of life.

11.3.1 The "Drake Equation"

This "equation" was conceived in 1961 by American astronomer and astrophysicist Frank Drake, and it estimates the number of possibly communicative civilizations N_T in our Galaxy:

$$N_T = R \times f_p \times n_E \times f_l \times f_i \times f_c \times L \,,$$

where:

- R is the yearly rate at which suitable stars are born;
- f_p is the fraction of stars with a planetary system;
- n_E is the number of Earth-like planets per planetary system;
- f_l is the fraction of those Earth-like planets where life can develop;
- f_i is the fraction of these planets on which intelligent life can develop;
- f_c is the fraction of planets with intelligent life that could develop technology;
- L is the lifetime of a civilization with communicating capability.

Let us examine each factor in it. We can distinguish among astronomical, planetary, and biological factors.

Astronomical Factors. The astronomical factors are the star formation rate R in our Galaxy, and the fraction f_p which develop a planetary system. The star formation rate R is estimated to be 2–5 stars/yr. The current estimate of f_p is about (0.2–0.5): thanks to technological innovation in the search for extraterrestrial planets, we discovered that a large fraction of stars have a planetary system.

Planetary Factors. The planetary factor of the equation is n_E, which depends on the "habitable zone" that correspond to the zone of the solar system where the temperature liquid water. In the solar system, the habitable zone (Fig. 11.5) lies between Venus and Mars: the Earth is the only planet located in the solar system's habitable zone today. As we discussed before, this zone is not fixed because conditions change: planets can get hotter or colder, and so they may not be forever habitable. We estimate, based also on the recent results on searches for extrasolar planets, that $n_E \sim 1$.

Biological Factors. These are the most difficult to estimate, and the values we assume here are guesses. f_l the fraction of the planets where life can develop, f_i the fraction of planets where intelligent life can develop, f_c the fraction of intelligent beings who can develop communication technology, and L the lifetime of civilization. Even if it is very difficult to give a range to these factors because we do not know the probability to find life based on liquid water, Drake estimated f_l range from 0.1 to 1; more recent studies suggest $f_l \sim 1$. As for the other factors: $f_i \sim 0.01 - 1$, $f_c \sim 0.1 - 1$, while L is valued to have a range from 10^3 to 10^6 years, being 10000 years a conservative estimate.

The number of communicative civilization in our Galaxy can be estimated to be:

$$N_T \simeq (2 \times 0.5) \times 1 \times (1 \times 0.1 \times 0.1 \times 10\,000) \simeq 100\,.$$

As one can see, due to the large uncertainties one cannot exclude that the value is just one (we know it has to be at least one). On the contrary, it is very unlikely that the chance to have intelligent life in a galaxy like the Milky Way is smaller than 0.01, which, given the fact that there are 10^{11} galaxies in the Universe, makes it very likely that life exists in some other galaxies. However, communication with these forms of civilization are, at the present state of technology, unthinkable.

The Drake equation is based on the idea of development of life:

$$star \rightarrow planet \rightarrow water \rightarrow life \rightarrow intelligence$$

however we cannot give a definition of life, and we cannot rule out that some form of life could be based, for example, on silicon instead of carbon, so all the factors in Drake's equation could take different values if we assume that life could also develop in extreme condition, for example where no liquid water exists.

11.3.2 The Search for Habitable Planets

One of the critical factors in Drake's equation is n_E: its estimated value is influenced by the number of habitable planets that we discovered in a planetary system. Our estimate of n_E is continuously changing due to the technological evolution that allows us to discover new extrasolar planets.

When scientists search for new habitable planets orbiting a star, they first want to determine the position of the star's habitable zone, and to do that they study the radiations emitted by the star: in fact, bigger stars are hotter than the Sun and so their habitable zone is farther out; otherwise, the habitable zone of smaller stars is tighter.

Planets are very small and dark compared to the stars, and so how can they be detected? If scientists cannot look at the planets, they study the stars and the effects that orbiting planets have on them.

There are several methods to detect planets orbiting a star. There of the main methods are listed below.

- Doppler spectroscopy. This method is the most effective to detect extrasolar planets. It relies on the fact that a star moves, responding to the gravitational force of the planet. These movements affect the starlight spectrum, via a periodic Doppler shift of the wavelengths of the emission.
- Transit photometry. With this method scientists can detect planets by measuring the dimming of the star as the planet that orbits it passes between the star and the observer on the Earth: if this dimming is periodic, and it lasts a fixed length of time, likely there is a planet orbiting the star.
- Microlensing. This is the method to detect planets at the largest distances from the Earth. The gravitational field of a host star acts like a lens, magnifying the light of a distant background star. This effect occurs only when the two stars are aligned. If the foreground lensing star hosts a planet, then that planet's own gravitational field can contribute in an appreciable way to the lensing (Fig. 11.9). Since such a precise alignment is not very likely, a large number of distant stars must be monitored in order to detect such effect. The galactic center region has a large number of stars, and thus this method is effective for planets lying between Earth and the center of the galaxy.

Looking for habitable planets, scientists want to find planets with mass, density and composition similar to the Earth: in large planets like Jupiter, the gravity force would be too strong for life; too small planets could never trap an atmosphere.

In Fig. 11.10 (left) we show the distance from their sun of the extrasolar planets discovered up to now, and the energy flux of their host star; in Fig. 11.10 (right) we can see how the research evolved in the years, and how, thanks to the scientific and technological innovation, we can find now planets with mass similar to the Earth.

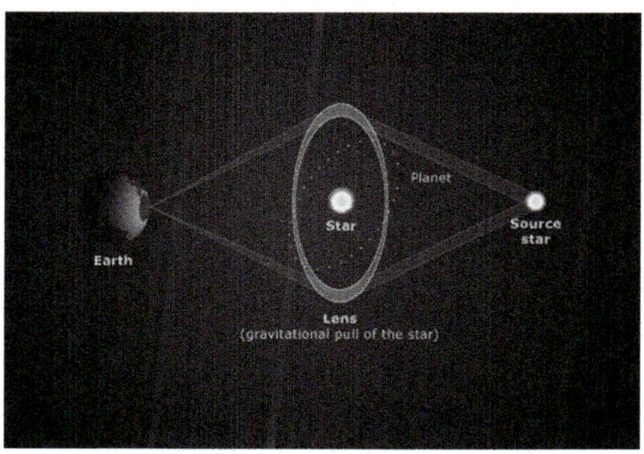

Fig. 11.9 A massive object will bend the light of a bright background object (the source). This can generate distorted, magnified, and brightened images of the background source. The effect can be enhanced if a planet orbiting the lens is aligned. http://www.teara.govt.nz/en/diagram/8008/gravitational-microlensing

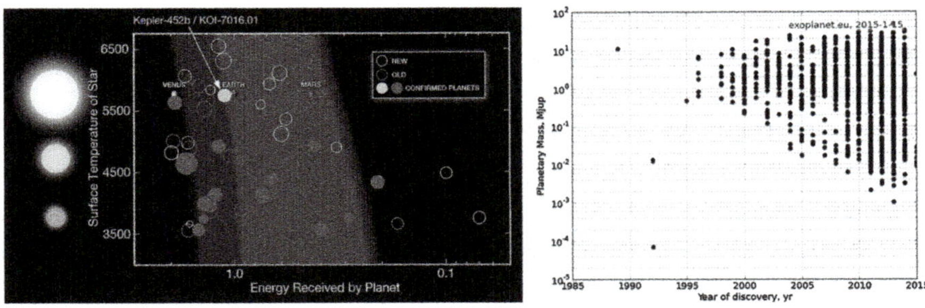

Fig. 11.10 *Left* discovered extrasolar planets in function of the distance from their sun, and the stellar flux. The Earth, Venus and Mars are included as a reference. Credit: NASA. *Right* plot of year of discovery of extrasolar planets versus their mass (normalized to Jupiter's mass). Credit: Wikimedia Commons

The first extrasolar planet (HD114762b) was discovered in 1989, its mass is 10 times Jupiter's mass.

The most important mission to detect Earth-like planets outside our solar system is nowadays the Kepler Mission; the spacecraft was launched on March 7th, 2009. A photometer analyzed over 145,000 stars in the Cygnus, Lyra, and Draco constellations, to detect a dimming of brightness which could be the proof of the existence of an orbiting planet.

The most important result of the Kepler mission is the discovery of the first Earth-like planet, in the habitable zone of the Cygnus Constellation: Kepler-186f–and more recently, of a possible Earth-like planet orbiting a Sun-like star: Kepler-452b.

11.3.2.1 Kepler-186f

On April 17, 2014, NASA announced the discovery of the first extrasolar Earth-like planet, orbiting a M-dwarf in the first habitable zone discovered outside our solar system: Kepler-186f (Fig. 11.11), in the constellation Cygnus, 500 lightyears from us. This planet was found by the NASA's Kepler Space Telescope, and many of its characteristics, composition and mass, make it similar to our planet. M-dwarfs, which have masses in the range of 0.1–0.5 solar masses, make up about 75 % of the stars within our galaxy.

Kepler-186f has a period of revolution around its sun of 130 days, it is likely to be rocky, and it is the first new discovered planet with dimensions similar to the Earth: in fact its radius is 1.1 times the Earth's one, and its estimated mass is 0.32 the Earth's one. It receives from its star one third the energy that Earth gets from the Sun, although it is much closer (just 0.36 astronomical units AU, where 1 AU is average distance between the Earth and the Sun): it is thus a cold planet, and it could not host human life.

Kepler-186f is located in a five-planet system; the other four planets in this system, Kepler-186b, Kepler-186c, Kepler-186d, and Kepler-186e, orbit around their sun

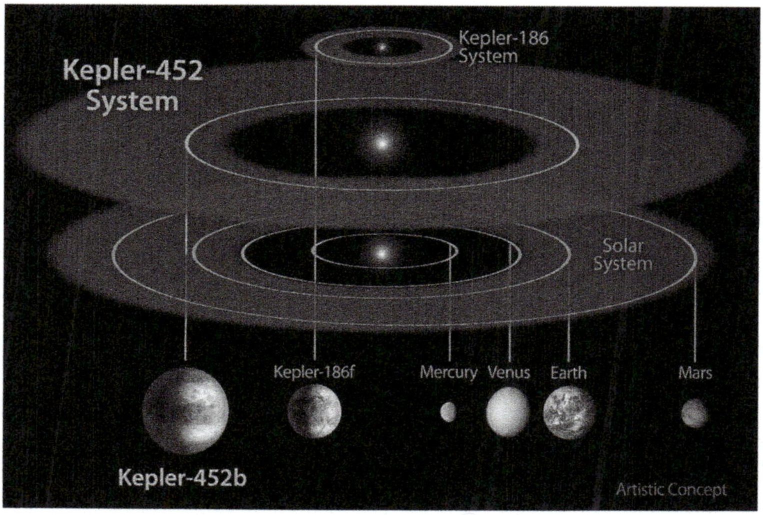

Fig. 11.11 Kepler-186f and Kepler-452b in their solar systems: comparison with the habitable zone of our Solar System. Credit: NASA

with periods of 4, 7, 13, and 22 days, respectively, they are too hot for life as we know it to develop.

11.3.2.2 Kepler-452b

On 23 July 2015, NASA announced the discovery of the first extrasolar Earth-like planet (potentially rocky) within the habitable zone of a Sun-like star (G star). At a distance of 1400 ly from the Earth and located in the constellation Cygnus, it has a revolution period of 385 days. The star is six billion years old, i.e.,1.5 billion years older than our Sun; Kepler-452b is receiving a power close to the one we receive from our Sun.

The similarities with the Earth are amazing; and the future promises more candidates possibly able to host a carbon-based life.

11.3.3 The Fermi Paradox

Enrico Fermi in 1951 tried to give an explanation to this lack of detection of alien communication; this is called the "Fermi paradox": where is everybody? Indeed, given the Drake's equation, we could expect that a contact could have been already established. Several possible answers have been suggested.

- We are alone. We have not received any signal just because nobody sent it, and life needs some proprieties, like liquid water, carbon, right temperature, that we can find just on Earth. But this opinion is difficult to accept—also because we cannot just give a univocal definition of life since life can evolve and develop also in places where, for example, water can be found only in frozen state.
- The evolution of civilizations able to communicate not last for long. There are two main reasons why a civilization can fall:
  1. Cultural reasons: populations evolute enough destroys themselves.
  2. Natural reasons: catastrophic events, like meteorites or cometary impacts.
- Communicative extraterrestrial civilizations do exist, but they are too far away from us. The galaxy is so extended (30 kpc) that any signal would take millions of years to get from a planet to another, and in this time a civilization could even become extinct.
- They do not want to communicate with us, maybe because they are afraid of our possible reaction. If we knew there are civilizations weaker than us in our Galaxy, would we attack them?
- We cannot understand their signals. All our attempts of communication are based on electromagnetic waves, but maybe they have already sent us signal based on neutrinos, or gravitational waves that we are not able to detect yet.

11.3.4 Listening to Messages from Space

One of the main unknowns is how could aliens communicate with us, and how can we receive and decrepit their signals. In this section we will describe what kind of signals we are trying to detect.

Let us first examine how far a signal can reach. First of all, we must know ho much energy a civilization can use for transmitting.

In 1964, Kardashev defined three levels of civilizations, based on the order of magnitude of power available to them:

- Type 1. Technological level close to the level presently attained on Earth, with energy consumption at $\simeq 4 \times 10^{12}$ W (four order of magnitude less than the total solar insulation).
- Type 2. A civilization capable of harnessing the energy radiated by its own star (if the host star is Sun-like, $\simeq 4 \times 10^{26}$ W).
- Type 3. A civilization in possession of energy on the scale of its own galaxy, (for the Milky Way, about 4×10^{37} W).

The above jumps might look too steep. Carl Sagan suggested defining intermediate values by interpolating the values given above:

$$K = \frac{\log_{10} P - 6}{10}$$

where value K is a civilization's rating and P is the power it controls. Using this extrapolation, humanity's civilization in 2011—average power was 17.4 TW—was of 0.72.

In general, the inverse square law for intensity applies here: $I = P/4\pi d^2$, with P the power of the signal. A rule of thumb for the distance which can be reached with a radio signal (most economic within the electromagnetic spectrum), a rule of thumb of the distance that can be reached with top technological devices at present technology (50 m dish size both for transmitting and receiving, best gains) is:

$$d \simeq 1\,\mathrm{kpc}\sqrt{\frac{P}{1\,\mathrm{GW}}}\,. \tag{11.1}$$

It seems thus difficult for a Type 1 civilization to reach beyond the scale of a galaxy based on radio communication.

11.3.4.1 SETI@home

The term SETI (search for extraterrestrial intelligence) refers to a number of activities to search for intelligent extraterrestrial life. As already discussed, communicating in space can be quite prohibitive, the main reason being cosmic distances. Receiving the visit of a spacecraft is extremely unlikely, so SETI is looking for radio waves that might have been sent by extraterrestrial intelligent civilizations (radio waves since this is the most economical way of communicating from the energetic point of view). SETI looks for "narrow band transmissions" which can be produced only by artificial equipment: the problem with these communications is that they are very difficult to single out from the many of them produced on Earth. Not even the world's biggest supercomputers could manage the task of studying all these noises of the Universe. So, a Internet-based, public-volunteer computing project, SETI@home, was set up: after downloading and installing an appropriate software on a personal computer, the executable gets switched on when the computer is not in use, receives 300 kb dates by Internet from the Arecibo radio telescope in Puerto Rico, and tries to find regularities in these data.

The rationale in the search is that we expect that the communication will be narrow-band, and periodical; thus we can isolate them with Fourier analysis or autocorrelation studies, tested at different wavelengths.

If aliens ever sent us messages, the real problem is if we can receive them or not, and distance being the main thing that could prevent signals from reaching us: as seen in the previous subsection, the distance from which a telescope could detect an extraterrestrial transmission depends on the sensitivity of the receiver and on the strength and type of the signal.

Most of the SETI searches are based on radio waves, but it is possible that the aliens would try to communicate using visible light, or other forms of energy/particles. Some SETI efforts are indeed addressed to search such signals. Distant civilizations might choose to communicate in our ways, like with ultraviolet light or X-rays—in particular infrared light has a potential value because it can penetrate interstellar dust—but all these forms of light are much more expensive in terms of energy cost. Some people have suggested, that extraterrestrial civilizations might use neutrinos, or gravitational waves but the problem with these kind of communication technologies is that they might involve physics we are not able to manage, yet. Some carriers of information offer the possibility to beam the emission; lasers in the visible range for example. Using current technology available on Earth (10 m reflectors as the transmitting and receiving apertures and a 4 MJ pulsed laser source), a 3 ns optical pulse could be produced which would be detectable at a distance of 1000 ly, outshining starlight from the host system by a factor of 10^4. It is not unlikely that our civilization could reach the full Galaxy with a beamed signal—which means that we should be able to choose our targets. In this case, Cherenkov telescopes, being equipped with the largest mirrors, would be the ideal target for aliens. Indeed, some effort has been done to look for extraterrestrial signals in Cherenkov telescopes, with no success.

11.3.5 Sending Messages to the Universe

We also try to communicate with alien civilizations, hoping that they will decrypt our signal and answer.

In this section, we outline how we try to communicate with technologically advanced civilizations outside the solar system. The most effective way possible for our technology is to broadcast a radio signal; however, several other attempts were done, in particular sending a "message in a bottle" (a handcraft plate in a satellite).

11.3.5.1 Arecibo

This message, sent into space by the Arecibo radio telescope in 1974, is composed by 1679 symbols. This number is the product of two prime numbers, in fact $1679 = 23 \times 73$. The message was aimed at the current location of globular star cluster M13 some 25,000 light-years away because M13 is a large and close collection of stars. We anyway welcome possible decoders from different galaxies.

Decrypting the binary message (Fig. 11.12), translating the number 0 with a black square and the number 1 with a colored square, the result is a matrix 23×73 (Fig. 11.13), that contains some messages about our world:

```
0000001010101000000000000101000001010000000100100010001000100101100 10
1010101010101010010010000000000000000000000000000000000000011000000000
0000000000110100000000000000000000011010000000000000000000101010000000000
000000000111110000000000000000000000000000000000110000111000110000011000
10000000000000110010000110100011000110000110101111101111101111101111 1
00000000000000000000000000100000000000000000001000000000000000000000000
0000100000000000000000011111100000000000000111110000000000000000000000000
110000110000111000110001000000010000000001000011010000110001110011010
111110111110111110111110000000000000000000000000000100000011000000000010
00000000001100000000000000000100000110000000000111111000001100000011111
00000000001100000000000000100000000100000000010000010000001100000001000
0000110000110000001000000000001100010000110000000000000000011001100000000
000000110001000011000000000110000110000001000000010000001000000001000
0010000000110000000010001000000001100000000100010000000001000000001000
0010000000100000001000000010000000000001100000000011000000001100000000
00100011101011000000000000100000001000000000000001000001111100000000000
0010000010111010010110110000001001110010011111101110000111000000110111
000000000101000001110110010000001010000011111100100000010100000110000
0010000011011000000000000000000000000000000000011100000100000000000000
0011101010001010101010100111000000000101010100000000000000000101000000
000000000111110000000000000000011111111100000000000011100000001110000000
000110000000000011000000011010000000000101100000110011000000011001 1000
01000101000001010001000010001001000100100010000000001000101000100000000
00000100001000010000000000000100000000010000000000000001001010000000000
01111001111101001111000
```

Fig. 11.12 The Arecibo message in binary form. *Source* Wikimedia Commons

Now, we will analyze the part that compose the message in the Fig. 11.13, from the top to the bottom set by set.

1. Reading from left to right, vertically from the bottom to the top, there are the numbers from 1 to 10 written in binary form, where in each column the black square at the bottom signs the beginning of the number: for example, the first number written in binary form of the left is $1 = 1 \times 2^0$ which is 1 in decimal form; then, we can find the number written in binary form 10 which is $0 \times 2^0 + 1 \times 2^1 = 2$ in decimal form; then, the number 111 is written in binary form, that correspond to $1 \times 2^0 + 1 \times 2^1 + 1 \times 2^2 = 3$, and so on. The numbers 8, 9, 10 are written on two columns.

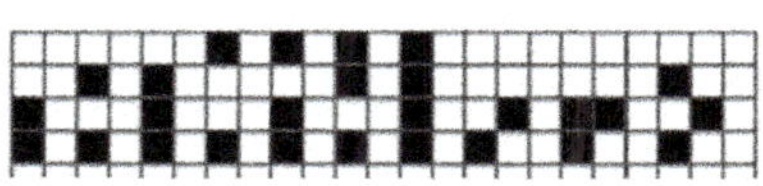

2. The atomic numbers 1, 6, 7, 8, and 15 of , respectively, hydrogen, carbon, nitrogen, oxygen and phosphorus, i.e., the component of the DNA.

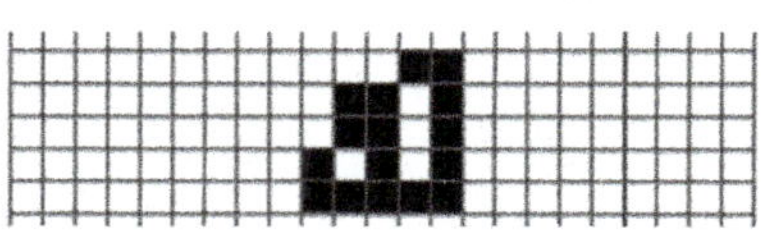

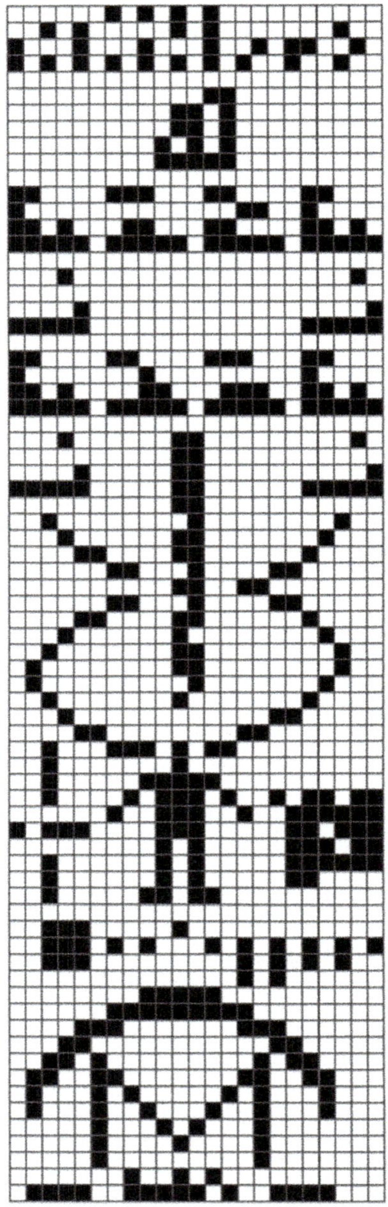

Fig. 11.13 Decrypting the binary message: we start with a rectangle 23 × 73. By Pengo (based on public information) (GFDL http://www.gnu.org/copyleft/fdl.html), via Wikimedia Commons

3. Nucleotides present in the DNA: deoxyribose (C_5H_7O), adenine ($C_5H_4N_5$), thymine ($C_5H_5N_2O_2$), phosphate (PO_4), cytosine ($C_4H_4N_3O$) and guanine ($C_5H_4N_5O$).
   They are described as a sequence of the five atoms that appear on the preceding line. For example, on the top left the number 75010 is written in binary form,

that matches the deoxyribose C_5H_7O: 7 atoms of hydrogen, 5 atoms of carbon, 0 atoms of nitrogen, 1 atom of oxygen, and 0 atoms of phosphorus.

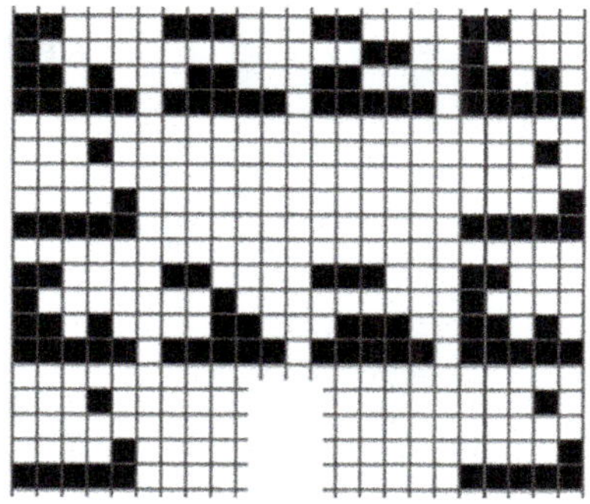

4. The helix structure of the DNA, and the central bar, read from the bottom to the top, where the lower black square on the left marks the beginning, represents the number of the nucleotides: in fact the bar is the number in binary form 11111111111101111111101101011110, that is in decimal form 4294441822 which was believed to be the case in 1974, when the message was sent—we think now that there are about 3.2 billion nucleotides that form the DNA.

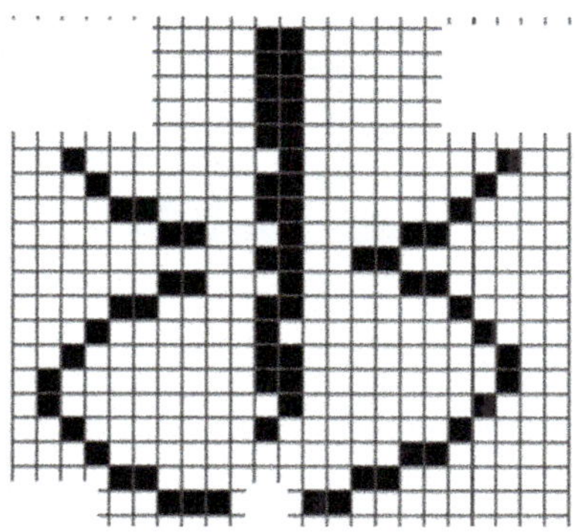

5. In the center a figure of a man; on the left the medium height of a man, i.e., 1.764 m, which is the product of 14, written in binary form by the horizontal line where the black square on the left marks the beginning, times the wavelength of the message (126 mm); on the right, the size of human population in binary form, and in this case it has to be read horizontally where the upper black square of the left marks the beginning: the number is 000011111111110111111011111111110110 (4292853750 in decimal form).

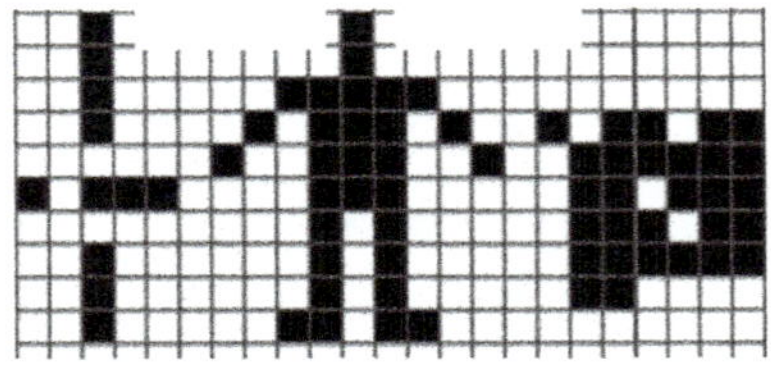

6. Our solar system, where the Earth is offset and the human figure is shown "standing on" the figure of the Earth.

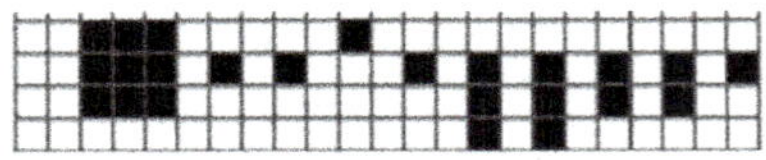

7. A drawing of the Arecibo Telescope, in Puerto Rico, with below the dimension of the telescope, 306.18 m, which is the product of the number 2.430 written in binary form (100101111110) in the two bottom rows, read horizontally and the black square on the low right in the central block marks the beginning of the number, multiplied by the wavelength of the message.

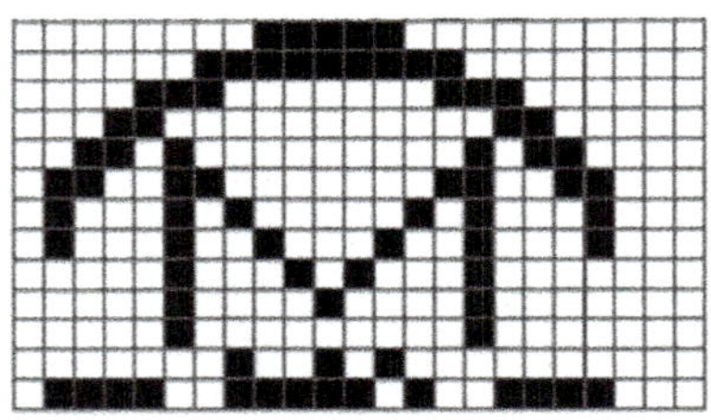

11.3.5.2 Pioneer 10

Pioneer 10 is an American space probe launched in 1972. A gold plaque (Fig. 11.14) was placed onboard (and also onboard the subsequent mission, Pioneer 11), containing information about the space mission and mankind. The information that we can find on the plaque are:

1. Hyperfine transition for neutral hydrogen, the most abundant element. The interaction between the proton and the neutron magnetic dipole moments in the ground state of neutral hydrogen results in a slight increase in energy when the spins are parallel, and a decrease when antiparallel. The transition between the two states causes the emission of a photon at a frequency about 1420 MHz, which means a period of about 7.04×10^{-10} s, and a wavelength of $\sim$21 cm. This is the key to read the message.

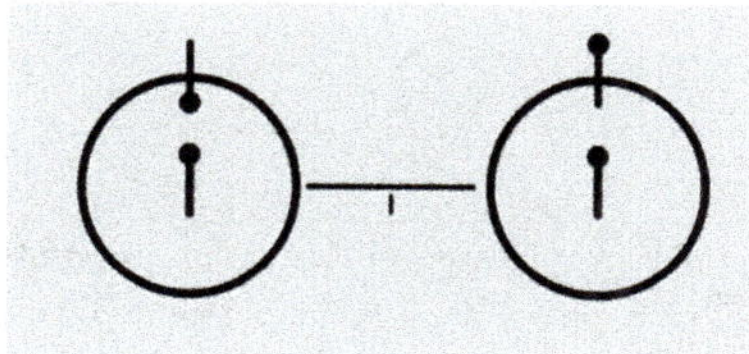

2. The figures of a man and a woman; between the vertical column brackets that indicate the height of the woman, the number eight can be seen in binary form 1000, where the vertical line means 1 and the horizontal lines mean 0: in unit of the wavelength of the hyperfine transition of the hydrogen, the result is 8×21 cm $= 168$ cm, which was at the time the average height of a woman. The right hand of the man is raised, as a good will sign, and it can even be a way to show the opposable thumb and how the limbs can be moved.

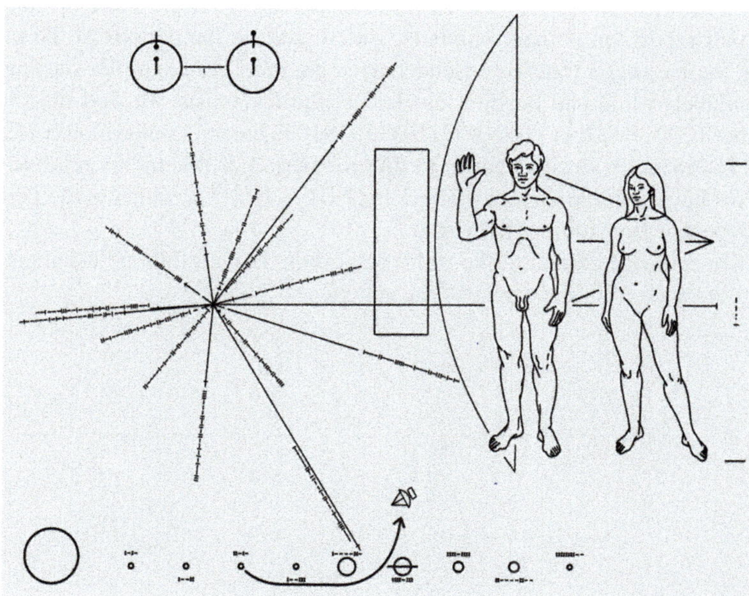

Fig. 11.14 The plaque onboard the Pioneer 10. *Source* Wikimedia Commons

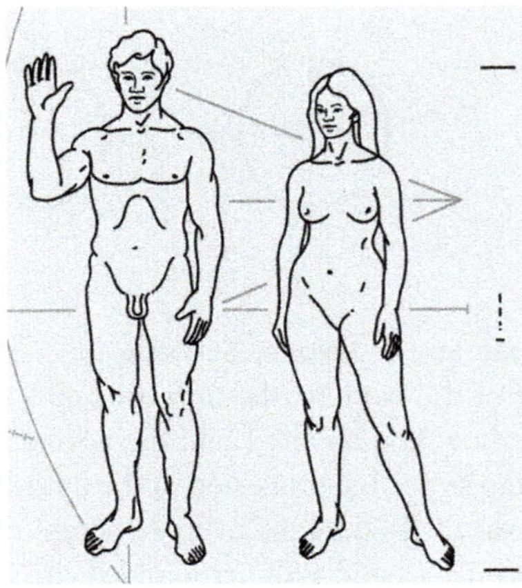

3. Relative position of the Sun to the center of the Galaxy, and 14 pulsars with their period; on the left, we can see 15 lines emanating from the same origin, 14 of the lines report long binary numbers, which indicate the periods of the pulsars, using the hydrogen transition frequency as the unit. For example, starting from the unlabeled line and heading clockwise, the fist pulsar we find matches the number 1000110001111100100011011101010 in binary form, which correspond to 1178486506 in decimal form: to find the period of this pulsar relative to the Sun we have to multiply this number by 7.04×10^{-10} s, which is the period of the hyperfine transition of hydrogen.
   The fifteenth line extend to the right, behind the human figures: it indicates the Sun's distance from the center of the galaxy.

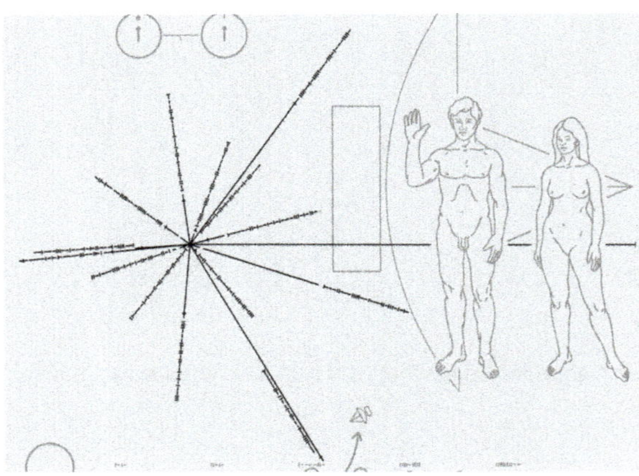

4. The solar system with the trajectory of Pioneer. In this section, it is indicated the distances of every single planet from the Sun, and they are relative to Mercury's distance from the Sun, which has a unit of 1010 in binary form, that is 10 in decimal form: for example the number relative to Saturn is 11110111, that is 247

in decimal from and this means that Saturn is 247 times further from the Sun than Mercury.

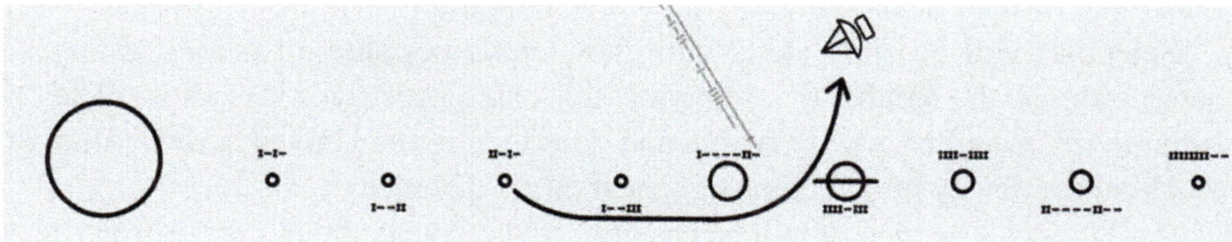

5. Silhouette of Pioneer relative to the size of the humans.

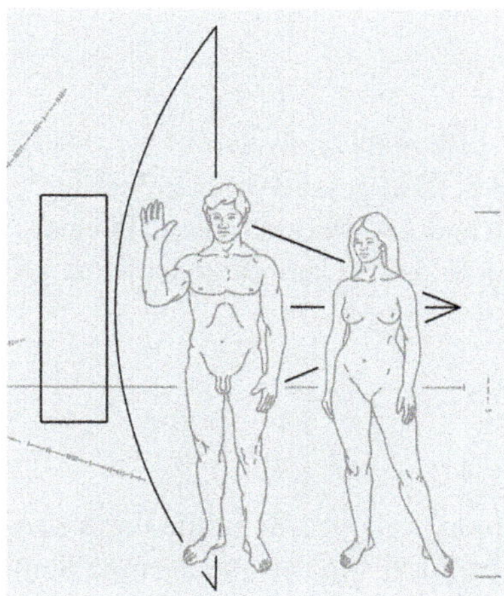

Given Eq. 11.1, the possibility to reach a distance of 25,000 light-years is very optimistic.

11.4 Conclusions

We discussed how technological and scientific innovation is contributing to discover new Earth-like planets where life could develop. But, what will happen in 30 years? What will we be able to discover? Where will the next mission take us? What will scientists study?

Scientists will analyze the light of planets around their stars to detect oxygen and other complex molecules that suggest the presence of an atmosphere, map other Earth-like planets, and study the presence of liquid water and volcanic activity on them. The next mission after Kepler will be the ESA satellite PLATO, foreseen for the year 2024. With an array of 34 telescopes mounted on a sun-shield, PLATO will allow 5 % of the sky to be monitored to incredible accuracy at any one time. Thanks to PLATO more than a million stars could be scrutinized for Earth-sized planets, providing a sensitivity an order of magnitude higher than Kepler. Plato will also not be tied down into staring at the same stars, instead monitoring 50 % of the sky on eight 30-day positions and two longer 3-year fields. This might allow a hundred of Earth-like planets with potentially habitable temperatures to be discovered.

Scientists will possibly study with new telescopes stars of nearby galaxies, to better estimate the number of communicative extraterrestrial civilizations. They will listen to the sound of gravity waves and neutrinos in the Universe: this will give to mankind the ability of detecting signals at larger distances.

If all this will happen, it will be easier to find evidence of extraterrestrial life, but this will require innovation, investment and perseverance.

Further Reading

[F11.1] L. Dartnell, "Life in the Universe", Oneworld 2007.
[F11.2] J. Chela-Flores, "The science of astrobiology", Springer 2011.
[F11.3] W.T. Sullivan and J.A. Baross (eds.), "Planets and Life: The Emerging Science of Astrobiology", Cambridge University Press 2007.

Exercises

1. The mass of the Moon is about 1/81 of the Earth's mass, and the mass of the Sun is 333,000 times the Earth's mass. The average Sun-Earth distance is 150×10^6 km, while the average Moon–Earth distance is 0.38×10^6 km (computed from center to center).
   (a) What is the ratio between the gravitational forces by the Moon and by the Sun?
   (b) What is the ratio between the tidal forces (i.e., between the differences of the forces at two opposite sides of the Earth along the line joining the two bodies)?
2. What is the maximum temperature for which the Earth could trap an atmosphere containing molecular oxygen O_2?
3. Assuming that the Sun is a blackbody emitting at a temperature of 6000 K (approximately the temperature of the photosphere), what is the temperature of Earth at equilibrium due to the radiation exchange with the Sun? Assume the Sun's radius to be 7000 km, i.e., 110 times the Earth's radius.

Appendix A
Periodic Table of the Elements

PERIODIC TABLE OF THE ELEMENTS

1 IA	2 IIA	3 IIIB	4 IVB	5 VB	6 VIB	7 VIIB	8	9 VIII	10	11 IB	12 IIB	13 IIIA	14 IVA	15 VA	16 VIA	17 VIIA	18 VIIIA
1 H Hydrogen 1.00794																	2 He Helium 4.002602
3 Li Lithium 6.941	4 Be Beryllium 9.012182											5 B Boron 10.811	6 C Carbon 12.0107	7 N Nitrogen 14.0067	8 O Oxygen 15.9994	9 F Fluorine 18.9984032	10 Ne Neon 20.1797
11 Na Sodium 22.98976928	12 Mg Magnesium 24.3050											13 Al Aluminum 26.9815386	14 Si Silicon 28.0855	15 P Phosph. 30.973762	16 S Sulfur 32.065	17 Cl Chlorine 35.453	18 Ar Argon 39.948
19 K Potassium 39.0983	20 Ca Calcium 40.078	21 Sc Scandium 44.955912	22 Ti Titanium 47.867	23 V Vanadium 50.9415	24 Cr Chromium 51.9961	25 Mn Manganese 54.938045	26 Fe Iron 55.845	27 Co Cobalt 58.933195	28 Ni Nickel 58.6934	29 Cu Copper 63.546	30 Zn Zinc 65.38	31 Ga Gallium 69.723	32 Ge German. 72.64	33 As Arsenic 74.92160	34 Se Selenium 78.96	35 Br Bromine 79.904	36 Kr Krypton 83.798
37 Rb Rubidium 85.4678	38 Sr Strontium 87.62	39 Y Yttrium 88.90585	40 Zr Zirconium 91.224	41 Nb Niobium 92.90638	42 Mo Molybd. 95.96	43 Tc Technet. (97.90722)	44 Ru Ruthen. 101.07	45 Rh Rhodium 102.90550	46 Pd Palladium 106.42	47 Ag Silver 107.8682	48 Cd Cadmium 112.411	49 In Indium 114.818	50 Sn Tin 118.710	51 Sb Antimony 121.760	52 Te Tellurium 127.60	53 I Iodine 126.90447	54 Xe Xenon 131.293
55 Cs Cesium 132.9054519	56 Ba Barium 137.327	57–71 Lantha-nides	72 Hf Hafnium 178.49	73 Ta Tantalum 180.94788	74 W Tungsten 183.84	75 Re Rhenium 186.207	76 Os Osmium 190.23	77 Ir Iridium 192.217	78 Pt Platinum 195.084	79 Au Gold 196.966569	80 Hg Mercury 200.59	81 Tl Thallium 204.3833	82 Pb Lead 207.2	83 Bi Bismuth 208.98040	84 Po Polonium (208.98243)	85 At Astatine (209.98715)	86 Rn Radon (222.01758)
87 Fr Francium (223.01974)	88 Ra Radium (226.02541)	89–103 Actinides	104 Rf Rutherford. (267.122)	105 Db Dubnium (268.125)	106 Sg Seaborg. (271.133)	107 Bh Bohrium (270.134)	108 Hs Hassium (269.134)	109 Mt Meitner. (276.151)	110 Ds Darmstadt. (281.162)	111 Rg Roentgen. (280.164)	112 Cn Copernicium (277)		114 Fl Flerovium (289)		116 Lv Livermorium (288)		

Lanthanide series	57 La Lanthan. 138.90547	58 Ce Cerium 140.116	59 Pr Praseodym. 140.90765	60 Nd Neodym. 144.242	61 Pm Prometh. (144.91275)	62 Sm Samarium 150.36	63 Eu Europium 151.964	64 Gd Gadolin. 157.25	65 Tb Terbium 158.92535	66 Dy Dyspros. 162.500	67 Ho Holmium 164.93032	68 Er Erbium 167.259	69 Tm Thulium 168.93421	70 Yb Ytterbium 173.054	71 Lu Lutetium 174.9668
Actinide series	89 Ac Actinium (227.02775)	90 Th Thorium 232.03806	91 Pa Protactin. 231.03588	92 U Uranium 238.02891	93 Np Neptunium (237.04817)	94 Pu Plutonium (244.06420)	95 Am Americ. (243.06138)	96 Cm Curium (247.07035)	97 Bk Berkelium (247.07031)	98 Cf Californ. (251.07959)	99 Es Einstein. (252.0830)	100 Fm Fermium (257.09510)	101 Md Mendelev. (258.09843)	102 No Nobelium (259.1010)	103 Lr Lawrenc. (262.110)

Source From K.A. Olive et al. (Particle Data Group), Chin. Phys. C 38 (2014) 090001

A. De Angelis and M.J.M. Pimenta, *Introduction to Particle and Astroparticle Physics*, Undergraduate Lecture Notes in Physics,
DOI 10.1007/978-88-470-2688-9

Appendix B
Properties of Materials

Material	Z	A	$\langle Z/A \rangle$	Nucl.coll. length λ_T {g cm^{-2}}	Nucl.inter. length λ_I {g cm^{-2}}	Rad.len. X_0 {g cm^{-2}}	$dE/dx\vert_{min}$ { MeV g^{-1}cm^2}	Density {g cm^{-3}} ({gℓ^{-1}})
H_2	1	1.00794(7)	0.99212	42.8	52.0	63.04	(4.103)	0.071(0.084)
D_2	1	2.01410177803(8)	0.49650	51.3	71.8	125.97	(2.053)	0.169(0.168)
He	2	4.002602(2)	0.49967	51.8	71.0	94.32	(1.937)	0.125(0.166)
Li	3	6.941(2)	0.43221	52.2	71.3	82.78	1.639	0.534
Be	4	9.012182(3)	0.44384	55.3	77.8	65.19	1.595	1.848
C diamond	6	12.0107(8)	0.49955	59.2	85.8	42.70	1.725	3.520
C graphite	6	12.0107(8)	0.49955	59.2	85.8	42.70	1.742	2.210
N_2	7	14.0067(2)	0.49976	61.1	89.7	37.99	(1.825)	0.807(1.165)
O_2	8	15.9994(3)	0.50002	61.3	90.2	34.24	(1.801)	1.141(1.332)
F_2	9	18.9984032(5)	0.47372	65.0	97.4	32.93	(1.676)	1.507(1.580)
Ne	10	20.1797(6)	0.49555	65.7	99.0	28.93	(1.724)	1.204(0.839)
Al	13	26.9815386(8)	0.48181	69.7	107.2	24.01	1.615	2.699
Si	14	28.0855(3)	0.49848	70.2	108.4	21.82	1.664	2.329
Cl_2	17	35.453(2)	0.47951	73.8	115.7	19.28	(1.630)	1.574(2.980)
Ar	18	39.948(1)	0.45059	75.7	119.7	19.55	(1.519)	1.396(1.662)
Ti	22	47.867(1)	0.45961	78.8	126.2	16.16	1.477	4.540
Fe	26	55.845(2)	0.46557	81.7	132.1	13.84	1.451	7.874
Cu	29	63.546(3)	0.45636	84.2	137.3	12.86	1.403	8.960
Ge	32	72.64(1)	0.44053	86.9	143.0	12.25	1.370	5.323
Sn	50	118.710(7)	0.42119	98.2	166.7	8.82	1.263	7.310
Xe	54	131.293(6)	0.41129	100.8	172.1	8.48	(1.255)	2.953(5.483)
W	74	183.84(1)	0.40252	110.4	191.9	6.76	1.145	19.300
Pt	78	195.084(9)	0.39983	112.2	195.7	6.54	1.128	21.450
Au	79	196.966569(4)	0.40108	112.5	196.3	6.46	1.134	19.320
Pb	82	207.2(1)	0.39575	114.1	199.6	6.37	1.122	11.350
U	92	[238.02891(3)]	0.38651	118.6	209.0	6.00	1.081	18.950
Air (dry, 1 atm)			0.49919	61.3	90.1	36.62	(1.815)	(1.205)
Shielding concrete			0.50274	65.1	97.5	26.57	1.711	2.300
Borosilicate glass (Pyrex)			0.49707	64.6	96.5	28.17	1.696	2.230
Lead glass			0.42101	95.9	158.0	7.87	1.255	6.220
Standard rock			0.50000	66.8	101.3	26.54	1.688	2.650
Methane (CH_4)			0.62334	54.0	73.8	46.47	(2.417)	(0.667)
Ethane (C_2H_6)			0.59861	55.0	75.9	45.66	(2.304)	(1.263)
Propane (C_3H_8)			0.58962	55.3	76.7	45.37	(2.262)	0.493(1.868)
Butane (C_4H_{10})			0.59497	55.5	77.1	45.23	(2.278)	(2.489)
Octane (C_8H_{18})			0.57778	55.8	77.8	45.00	2.123	0.703
Paraffin ($CH_3(CH_2)_{n\approx 23}CH_3$)			0.57275	56.0	78.3	44.85	2.088	0.930
Nylon (type 6, 6/6)			0.54790	57.5	81.6	41.92	1.973	1.18
Polycarbonate (Lexan)			0.52697	58.3	83.6	41.50	1.886	1.20
Polyethylene ($[CH_2CH_2]_n$)			0.57034	56.1	78.5	44.77	2.079	0.89
Polyethylene terephthalate (Mylar)			0.52037	58.9	84.9	39.95	1.848	1.40
Polyimide film (Kapton)			0.51264	59.2	85.5	40.58	1.820	1.42
Polymethylmethacrylate (acrylic)			0.53937	58.1	82.8	40.55	1.929	1.19
Polypropylene			0.55998	56.1	78.5	44.77	2.041	0.90
Polystyrene ($[C_6H_5CHCH_2]_n$)			0.53768	57.5	81.7	43.79	1.936	1.06
Polytetrafluoroethylene (Teflon)			0.47992	63.5	94.4	34.84	1.671	2.20
Polyvinyltoluene			0.54141	57.3	81.3	43.90	1.956	1.03
Aluminum oxide (sapphire)			0.49038	65.5	98.4	27.94	1.647	3.970
Barium flouride (BaF_2)			0.42207	90.8	149.0	9.91	1.303	4.893
Bismuth germanate (BGO)			0.42065	96.2	159.1	7.97	1.251	7.130
Carbon dioxide gas (CO_2)			0.49989	60.7	88.9	36.20	1.819	(1.842)
Solid carbon dioxide (dry ice)			0.49989	60.7	88.9	36.20	1.787	1.563
Cesium iodide (CsI)			0.41569	100.6	171.5	8.39	1.243	4.510
Lithium fluoride (LiF)			0.46262	61.0	88.7	39.26	1.614	2.635
Lithium hydride (LiH)			0.50321	50.8	68.1	79.62	1.897	0.820
Lead tungstate ($PbWO_4$)			0.41315	100.6	168.3	7.39	1.229	8.300
Silicon dioxide (SiO_2, fused quartz)			0.49930	65.2	97.8	27.05	1.699	2.200
Sodium chloride (NaCl)			0.55509	71.2	110.1	21.91	1.847	2.170
Sodium iodide (NaI)			0.42697	93.1	154.6	9.49	1.305	3.667
Water (H_2O)			0.55509	58.5	83.3	36.08	1.992	1.000(0.756)
Silica aerogel			0.50093	65.0	97.3	27.25	1.740	0.200

Source From K.A. Olive et al. (Particle Data Group), Chin. Phys. C 38 (2014) 090001

A. De Angelis and M.J.M. Pimenta, *Introduction to Particle and Astroparticle Physics*, Undergraduate Lecture Notes in Physics,
DOI 10.1007/978-88-470-2688-9

Appendix C
Physical and Astrophysical Constants

Quantity	Symbol, equation	Value	Uncertainty (ppb)
speed of light in vacuum	c	299 792 458 m s^{-1}	exact*
Planck constant	h	6.626 069 57(29)$\times 10^{-34}$ J s	44
Planck constant, reduced	$\hbar \equiv h/2\pi$	1.054 571 726(47)$\times 10^{-34}$ J s	44
		= 6.582 119 28(15)$\times 10^{-22}$ MeV s	22
electron charge magnitude	e	1.602 176 565(35)$\times 10^{-19}$ C = 4.803 204 50(11)$\times 10^{-10}$ esu	22, 22
conversion constant	$\hbar c$	197.326 9718(44) MeV fm	22
conversion constant	$(\hbar c)^2$	0.389 379 338(17) GeV^2 mbarn	44
electron mass	m_e	0.510 998 928(11) MeV/c^2 = 9.109 382 91(40)$\times 10^{-31}$ kg	22, 44
proton mass	m_p	938.272 046(21) MeV/c^2 = 1.672 621 777(74)$\times 10^{-27}$ kg	22, 44
		= 1.007 276 466 812(90) u = 1836.152 672 45(75) m_e	0.089, 0.41
deuteron mass	m_d	1875.612 859(41) MeV/c^2	22
unified atomic mass unit (u)	(mass ^{12}C atom)/12 = (1 g)/(N_A mol)	931.494 061(21) MeV/c^2 = 1.660 538 921(73)$\times 10^{-27}$ kg	22, 44
permittivity of free space	$\epsilon_0 = 1/\mu_0 c^2$	8.854 187 817 … $\times 10^{-12}$ F m^{-1}	exact
permeability of free space	μ_0	$4\pi \times 10^{-7}$ N A^{-2} = 12.566 370 614 … $\times 10^{-7}$ N A^{-2}	exact
fine-structure constant	$\alpha = e^2/4\pi\epsilon_0 \hbar c$	7.297 352 5698(24)$\times 10^{-3}$ = 1/137.035 999 074(44)†	0.32, 0.32
classical electron radius	$r_e = e^2/4\pi\epsilon_0 m_e c^2$	2.817 940 3267(27)$\times 10^{-15}$ m	0.97
(e^- Compton wavelength)/2π	$\lambda\!\!\!^{-}_e = \hbar/m_e c = r_e \alpha^{-1}$	3.861 592 6800(25)$\times 10^{-13}$ m	0.65
Bohr radius ($m_{\text{nucleus}} = \infty$)	$a_\infty = 4\pi\epsilon_0 \hbar^2/m_e e^2 = r_e \alpha^{-2}$	0.529 177 210 92(17)$\times 10^{-10}$ m	0.32
wavelength of 1 eV/c particle	$hc/(1\text{ eV})$	1.239 841 930(27)$\times 10^{-6}$ m	22
Rydberg energy	$hcR_\infty = m_e e^4/2(4\pi\epsilon_0)^2\hbar^2 = m_e c^2\alpha^2/2$	13.605 692 53(30) eV	22
Thomson cross section	$\sigma_T = 8\pi r_e^2/3$	0.665 245 8734(13) barn	1.9
Bohr magneton	$\mu_B = e\hbar/2m_e$	5.788 381 8066(38)$\times 10^{-11}$ MeV T^{-1}	0.65
nuclear magneton	$\mu_N = e\hbar/2m_p$	3.152 451 2605(22)$\times 10^{-14}$ MeV T^{-1}	0.71
electron cyclotron freq./field	$\omega^e_{\text{cycl}}/B = e/m_e$	1.758 820 088(39)$\times 10^{11}$ rad s^{-1} T^{-1}	22
proton cyclotron freq./field	$\omega^p_{\text{cycl}}/B = e/m_p$	9.578 833 58(21)$\times 10^{7}$ rad s^{-1} T^{-1}	22
gravitational constant‡	G_N	6.673 84(80)$\times 10^{-11}$ m^3 kg^{-1} s^{-2}	1.2×10^5
		= 6.708 37(80)$\times 10^{-39}$ $\hbar c$ (GeV/c^2)$^{-2}$	1.2×10^5
standard gravitational accel.	g_N	9.806 65 m s^{-2}	exact
Avogadro constant	N_A	6.022 141 29(27)$\times 10^{23}$ mol^{-1}	44
Boltzmann constant	k	1.380 6488(13)$\times 10^{-23}$ J K^{-1}	910
		= 8.617 3324(78)$\times 10^{-5}$ eV K^{-1}	910
molar volume, ideal gas at STP	$N_A k$(273.15 K)/(101 325 Pa)	22.413 968(20)$\times 10^{-3}$ m^3 mol^{-1}	910
Wien displacement law constant	$b = \lambda_{\max} T$	2.897 7721(26)$\times 10^{-3}$ m K	910
Stefan-Boltzmann constant	$\sigma = \pi^2 k^4/60\hbar^3 c^2$	5.670 373(21)$\times 10^{-8}$ W m^{-2} K^{-4}	3600
Fermi coupling constant**	$G_F/(\hbar c)^3$	1.166 378 7(6)$\times 10^{-5}$ GeV^{-2}	500
weak-mixing angle	$\sin^2\hat{\theta}(M_Z)$ ($\overline{\text{MS}}$)	0.231 26(5)††	2.2×10^5
$W^\pm$ boson mass	m_W	80.385(15) GeV/c^2	1.9×10^5
Z^0 boson mass	m_Z	91.1876(21) GeV/c^2	2.3×10^4
strong coupling constant	$\alpha_s(m_Z)$	0.1185(6)	5.1×10^6

π = 3.141 592 653 589 793 238 e = 2.718 281 828 459 045 235 γ = 0.577 215 664 901 532 861

1 in ≡ 0.0254 m	1 G ≡ 10^{-4} T	1 eV = 1.602 176 565(35) $\times 10^{-19}$ J	kT at 300 K = [38.681 731(35)]$^{-1}$ eV
1 Å ≡ 0.1 nm	1 dyne ≡ 10^{-5} N	1 eV/c^2 = 1.782 661 845(39) $\times 10^{-36}$ kg	0 °C ≡ 273.15 K
1 barn ≡ 10^{-28} m^2	1 erg ≡ 10^{-7} J 2.997 924 58 $\times 10^9$ esu = 1 C		1 atmosphere ≡ 760 Torr ≡ 101 325 Pa

* The meter is the length of the path traveled by light in vacuum during a time interval of 1/299 792 458 of a second.
† At $Q^2 = 0$. At $Q^2 \approx m_W^2$ the value is ~ 1/128.
‡ Absolute lab measurements of G_N have been made only on scales of about 1 cm to 1 m.
** See the discussion in Sec. 10, "Electroweak model and constraints on new physics."
†† The corresponding $\sin^2\theta$ for the effective angle is 0.23155(5).

Source From K.A. Olive et al. (Particle Data Group), Chin. Phys. C 38 (2014) 090001

A. De Angelis and M.J.M. Pimenta, *Introduction to Particle and Astroparticle Physics*, Undergraduate Lecture Notes in Physics,
DOI 10.1007/978-88-470-2688-9

Quantity	Symbol, equation	Value
speed of light	c	$299\,792\,458$ m s^{-1}
Newtonian gravitational constant	G_N	$6.673\,8(8) \times 10^{-11}$ m^3 kg^{-1} s^{-2}
Planck mass	$\sqrt{\hbar c/G_N}$	$1.220\,93(7) \times 10^{19}$ GeV/c^2
		$= 2.176\,51(13) \times 10^{-8}$ kg
Planck length	$\sqrt{\hbar G_N/c^3}$	$1.616\,20(10) \times 10^{-35}$ m
standard gravitational acceleration	g_N	$9.806\,65$ m s
jansky (flux density)	Jy	10^{-26} W m^{-2} Hz^{-1}
tropical year (equinox to equinox) (2011)	yr	$31\,556\,925.2$ s $\approx \pi \times 10^7$ s
sidereal year (fixed star to fixed star) (2011)		$31\,558\,149.8$ s $\approx \pi \times 10^7$ s
mean sidereal day (2011) (time between vernal equinox transits)		$23^{\mathrm{h}}\,56^{\mathrm{m}}\,04\overset{\mathrm{s}}{.}090\,53$
astronomical unit	au	$149\,597\,870\,700$ m
parsec (1 au/1 arc sec)	pc	$3.085\,677\,581\,49 \times 10^{16}$ m $= 3.262\ldots$ ly
light year (deprecated unit)	ly	$0.306\,6\ldots$ pc $= 0.946\,053\ldots \times 10^{16}$ m
Schwarzschild radius of the Sun	$2G_N M_\odot/c^2$	$2.953\,250\,077(2)$ km
Solar mass	$M_\odot$	$1.988\,5(2) \times 10^{30}$ kg
Solar equatorial radius	$R_\odot$	$6.9551(4) \times 10^8$ m
Solar luminosity	$L_\odot$	3.828×10^{26} W
Schwarzschild radius of the Earth	$2G_N M_\oplus/c^2$	$8.870\,055\,94(2)$ mm
Earth mass	$M_\oplus$	$5.972\,6(7) \times 10^{24}$ kg
Earth mean equatorial radius	$R_\oplus$	$6.378\,137 \times 10^6$ m
Solar angular velocity around the Galactic center	Θ_0/R_0	30.3 ± 0.9 km s^{-1} kpc^{-1}
Solar distance from Galactic center	R_0	$8.4(6)$ kpc
circular velocity at R_0	v_0 or Θ_0	$254(16)$ km s^{-1}
local disk density	ρ_{disk}	3–12 $\times 10^{-24}$ g cm^{-3} $\approx$ 2–7 GeV/c^2 cm^{-3}
local dark matter density	ρ_χ	canonical value 0.3 GeV/c^2 cm^{-3} within factor 2–3
escape velocity from Galaxy	v_{esc}	498 km/s $< v_{\mathrm{esc}} <$ 608 km/s
present day CMB temperature	T_0	$2.7255(6)$ K
present day CMB dipole amplitude		$3.355(8)$ mK
Solar velocity with respect to CMB		369(1) km/s towards $(\ell, b) = (263.99(14)^\circ, 48.26(3)^\circ)$
Local Group velocity with respect to CMB	v_{LG}	627(22) km/s towards $(\ell, b) = (276(3)^\circ, 30(3)^\circ)$
entropy density/Boltzmann constant	s/k	$2\,891.2\,(T/2.7255)^3$ cm^{-3}
number density of CMB photons	n_γ	$410.7(T/2.7255)^3$ cm^{-3}
baryon-to-photon ratio	$\eta = n_{\mathrm{b}}/n_\gamma$	$6.05(7) \times 10^{-10}$ (CMB)
		$5.7 \times 10^{-10} \leq \eta \leq 6.7 \times 10^{-10}$ (95% CL)
present day Hubble expansion rate	H_0	$100\,h$ km s^{-1} Mpc^{-1} $= h \times (9.777\,752$ Gyr$)^{-1}$
scale factor for Hubble expansion rate	h	$0.673(12)$
Hubble length	c/H_0	$0.925\,0629 \times 10^{26}\,h^{-1}$ m $= 1.37(2) \times 10^{26}$ m
scale factor for cosmological constant	$c^2/3H_0^2$	$2.85247 \times 10^{51}\,h^{-2}$ m^2 $= 6.3(2) \times 10^{51}$ m^2
critical density of the Universe	$\rho_{\mathrm{crit}} = 3H_0^2/8\pi G_N$	$2.775\,366\,27 \times 10^{11}\,h^2\,M_\odot$Mpc^{-3}
		$= 1.878\,47(23) \times 10^{-29}\,h^2$ g cm^{-3}
		$= 1.053\,75(13) \times 10^{-5}\,h^2$ (GeV/c^2) cm^{-3}
number density of baryons	n_{b}	$2.482(32) \times 10^{-7}$ cm^{-3}
		$(2.1 \times 10^{-7} < n_{\mathrm{b}} < 2.7 \times 10^{-7})$ cm^{-3} (95% CL)
baryon density of the Universe	$\Omega_{\mathrm{b}} = \rho_{\mathrm{b}}/\rho_{\mathrm{crit}}$	‡$0.02207(27)\,h^{-2}$ = †$0.0499(22)$
cold dark matter density of the universe	$\Omega_{\mathrm{cdm}} = \rho_{\mathrm{cdm}}/\rho_{\mathrm{crit}}$	‡$0.1198(26)\,h^{-2}$ = †$0.265(11)$
100 × approx to r_*/D_{A}	$100 \times \theta_{\mathrm{MC}}$	‡$1.0413(6)$
reionization optical depth	τ	‡$0.091^{+0.013}_{-0.014}$
scalar spectral index	n_s	‡$0.958(7)$
ln pwr primordial curvature pert. (k_0=0.05 Mpc^{-1})	$\ln(10^{10}\Delta^2_{\mathcal{R}})$	‡$3.090(25)$
dark energy density of the ΛCDM Universe	Ω_Λ	$0.685^{+0.017}_{-0.016}$
pressureless matter density of the Universe	$\Omega_{\mathrm{m}} = \Omega_{\mathrm{cdm}} + \Omega_{\mathrm{b}}$	$0.315^{+0.016}_{-0.017}$ (From Ω_Λ and flatness constraint)
dark energy equation of state parameter	w	♯$-1.10^{+0.08}_{-0.07}$ (*Planck*+WMAP+BAO+SN)
CMB radiation density of the Universe	$\Omega_\gamma = \rho_\gamma/\rho_c$	$2.473 \times 10^{-5}(T/2.7255)^4\,h^{-2} = 5.46(19) \times 10^{-5}$
effective number of neutrinos	N_{eff}	†3.36 ± 0.34
sum of neutrino masses	$\sum m_\nu$	< 0.23 eV (95% CL; CMB+BAO) $\Rightarrow \Omega_\nu h^2 < 0.0025$
neutrino density of the Universe	Ω_ν	$< 0.0025\,h^{-2} \Rightarrow\ < 0.0055$ (95% CL; CMB+BAO)
curvature	$\Omega_{\mathrm{tot}} = \Omega_{\mathrm{m}} + \ldots + \Omega_\Lambda$	♯$0.96^{+0.4}_{-0.5}$ (95%CL)
		♯$1.000(7)$ (95% CL; CMB+BAO)
fluctuation amplitude at $8\,h^{-1}$ Mpc scale	σ_8	†0.828 ± 0.012
running spectral index slope, $k_0 = 0.002$ Mpc^{-1}	$dn_s/d\ln k$	♯$-0.015(9)$
tensor-to-scalar field perturbations ratio, k_0=0.002 Mpc^{-1}	$r = T/S$	♯< 0.11 at 95% CL; no running
redshift at decoupling	z_{dec}	†1090.2 ± 0.7
age at decoupling	t_*	†3.72×10^5 yr
sound horizon at decoupling	$r_s(z_*)$	†147.5 ± 0.6 Mpc (*Planck* CMB)
redshift of matter-radiation equality	z_{eq}	†3360 ± 70
redshift at half reionization	z_{reion}	†11.1 ± 1.1
age at half reionization	t_{reion}	†462 Myr
age of the Universe	t_0	†13.81 ± 0.05 Gyr

‡ Parameter in six-parameter ΛCDM fit [2].
† Derived parameter in six-parameter ΛCDM fit [2].
♯ Extended model parameter [2].

Index

Symbols
$K^0 - \overline{K}^0$ Mixing, 310
ΛCDM model, 468, 477, 486
α_i matrices, 260
β matrix, 260
τ-θ puzzle, 95
γ^μ matrices, 260

A
Abelian symmetry groups, 324
Accelerators, 189
ACE experiment, 157
Acoustic peaks, 437
Active galactic nuclei (AGN), 576
AGASA experiment, 159
AGILE telescope, 166, 168
Aharonov, Yakir, 255
AMS-02 experiment, 157, 545
Andromeda galaxy, 9
Angular diameter distance, 428
ANTARES detector, 182
Anthropic principle, 409, 417
Antimatter, 87, 500, 547
  discovery, 88
Antiproton, 191, 547
Antiquark, 214
Arecibo radio telescope, 637
ARGO-YBJ detector, 169
Askar'yan effect, 108
Astrobiology, 617
Astronomical unit (AU), 7
Asymptotic freedom, 331, 340
Atacama Cosmology Telescope, 438
ATLAS experiment, 152
Aurorae, 624
Avalanche photodiodes (APD), 131
Axion, 385, 491
Axion-like-particle (ALP), 491, 610

B
Baikal NT-200 detector, 181
Barn, 28
Baryon Acoustic Oscillations, 428, 485
Baryons, 190, 214
  baryonic quantum number, 191
Beauty, *see* Bottom
BESS experiment, 157
β decay, 24, 440, 529, 530
Bethe formula, 102
Bethe, Hans, 102
Bi-spinor, 261
Big bang, 430, 473
Big crunch, 472
Big European Bubble Chamber (BEBC), 122
Bilinear covariants, 265
Binary systems, 556, 575
Binding energy, 13
BL lacertae objects (BL Lac), 578
Black disk, 350
Black hole, 473
  Kerr, 473
  Schwarzschild, 473
Blazar, 577, 578
Block, Martin, 289
Bohm, David, 255

A. De Angelis and M.J.M. Pimenta, *Introduction to Particle and Astroparticle Physics*, Undergraduate Lecture Notes in Physics,
DOI 10.1007/978-88-470-2688-9

Bohm–Aharonov effect, 255
Bose–Einstein statistics, 5
Boson, 5
  Goldstone, 367
  Higgs, 371, 384
Bothe, Walther, 90
Bottom
  Υ, 221
  bottomium, 221
Bragg curve, 110
Branching ratio, 34
Branes, 416
Breakdown potential, 123
Breit frame, 230
Bremsstrahlung, 105
Bubble chamber, 121

C
Cabibbo
  angle, 304
Cabibbo, Nicola, 305
Callan–Gross relation, 231
Calorimeter, 133
  electromagnetic, 133, 134
  hadronic, 134
  homogeneous, 133
  sampling, 133
CANGAROO telescope, 170
Casimir
  effect, 284
  operator, 201
Casimir, Hendrick, 284
Cepheids, 425
CERN, 15, 95, 145, 390
Chadwick, James, 25
Chamberlain, Owen, 67
Charge conjugation, 205
Charged weak currents, 296
Charm, 218, 307
  charmonium, 220
Charpak, George, 124
Cherenkov
  detector, 131
  emission in air showers, 156
  radiation, 107
  telescope, 170, 551
Cherenkov, Pavel, 107
Chirality, 292
CKM matrix, 308
Clebsch–Gordan coefficients, 207
Cloud chamber, 88, 121
Cluster model, 332
CMB polarization, 437
CMS experiment, 149
CNO cycle, 507
COBE, 434
Collection efficiency, 129
Color, 216
Color factors, 328
Compositeness, 417
Compton scattering, 111
Compton wavelength, 64
Confinement, 332, 333, 340
Contravariant representation, 46
Conversi, Pancini and Piccioni experiment, 91
Cosmic microwave background (CMB), 11, 429, 549
Cosmic rays, 15, 77
  acceleration, 562
    Hillas plot, 571
  arrival directions of charged, 543
  charged, 538
  discovery, 71
  electrons and positrons, 545
  gamma rays, 550
  GZK suppression, 540
  life and, 622
  photons, 549
  propagation of Galactic, 589
  spectrum, 539
Cosmological constant, 459
Cosmological principle, 461
Coulomb gauge, 253
Covariant derivative, 49, 255
Covariant representation, 46
CP violation, 310
  direct, 315
  in 2π modes, 312
  in the B sector, 317
  in the standard model, 319
  indirect, 316
  strong CP problem, 385
Crab Nebula, 16, 551, 573, 574
Crane, Richard, 267
Critical density, 465
Critical energy, 107
Cronin, Jim, 312
Cross-section, 27
  WW cross-section at LEP, 397
  ZZ cross section at LEP, 397

CTA observatory, 177, 599
CUORE, 534

D
Δ resonances, 211
  Δ^{++} puzzle, 216
D'Alembert operator, 49, 252
DAMA experiment, 498
DAMPE mission, 176
Dark energy, 14, 428, 472
Dark matter, 14, 443, 444, 488
  baryonic, 489
  cold, 463
  distributions, 446
  freeze-out, 477
  hot, 490
Dark noise, 129
DASI, 437
Davis, Ray, 178
Daya Bay experiment, 521, 522
de Sitter universe, 464
Deep inelastic scattering, 223
DELPHI experiment, 147
Density of final states, 35
Differential cross-section, 29
Dipole formula, 41
Diquark, 218, 335
Dirac
  equation, 86, 259
  notation, 80
Dirac, Paul Adrien Maurice, 85
Distance
  comoving, 424
  luminosity, 426
  proper, 426
  scale, 425
Drell-Yan process, 328, 343
Drift chamber, 125
Dwarf spheroidals, 447, 605

E
EAS, 154
Eddington luminosity, 577
EGRET instrument, 166
EHS, 142
Eightfold way, 211
Eikonal function, 350
Einasto profile of DM, 446
Einstein
  equations, 459
Einstein, Albert, 4
Elastic process, 30
Electromagnetic interaction, 6
Electromagnetic tensor, 57
Electroweak precision measurements, 391
Energy–momentum tensor, 459
Equations
  Dirac, 259
  Drake, 631
  Euler–Lagrange, 247
  Klein-Gordon, 270
  Maxwell, 250
Equivalence principle, 452
Euclidean space, 455
Euler–Lagrange equations, 247
EUSO mission, 164
EXO, 534
Extra dimensions, 415
Extragalactic background light (EBL), 592
Extraterrestrial life, *see also* Life
  messages from space, 636
  search for, 637
Extremophiles, 624

F
FCNC, 307
Fermi
  bubbles, 556
  constant, 286
  golden rule, 35, 280
  LAT, 166
  mechanism for CR acceleration, 562
  model of weak interactions, 285, 290
  observatory, 167
  paradox (extraterrestrial life), 635
Fermi, Enrico, 34
Fermi-Dirac statistics, 4
Fermion, 4
Feynman
  diagrams, 245, 277
Feynman, Richard, 39
Fine structure constant, 278
Fireball model, 581
Fitch, Val, 312
Flash chamber, 125
Flavor, 214
Fluorescence, 155
Form factor, 41
Four-scalar, 47
Fragmentation
  functions, 335
  string and cluster model, 332
Free streaming distance, 529

Freeze-out temperature, 440
Friedman, Jerome I., 228
Friedmann, Alexander, 461
Froissart bound, 350
FSRQ, 578

G
G-2 experiments, 267
Gain, 129
Galactic center, 606
Galactic coordinates, 18
Galaxy clusters, 605
Galilei, Galileo, 9
GALLEX, 508
GALPROP, 590
Gamma ray, 17, 550
 bursts, 553, 580
 emission from Fermi bubbles, 556
 flares, 552
 horizon, 597
 transients, 552
GAMMA400 mission, 176
Gargamelle, 297
Gaseous photomultipliers (GPM), 130
Gauge
 invariance, 361
 transformation, 251, 257
Geiger mode, 123
Geiger-Müller counter, 78, 123
General relativity, 452
Geodesic, 426, 455
 equations, 461
GIM mechanism, 304
Glashow, Sheldon Lee, 370
Glasma, 357
Glauber model, 352
Gluon, 7, 216, 326
Gottfried sum rule, 235
Grand unification, 408
Grand Unified Theories (GUT), 410, 484
Gravitational
 interaction, 6
 lensing, 444
 radiation, 559
  graviton, 559
 waves, 182
Gray disk, 350
GRB, *see* Gamma ray bursts
Greenhouse effect, 622
Gross, David J., 323
Group, 193
 abelian, 193
 SO(3), 200
 spin rotations, 199
 SU(2), 199
 SU(3), 202
 SU(n), 199
 U(n), 199
Gyromagnetic ratio, 266
GZK
 cutoff, 16, 593
 mechanism, 540
 neutrinos, 557

H
H.E.S.S. telescope, 168, 170, 172, 608
Hadrons, 191
Hamiltonian, 191, 247
HAWC observatory, 168, 170
HEGRA telescope, 170
Heisenberg
 principle, 2
Heisenberg, Werner, 2
Helicity, 43, 264
Hermitian conjugate, 194
Hess, Victor, 75
Higgs
 boson, 371, 384, 386
  discovery, 403
 mechanism, 368
Higgs, Peter, 363
High-energy gamma rays, 566
 attenuation edge, 597
 hadronic photoproduction, 569
 inverse Compton scattering, 567
 optical depth, 596
 self-synchrotron Compton mechanism, 567
High-z Supernova Search Team, 427
Hillas plot, 571
Hofstadter, Robert, 223
Homestake mine, 508
Hubble
 constant, 10, 422, 462
 law, 10, 422
Hubble, Erwin, 9
Hypercharge, 203, 212, 374

I
IceCube, 181, 559
 neutrino observations, 585
Imaging Atmospheric Cherenkov Telescope (IACT), 170
Indirect CP violation, 316

Inelastic interaction length, 114
Inflation, 437, 481
Inflaton, 481
Interaction length, 32
Inverse Compton, 112
Ionization counter, 123
Isospin, 203, 207
  weak, 373

J
J/ψ, 218, 307
  anomalous suppression, 356
  discovery, 218
  suppression in QGP, 357
Jarlskog invariant, 320
JEM-EUSO experiment, 163
Jet quenching, 355

K
K2K experiment, 522
KamLAND, 515–517, 525
KamLAND-Zen, 534
KATRIN, 531
Kendall, Henry W., 228
Kepler mission, 634
Klein, Oscar, 84
Klein-Gordon equation, 84, 252, 270
Km3NeT telescope, 182
Kobayashi, Makoto, 308
Koshiba, Masatoshi, 178
Kurie plot, 530

L
Lagrangian, 247
  density, 249
  mass term, 250
Landau-Pomeranchuk-Migdal effect, 114
Last scattering surface, 434
Leaky-box model, 590
Lederman, Leon, 506
Lemaitre, Georges, 461
Length contraction, 46
Leptons, 190, 237
  τ, *see* τ lepton
LHC, 15, 403
LHCb, 143
Lie algebra, 198
Life, 618
  abiogenesis, 626
  ammonia and, 621
  carbon-based, 618
  cosmic rays and, 622
  Drake equation, 631
  in the solar system, 628
  panspermia, 626
  recipe of, 619
  Schrödinger's definition, 619
  silicon-based, 619
LIGO observatory, 184
LISA project, 185
Livingston plot, 190
LNGS, 497
Local group, 9
Lorentz
  boost, 51
  force, 44, 253
  invariance violation, 607
  Lorentz–Heaviside convention, 250
  Lorentz-Invariant Fermi rule, 55
  transformations, 45, 51
Lorentz, Hendrik, 44
Lorenz gauge, 59, 252
Lorenz, Ludvig, 59
LPM effect, 114
LSND, 526
Luminosity, 29, 139
  differential, 139
  integrated, 139
Lund symmetric fragmentation function, 335

M
MACHO, 489
MAGIC telescope, 168, 170, 172, 174, 608
Magnetic field, 588
  Earth's, 623
  extragalactic, 588
  galactic, 588
Magnetosphere, 623
Magnitude, 585
Majorana, Ettore, 532
Mandelstam variables, 53
Maskawa, Toshihide, 308
Mather, John, 434
Maxwell equations, 44, 250
Maxwell, James Clerk, 44, 250
Mendeleev, Dimitri, 21
Mesons, 190
Metallicity, 451
Metric
  Kerr, 459
  Minkowski, 459
  Robertson–Walker, 458, 459

Schwarzschild, 459, 472
Metric tensor, 48
Microlensing, 434, 438, 490
MILAGRO, 169
Milky Way, 8
Millikan, Robert A., 77
Mills, Robert, 257
Minimal coupling, 255
Minkowski tensor, 47
MINOS, 523, 524
Mip or minimum ionizing particle, 102
Molière radius, 119
MOND, 449
Monte Carlo methods, 118
Mott cross-section, 42
Multijet production, 346
Multiple scattering, 109
Muon, 92
decay, 296
detection at accelerators, 141
in cosmic-ray showers, 159, 547

N

Natural units (NU), 62
Navarro, Frenk, and White (NFW) profile of DM, 446
NEMO experiment, 182
NESTOR project, 182
Neutral weak currents, 297
Neutralino, 413, 494
Neutrinos, 505, 557
appearance experiments, 513
disappearance experiments, 513
accelerator, 522
and hot dark matter, 528
astrophysical, 557
atmospheric, 516, 557, 585
beams, 297
CNO cycle, *see* CNO cycle
cosmogenic, 557
cosmological, 481
decoupling, 480
Dirac, 532
from SN1987A, *see* SN1987A
from the collapse of supernovae, 585
geo-neutrinos, 525
IceCube, *see* IceCube
Majorana, 532, 533
mass, 505
mass limits, 529
MSW effect, 514
number of light species, 395
oscillation in a two-flavor system, 511
oscillation length, 513
oscillations, 505
resonant oscillations, 515
solar, 506, 585
neutrino problem, 508
very-high-energy, 559
Neutron, 25, 207
Neutron star, 572
Newton, Isaac, 4
Nishimura-Kamata-Greisen function, 119
NKG function, 119
No-hair theorem, 473
Noether
theorem, 192, 194, 248
Noether, Emmy, 192
Non-Abelian symmetry groups, 324
Nuclear emulsion, 122
Nuclear suppression factor, 355
Nucleosynthesis
in stars, 439
primordial, 438

O

Ω^-, 213
Olbers' paradox, 451
OPERA, 525
Optical depth, 596
Optical theorem, 348
OZI rule, 217
Ozone layer, 623

P

Pacini, Domenico, 74
Pair production, 113
PAMELA
anomaly, 546, 604
PAMELA experiment, 157
Parity, 195, 204, 288
Parity violation, 288
Parsec, 9
Partial cross-section, 30
Partial wave amplitudes, 348
Partial width, 34
Particle accelerator, 136
Particle Data Book, 240
Particle lifetime, 33
Parton density function, 232
Partons, 229
Pauli
exclusion principle, 4

matrices, 200, 260
Pauli, Wolfgang, 4
PDF, *see* Parton density function
PDG, *see* Particle Data Book
Penning traps, 268
Penzias, Arno, 429
Periodic table, 21
Perlmutter, Saul, 427
Photodetector, 129
Photodisintegration, 593
Photoelectric effect, 111
Photoluminescence, 104
Photomultiplier tubes (PMT), 129
Photon decoupling, 431
Photonuclear effect, 113
Photosensors, 129
Pierre Auger Observatory, 162
Pioneer 10, 642
Planck
length, 64
mass, 65
mission, 434
scale, 3
time, 64
Planck, Max, 3
Planets, *see* Solar system
extrasolar, 633, 645
POLARBEAR, 438
Politzer, David , 323
Pontecorvo, Bruno, 519
Pontecorvo-Maki-Nakagawa-Sakata (PMNS) matrix, 519
Preons, 418
Primordial abundance of ^{4}He, 441
Propagator, 245, 277, 296, 382
Proper time, 45
Proportional counter, 123
Pseudo-rapidity, 145, 340
PSR 1913+16, 560
Pulsar, 572
Pulsar wind nebulae (PWN), 574

Q
QCD, 323
tests at LEP, 398
QED, 250
Quantum chromodynamics, *see* QCD
Quantum efficiency, 129
Quantum electrodynamics, *see* QED
Quark, 214
family, 222
hadronization, 220
quark-parton model, 229
top, 402
Quark–Gluon Plasma (QGP), 341, 353
Quasar, 577, *see also* Active Galactic Nuclei

R
Radiation length, 106
Radio emission, 156
Range, 110
Rapidity, 339
Rayleigh scattering, 113
Recombination
phase, 431
temperature, 432
Redshift, 422
Reines, Frederick, 505
Reionization
epoch, 433
optical depth parameter τ, 434
Renormalization, 281, 282
Resistive plate chamber, 128
RICH, 132
Riess, Adam, 427
Ring imaging Cherenkov detector, 132
Rosenbluth formula, 223, 281
Rossi, Bruno, 90
Rubbia, Carlo, 137, 300
Running coupling
and unification of forces, 409
for α, 283
for α_s, 329
Rutherford experiment, 23
Rutherford, Ernest, 23

S
SAGE, 508
Saha equation, 431
Sakharov conditions, 322
Salam, Abdus, 370
Schmidt, Brian, 427
Schrödinger's
definition of life, 619
equation, 79
Schrödinger, Erwin, 79
Schwartz, Melvin, 506
Schwarzschild
radius, 65, 472
Schwarzschild, Karl, 472
Scintillators, 127
Segrè, Emilio, 67
SETI, *see* Extraterrestrial life

Showers
electromagnetic, 115
extensive air showers, 154
depth of shower maximum X_{max}, 541
muons in, 547
hadronic, 119
SI, 63
Silicon detectors, 126
SiPM, 131
Smoot, George, 434
SN1987A, 558
SNO, 509, 511
SO(3), 200
Soft processes in QCD, 347
Solar system, 628
South Atlantic Anomaly, 624
South Pole Telescope, 438
Space-time, 49
Special Unitary group, *see* SU(n)
Spin, 5, 86, 260, 266
Spin rotations, 199
Spin-statistic theorem, 5
Spinor, 199
Spontaneous symmetry breaking, 363, 376
Standard candle, 425
Standard model, 3, 285, 383
Standard ruler, 428
Standard Solar Model, 506
Star
hypergiant, 12
main sequence, 12
red dwarf, 12
white dwarf, 13
Steinberger, Jack, 506
Strange particle, 94
Strangeness, 212
Streamer chamber, 125
String effect, 338
String model, 332, 333
String theories, 415
Strong CP problem, 491
Strong interaction, 6, 91
Structure functions, 227, 235
SU(2), 199, 285
SU(3), 202
$SU(3)_{flavor}$, 214
SU(n), 199
Super-K, *see* Super-Kamiokande
Super-Kamiokande, 508
Supernova, 13, 572
type Ia, 427
Supernova Cosmology Project, 427
Supernova remnant, 561, 572, 574
particle acceleration in, 561
shell, 574
Superstring theories, 416
SUSY, 412, 494
Symmetries
continuum, 195
CPT invariance, 206
discrete, 195
flavor, 214
Noether's theorem and, 191
phase shift invariance, 196
rotation invariance, 197
space Translation invariance, 196
time reversal, 206
Synchrotron, 142

T
τ lepton, 222, 237
T2K, 522
TA experiment, 163
Taylor, Richard E., 228
Telescope array, 163
Temperature power spectrum, 437
Tensor, 48
Thomson cross-section, 64
Tibet-AS, 169
Tides, 620
Time dilation, 45
Top, 222
quark, 402
Total cross-section, 27
Total width, 34
Tracking detector, 120
Transition amplitude, 35
Transition radiation detector (TRD), 133
Trigger system, 90, 127

U
U(1), 285, 324
U(n), 199
UA1, 145, 301
UA2, 301
Uncertainty principle, 26
Underlying event, 339
Unitarity
triangle, 309, 320, 321
Unitary group, *see* U(n)
Universe
age, 10, 450, 468
early, 473
expansion of the, 422

V
V-A, 290
Vacuum, 7
Vacuum polarization, 281
Van Allen belts, 623
van der Meer, Simon, 137, 300
Vela, 573
Veltman, Martinus, 363
VERITAS telescope, 168, 170, 172
Vertices in QCD, 327
VIRGO observatory, 184
Virial theorem, 443

W
W boson, 7, 285, 370, 378
Wavelength shifter, 127
Weak interaction, 6, 285
Weak isospin, 373
Weinberg
  angle, 377
Weinberg, Steven, 370
WHIPPLE telescope, 170
Wilczek, Frank, 323
Wilson, Charles Thomson Rees, 72
Wilson, Robert, 429
WIMP, 478, 490, 492, 603
  miracle, 479, 492
Wire chamber, 124
WISP, 490, 610
WMAP, 434
Wolfenstein parametrization (of the CKM matrix), 309

X
X-ray transition radiation, 108

Y
Yang, Chen, 257
Yang–Mills theories, 324
Yukawa potential, 41
Yukawa, Hideki, 42

Z
Z boson, 7, 285, 370, 378
  asymmetries, 393
  lineshape, 395
  partial widths of the, 392
Zero point energy, 284

Printed by Printforce, the Netherlands